Lecture Notes in Computational Science and Engineering 138

Editors:

More information about this series at http://www.springer.com/series/3527

Ronald Haynes • Scott MacLachlan
Xiao-Chuan Cai • Laurence Halpern
Hyea Hyun Kim • Axel Klawonn
Olof Widlund
Editors

Domain Decomposition Methods in Science and Engineering XXV

Springer

Editors
Ronald Haynes
Department of Mathematics
and Statistics
Memorial University of Newfoundland
St. John's, NL, Canada

Scott MacLachlan
Department of Mathematics
and Statistics
Memorial University of Newfoundland
St. John's, NL, Canada

Xiao-Chuan Cai
Department of Computer Science
University of Colorado
Boulder, CO, USA

Laurence Halpern
Laboratoire Analyse, Geometrie
Université Paris XIII
Villetaneuse, France

Hyea Hyun Kim
Department of Applied Mathematics
Kyung Hee University
Yongin, Republic of Korea

Axel Klawonn
Mathematisches Institut
Universität zu Köln
Köln, Germany

Olof Widlund
New York University
New York, NY, USA

ISSN 1439-7358 ISSN 2197-7100 (electronic)
Lecture Notes in Computational Science and Engineering
ISBN 978-3-030-56749-1 ISBN 978-3-030-56750-7 (eBook)
https://doi.org/10.1007/978-3-030-56750-7

Mathematics Subject Classification (2020): 65F10, 65N30, 65N55

Cover illustration: Page 92, Fig. 1.

This Springer imprint is published by the registered company Springer Nature Switzerland AG
The registered company address is: Gewerbestrasse 11, 6330 Cham, Switzerland

Preface

This volume contains a selection of 58 papers submitted to the 25th International Conference on Domain Decomposition Methods, hosted by Memorial University of Newfoundland, in St. John's, Newfoundland and Labrador, Canada, from July 23–27, 2018.

Background of the Conference Series

With its first meeting in Paris in 1987, the International Conference on Domain Decomposition Methods has been held in 15 countries in Asia, Europe, and North America, and now for the first time in Canada. The conference is held at roughly 18-months intervals. A complete list of 25 meetings appears below.

Domain decomposition is often seen as a form of divide-and-conquer for mathematical problems posed over a physical domain, reducing a large problem into a collection of smaller problems, each of which is much easier to solve computationally than the undecomposed problem, and most or all of which can be solved independently and concurrently, and then solving them iteratively in a consistent way. Much of the theoretical interest in domain decomposition algorithms lies in ensuring that the number of iterations required to converge is very small. Domain decomposition algorithms can be tailored to the properties of the physical system as reflected in the mathematical operators, to the number of processors available, and even to specific architectural parameters, such as cache size and the ratio of memory bandwidth to floating point processing rate, proving it to be an ideal paradigm for large-scale simulation on advanced architecture computers.

The principle technical content of the conference has always been mathematical, but the principle motivation has been to make efficient use of distributed memory computers for complex applications arising in science and engineering. While research in domain decomposition methods is presented at numerous venues, the International Conference on Domain Decomposition Methods is the only regularly occurring international forum dedicated to interdisciplinary technical interactions between theoreticians and practitioners working in the development, analysis, software implementation, and application of domain decomposition methods.

As we approach the dawn of exascale computing, where we will command 10^{18} floating point operations per second, clearly efficient and mathematically well-founded methods for the solution of large-scale systems become more and more important-as does their sound realization in the framework of modern HPC architectures. In fact, the massive parallelism, which makes exascale computing possible, requires the development of new solutions methods, which are capable of efficiently exploiting this large number of cores as well as the connected hierarchies for memory access. Ongoing developments such as parallelization in time asynchronous iterative methods, or nonlinear domain decomposition methods show that this massive parallelism does not only demand for new solution and discretization methods, but also allows to foster the development of new approaches.

Here is a list of the 25 first conferences on Domain Decomposition:

1. Paris, France, January 7-9, 1987
2. Los Angeles, USA, January 14-16, 1988
3. Houston, USA, March 20-22, 1989
4. Moscow, USSR, May 21-25, 1990
5. Norfolk, USA, May 6-8, 1991
6. Como, Italy, June 15-19, 1992
7. University Park, Pennsylvania, USA, October 27-30, 1993
8. Beijing, China, May 16-19, 1995
9. Ullensvang, Norway, June 3-8, 1996
10. Boulder, USA, August 10-14, 1997
11. Greenwich, UK, July 20-24, 1998
12. Chiba, Japan, October 25-20, 1999
13. Lyon, France, October 9-12, 2000
14. Cocoyoc, Mexico, January 6-11, 2002
15. Berlin, Germany, July 21-25, 2003
16. New York, USA, January 12-15, 2005
17. St. Wolfgang-Strobl, Austria, July 3-7, 2006
18. Jerusalem, Israel, January 12-17, 2008
19. Zhangjiajie, China, August 17-22, 2009
20. San Diego, California, USA, February 7-11, 2011
21. Rennes, France, June 25-29, 2012
22. Lugano, Switzerland, September 16-20, 2013
23. Jeju Island, Korea, July 6-10, 2015
24. Spitsbergen, Svalbard, Norway, February 6-10, 2017
25. St. John's, Newfoundland, Canada, July 23-27, 2018

International Scientific Committee on Domain Decomposition Methods

- Petter Bjørstad, University of Bergen, Norway
- Susanne Brenner, Louisiana State University, USA
- Xiao-Chuan Cai, CU Boulder, USA
- Martin Gander, University of Geneva, Switzerland
- Laurence Halpern, University Paris 13, France
- David Keyes, KAUST, Saudi Arabia
- Hyea Hyun Kim, Kyung Hee University, Korea
- Axel Klawonn, Universität zu Köln, Germany
- Ralf Kornhuber, Freie Universität Berlin, Germany
- Ulrich Langer, University of Linz, Austria
- Luca Pavarino, University of Pavia, Italy
- Olof Widlund, Courant Institute, USA
- Jinchao Xu, Penn State, USA
- Jun Zou, Chinese University of Hong Kong, Hong Kong

About the 25th Conference

The twenty-fifth International Conference on Domain Decomposition Methods had 187 participants from 20 different countries. The conference contained 12 invited presentations selected by the International Scientific Committee, fostering both experienced and younger scientists, 19 minisymposia around specific topics (including an industrial minisymposium), 6 contributed sessions, and a poster session. The present proceedings contain a selection of 58 papers grouped into three separate groups: 5 plenary papers, 39 minisymposia papers, and 14 contributed papers.

Sponsoring Organizations

- Memorial University
- Faculty of Engineering, Memorial University
- Department of Mathematics and Statistics, Memorial University
- Atlantic Association for Research in the Mathematical Sciences
- Fields Institute
- Centre de Recherches Mathématiques
- Pacific Institute for the Mathematical Sciences
- Canadian Applied and Industrial Mathematics Society
- National Science Foundation
- Lawrence Livermore National Laboratory
- Los Alamos National Laboratory

Local Organizing/Program Committee Members

- Ronald D. Haynes, Memorial University of Newfoundland
- Scott MacLachlan, Memorial University of Newfoundland
- Hermann Brunner, Memorial University of Newfoundland / Hong Kong Baptist University
- Sue Brenner, Louisiana State University
- Shaun Lui, University of Manitoba
- Emmanual Lorin, Carleton University

Plenary Presentations

- *A review of the computational aspects in Seismic Imaging and Reservoir Simulation*, Henri Calendra (Total E&P, France)
- *Computation of High Frequency Waves in Unbounded Domains: Perfectly Matched Layer and Source Transfer*, Zhiming Chen (Academy of Mathematics and Systems Science, China)
- *Anisotropic mesh adaptation using enriched reconstructed solutions*, André Fortin (Université Laval, Canada)
- *On Scalability, Optimal and Optimized Coarse Spaces*, Martin Gander (University of Geneva, Switzerland)
- *Hierarchical Algorithms on Hierarchical Architectures*, David Keyes (King Abdullah University of Science and Technology, Saudi Arabia)
- *Thoughts on Composing of Nonlinear Solvers*, Matthew Knepley (University at Buffalo, USA)
- *Finite Elements Methods for multiscale problems, and related issues*, Claude Le Bris (Ecole Nationale des Ponts et Chaussées and INRIA, France)
- *Reducing flops, communication and synchronization in sparse factorizations*, Xiaoye (Sherry) Li (Lawrence Berkeley National Laboratory, USA)
- *An algebraic view on BDDC - from local estimates to eigenvalue problems, parallel sums and deluxe scaling*, Clemens Pechstein (CST GmbH, Germany)
- *Domain decomposition tips the scales: From additive Schwarz methods to homogenization*, Daniel Peterseim (University of Augsburg, Germany)
- *Discretizations based on BDDC/FETI-DP Techniques*, Marcus Sarkis (Worcester Polytechnic Institute, USA)
- *Unified Analysis of Iterative Methods Based on One-Way Domain Decomposition*, Hui Zhang (Zhejiang Ocean University, China)

Acknowledgments

The organizers would like to thank all the participants for their enthusiasm and carefully prepared contributions that made this meeting a very successful event, both scientifically and socially. A warm thanks also to our sponsors that made the budget come together.

St. John's, June 2020.

Xiao-Chuan Cai
University of Colorado, Boulder, USA

Laurence Halpern
Université Paris 13, France

Ronald D. Haynes
Memorial University, Canada

Hyea Hyun Kim
Kyung Hee University, Korea

Axel Klawonn
University of Cologne, Germany

Scott MacLachlan
Memorial University, Canada

Olof Widlund
New York University, USA

Contents

List of Contributors

Elyes Ahmed
Inria, 2 rue Simone iff, 75589 Paris, France; (current address) Department of Mathematics, University of Bergen, P. O. Box 7800, N-5020 Bergen, Norway, e-mail: Elyes.Ahmed@uib.no

Katia Ait Ameur
Laboratoire Jacques Louis Lions (LJLL), Sorbonne Université, 75005 Paris, France C.E.A, CEA Saclay - DEN/DANS/DM2S/STMF/LMES - 91191 Gif-Sur-Yvette Cedex, France, e-mail: aitameur.katia@gmail.com

Blanca Ayuso de Dios
Dipartimento di Matematica e Applicazioni, Università di Milano-Bicocca, Milan, Italy, e-mail: blanca.ayuso@unimib.it

Philipp Birken
Centre for Mathematical Sciences, Lund University, Box 118, 22100, Lund, Sweden, e-mail: philipp.birken@na.lu.se

O. Bodart
The Lyon University, Université Jean Monnet Saint-Étienne, CNRS UMR 5208, Institut Camille Jordan, F-42023 Saint-Etienne, France, e-mail: olivier.bodart@univ-st-etienne.fr

Erik G. Boman
Center for Computing Research, Sandia National Laboratories, Albuquerque, New Mexico 87185, USA, e-mail: egboman@sandia.gov

Susanne C. Brenner
Department of Mathematics and Center for Computation and Technology, Louisiana State University, Baton Rouge, LA 70803, USA, e-mail: brenner@math.lsu.edu

Romain Brunet
Department of Mathematics and Statistics, University of Strathclyde, United Kingdom, e-mail: romain.brunet@strath.ac.uk

Xiao-Chuan Cai
Department of Computer Science, University of Colorado Boulder, Boulder, USA, e-mail: cai@cs.colorado.edu

J. G. Calvo
Centro de Investigación en Matemática Pura y Aplicada – Escuela de Matemática, Universidad de Costa Rica, San José, Costa Rica, e-mail: juan.calvo@ucr.ac.cr

Valérie Cayol
Laboratoire Magmas et Volcans, Université Jean Monnet-CNRS-IRD, Saint-Etienne F-42023, France, e-mail: v.cayol@opgc.fr

Faycal Chaouqui
Departement of Mathematics, Temple University, USA, e-mail: Faycal.Chaouqui@temple.edu

Rongliang Chen
Shenzhen Institutes of Advanced Technology, Shenzhen, China, e-mail: rl.chen@siat.ac.cn

A. Chorfi
LIMOS, Université Clermont-Auvergne – CNRS UMR 6158, F-63000 Clermont-Ferrand, France, e-mail: chorfi@isima.fr

Edmond Chow
School of Computational Science and Engineering, College of Computing, Georgia Institute of Technology, Atlanta, Georgia, USA, e-mail: echow@cc.gatech.edu

Eric T. Chung
Department of Mathematics, The Chinese University of Hong Kong, Hong Kong SAR, tschung@math.cuhk.edu.hk

Gabriele Ciaramella
Universität Konstanz, Germany, e-mail: gabriele.ciaramella@uni-konstanz.de

Xavier Claeys
Sorbonne Université, Université Paris-Diderot SPC, CNRS, INRIA, Laboratoire Jacques-Louis Lions, équipe Alpines, 75005 Paris, France, e-mail: claeys@ann.jussieu.fr

P-H. Cocquet
Université de la Réunion, France, e-mail: pierre-henri.cocquet@univ-reunion.fr

Francis Collino
POEMS, CNRS, INRIA, ENSTA Paris, Institut Polytechnique de Paris, 91120 Palaiseau, France, e-mail: francis.collino@orange.fr

Farshid Dabaghi
Laboratoire Magmas et Volcans, Université Jean Monnet-CNRS-IRD, Saint-Etienne F-42023, France e-mail: farshid.dabaghi@univ-st-etienne.fr

Christopher B. Davis,
Foundation Hall 250, Department of Mathematics, Tennessee Tech University, Cookeville, TN 38505, e-mail: CBDavis@tntech.edu

Stefan Dohr
Institute of Applied Mathematics, Graz University of Technology, Steyrergasse 30, 8010 Graz, Austria, e-mail: dohr@math.tugraz.at

Clark R. Dohrmann
Sandia National Laboratories, Albuquerque, New Mexico, U.S.A., e-mail: crdohrm@sandia.gov

Victorita Dolean
Université Côte d'Azur, France, Department of Mathematics and Statistics, University of Strathclyde, United Kingdom, e-mail: victorita.dolean@strath.ac.uk

Fabrizio Donzelli
Memorial University of Newfoundland, Canada, e-mail: fdonzelli@mun.ca

Maksymilian Dryja
Warsaw University,Banacha 2, 00-097 Warsaw, Poland e-mail: dryja@mimuw.edu.pl

Kyle Dunn
Cold Regions Research and Engineering Laboratory, ERDC - U.S. Army, Hanover, NH, e-mail: Kyle.G.Dunn@usace.army.mil

Oliver Fuhrer
Vulcan Inc., e-mail: oliverf@vulcan.com

Martin J. Gander
Université de Genève, Switzerland, e-mail: Martin.Gander@unige.ch

Longfei Gao
Division of Computer, Electrical and Mathematical Sciences and Engineering, King Abdullah University of Science and Technology, Thuwal 23955-6900, Saudi Arabia, e-mail: longfei.gao@kaust.edu.sa

Christian Glusa
Center for Computing Research, Sandia National Laboratories, Albuquerque, New Mexico 87185, USA, e-mail: caglusa@sandia.gov

Yaguang Gu
Department of Mathematics, Hong Kong Baptist University, Kowloon Tong, Hong Kong, e-mail: 16482980@life.hkbu.edu.hk

Laurence Halpern
Université Paris 13, France, e-mail: halpern@math.univ-paris13.fr

Muhammad Hassan
RWTH Aachen University, Germany, e-mail: hassan@mathcces.rwth-aachen.de

Ronald D. Haynes
Memorial University, St. John's, Newfoundland, Canada, e-mail: rhaynes@mun.ca

Alexander Heinlein
Department of Mathematics and Computer Science and Center for Data and Simulation Science, University of Cologne, Weyertal 86-90, 50931 Köln, Germany, e-mail: alexander.heinlein@uni-koeln.de

Julian Hennicker
Université de Genève, 2-4 rue du Lièvre, CP 64, CH-1211 Genève, Switzerland, e-mail: julian.hennicker@unige.ch

Richard M. Höfer
University of Bonn, Germany, e-mail: hoefer@iam.uni-bonn.de

Caroline Japhet
Université Paris 13, Sorbonne Paris Cité, LAGA, CNRS(UMR 7539), 93430, Villetaneuse, France, e-mail: japhet@math.univ-paris13.fr

Patrick Joly
POEMS, CNRS, INRIA, ENSTA Paris, Institut Polytechnique de Paris, 91120 Palaiseau, France, e-mail: patrick.joly@inria.fr

Michel Kern
Inria, 2 rue Simone iff, 75589 Paris, France, and Université Paris-Est, CERMICS (ENPC), 77455 Marne-la-Vallée 2, France, e-mail: michel.kern@inria.fr,

David Keyes
Division of Computer, Electrical and Mathematical Sciences and Engineering, King Abdullah University of Science and Technology, Thuwal 23955-6900, Saudi Arabia, e-mail: david.keyes@kaust.edu.sa

Hyea Hyun Kim
Department of Applied Mathematics and Institute of Natural Sciences, Kyung Hee University, Korea, hhkim@khu.ac.kr

Axel Klawonn
Department of Mathematics and Computer Science and Center for Data and Simulation Science, University of Cologne, Weyertal 86-90, 50931 Köln, Germany, e-mail: axel.klawonn@uni-koeln.de

Jonas Koko
LIMOS, UMR 6158, Université Clermont Auvergne, BP 10448, F-63173 Aubière Cedex, France e-mail: jonas.koko@uca.fr

Piotr Krzyżanowski
University of Warsaw, Poland, e-mail: p.krzyzanowski@mimuw.edu.pl

Martin J. Kühn
CERFACS (Centre Européen de Recherche et de Formation Avancée en Calcul Scientifique), 42 Avenue Gaspard Coriolis, 31057 Toulouse Cedex 01, France, e-mail: martin.kuehn@cerfacs.fr

Iryna Kulchytska-Ruchka
Institut für Teilchenbeschleunigung und Elektromagnetische Felder, Technische Universität Darmstadt, Schlossgartenstrasse 8, D-64289 Darmstadt, Germany, e-mail: kulchytska@gsc.tu-darmstadt.de

Pratik M. Kumbhar
Section de Mathématiques, Université de Genève, Switzerland, e-mail: pratik.kumbhar@unige.ch

Felix Kwok
Department of Mathematics, Hong Kong Baptist University, Kowloon Tong, Hong Kong e-mail: felix_kwok@hkbu.edu.hk

Ulrich Langer
Johann Radon Institute, Altenberg Strasse 69, 4040 Linz, Austria, e-mail: ulrich.langer@ricam.oeaw.ac.at

Martin Lanser
Department of Mathematics and Computer Science and Center for Data and Simulation Science, University of Cologne, Weyertal 86-90, 50931 Köln, Germany, e-mail: martin.lanser@uni-koeln.de

Michael C. Leistner,
Technical University of Munich, Department of Mechanical Engineering, Boltzmannstr. 15 D - 85748 Garching, Germany, e-mail: m.leistner@tum.de

Yingjie Liu
School of Mathematics, Georgia Institute of Technology, Atlanta, GA 30332, USA, e-mail: yingjie@math.gatech.edu

Dalibor Lukas
Department of Applied Mathematics, VŠB – Technical University of Ostrava, 17. listopadu 2172/15, 708 00 Ostrava-Poruba, Czech Republic, e-mail: dalibor.lukas@vsb.cz

Thibaut Lunet
University of Geneva, 2-4 rue du Lièvre, 1211 Genève 4, Suisse, e-mail: thibaut.lunet@unige.ch

Li Luo
Extreme Computing Research Center, King Abdullah University of Science and Technology, Thuwal, Saudi Arabia, e-mail: li.luo@kaust.edu.sa

Yvon Maday
Laboratoire Jacques Louis Lions (LJLL), Sorbonne Université, 75005 Paris, France, and Institut Universitaire de France, e-mail: maday@ann.jussieu.fr

Alexandre Madureira
Laboratório Nacional Computação Científica, Petrópolis, Brazil, e-mail: alm@lncc.br

Lukas Maly
IT4Innovations and Department of Applied Mathematics, VŠB – Technical University of Ostrava, 17. listopadu 2172/15, 708 00 Ostrava-Poruba, Czech Republic, e-mail: lukas.maly@vsb.cz

Parisa Mamooler
Université de Genève, Section de mathématiques, Switzerland, e-mail: Parisa.Mamooler@unige.ch

Leszek Marcinkowski
Faculty of Mathematics, University of Warsaw, Banacha 2, 02-097 Warszawa, Poland, e-mail: Leszek.Marcinkowski@mimuw.edu.pl

Roland Masson
Université Côte d'Azur, CNRS, Inria team COFFEE, LJAD, France, e-mail: roland.masson@unice.fr

Ian May
Simon Fraser University, 8888 University Dr, Burnaby, BC V5A 1S6, Canada, e-mail: mayianm@sfu.ca

Michal Merta
IT4Innovations and Department of Applied Mathematics, VŠB – Technical University of Ostrava, 17. listopadu 2172/15, 708 33 Ostrava-Poruba, Czech Republic, e-mail: michal.merta@vsb.cz

Khaled Mohammad
Memorial University, St. John's, Newfoundland, Canada, e-mail: km2605@mun.ca

Azahar Monge
Centre for Mathematical Sciences, Lund University, Box 118, 22100, Lund, Sweden, and DeustoTech, University of Deusto, Avenida Universidades 24, 48007, Bilbao, Spain, e-mail: azahar.monge@deusto.es

Sandie Moody
University of Geneva, Switzerland, e-mail: Sandie.Moody@unige.ch

Günther Of
Institute of Applied Mathematics, Graz University of Technology, Steyrergasse 30, 8010 Graz, Austria, e-mail: of@tugraz.at

Eun-Hee Park
School of General Studies, Kangwon National University, Samcheok, Gangwon 25913, Republic of Korea, e-mail: eh.park@kangwon.ac.kr

Emile Parolin
POEMS, CNRS, INRIA, ENSTA Paris, Institut Polytechnique de Paris, 91120 Palaiseau, France, e-mail: emile.parolin@inria.fr

Luca F. Pavarino
Dipartimento di Matematica, Università degli Studi di Pavia, Via Ferrata 5, 27100 Pavia, Italy, e-mail: luca.pavarino@unipv.it

Clemens Pechstein
Dassault Systèmes Austria GmbH, Semmelweisstraße 15, 4600 Wels, Austria, e-mail: clemens.pechstein@3ds.com

Daniel Peterseim
Universität Augsburg, Germany, e-mail: daniel.peterseim@math.uni-augsburg.de

Kendall H. Pierson
Sandia National Laboratories, Albuquerque, New Mexico, U.S.A., e-mail: khpiers@sandia.gov

Talal Rahman
Faculty of Engineering and Science, Western Norway University of Applied Sciences, Inndalsveien 28, 5063 Bergen, Norway, e-mail: Talal.Rahman@hvl.no

Sivasankaran Rajamanickam
Center for Computing Research, Sandia National Laboratories, Albuquerque, New Mexico 87185, USA, e-mail: srajama@sandia.gov

Paritosh Ramanan
School of Computational Science and Engineering, College of Computing, Georgia Institute of Technology, Atlanta, Georgia, USA, e-mail: paritoshpr@gatech.edu

Oliver Rheinbach
INMO, Technische Universität Bergakademie Freiberg, Akademiestr. 6, 09599 Freiberg, Germany, e-mail: oliver.rheinbach@math.tu-freiberg.de

Daniel J. Rixen,
Technical University of Munich, Department of Mechanical Engineering, Boltzmannstr. 15 D - 85748 Garching, Germany, e-mail: rixen@tum.de

Friederike Röver
Institut für Numerische Mathematik und Optimierung, Technische Universität Bergakademie Freiberg, Akademiestr. 6, 09599 Freiberg, Germany, e-mail:

friederike.roever@math.tu-freiberg.de

Albert E. Ruehli
EMC Laboratory, Missouri University of Science And Technology, U.S, e-mail: ruehlia@mst.edu

Steven Ruuth
Simon Fraser University, 8888 University Dr, Burnaby, BC V5A 1S6, Canada, e-mail: sruuth@sfu.ca

Kévin Santugini-Repiquet
Bordeaux INP, IMB, UMR 5251, F-33400,Talence, France, e-mail: Kevin.Santugini-Repiquet@bordeaux-inp.fr

Marcus Sarkis
Mathematical Sciences Department, Worcester Polytechnic Institute, MA, USA, e-mail: msarkis@wpi.edu

Simone Scacchi
Dipartimento di Matematica, Università degli Studi di Milano, Via Saldini 50, 20133 Milano, Italy, e-mail: simone.scacchi@unimi.it

Sebastian Schöps
Institut für Teilchenbeschleunigung und Elektromagnetische Felder, Technische Universität Darmstadt, Schlossgartenstrasse 8, D-64289 Darmstadt, Germany, e-mail: schoeps@temf.tu-darmstadt.de

Andreas S. Seibold,
Technical University of Munich, Department of Mechanical Engineering, Boltzmannstr. 15 D - 85748 Garching, Germany, e-mail: andreas.seibold@tum.de

Benjamin Stamm
RWTH Aachen University, Germany, e-mail: stamm@mathcces.rwth-aachen.de

Olaf Steinbach
Institute of Applied Mathematics, Graz University of Technology, Steyrergasse 30, 8010 Graz, Austria, e-mail: o.steinbach@tugraz.at

Li-Yeng Sung
Department of Mathematics and Center for Computation and Technology, Louisiana State University, Baton Rouge, LA 70803, USA, e-mail: sung@math.lsu.edu

Marc Tajchman
C.E.A, CEA Saclay - DEN/DANS/DM2S/STMF/LMES - 91191 Gif-Sur-Yvette Cedex, France, e-mail: marc.tajchman@cea.fr

Hansong Tang
Department of Civil Engineering, City College, City University of New York, NY 10031, USA, e-mail: htang@ccny.cuny.edu

Xuemin Tu
Department of Mathematics, University of Kansas, 1460 Jayhawk Blvd, Lawrence, KS 66045, U.S.A., e-mail: xuemin@ku.edu

Serge Van Criekingen
CNRS/IDRIS, France, e-mail: serge.van.criekingen@idris.fr

Tommaso Vanzan
Section de mathématiques, Université de Genève, 2-4 rue du Lièvre, Genève, Switzerland, e-mail: tommaso.vanzan@unige.ch

Dora Varga
Universität Augsburg, Germany, e-mail: dora.varga@math.uni-augsburg.de

Barbara Verfürth
Universität Augsburg, Germany, e-mail: barbara.verfuerth@math.uni-augsburg.de

Bin Wang
Department of Mathematical Sciences, Hood College, 401 Roesmont Ave., Frederick, MD 21701, U.S.A., e-mail: wang@hood.edu

Junxian Wang
School of Mathematics and Computational Science, Xiangtan University, China, wangjunxian@xtu.edu.cn

Kening Wang
Department of Mathematics and Statistics, University of North Florida, Jacksonville, FL 32224, USA, e-mail: kening.wang@unf.edu

Janine Weber
Department of Mathematics and Computer Science, University of Cologne, Weyertal 86-90, 50931 Köln, Germany, e-mail: janine.weber@uni-koeln.de

Olof B. Widlund
Courant Institute, 251 Mercer Street, New York, NY 10012, USA, e-mail: widlund@cims.nyu.edu

X. Xiang
Qian Xuesen Laboratory of Space Technology, Beijing, China, e-mail: xiangxueshuang@qxslab.cn

Huidong Yang
Johann Radon Institute, Altenberg Strasse 69, 4040 Linz, Austria, e-mail: huidong.yang@oeaw.ac.at

Yi Yu
Worcester Polytechnic Institute, 100 Institute Rd, Worcester, USA, e-mail: yyu5@wpi.edu

Stefano Zampini
Extreme Computing Research Center, King Abdullah University of Science and Technology, Thuwal 23955, Saudi Arabia, e-mail: stefano.zampini@kaust.edu.sa

Jan Zapletal
IT4Innovations and Department of Applied Mathematics, VŠB – Technical University of Ostrava, 17. listopadu 2172/15, 708 33 Ostrava-Poruba, Czech Republic, e-mail: jan.zapletal@vsb.cz

Hui Zhang
Xi'an Jiaotong-Liverpool University, Department of Mathematical Sciences, Suzhou 215123, China; Zhejiang Ocean University, Key Laboratory of Oceanographic Big Data Mining & Application of Zhejiang Province, Zhoushan 316022, China, e-mail: mike.hui.zhang@hotmail.com

Yunrong Zhu
Department of Mathematics & Statistics, Idaho State University, 921 S. 8th Ave., Stop 8085 Pocatello, ID 83209, USA. e-mail: zhuyunr@isu.edu

Part I
Plenary Talks (PT)

Does the Partition of Unity Influence the Convergence of Schwarz Methods?

Martin J. Gander

1 Which Schwarz Methods Need a Partition of Unity?

The classical alternating Schwarz method does not need a partition of unity in its definition [3]: one solves one subdomain after the other, stores subdomain solutions, and always uses the newest data available from the neighboring subdomains. In the parallel Schwarz method introduced by Lions in [10], where all subdomains are solved simultaneously, one also stores subdomain solutions, but one has to distinguish two cases: if in the decomposition there are never more than two subdomains that intersect, which we call the *no crosspoint assumption*, then one also does not need a partition of unity to define the method, one simply takes data from the neighboring subdomains with which the subdomain intersects, and in that case the parallel Schwarz method has a variational interpretation [10]. If however points of the boundary of a subdomain are contained in more than one neighboring subdo main, then one has to decide from which neighboring subdomain to take data, or one can use a linear combination. In this case, the parallel Schwarz method does not have a variational interpretation [10], for an example, see Figure 2.2 in [3]. The decision from which of the neighboring subdomains data should be taken has to be made only on the boundary of each subdomain, and by the maximum principle, it is better to take data as far away from the boundary of the neighboring subdomains as possible to benefit from the largest error decay. This can be achieved if the overlapping domain decomposition is obtained from a non-overlapping one by enlarging the non-overlapping subdomains equally, and then using the subdomain solutions restricted to the non-overlapping subdomains to define a global approximation from which data is taken for the next iteration, see [3] for more details.

The situation for the algebraic Schwarz methods is more delicate, since these methods define approximate iterates over the entire domain only, so in the overlap, where necessarily more than one iterate is available, one has to decide which one or

M. J. Gander
Université de Genève, Switzerland e-mail: Martin.Gander@unige.ch

R. Haynes et al. (eds.), *Domain Decomposition Methods in Science and Engineering XXV*, Lecture Notes in Computational Science and Engineering 138,
https://doi.org/10.1007/978-3-030-56750-7_1

which combination should be stored. For the multiplicative Schwarz method, which was proved to be equivalent to the discretization of the alternating Schwarz method, see [3], one also does not need a partition of unity, since the method stores the most recently updated values in the overlaps. Additive Schwarz does also not use a partition of unity, it adds all contributions in the overlap, which however leads to a non convergent stationary iterative method [3]. Additive Schwarz is thus not equivalent to a discretization of the parallel Schwarz method of Lions [3], it has to be used as a preconditioner for a Krylov method, which corrects the error made by Additive Schwarz adding all contributions in the overlap. In RAS [1], implicitly a partition of unity was defined by "neglecting part of the communication routine", but any other partition of unity could be used as well. A natural question is if the choice of the partition of unity influences the convergence properties of RAS. It was proved in [3] that RAS is equivalent to the discretization of the parallel Schwarz method of Lions, if the parallel Schwarz method of Lions uses as partition of unity the restriction to the non-overlapping domain decomposition, as described above. Similarly, it was shown in [9] that Additive Schwarz with Harmonic extension [1] is also equivalent to the discretization of the parallel Schwarz method of Lions, but only under certain restrictions on the decomposition. Finally, also a variant called Restricted Additive Schwarz with Harmonic extension (RASH) was introduced in [1], but it was found to have less good convergence properties, even though RASH is symmetric for symmetric problems, while RAS and ASH are not.

Optimized transmission conditions [2] were introduced for RAS in [12, 11] leading the Optimized Restricted Additive Schwarz method (ORAS), and also for ASH leading to OASH [9], and in both cases a direct equivalence to discretized optimized Schwarz methods was proved. A symmetric variant ORASH was proposed in [5] (under the name SORAS), which needs a special coarse correction to permit a convergence analysis of the method using the abstract Schwarz framework. The symmetrized version ORASH has also been studied again with radiation transmission conditions for the Helmholtz case in [4], see also the earlier work for Helmholtz in [7, 8] for a BDD variant with overlap.

We are interested in understanding if the choice of partition of unity influences the convergence of RAS and RASH and their optimized variants ORAS and ORASH. We will prove that in the two subdomain case the choice of partition of unity has no influence on the convergence properties of RAS, and ORAS under an additional condition on the partition of unity, while RASH and ORASH are extremely sensitive to the choice of the partition of unity. The main reason for this is that RAS and ORAS are equivalent to classical and optimized parallel Schwarz methods, while RASH and ORASH have no such interpretation as iterative domain decomposition methods, and generate an extra residual term which we compute explicitly. We also investigate the many subdomain case, including cross points, and show numerically that the partition of unity in the presence of cross points has the same weak influence on the functioning of RAS as on the equivalent discretized parallel Schwarz method of Lions. RASH however is extremely sensitive, and its convergence properties are much less favorable than the convergence properties of RAS.

2 Partitions of Unity for RAS, RASH, ORAS and ORASH

To get a better understanding on how information is transmitted between subdomains in RAS, ORAS, RASH and ORASH, and how this is influenced by the partition of unity used, we consider as our model problem

$$(\eta - \Delta)u = f \text{ in } \Omega, \quad u = 0 \text{ on } \partial\Omega, \tag{1}$$

where $\eta \geq 0$ is a parameter, $\Omega \subset \mathbb{R}^2$ and f is some given source function $f : \Omega \to \mathbb{R}$. We discretize (1) using a finite element or finite difference method and obtain the linear system

$$A\mathbf{u} = \mathbf{f}, \tag{2}$$

where $A \in \mathbb{R}^{m\times m}$ is the system matrix, $\mathbf{f} \in \mathbb{R}^m$ is the discretization of the source term, and $\mathbf{u} \in \mathbb{R}^m$ is an approximation of the solution at the grid points. Schwarz methods are based on a decomposition of the domain Ω into overlapping subdomains Ω_j, $j = 1, 2, \ldots, J$. At the discrete level this decomposition can be identified with a decomposition of the degrees of freedom of the discrete system (2) into a set of overlapping or non-overlapping subsets[1], and is most easily represented by restriction matrices R_j of size $m_j \times m$ which are restrictions of the identity matrix to the rows corresponding to the degrees of freedom in subdomain Ω_j. The restriction operators R_j can also be used to define the subdomain matrices $A_j := R_j A R_j^T$, which correspond to subdomain problems with Dirichlet transmission conditions. We next define a discrete partition of unity represented by diagonal matrices $\chi_j \in \mathbb{R}^{m\times m}$ such that $\sum_{j=1}^J \chi_j = I$, the identity, and which equal one on the diagonal for degrees of freedom that belong to one subdomain only. This discrete partition of unity can conveniently be used to define also modified restriction matrices $\widetilde{R}_j := R_j \chi_j$, and the classical choice we have seen in Section 1 is to use a non-overlapping partition to define the χ_j, which leads to $\widetilde{R}_j$ matrices that still only contain zeros and ones, see [1]. There are however also other possibilities, and we define in particular the five partition of unity functions χ_j^ℓ, $\ell = 1, 2, 3, 4, 5$ shown in Figure 1. The first one is the one used in RAS. The second one computes the average of the two subdomain solutions in the overlap. The third one takes a linear combination, weighted by a linear function depending on the distance from the interfaces, and the fourth and fifth one are spline functions, the last one staying longer close to one on the boundary then the former. Each partition of unity function χ_j^ℓ leads to an associated restriction matrix $\widetilde{R}_j^\ell := R_j \chi_j^\ell$ for $\ell = 1, 2, 3, 4, 5$.

We can now define the discrete Schwarz methods RAS, RASH, ORAS and ORASH by defining the preconditioning matrix M^{-1} in the stationary iterative method

$$\mathbf{u}^n = \mathbf{u}^{n-1} + M^{-1}(\mathbf{f} - A\mathbf{u}^{n-1}), \tag{3}$$

[1] For a detailed explanation why a non-overlapping decomposition at the algebraic level still implies an overlapping decomposition at the continuous level for classical finite element and finite difference methods, see [3, Section 3]

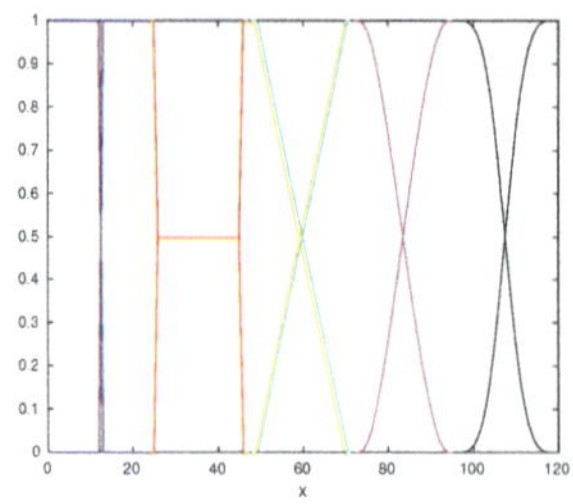

Fig. 1: Five partitions of unity functions we will test, shown in one dimension across a typical overlap size.

$$M_{RAS_\ell}^{-1} = \sum_{j=1}^{J} (\widetilde{R}_j^\ell)^T A_j^{-1} R_j, \quad M_{RASH_\ell}^{-1} = \sum_{j=1}^{J} (\widetilde{R}_j^\ell)^T A_j^{-1} \widetilde{R}_j^\ell,$$

$$M_{ORAS_\ell}^{-1} = \sum_{j=1}^{J} (\widetilde{R}_j^\ell)^T \widetilde{A}_j^{-1} R_j, \text{ and } M_{ORASH_\ell}^{-1} = \sum_{j=1}^{J} (\widetilde{R}_j^\ell)^T \widetilde{A}_j^{-1} \widetilde{R}_j^\ell,$$

where the subdomain matrices $\widetilde{A}_j$ correspond to subdomain problems with Robin transmission conditions, see [11].

3 Influence of the Partition of Unity on RAS and RASH in 1D

We start by a numerical experiment in one spatial dimension: we use $\Omega := (0, 1)$, two subdomains $\Omega_1 := (0, \beta)$ and $\Omega_2 := (\alpha, 1)$ with $\alpha < \beta$ and solve the model problem in (1) for $\eta = 0$ with boundary conditions $u(0) = 0$ and $u(1) = 1$, so that the solution is a straight line going from zero to one. We discretize the problem using centered finite differences with $m = 100$ interior mesh points, which leads to the mesh size $h = \frac{1}{m+1}$, and we assign the first b mesh points to the first subdomain matrix, and the last $m-a$ mesh points to the second subdomain matrix, which implies that $A_1 \in \mathbb{R}^{b\times b}$, $A_2 \in \mathbb{R}^{m-a\times m-a}$, $R_1^\ell \in \mathbb{R}^{b\times m}$ $R_2^\ell \in \mathbb{R}^{m-a\times m}$ and that $\alpha = ah$ and $\beta = (b+1)h$. We choose $a = 40$ and $b = 60$. The parallel Schwarz method of Lions, which does not need a partition of unity in the case of two subdomains, would then compute

$$A_1 \begin{bmatrix} u_{1,1}^n \\ \vdots \\ u_{1,b-1}^n \\ u_{1,b}^n \end{bmatrix} = \begin{bmatrix} f_1 \\ \vdots \\ f_{b-1} \\ f_b - \frac{1}{h^2} u_{2,b+1}^{n-1} \end{bmatrix}, \quad A_2 \begin{bmatrix} u_{2,a+1}^n \\ u_{2,a+2}^n \\ \vdots \\ u_{2,m}^n \end{bmatrix} = \begin{bmatrix} f_{a+1} - \frac{1}{h^2} u_{1,a}^{n-1} \\ f_{a+2} \\ \vdots \\ f_m \end{bmatrix}. \tag{4}$$

We show in Figure 2 in the first five panels the iterates of RAS and RASH when using the five partition of unity functions χ^ℓ. Note that the iterates of the parallel Schwarz

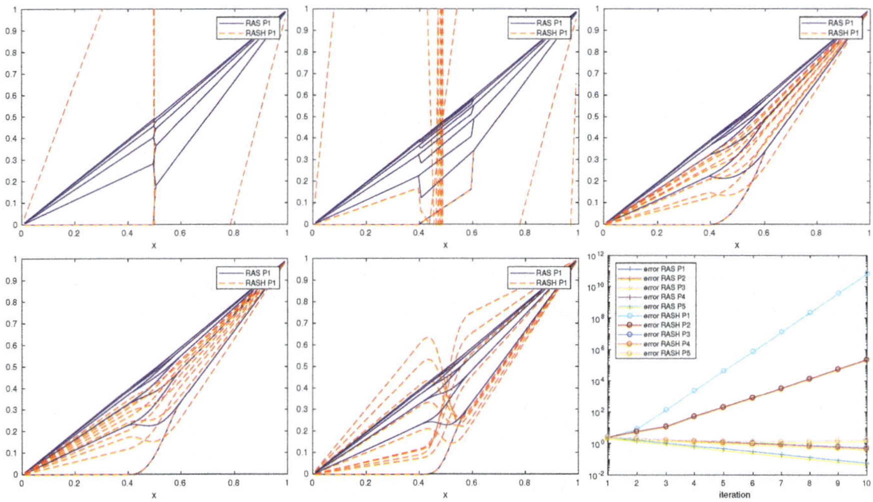

Fig. 2: Iterates of RAS (blue) and RASH (red) for the five partitions of unity, and corresponding convergence curves.

method would just converge monotonically from below to the solution which is a straight line from zero to one. We see that RAS (in solid blue) is converging in the same way for all five partition of unity functions, we only see a difference in the overlap depending on the partition of unity used. RASH however (in dashed red) is diverging violently for the first two partition of unity functions χ^1 and χ^2, converging, albeit more slowly than RAS for the partition of unity functions χ^3 and χ^4, and diverging again for the partition of unity function χ^5. The corresponding convergence curves are shown in the last panel in Figure 2, and we see indeed that RAS converges at the same rate for all partition of unity functions, while RASH only converges for two, and is substantially slower than RAS.

We now prove that the convergence of RAS does not depend on the choice of the partition of unity function, and that RAS is a faithful implementation of the parallel Schwarz method of Lions.

Theorem 1 (The convergence of RAS$_\ell$ does not depend on the partition of unity used) *If the initial iterate* $\mathbf{u}^0$ *of RAS$_\ell$ satisfies* $u_a^0 = u_{2,a}^0$ *and* $u_{b+1}^0 = u_{1,b+1}^0$, *where* $u_{2,a}^0$ *and* u_{b+1}^0 *are the initial guess of the parallel Schwarz method of Lions* (4), *then the iterates of RAS$_\ell$ outside the overlap coincide with the iterates of the discretized parallel Schwarz method of Lions* (4), $u_j^n = u_{1,j}^n$, $j \in \{1, 2, \ldots a\} \cup \{b+1, b+2, \ldots, m\}$, *independently of the partition of unity* χ^ℓ *used in RAS$_\ell$.*

Proof The proof is by induction: according to the iteration formula for RAS$_\ell$ in (3), one first computes the residual $\mathbf{f} - A\mathbf{u}^0$, which can be written partitioned into two parts in two different ways,

$$\begin{bmatrix} f_1 \\ \vdots \\ f_b \\ f_{b+1} \\ \vdots \\ f_m \end{bmatrix} - \begin{bmatrix} A_1 \begin{bmatrix} u_1^0 \\ \vdots \\ u_b^0 \end{bmatrix} + \begin{bmatrix} 0 \\ \vdots \\ \frac{1}{h^2} u_{b+1}^0 \end{bmatrix} \\ B_1 \begin{bmatrix} u_{b+1}^0 \\ \vdots \\ u_m^0 \end{bmatrix} + \begin{bmatrix} \frac{1}{h^2} u_b^0 \\ 0 \\ \vdots \end{bmatrix} \end{bmatrix} = \begin{bmatrix} f_1 \\ \vdots \\ f_a \\ f_{a+1} \\ \vdots \\ f_m \end{bmatrix} - \begin{bmatrix} B_2 \begin{bmatrix} u_1^0 \\ \vdots \\ u_a^0 \end{bmatrix} + \begin{bmatrix} 0 \\ \vdots \\ \frac{1}{h^2} u_{a+1}^0 \end{bmatrix} \\ A_2 \begin{bmatrix} u_{a+1}^0 \\ \vdots \\ u_m^0 \end{bmatrix} + \begin{bmatrix} \frac{1}{h^2} u_a^0 \\ 0 \\ \vdots \end{bmatrix} \end{bmatrix},$$

where B_j is the remaining diagonal block for the subdomain matrix A_j and of no importance, since the following restriction step in RAS$_\ell$ removes it,

$$R_1(\mathbf{f} - A\mathbf{u}^0) = \begin{bmatrix} f_1 \\ \vdots \\ f_b - \frac{1}{h^2} u_{b+1}^0 \end{bmatrix} - A_1 \begin{bmatrix} u_1^0 \\ \vdots \\ u_b^0 \end{bmatrix}, \quad R_2(\mathbf{f} - A\mathbf{u}^0) = \begin{bmatrix} f_{a+1} - \frac{1}{h^2} u_a^0 \\ \vdots \\ f_m \end{bmatrix} - A_2 \begin{bmatrix} u_{a+1}^0 \\ \vdots \\ u_m^0 \end{bmatrix}. \tag{5}$$

Next the subdomain solves A_j^{-1} are applied in parallel, which cancel the remaining A_j matrices,

$$A_1^{-1} R_1(\mathbf{f} - A\mathbf{u}^0) = A_1^{-1} \begin{bmatrix} f_1 \\ \vdots \\ f_b - \frac{1}{h^2} u_{b+1}^0 \end{bmatrix} - \begin{bmatrix} u_1^0 \\ \vdots \\ u_b^0 \end{bmatrix}, \quad A_2^{-1} R_2(\mathbf{f} - A\mathbf{u}^0) = A_2^{-1} \begin{bmatrix} f_{a+1} - \frac{1}{h^2} u_a^0 \\ \vdots \\ f_m \end{bmatrix} - \begin{bmatrix} u_{a+1}^0 \\ \vdots \\ u_m^0 \end{bmatrix}.$$

We now see that due to the assumption of identical starting values, $u_a^0 = u_{2,a}^0$ and $u_{b+1}^0 = u_{1,b+1}^0$, precisely the subdomain solves of the parallel Schwarz method of Lions (4) appeared,

$$\begin{bmatrix} u_{1,1}^1 \\ \vdots \\ u_{1,b}^1 \end{bmatrix} = A_1^{-1} \begin{bmatrix} f_1 \\ \vdots \\ f_b - \frac{1}{h^2} u_{b+1}^0 \end{bmatrix}, \quad \begin{bmatrix} u_{2,a+1}^1 \\ \vdots \\ u_{2,m}^1 \end{bmatrix} = A_2^{-1} \begin{bmatrix} f_{a+1} - \frac{1}{h^2} u_a^0 \\ \vdots \\ f_m \end{bmatrix},$$

and we therefore obtain in the last combination step of RAS$_\ell$

$$\begin{aligned} \mathbf{u}^1 &= \begin{bmatrix} u_1^0 \\ \vdots \\ u_{m-1}^0 \end{bmatrix} + (\widetilde{R}_1^\ell)^T \left(\begin{bmatrix} u_{1,1}^1 \\ \vdots \\ u_{1,b}^1 \end{bmatrix} - \begin{bmatrix} u_1^0 \\ \vdots \\ u_{b-1}^0 \end{bmatrix} \right) + (\widetilde{R}_2^\ell)^T \left(\begin{bmatrix} u_{2,a+1}^1 \\ \vdots \\ u_{2,m}^1 \end{bmatrix} - \begin{bmatrix} u_{a+1}^0 \\ \vdots \\ u_m^0 \end{bmatrix} \right) \\ &= (\widetilde{R}_1^\ell)^T \begin{bmatrix} u_{1,1}^1 \\ \vdots \\ u_{1,b}^1 \end{bmatrix} + (\widetilde{R}_2^\ell)^T \begin{bmatrix} u_{2,a+1}^1 \\ \vdots \\ u_{2,m}^1 \end{bmatrix}, \end{aligned} \tag{6}$$

because the old iterate cancels due to the partition of unity used in the $\widetilde{R}_j^\ell$, and the same property also shows that the new iterate $\mathbf{u}^1$ of RAS$_\ell$ coincides outside the overlap with the parallel Schwarz iterates from (4), and this independently of the partition of unity used. Induction now concludes the proof. □

So why does RASH$_\ell$ fail? This can be seen in step (5), where in RASH$_\ell$ the $\widetilde{R}_j^\ell$ operators would be applied containing the partition of unity. Adding and subtracting $R_j(\mathbf{f} - A\mathbf{u}^0)$, we obtain in RASH$_\ell$

$$\widetilde{R}_1(\mathbf{f} - A\mathbf{u}^0) = \begin{bmatrix} f_1 \\ \vdots \\ f_b - \frac{1}{h^2}u_{b+1}^0 \end{bmatrix} - A_1 \begin{bmatrix} u_1^0 \\ \vdots \\ u_b^0 \end{bmatrix} + (\widetilde{R}_1 - R_1)(\mathbf{f} - A\mathbf{u}^0), \tag{7}$$

$$\widetilde{R}_2(\mathbf{f} - A\mathbf{u}^0) = \begin{bmatrix} f_{a+1} - \frac{1}{h^2}u_a^0 \\ \vdots \\ f_m \end{bmatrix} - A_2 \begin{bmatrix} u_{a+1}^0 \\ \vdots \\ u_m^0 \end{bmatrix} + (\widetilde{R}_2 - R_2)(\mathbf{f} - A\mathbf{u}^0). \tag{8}$$

This implies that in the last combination step of RASH$_\ell$ artificial source terms are left,

$$\mathbf{u}^1 = (\widetilde{R}_1^\ell)^T \begin{bmatrix} u_{1,1}^1 \\ \vdots \\ u_{1,b}^1 \end{bmatrix} + (\widetilde{R}_2^\ell)^T \begin{bmatrix} u_{2,a+1}^1 \\ \vdots \\ u_{2,m}^1 \end{bmatrix} + \left(\sum_{j=1}^{2} \widetilde{R}_j^T A_j^{-1} (\widetilde{R}_j - R_j) \right) (\mathbf{f} - A\mathbf{u}^0). \tag{9}$$

These source terms modify the correct Schwarz iterates, and even though these artificial source terms go to zero when the residual goes to zero, they greatly affect the convergence, and can even lead to divergence, see Figure 2.

4 RAS, RASH, ORAS and ORASH in 2D

The generalization of Theorem 1 to higher spatial dimensions and more than two subdomains does not present any difficulties under the no-crosspoint assumption[2]. As an illustration, we show numerical experiments on the unit square, solving the model problem (1) for $\eta = 0$ using a uniform mesh size $h = \frac{1}{40}$ and two equal subdomains which overlap by $11h$ and $\mathbf{f} = 0$, which means that we simulate directly the error equations. We show in Figure 3 the third iteration starting with the same random initial guess for RAS$_\ell$ (left) and the corresponding results for RASH$_\ell$ (right). As in one spatial dimension, RAS$_\ell$ converges outside of the overlap like the parallel Schwarz method of Lions, only in the overlap one can see the influence of the partition of unity, which does not affect the convergence. This is very different for RASH$_\ell$, where convergence can be completely destroyed by the partition of unity.

[2] Even in the presence of cross points, the equivalence of the discretized parallel Schwarz method of Lions and RAS is proved in [3, Theorem 3.5] for a partition of unity of the form χ^1, the proof for other partitions of unity can be obtained following the arguments in the proof of Theorem 1.

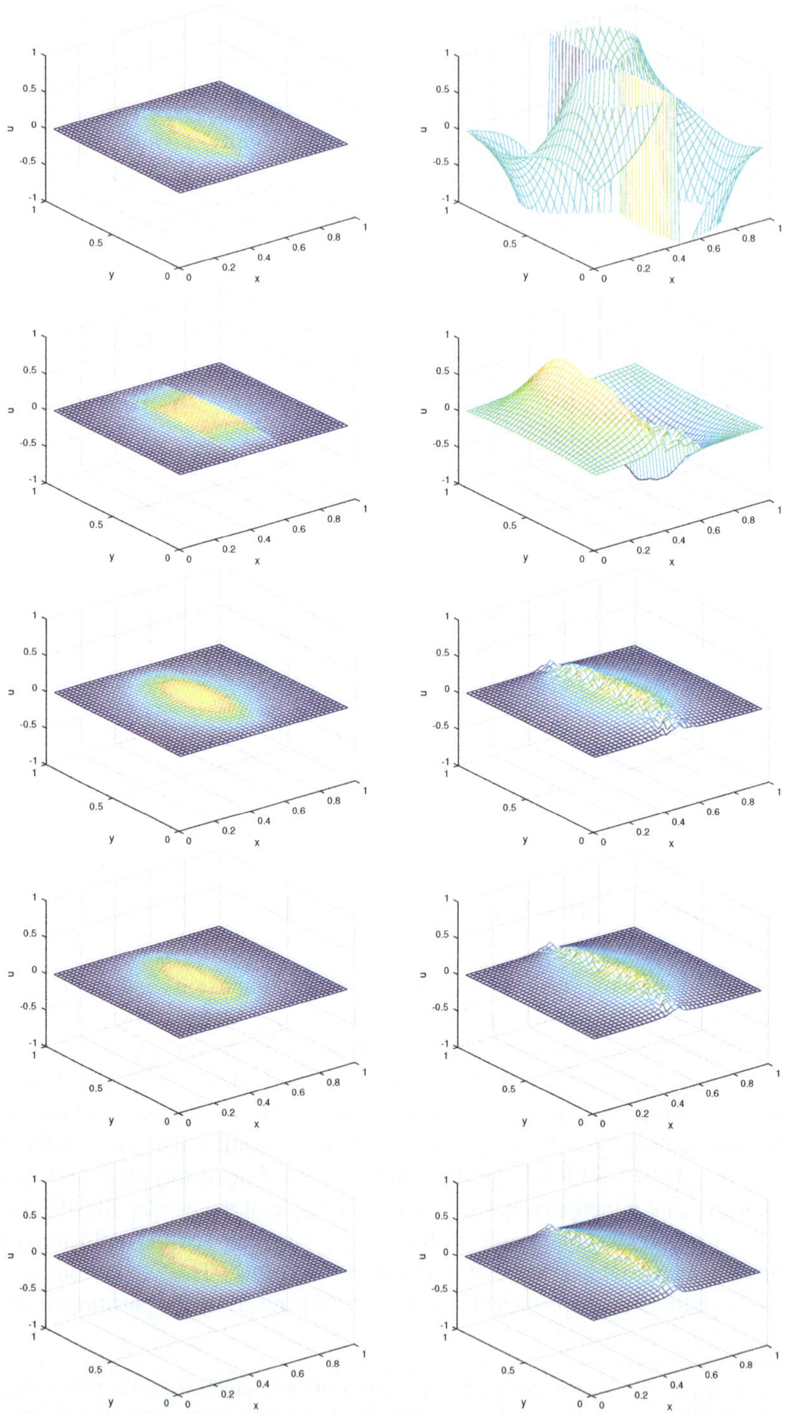

Fig. 3: Third iterate of RAS_ℓ (left) and RASH_ℓ (right) for $\ell = 1, 2, 3, 4, 5$ corresponding to the five different partitions of unity.

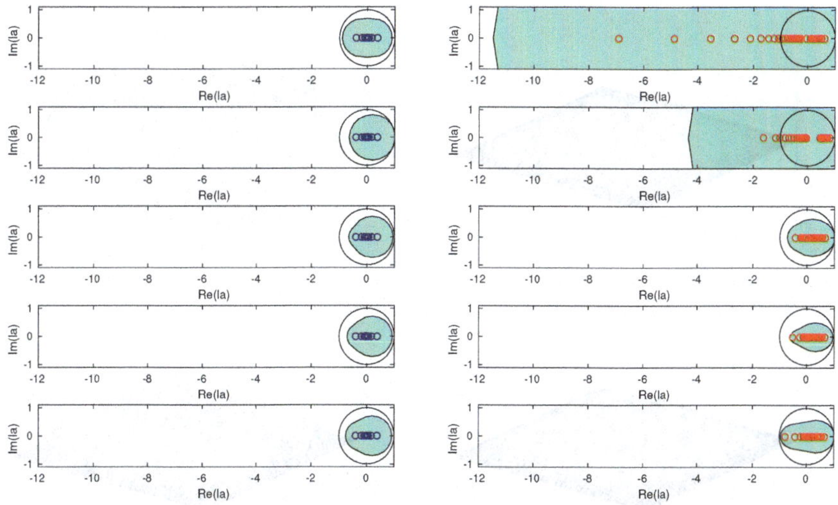

Fig. 4: Spectra and numerical range of the preconditioned operators with RAS_ℓ (left) and RASH_ℓ (right) for $\ell = 1, 2, 3, 4, 5$ corresponding to the five different partitions of unity.

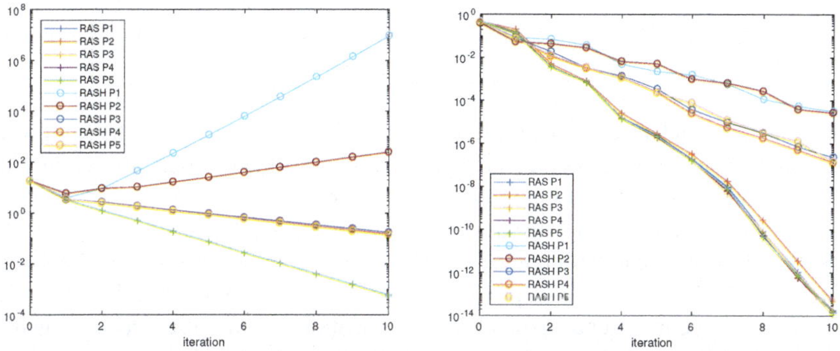

Fig. 5: Convergence of RAS_ℓ and RASH_ℓ as iterative solvers (left) and when used as preconditioners for GMRES (right), for for $\ell = 1, 2, 3, 4, 5$ corresponding to the five different partitions of unity.

In Figure 4 we show the spectra and numerical range of the preconditioned operators. As expected, we see that the spectra of RAS_ℓ are not affected by the partition of unity, while the spectra of RASH_ℓ are: the first two partitions of unity cause large negative eigenvalues, which explain the divergence of the iterative method in this case. The smoother partitions of unity lead to convergent methods, but the spectra are clearly less favorable for convergence. A similar observation holds also for the numerical range which can be related to the convergence of preconditioned GMRES: for RAS_ℓ, the numerical range is very similar, which indicates similar convergence for GMRES, whereas for RASH_ℓ, the first two partitions of unity lead to a much larger numerical range which is unfavorable for GMRES. This is illustrated in Figure 5, where we see on the left that as iterative solver, RAS_ℓ faithfully produces the same convergence behavior of the parallel Schwarz method

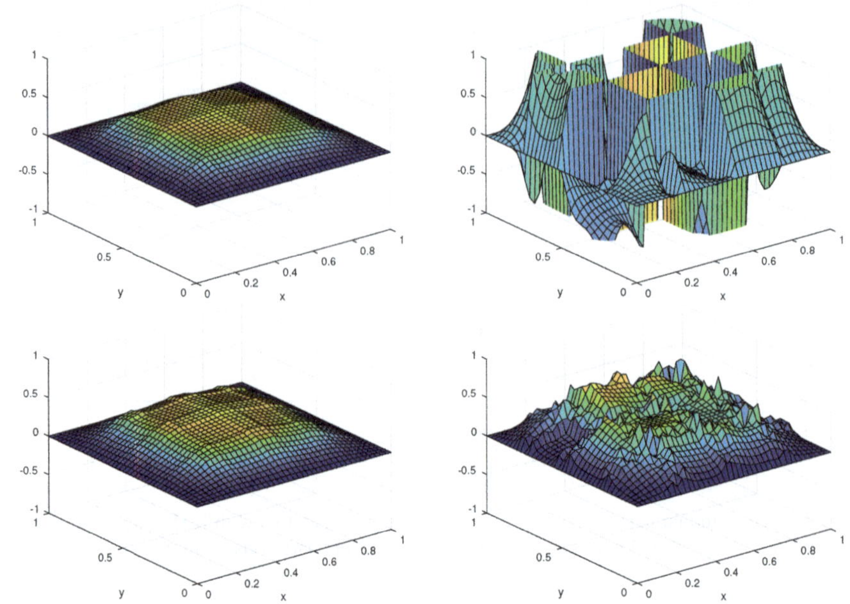

Fig. 6: Third iterate of RAS_ℓ (left) and RASH_ℓ (right) for $\ell = 1, 2$ and the 4×4 subdomain case.

of Lions independently of the partition of unity used, which leads on the right when used as preconditioner to rapid convergence of the residuals, not identical, since the residuals are also minimized in the overlap, where the partition of unity has a slight influence on the numerical range as seen in Figure 4 on the left. This is very different for RASH_ℓ, which can both converge and diverge as an iterative solver, see Figure 5 on the left. When used as a preconditioner, RASH_ℓ is much less effective than RAS_ℓ, and the convergence depends on the partition of unity used: as indicated by the numerical range in Figure 4 on the right, the first two partitions of unity lead to worse convergence of GMRES for RASH_ℓ, see Figure 5 on the right.

We next investigate for the first two partitions of unity the case where cross points are present, namely a decomposition of the unit square into 4×4 subdomains, using the same mesh size $h = \frac{1}{40}$ but a smaller overlap $3h$ to still clearly see the subdomains, see Figure 6. Like the parallel Schwarz method of Lions, RAS depends only little on the partition of unity used, while RASH depends very strongly. In Figure 7 we show the spectra and numerical range of the preconditioned operators, and we see that while RAS_ℓ also is convergent in the presence of cross points, RASH_ℓ is not for the two partitions of unity. We show in Figure 8 (left) the corresponding convergence plots. We observe that in the presence of cross points, the convergence of RAS depends a little on the partition of unity, exactly like the parallel Schwarz method of Lions: the first partition of unity is better than the second one, since it takes data further away from the interfaces, which is better for the Schwarz method by the maximum principle. The dependence of RASH is however very strong: we see violent divergence for the first partition of unity, and also slow divergence for the

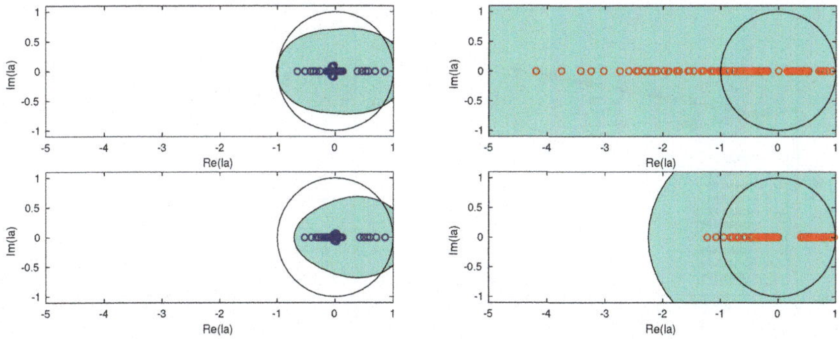

Fig. 7: Spectra and numerical range of the preconditioned operators with RAS_ℓ (left) and RASH_ℓ (right) for $\ell = 1, 2$ and the 4×4 subdomain case.

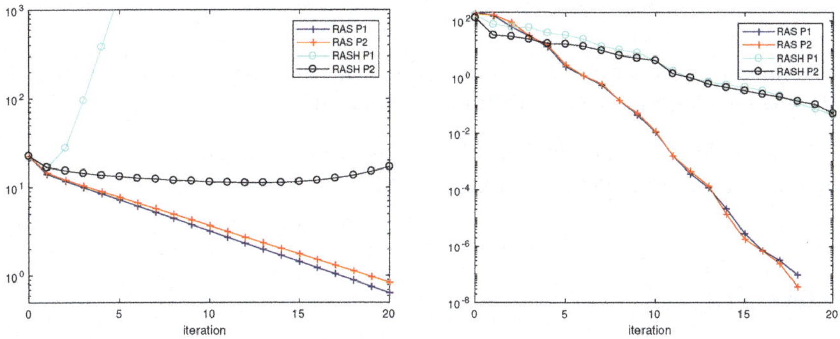

Fig. 8: Convergence of RAS_ℓ and RASH_ℓ as iterative solvers (left) and when used as preconditioners for GMRES (right), for $\ell = 1, 2$ and the 4×4 subdomain case.

second one. The spectrum and numerical range in Figure 7 (right) explains their less favorable properties as preconditioners, see Figure 8 (right).

We finally test the optimized variants ORAS and ORASH: it was shown in [11] that ORAS is a discretization of the optimized Schwarz method with Robin transmission conditions for the partition of unity function χ^1, and this result holds provided the partition of unity equals one at least for the first layer inside the overlap, which is almost satisfied by χ^5 as well, but not for the other partitions of unity. We show in Figure 9 the results corresponding to Figure 5 but now using ORAS_ℓ and ORASH_ℓ. We see that ORAS_1 performs indeed best, like an optimized Schwarz method and much better than RAS_ℓ. ORAS_5 also works, but ORAS_2, ORAS_3 and ORAS_4 are now not functioning properly, since the partition of unity overwrites the location where derivative information needs to be extracted. ORASH_ℓ never works properly, which then also leads to very poor performance when used as a preconditioner, see Figure 9 on the right, even worse than RAS_ℓ, and only marginally better than RASH_ℓ. It is therefore delicate to use the symmetrized versions RASH_ℓ and ORASH_ℓ, and for ORAS_ℓ the partition of unity needs to satisfy a constraint. Note

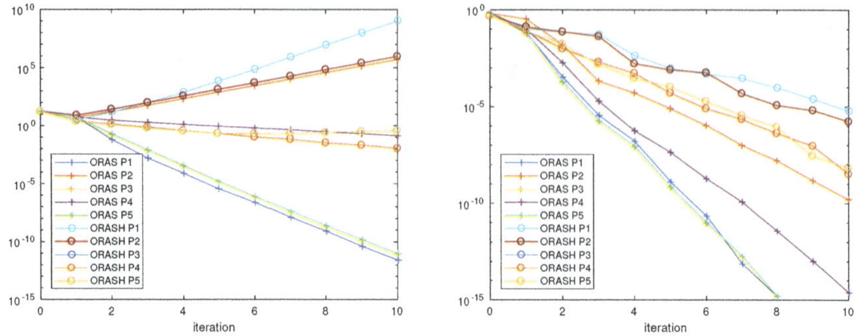

Fig. 9: Convergence of ORAS_ℓ and ORASH_ℓ as iterative solvers (left) and when used as preconditioners for GMRES (right), for $\ell = 1, 2, 3, 4, 5$ corresponding to the five different partitions of unity.

that similar problems were also observed in an alternating version in [6, subsection 6.1] when not keeping the correct Robin interface data.

Acknowledgments: The author would like to thank an anonymous referee for pointing out the important references [6, 7, 8].

References

1. Cai, X.C., Sarkis, M.: A restricted additive Schwarz preconditioner for general sparse linear systems. SIAM Journal on Scientific Computing **21**(2), 792–797 (1999)
2. Gander, M.J.: Optimized Schwarz methods. SIAM Journal on Numerical Analysis **44**(2), 699–731 (2006)
3. Gander, M.J.: Schwarz methods over the course of time. Electron. Trans. Numer. Anal **31**(5), 228–255 (2008)
4. Graham, I.G., Spence, E.A., Zou, J.: Domain decomposition with local impedance conditions for the Helmholtz equation (2018). Https://arxiv.org/abs/1806.03731, Presentation at DD25
5. Haferssas, R.M.: Espaces grossiers pour les méthodes de décomposition de domaine avec conditions d'interface optimisées. Ph.D. thesis, Université Pierre et Marie Curie, Paris VI (2016)
6. Kimn, J.H.: A convergence theory for an overlapping Schwarz algorithm using discontinuous iterates. Numer. Math. **100**, 117–139 (2005)
7. Kimn, J.H., Sarkis, M.: OBDD: overlapping balancing domain decomposition methods and generalizations to Helmholtz equations. In: Domain Decomposition Methods in Science and Engineering XVI, pp. 317–324. Springer (2006)
8. Kimn, J.H., Sarkis, M.: Restricted overlapping balancing domain decomposition and restricted coarse problem for the Helmholtz equation. Computer Methods in Applied Mechanics and Engineering **196**, 1507–1514 (2007)
9. Kwok, F.: Optimized additive Schwarz with harmonic extension as a discretization of the continuous parallel Schwarz method. SIAM Journal on Numerical Analysis **49**(3), 1289–1316 (2011)
10. Lions, P.L.: On the Schwarz alternating method. I. In: First international symposium on domain decomposition methods for partial differential equations, pp. 1–42. Paris, France (1988)

11. St-Cyr, A., Gander, M.J., Thomas, S.J.: Optimized multiplicative, additive, and restricted additive Schwarz preconditioning. SIAM Journal on Scientific Computing **29**(6), 2402–2425 (2007)
12. St-Cyr, A., Gander, M.J., Thomas, S.J.: Optimized restricted additive Schwarz methods. In: Domain Decomposition Methods in Science and Engineering XVI, pp. 213–220. Springer (2007)

Adaptive BDDC Based on Local Eigenproblems

Clemens Pechstein

1 Introduction

FETI-DP (dual-primal finite element tearing and interconnecting) and BDDC (balancing domain decomposition by constraints) are among the leading non-overlapping domain decomposition preconditioners. For standard symmetric positive definite (SPD) problems and standard discretizations, the spectral condition number κ of the preconditioned systems of either FETI-DP or BDDC can be bounded from above by $C\,(1 + \log(H/h))^2$. Here, H is the subdomain diameter and h the discretization parameter, and C is a constant independent of H, h, and the number of subdomains; for more details see e.g. [19]. In the past decade, there has been significant effort in analyzing the dependence of the constant C on problem parameters, such as coefficient values. This research has also led to new parameter choices of the preconditioners themselves, such as more sophisticated scalings and primal constraints. In particular, *adaptive* choices of primal constraints have been studied, starting with the pioneering work by Mandel and Sousedík [13] and later with Šístek [15], continued from different angles by Spillane and Rixen [18] as well as Klawonn, Radtke, and Rheinbach [8], and meanwhile pursued by various researchers [2, 5, 6, 9, 10, 20].

Our own work started with talks and slides [4, 16] and led to the comprehensive paper [17], where we present a rigorous and quite general theoretical framework for adaptive BDDC preconditioners and show the connections and differences between various existing methods. The paper [17] appears to be rather long and technical. In the contribution at hand, we would like to summarize the big picture from a less detailed perspective in favor of simplicity.

Clemens Pechstein
Dassault Systèmes Austria GmbH, Semmelweisstraße 15, 4600 Wels, Austria, e-mail: clemens.pechstein@3ds.com

R. Haynes et al. (eds.), *Domain Decomposition Methods in Science and Engineering XXV*,
Lecture Notes in Computational Science and Engineering 138,
https://doi.org/10.1007/978-3-030-56750-7_2

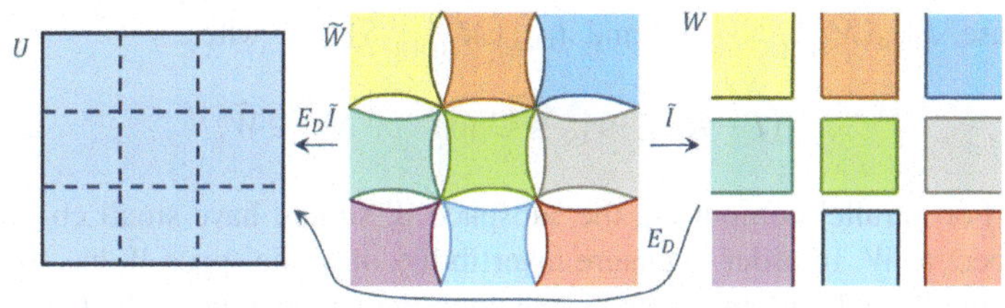

Fig. 1: Sketch of the spaces U, $\widetilde{W}$, and W for primal dofs on subdomain vertices in 2D.

2 BDDC Basics

The original problem to be solved reads $Au^* = f$, with the SPD matrix $A \in \mathbb{R}^{n \times n}$, originating from a PDE on the global domain Ω and projected to a finite-dimensional space via a finite element method (FEM), discontinuous Galerkin (DG) method, or isogeometric analysis (IGA). Using a decomposition of the domain Ω into non-overlapping subdomains $\{\Omega_i\}_{i=1}^N$, each degree of freedom (dof) is associated with one or several subdomains. After formal static condensation of the *inner* dofs (those owned only by one subdomain), we are left with the interface system

$$\widehat{S}\widehat{u} = \widehat{g}, \tag{1}$$

where $\widehat{S} : U \to U$ is again SPD. The global Schur complement $\widehat{S}$ can be assembled from subdomain contributions S_i in the following way,

$$\widehat{S} = R^T S R = \sum_{i=1}^N R_i^T S_i R_i\,, \tag{2}$$

where $R_i : U \to W_i$ is the restriction matrix that selects from all global dofs those of subdomain i, and $S = \operatorname{diag}(S_i)_{i=1}^N : W \to W := \prod_{i=1}^N W_i$. Throughout the paper we assume that each matrix S_i is symmetric positive semidefinite (SPSD).

The space U of interface dofs can be visualized as the global *continuous* space, whereas W can be visualized as a *discontinuous* space, see Fig. 1. The subspace of W containing *continuous* functions is $\widehat{W} := \operatorname{range}(R)$.

The balancing domain decomposition by constraints (BDDC) preconditioner [3, 12] can be seen as a fictitious space preconditioner [17, Appendix A]: Selecting a subspace $\widetilde{W} \subset W$ such that $\widetilde{W} \supset \widehat{W}$, the preconditioner reads

$$M_{\text{BDCC}}^{-1} := E_D \tilde{I} \widetilde{S}^{-1} \tilde{I}^T E_D^T\,, \tag{3}$$

where $\widetilde{S}$ is the restriction of S to $\widetilde{W}$, $\tilde{I}\colon \widetilde{W} \to W$ is the natural embedding operator, and $E_D\colon W \to U$ is a linear averaging operator mapping back to the original space U. In the following, we assume that $E_D R = I$, such that RE_D becomes a projection.

In this case, $\lambda_{\min}(M_{\text{BDDC}}^{-1}\widehat{S}) \geq 1$ and $\lambda_{\max}(M_{\text{BDDC}}^{-1}\widehat{S}) \leq C$ where[1]

$$|(I - RE_D)w|_{\widetilde{S}}^2 \leq C|w|_{\widetilde{S}}^2 \qquad \forall w \in \widetilde{W}, \tag{4}$$

cf. [12]. For parallel computing, the subspace $\widetilde{W}$ should have small co-dimension with respect to W. In order to ensure invertibility of $\widetilde{S}$, the space $\widetilde{W}$ has to be made smaller than W and with that allow some coupling between individual subdomains. In case of highly varying coefficients, one typically needs even more coupling. For some motivating numerical results obtained by adaptively chosen spaces $\widetilde{W}$ see e.g. [6, Sect. 8] (using the pair-based approach).

We decompose the global set of interface dofs into equivalence classes called *globs* such that dofs within a glob are shared by the same set of subdomains. In the sequel, we refer to globs shared by two subdomains simply as *faces*.[2] Next, we define a *primal dof* as a linear combination of regular dofs within the same glob.

For the following investigation we make two assumptions:

1. The space $\widetilde{W}$ is based on *primal constraints*, i.e., it is the subspace of W where on each glob, the associated primal dofs are continuous across the subdomains.
2. The averaging operator is block-diagonal w.r.t. the glob partition, i.e., the global dofs of $E_D w$ associated with a glob only depend on the values of w_i on that glob.

To formulate these assumptions more precisely, we need some notation. Let R_{iG} extract the dofs of W_i that belong to glob G and let $\mathcal{N}_G$ denote the set of subdomains sharing glob G. Then Assumption 1 reads

$$\widetilde{W} = \{w \in W : Q_G^T(R_{iG}w_i - R_{jG}w_j) = 0 \quad \forall i, j \in \mathcal{N}_G\}, \tag{5}$$

where Q_G^T is the matrix evaluating all primal dofs on G. Assumption 2 reads $\widehat{R}_G E_D w = \sum_{i \in \mathcal{N}_G} D_{iG} R_{iG} w_i$, where $\widehat{R}_G$ extracts the dofs of U that belong to glob G and $\{D_{jG}\}_{j \in \mathcal{N}_G}$ are local weighing matrices, not necessarily diagonal. To ensure that RE_D is a projection, we assume the glob-wise partition of unity property

$$\sum\nolimits_{j \in \mathcal{N}_G} D_{jG} = I. \tag{6}$$

There are several ways to realize the application of $\widetilde{I}\widetilde{S}^{-1}\widetilde{I}^T$ in practice (see [3, 11] and [17, Appendix C]), but all essentially boil down to block factorization where a sparse matrix on the space of primal dofs forms the coarse problem, whereas independent subdomain problems with the primal dofs being fixed form the remainder.

[1] assuming that RE_D is different from zero and identity

[2] In simple setups, one may visualize globs as open faces, open edges, and vertices, but this can change due to the geometry and/or the particular discretization.

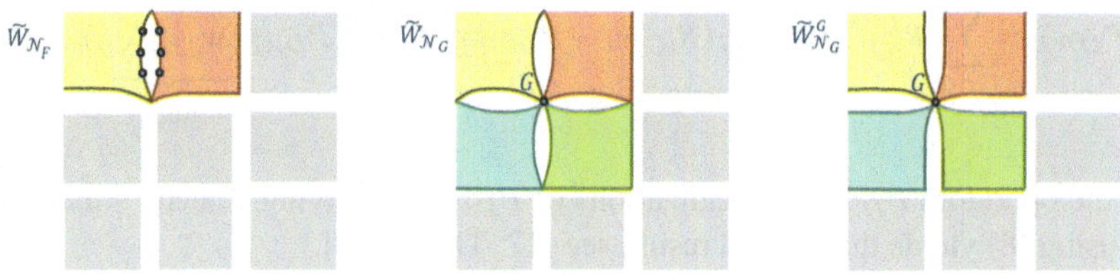

Fig. 2: Sketch of spaces and support (indicated by dots) of local operator $P_{D,G}$ for glob-based approach. *Left:* $\widetilde{W}_{N_F}$ for face $G = F$. *Middle:* $\widetilde{W}_{N_G}$ for vertex G. *Right:* $\widetilde{W}^G_{N_G}$ for vertex G.

3 Localization, Eigenproblems, and Adaptivity

Under the assumptions from the previous section, the global estimate (4) can be localized. In the following, we consider two kinds of localizations and work out the associated generalized eigenproblem and adaptive coarse space enrichment.

3.1 Glob-based approach

Let $P_{D,G}\colon W_{N_G} \to W_{N_G} := \prod_{i\in N_G} W_i$ be given by

$$(P_{D,G}w)_i = R_{iG}^T \sum\nolimits_{j\in N_G} D_{jG}(R_{iG}w_i - R_{jG}w_j), \tag{7}$$

and let $\widetilde{W}_{N_G} \subset W_{N_G}$ denote the subspace of functions where on all *neighboring* globs of G, the primal constraints of the global problem are enforced (see Fig. 2).[3] Two globs G and G' are *neighbors* if they share at least two common subdomains.

Theorem 1 *If for each glob G the inequality*

$$\sum\nolimits_{i\in N_G} |(P_{D,G}w)_i|^2_{S_i} \le \omega_G \sum\nolimits_{i\in N_G} |w_i|^2_{S_i} \qquad \forall w \in \widetilde{W}_{N_G} \tag{8}$$

holds, then

$$\kappa(M_{\text{BDDC}}^{-1}\widehat{S}) \le \big(\max_{i=1,\dots,N} |\mathcal{G}_i|^2\big)\big(\max_G \omega_G\big), \tag{9}$$

where $|\mathcal{G}_i|$ is the number of globs associated with subdomain i.

Proof We only have to show estimate (4), i.e., $|P_D w|^2_{\widetilde{S}} \le C|w|^2_{\widetilde{S}}$ for all $w \in \widetilde{W}$ where $P_D := (I - RE_D)$, cf. [12, Thm. 5] and [17, Sect. 2.3]. Under our assumptions, for any $w \in \widetilde{W}$,

[3] Precisely, $\widetilde{W}_{N_G} = \{w \in W_{N_G} : \forall i \ne j \in N_G \forall G', \{i,j\} \subset N_{G'} : Q_{G'}^T(R_{iG'}w_i - R_{jG'}w_j) = 0\}$.

$$(P_D w)_i = \sum_{G\in\mathcal{G}_i} R_{iG}^T \sum_{j\in\mathcal{N}_G} D_{jG}(R_{iG}w_i - R_{jG}w_j) = \sum_{G\in\mathcal{G}_i} (P_{D,G}\underbrace{[w_j]_{j\in\mathcal{N}_G}}_{\in\widetilde{W}_{\mathcal{N}_G}})_i\,. \tag{10}$$

I.e., the operators $P_{D,G}$ are localizations of P_D, and applying Cauchy's inequality and using (8) yields the desired result, see [17, Thm. 3.10]. □

Remark 1 If all dofs of glob G are primal dofs ($Q_G = I$) then $P_{D,G} = 0$. For such globs, estimate (8) holds with $\omega_G = 0$ and need not be accounted for in $|\mathcal{G}_i|$ in (9).

Generalized Eigenproblem. With $S_{\mathcal{N}_G} := \operatorname{diag}(S_j)_{j\in\mathcal{N}_G}$, estimate (8) is linked to the generalized eigenproblem, find $(w,\lambda)\in\widetilde{W}_{\mathcal{N}_G}\times\mathbb{R}$ such that

$$z^T S_{\mathcal{N}_G} w = \lambda\, z^T P_{D,G}^T S_{\mathcal{N}_G} P_{D,G} w \qquad \forall z\in\widetilde{W}_{\mathcal{N}_G}\,, \tag{11}$$

Since the local operator $P_{D,G}$ is a projection [17, Lemma 3.8], (11) is of the form $Ax = \lambda P^T A P x$ where A is SPSD and P a projection. Therefore all finite eigenvalues λ of (11) fulfill $\lambda\le 1$, see Lemma 1 in the Appendix. Moreover, if the smallest eigenvalue λ_1 is positive, then (8) holds with $\omega_G = \lambda_1^{-1}$. We further obtain the improved bound

$$|P_{D,G}w|^2_{S_{\mathcal{N}_G}} \le \lambda_{k+1}^{-1}|w|^2_{S_{\mathcal{N}_G}} \tag{12}$$

for all $w\in\widetilde{W}_{\mathcal{N}_G}$ such that

$$(y^{(\ell)})^T P_{D,G}^T S_{\mathcal{N}_G} P_{D,G} w = 0 \qquad \forall \ell = 1,\ldots,k, \tag{13}$$

where $y^{(1)},\ldots,y^{(k)}$ are the eigenvectors corresponding to the k smallest eigenvalues of (11), cf. [13]. As a viable alternative, we can replace the space $\widetilde{W}_{\mathcal{N}_G}$ in (11) by the space $\widetilde{W}_{\mathcal{N}_G}^G$ where just the primal constraints on G are enforced (but not on its neighbors),[4] see Fig. 2, cf. [17, Strategy 4]. This discards any (good) influence of primal constraints on neighboring globs but makes the underlying operator much more simple to implement.

Adaptive enrichment. We show now how to realize (13) by primal constraints.

If $G = F$ is a face shared by two subdomains i and j then (13) reads

$$\underbrace{(R_{iF}y_i^{(\ell)} - R_{jF}y_j^{(\ell)})^T \begin{bmatrix} D_{jF} \\ -D_{iF}\end{bmatrix}^T \begin{bmatrix} S_{iF} & 0 \\ 0 & S_{jF}\end{bmatrix}\begin{bmatrix} D_{jF} \\ -D_{iF}\end{bmatrix}}_{=:(Q_F^*)^T}(R_{iF}w_i - R_{jF}w_j) = 0, \tag{14}$$

where $S_{kF} := R_{kF}^T S_k R_{kF}$ is the principal minor of S_k associated with the dofs on F. Apparently, the columns of Q_F^* make up the new primal dofs.

For globs shared by more than just two subdomains, it has turned out to be a challenge to enforce (13) in terms of primal constraints. For simplicity we assume

[4] Precisely, $\widetilde{W}_{\mathcal{N}_G}^G := \{w\in W_{\mathcal{N}_G} : Q_G^T(R_{iG}w_i - R_{jG}w_j) = 0 \quad \forall i,j\in\mathcal{N}_G\}$.

that G is a glob shared by three subdomains i, j, and k (the general case follows the same idea, cf. [17, Sect. 5.4]). Then the constraints (13) take the form

$$\begin{bmatrix} c_i^{(\ell)} \\ c_j^{(\ell)} \\ c_k^{(\ell)} \end{bmatrix}^T \begin{bmatrix} (I - D_{iG})w_{iG} - D_{jG}w_{jG} - D_{kG}w_{kG} \\ -D_{iG}w_{iG} + (I - D_{jG})w_{jG} - D_{kG}w_{kG} \\ -D_{iG}w_{iG} - D_{jG}w_{jG} + (I - D_{kG})w_{kG} \end{bmatrix} = 0, \tag{15}$$

where $c^{(\ell)} = S_{N_G} P_{D,G} y^{(\ell)}$ and w_{iG} is a short hand for $R_{iG} w_i$. We introduce

$$\widehat{w}_G := \tfrac{1}{3}(w_{iG} + w_{jG} + w_{kG}), \quad \check{w}_{2G} := w_{iG} - w_{jG}, \quad \check{w}_{3G} := w_{iG} - w_{kG}, \tag{16}$$

together with the corresponding inverse transformation

$$w_{iG} = \widehat{w}_G - \tfrac{1}{3}\check{w}_{2G} - \tfrac{1}{3}\check{w}_{3G}, \qquad w_{jG} = \widehat{w}_G + \tfrac{2}{3}\check{w}_{2G} - \tfrac{1}{3}\check{w}_{3G}, \tag{17}$$
$$w_{kG} = \widehat{w}_G - \tfrac{1}{3}\check{w}_{2G} + \tfrac{2}{3}\check{w}_{3G}.$$

We apply this transformation to (15) and find out that due to the partition of unity property (6), the whole expression is independent of $\widehat{w}_G$, and so (15) is of form

$$(\check{c}_2^{(\ell)})^T \check{w}_{2G} + (\check{c}_3^{(\ell)})^T \check{w}_{3G} = 0. \tag{18}$$

In [17] we enforce this constraint by the two stronger constraints $(\check{c}_2^{(\ell)})^T \check{w}_{2G} = 0$ and $(\check{c}_3^{(\ell)})^T \check{w}_{3G} = 0$, rewritten in the original variables,

$$(\check{c}_2^{(\ell)})^T (R_{iG}w_i - R_{jG}w_j) = 0, \qquad (\check{c}_3^{(\ell)})^T (R_{iG}w_i - R_{kG}w_k) = 0. \tag{19}$$

With the enforcement of an even stronger set, namely,

$$\left.\begin{aligned} (\check{c}_2^{(\ell)})^T (R_{mG}w_m - R_{nG}w_n) = 0 \\ (\check{c}_3^{(\ell)})^T (R_{mG}w_m - R_{nG}w_n) = 0 \end{aligned}\right\} \qquad \forall m, n \in \{i, j, k\}, \tag{20}$$

we see that $\check{c}_2^{(\ell)}$, $\check{c}_3^{(\ell)}$ define the new primal dofs. In [17, Thm. 5.18], we show that it is more favorable to use the stronger primal constraints (20) than using so-called generalized primal constraints realizing exactly (15).

3.2 Pair-based approach

A different way of writing the P_D operator (compared to (10)) is

$$(P_D w)_i = \sum_{j \in \mathcal{N}_i} \sum_{G:\{i,j\} \subset \mathcal{N}_G} R_{iG}^T D_{jG}(R_{iG}w_i - R_{jG}w_j), \tag{21}$$

Fig. 3: Sketch of spaces and support of $P_{D,\Gamma_{ij}}$ for pair-based approach. Dots indicate support of the local operator $P_{D,\Gamma_{ij}}$. *Left:* Sketch of the space $\widetilde{W}_{ij}$. *Right:* Sketch of $\widetilde{W}_{ik}$.

where $\mathcal{N}_i$ denotes the set of subdomains that share a non-trivial set of globs with subdomain i. It was used in the early works [14, 15] and put on solid ground in [6].

Defining for $i \neq j$ the *generalized facet*

$$\Gamma_{ij} := \bigcup\nolimits_{G:\ \{i,j\}\subset \mathcal{N}_G} G, \tag{22}$$

and collecting only the non-trivial ones into the set Υ, we obtain

$$(P_D w)_i = \sum_{j:\,\Gamma_{ij}\in\Upsilon} \underbrace{R_{i\Gamma_{ij}}^T D_{j\Gamma_{ij}}(R_{i\Gamma_{ij}} w_i - R_{j\Gamma_{ij}} w_j)}_{=:(P_{D,\Gamma_{ij}} w_{ij})_i}, \tag{23}$$

where $R_{i\Gamma_{ij}}$ extracts the dofs on Γ_{ij}, the matrix $D_{j\Gamma_{ij}}$ is block-diagonal with blocks $\{D_{jG}\}_{G\subset\Gamma_{ij}}$, and $P_{D,\Gamma_{ij}} : W_{ij} \to W_{ij} := W_i \times W_j$. Before we can formulate the counterpart of Theorem 1, we have to introduce the subspace $\widetilde{W}_{ij}$ of W_{ij} where all primal constraints between subdomain i and j are enforced,[5] see Fig. 3.

Theorem 2 *If for each generalized facet $\Gamma_{ij} \in \Upsilon$ the inequality*

$$|(P_{D,\Gamma_{ij}} w)_i|_{S_i}^2 + |(P_{D,\Gamma_{ij}} w)_j|_{S_j}^2 \leq \omega_{ij}\big(|w_i|_{S_i}^2 + |w_j|_{S_i}^2\big) \qquad \forall w \in \widetilde{W}_{ij} \tag{24}$$

holds, then

$$\kappa(M_{\mathrm{BDDC}}^{-1}\widehat{S}) \leq \big(\max_{i=1,\ldots,N} n_i^2\big) \max_{\Gamma_{ij}\in\Upsilon} \omega_{ij}\,,$$

with $n_i := |\{j : \Gamma_{ij} \in \Upsilon\}|$ the number of pairs associated with subdomain i.

Proof The proof is similar to that of Theorem 1, see also [17, Lemma 3.16]. □

Generalized eigenproblem. The generalized eigenproblem associated with estimate (24) is finding $(w, \lambda) \in \widetilde{W}_{ij} \times \mathbb{R}$ such that

$$z^T S_{ij} w = \lambda\, z^T P_{D,\Gamma_{ij}}^T S_{ij} P_{D,\Gamma_{ij}} w \qquad \forall z \in \widetilde{W}_{ij}\,, \tag{25}$$

where $S_{ij} := \mathrm{diag}(S_i, S_j)$.

[5] Precisely, $\widetilde{W}_{ij} := \{(w_i, w_j) \in W_i \times W_j : \forall G, \{i,j\} \subset \mathcal{N}_G : Q_G^T(R_{iG} w_i - R_{jG}) = 0\}$.

Remark 2 Unlike the operator $P_{D,G}$ from Sect. 3.1, the operator $P_{D,\Gamma_{ij}}$ in general *fails* to be a projection,[6] so Lemma 1 from the appendix (and some other tools from [17]) cannot be applied.

We obtain the improved bound

$$|P_{D,\Gamma_{ij}} w|_{S_{ij}}^2 \le \lambda_{k+1}^{-1} |w|_{S_{ij}}^2 \tag{26}$$

for all $w \in \widetilde{W}_{ij}$ such that

$$(y^{(\ell)})^T P_{D,\Gamma_{ij}}^T S_{ij} P_{D,\Gamma_{ij}} w = 0 \qquad \forall \ell = 1, \ldots, k, \tag{27}$$

where $y^{(1)}, \ldots, y^{(k)}$ are the first k eigenvectors of (25).

Adaptive enrichment. We wish to enforce condition (27) by primal constraints and follow [15, 10, 6]. Because of its particular form, $P_{D,\Gamma_{ij}} w$ only depends on the *difference* of w_i and w_j on Γ_{ij}, and so for fixed ℓ, (27) can be written as

$$(c^{(\ell)})^T (R_{i\Gamma_{ij}} w_i - R_{j\Gamma_{ij}} w_j) = 0. \tag{28}$$

Splitting the dofs of Γ_{ij} into globs, we can express the latter as

$$\sum_{G\colon \{i,j\} \subset \mathcal{N}_G} (c_G^{(\ell)})^T (R_{iG} w_i - R_{jG} w_j) = 0, \tag{29}$$

where $c^{(\ell)} = [c_G^{(\ell)}]_{G\colon \{i,j\} \subset \mathcal{N}_G}$, up to possible renumbering. Apparently, (29) holds if we enforce the stronger conditions

$$(c_G^{(\ell)})^T (R_{kG} w_k - R_{\ell G} w_\ell) = 0 \qquad \forall k, \ell \in \mathcal{N}_G \tag{30}$$

for each glob G such that $\{i, j\} \subset \mathcal{N}_G$. Conditions (30) have exactly the form of primal constraints and will imply (27).

Remark 3 Apparently, for glob G shared by $m > 2$ subdomains, we have to collect the adaptive primal constraints originating from $(m-1)(m-2)$ pairs. E.g., for an edge (in three dimensions) shared by three subdomains, these are three pairs; if it is shared by four subdomains, six pairs. In order to avoid redundancy, an orthonormalization procedure should be applied, e.g., modified Gram-Schmidt. For the typical 3D mesh decompositions created by METIS, it is very unlikely that an edge will be shared by more than three subdomains [6]. Nevertheless, there can be a large number of short edges or thin faces, see also [6, Sect. 7].

---

[6] A simple counterexample can be constructed by looking at an edge shared by two subdomains with its endpoints shared by four and using the multiplicity scaling. At an interior dofs x of the edge $P_{D,\Gamma_{ij}}$ evaluates as $\pm\frac{1}{2}(w_i(x) - w_j(x))$ whereas at an endpoint x as $\pm\frac{1}{4}(w_i(x) - w_j(x))$.

3.3 Three different approaches

We basically have three approaches:

1. The glob-based approach with the original space $\widetilde{W}_{N_G}$ where primal constraints on neighboring globs are enforced ([17, Strategy 1–3]),
2. the glob-based approach with $\widetilde{W}^G_{N_G}$ where no constraints are enforced on neighboring globs ([17, Strategy 4]),
3. the pair-based approach (where constraints on neighboring globs are sometimes enforced, sometimes not, see also [6]).

The difference between the glob- and pair-based approach is not only the space but also the localized P_D operator, see Fig. 2 and Fig. 3. A priori, there is no theoretical argument on which of the three approaches is better, and dedicated numerical studies will be necessary to find out more. For typical METIS partitions, the pair-based approach involves a smaller number of eigenproblems, while potentially creating some unnecessary constraints.[7] Each of the approaches can be hard to load balance, the glob-based approach likely more difficult (if one thinks of subdomain edges).

4 Simplification of the generalized eigenproblems

In this subsection, we pursue only the glob-based localization from Sect. 3.1 with the space $\widetilde{W}_{N_G}$ in (11) being replaced by the space $\widetilde{W}^G_{N_G}$ where just the primal constraints on G are enforced (but not on its neighbors), see Fig. 2.

In the following, suppose that F is a face shared by subdomains i and j. The eigenproblem (11) involves the dofs on the subdomain (boundary) i *and* j, so more than twice as many dofs as on F. However, the matrix $P_{D,F}$ has a large kernel with co-dimension equal to the number of dofs on F. Hence, there are many infinite eigenvalues that are irrelevant to our consideration (recall that we are after the first few smallest eigenvalues and their associated eigenvectors). It turns out that using Schur complement techniques, the eigenproblem (11) can be reduced to an equivalent one in the sense that the number of infinite eigenvalues is reduced, the rest of the spectrum is untouched, and the full eigenvectors can easily be reconstructed from the reduced ones, see [17, Principle 4.4]. For the face, (11) (on $\widetilde{W}^G_{N_G}$) is equivalent (up to infinite eigenvalues) to

$$\check{z}_F^T (S^\star_{iF} : S^\star_{jF}) \check{w}_F = \lambda\, \check{z}_F^T M_F \check{w}_F \tag{31}$$

where $\check{w}_F = w_{iF} - w_{jF}$, and so the initially chosen primal dofs on F of $\check{w}_F$, $\check{z}_F$ vanish. Above, $S^\star_{kF}$ is the Schur complement of S_k eliminating all dofs except those

[7] Let us note that the sizes of the corresponding eigenproblems will not differ much, provided that one applies the same reduction technique. A comparison of approach 2. and 3. with *different* reduction techniques can be found in [10].

on F, $S^{\star}_{iF} : S^{\star}_{jF}$ is the *parallel sum* [1], defined by $A : B = A(A+B)^{\dagger}B$ (see [17, Sect. 5.1]), and $M_F = D_{iF}^T S_{jF} D_{iF} + D_{jF}^T S_{iF} D_{jF}$, cf. [17, Sect. 5.2].

For the choice of the deluxe scaling, $D_{iF} = (S_{iF} + S_{jF})^{-1} S_{iF}$ it can be shown [17, Sect. 6.1] that $M_F = S_{iF} : S_{jF}$, such that the eigenproblem (here with no initial primal constraints) takes the form

$$(S^{\star}_{iF} : S^{\star}_{jF}) \check{w}_F = \lambda\, (S_{iF} : S_{jF}) \check{w}_F \,. \tag{32}$$

This has been implemented in a PETSc version of BDDC by Zampini [20].

For globs G shared by more than two subdomains, one can easily eliminate the dofs not on G in a first step such that one is left with an eigenproblem of size $\#\mathcal{N}_G \times \#\mathsf{dofs}(G)$ and with kernel dimension $\#\mathsf{dofs}(G)$. Getting rid of the kernel completely is possible but more tricky. But this is not so severe since the number of dofs on such a glob (e.g. an edge) is typically much less than on a face, and so one can usually afford computing with the eigenproblem from the first reduction step. Even so, some decoupling approaches have been suggested, see [2, 5] and [17, Sect. 5.5, Sect. 5.6, and Sect. 6.4].

5 Optimality of the deluxe scaling

Let F be a face and consider the reduced eigenproblem from the previous section,

$$\check{z}_F^T T_F \check{w}_F = \lambda\, \check{z}_F^T \underbrace{\left[X^T S_{jF} X + (I-X)^T S_{iF} (I-X)\right]}_{M_F(X)} \check{w}_F \,, \tag{33}$$

where $T_F = S^{\star}_{iF} : S^{\star}_{jF}$ and where we have set $D_{iF} = X$ and $D_{jF} = I - X$ in order to obtain the partition of unity. The choice of the weighting matrix, here X, can have quite an influence on the spectrum of (33). It is of course desirable to have a spectrum that has as few small eigenvalues (lower outliers) as possible. Depending on the problem, outliers can often not be avoided, but at least its number could be minimized, because this number will be the number of new primal constraints on face F if one aims at a robust method, cf. [2].

For simplicity, we first look at the case where F consists of a single dof and so S_{iF}, S_{jF}, and X are scalars. Since T_F is fixed, we see that minimizing the quadratic expression $(S_{iF} + S_{jF})X^2 - 2S_{iF}X$ is favorable, because then the (only) eigenvalue λ is maximized (such that the local bound $\omega_F = \lambda^{-1}$ is small). Minimization is achieved with the choice $X^* = \frac{S_{iF}}{S_{iF}+S_{iF}}$ which is the well-known weighted counting function (with exponent $\gamma = 1$, see [19, Sect. 6]).

If F has more than one dof, no initial primal constraints, and if T_F is non-singular,[8] then it is favorable to minimize the *trace* of the matrix on the right-hand side of (33)

[8] If one of the subdomain "Neumann" matrices $S^{\star}_{kF}$ is singular (e.g., corresponding to the Laplace operator on a floating subdomain), then also $T_F = S^{\star}_{iF} : S^{\star}_{jF}$ is singular.

because of the following (for details see [17, Sect. 6.2]). Firstly, the trace of a matrix equals the sum of its eigenvalues and is similarity-invariant, i.e.,

$$\mathrm{tr}(M_F(X)) = \mathrm{tr}\big(T_F^{-1/2} M_F(X) T_F^{-1/2}\big) = \sum\nolimits_{k=1}^{n} \lambda_k^{-1}, \tag{34}$$

where $\lambda_1, \ldots, \lambda_n$ are the (generalized) eigenvalues of (33). Secondly, minimizing $\sum_{k=1}^n \lambda_k^{-1}$ means that it is less likely that the smallest eigenvalues are very small. At the minimum, we obtain $X^* = (S_{iF} + S_{jF})^{-1} S_{iF}$, the deluxe scaling. For numerical studies comparing different scalings (including deluxe) see, e.g., [8, 7].

Acknowledgements The author would like to thank Clark Dohrmann (Sandia National Laboratories) for many fruitful discussions on BDDC, especially during the week of the DD25 conference.

Appendix

We consider the generalized eigenproblem $Ax = \lambda Bx$ with SPSD matrices A, B and call (λ, x) a *genuine* eigenpair if $\lambda \in \mathbb{R}$ and $x \notin \ker(A) \cap \ker(B)$. We call (∞, x) an eigenpair with *infinite* eigenvalue if $x \in \ker(B) \setminus \{0\}$, and (λ, x) an *ambiguous* eigenpair if $x \in \ker(A) \cap \ker(B)$.

Lemma 1 *Let us consider the generalized eigenproblem*

$$Ax = \lambda P^T A P x, \tag{35}$$

where $A \in \mathbb{R}^{n \times n}$ *is SPSD and* $P \in \mathbb{R}^{n \times n}$ *a projection. Then all genuine eigenvalues* λ *of* (35) *fulfill* $\lambda \leq 1$.

Proof [17, Lemma 4.12] yields that the infinite eigenspace is

$$V_\infty = \ker(P^T A P) = \ker(P) \oplus (\ker(A) \cap \mathrm{range}(P))$$

and the ambiguous eigenspace turns out to be

$$V_{\mathrm{amb}} := \ker(A) \cap V_\infty = (\ker(A) \cap \ker(P)) \oplus (\ker(A) \cap \mathrm{range}(P)).$$

The latter is a subspace of the above: $V_{\mathrm{amb}} \subset V_\infty$. Since $V_\infty \subset \ker(P^T A P)$, we can eliminate V_∞ and obtain an eigenproblem that has the same finite and non-ambigious eigenvalues as (35). For such an elimination, we need a space splitting $\mathbb{R}^n = V_\infty \oplus V_c$. Here we use some complementary space V_c with the property $V_c \subset \mathrm{range}(P)$ (that is feasible because $\mathbb{R}^n = \ker(P) \oplus \mathrm{range}(P)$). We have the property that $Py = y$ on V_c because P is a projection. Following [17, Principle 4.4], the reduced eigenproblem reads: find $(\lambda, y) \in \mathbb{R} \times V_c$ such that

$$z^T S y = \lambda z^T A y \qquad \forall z \in V_c\,,$$

where S is the Schur complement w.r.t. V_∞ such that $y^T S y \leq (x+y)^T A(x+y)$ for all $x \in V_\infty$ and $y \in V_c$. Note that A is definite on V_c because $V_c \subset \mathrm{range}(P)$ but $\ker(A) \cap V_c = \{0\}$ since $V_\infty \supset \ker(A) \cap \mathrm{range}(P)$. Since A is definite on V_c, the right-hand side matrix of the reduced eigenproblem is definite and so we can express the maximal eigenvalue $\lambda_{\max}$ in terms of the Rayleigh quotient. The proof is completed by using the minimizing property of the Schur complement. □

References

1. Anderson, Jr., W.N., Duffin, R.J.: Series and parallel addition of matrices. J. Math. Anal. Appl. **26**(3), 576–594 (1969)
2. Calvo, J.G., Widlund, O.B.: Adaptive choice of primal constraints for BDDC domain decomposition algorithms. Electron. Trans. Numer. Anal. **45**, 524–544 (2016)
3. Dohrmann, C.R.: A preconditioner for substructuring based on constrained energy minimization. SIAM J. Sci. Comput. **25**(1), 246–258 (2003)
4. Dohrmann, C.R., Pechstein, C.: Constraint and weight selection algorithms for BDDC (2012). Talk by Dohrmann at the Domain Decomp. Meth. Sci. Engrg. XXI, Rennes, France, `http://www.numa.uni-linz.ac.at/~clemens/dohrmann-pechstein-dd21-talk.pdf`
5. Kim, H.H., Chung, E.T., Wang, J.: BDDC and FETI-DP preconditioners with adaptive coarse spaces for three-dimensional elliptic problems with oscillatory and high contrast coefficients. J. Comput. Phys. **349**, 191–214 (2017)
6. Klawonn, A., Kühn, M., Rheinbach, O.: Adaptive coarse spaces for FETI-DP in three dimensions. SIAM J. Sci. Comput. **38**(5), A2880–A2911 (2016)
7. Klawonn, A., Kühn, M., Rheinbach, O.: Adaptive FETI-DP and BDDC methods with a generalized transformation of basis for heterogeneous problems. Electron. Trans. Numer. Anal. **49**, 1–27 (2018)
8. Klawonn, A., Radtke, P., Rheinbach, O.: FETI-DP with different scalings for adaptive coarse spaces. PAMM Proc. Appl. Math. Mech. **14**, 835–836 (2014)
9. Klawonn, A., Radtke, P., Rheinbach, O.: FETI-DP methods with an adaptive coarse space. SIAM J. Numer. Anal. **53**(1), 297–320 (2015)
10. Klawonn, A., Radtke, P., Rheinbach, O.: A comparison of adaptive coarse spaces for iterative substructuring in two dimensions. Electron. Trans. Numer. Anal. **45**, 75–106 (2016)
11. Klawonn, A., Widlund, O.B.: Dual-primal FETI methods for linear elasticity. Comm. Pure Appl. Math. **59**(11), 1523–1572 (2006)
12. Mandel, J., Dohrmann, C.R., Tezaur, R.: An algebraic theory for primal and dual substructuring methods by constraints. Appl. Numer. Math. **54**(2), 167–193 (2005)
13. Mandel, J., Sousedík, B.: Adaptive selection of face coarse degrees of freedom in the BDDC and FETI-DP iterative substructuring methods. Comput. Methods Appl. Mech. Engrg. **196**(8), 1389–1399 (2007)
14. Mandel, J., Sousedík, B.: BDDC and FETI-DP under minimalist assumptions. Computing **81**(4), 269–280 (2007)
15. Mandel, J., Sousedík, B., Šístek, J.: Adaptive BDDC in three dimensions. Math. Comput. Simulation **82**(10), 1812–1831 (2012)
16. Pechstein, C., Dohrmann, C.R.: Modern domain decomposition solvers – BDDC, deluxe scaling, and an algebraic approach (2013). Talk by Pechstein at RICAM, Linz, Austria, `http://people.ricam.oeaw.ac.at/c.pechstein/pechstein-bddc2013.pdf`
17. Pechstein, C., Dohrmann, C.R.: A unified framework for adaptive BDDC. Electron. Trans. Numer. Anal. **46**, 273–336 (2017)
18. Spillane, N., Rixen, D.: Automatic spectral coarse spaces for robust FETI and BDD algorithms. Int. J. Numer. Meth. Engng. **95**(11), 953–990 (2013)

19. Toselli, A., Widlund, O.B.: Domain Decomposition Methods – Algorithms and Theory, *Springer Series in Computational Mathematics*, vol. 34. Springer-Verlag, Berlin (2005)
20. Zampini, S.: PCBDDC: a class of robust dual-primal preconditioners in PETSc. SIAM J. Sci. Comput. **38**(5), S282–S306 (2016)

From Domain Decomposition to Homogenization Theory

Daniel Peterseim, Dora Varga, and Barbara Verfürth

1 Introduction

Elliptic boundary value problems with oscillatory coefficients play a key role in the mathematical modelling and simulation of complex multiscale problems, for instance transport processes in porous media or the mechanical analysis of composite and multifunctional materials. The characteristic properties of such processes are determined by a complex interplay of effects on multiple non-separable length and time scales. The challenge is that the resolution of all details on all relevant scales may easily lead to a number of degrees of freedom and computational work in a direct numerical simulation which exceed today's computing resources by multiple orders of magnitude. The observation and prediction of physical phenomena from multiscale models, hence, requires insightful methods that effectively represent unresolved scales, i.e., multiscale methods.

Homogenization is such a multiscale method. It seeks a simplified model that is able to capture the macroscopic responses of the process adequately by a few localized computations on the microscopic scale. Consider, e.g., prototypical second order linear elliptic model problems with highly oscillatory periodic diffusion coefficients that oscillate at frequency ε^{-1} for some small parameter $0 < \varepsilon \ll 1$. Then, the theory of homogenization shows that there exists a constant coefficient such that the corresponding diffusion process represents the macroscopic behaviour correctly. In practice, this yields a two- or multi-scale method that first computes the effective coefficient which is implicitly given through some PDE on the microscopic periodic cell, and then solves the macroscopic effective PDE. This is done for instance in

Daniel Peterseim
Universität Augsburg e-mail: daniel.peterseim@math.uni-augsburg.de

Dora Varga
Universität Augsburg e-mail: dora.varga@math,uni-augsburg.de

Barbara Verfürth
Universität Augsburg e-mail: barbara.verfuerth@math.uni-augsburg.de

R. Haynes et al. (eds.), *Domain Decomposition Methods in Science and Engineering XXV*,
Lecture Notes in Computational Science and Engineering 138,
https://doi.org/10.1007/978-3-030-56750-7_3

the Multiscale Finite Element Method [8] or the Heterogeneous Multiscale Method [2]. In certain cases, the error of such procedures can be quantified in terms of the microscopic length scale ε. The approach and its theoretical foundation can be generalized to certain classes of non-periodic problems. However, the separation of scales, i.e., the separation of the characteristic frequencies of the diffusion coefficient and macroscopic frequencies of interest, seems to be essential for both theory and computation.

There is a more recent class of numerical homogenization methods that can deal with arbitrarily rough diffusion coefficients beyond scale separation [31, 23]. While, at first glance, these methods seemed to be only vaguely connected to classical homogenization theory, the recent paper [14] identifies them as a natural generalization of some new characterization of classical homogenization. Another deep connection, which was always believed to exist in the community of domain decomposition methods, is the one between homogenization and domain decomposition. This one was made precise only recently by Kornhuber and Yserentant [28, 26, 27]. By combining their iterative approach to homogenization and the results of [14], the present paper illuminates the role of domain decomposition in the theory of homogenization and provides homogenization limits without any advanced compactness arguments or two scale limits. In addition, compared with [14], we are able to drop a technical assumption on some artificial symmetries of the diffusion coefficient with respect to the periodic cell.

Our new construction of effective coefficients (see Sections 3–4) is not necessarily any easier than the classical one. For the simple diffusion model problem, this is merely an instance of mathematical curiosity and we do not mean to rewrite homogenization theory. However, the connection between homogenization theory and domain decomposition and, in particular, the method of proof turn out to be very interesting and, moreover, they unroll their striking potential for problems beyond scale separation and periodicity. Using this approach, new theoretical results could be derived and some of them are briefly discussed in Section 5.

2 Model problem and classical homogenization

For the sake of illustration we restrict ourselves to the simplest possible yet representative and relevant setting. Let $\Omega = [0, 1]^2$ be the unit square and $\varepsilon\Omega := [0, \varepsilon]^2$. Moreover, let $A_1 \in L^\infty(\Omega; \mathbb{R}^{2\times 2})$ be a symmetric, uniformly elliptic, Ω-periodic (matrix-valued) coefficient and let $A_\varepsilon(x) := A_1(\frac{x}{\varepsilon})$, $x \in \Omega$. The extension to a cuboidal domain in $3D$ is straight forward. We denote by $V := H^1_\#(\Omega)_{/\mathbb{R}}$ the equivalence class of Ω-periodic functions in $H^1(\Omega)$ factorised by constants, and similarly, by $V_\varepsilon := H^1_\#(\varepsilon\Omega)_{/\mathbb{R}}$ for their ε-periodic counterparts. The model problem under consideration then reads: given $f \in L^2(\Omega)$, find a function $u_\varepsilon \in V$ such that

$$\int_\Omega A_\varepsilon(x)\nabla u_\varepsilon(x) \cdot \nabla v(x)\, \mathrm{d}x = \int_\Omega f(x)v(x)\, \mathrm{d}x, \tag{1}$$

for all $v \in V$. In order to ensure the well-posedness of the problem, we assume that $A_\varepsilon \in \mathcal{M}_{\alpha\beta}$, where $\mathcal{M}_{\alpha\beta}$ is defined as

$$\mathcal{M}_{\alpha\beta} := \{A \in L^\infty(\Omega) \mid \alpha|\xi|^2 \le \xi \cdot A(x)\xi \le \beta|\xi|^2 \text{ for all } \xi \in \mathbb{R}^2 \text{ and a.a. } x \in \Omega\}.$$

The idea behind classical homogenization is to look for a so-called effective (homogenized) coefficient $A_0 \in \mathcal{M}_{\alpha\beta}$ so that the solution $u_0 \in V$ of the problem

$$\int_\Omega A_0 \nabla u_0 \cdot \nabla v \,\mathrm{d}x = \int_\Omega f v \,\mathrm{d}x, \tag{2}$$

for all $v \in V$, represents the limit of the sequence $\{u_\varepsilon\}_{\varepsilon>0}$ of solutions of the problem (1). In general, explicit representations of effective coefficients are not known, except for the simple case of the one-dimensional or (locally) periodic setting. However, the so-called energy method of Murat and Tartar ([34]) or the two-scale convergence ([3]) provide us with the following form

$$\Big(A_0\Big)_{kj} = \int_\Omega \Big(A_1(x)(e_j + \nabla w_j(x))\Big) \cdot \Big(e_k + \nabla w_k(x)\Big) \,\mathrm{d}x, \tag{3}$$

where w_j are defined as the unique solutions in V of the so-called cell problems

$$\int_\Omega A_1(x)\left(\nabla w_j(x) - e_j\right) \cdot \nabla v(x) \,\mathrm{d}x = 0,$$

for all $v \in V$, with the canonical basis $(e_j)_{j=1}^2$ of $\mathbb{R}^2$. The substitution $x \mapsto \frac{x}{\varepsilon}$ yields

$$\begin{aligned} 0 &= \varepsilon^{-2} \int_{\varepsilon\Omega} A_1\Big(\frac{x}{\varepsilon}\Big)\Big(\nabla \underbrace{w_j\Big(\frac{x}{\varepsilon}\Big)}_{=:\hat{q}_j(x)} - e_j\Big) \cdot \nabla \underbrace{v\Big(\frac{x}{\varepsilon}\Big)}_{=:v_\varepsilon(x)} \,\mathrm{d}x \\ &= \int_\Omega A_\varepsilon(x)(\nabla \hat{q}_j(x) - e_j) \cdot \nabla v_\varepsilon(x) \,\mathrm{d}x. \end{aligned} \tag{4}$$

Since all functions v_ε in V_ε can be written as $v(\frac{x}{\varepsilon})$ for a certain function $v \in V$, equation (4) yields that the function $\hat{q}_j \in V_\varepsilon$ solves

$$\int_\Omega A_\varepsilon(x)(\nabla \hat{q}_j(x) - e_j) \cdot \nabla v_\varepsilon(x) \,\mathrm{d}x = 0, \tag{5}$$

for all $v_\varepsilon \in V_\varepsilon$. Moreover, $\hat{q}_j \in V_\varepsilon \subset V$ solves the same problem in the space V, i.e.,

$$\int_\Omega A_\varepsilon(x)(\nabla \hat{q}_j(x) - e_j) \cdot \nabla v(x) \,\mathrm{d}x = 0,$$

for all $v \in V$, since the solution of an elliptic model problem with periodic data (coefficient, source function) is also periodic, with the same period.

3 Novel characterization of the effective coefficient

In order to define the effective coefficient from the alternative perspective of finite elements, we first introduce the necessary notation on meshes, spaces, and interpolation operators.

We consider structured triangulations of $\Omega = [0, 1]^2$ as depicted in Figure 1, where the triangles T form the triangulation $\mathcal{T}_H$ and the boldface squares Q are part of the square mesh Q_H. The theoretical arguments below require the triangulation to be aligned with the periodicity cells of the coefficient represented by the elements of Q_H. Moreover, $\mathcal{T}_H$ should not introduce any nodes in the interiors of those cells. We shall emphasize that the general numerical homogenization method of Section 5 can deal with fairly general meshes. Denote the set of nodes by $\mathcal{N}_{\mathcal{T}_H} = \mathcal{N}_{Q_H}$. Since we are working with periodic boundary conditions, we will frequently understand Q_H and $\mathcal{T}_H$ as periodic partitions (or partitions of the torus or partitions of the whole $\mathbb{R}^2$), i.e., we identify opposite faces of the unit square. The parameter H denotes the length of the quadrilaterals and is supposed to be not smaller than the microscopic length scale ε of the model problem.

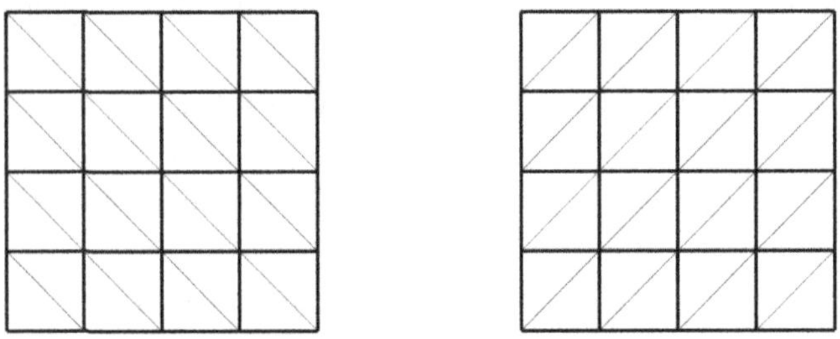

Fig. 1: Admissible triangulations.

Let $\mathcal{P}_1(\mathcal{T}_H)$ denote the space of globally continuous piecewise affine functions on Ω with periodic boundary conditions. As in the continuous case with V, we also factor out the constants here, i.e., in fact we consider $(\mathcal{P}_1(\mathcal{T}_H))_{/\mathbb{R}}$, but still write $\mathcal{P}_1(\mathcal{T}_H)$ for simplicity. Since $\varepsilon \leq H$ is assumed, the finite element method with the space $\mathcal{P}_1(\mathcal{T}_H)$ does not yield faithful approximations of the solution u_ε to (1); see, e.g., [37, Sec. 1]. We introduce a bounded local linear projection operator I_H: $V \to \mathcal{P}_1(\mathcal{T}_H)$, which can be seen as a composition $I_H := E_H \circ \Pi_H$, where a function $v \in V$ is first approximated on every element $T \in \mathcal{T}_H$ by its L^2-orthogonal projection Π_H onto the space of affine functions. Hence, a possibly globally discontinuous function $\Pi_H v$ is obtained. In the second step E_H, the values at the inner vertices of the triangulation are averages of the respective contributions from the single elements, i.e.,

$$E_H \circ \Pi_H(v)(z) := \frac{1}{\#\{T \in \mathcal{T}_H, z \in T\}} \sum_{\substack{T \in \mathcal{T}_H \\ z \in T}} \Pi_H(v)|_T(z)$$

for all vertices z, where the triangulation is understood in a periodic manner, see [35].

Let $W := \mathrm{kern} I_H$ be the kernel of the quasi-interpolation operator I_H. It can be seen as the set of rapidly oscillating functions, which cannot be captured by standard finite elements functions on the (coarse) mesh $\mathcal{T}_H$. Motivated by the reformulation (5) of the cell problems and the interpretation of W as rapidly oscillating functions, we define now the correctors $q^\infty_{Q,j}$ as the unique solutions in W of the following variational problems

$$\int_\Omega A_\varepsilon(x)\nabla q^\infty_{Q,j}(x) \cdot \nabla w(x)\,\mathrm{d}x = \int_Q A_\varepsilon(x)e_j \cdot \nabla w(x)\,\mathrm{d}x, \tag{6}$$

for all $w \in W$, and the correctors are defined for every $Q \in Q_H$, $j = 1, 2$. We define the following w.r.t. Q_H piecewise constant numerical coefficient A^∞_H which will play the main role in Proposition 1:

$$\left[A^\infty_{H|Q}\right]_{kj} = \frac{1}{|Q|}\int_Q A_\varepsilon(x)e_j \cdot e_k\,\mathrm{d}x - \frac{1}{|Q|}\int_\Omega A_\varepsilon(x)\nabla q^\infty_{Q,j}(x) \cdot e_k\,\mathrm{d}x, \tag{7}$$

for all $Q \in Q_H$, $k, j = 1, 2$.

Proposition 1 *In the case that the mesh size H is an integer multiple of ε, the coefficient A^∞_H coincides with the homogenized coefficient A_0 from classical homogenization defined in* (3).

Proof We will first show that the function $q_j := \sum_{Q \subset Q_H} q^\infty_{Q,j}$ coincides with the corrector $\hat{q}_j \in V_\varepsilon$, the unique solution of the problem (5). The crucial observation needed for the proof is the fact that the space of ε-periodic functions is contained in the kernel W of the quasi-interpolation operator I_H, in the case of the present setting with the triangulations $\mathcal{T}_H$ and Q_H. To see this we observe that, for an ε-periodic function $v_\varepsilon \in V_\varepsilon$, the values $I_H(v_\varepsilon)(z)$ coincide for all $z \in \mathcal{N}_{\mathcal{T}_H}$. That is, $I_H(v_\varepsilon) \in \mathcal{P}_1(\mathcal{T}_H)$ is a global constant. As we factored out the constants, we can take the zero function as representative, i.e., $I_H(v_\varepsilon) = 0$.

Moreover, summing up the equations (6) over all $Q \in Q_H$ and taking advantage of the symmetry of A_ε, we get that $q_j := \sum_{Q \in Q_H} q^\infty_{Q,j}$ solves

$$\int_\Omega A_\varepsilon(x)\nabla q_j(x) \cdot \nabla w(x)\,\mathrm{d}x = \int_\Omega A_\varepsilon(x)e_j \cdot \nabla w(x)\,\mathrm{d}x \tag{8}$$

$$= \int_\Omega A_\varepsilon(x)\nabla w(x) \cdot e_j\,\mathrm{d}x, \tag{9}$$

for all $w \in W$, and in particular for all $w \in V_\varepsilon$. The combination of (5) and (9) readily yields that $q_j \equiv \hat{q}_j$, $j = 1, 2$. Moreover, (9) with $w = q^\infty_{Q,k}$ implies

$$\begin{aligned}\int_\Omega A_\varepsilon(x)\nabla q^\infty_{Q,k}(x)\cdot\nabla e_j\,\mathrm{d}x &= \int_\Omega A_\varepsilon(x)\nabla q_j(x)\cdot\nabla q^\infty_{Q,k}(x)\,\mathrm{d}x\\ &= \int_\Omega A_\varepsilon(x)\nabla q^\infty_{Q,k}(x)\cdot\nabla q_j(x)\,\mathrm{d}x\\ &= \int_Q A_\varepsilon(x)e_k\cdot\nabla q_j(x)\,\mathrm{d}x.\end{aligned}$$

Hence, in the definition of A^∞_H we can replace the second term, namely

$$\begin{aligned}\left(A^\infty_{H|Q}\right)_{kj} &= \frac{1}{|Q|}\int_Q A_\varepsilon(x)e_j\cdot e_k\,\mathrm{d}x - \frac{1}{|Q|}\int_Q A_\varepsilon(x)e_j\cdot\nabla q_k(x)\,\mathrm{d}x\\ &= \frac{1}{|Q|}\int_Q A_\varepsilon(x)e_j\cdot(e_k-\nabla\hat{q}_k(x))\,\mathrm{d}x\\ &= \left(A_0\right)_{kj},\end{aligned}$$

for $j,k=1,2$. □

4 Numerical effective coefficient by domain decomposition

The correctors $q^\infty_{Q,j}$ defined in the previous section require the solution of a global problem involving the oscillating coefficient A_ε. Employing domain decomposition, we introduce localized variants and then use arguments from the theory of iterative (domain decomposition) methods as presented in [26, 28] to show that the error decays exponentially in the number of iterations. With the localized correctors, we then introduce an effective localized coefficient A^ℓ_H which is piecewise constant on $\mathcal{Q}_H$.

Let ω_i be the union of all squares $Q\in\mathcal{Q}_H$ having the vertex z_i as a corner and let

$$W_i=\{v-I_Hv\mid v\in H^1_0(\omega_i)\}.$$

We emphasize that ω_i is understood as a subset of $\mathbb{R}^2$, i.e., it is continued over the periodic boundary. The functions in W_i vanish outside a small neighbourhood of the vertex z_i. The W_i are closed subspaces of the kernel W of I_H, see [26]. Let P_i be the a_ε-orthogonal projection from V to W_i, defined via the equation

$$a_\varepsilon(P_iv,w_i)=a_\varepsilon(v,w_i),\quad \forall w_i\in W_i.$$

Introducing the with respect to the bilinear form $a_\varepsilon(\cdot,\cdot)$ symmetric operator

$$P=P_1+P_2+\cdots+P_n,$$

the following properties are proved in [26]:

Lemma 1 *There are constants K_1 and K_2, independent of H and ε, such that*

$$K_1^{-1} a_\varepsilon(v,v) \le a_\varepsilon(Pv,v) \le K_2 a_\varepsilon(v,v)$$

for all $v \in V$. Moreover, for an appropriate scaling factor ϑ only depending on K_1 and K_2, there exists a positive constant $\gamma < 1$ such that

$$\| \mathrm{id} - \vartheta P \|_{\mathcal{L}(V,V)} \le \gamma. \tag{10}$$

Starting from $q_{Q,j}^0 = 0$, $j = 1,2$, the localized correctors $q_{Q,j}^\ell$ are defined for all $Q \in \mathcal{Q}_H$ via

$$q_{Q,j}^{\ell+1} = q_{Q,j}^\ell + \vartheta P(x_j 1_Q - q_{Q,j}^\ell), \qquad j = 1,2, \tag{11}$$

where 1_Q denotes the characteristic function of Q and x_j denotes the j-th component of the (vector-valued) function $x \mapsto x$. The scaling factor ϑ is chosen as discussed in Lemma 1. The correction $P(x_j 1_Q - q_{Q,j}^\ell)$ is the sum of its components $C_i^\ell = P_i(x_j 1_Q - q_{Q,j}^\ell)$ in the subspaces W_i of W, where the C_i^ℓ solve the local equations

$$a_\varepsilon(C_i^\ell, w_i) = a_\varepsilon(x_j 1_Q, w_i) - a_\varepsilon(q_{Q,j}^\ell, w_i), \qquad \forall w_i \in W_i. \tag{12}$$

The sloppy notation using 1_Q as argument in a_ε is to denote that the integration is over the element Q only, i.e., $a_\varepsilon(x_j 1_Q, w_i) = \int_Q A_\varepsilon e_j \cdot \nabla w_i \, dx$. Since the local projections P_i only slightly increase the support of a function, we deduce inductively that the support of $q_{Q,j}^\ell$ is contained in an ℓH-neighbourhood of Q. In particular, in each step of (11) only a few local problems of type (12) have to be solved.

We now replace $q_{Q,j}^\infty$ by its localized variant $q_{Q,j}^\ell$ in the definition of the numerical effective coefficient. This procedure is justified by an exponential error estimate in Proposition 2. We define the piecewise constant (on the mesh $\mathcal{Q}_H$) (localized) effective matrix A_H^ℓ via

$$\left(A_H^\ell|_Q\right)_{kj} = \frac{1}{|Q|}\int_Q A_\varepsilon(x) e_j \cdot e_k \, dx - \frac{1}{|Q|}\int_\Omega A_\varepsilon \nabla q_{Q,j}^\ell(x) \cdot e_k \, dx. \tag{13}$$

Since the numerical effective coefficient (7) is the "true" one in the sense that $A_H^\infty = A_0$, we simply need to estimate the error of the iterative approximation.

Proposition 2 *Let H be an integer multiple of ε and let the localization parameter ℓ be chosen of order $\ell \approx |\log H|$. Then,*

$$\|A_H^\infty - A_H^\ell\|_{L^\infty(\Omega)} \lesssim H. \tag{14}$$

Proof We first estimate the error between the correctors $q_{Q,j}^\infty$ and $q_{Q,j}^\ell$. Using the definition of $q_{Q,j}^\infty$ in (6), we deduce that $P(x_j 1_Q) = P(q_{Q,j}^\infty)$. Hence, we can characterize the error between the correctors $q_{Q,j}^\infty$ and their localized approximations $q_{Q,j}^\ell$ via

$$q_{Q,j}^\infty - q_{Q,j}^\ell = (\mathrm{id} - \vartheta P)^\ell q_{Q,j}^\infty.$$

Using (10), this yields the exponential convergence of $q^{\ell}_{Q,j}$ towards $q^{\infty}_{Q,j}$, i.e.,

$$\|\nabla(q^{\infty}_{Q,j} - q^{\ell}_{Q,j})\| \lesssim \gamma^{\ell} \|\nabla q^{\infty}_{Q,j}\| \lesssim \gamma^{\ell} |Q|^{1/2}. \tag{15}$$

By the definitions of A^{∞}_{H} in (7) and A^{ℓ}_{H} in (13), we obtain

$$\begin{aligned}\left|(A^{\infty}_{H}|_{Q})_{jk} - (A^{\ell}_{H}|_{Q})_{jk}\right| &= |Q|^{-1}\left|\int_{\Omega} A_{\varepsilon}\nabla(q^{\ell}_{Q,j} - q^{\infty}_{Q,j}) \cdot e_k \, dx\right| \\ &\lesssim |Q|^{-1}\|e_k\|_{L^2(\Omega)}\|\nabla(q^{\ell}_{Q,j} - q^{\infty}_{Q,j})\|_{L^2(\Omega)}.\end{aligned}$$

Estimate (15) and the choice $\ell \approx |\log H|$ readily imply the assertion. □

The same estimate was previously derived in [14] with a slightly different localization strategy and with more restrictive conditions on the triangulation. There, the homogenization error in the L^2-norm is quantified as follows. Let Ω be convex. Let $u_{\varepsilon} \in V$ solve (1) and let $u_0 \in V$ be the solution to (2). For sufficiently small ε, it holds that

$$\|u_{\varepsilon} - u_0\|_{L^2(\Omega)} \lesssim \varepsilon |\log \varepsilon|^2 \|f\|_{L^2(\Omega)}.$$

This estimate recovers the classical result that $u_{\varepsilon} \to u_0$ strongly in L^2 and furthermore states that the convergence is almost linear for right-hand sides $f \in L^2(\Omega)$. We shall emphasize that the proofs of [14] are solely based on standard techniques of finite elements. The authors believe that such a result is also possible in the slightly more general setup of this paper. However, it seems that there is no simple argument but the generalization requires to revise the analysis of [14] step by step which is far beyond the scope of this paper.

5 Beyond periodicity and scale separation

The numerical approach presented in Section 4 does not essentially rely on the assumption of periodicity or separation of scales (between the length scales of the computational domain and the material structures). Of course, in such general situations, one cannot identify a constant effective coefficient. Instead the goal is to faithfully approximate the analytical solution by a (generalized) finite element method based on a (coarse) mesh, which does not need to resolve the fine material structures and thereby is computationally efficient.

For this generalization, note that the definition (11) can be formulated verbatim for any boundary value problem involving a potentially rough, but not necessarily periodic diffusion tensor $A \in L^{\infty}(\Omega)$. Moreover, the choice of the function $x_j 1_Q$ in the definition of $q^{\ell}_{Q,j}$ can be generalized to any function $v \in V$ in the following way. Define the operator $C^{\ell}_{T} : V \to W$ inductively via $C^{0}_{T} = 0$ and

$$C^{\ell+1}_{T} = C^{\ell}_{T} + \vartheta P(\mathrm{id}|_{T} - C^{\ell}_{T})$$

for all $T \in \mathcal{T}_H$, see [26]. Instead of modifying the diffusion tensor as in the previous sections, we then modify the basis functions and define a generalized finite element method using the test and ansatz spaces $V_H^\ell := (\mathrm{id} + C^\ell)\mathcal{P}_1(\mathcal{T}_H)$ with $C^\ell := \sum_{T \in \mathcal{T}_H} C_T^\ell$. This method is known as the Localized Orthogonal Decomposition (LOD) [23, 31, 19, 37] and originally arose from the concept of the Variational Multiscale Method [24, 25]. Note that mostly a slightly different definition of the correctors C_T^ℓ based on patches of diameter ℓH around the element T is used. The present approach via domain decomposition and iterative solvers was developed recently in [28, 26]. It has been shown in [31, 19] for instance, that the method approximates the analytical solution with an energy error of the order H even in the pre-asymptotic regime if the localization parameter ℓ is chosen of the order $\ell \approx |\log H|$ as in Proposition 2. Hence, the Localized Orthogonal Decomposition can efficiently treat general multiscale problems. Besides the above mentioned Galerkin-type ansatz with modified ansatz and test functions, Petrov-Galerkin formulations of the method [9] may have computational advantages [10] and even meshless methods are possible [21].

The Localized Orthogonal Decomposition is not restricted to elliptic diffusion problems and has underlined its potential in various applications and with respect to different (computational) challenges. Starting from the already mentioned application in the geosciences, we underline that the material coefficients are often characterized not only by rapid oscillations but also by a high contrast, i.e., the ratio β/α is large. Many error estimates, also for the standard LOD, are contrast-dependent, but a careful choice of the interpolation operator, see [17, 40], can overcome this effect. Apart from simple diffusion problems, porous media [7], elasticity problems [22] or coupling of those such as in poroelasticity [4] play important roles in these (and many other) applications. For instance in elasticity theory, not only heterogeneous materials are treated, but also the effect of locking can be reduced by the multiscale method in [22].

Another important area of research are acoustic and electromagnetic wave propagation problems, where the considered prototypical equations are the Helmholtz and Maxwell's equations. It is well known that standard finite element discretizations of the (indefinite) Helmholtz equation are only well-posed and converging under a rather restrictive resolution condition between the mesh size and the wavenumber. In a series of paper [6, 13, 38], it was analysed that the LOD can relax this resolution condition if the localization parameter grows logarithmically with the wavenumber. For large wavenumbers, this is a great computational gain in comparison to standard numerical methods that even allows the simulation of physical phenomena in high contrast regimes [41]. Maxwell's equations, studied in [12, 42], on the other hand, pose a challenge as the involved curl-operator has a large kernel. Moreover, the natural finite element space are Nédélec's edge elements, for which stable interpolation operators are much less developed than for Lagrange finite elements. In the context of problems not based on standard Lagrange spaces, we also mention the mixed problem utilizing Raviart-Thomas spaces in [16]. Considering wave problems, the time-dependent wave equation with different time discretizations was studied in [1, 30]. Concerning time-dependency, an important question for the LOD construction is how to deal with time-dependent diffusion tensors. [18] presents an a posteriori error estimator in order to adaptively decide which correction to recompute in the next time step.

Apart from the treatment of multiscale coefficients in a variety of partial differential equations, the methodology can also be seen as a stabilization scheme similar as its origin the variational multiscale methods. This has been exploited to deal with the pollution effect in Helmholtz problems mentioned above, for convection dominated diffusion problems [29] and, more importantly, to bypass CFL conditions in the context of explicit wave propagation on adaptive meshes [39].

Further unexpected applications are linear and nonlinear eigenvalue problems [32, 33], in particular the quantum-physical simulation based on the Gross-Pitaevskii equation. While the LOD can be employed to speed-up ground state computations for rather rough potentials [20], the underlying technique of localization by domain decomposition turned out to be of great value to provide (analytical) insight into the phenomenon of Anderson localization in this context. The recent paper [5] predicts and quantifies the emergence of localized eigenstates and might inspire progress regarding the understanding of localization effects which are observed for many other problems as well.

The present contribution aimed at unifying the view of the LOD and classical homogenization and domain decomposition. As already mentioned, close connections exist with [14] and its extension to stochastic homogenization [15]. Further applications involve a multilevel generalization of LOD named gamblets [36] (due to a possible game-theoretic interpretation). This multilevel variant allows surprising results such as a sparse representation of the expected solution operator for random elliptic boundary value problems [11] which may inspire new computational strategies for uncertainty quantification in the future.

Acknowledgements This work was supported in part by the German Research Foundation (DFG) through the Priority Programme 1648 "Reliable simulation techniques in solid mechanics" under (PE2143/2-2).

References

1. Abdulle, A., Henning, P.: Localized orthogonal decomposition method for the wave equation with a continuum of scales. Math. Comp. **86**(304), 549–587 (2017). DOI:10.1090/mcom/3114
2. Abdulle, A., Weinan, E., Engquist, B., Vanden-Eijnden, E.: The heterogeneous multiscale method. Acta Numer. **21**, 1–87 (2012). DOI:10.1017/S0962492912000025
3. Allaire, G.: Homogenization and two-scale convergence. SIAM J. Math. Anal. **23**(6), 1482–1518 (1992). DOI:10.1137/0523084
4. Altmann, R., Chung, E., Maier, R., Peterseim, D., Pun, S.: Computational multiscale methods for linear heterogeneous poroelasticity. ArXiv preprint **1801.00615** (2019). Accepted for publication in J. Comput. Math.
5. Altmann, R., Henning, P., Peterseim, D.: Quantitative anderson localization of schrödinger eigenstates under disorder potentials. ArXiv preprint **1803.09950** (2018)

6. Brown, D.L., Gallistl, D., Peterseim, D.: Multiscale Petrov-Galerkin method for high-frequency heterogeneous Helmholtz equations. In: Meshfree methods for partial differential equations VIII, *Lect. Notes Comput. Sci. Eng.*, vol. 115, pp. 85–115. Springer, Cham (2017)
7. Brown, D.L., Peterseim, D.: A multiscale method for porous microstructures. Multiscale Model. Simul. **14**(3), 1123–1152 (2016). DOI:10.1137/140995210
8. Efendiev, Y., Hou, T.Y.: Multiscale finite element methods, *Surveys and Tutorials in the Applied Mathematical Sciences*, vol. 4. Springer, New York (2009). Theory and applications
9. Elfverson, D., Ginting, V., Henning, P.: On multiscale methods in Petrov-Galerkin formulation. Numer. Math. **131**(4), 643–682 (2015). DOI:10.1007/s00211-015-0703-z
10. Engwer, C., Henning, P., Målqvist, A., Peterseim, D.: Efficient implementation of the localized orthogonal decomposition method. Comput. Methods Appl. Mech. Engrg. **350**, 123–153 (2019). DOI:10.1016/j.cma.2019.02.040
11. Feischl, M., Peterseim, D.: Sparse compression of expected solution operators. ArXiv preprint **1807.01741** (2018)
12. Gallistl, D., Henning, P., Verfürth, B.: Numerical homogenization of $\mathbf{H}(\mathrm{curl})$-problems. SIAM J. Numer. Anal. **56**(3), 1570–1596 (2018). DOI:10.1137/17M1133932
13. Gallistl, D., Peterseim, D.: Stable multiscale Petrov-Galerkin finite element method for high frequency acoustic scattering. Comput. Methods Appl. Mech. Engrg. **295**, 1–17 (2015). DOI:10.1016/j.cma.2015.06.017. URL https://doi.org/10.1016/j.cma.2015.06.017
14. Gallistl, D., Peterseim, D.: Computation of quasi-local effective diffusion tensors and connections to the mathematical theory of homogenization. Multiscale Model. Simul. **15**(4), 1530–1552 (2017). DOI:10.1137/16M1088533
15. Gallistl, D., Peterseim, D.: Numerical stochastic homogenization by quasi-local effective diffusion tensors. ArXiv preprint **1702.08858** (2019). Accepted for publication in Communications in Mathematical Sciences
16. Hellman, F., Henning, P., Målqvist, A.: Multiscale mixed finite elements. Discrete Contin. Dyn. Syst. Ser. S **9**(5), 1269–1298 (2016). DOI:10.3934/dcdss.2016051
17. Hellman, F., Målqvist, A.: Contrast independent localization of multiscale problems. Multiscale Model. Simul. **15**(4), 1325–1355 (2017). DOI:10.1137/16M1100460
18. Hellman, F., Målqvist, A.: Numerical homogenization of elliptic PDEs with similar coefficients. Multiscale Model. Simul. **17**(2), 650–674 (2019). DOI:10.1137/18M1189701
19. Henning, P., Målqvist, A.: Localized orthogonal decomposition techniques for boundary value problems. SIAM J. Sci. Comput. **36**(4), A1609–A1634 (2014). DOI:10.1137/130933198
20. Henning, P., Målqvist, A., Peterseim, D.: Two-level discretization techniques for ground state computations of Bose-Einstein condensates. SIAM J. Numer. Anal. **52**(4), 1525–1550 (2014). DOI:10.1137/130921520
21. Henning, P., Morgenstern, P., Peterseim, D.: Multiscale partition of unity. In: Meshfree methods for partial differential equations VII, *Lect. Notes Comput. Sci. Eng.*, vol. 100, pp. 185–204. Springer, Cham (2015)
22. Henning, P., Persson, A.: A multiscale method for linear elasticity reducing Poisson locking. Comput. Methods Appl. Mech. Engrg. **310**, 156–171 (2016). DOI:10.1016/j.cma.2016.06.034
23. Henning, P., Peterseim, D.: Oversampling for the multiscale finite element method. Multiscale Model. Simul. **11**(4), 1149–1175 (2013). DOI:10.1137/120900332
24. Hughes, T.J.R., Feijóo, G.R., Mazzei, L., Quincy, J.B.: The variational multiscale method—a paradigm for computational mechanics. Comput. Methods Appl. Mech. Engrg. **166**(1-2), 3–24 (1998). DOI:10.1016/S0045-7825(98)00079-6
25. Hughes, T.J.R., Sangalli, G.: Variational multiscale analysis: the fine-scale Green's function, projection, optimization, localization, and stabilized methods. SIAM J. Numer. Anal. **45**(2), 539–557 (2007). DOI:10.1137/050645646
26. Kornhuber, R., Peterseim, D., Yserentant, H.: An analysis of a class of variational multiscale methods based on subspace decomposition. Math. Comp. **87**(314), 2765–2774 (2018). DOI:10.1090/mcom/3302

27. Kornhuber, R., Podlesny, J., Yserentant, H.: Direct and iterative methods for numerical homogenization. In: Domain decomposition methods in science and engineering XXIII, *Lect. Notes Comput. Sci. Eng.*, vol. 116, pp. 217–225. Springer, Cham (2017)
28. Kornhuber, R., Yserentant, H.: Numerical homogenization of elliptic multiscale problems by subspace decomposition. Multiscale Model. Simul. **14**(3), 1017–1036 (2016). DOI:10.1137/15M1028510
29. Li, G., Peterseim, D., Schedensack, M.: Error analysis of a variational multiscale stabilization for convection-dominated diffusion equations in two dimensions. IMA J. Numer. Anal. **38**(3), 1229–1253 (2018). DOI:10.1093/imanum/drx027
30. Maier, R., Peterseim, D.: Explicit computational wave propagation in micro-heterogeneous media. BIT **59**(2), 443–462 (2019). DOI:10.1007/s10543-018-0735-8
31. Målqvist, A., Peterseim, D.: Localization of elliptic multiscale problems. Math. Comp. **83**(290), 2583–2603 (2014). DOI:10.1090/S0025-5718-2014-02868-8
32. Målqvist, A., Peterseim, D.: Computation of eigenvalues by numerical upscaling. Numer. Math. **130**(2), 337–361 (2015). DOI:10.1007/s00211-014-0665-6
33. Målqvist, A., Peterseim, D.: Generalized finite element methods for quadratic eigenvalue problems. ESAIM Math. Model. Numer. Anal. **51**(1), 147–163 (2017). DOI:10.1051/m2an/2016019
34. Murat, F., Tartar, L.: H-convergence. Séminaire d'Analyse Fonctionnelle et Numérique de l'Université d'Alger (1978)
35. Ohlberger, M., Verfürth, B.: Localized Orthogonal Decomposition for two-scale Helmholtz-type problems. AIMS Mathematics **2**(3), 458–478 (2017)
36. Owhadi, H.: Multigrid with rough coefficients and multiresolution operator decomposition from hierarchical information games. SIAM Rev. **59**(1), 99–149 (2017). DOI:10.1137/15M1013894
37. Peterseim, D.: Variational multiscale stabilization and the exponential decay of fine-scale correctors. In: Building bridges: connections and challenges in modern approaches to numerical partial differential equations, *Lect. Notes Comput. Sci. Eng.*, vol. 114, pp. 341–367. Springer, [Cham] (2016)
38. Peterseim, D.: Eliminating the pollution effect in Helmholtz problems by local subscale correction. Math. Comp. **86**(305), 1005–1036 (2017). DOI:10.1090/mcom/3156
39. Peterseim, D., Schedensack, M.: Relaxing the CFL condition for the wave equation on adaptive meshes. J. Sci. Comput. **72**(3), 1196–1213 (2017). DOI:10.1007/s10915-017-0394-y
40. Peterseim, D., Scheichl, R.: Robust numerical upscaling of elliptic multiscale problems at high contrast. Comput. Methods Appl. Math. **16**(4), 579–603 (2016). DOI:10.1515/cmam-2016-0022
41. Peterseim, D., Verfürth, B.: Computational high frequency scattering from high contrast heterogeneous media. ArXiv preprint **1902.09935** (2019)
42. Verfürth, B.: Numerical homogenization for indefinite H(curl)-problems. In: Proceedings of Equadiff 2017 conference, pp. 137–146. Slovak University of Technology, Bratislava (2017)

Robust Model Reduction Discretizations Based on Adaptive BDDC Techniques

Alexandre Madureira and Marcus Sarkis

1 Introduction

Consider the problem of finding the weak solution $u : \Omega \to \mathbb{R}$ of

$$\begin{aligned} -\operatorname{div} \mathcal{A}\nabla u &= f \quad \text{in } \Omega, \\ u &= 0 \quad \text{on } \partial\Omega. \end{aligned} \tag{1}$$

Here $\Omega \subset \mathbb{R}^2$ is an open bounded domain with polygonal boundary $\partial\Omega$, the symmetric tensor $\mathcal{A} \in [L^\infty(\Omega)]^{2\times 2}_{\text{sym}}$ is uniformly positive definite and bounded.

For almost all $\boldsymbol{x} \in \Omega$ let the positive constants c_1 and c_2 be such that

$$c_1|\boldsymbol{v}|^2 \le a_{\min}(\boldsymbol{x})|\boldsymbol{v}|^2 \le \mathcal{A}(\boldsymbol{x})\,\boldsymbol{v}\cdot\boldsymbol{v} \le a_{\max}(\boldsymbol{x})|\boldsymbol{v}|^2 \le c_2|\boldsymbol{v}|^2 \text{ for all } \boldsymbol{v} \in \mathbb{R}^2, \text{ a.e. } \boldsymbol{x} \in \Omega.$$

The associated variational formulation is given by: Find $u \in H^1_0(\Omega)$ such that

$$a(u,v) := \int_\Omega \mathcal{A}\nabla u \cdot \nabla v\, d\boldsymbol{x} = \int_\Omega f v\, d\boldsymbol{x} =: (f,v) \quad \forall v \in H^1_0(\Omega).$$

Recently, methods that do not rely on the regularity of the solution were introduced: generalized finite element methods [1], the rough polyharmonic splines [22], the variational multiscale method (VMS) [13], and the Localized Orthogonal Decomposition (LOD) [16, 10]. These methods are based on splitting approximation spaces into *fine* and *multiscale* subspaces, and the numerical solution of (1) is sought in the latter. We note that these works were designed for the low-contrast case, that is, c_2/c_1 not large. We note that for a class of coefficients $\mathcal{A}$, that is, when local Poincaré inequality constants are not large, the LOD methodology works [24].

Alexandre Madureira
Laboratório Nacional Computação Científica, Petrópolis, Brazil, e-mail: alm@lncc.br

Marcus Sarkis
Worcester Polytechnic Institute, e-mail: msarkis@wpi.edu

R. Haynes et al. (eds.), *Domain Decomposition Methods in Science and Engineering XXV*, Lecture Notes in Computational Science and Engineering 138,
https://doi.org/10.1007/978-3-030-56750-7_4

On the other side, there exist several domain decomposition solvers which are optimal with respect to mesh and contrast. All of them are based on extracting coarse basis functions from local generalized eigenvalue problems. For non-overlapping domain decomposition based on the technique named *adaptive choice of primal constraints* was introduced in [19], revisited in [23, 15]; see also [6, 21] and references therein. We note that earlier ideas were also introduced in [3]. This robustness also was developed for overlapping domain decomposition methods and we refer the earlier works in [7, 20].

In this paper we consider *Approximate Component Mode Synthesis*–ACMS methods [5, 4, 2, 9, 12, 11, 14]; these methods require extra solution regularity and do not work for high contrast. The goal here is to develop a discretization that has optimal energy a priori error approximation, assuming no regularity on the solution and on $\mathcal{A}$. To do that we combine adaptive BDDC and LOD techniques; see also [17] for a similar combination however for mixed finite element discretizations.

The remainder of the this paper is organized as follows. Section 2 describes the substructuring decomposition into interior and interface unknowns and in Section 3 we present the goal of the paper. In Section 4 the model reduction method via adaptive BDDC is proposed and the results are discussed. In Section 5 we consider how to deal with elementwise problems. In Section 6 numerical results are presented.

2 Discrete Substructuring Formulation

We start by defining a partition of Ω by a triangular finite element regular mesh $\mathcal{T}_H$ with elements of characteristic length $H > 0$. Let $\partial\mathcal{T}_h$ be the mesh skeleton, and $\mathcal{N}_H$ the set of nodes on $\partial\mathcal{T}_h\backslash\partial\Omega$. Consider $\mathcal{T}_h$, a refinement of $\mathcal{T}_H$, in the sense that every (coarse) edge of the elements in $\mathcal{T}_H$ can be written as a union of edges of $\mathcal{T}_h$. We assume that $h < H$. Let $\mathcal{N}_h$ be the set of nodes of $\mathcal{T}_h$ on the skeleton $\partial\mathcal{T}_h\backslash\partial\Omega$; thus all nodes in $\mathcal{N}_h$ belong to edges of elements in $\mathcal{T}_H$.

For $v \in H^1(\Omega)$ let

$$|v|^2_{H^1_{\mathcal{A}}(\Omega)} = \|\mathcal{A}^{1/2}\nabla v\|^2_{L^2(\Omega)}, \qquad |v|^2_{H^1_{\mathcal{A}}(\mathcal{T})} = \sum_{\tau\in\mathcal{T}} \|\mathcal{A}^{1/2}\nabla v\|^2_{L^2(\tau)},$$

where $\mathcal{T} \subset \mathcal{T}_H$ denotes a given set of elements. Let $V_h \subset H^1_0(\Omega)$ be the space of continuous piecewise linear functions related to $\mathcal{T}_h$. Let $u_h \in V_h$ such that

$$a(u_h, v_h) = (f, v_h) \quad \text{for all } v_h \in V_h.$$

We assume that u_h approximates u well, but we remark that u_h is never computed; the goal here is to develop numerical schemes which yield good approximations for u_h, therefore, the schemes proposed can be viewed as a model reduction method.

We can decompose $u_h = u_h^{\mathbb{B}} \oplus u_h^{\mathbb{H}}$ in its bubble (belonging to $V_h^{\mathbb{B}}$) and a-discrete harmonic components (belonging to $V_h^{\mathbb{H}}$), respectively, where

$$
\begin{aligned}
V_h^{\mathbb{B}} &= \{v_h \in V_h : v_h = 0 \text{ on } \partial\tau,\ \tau \in \mathcal{T}_H\},\\
V_h^{\mathbb{H}} &= \{u_h^{\mathbb{H}} \in V_h : a(u_h^{\mathbb{H}}, v_h^{\mathbb{B}}) = 0 \text{ for all } v_h^{\mathbb{B}} \in V_h^{\mathbb{B}}\},
\end{aligned}
$$

i.e., $V_h^{\mathbb{H}} = (V_h^{\mathbb{B}})^{\perp_a}$. It follows immediately from the definitions that

$$
a(u_h^{\mathbb{H}}, v_h^{\mathbb{H}}) = (f, v_h^{\mathbb{H}}) \quad \text{for all } v_h^{\mathbb{H}} \in V_h^{\mathbb{H}}, \qquad a(u_h^{\mathbb{B}}, v_h^{\mathbb{B}}) = (f, v_h^{\mathbb{B}}) \quad \text{for all } v_h^{\mathbb{B}} \in V_h^{\mathbb{B}}.
$$

Although the problem related to $u_h^{\mathbb{B}}$ is global, it can be decomposed in local uncoupled problems, as discussed in Section 5.

Note that any function in $V_h^{\mathbb{H}}$ is uniquely determined by its trace on the boundary of elements in $\mathcal{T}_H$. Let us define

$$
\Lambda_h = \{v_h|_{\partial\mathcal{T}_h} : v_h \in V_h^{\mathbb{H}}\} \subset H^{1/2}(\partial\mathcal{T}_h),
$$

and $T : \Lambda_h \to V_h^{\mathbb{H}}$ be the *local* discrete-harmonic extension operator given by

$$
(T\mu_h)|_{\partial\mathcal{T}_h} = \mu_h, \qquad \text{and} \qquad a(T\mu_h, v_h^{\mathbb{B}}) = 0 \quad \text{for all } v_h^{\mathbb{B}} \in V_h^{\mathbb{B}}.
$$

For $\tau \in \mathcal{T}_H$, let $\Lambda_h^\tau = \Lambda_h|_{\partial\tau}$, that is, the restriction of functions on Λ_h to $\partial\tau$. Define the bilinear forms $s : \Lambda_h \times \Lambda_h \to \mathbb{R}$ and $s^\tau : \Lambda_h^\tau \times \Lambda_h^\tau \to \mathbb{R}$ such that, for μ_h, $v_h \in \Lambda_h$,

$$
s(\mu_h, v_h) = \sum_{\tau\in\mathcal{T}_H} s_\tau(\mu_h^\tau, v_h^\tau) \quad \text{where} \quad s_\tau(\mu_h^\tau, v_h^\tau) = \int_\tau \mathcal{A}\nabla T^\tau \mu_h^\tau \cdot \nabla T^\tau v_h^\tau \, d\boldsymbol{x}
$$

where T^τ is the restriction of T to τ. Let $\lambda_h = u_h|_{\partial\mathcal{T}_h}$. Then $u_h^{\mathbb{H}} = T\lambda_h$ and

$$
s(\lambda_h, \mu_h) = (f, T\mu_h) \quad \text{for all } \mu_h \in \Lambda_h. \tag{2}
$$

3 Main Goal of the Paper

Let us introduce $\rho \in L^\infty(\Omega)$ such that $\rho(\boldsymbol{x}) \in [\rho_{\min}, \rho_{\max}]$ almost everywhere for some positive constants $\rho_{\min}$ and $\rho_{\max}$, and define $g = f/\rho$ and the spaces for g or f such that

$$
\|g\|_{L^2_\rho(\Omega)} = \|\rho^{1/2} g\|_{L^2(\Omega)} = \|f\|_{L^2_{1/\rho}(\Omega)} < \infty
$$

The main goal of this paper is the following: Given a threshold δ, construct a lower-dimensional subspace $\Lambda_h^{ms} \subset \Lambda_h$, such that for any $g \in L^2_\rho(\Omega)$ (or equivalently $f \in L^2_{1/\rho}(\Omega)$) the multiscale solution $\lambda_h^{ms}(g) \in \Lambda_h^{ms}$ of

$$
s(\lambda_h^{ms}, \mu_h^{ms}) = (\rho g, T\mu_h^{ms}) \quad \text{for all } \mu_h^{ms} \in \Lambda_h^{ms}
$$

satisfies

$$|u_h^{\mathbb{H}} - T\lambda_h^{ms}|^2_{H^1_{\mathcal{A}}(\Omega)} = s(\lambda_h - \lambda_h^{ms}, \lambda_h - \lambda_h^{ms}) \leq C\delta^2 \|g\|^2_{L_\rho(\Omega)}. \tag{3}$$

where the constant C does not depend on g, $\mathcal{A}$ or ρ.

The reason for introducing the weight function ρ is to normalize the equation (1). For example, assume that $\mathcal{A}(\boldsymbol{x}) = 10^{-6}$. Then the solution of (1) satisfies $-\Delta u = f/10^{-6}$. This means that if we want to obtain an approximation like (3) with $\rho = 1$ and with C independently of $\mathcal{A}$, it would require a large space Λ_h^{ms}, maybe as large as the fine space Λ_h. So, it is natural for this case to choose $\rho = 10^{-6}$. And vice-versa, if $\mathcal{A}(\boldsymbol{x}) = 10^6$, an estimate like (3) with $\rho = 1$ would be too easy, it would not give a good relative energy approximation. We think that a judicious choice is $\rho(\boldsymbol{x}) = a_{\min}(\boldsymbol{x})$ since the approximation also will capture the anisotropy of $\mathcal{A}(\boldsymbol{x})$. Another reason is that similarly as discussed in [8], the dimension of the space Λ_h^{ms} is related to the number of highly conductive fingerings crossing the edges of the coarse triangulation $\mathcal{T}_H$.

4 Model Reduction via BDDC

We now propose a scheme to approximate λ_h in (2) based on LOD and BDDC techniques. Decompose $\Lambda = \Lambda^0 \oplus \widetilde{\Lambda}_h$ by

$$\widetilde{\Lambda}_h = \{\lambda \in \Lambda_h : \lambda(\boldsymbol{x}_i) = 0 \text{ for all } \boldsymbol{x}_i \in \mathcal{N}_H\},$$
$$\Lambda^0 = \{\lambda \in \Lambda_h : \lambda(\boldsymbol{x}_i) = 0 \text{ for all } \boldsymbol{x}_i \in \mathcal{N}_h \backslash \mathcal{N}_H\}.$$

Let e be an edge of $\partial\mathcal{T}_H \backslash \partial\Omega$ shared by the elements τ and τ' of $\mathcal{T}_H$, and denote $\widetilde{\Lambda}_h^e = \widetilde{\Lambda}_h|_e$, that is, the restriction of functions on $\widetilde{\Lambda}_h$ to e. Note that a function $\widetilde{\mu}_h^e \in \widetilde{\Lambda}_h^e$ vanishes at the end-points of e; it is thus possible to extend continuously by zero to either $\partial\tau$ or $\partial\tau'$. Let us denote this extension by $R_{e,\tau}^T : \widetilde{\Lambda}_h^e \to \Lambda_h^\tau$.

Let us define $S_{ee}^\tau : \widetilde{\Lambda}_h^e \to (\widetilde{\Lambda}_h^e)'$, where $(\widetilde{\Lambda}_h^e)'$ is the dual space of $\widetilde{\Lambda}_h^e$, by

$$(\widetilde{\mu}_h^e, S_{ee}^\tau \widetilde{v}_h^e)_e = (R_{e,\tau}^T \widetilde{\mu}_h^e, S^\tau R_{e,\tau}^T \widetilde{v}_h^e)_{\partial\tau} \quad \text{for all } \widetilde{\mu}_h^e, \widetilde{v}_h^e \in \widetilde{\Lambda}_h^e,$$

where $(\cdot,\cdot)_e$ is the $L^2(e)$ inner product and

$$(\mu_h^\tau, S^\tau v_h^\tau)_{\partial\tau} = \int_\tau \mathcal{A}\nabla T^\tau \mu_h^\tau \cdot \nabla T^\tau v_h^\tau \, d\boldsymbol{x} \qquad \text{for all } \mu_h^\tau, v_h^\tau \in \Lambda_h^\tau.$$

In a similar fashion, define $S_{e^ce}^\tau$, $S_{ee^c}^\tau$ and $S_{e^ce^c}^\tau$, related to the degrees of freedom on $e^c = \partial\tau \backslash e$. We remind that e is an open edge, not containing its endpoints.

Let us introduce M_{ee}^τ by

$$(\tilde{\mu}_h^e, M_{ee}^\tau \tilde{v}_h^e)_e = \int_\tau \rho\, (T^\tau R_{e,\tau}^T \tilde{\mu}_h^e)\, (T^\tau R_{e,\tau}^T \tilde{v}_h^e) \, d\boldsymbol{x}$$

and define $\widehat{S}_{ee}^{\tau} = \delta^{-2} M_{ee}^{\tau} + S_{ee}^{\tau}$, where δ is the target precision of the method, that can be set by the user.

Define also

$$\widetilde{S}_{ee}^{\tau} = S_{ee}^{\tau} - S_{ee^c}^{\tau}(S_{e^ce^c}^{\tau})^{-1}S_{e^ce}^{\tau},$$

and it is easy to show that

$$(\widetilde{v}_h^e, \widetilde{S}_{ee}^{\tau}\widetilde{v}_h^e) \leq (v_h, S^{\tau} v_h) \quad \text{for all } v_h \in \Lambda_h^{\tau} \text{ so that } R_{e,\tau}v_h = \widetilde{v}_h^e, \tag{4}$$

where the restriction operator $R_{e,\tau} : \Lambda_h \to \widetilde{\Lambda}_h^e$ is so that $R_{e,\tau}v_h(\boldsymbol{x}_i) = \widetilde{v}_h^e(\boldsymbol{x}_i)$ for all nodes $\boldsymbol{x}_i \in \mathcal{N}_e := (\mathcal{N}_h \backslash \mathcal{N}_H) \cap e$.

In what follows, to take into account high contrast coefficients, we consider the following generalized eigenvalue problem: Find eigenpairs $(\alpha_i^e, \tilde{\mu}_{h,i}^e) \in (\mathbb{R}, \widetilde{\Lambda}_h^e)$, where $\alpha_1^e \geq \alpha_2^e \geq \alpha_3^e \geq \cdots \geq \alpha_{\mathcal{N}_e}^e > 1$, such that if the edge e of $\partial\mathcal{T}_H \backslash \partial\Omega$ is shared by elements τ and τ' of $\mathcal{T}_H$, we solve

$$(\widehat{S}_{ee}^{\tau} + \widehat{S}_{ee}^{\tau'})\tilde{\mu}_{h,i}^e = \alpha_i^e(\widetilde{S}_{ee}^{\tau} + \widetilde{S}_{ee}^{\tau'})\tilde{\mu}_{h,i}^e. \tag{5}$$

The eigenfunctions $\tilde{\mu}_{h,i}^e$ are chosen to be orthonormal with respect to the norm $(\cdot, (\widehat{S}_{ee}^{\tau} + \widehat{S}_{ee}^{\tau'})\cdot)_e$.

Now we decompose $\widetilde{\Lambda}_h^e := \widetilde{\Lambda}_h^{e,\triangle} \oplus \widetilde{\Lambda}_h^{e,\Pi}$ where for a given $\alpha_{\text{stab}} > 1$,

$$\widetilde{\Lambda}_h^{e,\triangle} := \text{span}\{\tilde{\mu}_{h,i}^e : \alpha_i^e < \alpha_{\text{stab}}\}, \qquad \widetilde{\Lambda}_h^{e,\Pi} := \text{span}\{\tilde{\mu}_{h,i}^e : \alpha_i^e \geq \alpha_{\text{stab}}\}.$$

The value of α_{stab} is tuned with $\mathcal{A}(\boldsymbol{x}) = \rho(\boldsymbol{x}) = 1$ so that the dimension of $\widetilde{\Lambda}_h^{e,\Pi}$ is small. Hence, for general $\mathcal{A}(\boldsymbol{x})$ and $\rho(\boldsymbol{x})$, the space $\widetilde{\Lambda}_h^{e,\Pi}$ will consist mostly of eigenvectors associated to the heterogeneities of $\mathcal{A}(\boldsymbol{x})$ with respect to $\rho(\boldsymbol{x})$.

For adaptive BDDC preconditioners, in general the generalized eigenvalue problem is defined by

$$(S_{ee}^{\tau} + S_{ee}^{\tau'})\tilde{\mu}_{h,i}^e = \alpha_i^e(\widetilde{S}_{ee}^{\tau} + \widetilde{S}_{ee}^{\tau'})\tilde{\mu}_{h,i}^e \tag{6}$$

We note that this generalized eigenvalue problem would be enough for establishing exponential decay for the multiscale basis functions. In (5), the term $\delta^{-2}M_{ee}^{\tau}$ was added to S_{ee}^{τ}. This is needed when dealing with approximation results such as Theorem 4 since in the proof it is required that $\|v\|_{L_\rho^2(\Omega)} \leq \delta |v|_{H_{\mathcal{A}}^1(\Omega)}$ for $v \in T\widetilde{\Lambda}_h^{\triangle}$ defined below.

To define our ACMS–NLSD (*Approximate Component Mode Synthesis Non-Localized Spectral Decomposition*) method for high-contrast coefficients, let

$$\widetilde{\Lambda}_h^{\Pi} = \{\tilde{\mu}_h \in \widetilde{\Lambda}_h : \tilde{\mu}_h|_e \in \widetilde{\Lambda}_h^{e,\Pi} \text{ for all } e \in \partial\mathcal{T}_H\},$$
$$\widetilde{\Lambda}_h^{\triangle} = \{\tilde{\mu}_h \in \widetilde{\Lambda}_h : \tilde{\mu}_h|_e \in \widetilde{\Lambda}_h^{e,\triangle} \text{ for all } e \in \partial\mathcal{T}_H\}.$$

Note that $\Lambda_h = \Lambda_h^{\Pi} \oplus \widetilde{\Lambda}_h^{\triangle}$, where

$$\Lambda_h^{\Pi} = \Lambda_h^0 \oplus \widetilde{\Lambda}_h^{\Pi}$$

and Λ_h^0 is the set of functions on Λ_h which vanish on all nodes of $\mathcal{N}_h \backslash \mathcal{N}_H$. Denote

$$(v_h, S\mu_h)_{\partial\mathcal{T}_h} = \sum_{\tau\in\mathcal{T}_H} (v_h^\tau, S^\tau \mu_h^\tau)_{\partial\tau}.$$

We now introduce the ACMS–NLSD multiscale functions. For $\tau \in \mathcal{T}_H$, consider the operator $P^{\tau,\Delta} : \Lambda_h \to \widetilde{\Lambda}_h^\Delta$ as follows: Given $\mu_h \in \Lambda_h$, find $P^{\tau,\Delta}\mu_h \in \widetilde{\Lambda}_h^\Delta$ solving

$$(\widetilde{v}_h^\Delta, SP^{\tau,\Delta}\mu_h)_{\partial\mathcal{T}_h} = (\widetilde{v}_h^\Delta, S^\tau \mu_h)_{\partial\tau} \quad \text{for all } \widetilde{v}_h^\Delta \in \widetilde{\Lambda}_h^\Delta \tag{7}$$

and define $P^\Delta : \Lambda_h \to \widetilde{\Lambda}_h^\Delta$ given by $P^\Delta = \sum_{\tau\in\mathcal{T}_H} P^{\tau,\Delta}$. It is easy to see that P^Δ is an orthogonal projection on $\widetilde{\Lambda}_h^\Delta$ with respect to S.

Consider $\Lambda_h^{\text{ms}} = (I - P^\Delta)\Lambda_h^\Pi$. We note that $(I - P^\Delta)\Lambda_h^\Pi \neq \Lambda_h^\Pi$ since Λ_h^Π and $\widetilde{\Lambda}_h^\Delta$ are not orthogonal with respect to S. What we have is that $\widetilde{\Lambda}_h^{e,\Delta}$ and $\widetilde{\Lambda}_h^{e,\Pi}$ are orthogonal with respect to $\widetilde{S}_{ee}^\tau + \widetilde{S}_{ee}^{\tau'}$. If $\mu_h^\Pi \in \Lambda_h^\Pi$ is a local function, $(I - P^\Delta)\mu_h^\Pi$ will not be necessarily local. However, we can show its exponential decay.

The ACMS–NLSD method is defined by: Find $\lambda_h^{\text{ms}} \in \Lambda_h^{\text{ms}}$ such that

$$(v_h^{\text{ms}}, S\lambda_h^{\text{ms}})_{\partial\mathcal{T}_h} = (\rho g, Tv_h^{\text{ms}}) \quad \text{for all } v_h^{\text{ms}} \in \Lambda_h^{\text{ms}}. \tag{8}$$

Note that

$$(v_h^{\text{ms}}, S\lambda_h^{\text{ms}})_{\partial\mathcal{T}_h} = \int_\Omega \mathcal{A}\nabla Tv_h^{\text{ms}} \cdot \nabla T\lambda_h^{\text{ms}}\, d\boldsymbol{x} = \int_\Omega \rho g T v_h^{\text{ms}}\, d\boldsymbol{x}.$$

Remark 1 In [12, 11], different but still local eigenvalue problems are introduced, aiming to build approximation spaces. Their analysis however requires extra regularity of the coefficients, and the error estimate is not robust with respect to contrast.

Below we present several results where proofs will be published in [18].

Using local arguments, the next lemma states that a weighted Poincaré inequality can be obtained on the space $\widetilde{\Lambda}_h^\Delta$.

Lemma [18] Let $\widetilde{\mu}_h^\Delta \in \widetilde{\Lambda}_h^\Delta$. Then

$$\|T\widetilde{\mu}_h^\Delta\|_{L^2_\rho(\Omega)} \leq (L^2\alpha_{\text{stab}})^{1/2}\delta|T\widetilde{\mu}_h^\Delta|_{H^1_\mathcal{A}(\Omega)}, \tag{9}$$

where L is the maximum number of edges that an element of $\mathcal{T}_H$ can have. □

The next lemma states that the energy stability of the interpolation onto the primal space Λ^Π.

Lemma [18] Let $\mu_h \in \Lambda_h$ and let $\mu_h = \mu_h^\Pi + \widetilde{\mu}_h^\Delta$. Then

$$|T\mu_h^\Pi|_{H^1_\mathcal{A}(\Omega)} \leq (2 + 2L^2\alpha_{\text{stab}})^{1/2}|T\mu_h|_{H^1_\mathcal{A}(\Omega)}.$$

The next lemma follows directly from the definition of the generalized eigenvalue problem and properties of $\widetilde{\Lambda}_h^{e,\Delta}$ and (4).

Lemma [18] Let e be a common edge of τ, $\tau' \in \mathcal{T}_H$, and $\widetilde{\mu}_h^{\triangle} \in \widetilde{\Lambda}_h^{\triangle}$. Then, defining $\widetilde{\mu}_h^{e,\triangle} = \widetilde{\mu}_h^{\triangle}|_e$ and $\widetilde{\mu}_h^{\tau,\triangle} = \widetilde{\mu}_h^{\triangle}|_{\partial\tau}$ it follows that

$$|T^\tau R_{e,\tau}^T \widetilde{\mu}_h^{e,\triangle}|^2_{H^1_{\mathcal{A}}(\tau)} + |T^\tau R_{e,\tau'}^T \widetilde{\mu}_h^{e,\triangle}|^2_{H^1_{\mathcal{A}}(\tau')} \le \alpha_{\text{stab}}\left(|T^\tau \widetilde{\mu}_h^{\tau,\triangle}|^2_{H^1_{\mathcal{A}}(\tau)} + |T^\tau \widetilde{\mu}_h^{\tau',\triangle}|^2_{H^1_{\mathcal{A}}(\tau')}\right)$$

Our main theorem follows.

Theorem [18] Let $\lambda_h = u_h|_{\partial\mathcal{T}_h}$, and λ_h^{ms} solution of (8). Then $\lambda_h - \lambda_h^{ms} \in \widetilde{\Lambda}_h^{\triangle}$ and

$$|u_h^{\mathbb{H}} - T\lambda_h^{ms}|^2_{H^1_{\mathcal{A}}(\Omega)} \le L^2 \alpha_{\text{stab}} \delta^2 \|g\|^2_{L^2_\rho(\Omega)}.$$

4.1 Decaying for the High-Contrast Case

In the next two lemmas, we show first that we can control the energy on the exterior region outside the patch of j-neighbor elements $\mathcal{T}_{j+1}(\tau)$ by the energy on the strip $\mathcal{T}_{j+2}(\tau)\backslash\mathcal{T}_j(\tau)$. Next, we state the exponential decay of $P^{\tau,\triangle}v_h$.

Lemma [18] Let $\mu_h \in \Lambda_h$ and let $\tilde{\phi}_h^{\triangle} = P^{\tau,\triangle}\mu_h$ for some fixed element $\tau \in \mathcal{T}_H$. Then, for any integer $j \ge 1$,

$$|T\tilde{\phi}_h^{\triangle}|^2_{H^1_{\mathcal{A}}(\mathcal{T}_H\backslash\mathcal{T}_{j+1}(\tau))} \le L^2 \alpha_{\text{stab}} |T\tilde{\phi}_h^{\triangle}|^2_{H^1_{\mathcal{A}}(\mathcal{T}_{j+2}(\tau)\backslash\mathcal{T}_j(\tau))}.$$

The next lemma states the exponential decay of $P^{\tau,\triangle}v_h$.

Corollary [18] Assume that $\tau \in \mathcal{T}_H$ and $v_h \in \Lambda_h$ and let $\tilde{\phi}_h^{\triangle} = P^{\tau,\triangle}v_h \in \widetilde{\Lambda}_h^{\triangle}$. For any integer $j \ge 1$,

$$|T\tilde{\phi}_h^{\triangle}|^2_{H^1_{\mathcal{A}}(\mathcal{T}_H\backslash\mathcal{T}_{j+1}(\tau))} \le e^{-\frac{[(j+1)/2]}{1+L^2\alpha_{\text{stab}}}} |T\tilde{\phi}_h^{\triangle}|^2_{H^1_{\mathcal{A}}(\mathcal{T}_H)}.$$

where $[s]$ is the integer part of s. □

Inspired by the exponential decay stated in Corollary 6, we define the operator $P^{\triangle,j}$ as follows. First, for a fixed $\tau \in \mathcal{T}_H$, let

$$\widetilde{\Lambda}_h^{\triangle,\tau,j} = \{\widetilde{\mu}_h \in \widetilde{\Lambda}_h^{\triangle} : T\widetilde{\mu}_h = 0 \text{ on } \mathcal{T}_H\backslash\mathcal{T}_j(\tau)\},$$

i.e., the support of $\widetilde{\Lambda}_h^{\triangle,\tau,j}$ is just a patch of size j elements around the element τ.

For $\mu_h \in \Lambda_h$, define $P^{\triangle,\tau,j}\mu_h \in \widetilde{\Lambda}_h^{\tau,j}$ such that

$$s(P^{\triangle,\tau,j}\mu_h, \widetilde{\mu}_h) = s_\tau(\mu_h, \widetilde{\mu}_h) \quad \text{for all } \widetilde{\mu}_h \in \widetilde{\Lambda}_h^{\triangle,\tau,j},$$

and let

$$P^{\triangle,j}\mu_h = \sum_{\tau\in\mathcal{T}_H} P^{\triangle,\tau,j}\mu_h. \tag{10}$$

Finally, define the approximation $\lambda_H^{\Pi,j} \in \Lambda_H^{\Pi}$ such that

$$s\big((I - P^{\Delta,j})\lambda_H^{\Pi,j}, (I - P^{\Delta,j})\mu_H^{\Pi}\big) = \big(\rho g, T(I - P^{\Delta,j})\mu_H^{\Pi}\big) \quad \text{for all } \mu_H^{\Pi} \in \Lambda_H^{\Pi}, \tag{11}$$

and then let $\lambda_h^{ms,j} = (I - P^{\Delta,j})\lambda_H^{\Pi,j}$. We name as ACMS–LSD (*Approximate Component Mode Synthesis Localized Spectral Decomposition*) method.

We now state the approximation error of the method, starting by a technical result essential to obtain the final estimate.

Lemma [18] Consider $v_h \in \Lambda_h$ and the operators P^{Δ} defined by (7) and $P^{\Delta,j}$ by (10) for $j > 1$. Then

$$|T(P^{\Delta} - P^{\Delta,j})v_h|^2_{H^1_{\mathcal{A}}(\mathcal{T}_H)} \le (c_\gamma j)^2 (L^2 \alpha_{\text{stab}})^2 e^{-\frac{\lfloor (j-1)/2 \rfloor}{1+L^2\alpha_{\text{stab}}}} |T v_h|^2_{H^1_{\mathcal{A}}(\mathcal{T}_H)},$$

where c_γ is a constant depending only on the shape of $\mathcal{T}_H$ such that

$$\sum_{\tau \in \mathcal{T}_H} |v|^2_{H^1(\mathcal{T}_j(\tau))} \le (c_\gamma j)^2 |v|^2_{H^1(\mathcal{T}_H)} \qquad \forall v \in H^1(\mathcal{T}_H). \tag{12}$$

Theorem [18] Define $u_h^{\mathbb{H}}$ by (2) and let $\lambda_h^{ms,j} = (I - P^{\Delta,j})\lambda_H^{\Pi,j}$, where $\lambda_H^{\Pi,j}$ is as in (11). Then

$$|u_h^{\mathbb{H}} - T\lambda^{ms,j}|_{H^1_{\mathcal{A}}(\mathcal{T}_H)} \le \delta L (2\alpha_{\text{stab}})^{1/2} \|g\|_{L^2_\rho(\Omega)} + c_\gamma j L^2 \alpha_{\text{stab}} e^{-\frac{\lfloor (j-1)/2 \rfloor}{2(1+L^2\alpha_{\text{stab}})}} |u_h^{\mathbb{H}}|_{H^1_{\mathcal{A}}(\mathcal{T}_H)}.$$

5 Spectral Multiscale Problems inside Substructures

To approximate $u_h^{\mathbb{B}}$ on an element $\tau \in \mathcal{T}_H$, we introduce a multiscale method by first building the approximation space $V^{\mathbb{B},\text{ms}}(\tau) := \text{Span}\{\psi^\tau_{h,1}, \psi^\tau_{h,2}, \cdots, \psi^\tau_{h,N_\tau}\}$ generated by the following generalized eigenvalue problem: Find the eigenpairs $(\alpha_i^\tau, \psi_{h,i}^\tau) \in (\mathbb{R}, V_h^{\mathbb{B}}(\tau))$ such that

$$a_\tau(v_h, \psi^\tau_{h,i}) = \alpha_i^\tau (\rho v_h, \psi^\tau_{h,i}) \qquad \text{for all} \quad v_h \in V_h^{\mathbb{B}}(\tau)$$

where

$$a_\tau(v_h, \psi^\tau_{h,i}) = \int_\tau \mathcal{A}\boldsymbol{\nabla} v_h \cdot \boldsymbol{\nabla}\psi^\tau_{h,i}\, d\boldsymbol{x} \qquad \text{and} \qquad (\rho v_h, \psi^\tau_{h,i})_\tau = \int_\tau \rho v_h \psi^\tau_{h,i}\, d\boldsymbol{x},$$

and $0 < \alpha_1^\tau \le \alpha_2^\tau \le \cdots \le \alpha^\tau_{N_\tau} < 1/\delta^2$ and $\alpha^\tau_{N_\tau+1} \ge 1/\delta^2$. The local multiscale problem is defined by: Find $u_h^{\mathbb{B},\text{ms}} \in V_h^{\mathbb{B},\text{ms}}$ such that

$$a(u_h^{\mathbb{B},\text{ms}}, v_h) = (\rho g, v_h) \qquad \text{for all} \qquad v_h \in V_h^{\mathbb{B},\text{ms}}.$$

We obtain

$$|u_h^{\mathbb{B}} - u_h^{\mathbb{B},\text{ms}}|^2_{H^1_{\mathcal{A}}(\Omega)} = (\rho g, u_h^{\mathbb{B}} - u_h^{\mathbb{B},\text{ms}}) \leq \delta |u_h^{\mathbb{B}} - u_h^{\mathbb{B},\text{ms}}|_{H^1_{\mathcal{A}}(\Omega)} \|g\|_{L^2_\rho(\Omega)},$$

and therefore,

$$|u_h^{\mathbb{B}} - u^{\mathbb{B},\text{ms}}|_{H^1_{\mathcal{A}}(\Omega)} \leq \delta \|g\|_{L^2_\rho(\Omega)}.$$

6 Numerical Experiments

Let $\Omega = (0,1) \times (0,1)$. We consider a Cartesian coarse mesh made of $2^M \times 2^M$ squares subdomains. We next subdivide each square subdomain into $2^{N-M} \times 2^{N-M}$ equal fine squares and then subdivide further into two 45-45-90 triangular elements. Denote $H = 2^{-M}$ and $h = 2^{-N}$ as the sizes of the subdomains and the fine elements, respectively.

The first numerical test is to examine the exponential decay of the multiscale basis functions. We assume that $\mathcal{A}(\boldsymbol{x})$ is scalar and $\rho(\boldsymbol{x}) = \mathcal{A}(\boldsymbol{x})$. The distribution of $\rho(\boldsymbol{x})$ is shown in the left Figure 1. The coefficient $\rho = 100$ inside the **H**-shape region and $\rho = 1$ outside. We assume that $N = 6$ and $M = 3$, that is, 8×8 subdomain distribution and 8×8 local mesh inside each subdomain. This distribution of the coefficients $\mathcal{A}(\boldsymbol{x})$ and subdomains has the property that $\mathcal{A}(\boldsymbol{x}) = 100$ at the subdomain corner node at $\boldsymbol{x} = (1/2, 1/2)$ and $\mathcal{A} = 1$ at the remaining subdomains corners nodes. Figure 1 on the right shows the decay of the multiscale basis function associated to the coarse node $\boldsymbol{x} = (1/2, 1/2)$ when $\Lambda_h^{\Pi} = \Lambda_h^0$ (equivalently $\widetilde{\Lambda}_h^{\Delta} = \widetilde{\Lambda}_h$), that is, with $\alpha_{stab} = \infty$ (without edges eigenfunctions). We can see that this multiscale basis function does not decay exponentially away from $\boldsymbol{x} = (1/2, 1/2)$. The white holes you see in the picture occurs because the value of the function is closed to zero. The reason for the non-decay is because this basis function wants to have small energy, that is, this basis function wants to have value near one on the **H**-shape region since $\mathcal{A}$ is large there. We now consider the adaptive case with $\alpha_{stab} = 1.5$. On the left and right of Figure 2 we show the exponential decay (in the log-normal scale) when $\delta = \infty$ and $\delta = H$, respectively. As expected from the theory, the eigenvalue problem (6) is enough to obtain the exponential decay, however, it is not enough for approximation.

In the second numerical test we keep the same distribution of coefficients in Figure 1 again choose $N = 6$ and $M = 3$. To make the problem a little more complicated, we multiply $\mathcal{A}$ and ρ in each element by independently uniformly random distributions between zero and one. Similarly, we let f to be constant in each element given by another independently uniformly random distributions between zero and one. In Table 1 we show the energy errors for different values of δ. We also include the total number of edges functions required by the ACMS–NLSD method (without localization) for a δ tolerance. We take $\alpha_{stab} = 1.5$. Just as a reference, there are 112 interior subdomain edges; see that we can obtain a 0.22% relative energy error using an average of one eigenvector per subdomain edge.

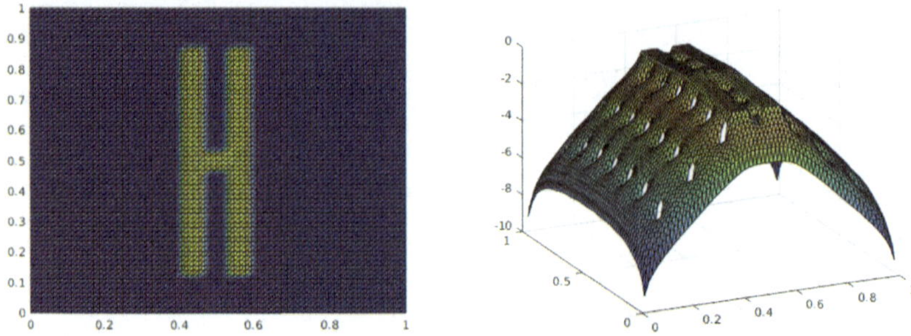

Fig. 1: On left, the distribution of the coefficient for a 8×8 subdomain decomposition. On the right, the plot of a multiscale basis functions without adaptivity. Note that there is no exponential decay whatsoever.

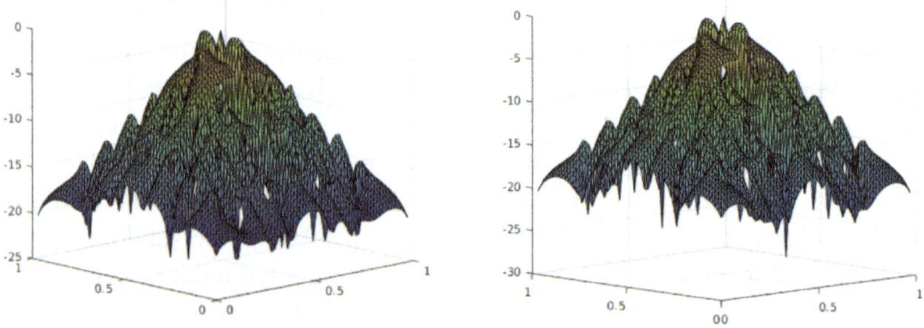

Fig. 2: Log-normal plot showing the decay of a multiscale basis functions with adaptivity, for $\delta = \infty$ (left figure) and $\delta = H$ (right figure)

δ	$\|u - u^{ms}\|_{H_a^1}$	$\frac{\|u-u^{ms}\|_{H_a^1}}{\|u\|_{H_a^1}}$	$\frac{\|u-u^{ms}\|_{H_a^1}}{\\|f\\|_{L_\rho^2}}$	Neigs
1/8	0.0095	0.0083	0.0079	78
1/16	0.0064	0.0056	0.0053	92
1/32	0.0025	0.0022	0.0021	112
1/64	0.0014	0.0012	0.0011	226

Table 1: The energy errors for different target accuracies δ. The last column shows Neigs (the total number of multiscale edges functions).

The last numerical test we investigate the dependence of the energy error $|u - u^{ms,j}|_{H_a^1}$ with respect to the localization j, that is, the ACMS–LSD method with localization j. We can see in Table 2 that the localization works really well.

$\delta\backslash j$	-1	0	1	2
1/8	0.43870	0.0095	0.0095	0.0095
1/16	0.0977	0.0064	0.0064	0.0064
1/32	0.1702	0.0025	0.0025	0.0025
1/64	0.0795	0.0014	0.0014	0.0014

Table 2: The energy errors for different target accuracies δ and localization j.

References

1. Babuska, I., Lipton, R.: Optimal local approximation spaces for generalized finite element methods with application to multiscale problems. Multiscale Model. Simul. **9**(1), 373–406 (2011). DOI:10.1137/100791051
2. Bampton, M.C.C., Craig JR., R.R.: Coupling of substructures for dynamic analyses. AIAA Journal **6**(7), 1313–1319 (1968). DOI:10.2514/3.4741
3. Bjørstad, P.E., Koster, J., Krzyżanowski, P.: Domain decomposition solvers for large scale industrial finite element problems. In: T. Sørevik, F. Manne, A.H. Gebremedhin, R. Moe (eds.) Applied Parallel Computing. New Paradigms for HPC in Industry and Academia, pp. 373–383. Springer Berlin Heidelberg, Berlin, Heidelberg (2001)
4. Bourquin, F.: Analysis and comparison of several component mode synthesis methods on one-dimensional domains. Numer. Math. **58**(1), 11–33 (1990). DOI:10.1007/BF01385608
5. Bourquin, F.: Component mode synthesis and eigenvalues of second order operators: discretization and algorithm. RAIRO Modél. Math. Anal. Numér. **26**(3), 385–423 (1992). DOI:10.1051/m2an/1992260303851
6. Calvo, J.G., Widlund, O.B.: An adaptive choice of primal constraints for BDDC domain decomposition algorithms. Electron. Trans. Numer. Anal. **45**, 524–544 (2016)
7. Galvis, J., Efendiev, Y.: Domain decomposition preconditioners for multiscale flows in high-contrast media. Multiscale Model. Simul. **8**(4), 1461–1483 (2010). DOI:10.1137/090751190
8. Galvis, J., Efendiev, Y.: Domain decomposition preconditioners for multiscale flows in high contrast media: reduced dimension coarse spaces. Multiscale Model. Simul. **8**(5), 1621–1644 (2010). DOI:10.1137/100790112
9. Heinlein, A., Hetmaniuk, U., Klawonn, A., Rheinbach, O.: The approximate component mode synthesis special finite element method in two dimensions: parallel implementation and numerical results. J. Comput. Appl. Math. **289**, 116–133 (2015). DOI:10.1016/j.cam.2015.02.053
10. Hellman, F., Henning, P., Må lqvist, A.: Multiscale mixed finite elements. Discrete Contin. Dyn. Syst. Ser. S **9**(5), 1269–1298 (2016). DOI:10.3934/dcdss.2016051
11. Hetmaniuk, U., Klawonn, A.: Error estimates for a two-dimensional special finite element method based on component mode synthesis. Electron. Trans. Numer. Anal. **41**, 109–132 (2014)
12. Hetmaniuk, U.L., Lehoucq, R.B.: A special finite element method based on component mode synthesis. M2AN Math. Model. Numer. Anal. **44**(3), 401–420 (2010). DOI:10.1051/m2an/2010007
13. Hughes, T.J.R., Feijóo, G.R., Mazzei, L., Quincy, J.B.: The variational multiscale method—a paradigm for computational mechanics. Comput. Methods Appl. Mech. Engrg. **166**(1-2), 3–24 (1998). DOI:10.1016/S0045-7825(98)00079-6
14. Hurty, W.C.: Vibrations of structural systems by component mode synthesis. Journal of the Engineering Mechanics Division **86**(4), 51–70 (1960)
15. Klawonn, A., Radtke, P., Rheinbach, O.: FETI-DP methods with an adaptive coarse space. SIAM J. Numer. Anal. **53**(1), 297–320 (2015). DOI:10.1137/130939675

16. Må lqvist, A., Peterseim, D.: Localization of elliptic multiscale problems. Math. Comp. **83**(290), 2583–2603 (2014). DOI:10.1090/S0025-5718-2014-02868-8
17. Madureira, A., Sarkis, M.: Adaptive deluxe bddc mixed and hybrid primal discretizations. In: P.E. Bjørstad, S.C. Brenner, L. Halpern, H.H. Kim, R. Kornhuber, T. Rahman, O.B. Widlund (eds.) Domain Decomposition Methods in Science and Engineering XXIV, pp. 465–473. Springer International Publishing, Cham (2018)
18. Madureira, A.L., Sarkis, M.: Adaptive ACMS: A robust localized Approximated Component Mode Synthesis Method. arXiv e-prints arXiv:1709.04044 (2017)
19. Mandel, J., Sousedík, B.: Adaptive selection of face coarse degrees of freedom in the BDDC and the FETI-DP iterative substructuring methods. Comput. Methods Appl. Mech. Engrg. **196**(8), 1389–1399 (2007). DOI:10.1016/j.cma.2006.03.010
20. Nataf, F., Xiang, H., Dolean, V., Spillane, N.: A coarse space construction based on local Dirichlet-to-Neumann maps. SIAM J. Sci. Comput. **33**(4), 1623–1642 (2011). DOI:10.1137/100796376
21. Oh, D.S., Widlund, O.B., Zampini, S., Dohrmann, C.R.: BDDC algorithms with deluxe scaling and adaptive selection of primal constraints for Raviart-Thomas vector fields. Math. Comp. **87**(310), 659–692 (2018). DOI:10.1090/mcom/3254
22. Owhadi, H., Zhang, L., Berlyand, L.: Polyharmonic homogenization, rough polyharmonic splines and sparse super-localization. ESAIM Math. Model. Numer. Anal. **48**(2), 517–552 (2014). DOI:10.1051/m2an/2013118
23. Pechstein, C., Dohrmann, C.R.: A unified framework for adaptive BDDC. Electron. Trans. Numer. Anal. **46**, 273–336 (2017)
24. Peterseim, D., Scheichl, R.: Robust numerical upscaling of elliptic multiscale problems at high contrast. Comput. Methods Appl. Math. **16**(4), 579–603 (2016). DOI:10.1515/cmam-2016-0022

Analysis of Double Sweep Optimized Schwarz Methods: the Positive Definite Case

Martin J. Gander and Hui Zhang

1 Introduction

Over the last decade, substantial research efforts have gone into developing preconditioners for time harmonic wave propagation problems, like the Helmholtz and the time harmonic Maxwell's equations. Such equations are much harder to solve than diffusive problems like Laplace's equation, because of two main reasons: first, the pollution effect [1] requires much finer meshes than would be necessary just to resolve the signal computed, and second, classical iterative methods all exhibit severe convergence problems when trying to solve the very large discrete linear systems obtained [9]. These research efforts have led to innovative new preconditioners, like optimized Schwarz methods (OSM) [5, 11], Analytic Incomplete LU (AILU) [12], the sweeping preconditioner [7, 8], the source transfer domain decomposition [3, 4], the method based on single layer potentials [14], and the method of polarized traces [15], for a more complete treatment, see [13] and references therein. In [13], it was shown that all these methods can be written as alternating optimized Schwarz methods called Double Sweep Optimized Schwarz Methods (DOSMs). We study here analytically the contraction properties of DOSMs for the model problem

$$\begin{aligned} &\eta u - u_{xx} - u_{yy} = f \quad \text{in } \Omega := (-\tfrac{L}{2}, 1 + \tfrac{L}{2}) \times (0, \pi), \\ &u = 0 \quad \text{at } y \in \{0, \pi\}, \quad \mathcal{B}_1^l u = 0 \quad \text{at } x = -\tfrac{L}{2}, \quad \mathcal{B}_N^r u = 0 \quad \text{at } x = 1 + \tfrac{L}{2}, \end{aligned} \tag{1}$$

Martin J. Gander
University of Geneva, Section of Mathematics, Rue du Lievre 2-4, CP 64, 1211 Geneva 4, e-mail: martin.gander@unige.ch

Hui Zhang (corresponding author)
Xi'an Jiaotong-Liverpool University, Department of Mathematical Sciences & Laboratory for Intelligent Computing and Financial Technology (KSF-P-02), Suzhou 215123, China; Zhejiang Ocean University, Key Laboratory of Oceanographic Big Data Mining & Application of Zhejiang Province, Zhoushan 316022, China, e-mail: mike.hui.zhang@hotmail.com

R. Haynes et al. (eds.), *Domain Decomposition Methods in Science and Engineering XXV*, Lecture Notes in Computational Science and Engineering 138,
https://doi.org/10.1007/978-3-030-56750-7_5

where $L \geq 0$ is a parameter which will be related to the overlap, $\mathcal{B}_1^l$ and $\mathcal{B}_N^r$ are linear trace operators, and $N \geq 2$ is an integer number to be defined below. While in our derivations in Section 2 and 3 we consider $\eta \in \mathbb{C}$, and we thus also include the Helmholtz case of interest, we will focus for our results then in Section 4 on the positive definite case $\eta > 0$.

2 Iteration Matrix of DOSM

Using a Fourier sine expansion of u solution to (1) in the y direction with Fourier parameter k, we obtain for the Fourier coefficients [1] for $k = 1, 2, \dots$ the problem

$$\begin{aligned} &(k^2 + \eta)u - u_{xx} = f \quad \text{in } (-\tfrac{L}{2}, 1 + \tfrac{L}{2}),\ k \in \mathbb{N}^+, \\ &\mathcal{B}_1^l u = 0 \quad \text{at } x = -\tfrac{L}{2}, \quad \mathcal{B}_N^r u = 0 \quad \text{at } x = 1 + \tfrac{L}{2}. \end{aligned} \tag{2}$$

We decompose the domain $(-\frac{L}{2}, 1 + \frac{L}{2})$ into N overlapping subdomains of equal width $H + L := \frac{1}{N} + L$, denoted by $\Omega_j := ((j-1)H - \frac{L}{2}, jH + \frac{L}{2})$, and we denote the restricted solution by $u_j := u|_{\Omega_j}$, $j = 1, .., N$. In DOSM, (2) is reformulated as transmission problems on the Ω_j for $j = 1, .., N$,

$$\begin{aligned} &(k^2 + \eta)u_j - (u_j)_{xx} = f \quad \text{in } \Omega_j, \\ &\mathcal{B}_j^l(u_j - u_{j-1}) = 0 \text{ at } x = (j-1)H - \tfrac{L}{2}, \mathcal{B}_j^r(u_j - u_{j+1}) = 0 \text{ at } x = jH + \tfrac{L}{2}, \end{aligned} \tag{3}$$

where $\mathcal{B}_j^l$ and $\mathcal{B}_j^r$ are linear trace operators and u_0, u_{N+1} are identically zero. Let $g_j^l := \mathcal{B}_j^l u_j$ at $x = x_j^l := (j-1)H - \frac{L}{2}$ and $g_j^r := \mathcal{B}_j^r u_j$ at $x = x_j^r := jH + \frac{L}{2}$. To rewrite (3) in terms of the interface data $[g_2^l; ..; g_N^l; g_1^r; ..; g_{N-1}^r]$, we define the trace-to-trace operators (see also Figure 1)

$$\begin{aligned} a_j &: \left(\ell_j \text{ at } x = x_j^l\right) \to \left(\mathcal{B}_{j+1}^l v_j \text{ at } x = x_{j+1}^l\right) \text{ with } v_j \text{ solving} \\ &\quad (k^2 + \eta)v_j - (v_j)_{xx} = 0 \quad \text{in } \Omega_j, \quad \mathcal{B}_j^l v_j = \ell_j \text{ at } x = x_j^l, \quad \mathcal{B}_j^r v_j = 0 \text{ at } x = x_j^r, \\ b_j &: \left(\gamma_j \text{ at } x = x_j^r\right) \to \left(\mathcal{B}_{j+1}^l v_j \text{ at } x = x_{j+1}^l\right) \text{ with } v_j \text{ solving} \\ &\quad (k^2 + \eta)v_j - (v_j)_{xx} = 0 \quad \text{in } \Omega_j, \quad \mathcal{B}_j^l v_j = 0 \text{ at } x = x_j^l, \quad \mathcal{B}_j^r v_j = \gamma_j \text{ at } x = x_j^r, \\ c_j &: \left(\gamma_j \text{ at } x = x_j^r\right) \to \left(\mathcal{B}_{j-1}^r v_j \text{ at } x = x_{j-1}^r\right) \text{ with } v_j \text{ solving} \\ &\quad (k^2 + \eta)v_j - (v_j)_{xx} = 0 \quad \text{in } \Omega_j, \quad \mathcal{B}_j^l v_j = 0 \text{ at } x = x_j^l, \quad \mathcal{B}_j^r v_j = \gamma_j \text{ at } x = x_j^r, \\ d_j &: \left(\ell_j \text{ at } x = x_j^l\right) \to \left(\mathcal{B}_{j-1}^r v_j \text{ at } x = x_{j-1}^r\right) \text{ with } v_j \text{ solving} \\ &\quad (k^2 + \eta)v_j - (v_j)_{xx} = 0 \quad \text{in } \Omega_j, \quad \mathcal{B}_j^l v_j = \ell_j \text{ at } x = x_j^l, \quad \mathcal{B}_j^r v_j = 0 \text{ at } x = x_j^r. \end{aligned}$$

[1] We still denote the Fourier transformed quantities for simplicity by the same symbols u, $\mathcal{B}_1^l$ and $\mathcal{B}_N^r$ to avoid a more complicated notation.

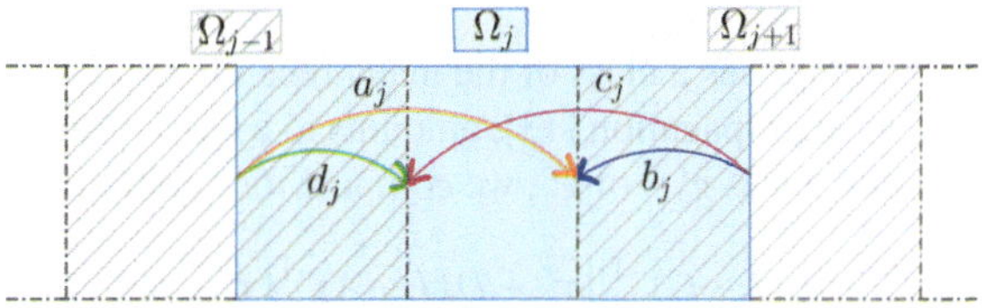

Fig. 1: Illustration of the interface-to-interface operators.

In the Fourier basis, the operators a_j, b_j, c_j, d_j reduce to scalars which we compute now explicitly. To simplify the notation, let $s := \sqrt{k^2+\eta}$ with $\operatorname{Re} s > 0$ if $k^2+\eta$ is not exactly on the negative real axis, $s := c_{\text{sgn}}\mathrm{i}\sqrt{-k^2-\eta}$ if $k^2+\eta < 0$ with $c_{\text{sgn}} \in \{1,-1\}$ a conventional sign-value from the time-dependence $\mathrm{e}^{c_{\text{sgn}}\mathrm{i}\sqrt{-\eta}t}$. For typical OSM transmission conditions (not just DOSM) of the form $\mathcal{B}_j^l = -q_j^l\partial_x + p_j^l$ and $\mathcal{B}_j^r = q_j^r\partial_x + p_j^r$, we define

$$R_j^l := \frac{p_j^l - q_j^l s}{p_j^l + q_j^l s}\mathrm{e}^{-Ls}, \qquad R_j^{lr} := \frac{p_{j-1}^r - q_{j-1}^r s}{p_j^l + q_j^l s}\mathrm{e}^{-Ls}, \qquad Q_j^{lr} := \frac{p_{j-1}^r + q_{j-1}^r s}{p_j^l + q_j^l s},$$

$$R_j^r := \frac{p_j^r - q_j^r s}{p_j^r + q_j^r s}\mathrm{e}^{-Ls}, \qquad R_j^{rl} := \frac{p_{j+1}^l - q_{j+1}^l s}{p_j^r + q_j^r s}\mathrm{e}^{-Ls}, \qquad Q_j^{rl} := \frac{p_{j+1}^l + q_{j+1}^l s}{p_j^r + q_j^r s},$$

$$R_j^{ll} := \frac{p_{j+1}^l - q_{j+1}^l s}{p_j^l + q_j^l s}\mathrm{e}^{-Ls}, \qquad R_j^{rr} := \frac{p_{j-1}^r - q_{j-1}^r s}{p_j^r + q_j^r s}\mathrm{e}^{-Ls}, \qquad Q_j^{ll(rr)} := \frac{p_{j\pm1}^{l(r)} + q_{j\pm1}^{l(r)} s}{p_j^{l(r)} + q_j^{l(r)} s},$$

and we have for $L \geq 0$

$$a_j = \frac{(Q_j^{ll} - R_j^r R_j^{ll})\mathrm{e}^{-Hs}}{1 - R_j^l R_j^r \mathrm{e}^{-2Hs}}, \qquad b_j = \frac{R_j^{rl} - R_j^l Q_j^{rl}\mathrm{e}^{-2Hs}}{1 - R_j^l R_j^r \mathrm{e}^{-2Hs}},$$

$$c_j = \frac{(Q_j^{rr} - R_j^l R_j^{rr})\mathrm{e}^{-Hs}}{1 - R_j^r R_j^l \mathrm{e}^{-2Hs}}, \qquad d_j = \frac{R_j^{lr} - R_j^r Q_j^{lr}\mathrm{e}^{-2Hs}}{1 - R_j^r R_j^l \mathrm{e}^{-2Hs}}.$$

When $\mathcal{B}_j^l = \mathcal{B}_j^r = 1$, i.e. the classical alternating Schwarz case, we have

$$a_j = c_j = \frac{(1-\mathrm{e}^{-2Ls})\mathrm{e}^{-Hs}}{1-\mathrm{e}^{-2Hs-2Ls}}, \qquad b_j = d_j = \frac{(1-\mathrm{e}^{-2Hs})\mathrm{e}^{-Ls}}{1-\mathrm{e}^{-2Hs-2Ls}}.$$

Using the operators a_j, b_j, c_j and d_j, we can rewrite (3) as the linear system

$$\begin{pmatrix} I - A & -B \\ -D & I - C \end{pmatrix} \begin{pmatrix} g^l \\ g^r \end{pmatrix} = \begin{pmatrix} \tau^l \\ \tau^r \end{pmatrix}. \tag{4}$$

where A (C) has all its non-zero entries on the subdiagonal (superdiagonal) as $(a_2, \ldots, a_{N-1})$ $((c_2, \ldots, c_{N-1}))$, $B = \operatorname{diag}(b_1, \ldots, b_{N-1})$, $D = \operatorname{diag}(d_2, \ldots, d_N)$, and $\tau_j^l := \mathcal{B}_j^l v_j$, $\tau_j^r := \mathcal{B}_j^r v_j$ with v_j satisfying

$$(k^2+\eta)v_j - (v_j)_{xx} = f \quad \text{in } \Omega_j,$$
$$\mathcal{B}^l_j v_j = 0 \quad \text{at } x = x^l_j, \quad \mathcal{B}^r_j v_j = 0 \quad \text{at } x = x^r_j.$$

The DOSM, which comprises a class of the many recently invented preconditioners for time harmonic wave propagation [13], amounts to a block Gauss-Seidel iteration for (4): given an initial guess $g^{r,0}$ of g^r, we compute for iteration index $m = 0, 1, \ldots$

$$g^{r,m+1} := (I-C)^{-1}\left[\tau^r + D(I-A)^{-1}(\tau^l + Bg^{r,m})\right]. \tag{5}$$

We denote by $\epsilon^{r,m} := g^{r,m} - g^r$ the error, which then by (5) satisfies a recurrence relation with iteration matrix T,

$$\epsilon^{r,m+1} = T\epsilon^{r,m} := (I-C)^{-1}D(I-A)^{-1}B\epsilon^{r,m}. \tag{6}$$

3 Eigenvalues of the iteration matrix T

To understand the convergence properties of these methods, we need to study the spectral radius of T. We first compute the inverse of T if B and D are invertible. For simplicity, we assume from now on that $\mathcal{B}^l_j = \mathcal{B}^l$, $j = 2, .., N$ are the same and $\mathcal{B}^r_j = \mathcal{B}^r$, $j = 1, .., N-1$ are the same. Therefore, b_j, d_j, $j = 2, .., N$ have the same value and we denote them by b and d. In addition $a_j = c_j$, $j = 2, .., N-1$ which also have the same value denoted by a. We thus obtain

$$\begin{aligned} T^{-1} &= B^{-1}(I-A)D^{-1}(I-C) \\ &= b^{-1}d^{-1}\begin{pmatrix} b_1^{-1}b & -b_1^{-1}ba & & & \\ -a & a^2+1 & -a & & \\ & \ddots & \ddots & \ddots & \\ & & \ddots & a^2+1 & -a \\ & & & -a & a^2+d_N^{-1}d \end{pmatrix} =: b^{-1}d^{-1}\tilde{T}. \end{aligned} \tag{7}$$

Let $\lambda_{\max}(\cdot)$ and $\lambda_{\min}(\cdot)$ denote the largest and smallest eigenvalue in modulus, and $\rho := \lambda_{\max}(T)$. From (7), we have $\rho = bd\lambda_{\min}^{-1}(\tilde{T})$. Let λ be an eigenvalue of $\tilde{T}$ and $\mathbf{v} = (v_j)_{j=1}^{N-1} \in \mathbb{C}^{N-1}$ the associated eigenvector. It follows that

$$-av_{j-1} + (a^2+1-\lambda)v_j - av_{j+1} = 0, \tag{8a}$$
$$(b_1^{-1}b - \lambda)v_1 - b_1^{-1}bav_2 = 0, \tag{8b}$$
$$-av_{N-2} + \left(a^2 + d_N^{-1}d - \lambda\right)v_{N-1} = 0. \tag{8c}$$

Note that $v_j = \xi_1\mu^j + \xi_2\mu^{-j}$, $j \in \mathbb{Z}$ is the general solution of (8a) if $\mu \neq \pm 1$ satisfies

$$-a + (a^2+1-\lambda)\mu - a\mu^2 = 0, \text{ or } \lambda = 1 + a^2 - a(\mu + \mu^{-1}). \tag{9}$$

Subtracting (8a) at $j = 1$ and $j = N - 1$ from (8b) and (8c) gives the equivalent boundary conditions

$$-av_0 + (a^2 + 1 - b_1^{-1}b)v_1 - a(1 - b_1^{-1}b)v_2 = 0,$$
$$(1 - d_N^{-1}d)v_{N-1} - av_N = 0.$$

Further substituting $v_j = \xi_1\mu^j + \xi_2\mu^{-j}$ into the above equations leads to

$$\left[-a + (a^2 + 1 - b_1^{-1}b)\mu - a(1 - b_1^{-1}b)\mu^2\right]\xi_1 +$$
$$\left[-a + (a^2 + 1 - b_1^{-1}b)\mu^{-1} - a(1 - b_1^{-1}b)\mu^{-2}\right]\xi_2 = 0,$$
$$\left[(1 - d_N^{-1}d) - a\mu\right]\mu^{N-1}\xi_1 + \left[(1 - d_N^{-1}d) - a\mu^{-1}\right]\mu^{1-N}\xi_2 = 0.$$

Since $\mathbf{v} \neq 0$, the determinant of the above linear system for $[\xi_1; \xi_2]$ must vanish, i.e.

$$\begin{aligned}&\left[-a + (a^2 + 1 - b_1^{-1}b)\mu - a(1 - b_1^{-1}b)\mu^2\right]\left[(1 - d_N^{-1}d) - a\mu^{-1}\right]\mu^{1-N}\\ &= \left[-a + (a^2 + 1 - b_1^{-1}b)\mu^{-1} - a(1 - b_1^{-1}b)\mu^{-2}\right]\left[(1 - d_N^{-1}d) - a\mu\right]\mu^{N-1}.\end{aligned} \tag{10}$$

Assume $a \neq 0$ and let $\beta_1 := a^{-1}(1 - b_1^{-1}b)$, $\beta_N := a^{-1}(1 - d_N^{-1}d)$. We can rewrite (10) as

$$\mu^{2N} = \frac{(1 - a\mu)(1 - \beta_1\mu)(1 - \beta_N\mu)}{(1 - a\mu^{-1})(1 - \beta_1\mu^{-1})(1 - \beta_N\mu^{-1})}. \tag{11}$$

In the special case when $\mathcal{B}_1^l = \mathcal{B}^l$ and $\mathcal{B}_N^r = \mathcal{B}^r$, we have $\beta_1 = \beta_N = 0$ so that

$$\mu^{2N} = \frac{1 - a\mu}{1 - a\mu^{-1}}. \tag{12}$$

Remark 1 The value $\lambda - (1 \mp a)^2$ corresponding to $\mu = \pm 1$ in (9) is an eigenvalue of T^{-1} if and only if $v_j = (\xi_1 + \xi_2 j)(\pm 1)^j$ is a non-zero solution of (8) or equivalently

$$\begin{aligned}&(1 \mp a)(\pm 1 \mp b_1^{-1}b - a)[(1 - d_N^{-1}d)(N - 1) \mp aN]\\ &\qquad = [\pm a^2 + (\pm 1 - 2a)(1 - b_1^{-1}b)][1 - d_N^{-1}d \mp a].\end{aligned}$$

In the special case of $\mathcal{B}_1^l = \mathcal{B}^l$ and $\mathcal{B}_N^r = \mathcal{B}^r$, the above condition becomes $\pm a^2 N(1 \mp a) = -a^3$, that is, $a = 0$ or $\pm\frac{N}{N-1}$.

4 Roots of the Polynomial Equation for μ

We first observe the following facts: $\mu = \pm 1$ are two roots of (11), and the other roots appear in pairs as μ, μ^{-1}. Our goal in this section is to locate all the roots in the complex plane. We assume from now on that $\eta \geq 0$ and thus $a, \beta_1, \beta_N \in \mathbb{R}$. Hence complex roots of (11) appear in conjugate pairs. We begin with the simplest

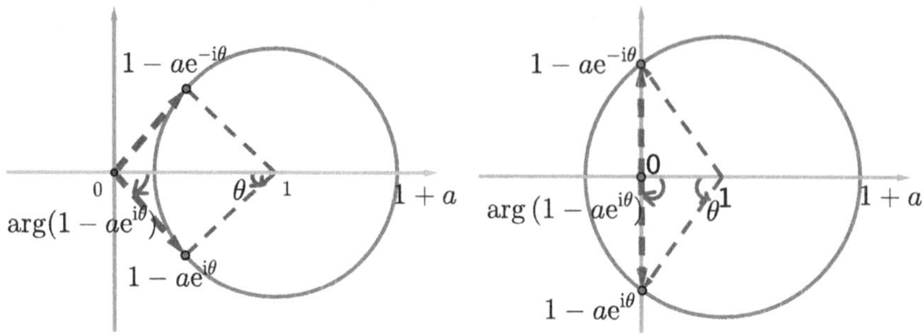

Fig. 2: Image of $1 - a\mathrm{e}^{\mathrm{i}\theta}$. Left: $0 < a < 1$. Right: $a > 1$.

case (12). We assume the argument $\arg z$ of a complex number z to take values in $(-\pi, \pi]$.

Lemma 1 *If* $a \in [-1, 1]\backslash\{0\}$, *then the roots of* (12) *are* ± 1 *and* $(\operatorname{sign} a)\, \mathrm{e}^{\pm\mathrm{i}\,\theta_j}$ *for some* $\theta_j \in \left[(j - \frac{1}{2})\pi/N, j\pi/N\right)$, $j = 1, .., N-1$.

Proof Since (12) is invariant under the transform $a \longrightarrow -a, \mu \longrightarrow -\mu$, we can assume $a > 0$. Substituting the ansatz $\mu = \mathrm{e}^{\mathrm{i}\theta}$, $\theta \in (-\pi, \pi]$ into (12), we obtain for θ the equation

$$w(\theta) := \mathrm{e}^{\mathrm{i}2N\theta} = \frac{1 - a\mathrm{e}^{\mathrm{i}\theta}}{1 - a\mathrm{e}^{-\mathrm{i}\theta}} =: z(\theta). \tag{13}$$

Since (13) is invariant under the transform $\theta \longrightarrow -\theta$, we need only to show that $w(\theta) = z(\theta)$ has $N-1$ roots for $\theta \in (0, \pi)$. On the one hand, we note that $1 - a\mathrm{e}^{\mathrm{i}\theta}$ turns around 1 with radius a, see Figure 2. It follows that $z(\theta)$ moves on the unit circle: first from $z(0) = 1$ clockwise to the extremal point $z(\arccos a)$ with $\arg z = -2 \arcsin a$, and then back counter-clockwise to $z(\pi) = 1$. On the other hand, $w(\theta)$ starts from $w(0) = 1$ and turns counter-clockwise along the unit circle N times. Hence, in each lower semi-cycle $\theta \in \left[(j - \frac{1}{2})\pi/N, j\pi/N\right)$, $j = 1, .., N-1$ there must exist a value of θ such that $w(\theta) = z(\theta)$. □

5 Numerical Study of the Convergence Factor

As before, we focus on the regime $k^2 + \eta > 0$, and therefore $s = \sqrt{k^2 + \eta}$ varies in $[s_{\min}, s_{\max}]$. Typically, $s_{\max}$ is linked to N, for example, if H is proportional to the mesh size and a second-order discretization is used, we have $s_{\max} = O(N)$. On the other hand, $s_{\min}$ is in this case a constant, which for our sine expansion has the value $s_{\min} = \sqrt{1 + \eta}$ stemming from the lowest Fourier mode $k = 1$.

In the special case of the classical alternating Schwarz methods, $\mathcal{B}_j^l = \mathcal{B}_j^r = 1$, we have $a \in (0, 1)$, $b \in (0, 1)$. By (9) and Lemma 1, we get $\lambda_{\min} = 1 + a^2 - 2a \cos \theta_1 > 0$ for some $\theta_1 \in \left[\frac{\pi}{2N}, \frac{\pi}{N}\right)$. Therefore, the convergence factor $\rho = b^2 \lambda_{\min}^{-1}$ becomes

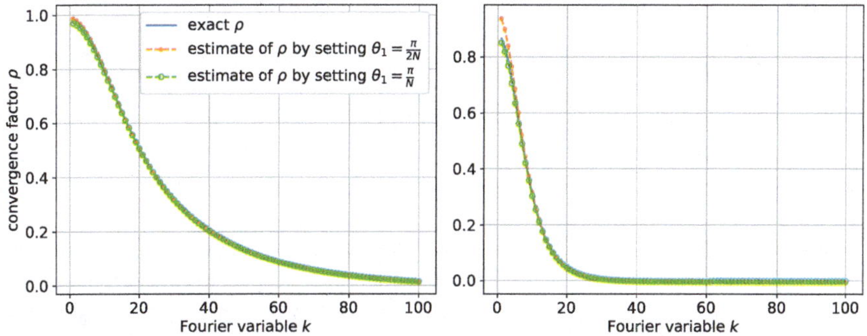

Fig. 3: Convergence factor of alternating Schwarz with $N = 10$, $L = \frac{1}{5N}$ (left), $\frac{4}{5N}$ (right) and $s \in [\sqrt{2}, \sqrt{10^4+1}]$.

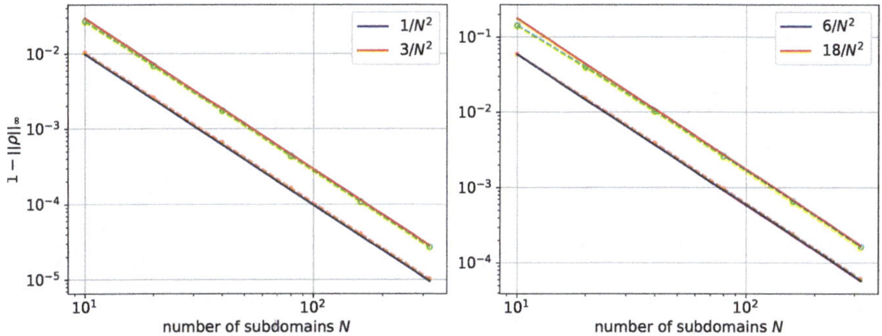

Fig. 4: Scaling of alternating Schwarz with $L = \frac{1}{5N}$ (left), $\frac{4}{5N}$ (right) and $s \in [\sqrt{2}, \sqrt{100N^2+1}]$. The dashed lines correspond to the upper and lower bounds of $1 - \|\rho\|_\infty$.

$$0 < \rho = \frac{e^{-2Ls}(1 - e^{-2Hs})^2}{r_1^2 + r_2^2 - 2r_1 r_2 \cos\theta_1}, \quad r_1 := 1 - e^{-2Hs-2Ls}, \quad r_2 := (1 - e^{-2Ls})e^{-Hs}.$$

Substituting θ_1 with its lower (upper) bound into the above expression yields an upper (lower) bound of ρ. In Figure 3, we compare these bounds with the exact value of ρ computed numerically. We see that the bounds are quite sharp. Then using these bounds we get the scaling of $\|\rho\|_\infty := \max_s |\rho(s)|$ with the number of subdomains N, and the convergence deteriorates, see Figure 4.

For optimized Schwarz, since $\eta \geq 0$, it is natural to use positive p_j^l, p_j^r [10]. In the special case of $p_j^l = p_j^r = p(k^2) > 0$, we find that $R \in (-1,1)$, $a \in (0,1)$, $b = d \in (-1,1)$. Again, by (9) and Lemma 1, we have $\lambda_{\min} = 1 + a^2 - 2a\cos\theta_1 > 0$ for some $\theta_1 \in [\frac{\pi}{2N}, \frac{\pi}{N})$. Therefore, the convergence factor $\rho = b^2\lambda_{\min}^{-1}$ becomes

$$0 < \rho = \frac{R^2(1 - e^{-2Hs})^2}{r_1^2 + r_2^2 - 2r_1 r_2 \cos\theta_1}, \quad r_1 := 1 - R^2 e^{-2Hs}, \quad r_2 := (1 - R^2)e^{-Hs}.$$

We first take $p > 0$ a constant. In Figure 5, we show how good the bounds of ρ are,

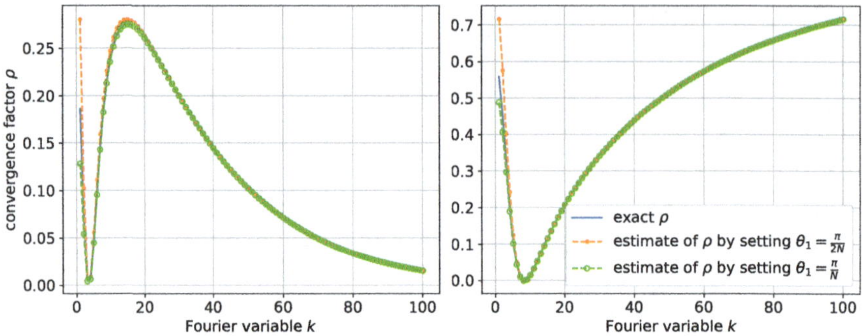

Fig. 5: Convergence factor of DOSM with p a constant obtained by numerically minimizing the upper bound of $\|\rho\|_\infty$, $N = 10$, $L = \frac{1}{5N}$ (left), $L = 0$ (right) and $s \in [\sqrt{2}, \sqrt{10^4+1}]$.

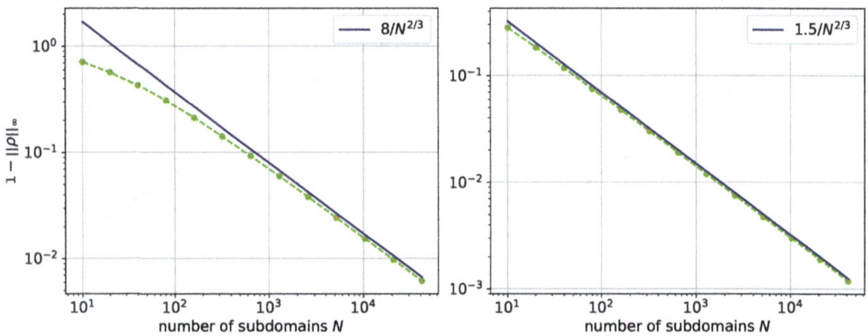

Fig. 6: Scaling of DOSM with p a constant obtained by numerically minimizing the upper bound of $\|\rho\|_\infty$, $L = \frac{1}{5N}$ (left), $L = 0$ (right) and $s \in [\sqrt{2}, \sqrt{100N^2+1}]$. The dashed lines correspond to the upper and lower bounds of $1 - \|\rho\|_\infty$ which are too close to be distinguished.

and also the results of minimizing the upper bound of $\|\rho\|_\infty$. Given p optimized in this way (dependent on N), we find that $\|\rho\|_\infty \approx 1 - O(N^{-2/3})$, both with minimal overlap and without overlap; see Figure 6. Next, we take $p = \tilde{p}_0 + \tilde{p}_2 k^2$ corresponding to the second-order boundary condition $\mathcal{B}_{l,r} = \mp\tilde{p}_0\partial_x - \tilde{p}_2\partial_{yy}$. We show the upper and lower bounds of ρ in Figure 7. Using numerically optimized parameters $\tilde{p}_0$ and $\tilde{p}_2$ (dependent on N), we find that $\|\rho\|_\infty \approx 1 - O(N^{-1/3})$ with minimal overlap and $\|\rho\|_\infty \approx 1 - O(N^{-2/5})$ without overlap; see Figure 8.

We can also choose $p(k^2)$ to be a more accurate approximation of $s = \sqrt{k^2 + \eta}$ to obtain an even smaller reflection coefficient $R = \exp(-Ls)(p-s)/(p+s)$. In the recently invented methods [13], Perfectly Matched Layers (PMLs; see [2, 6]) are most commonly used. Starting from a boundary $x = x_0$, a PML $[x_0, x_0 + D]$ is added outside a domain, and a new variable

$$\tilde{x} := \begin{cases} x + \int_0^{x-x_0} \sigma(|t|)\,\mathrm{d}t, & x \in [x_0, x_0 + D], \\ x, & \text{inside the domain,} \end{cases}$$

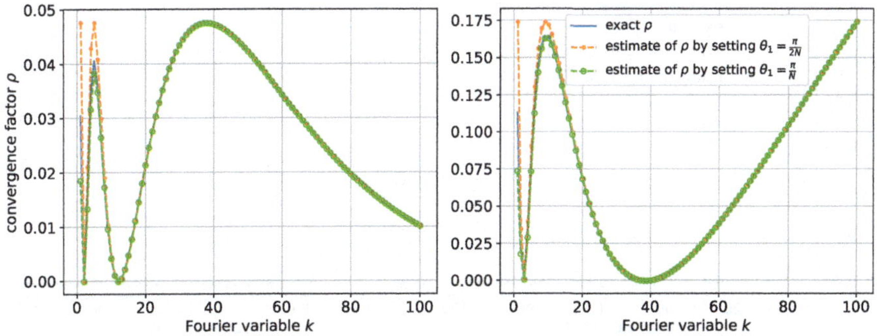

Fig. 7: Convergence factor of DOSM with $p = \tilde{p}_0 + \tilde{p}_2 k^2$ obtained by numerically minimizing the upper bound of $\|\rho\|_\infty$, $N = 10$, $L = \frac{1}{5N}$ (left), $L = 0$ (right) and $s \in [\sqrt{2}, \sqrt{10^4 + 1}]$.

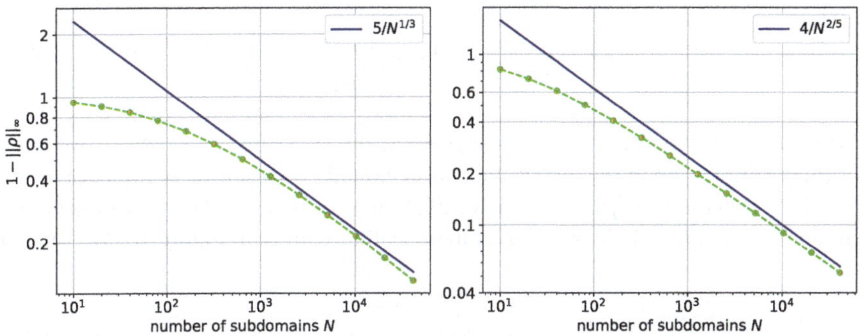

Fig. 8: Scaling of DOSM with $p = \tilde{p}_0 + \tilde{p}_2 k^2$ obtained by numerically minimizing the upper bound of $\|\rho\|_\infty$, $L = \frac{1}{5N}$ (left), $L = 0$ (right) and $s \in [\sqrt{2}, \sqrt{100N^2 + 1}]$. The dashed lines correspond to the upper and lower bounds of $1 - \|\rho\|_\infty$ which are too close to be distinguished.

is used for the model on the augmented domain $(k^2 + \eta)u - u_{\tilde{x}\tilde{x}} = \tilde{f}$, where $\tilde{f}$ is the zero extension of f and a homogeneous Dirichlet condition is put on the augmented boundary $x = x_0 + D$. This model amounts to imposing on $x = x_0$ the boundary condition $\mathrm{sign}(D)\partial_x u + \mathrm{DtN}_D u = 0$, where DtN_D is the Dirichlet-to-Neumann operator defined by

$$\begin{aligned}\mathrm{DtN}_D : (\gamma \text{ at } x = x_0) \longrightarrow (-\mathrm{sign}(D)\partial_x v \text{ at } x = x_0) \text{ with } v \text{ solving}\\ (k^2 + \eta)v - v_{\tilde{x}\tilde{x}} = 0 \quad \text{for } x \in [x_0, x_0 + D],\\ v = 0 \quad \text{at } x = x_0 + D, \quad v = \gamma \quad \text{at } x = x_0.\end{aligned}$$

In our case, DtN_D reduces to a scalar. Note that $\tilde{x}(x = x_0) = x_0$, $\tilde{x}(x = x_0 + D) = x_0 + D + \int_0^D \sigma(|t|)\,\mathrm{d}t =: x_0 + D + \bar{\sigma}$. From the above definition, we have

$$v = \xi_1 \mathrm{e}^{-s(\tilde{x}-x_0)} + \xi_2 \mathrm{e}^{s(\tilde{x}-x_0)}, \quad v(\tilde{x} = x_0 + D + \bar{\sigma}) = 0, \quad v(\tilde{x} = x_0) = \gamma.$$

Hence, we obtain in our case

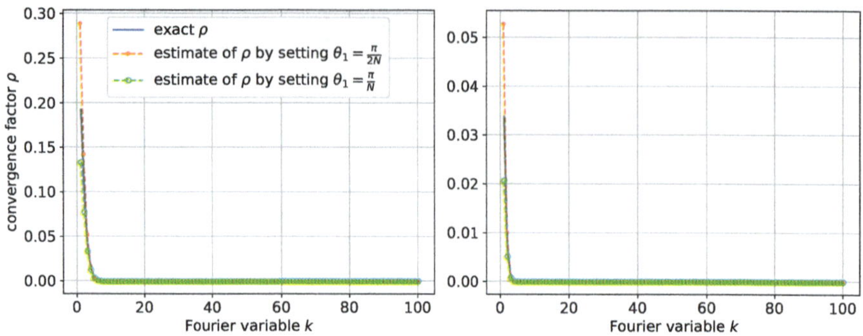

Fig. 9: Convergence factor of DOSM with p = DtN_D from PMLs, $N = 10$, $L = 0$, $\bar{\sigma} = 5D$, $D = 0.05$ (left), 0.1 (right) and $s \in [\sqrt{2}, \sqrt{10^4+1}]$.

$$\mathrm{DtN}_D = s \cdot \frac{1 + \mathrm{e}^{-2(D+\bar{\sigma})s}}{1 - \mathrm{e}^{-2(D+\bar{\sigma})s}}.$$

Typically, one chooses $\bar{\sigma}$ linearly dependent on D. Using $p = \mathrm{DtN}_D$, we show in Figure 9 how good our upper and lower bounds of ρ are. It is impressive that doubling D decreases $\|\rho\|_\infty$ by a factor of about six. Then, for D proportional to the subdomain size $H = 1/N$, we look at their scaling with N in Figure 10. We see that

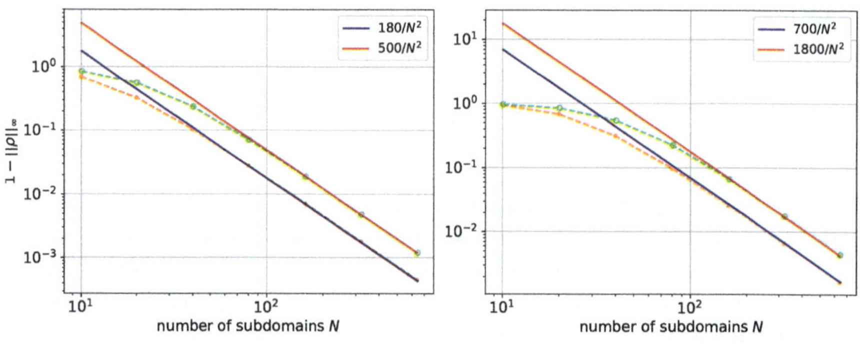

Fig. 10: Scaling of DOSM with $p = \mathrm{DtN}_D$ from PMLs, $L = 0$, $\bar{\sigma} = 5D$, $D = \frac{1}{2N}$ (left), $D = \frac{1}{N}$ (right) and $s \in [\sqrt{2}, \sqrt{100N^2+1}]$. The dashed lines correspond to the upper and lower bounds of $1 - \|\rho\|_\infty$.

the improvement by doubling D is only on the constant factor, and the deterioration $\|\rho\|_\infty \approx 1 - O(N^{-2})$ is the same as for the alternating Schwarz method. Hence, to have convergence independent of N, we must let the relative PML width D/H grow with N. To see how big a D is necessary, we test a range of D in Figure 11, where we can read for which size D and which N the bounds of $\|\rho\|_\infty$ equal to 0.2. We then plot these pairs in Figure 12, which indicates that a constant PML size D, independent of the number of subdomains N, is necessary and sufficient. The sufficiency is further

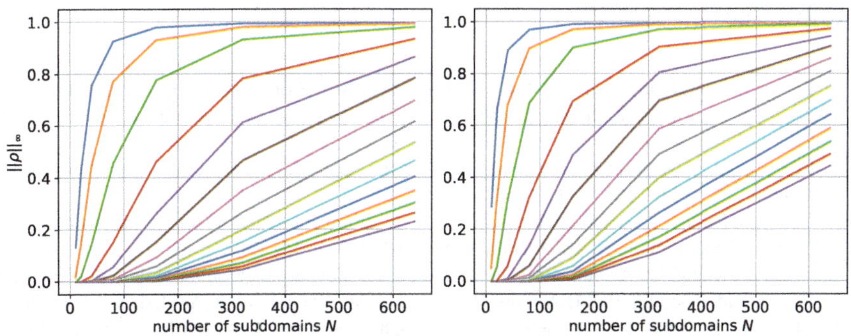

Fig. 11: Scaling of DOSM with different sizes D of PMLs, $L = 0$, $\bar{\sigma} = 5D$ and $s \in [\sqrt{2}, \sqrt{100N^2+1}]$. Left: lower bound. Right: upper bound. From top to bottom: $D/H = 0.5, 1, 2, 4, 6, .., 26$ where $H = 1/N$ is the subdomain width.

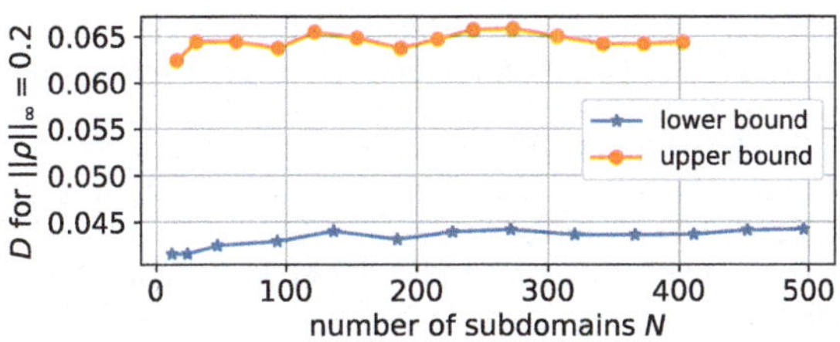

Fig. 12: Necessary PML size D to let DOSM converge independently of N, when $L = 0$, $\bar{\sigma} = 5D$ and $s \in [\sqrt{2}, \sqrt{100N^2+1}]$.

shown in Figure 13. Note that in our setting a fixed physical PML size independent

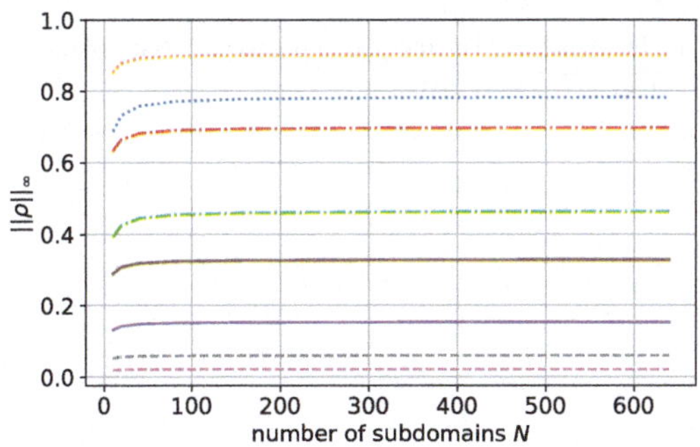

Fig. 13: Scaling of DOSM using fixed sizes of PMLs, $L = 0$, $\bar{\sigma} = 5D$ and $s \in [\sqrt{2}, \sqrt{100N^2+1}]$. From top to bottom each pair of lines correspond to the lower and upper bounds of $\|\rho\|_\infty$ for $D = 0.0125, 0.025, 0.05, 0.1$.

of the number of subdomains N means a linear growth of mesh points in the PMLs, not a logarithmic one.

References

1. Babuska, I.M., Sauter, S.A.: Is the pollution effect of the FEM avoidable for the Helmholtz equation considering high wave numbers? SIAM Journal on numerical analysis **34**(6), 2392–2423 (1997)
2. Berenger, J.P.: A perfectly matched layer for the absorption of electromagnetic waves. J. Comput. Phys. **114**, 185–200 (1994)
3. Chen, Z., Xiang, X.: A source transfer domain decomposition method for Helmholtz equations in unbounded domain. SIAM J. Numer. Anal. **51**, 2331–2356 (2013)
4. Chen, Z., Xiang, X.: A source transfer domain decomposition method for Helmholtz equations in unbounded domain Part II: Extensions. Numer. Math. Theor. Meth. Appl. **6**, 538–555 (2013)
5. Chevalier, P., Nataf, F.: Symmetrized method with optimized second-order conditions for the Helmholtz equation. In: J. Mandel, C. Farhat, X.C. Cai (eds.) Domain Decomposition Methods 10, pp. 400–407. AMS (1998)
6. Chew, W.C., Jin, J.M., Michielssen, E.: Complex coordinate stretching as a generalized absorbing boundary condition. Microw. Opt. Technol. Lett. **15**, 363–369 (1997)
7. Engquist, B., Ying, L.: Sweeping preconditioner for the Helmholtz equation: Hierarchical matrix representation. Comm. Pure Appl. Math. **LXIV**, 0697–0735 (2011)
8. Engquist, B., Ying, L.: Sweeping preconditioner for the Helmholtz equation: Moving perfectly matched layers. Multiscale Model. Sim. **9**, 686–710 (2011)
9. Ernst, O., Gander, M.J.: Why it is difficult to solve Helmholtz problems with classical iterative methods. In: I. Graham, T. Hou, O. Lakkis, R. Scheichl (eds.) Numerical Analysis of Multiscale Problems, pp. 325-363. Springer-Verlag, Berlin (2012)
10. Gander, M.J.: Optimized Schwarz methods. SIAM Journal on Numerical Analysis **44**(2), 699–731 (2006)
11. Gander, M.J., Magoules, F., Nataf, F.: Optimized Schwarz methods without overlap for the Helmholtz equatio n. SIAM J. Sci. Comput. **24**, 38–60 (2002)
12. Gander, M.J., Nataf, F.: An incomplete LU preconditioner for problems in acoustics. J. Comput. Acoust. **13**, 455–476 (2005)
13. Gander, M.J., Zhang, H.: A class of iterative solvers for the Helmholtz equation: Factorizations, sweeping preconditioners, source transfer, single layer potentials, polarized traces, and optimized Schwarz methods. SIAM Review **61**(1), 3–76 (2019)
14. Stolk, C.C.: A rapidly converging domain decomposition method for the Helmholtz equation. J. Comput. Phys. **241**, 240–252 (2013)
15. Zepeda-Núñez, L., Demanet, L.: The method of polarized traces for the 2D Helmholtz equation. J. Comput. Phys. **308**, 347–388 (2016)

Part II
Talks in Minisymposia (MT)

Dirichlet-Neumann Preconditioning for Stabilised Unfitted Discretization of High Contrast Problems

B. Ayuso de Dios, K. Dunn, M. Sarkis, and S. Scacchi

1 Introduction

Let $\Omega \subset \mathbb{R}^2$ be a polygonal domain with an immersed simple closed smooth interface $\Gamma \in C^2$, such that $\overline{\Omega} = \overline{\Omega}^- \cup \overline{\Omega}^+$, and $\Gamma := \overline{\Omega}^- \cap \overline{\Omega}^+$ is far away from $\partial\Omega$ (i.e, either Ω^+ or Ω^- is a *floating subdomain*; i.e., one of them does not touch $\partial\Omega$). Given $f \in L^2(\Omega)$ we set $f^\pm = f_{|\Omega^\pm}$ and consider the problem of finding u_* such that

$$\begin{cases} -\nabla \cdot (\rho_\pm \nabla u_*^\pm) = f^\pm \quad \text{in } \Omega^\pm, & u_*^\pm \quad = 0 \quad \text{on } \partial\Omega^\pm \backslash \Gamma \\ \qquad [u_*] = 0 \quad \text{on } \Gamma, & [\rho \nabla u_*] \quad = 0 \quad \text{on } \Gamma, \end{cases} \tag{1}$$

where $u_*^\pm = u_*|_{\Omega^\pm}$ and $\mathbf{n}^\pm$ denote the unit normal outward to $\Omega^\pm$. The jump conditions on Γ enforce the continuity of the solution and its flux across the interface. The jump operators are defined by

$$[\rho \nabla u_*] = \rho_+ \nabla u_*^+ \cdot \mathbf{n}^+ + \rho_- \nabla u_*^- \cdot \mathbf{n}^- \quad \text{and} \quad [u_*] = u_*^+ - u_*^-. \tag{2}$$

We also assume that the diffusion coefficients $\rho_\pm > 0$ are constant and satisfy $\rho_- \leq \rho_+$. Note that $u_*^\pm \in H^2(\Omega^\pm)$, but $u_* \in H^{1+\epsilon}(\Omega)$ with $\epsilon > 0$. To approximate

Blanca Ayuso de Dios
Dipartimento di Matematica e Applicazioni, Università di Milano-Bicocca, Milan, Italy, e-mail: blanca.ayuso@unimib.it

Kyle Dunn
Cold Regions Research and Engineering Laboratory, ERDC - U.S. Army, Hanover, NH, e-mail: Kyle.G.Dunn@usace.army.mil

Marcus Sarkis
Mathematical Sciences Department, Worcester Polytechnic Institute, MA, e-mail: msarkis@wpi.edu

Simone Scacchi
Dipartimento di Matematica, Università degli Studi di Milano, Milan, Italy, e-mail: simone.scacchi@unimi.it

R. Haynes et al. (eds.), *Domain Decomposition Methods in Science and Engineering XXV*,
Lecture Notes in Computational Science and Engineering 138,
https://doi.org/10.1007/978-3-030-56750-7_6

(1) we consider the stabilised unfitted FE approximation from [3].

A class of unfitted finite element methods were introduced in the seminal works of [1] and in recent years there has been a renewed interest in these type of approaches, giving rise to numerous novel methods; the immersed boundary method [2], XFEM [5], the finite cell method (FCM) [6], and CutFEM [4, 8]. The use of unfitted meshes is particularly relevant for interface problems. However, in spite of the upsurge in research for unfitted approaches, the design and analysis of robust solvers for the resulting linear and nonlinear systems still seem elusive. Simple preconditioning strategies are explored for finite cell discretizations in [10] and multigrid-type method are proposed in [9]. In the present contribution we focus on the construction of a simple Dirichlet-Neumann (DN) domain decomposition preconditioner for the CutFEM method introduced in [3] and demonstrate its robustness also in the hard inclusion case. Due to space restrictions, we focus on a very simple version and stick to the algebraic description of the solver. Details on the analysis as well as further tailored preconditioners will be found in [7].

2 Basic Notation and Unfitted Stabilized Discretization

Let $\{\mathcal{T}_h\}_{h>0}$ be a family of uniform partitions of Ω into squares T of diameter h. We assume that for each T, $\Gamma \cap \partial T$, is either empty or occurs at exactly two different edges of ∂T[1]. We also define:

$$\mathcal{T}_h^{\pm} := \{T \in \mathcal{T}_h : \overline{T} \cap \overline{\Omega}^{\pm} \neq \emptyset\}, \quad \mathcal{T}_h^{\Gamma} := \{T \in \mathcal{T}_h : \overline{T} \cap \Gamma \neq \emptyset\}.$$

For $T \in \mathcal{T}_h^{\Gamma}$ we denote $T_\Gamma = \overline{T} \cap \Gamma$. We also introduce the discrete domains

$$\Omega_h^{\pm} := \mathrm{Int}\left(\bigcup_{T \in \mathcal{T}_h^{\pm}} \overline{T}\right) \qquad \Omega_h^{\Gamma} := \mathrm{Int}\left(\bigcup_{T \in \mathcal{T}_h^{\Gamma}} \overline{T}\right), \quad \text{and} \quad \Omega_{h,0}^{\pm} = \Omega_h^{\pm} \backslash \overline{\Omega}_h^{\Gamma},$$

where $\mathrm{Int}(K)$ denotes the interior of the set K. Note that $\Omega_h^+ \cup \Omega_h^- = \Omega$ is an overlapping partition of Ω while a non-overlapping partition is given by $\Omega = \Omega_{h,0}^+ \cup \overline{\Omega}_h^{\Gamma} \cup \Omega_{h,0}^-$ (see Figure 1.) Finally we introduce the following subsets of edges of elements in $\mathcal{T}_h^{\Gamma}$:

$$\mathcal{E}_h^{\Gamma,\pm} := \{e = \mathrm{Int}(\partial T_1 \cap \partial T_2) : T_1 \neq T_2 \in \mathcal{T}_h^{\pm}, \text{ and } T_1 \in \mathcal{T}_h^{\Gamma} \text{ or/and } T_2 \in \mathcal{T}_h^{\Gamma}\}.$$

Note that $\mathcal{E}_h^{\Gamma,+}$ (resp. $\mathcal{E}_h^{\Gamma,-}$) does not contain any edges on $\partial\Omega_h^+$ (resp. $\partial\Omega_h^-$).

• *Finite Element Spaces:* We consider FE spaces of piecewise bilinear polynomials whose support is contained in $\Omega_h^{\pm}$, $\Omega_{h,0}^{\pm}$ and Ω_h^{Γ}, respectively:

[1] This assumption is only needed in the stability and error analysis of the method.

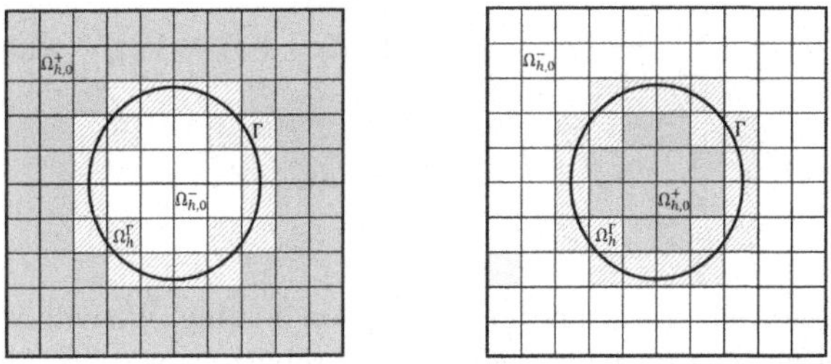

Fig. 1: Domain configurations: Soft inclusion (leftmost), Hard inclusion (center) cases

$$V^{\pm} = \{v \in C(\Omega_h^{\pm}) : v|_T \in \mathbb{Q}^1(T), \forall T \in \mathcal{T}_h^{\pm}, \text{ and } v|_{\partial\Omega_h^{\pm}\cap\partial\Omega} \equiv 0\},$$

$$V_0^{\pm} = \{v \in V^{\pm} \ : \ v|_T \equiv 0 \quad \forall\, T \in \Omega_h^{\Gamma}\}, \qquad W^{\pm} = \{v \text{ restricted to } \Omega_h^{\Gamma}, \quad v \in V^{\pm}\}\ .$$

With a small abuse of notation, we set $V_h = V^+ \times V^-$ where it is understood

$$u_h \in V_h = V^+ \times V^- \qquad u_h = (u^+, u^-) \text{ with } u^+ \in V^+\ , \quad u^- \in V^-\ .$$

That is, the FE space V_h is defined by *a copy of two FE piecewise functions*: one from V^+ defined on Ω_h^+ and another from V^- defined over Ω_h^-.

• *The stabilised unfitted Nitsche approximation:* the method reads: find $u_h = (u^+, u^-) \in V_h = V^+ \times V^-$, such that:

$$a_h(u_h, v_h) = (f^+, v^+)_{\Omega^+} + (f^-, v^-)_{\Omega^-}, \qquad \text{for all } v_h = (v^+, v^-) \in V^+ \times V^-\ , \tag{3}$$

where $(\cdot,\cdot)_{\Omega^{\pm}}$ denotes the $L^2(\Omega^{\pm})$ inner product and $a_h : V_h \times V_h \longrightarrow \mathbb{R}$ is given as:

$$a_h(u_h, v_h) = \int_{\Omega^-} \rho_- \nabla u^- \cdot \nabla v^- dx + \int_{\Omega^+} \rho_+ \nabla u^+ \cdot \nabla v^+ dx \tag{4}$$

$$+ \int_{\Gamma} \left(\{\rho\nabla v_h\}_w \cdot \mathbf{n}^- \,[u_h] + \{\rho \nabla u_h\}_w \cdot \mathbf{n}^- \,[v_h]\right) ds + \sum_{T\in\mathcal{T}_h^{\Gamma}} \frac{\gamma_\Gamma}{h_T} \{\rho\}_H \int_{T_\Gamma} [u_h]\,[v_h]\, ds$$

$$+ \sum_{e\in\mathcal{E}_h^{\Gamma,-}} \gamma_- |e| \int_e \rho_- [\nabla u^-]\,[\nabla v^-]\, ds + \sum_{e\in\mathcal{E}_h^{\Gamma,+}} \gamma_+ |e| \int_e \rho_+ [\nabla u^+]\,[\nabla v^+]\, ds,$$

where $\gamma^{\Gamma}, \gamma^-$, and γ^+ are positive (moderate) constants and $|e|$ is the diameter of the edge e. Here, $[\cdot]$ refers to the jump operator as in (2) while $\{\cdot\}_H$ and $\{\cdot\}_\omega$ denote the harmonic and weighted averages defined by:

$$\{\rho\}_H = \frac{2\rho^+\rho^-}{\rho^+ + \rho^-}\ , \qquad \{\rho\nabla v_h\}_\omega := (\omega_- \rho^- \nabla v^- + \omega_+ \rho^+ \nabla v^+), \quad \omega_\mp = \frac{\rho^\pm}{\rho^+ + \rho^-}\ .$$

Continuity and coercivity of $a_h(\cdot,\cdot)$ in (4) can be shown with respect to the norm:

$$\|v_h\|_{V_h}^2 := |v^+|_{V^+}^2 + |v^-|_{V^-}^2 + \sum_{T\in\mathcal{T}_h^\Gamma} \frac{\gamma_\Gamma}{h_T}\{\rho\}_H \int_{T_\Gamma} [v_h]^2\, ds \quad \forall v_h \in V_h\,, \quad \text{with}$$

$$|v^\pm|_{V^\pm}^2 := \int_{\Omega^\pm} \rho_\pm |\nabla v^\pm|^2\, dx + \sum_{e\in\mathcal{E}_h^{\Gamma,\pm}} \gamma_\pm |e| \int_e \rho_\pm [\nabla v^\pm]^2\, ds, \quad \forall v^\pm \in V^\pm\,. \tag{5}$$

We remark that the semi-norm $|\cdot|_{V^+}$ is a norm if Ω^+ is non floating. We will denote by $(\cdot,\cdot)_{V^+}$ to its originating inner product. Optimal and robust error estimates are proved in [3].

3 Dirchlet-Neumann preconditioner

We describe now a preconditioner for the linear system resulting from (3) based on the non-overlapping decomposition $\Omega_{h,0}^+ \cup \overline{\Omega}_h^\Gamma \cup \Omega_{h,0}^-$. Associated with such a decomposition, and owing to the *fat interface* we consider the somehow asymmetric splitting of the space $V_h = (V_0^+, W^+) \times V^-$, we first introduce some notation. We denote by $\mathcal{R}_\pm : V_h \longrightarrow V^\pm$ the restriction operators to $\Omega_h^\pm$ such that $\mathcal{R}_\pm u_h = u^\pm$. The corresponding prolongation operators $\mathcal{R}_\pm^T : V^\pm \longrightarrow V_h$ are defined as the extension to V_h by zero, i.e., $\mathcal{R}_+^T u^+ = (u^+, 0)$ and $\mathcal{R}_-^T u^- = (0, u^-)$. Similarly, we introduce the restriction and prolongation operators

$$\begin{array}{lll} \mathcal{R}_{W^\pm} : V_h \longrightarrow W^\pm & \mathcal{R}_{0^\pm} : V_h \longrightarrow V_0^\pm & \mathcal{R}_W : V_h \longrightarrow W_h \\ \mathcal{R}_{W^\pm}^T : W^\pm \longrightarrow V_h & \mathcal{R}_{0^\pm}^T : V_0^\pm \longrightarrow V_h & \mathcal{R}_W^T : W_h \longrightarrow V_h \end{array}$$

We define the bilinear forms $a_0^+ : V_0^+ \times V_0^+ \longrightarrow \mathbb{R}$ and $a^- : V^- \times V^- \longrightarrow \mathbb{R}$

$$\begin{array}{lll} a_0^+ : V_0^+ \times V_0^+ \longrightarrow \mathbb{R} & a_0^+(u_0^+, v_0^+) := a_h(\mathcal{R}_{0^+}^T u_0^+, \mathcal{R}_{0^+}^T v_0^+) & \forall u_0^+, v_0^+ \in V_0^+ \\ a^- : V^- \times V^- \longrightarrow \mathbb{R} & a^-(u^-, v^-) := a_h(\mathcal{R}_-^T u^-, \mathcal{R}_-^T v^-) & \forall u^-, v^- \in V^- \end{array}$$

We now introduce the *local solvers*. Let $u_{f,0}^+ \in V_0^+$ and $u_f^- \in V^-$ be the local solutions with support in $\Omega_{h,0}^+$ and Ω_h^-, respectively, defined by:

$$a_0^+(u_{f,0}^+, v_0^+) = (f^+, v_0^+)_{\Omega^+} \;\; \forall v_0^+ \in V_0^+ \qquad a^-(u_f^-, v^-) = (f^-, v^-)_{\Omega^-} \quad v^- \in V^-\,.$$

We set $\mathcal{P}_h u_h = \mathcal{R}_{0^+}^T u_{f,0}^+ + \mathcal{R}_-^T u_f^-$ and note that $u_h - \mathcal{P}_h u_h$ lies in the orthogonal complement of $\mathcal{R}_{0^+}^T V_0^+ + \mathcal{R}_-^T V^-$ in V_h with respect to the inner product $a_h(\cdot,\cdot)$. This suggests the splitting $u_h = \mathcal{P}_h u_h + \mathcal{H}_h u_h$, with $\mathcal{H}_h u_h = (\mathcal{H}_+ u_h, \mathcal{H}_- u_h) \in V_h$ a suitable *discrete harmonic extension* of $(u_h^+)|_{\Omega_h^\Gamma}$ that we briefly sketch next. Recall that W^+ is the restriction of the space V^+ to Ω_h^Γ. Given $\eta^+ \in W^+$, we define $\mathcal{H}_\pm : W^+ \longrightarrow V^\pm$ to be the discrete harmonic extension of η^+ such that

$$a_h(\mathcal{R}_+^T \mathcal{H}_+ \eta^+, \mathcal{R}_{0^+}^T v_0^+) = 0 \quad \forall v_0^+ \in V_0^+ \qquad \text{and} \qquad \mathcal{R}_{W^+} \mathcal{R}_+^T \mathcal{H}_+ \eta^+ = \eta^+$$

and

$$a_h\big((\mathcal{R}_+\mathcal{R}_{W^+}^T\eta^+, \mathcal{H}_-\eta^+), \mathcal{R}_-^T v^-\big) = 0 \qquad \forall v^+ \in V^- .$$

Finally, we set $\mathcal{H}_h\eta^+ = (\mathcal{H}_+\eta^+, \mathcal{H}_-\eta^+)$ and introduce the Schur complement operator $\mathcal{S} : W^+ \longrightarrow W^+$:

$$< \mathcal{S}\eta, w >:= a_h(\mathcal{H}_h\eta^+, \mathcal{H}_h w^+) \qquad \forall \eta^+, w^+ \in W^+ . \tag{6}$$

From the definition of $\mathcal{P}_h u_h$ it follows

$$a_h(\mathcal{H}_h u_h, \mathcal{H}_h v_h) = (f, v_h)_\Omega - a_h(\mathcal{P}_h u_h, v_h) \qquad \forall v_h \in V_h . \tag{7}$$

We focus now on constructing preconditioners $\mathcal{B}^{-1}$ for the operator $\mathcal{S}$ and hence for the system (7). The basic guide to ensure robustness will be to use, when possible, the local Schur complement corresponding to the largest coefficient, ρ_+:

$$< \mathcal{S}_+\eta, w >:= (\mathcal{H}_+\eta, \mathcal{H}_+ w^+)_{V^+} \qquad \forall \eta, w \in W^+ , \tag{8}$$

where $(\cdot,\cdot)_{V+}$ is the originating inner-product for the norm $|\cdot|_{V^+}$ in (5). We need to distinguish two cases:

- Ω^+ is not floating subdomain and we set $\mathcal{B}^{-1} = \mathcal{S}_+^{-1}$.
- Ω^+ is a floating subdomain; since $\mathcal{S}_+$ is not invertible, we define $\mathcal{B}^{-1}$ as a suitable regularisation of $\mathcal{S}_+$. We propose one level and two level methods.

4 Algebraic formulation of the DN preconditioner

After choosing standard Lagrangian basis for $V^\pm$, problem (3) reduces to a linear algebraic system $\mathbb{A}\mathbb{U} = \mathbb{F}$. We consider the block structure of $\mathbb{A}$ that results from splitting the degrees of freedom (dofs) of the discrete space V_h into three sets:

- dofs associated with V_0^+ (in the interior of Ω^+) are indicated by I^+;
- dofs related to W^+, indicated by W^+;
- dofs associated with V^-(dofs related to V_0^- and W^-), indicated by V^-.

$$\begin{bmatrix} \mathbb{A}_{I^+I^+} & \mathbb{A}_{I^+W^+} & 0 \\ \mathbb{A}_{W^+I^+} & \boxed{\mathbb{A}_{W^+W^+}^+ + \mathbb{A}_{W^+W^+}^-} & \mathbb{A}_{W^+V^-} \\ 0 & \mathbb{A}_{V^-W^+} & \mathbb{A}_{V^-V^-} \end{bmatrix} \begin{bmatrix} \mathbb{U}_{I^+} \\ \mathbb{U}_{W^+} \\ \mathbb{U}_{V^-} \end{bmatrix} = \begin{bmatrix} \mathbb{F}_{I^+} \\ \mathbb{F}_{W^+} \\ \mathbb{F}_{V^-} \end{bmatrix} .$$

Here, we have highlighted that the stiffness block with dofs from W^+ in the *fat interface* has contributions from Ω_h^+ and Ω_h^-. Performing static condensation of the interior variables I^+ and V^- we obtain the Schur complement system

$$\mathbb{S}\mathbb{U}_{W^+} = \mathbb{G}_{W^+}, \qquad \mathbb{S} = \mathbb{S}_+ + \mathbb{S}_- ,$$

where $\mathbb{G}_{W^+} = \mathbb{F}_{W^+} - \mathbb{A}_{W^+I^+}\mathbb{A}_{I^+I^+}^{-1}\mathbb{F}_{I^+} - \mathbb{A}_{W^+V^-}\mathbb{A}_{V^-V^-}^{-1}\mathbb{F}_{V^-}$, and $\mathbb{S}$ is given by

$$\mathbb{S} = \mathbb{S}_+ + \mathbb{S}_- \qquad \text{with} \qquad \begin{cases} \mathbb{S}_+ = \mathbb{A}^+_{W^+W^+} - \mathbb{A}_{W^+I^+}\mathbb{A}^{-1}_{I^+I^+}\mathbb{A}_{I^+W^+} \\ \mathbb{S}_- = \mathbb{A}^-_{W^+W^+} - \mathbb{A}_{W^+V^-}\mathbb{A}^{-1}_{V^-V^-}\mathbb{A}_{V^-W^+}, \end{cases}$$

Soft inclusion: Ω_h^+ in Non-Floating Subdomain Case: In this case we set $\mathcal{B}^{-1} = \mathcal{S}_+^{-1}$ since the operator is invertible. At the algebraic level we arrive at $\mathbb{S}_+^{-1}\mathbb{S}\mathbb{U}_{W^+} = \mathbb{S}_+^{-1}\mathbb{G}_{W^+}$. The action of the DN preconditioner $\mathbb{S}_+^{-1}$ on a generic residual vector $\mathbb{R}_{W^+}$ consists of solving the linear system

$$\begin{bmatrix} \mathbb{A}_{I^+I^+} & \mathbb{A}_{I^+W^+} \\ \mathbb{A}_{W^+I^+} & \mathbb{A}^+_{W^+W^+} \end{bmatrix} \begin{bmatrix} \mathbb{V}_{I^+} \\ \mathbb{V}_{W^+} \end{bmatrix} = \begin{bmatrix} 0 \\ \mathbb{R}_{W^+} \end{bmatrix}.$$

and letting $\mathbb{V}_{W^+} := \mathbb{S}_+^{-1}\mathbb{R}_{W^+}$.

Hard inclusion: Ω_h^+ is the Floating Subdomain: Since $\mathbb{S}_+$ is not invertible we consider two different strategies: a regularisation and the use of a one dimensional coarse solver to account for the kernel of $\mathbb{S}_+$.

- *One-Level DN:* The action of the preconditioner amounts to solving

$$\left(\begin{bmatrix} \mathbb{A}_{I^+I^+} & \mathbb{A}_{I^+W^+} \\ \mathbb{A}_{W^+I^+} & \mathbb{A}^+_{W^+W^+} \end{bmatrix} + \frac{\{\rho\}_H}{D_+^2} \begin{bmatrix} \mathbb{M}^+_{I^+I^+} & \mathbb{M}^+_{I^+W^+} \\ \mathbb{M}^+_{W^+I^+} & \mathbb{M}^+_{W^+W^+} \end{bmatrix} \right) \begin{bmatrix} \mathbb{V}_{I^+} \\ \mathbb{V}_{W^+} \end{bmatrix} = \begin{bmatrix} 0 \\ \mathbb{R}_{W^+} \end{bmatrix},$$

and setting $\mathbb{S}^{-1}_{+,one}\mathbb{R}_{W^+} = \mathbb{V}_{W^+}$. Here, $\mathbb{M}^+$ stands for the mass matrix associated with V^+ (i.e., defined over Ω_h^+), and $D_+ :=\text{diam}(\Omega_h^+)$ and is used to regularise $\mathbb{S}_+$.

- *Two Level DN preconditioner:* The idea is to first solve in the space orthogonal to the (one-dimensional) kernel of $\mathbb{S}_+$ and then correct with a coarse solver that accounts for the contribution in $\ker(\mathbb{S}_+)$. Hence, the practical implementation of the two level solver $\mathbb{S}^{-1}_{+,two}$ amounts to first solving

$$\begin{bmatrix} \mathbb{A}_{I^+I^+} & \mathbb{A}_{I^+W^+} \\ \mathbb{A}_{W^+I^+} & \mathbb{A}^+_{W^+W^+} \end{bmatrix} \begin{bmatrix} \mathbb{V}_{I^+} \\ \mathbb{V}_{W^+} \end{bmatrix} + \begin{bmatrix} \mathbb{M}^+_{I^+I^+} & \mathbb{M}^+_{I^+W^+} \\ \mathbb{M}^+_{W^+I^+} & \mathbb{M}^+_{W^+W^+} \end{bmatrix} \begin{bmatrix} \mathbf{1}_{I^+} \\ \mathbf{1}_{W^+} \end{bmatrix} \lambda = \begin{bmatrix} 0 \\ \mathbb{R}_{W^+} \end{bmatrix},$$

with the constraint

$$\begin{bmatrix} \mathbf{1}_{I^+} \\ \mathbf{1}_{W^+} \end{bmatrix}^T \begin{bmatrix} \mathbb{M}^+_{I^+I^+} & \mathbb{M}^+_{I^+W^+} \\ \mathbb{M}^+_{W^+I^+} & \mathbb{M}^+_{W^+W^+} \end{bmatrix} \begin{bmatrix} \mathbb{V}_{I^+} \\ \mathbb{V}_{W^+} \end{bmatrix} = 0$$

and then define $\mathbb{S}_+^{\dagger}\mathbb{R}_{W^+} = \mathbb{V}_{W^+}$. Here $\mathbf{1}^+_{I^+}$ and $\mathbf{1}^+_{W^+}$ are vectors of ones in V_0^+ and W^+, respectively. The matrix representation of the two level preconditioner (with coarse space) is defined via $\mathbb{S}^{-1}_{+,two} = \mathbb{S}_+^{\dagger} + \mathbf{1}_{W^+}(\mathbf{1}_{W^+}, \mathbb{S}\mathbf{1}_{W^+})^{-1}\mathbf{1}^T_{W^+}$. Note that $\mathbb{S}\mathbf{1}_{W^+} = \mathbb{S}_-\mathbf{1}_{W^+}$.

5 Numerical Results

We consider the domain $\Omega = (0,1)^2$ and study the performance of the Dirichlet-Neumann (DN) preconditioner for the CutFEM approximation (3) to (1) with $\Omega^{\mp}$ a disk of radius 0.15 and $\Omega^{\pm} = (0,1)^2 \setminus \overline{\Omega}^{\mp}$ and always $\rho_- \leq \rho_+$. We use CG and PCG as a solver with zero initial guess and tolerance 10^{-6} for the relative residual. In the tables we report the estimated (via Lanzcos algorithm) condition numbers (denoted by κ_2) and the number of iterations (denoted by **it**) required by CG and PCG for convergence. Table 1 reports the results in the case where Ω^+ is

ρ_-	**full CG**		**schur NO precond.**		**schur DN preconditioned**	
	κ_2	it	κ_2	it	κ_2	it
1	3.32e+3	(218)	388.40	(75)	1.95	(14)
10^{-2}	2.06e+4	(575)	362.15	(91)	1.01	(15)
10^{-4}	2.01e+6	(2828)	361.71	(93)	1.00	(4)
10^{-6}	2.01e+8	(5418)	361.70	(93)	1.00	(3)

Table 1: Robustness with respect to ρ: Ω^- is the floating subdomain. Here, $\rho_+ = 1$ and $h = 1/64$.

non-floating, therefore using $\mathcal{S}_+^{-1}$ as a preconditioner. $\mathcal{S}_+^{-1}$ performs robustly when the ratio ρ_+/ρ_- increases. In the case where Ω^+ is the floating subdomain, we use one level and two level DN preconditioners. The results regarding optimality and robustness of these preconditioners are reported in Table 2 and 3, respectively. Notice that both preconditioners perform optimally and show robustness with respect to the jumping coefficient. In particular, the one-level DN preconditioner seems to be enough effective for the considered setting.

$1/h$	**full CG**		**schur NO precond.**		**DN Two-Level**		**DN one-level**	
	κ_2	it	κ_2	it	κ_2	it	κ_2	it
8	6.38e+3	252	4.09e+2	79	6.76	11	3.51	14
16	1.77e+4	520	8.60e+3	224	6.39	15	2.11	14
32	5.83e+4	863	1.09e+4	423	6.29	16	2.09	14
64	2.14e+4	1625	1.86e+4	551	6.34	16	2.08	14
128	8.19e+5	3163	3.79e+4	832	6.37	16	2.13	14
256	3.20e+6	6140	7.43e+4	1148	6.39	16	2.19	14

Table 2: Optimality with respect to h: floating circle Ω^+ embedded in $[0,1]^2$. $\rho_+ = \rho_- = 1$.

ρ_+	full CG		schur NO precond.		DN Two-Level		DN one-level	
	κ_2	it	κ_2	it	κ_2	it	κ_2	it
1	2.14e+5	1625	1.86e+4	539	6.37	16	2.13	14
10^2	2.00e+7	12906	1.81e+6	765	6.33	6	1.83	5
10^4	2.00e+9	>100000	1.81e+8	897	6.33	4	1.83	4
10^6	5.70e+10	>100000	1.81e+10	1026	6.33	3	1.83	3
10^8	4.20e+12	>100000	1.83e+12	1326	6.33	3	1.83	3

Table 3: Robustness with respect to ρ. Floating Ω^+ with jumping coefficients. Here, $\rho_- = 1$, $1/h = 64$.

References

1. Barrett, J.W., Elliott, C.M.: Fitted and unfitted finite-element methods for elliptic equations with smooth interfaces. IMA J. Numer. Anal. **7**(3), 283–300 (1987). DOI:10.1093/imanum/7.3.283
2. Boffi, D., Gastaldi, L.: A finite element approach for the immersed boundary method. Comput. & Structures **81**(8-11), 491–501 (2003). DOI:10.1016/S0045-7949(02)00404-2. In honour of Klaus-Jürgen Bathe
3. Burman, E., Guzmán, J., Sánchez, M.A., Sarkis, M.: Robust flux error estimation of an unfitted Nitsche method for high-contrast interface problems. IMA J. Numer. Anal. **38**(2), 646–668 (2018)
4. Burman, E., Hansbo, P.: Fictitious domain finite element methods using cut elements: II. A stabilized Nitsche method. Appl. Numer. Math. **62**(4), 328–341 (2012). DOI:10.1016/j.apnum.2011.01.008
5. Chessa, J., Smolinski, P., Belytschko, T.: The extended finite element method (XFEM) for solidification problems. Internat. J. Numer. Methods Engrg. **53**(8), 1959–1977 (2002). DOI:10.1002/nme.386
6. Dauge, M., Düster, A., Rank, E.: Theoretical and numerical investigation of the finite cell method. J. Sci. Comput. **65**(3), 1039–1064 (2015). DOI:10.1007/s10915-015-9997-3
7. Ayuso de Dios, B., Dunn, K., Sarkis, M., Scacchi, S.: Simple preconditioners for cutfem methods (2019). (work in preparation)
8. Hansbo, A., Hansbo, P.: An unfitted finite element method, based on Nitsche's method, for elliptic interface problems. Comput. Methods Appl. Mech. Engrg. **191**(47-48), 5537–5552 (2002). DOI:10.1016/S0045-7825(02)00524-8
9. Ludescher, T., Groβ, S., Reusken, A.: A multigrid method for unfitted finite element discretizations of elliptic interface problems. Techical report IGPM 481, RWTH Aachen (2018)
10. de Prenter, F., Verhoosel, C.V., van Zwieten, G.J., van Brummelen, E.H.: Condition number analysis and preconditioning of the finite cell method. Comput. Methods Appl. Mech. Engrg. **316**, 297–327 (2017). DOI:10.1016/j.cma.2016.07.006

Virtual Coarse Spaces for Irregular Subdomain Decompositions

Juan G. Calvo

1 Introduction

Consider the model problem: Find $u \in H^1(\Omega)$ such that

$$-\nabla \cdot (\rho(x)\nabla u) = f(x), \quad x \in \Omega, \tag{1}$$

for a given polygonal domain $\Omega \subset \mathbb{R}^2$ and $\rho(x) > 0$, along with homogeneous boundary conditions. A standard approach to solve (1) is to discretize with Finite Element Methods (FEM) for which there is vast literature on the construction of Domain Decomposition (DD) algorithms; see, e.g., [11] for a complete study. As usual, we will decompose the domain Ω into N non-overlapping subdomains $\{\Omega_i\}_{i=1}^N$, each of which is the union of elements of the triangulation $\mathcal{T}_h$ of Ω. Each Ω_i will be simply connected and will have a connected boundary $\partial\Omega_i$. We then construct overlapping subdomains Ω_i' by adding layers of elements to Ω_i.

One of the simplest DD algorithms consists in splitting the finite dimensional space V_h (associated with the fine triangulation of the domain) as

$$V_h = R_0^T V_0 + \sum_{i=1}^{N} R_i^T V_i,$$

where $V_1, \dots, V_N$ represent local spaces related to $\Omega_1', \dots, \Omega_N'$, respectively, with corresponding extension operators $R_i^T : V_i \to V_h$, and V_0 is a coarse space which is related to V_h by the operator $R_0^T : V_0 \to V_h$. Originally, these methods arose in the presence of regular decompositions where usual Finite Element spaces can be defined. In the past few years, there has been some efforts to study how to define coarse spaces if irregular subdomains as the ones obtained by mesh partitioners

J. G. Calvo
Centro de Investigación en Matemática Pura y Aplicada – Escuela de Matemática, Universidad de Costa Rica, San José, Costa Rica, e-mail: juan.calvo@ucr.ac.cr

R. Haynes et al. (eds.), *Domain Decomposition Methods in Science and Engineering XXV*, Lecture Notes in Computational Science and Engineering 138,
https://doi.org/10.1007/978-3-030-56750-7_7

are considered; see, e.g., [5, 14, 6], where a complete theory is developed for Jones subdomains and nodal elliptic problems. For Raviart-Thomas and Nédélec elements, see [9, 2]. These studies are based on energy minization, and require to obtain local discrete harmonic functions by solving Dirichlet problems on the subdomains. In a more general setting, adaptive coarse spaces can be defined as in [7, 10, 8].

On the other hand, Virtual Element Methods (VEM) [1, 12, 13] allow to handle general polygonal elements. In the case of triangular elements, VEM reduces to the usual FEM. Thus, VEM is a natural choice for constructing space of functions on irregular subdomains. As studied in [3, 4], considering a virtual space on an irregular decomposition allows us to avoid the computation of discrete harmonic functions, while we keep the typical bound for the condition number of the preconditioned system; see Theorem 1 below. In this setting, we can define general virtual functions for such irregular decompositions. However, virtual functions cannot be evaluated at interior nodes, and the operator R_0^T plays an essential role into approximating functions in V_0. Two different approaches have been studied so far: we can construct V_0 based on linear interpolants [4], or we can use projections onto polynomial spaces of degree of at least two [3], which we will discuss in this manuscript.

Instead of having a triangular mesh and a FEM discretization for problem (1), we could also consider a discretization based on VEM. There is a lack of literature on DD methods for such type of problems. At the DD25 Conference, held in Saint John's, Canada on July 2018, interesting talks by Yunrong Zhu (Auxiliary Space Preconditioners for Virtual Element Discretization) and Daniele Prada (FETI-DP for Three Dimensional VEM) addressed this problem with different approaches as ours. We note that the theory developed in [3, 4] is also useful for designing Schwarz operators for discretizations obtained by VEM, and it is possible to obtain similar bounds for the condition number of the preconditioned system.

2 Description of the preconditioner

In this section we describe the discretizacion of the model problem and the construction of the additive preconditioner. We refer [13] for general details on VEM, [3, 4] for a detailed explanation on the coarse space definition, and [11, Chapter 3] for a complete study of overlapping preconditioners.

The usual weak form for problem (1) is: Find $u \in H_0^1(\Omega)$ such that

$$a(u,v) := \int_\Omega \nabla u \cdot \nabla v \, d\boldsymbol{x} = (f,v) \quad \forall\, v \in H_0^1(\Omega), \tag{2}$$

where $(\cdot,\cdot)$ is the usual inner product in $L^2(\Omega)$. When using nodal Lagrange triangular elements, we consider the lowest-order finite-dimensional Lagrange space V_h, which consists of continuous piecewise-linear functions on each element, and Problem (2) becomes: Find $u_h \in V_h$ such that

$$a(u_h, v_h) = (f, v_h) \quad \forall\, v_h \in V_h. \tag{3}$$

When using VEM, we can consider a general triangulation $\mathcal{T}_h$ composed by general polygons as in Figure 1 (not necessarily similar or with the same number of edges), and V_h now contains piecewise-linear continuous functions on the boundary of each element that are harmonic in its interior. We omit further details on how to modify the bilinear form $a(\cdot,\cdot)$ and the right-hand side in Equation (3) when VEM are used; see, e.g., [13]. We then obtain a linear system $Au = f$, for which we will describe the construction of an additive preconditioner.

2.1 Virtual coarse space

We present the coarse spaces considered in [3, 4]. We first define the lowest-order virtual element space on the polygonal decomposition $\{\Omega_i\}_{i=1}^N$ of Ω. For each Ω_i, consider the set

$$\mathcal{B}(\partial\Omega_i) := \left\{ v \in C^0(\partial\Omega_i) : v|_e \in \mathcal{P}_1(e)\ \forall\, e \subset \partial\Omega_i \right\},$$

where e represents any straight segment of the boundary of the polygon Ω_i. The local virtual space is then defined as

$$V^{\Omega_i} := \left\{ v \in H^1(\Omega_i) : v|_{\partial\Omega_i} \in \mathcal{B}(\partial\Omega_i),\ \Delta v = 0 \right\}.$$

A natural choice for the coarse space of the two-level algorithm is the global virtual space

$$V_0 := \left\{ v \in H^1(\Omega) : v|_{\Omega_i} \in V^{\Omega_i} \right\}.$$

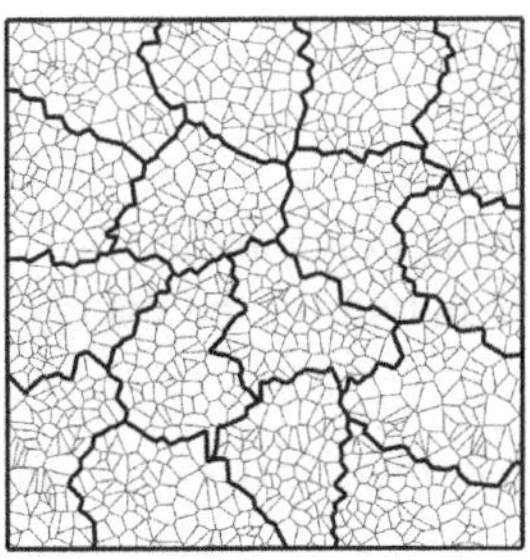

Fig. 1: General polygonal mesh for VEM with irregular subdomains.

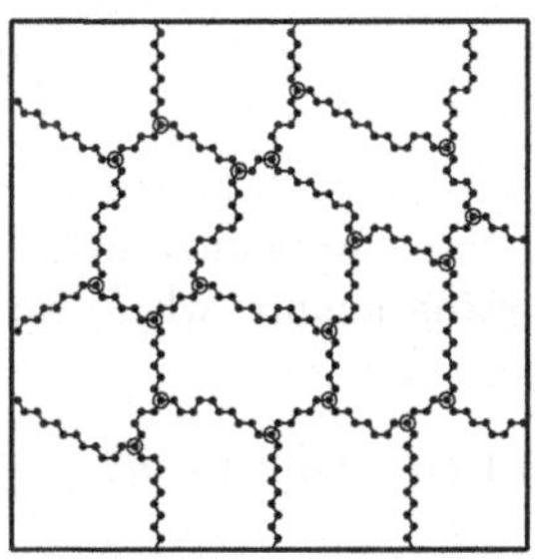

Fig. 2: Decomposition $\{\Omega_i\}$. The coarse space V_0 has one degree of freedom per polygonal vertex (black dots). The reduced coarse space V_0^R has only one degree of freedom per subdomain vertex (black circles)

Hence, a function in V_0 is continuous, piecewise-linear on the boundary of each Ω_i, and harmonic in the interior of each subdomain. Thus, it is completely determined by its values at the vertices of the polygonal domain Ω_i and the dimension of V_0 can be quite large; see Figure 2 for an example with an hexagonal mesh. Therefore, we define a reduced coarse space as follows.

For each subdomain vertex $\boldsymbol{x}_0$ we define a coarse function $\psi^H_{\boldsymbol{x}_0} \in V_0$ by choosing appropriately its degrees of freedom, a construction modified from [5]. First, we set $\psi^H_{\boldsymbol{x}_0}(\boldsymbol{x}) = 0$ for all the subdomain vertices $\boldsymbol{x}$, except at $\boldsymbol{x}_0$ where $\psi^H_{\boldsymbol{x}_0}(\boldsymbol{x}_0) = 1$. Second, we set the degrees of freedom related to the nodal values on each subdomain edge. If $\boldsymbol{x}_0$ is not an endpoint of $\mathcal{E}$, then $\psi^H_{\boldsymbol{x}_0}$ vanishes on that edge. If $\mathcal{E}$ has endpoints $\boldsymbol{x}_0$ and $\boldsymbol{x}_1$, let $\boldsymbol{d}_\mathcal{E}$ be the unit vector with direction from $\boldsymbol{x}_1$ to $\boldsymbol{x}_0$. For any node $\widetilde{\boldsymbol{x}} \in \mathcal{E}$ set

$$\psi^H_{\boldsymbol{x}_0}(\widetilde{\boldsymbol{x}}) = \begin{cases} 0, \text{ if} & (\widetilde{\boldsymbol{x}} - \boldsymbol{x}_1)\cdot \boldsymbol{d}_\mathcal{E} < 0 \\ \dfrac{(\widetilde{\boldsymbol{x}} - \boldsymbol{x}_1)\cdot \boldsymbol{d}_\mathcal{E}}{|\boldsymbol{x}_0 - \boldsymbol{x}_1|}, \text{ if} & 0 \le (\widetilde{\boldsymbol{x}} - \boldsymbol{x}_1)\cdot \boldsymbol{d}_\mathcal{E} \le |\boldsymbol{x}_0 - \boldsymbol{x}_1| \\ 1, \text{ if} & (\widetilde{\boldsymbol{x}} - \boldsymbol{x}_1)\cdot \boldsymbol{d}_\mathcal{E} > |\boldsymbol{x}_0 - \boldsymbol{x}_1| \end{cases}$$

It is clear that $\psi^H_{\boldsymbol{x}_0}(\boldsymbol{x}_0) = 1$, $\psi^H_{\boldsymbol{x}_0}(\boldsymbol{x}_1) = 0$, and that the function varies linearly in the direction of $\boldsymbol{d}_\mathcal{E}$ for such nodes. In this way, we define all the degrees of freedom of $\psi^H_{\boldsymbol{x}_0} \in V_0$. By construction, $0 \le \psi^H_{\boldsymbol{x}_0} \le 1$ and $\sum_{\boldsymbol{x}_0} \psi^H_{\boldsymbol{x}_0} \equiv 1$.

We then define the *reduced coarse space* as the span of $\{\psi^H_{\boldsymbol{x}_0}\}$, i.e.,

$$V_0^R := \left\{ v \in H_0^1(\Omega) : v = \sum_{\boldsymbol{x}_0} \alpha_{\boldsymbol{x}_0} \psi^H_{\boldsymbol{x}_0} \right\} \subset V_0$$

for some real coefficients $\alpha_{\boldsymbol{x}_0}$; see [4, Section 6]. We point out that in the case where the partition $\{\Omega_i\}$ is composed by triangles or squares, $V_0 = V_0^R$ and they reduce to the usual linear or bilinear finite element space, respectively. We can naturally define a linear interpolant $I^H : V_h \to V_0^R$ by

$$I^H u := \sum_{\boldsymbol{x}_0} u(\boldsymbol{x}_0)\psi^H_{\boldsymbol{x}_0},$$

and it is easy to deduce that I^H reproduces linear polynomials. We can prove the following lemma, where we present an upper bound for the energy of coarse functions:

Lemma 1 *Given $u \in V_h$, let $u_0 := I^H u \in V_0^R$. Then, there exists a constant C such that*

$$|u_0|^2_{H^1(\Omega_i)} \le C\left(1 + \log \frac{H_i}{h_i}\right) |u|^2_{H^1(\Omega_i)},$$

where H_i is the diameter of Ω_i and h_i is the smallest element diameter of the triangulation of Ω_i. Here, C depends only on the aspect ratio of Ω_i and the number of subdomain vertices on $\partial\Omega_i$.

Proof See [3, Lemma 4.4 and Theorem 6.1], [4, Lemma 5.6]. □

Since virtual coarse functions cannot be evaluated at internal nodes of the subdomains, we still need to define an appropriate operator $R_0^T : V_0^R \longrightarrow V_h$, such that each function in V_0^R is well-approximated in V_h. We could:

(1) Solve a Dirichlet problem on each subdomain in order to compute the discrete harmonic extension of the values on the boundary of each Ω_i, as it is done in [5, 14, 6].
(2) Triangulate each subdomain Ω_i and define R_0^T as a piecewise-linear interpolant onto such triangulations; see [4, Section 3.1] for further details and assumptions that are required.
(3) Construct a projection $\Pi^{\nabla}_{\Omega_i,k} u_0$ for a given function u_0 in V^{Ω_i}, onto the polynomial space defined on Ω_i of degree $k \geq 2$, and this operator can be constructed by knowing only the degrees of freedom of the virtual functions; see [3, Section 6.1] for implementation details. The main advantage in this approach is that in order to compute all the internal degrees of freedom, we only need to solve a linear system with $k(k-1)/2$ unknowns. Thus, in the interior of each subdomain we approximate u_0 by $\Pi^{\nabla}_{\Omega_i,k} u_0$, avoiding discrete harmonic extensions.

It can be shown that the following estimates hold:

Lemma 2 *Given $u \in V_h$, let $u_0 := I^H u \in V_0^R$. Then there exists a constant C such that*

$$\|u - R_0^T u_0\|^2_{L^2(\Omega_i)} \leq CH_i^2 \left(1 + \log \frac{H_i}{h_i}\right) |u|^2_{H^1(\Omega_i)},$$

$$|R_0^T u_0|^2_{H^1(\Omega_i)} \leq C \left(1 + \log \frac{H_i}{h_i}\right) |u|^2_{H^1(\Omega_i)},$$

where C is independent of H_i and h_i.

Proof See [4, Lemma 3, Lemma 4] and [3, Lemma 5.7] for cases (2) and (3), respectively. For case (1), similar estimates holds; see the proof in [5, Theorem 3.1], [6, Theorem 3.1] □

2.2 Local spaces and preconditioner

For each subdomain Ω_i, we construct the overlapping subdomain Ω_i' by adding layers of elements to Ω_i and denote by δ_i the size of the overlap. The local virtual space is then defined by

$$V_i := \left\{v \in H_0^1(\Omega_i') : v|_K \in \mathcal{B}(\partial K),\ \Delta v|_K = 0 \text{ in } K,\ \forall\, K \subset \Omega_i'\right\}.$$

Thus, the degrees of freedom are the values at all the nodes in the interior of Ω_i', and it is straightforward to define zero extension operators $R_i^T : V_i \longrightarrow V_h$. Consider the matrix representation of the operators R_i^T denoted again by R_i^T. We use exact local solvers and define $\tilde{A}_i = R_i A R_i^T$, $0 \leq i \leq N$. Schwarz projections are given by

$$P_i = R_i^T \tilde{A}_i^{-1} R_i A, \; 0 \le i \le N.$$

The additive preconditioned operator is defined by

$$P_{ad} := \sum_{i=0}^{N} P_i = A_{ad}^{-1} A, \text{ with } A_{ad}^{-1} = \sum_{i=0}^{N} R_i^T \tilde{A}_i^{-1} R_i. \tag{4}$$

Multiplicative and hybrid preconditioners can be considered as well; see [11, Section 2.2]. We can then prove the following result:

Theorem 1 *There exists a constant C, independent of H, h and ρ, such that the condition number of the preconditioned system $\kappa(A_{ad}^{-1}A)$ satisfies*

$$\kappa(A_{ad}^{-1}A) \le C\left(1 + \log\frac{H}{h}\right)\left(1 + \frac{H}{\delta}\right),$$

where the ratios H/h and H/δ denote their maximum value over all the subdomains.

Proof See [4, Theorem 6.1], [3, Theorem 4.1]. □

3 Some numerical results

We first provide a comparison of the running time when assembling R_0^T by discrete harmonic extensions and by quadratic and cubic polynomial approximations; see Figure 3 where we have used a serial implementation in MATLAB with $N = 4$ METIS subdomains and triangular elements.

We also include an experiment with a different application of the virtual coarse spaces. We approximate accurately harmonic functions with given Dirichlet boundary conditions in a domain Ω, by using the projector $\Pi_{\Omega,k}^{\nabla}$ for sufficiently large k. Instead of solving the resulting ill-conditioned linear system $Au = f$ that arises from FEM or VEM, we can approximate the nodal values in the interior nodes of $\mathcal{T}_h$ by evaluating $u_h := \Pi_{\Omega,k}^{\nabla} u$. In order to do so, we just need to solve a linear system with $k(k-1)/2$ unknowns. We remark that in the construction of the preconditioner (4),

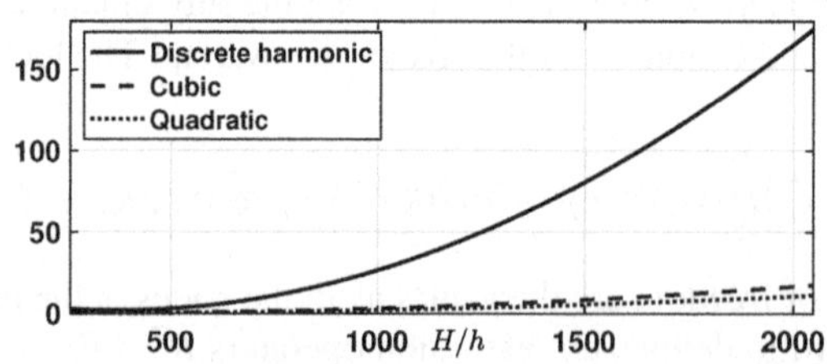

Fig. 3: Time (in seconds) required for computing R_0^T with discrete harmonic extensions, quadratic and cubic projections, as a function of H/h, with $N = 4$ irregular subdomains.

a competitive number of iterations can be obtained with just $k = 2$ or $k = 3$, since they provide good-enough approximations for functions in the virtual coarse space. Here instead, we construct the projection onto the domain Ω, obtaining directly u_h.

For simplicity, we consider the unit square $\Omega = [0, 1]^2$ with boundary conditions such that the exact solution is $u(x, y) = (e^{2x}+e^{-2x})\sin(2y)$. We consider a triangular partition for Ω; the inf-norm of the error in the approximation is shown in Figure 4, for different values of k and mesh size h. As we observe, for a fixed k, the error decreases quadratically as a function of h, and it reaches a minimum value that depends on k, for which $\Pi^{\nabla}_{\Omega,k}$ cannot improve the approximation. We remark that further exploration is required, and this approach is being studied for problems in two and three dimensions.

For further experiments on the performance of the preconditioner (4), we refer to the numerical experiments shown in [3, 4].

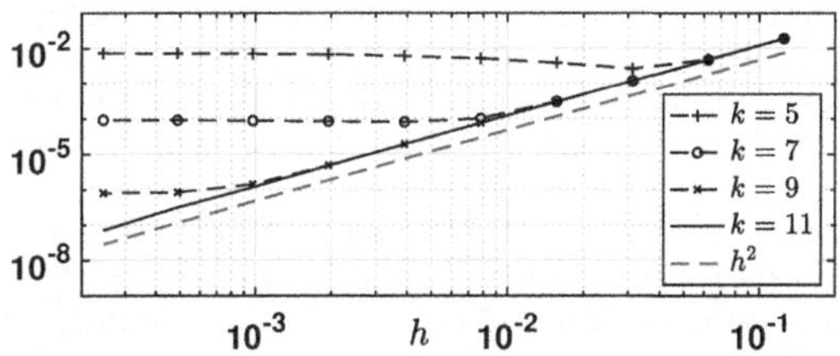

Fig. 4: Inf-norm of the error, $\|u - u_h\|_\infty$, as a function of h, in the approximation of the solution of Laplace's equation in the unit square by computing $\Pi^{\nabla}_{\Omega,k}u$. Convergence is quadratic as a function of h.

4 Conclusions

We note that the main advantage of our approach with respect to previous studies is that no discrete harmonic extensions are required in the algorithm, saving computational time. We also aim to contribute and enrich the literature related to iterative solvers for VEM discretizations, since there is a lack of theoretical analysis for such problems. Even though theory does not include the case of a discontinuous coefficient in the interior of each subdomain, a reasonable number of iterations is obtained even for extreme cases of discontinuities and high-contrast jumps across the elements; see [3, Section 6.2.4]. For higher values of k, we can directly obtain more accurate approximations of harmonic functions, as shown in Figure 4. For preconditioning, experimentally we have found that using quadratic or cubic polynomials is sufficient, but we can use higher degree spaces in order to improve accuracy in the approximation of harmonic functions.

References

1. Ahmad, B., Alsaedi, A., Brezzi, F., Marini, L.D., Russo, A.: Equivalent projectors for virtual element methods. Comput. Math. Appl. **66**(3), 376–391 (2013). DOI:10.1016/j.camwa.2013.05.015
2. Calvo, J.G.: A two-level overlapping Schwarz method for H(curl) in two dimensions with irregular subdomains. Electron. Trans. Numer. Anal. **44**, 497–521 (2015)
3. Calvo, J.G.: On the approximation of a virtual coarse space for domain decomposition methods in two dimensions. Math. Models Methods Appl. Sci. **28**(7), 1267–1289 (2018). DOI:10.1142/S0218202518500343
4. Calvo, J.G.: An overlapping Schwarz method for virtual element discretizations in two dimensions. Comput. Math. Appl. **77**(4), 1163–1177 (2019). DOI:10.1016/j.camwa.2018.10.043
5. Dohrmann, C.R., Klawonn, A., Widlund, O.B.: Domain decomposition for less regular subdomains: overlapping Schwarz in two dimensions. SIAM J. Numer. Anal. **46**(4), 2153–2168 (2008). DOI:10.1137/070685841
6. Dohrmann, C.R., Widlund, O.B.: An alternative coarse space for irregular subdomains and an overlapping Schwarz algorithm for scalar elliptic problems in the plane. SIAM J. Numer. Anal. **50**(5), 2522–2537 (2012). DOI:10.1137/110853959
7. Galvis, J., Efendiev, Y.: Domain decomposition preconditioners for multiscale flows in high contrast media: reduced dimension coarse spaces. Multiscale Model. Simul. **8**(5), 1621–1644 (2010). DOI:10.1137/100790112
8. Heinlein, A., Klawonn, A., Knepper, J., Rheinbach, O.: An adaptive GDSW coarse space for two-level overlapping schwarz methods in two dimensions. In: Domain decomposition methods in science and engineering XXIV, *Lect. Notes Comput. Sci. Eng.*, vol. 125, pp. 373–382. Springer, Berlin (2018)
9. Oh, D.S., Widlund, O.B., Zampini, S., Dohrmann, C.R.: BDDC algorithms with deluxe scaling and adaptive selection of primal constraints for Raviart-Thomas vector fields. Math. Comp. **87**(310), 659–692 (2018). DOI:10.1090/mcom/3254
10. Spillane, N., Dolean, V., Hauret, P., Nataf, F., Pechstein, C., Scheichl, R.: Abstract robust coarse spaces for systems of PDEs via generalized eigenproblems in the overlaps. Numer. Math. **126**(4), 741–770 (2014). DOI:10.1007/s00211-013-0576-y
11. Toselli, A., Widlund, O.: Domain decomposition methods—algorithms and theory, *Springer Series in Computational Mathematics*, vol. 34. Springer-Verlag, Berlin (2005). DOI:10.1007/b137868
12. Beirão da Veiga, L., Brezzi, F., Cangiani, A., Manzini, G., Marini, L.D., Russo, A.: Basic principles of virtual element methods. Math. Models Methods Appl. Sci. **23**(1), 199–214 (2013). DOI:10.1142/S0218202512500492
13. Beirão da Veiga, L., Brezzi, F., Marini, L.D., Russo, A.: The hitchhiker's guide to the virtual element method. Math. Models Methods Appl. Sci. **24**(8), 1541–1573 (2014). DOI:10.1142/S021820251440003X
14. Widlund, O.B.: Accomodating irregular subdomains in domain decomposition theory. In: Domain decomposition methods in science and engineering XVIII, *Lect. Notes Comput. Sci. Eng.*, vol. 70, pp. 87–98. Springer, Berlin (2009). DOI:10.1007/978-3-642-02677-5_8

A Local Coarse Space Correction Leading to a Well-Posed Continuous Neumann-Neumann Method in the Presence of Cross Points

Faycal Chaouqui, Martin J. Gander, and Kévin Santugini-Repiquet

1 Introduction

Neumann-Neumann methods (NNMs) are among the best parallel solvers for discretized partial differential equations, see [12] and references therein. Their common polylogarithmic condition number estimate shows their effectiveness for many discretized elliptic problems, see [9, 10, 5]. However, NNM was originally described in [1] as an iteration at the continuous level like the classical Schwarz method, but only for two subdomains, see also [11]. This is because in contrast to the Schwarz method, it does not converge for general decompositions into many subdomains when used as a stationary iteration [4, 3]. Furthermore, for decompositions presenting cross points, NNM is not well-posed in H^1 and has as a stationary iteration a convergence factor that deteriorates polylogarithmically in the mesh size h, see [4]. The iterates being discontinuous at the cross points also prevents NNM from being well-posed in H^2. We propose here a very specific local coarse space that leads to a well posed NNM at the continuous level for the model problem

$$-\Delta u = f \text{ in } \Omega, \quad u = 0 \text{ on } \partial\Omega, \tag{1}$$

where $f \in L^2(\Omega)$, and Ω can be decomposed as in Fig. 1, i.e. the decomposition can contain cross points. In Section 2 we present NNM at the continuous level for a 2×1 decomposition and show why it is always well-posed in H^1. In Section 3 we show why NNM for a 2×2 decomposition containing a cross point is not in general well-posed in H^1. To make it well-posed in H^2, we introduce a very specific

Faycal Chaouqui
Temple University, Departement of Mathematics, e-mail: Faycal.Chaouqui@temple.edu

Martin J. Gander
Université de Genève, Section de mathématiques, e-mail: Martin.Gander@unige.ch

Kévin Santugini-Repiquet
Bordeaux INP, IMB, UMR 5251, F-33400,Talence, France, e-mail: Kevin.Santugini-Repiquet@bordeaux-inp.fr

R. Haynes et al. (eds.), *Domain Decomposition Methods in Science and Engineering XXV*,
Lecture Notes in Computational Science and Engineering 138,
https://doi.org/10.1007/978-3-030-56750-7_8

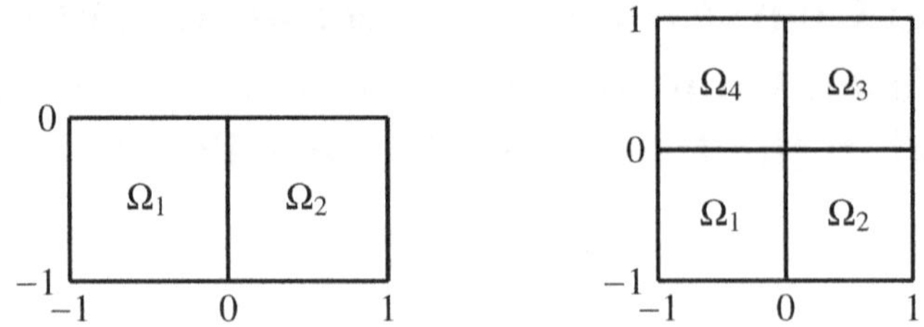

Fig. 1: Decomposition without a cross point (left) and with a cross point (right)

Algorithm 1: NNM for a 2×1 decomposition

---

1. Set g_{12}^0 to zero or any inexpensive initial guess.
2. For $n = 0, 1, \ldots$ until convergence

   a. Solve the Dirichlet problems

$$\begin{aligned} -\Delta u_1^n &= f \text{ in } \Omega_1, & -\Delta u_2^n &= f \text{ in } \Omega_2, \\ u_1^n &= g_{12}^n \text{ on } \Gamma, & u_2^n &= g_{12}^n \text{ on } \Gamma, \\ u_1^n &= 0 \text{ on } \partial\Omega_1 \cap \partial\Omega, & u_2^n &= 0 \text{ on } \partial\Omega_2 \cap \partial\Omega. \end{aligned}$$

   b. Solve the Neumann problems

$$\begin{aligned} -\Delta \psi_1^n &= 0 \text{ in } \Omega_1, & -\Delta \psi_2^n &= 0 \text{ in } \Omega_2, \\ \frac{\partial \psi_1^n}{\partial n_1} &= \frac{1}{2}\left(\frac{\partial u_1^n}{\partial n_1} + \frac{\partial u_2^n}{\partial n_2}\right) \text{ on } \Gamma, & \frac{\partial \psi_2^n}{\partial n_2} &= \frac{1}{2}\left(\frac{\partial u_1^n}{\partial n_1} + \frac{\partial u_2^n}{\partial n_2}\right) \text{ on } \Gamma, \\ \psi_1^n &= 0 \text{ on } \partial\Omega_1 \cap \partial\Omega, & \psi_2^n &= 0 \text{ on } \partial\Omega_2 \cap \partial\Omega, \end{aligned}$$

   where n_i is the outward pointing normal on $\partial\Omega_i$, $i = 1, 2$.

   c. Update the trace $g_{12}^{n+1} = g_{12}^n - \frac{1}{2}(\psi_1^n + \psi_2^n)$ on Γ.

---

local coarse space correction. Our new NNM then converges as a stationary iterative solver, also in the presence of cross points, and we show numerically that it is a better preconditioner than the classical NNM in the case of many cross points.

2 Existence of iterates for a 2×1 decomposition

For a decomposition as shown in Fig. 1 (left), let $\Gamma := \{0\} \times (0, 1)$ be the interface between Ω_1 and Ω_2. The NNM in Algorithm 1 is well-posed with iterates in H^1:

Theorem 1 *If* $g_{12}^0 \in H_{00}^{\frac{1}{2}}(\Gamma)$ *(the* Lions-Magenes *space defined in [7, Chapter 1]), then Algorithm 1 is well-posed and for all* $n \geq 0$ *we have* $u_i^n \in V_i$*, where* $V_i := \{v \in H^1(\Omega_i) : v = 0 \text{ on } \partial\Omega_i \cap \partial\Omega\}$ *for* $i = 1, 2$.

To prove Theorem 1, we first need to prove

Lemma 1 *Denote by* $\gamma\colon V_1 \mapsto H_{00}^{\frac{1}{2}}(\Gamma)$ *the restriction map on* Ω_1*. There exists* $C_1 > 0$ *such that for all* $v_1 \in V_1$

$$\|\gamma v_1\|_{H_{00}^{\frac{1}{2}}(\Gamma)} \leq C_1 \|v_1\|_{V_1}. \tag{2}$$

Moreover, there exists $C_2 > 0$ *such that for all* $g \in H_{00}^{\frac{1}{2}}(\Gamma)$*, there exists* $\widetilde{v}_2$ *such that* $\widetilde{\gamma}\widetilde{v}_2 = g$*, and*

$$\|\widetilde{v}_2\|_{V_2} \leq C_2 \|g\|_{H_{00}^{\frac{1}{2}}(\Gamma)}, \tag{3}$$

where $\widetilde{\gamma}\colon V_2 \mapsto H_{00}^{\frac{1}{2}}(\Gamma)$ *denotes the restriction map on* Ω_2.

Proof The continuity and surjectivity of $\gamma\colon V_1 \mapsto H_{00}^{\frac{1}{2}}(\Gamma)$ comes from [8, Chapter 4,Th 2.3] and the definition of $H_{00}^{\frac{1}{2}}(\Gamma)$. Let $g \in H_{00}^{\frac{1}{2}}(\Gamma)$. The surjectivity of $\widetilde{\gamma}\colon V_2 \mapsto H_{00}^{\frac{1}{2}}(\Gamma)$ ensures the existence of $\widetilde{v}_2 \in V_2$ such that the equality $\widetilde{\gamma}\widetilde{v}_2 = g$ holds. Using then the open mapping theorem for $\widetilde{\gamma}$; see e.g. [2, Chapter 2,Th 2.6], we know that there exists $C_2 > 0$ such that Eq. (3) holds, which concludes the proof. □

Proof (of Theorem 1) Since g_{12}^0 satisfies the H^1-compatibility relations, we know by the Lax-Millgram Lemma that $u_1^0 \in V_1$ and $u_2^0 \in V_2$. Now it suffices to show that ψ_1^0 and ψ_2^0 are also in V_1 and V_2. We know that ψ_1^0 and ψ_2^0 satisfy

$$\int_{\Omega_1} \nabla\psi_1^0 \nabla v_1 = \int_\Gamma \frac{1}{2}\left(\frac{\partial u_1^0}{\partial n_1} + \frac{\partial u_2^0}{\partial n_2}\right) v_1, \quad \text{for all } v_1 \in V_1,$$

$$\int_{\Omega_2} \nabla\psi_1^0 \nabla v_2 = \int_\Gamma \frac{1}{2}\left(\frac{\partial u_1^0}{\partial n_1} + \frac{\partial u_2^0}{\partial n_2}\right) v_2, \quad \text{for all } v_2 \in V_2.$$

In order to apply the Lax-Milgram Lemma, it suffices to show that $b_1(v_1) := \int_\Gamma \frac{1}{2}(\frac{\partial u_1^0}{\partial n_1} + \frac{\partial u_2^0}{\partial n_2})v_1$ and $b_2(v_2) := \int_\Gamma \frac{1}{2}(\frac{\partial u_1^0}{\partial n_1} + \frac{\partial u_2^0}{\partial n_2})v_2$, define a continuous map on V_1 and V_2. It suffices to prove this for b_1, and the same then holds for b_2. Indeed, we have for all $v_1 \in V_1$

$$\begin{aligned}
b_1(v_1) &= \left\langle \frac{\partial u_1^0}{\partial n_1}, v_1 \right\rangle_{H^{-\frac{1}{2}}(\Gamma), H_{00}^{\frac{1}{2}}(\Gamma)} + \left\langle \frac{\partial u_2^0}{\partial n_2}, v_1 \right\rangle_{H^{-\frac{1}{2}}(\Gamma), H_{00}^{\frac{1}{2}}(\Gamma)} \\
&= \left\langle \frac{\partial u_1^0}{\partial n_1}, \gamma v_1 \right\rangle_{H^{-\frac{1}{2}}(\Gamma), H_{00}^{\frac{1}{2}}(\Gamma)} + \left\langle \frac{\partial u_2^0}{\partial n_2}, \widetilde{\gamma v_2} \right\rangle_{H^{-\frac{1}{2}}(\Gamma), H_{00}^{\frac{1}{2}}(\Gamma)} \\
&= -\int_{\Omega_1} f v_1 + \int_{\Omega_1} \nabla u_1^0 \nabla v_1 - \int_{\Omega_2} f \widetilde{v}_2 + \int_{\Omega_2} \nabla u_2^0 \nabla \widetilde{v}_2.
\end{aligned}$$

Hence, $|b_1(v_1)| \le C(\|v_1\|_{V_1} + \|\widetilde{v}_2\|_{V_2}) \le C(1 + C_1C_2)\|v_1\|_{V_1}$. We deduce then that b_1 is a continuous map on V_1. In the same manner, we prove that b_2 is continuous on V_2, and by applying the Lax-Milgram Lemma, we obtain that ψ_1^0 and ψ_2^0 are in V_1 and V_2. Finally, we conclude that $g_{12}^1 \in H_{00}^{\frac{1}{2}}(\Gamma)$. Repeating then the same arguments, we conclude that $g_{12}^n \in H_{00}^{\frac{1}{2}}(\Gamma)$ for all $n \ge 0$. □

3 Existence of iterates for a 2 × 2 decomposition

We now study the well-posedness of NNM for a 2×2 decomposition, see Fig. 1 (right). The well-posedness in this case cannot be treated as in Section 2. In fact, let $\Gamma_{12} := \{0\} \times (-1, 0)$, $\Gamma_{23} := (0, 1) \times \{0\}$, $\Gamma_{34} := \{0\} \times (0, 1)$, $\Gamma_{41} := (-1, 0) \times \{0\}$ be the shared interfaces. Then $g_{12}^0 \in H^{\frac{1}{2}}(\Gamma_{12})$, $g_{23}^0 \in H^{\frac{1}{2}}(\Gamma_{23})$, $g_{34}^0 \in H^{\frac{1}{2}}(\Gamma_{34})$, $g_{41}^0 \in H^{\frac{1}{2}}(\Gamma_{41})$ is not sufficient for the first iterates to exist: the traces need to satisfy additional assumptions which are known as the H^1-compatibility relations ($\mathcal{CR}_1$) which are

$$\int_0^\varepsilon \left|g_{12}^0(-\sigma) - g_{41}^0(-\sigma)\right|^2 \frac{d\sigma}{\sigma} < \infty, \quad \int_0^\varepsilon \left|g_{12}^0(-\sigma) - g_{23}^0(\sigma)\right|^2 \frac{d\sigma}{\sigma} < \infty,$$

$$\int_0^\varepsilon \left|g_{23}^0(-\sigma) - g_{34}^0(-\sigma)\right|^2 \frac{d\sigma}{\sigma} < \infty, \quad \int_0^\varepsilon \left|g_{34}^0(-\sigma) - g_{41}^0(\sigma)\right|^2 \frac{d\sigma}{\sigma} < \infty,$$

for $\varepsilon > 0$ small enough; see [8, chapter 4,Th 2.3]. However, even if the initial iterates satisfy $\mathcal{CR}_1$, this does in general not hold for the following iterates. This explains why NNM is in general not well-defined for a 2×2 decomposition with a cross point. This is also the reason why NNM does not converge iteratively and has a convergence factor that grows logarithmically with respect to the mesh size after discretization as we mentioned in Section 1. We propose here to add a very specific local coarse space correction such that NNM becomes well-posed. Since the $\mathcal{CR}_1$ are global, it is not clear how to define a coarse space such that NNM with the additional coarse correction satisfies these conditions systematically. We thus look for a coarse space correction such the iterates are not in H^1 but rather in H^2. However, even the condition $g_{12}^0 \in H^{\frac{3}{2}}(\Gamma_{12})$, $g_{23}^0 \in H^{\frac{3}{2}}(\Gamma_{23})$, $g_{34}^0 \in H^{\frac{3}{2}}(\Gamma_{34})$, $g_{41}^0 \in H^{\frac{3}{2}}(\Gamma_{41})$ does not ensure the existence of H^2 iterates, and one needs to satisfy the so-called

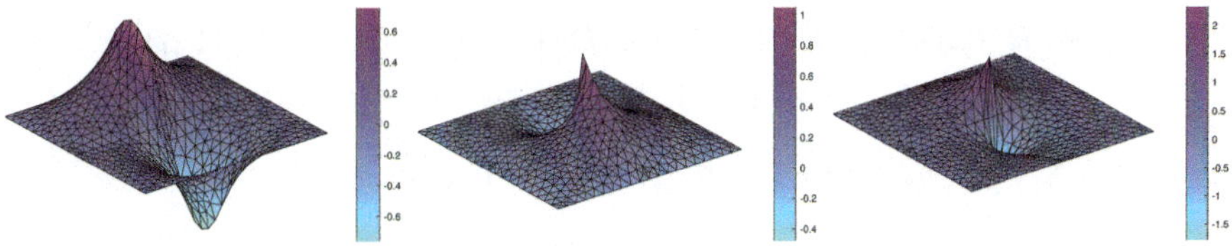

Fig. 2: Iterates 1,2,3 of NNM for the solution of Eq. (1) (note the different scale).

Algorithm 2: NNM for a 2×2 decomposition

1. Initialize $g_{12}^0, g_{23}^0, g_{34}^0, g_{41}^0$.
2. For $n = 0, 1, \ldots$ until convergence
   - Compute $g_{12}^{n+\frac{1}{2}}, g_{23}^{n+\frac{1}{2}}, g_{34}^{n+\frac{1}{2}}, g_{41}^{n+\frac{1}{2}}$ using NNM (which used superscript $n+1$).
   - Find $\varphi_{12}, \varphi_{23}, \varphi_{34}, \varphi_{41} \in H^{\frac{3}{2}}$ in a given local coarse space, e.g. in (4), s.t.

$$g_{12}^{n+1} := g_{12}^{n+\frac{1}{2}} + \varphi_{12},\ g_{23}^{n+1} := g_{23}^{n+\frac{1}{2}} + \varphi_{23},$$
$$g_{34}^{n+1} := g_{34}^{n+\frac{1}{2}} + \varphi_{34},\ g_{41}^{n+1} := g_{41}^{n+\frac{1}{2}} + \varphi_{41},$$

   satisfiy by solving (5) the compatibility conditions

$$g_{12}^{n+1}(0) = g_{23}^{n+1}(0) = g_{34}^{n+1}(0) = g_{41}^{n+1}(0).$$

H^2-compatibility relations ($\mathcal{CR}_2$). This can also be illustrated numerically: we show in Fig. 2 the first iterates of NNM for Eq. (1) with $f = 1$ discretized by P_1 finite elements with a mesh size $h = 0.1$, and starting with smooth traces along the shared edges. The iterates in Fig. 2 show that NNM does not converge iteratively and has a discontinuity that forms at the origin. This discontinuity cannot happen if the iterates are in H^2 since their traces are in $H^{\frac{3}{2}}$, hence continuous at the cross point. One can show that this is the only problem that needs to be fixed in order to have a well-posed method. We thus propose to add a coarse space correction consisting of functions that are in $H^{\frac{3}{2}}$ on the common edges such that we enforce the continuity of the iterates at the origin. The NNM with this local coarse space correction is given in Algorithm 2. The next theorem ensures the well-posedness of Algorithm 2.

Theorem 2 *If* $(g_{12}^0, g_{23}^0, g_{34}^0, g_{41}^0) \in H^{\frac{3}{2}}(\Gamma_{12}) \times H^{\frac{3}{2}}(\Gamma_{23}) \times H^{\frac{3}{2}}(\Gamma_{34}) \times H^{\frac{3}{2}}(\Gamma_{41})$ *satisfy* $g_{12}^0(0) = g_{23}^0(0) = g_{34}^0(0) = g_{41}^0(0)$, *then Algorithm 2 is well-posed and for all* $n \geq 0$ *we have* $u_i^n \in H^2(\Omega_i) \cap V_i$, *where* $V_i := \left\{v \in H^1(\Omega_i) : v = 0 \text{ on } \partial\Omega_i \cap \partial\Omega\right\}$ *for* $i = 1, \ldots, 4$.

We first state a result for the H^2 compatibility relations ($\mathcal{CR}_2$) which can be found in [8, chapter 4,Th 2.3].

Theorem 3 *Define the trace mapping*

$$\gamma : H^2(\Omega_1) \cap V_1 \mapsto H^{\frac{3}{2}}(\Gamma_{12}) \times H^{\frac{1}{2}}(\Gamma_{12}) \times H^{\frac{3}{2}}(\Gamma_{41}) \times H^{\frac{1}{2}}(\Gamma_{41})$$
$$u \mapsto \left(u(0,\cdot), \partial_x u(0,\cdot), u(\cdot,0), \partial_y u(\cdot,0)\right).$$

Then $(g_{12}, h_{12}, g_{41}, h_{41}) \in \mathrm{Im}\,(\gamma)$ *iff* $g_x(0) = g_y(0)$ *and*

$$\int_0^{\varepsilon} \left|g_{12}'(-\sigma) - h_{41}(-\sigma)\right|^2 \frac{\mathrm{d}\sigma}{\sigma} < \infty, \quad \int_0^{\varepsilon} \left|g_{41}'(-\sigma) - h_{12}(-\sigma)\right|^2 \frac{\mathrm{d}\sigma}{\sigma} < \infty,$$

for $\varepsilon > 0$ *sufficiently small.*

From Theorem 3, we obtain the corollaries

Corollary 1 *Define the mapping*

$$\gamma_D : H^2(\Omega_1) \cap V_1 \mapsto H^{\frac{3}{2}}(\Gamma_{12}) \times H^{\frac{3}{2}}(\Gamma_{41})$$
$$u \mapsto (u(0,\cdot), u(\cdot,0)).$$

Then $(g_{12}, g_{41}) \in \mathrm{Im}\,(\gamma_D)$ *iff* $g_{12}(0) = g_{41}(0)$.

Proof In fact, it suffices to define $h_{41} := g_{12}'(\sigma) \in H^{\frac{1}{2}}(\Gamma_{41})$ and $h_{12} := g_{41}'(\sigma) \in H^{\frac{1}{2}}(\Gamma_{12})$ and apply Theorem 3 to $(g_{12}, h_{12}, g_{41}, h_{41})$. □

Corollary 2 *Define the mapping*

$$\gamma_N : H^2(\Omega_1) \cap V_1 \mapsto H^{\frac{1}{2}}(\Gamma_{12}) \times H^{\frac{1}{2}}(\Gamma_{41})$$
$$u \mapsto \left(\partial_x u(0,\cdot), \partial_y u(\cdot,0)\right).$$

Then γ_N *is onto.*

Proof Here again, it suffices to define $g_{12} := -\psi(\sigma) \int_{\sigma}^{0} h_{41}(\sigma')\, d\sigma'$ and $g_{41} := -\psi(\sigma) \int_{\sigma}^{0} h_{12}(\sigma')\, d\sigma'$, where $\psi(\sigma) \in C^{\infty}[-1,0]$ such that $\psi(\sigma) = 1$ on $(-\varepsilon, 0]$ and $\psi(\sigma) = 0$ on $[-1, -2\epsilon)$, and apply Theorem 3 to $(g_{12}, h_{12}, g_{41}, h_{41})$. □

Proof (of Theorem 2) We start by showing that $u_i^0 \in H^2(\Omega_i) \cap V_i$ for $i = 1, \ldots, 4$. We prove it for u_1^0 and the proof for the remaining u_i^0 is exactly the same. We have that $g_{12}^0 \in H^{\frac{3}{2}}(\Gamma_{12})$ and $g_{14}^0 \in H^{\frac{3}{2}}(\Gamma_{14})$ and they satisfy $g_{12}^0(0) = g_{14}^0(0)$, hence using Corollary 1 we know that there exists $w_1 \in H^2(\Omega_1) \cap V_1$ such that $\widetilde{u}_1 := u_1^0 - w_1 \in H_0^1(\Omega_1)$ is the solution of the variational problem

$$\int_{\Omega_1} \nabla \widetilde{u}_1 \nabla v = \int_{\Omega_1} (f + \Delta w_1) v, \text{ for all } v \in H_0^1(\Omega_1),$$

which using the result in [6, Chapter 3, p 147] has a unique solution in $H^2(\Omega_1) \cap H_0^1(\Omega_1)$, and it follows that $u_1^0 = \widetilde{u}_1 + w_1 \in H^2(\Omega_1) \cap V_1$. In the same manner we can show that u_2^0, u_3^0, u_4^0 are in $H^2(\Omega_2) \cap V_2$, $H^2(\Omega_3) \cap V_3$ and $H^2(\Omega_4) \cap V_4$. Now, since

$$\frac{\partial\psi_1^0}{\partial n_1}_{|\Gamma_{12}} = \frac{1}{2}\left(\frac{\partial u_1^0}{\partial n_1} + \frac{\partial u_2^0}{\partial n_2}\right) \in H^{\frac{1}{2}}(\Gamma_{12}), \quad \frac{\partial\psi_1^0}{\partial n_1}_{|\Gamma_{14}} = \frac{1}{2}\left(\frac{\partial u_1^0}{\partial n_1} + \frac{\partial u_4^0}{\partial n_4}\right) \in H^{\frac{1}{2}}(\Gamma_{14}),$$

we know by Corollary 2 that there exists again a function $\widetilde{w}_1 \in H^2(\Omega_1) \cap V_1$ such that $\widetilde{\psi}_1 := \psi_1^0 - \widetilde{w}_1 \in H^2(\Omega_1) \cap V_1$ is the solution of the variational problem

$$\int_{\Omega_1} \nabla\widetilde{\psi}_1 \nabla v = \int_{\Omega_1} \Delta\widetilde{w}_1 v \text{ for all } v \in V_1,$$

which has a solution $\widetilde{\psi}_1 \in H^2(\Omega_1) \cap V_1$, hence $\psi_1^0 = \widetilde{\psi}_1 + \widetilde{w}_1 \in H^2(\Omega_1) \cap V_1$. The same conclusion can be drawn for ψ_2^0, ψ_3^0, ψ_4^0 using the same reasoning. It follows then that $g_{12}^1, g_{23}^1, g_{34}^1, g_{41}^1$ are in $H^{\frac{3}{2}}(\Gamma_{12})$, $H^{\frac{3}{2}}(\Gamma_{23})$, $H^{\frac{3}{2}}(\Gamma_{34})$ and $H^{\frac{3}{2}}(\Gamma_{41})$ respectively. Since the coarse functions $\phi_{12}, \phi_{23}, \phi_{34}, \phi_{41}$ in Algorithm 2 are chosen such that $g_{12}^1(0) = g_{23}^1(0) = g_{34}^1(0) = g_{41}^1(0)$, we can apply again Corollary 1. We proceed again as before to prove that the next iterates are well defined, and so on. Finally, we conclude that Algorithm 2 is well defined with iterates $u_i^n \in H^2(\Omega_i) \cap V_i$ for $i = 1, \ldots, 4$. This finishes the proof. □

It remains to choose the coarse basis, and a first idea is to use linear functions,

$$\begin{aligned} \varphi_{12} &:= \alpha_{12}(1+y), \; \varphi_{23} := \alpha_{23}(1-x), \\ \varphi_{34} &:= \alpha_{34}(1-y), \; \varphi_{41} := \alpha_{41}(1+x), \end{aligned} \tag{4}$$

where the coefficients α are determined using the pseudo inverse,

$$\begin{bmatrix} \alpha_{12} \\ \alpha_{23} \\ \alpha_{34} \\ \alpha_{41} \end{bmatrix} = \begin{bmatrix} 1 & -1 & 0 & 0 \\ 0 & 1 & -1 & 0 \\ 0 & 0 & 1 & 1 \end{bmatrix}^\dagger \begin{bmatrix} g_{23}^{n+\frac{1}{2}}(0) - g_{12}^{n+\frac{1}{2}}(0) \\ g_{34}^{n+\frac{1}{2}}(0) - g_{23}^{n+\frac{1}{2}}(0) \\ g_{41}^{n+\frac{1}{2}}(0) - g_{34}^{n+\frac{1}{2}}(0) \end{bmatrix}, \tag{5}$$

i.e. we compute the smallest correction to obtain continuous traces at the cross point. The plots in Fig. 3 (top) show that this local linear coarse correction is sufficient to obtain a convergent iterative method which does not form a singularity at the cross point any more. To investigate how the convergence depends on the basis chosen, we now use exponentially decaying functions of the form $e^{-\mu x}$ and $e^{-\mu y}$. Choosing $\mu := 3$, we obtain the results shown in Fig. 3 (bottom): convergence is much faster than with the linear coarse basis; see also Fig. 4 (left) for a comparison. The number of iterations required for NNM with our local coarse correction to reach a tolerance of 10^{-6} for mesh size $h = 0.4, 0.2, 0.1, 0.05, 0.03$ is 9, 15, 19, 23, 26 with the linear coarse functions, and 7, 7, 7, 7, 10 with the exponential ones. We finally test Algorithm 2 with Krylov acceleration (GMRES), for the case of nine cross points and the exponential coarse basis functions: the result is shown in Figure 3 (right), and we see that the fact to be well posed in function space leads to a more effective preconditioner.

We thus answered an interesting question in this short manuscript, namely why NNM only appears in the literature for two subdomains at the continuous level, and

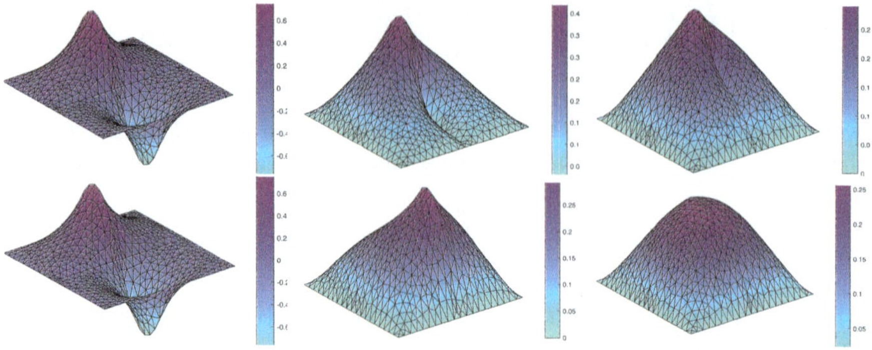

Fig. 3: Iterates 1,2,3 of Algorithm 2 for Eq. (1) using a linear coarse basis (top) and an exponentially decaying coarse basis (bottom)

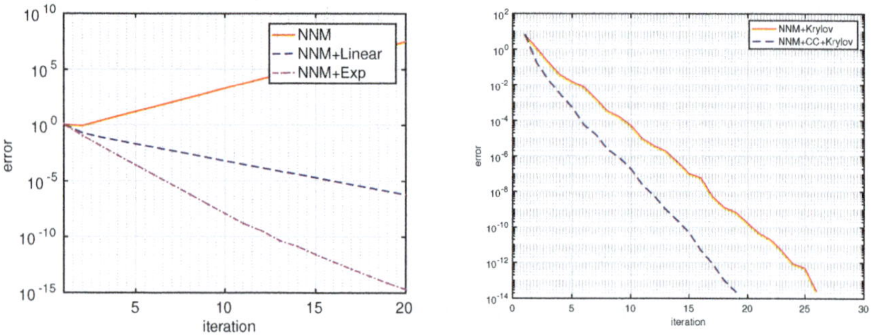

Fig. 4: Error curves of NNM with and without coarse correction for one cross point (left), and with Krylov acceleration for nine cross points (right)

otherwise only at the discrete level as a preconditioner: it is because it is not well posed at the continuous level in the many subdomain case with cross points. We then showed that a specific local coarse space can make NNM well posed at the continuous level, which both leads to a convergent iterative NNM algorithm, and a better preconditioner in the presence of cross points. We are currently investigating if coarse basis functions exist for which we can prove that the convergence factor of NNM becomes independent of the mesh size h like for 2 subdomains.

References

1. Bourgat, J.F., Glowinski, R., Le Tallec, P., Vidrascu, M.: Variational formulation and algorithm for trace operator in domain decomposition calculations. Institut National de Recherche en Informatique et en Automatique (1988)
2. Brezis, H.: Functional analysis, Sobolev spaces and partial differential equations. Springer Science & Business Media (2010)

3. Chaouqui, F., Ciaramella, G., Gander, M., Vanzan, T.: On the scalability of classical one-level domain-decomposition methods. Vietnam J. Math. **64**(4), 1053–1088 (2018)
4. Chaouqui, F., Gander, M.J., Santugini-Repiquet, K.: A Coarse Space to Remove the Logarithmic Dependency in Neumann-Neumann Methods. In: Domain Decomposition Methods in Science and Engineering XXIV. Springer (2018)
5. Dohrmann, C.R.: A preconditioner for substructuring based on constrained energy minimization. SIAM J. Sci. Comput. **25**(1), 246–258 (2003)
6. Grisvard, P.: Elliptic problems in nonsmooth domains, vol. 69. SIAM (2011)
7. Lions, J.L., Magenes, E.: Non-Homogeneous Boundary Value Problems and Applications, Vol I. Springer, New York (1972)
8. Lions, J.L., Magenes, E.: Non-homogeneous boundary value problems and applications, Vol II. Springer, New York (1972)
9. Mandel, J.: Balancing domain decomposition. Int. J. Numer. Methods Biomed. Eng. **9**(3), 233–241 (1993)
10. Mandel, J., Brezina, M.: Balancing domain decomposition for problems with large jumps in coefficients. Math. Comp. **65**(216), 1387–1401 (1996)
11. Quarteroni, A., Valli, A.: Domain decomposition methods for partial differential equations numerical mathematics and scientific computation. Oxford University Press (1999)
12. Toselli, A., Widlund, O.B.: Domain Decomposition Methods — Algorithms and Theory, *Springer Series in Computational Mathematics*, vol. 34. Springer-Verlag Berlin Heidelberg (2005)

Happy 25th Anniversary DDM! ... But How Fast Can the Schwarz Method Solve Your Logo?

Gabriele Ciaramella and Martin J. Gander

1 The ddm logo problem and the Schwarz method

"Vous n'avez vraiment rien à faire"![1] This was the smiling reaction of Laurence Halpern when the first author told her about our wish to accurately estimate the convergence rate of the Schwarz method for the solution of the ddm logo[2], see Figure 1 (left). Anyway, here we are: to honor the 25th anniversary of the domain decomposition conference, we study the convergence rate of the alternating Schwarz method for the solution of Laplace's equation defined on the ddm logo. This method was invented by H.A. Schwarz in 1870 [12] for the solution of the Laplace problem

$$\Delta u = 0 \text{ in } \Omega, \quad u = g \text{ on } \partial\Omega. \tag{1}$$

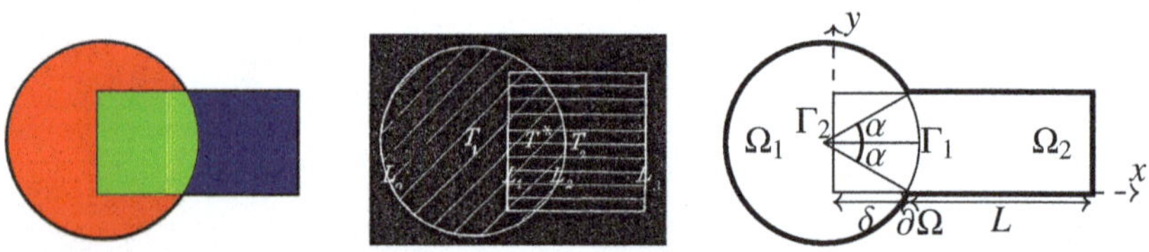

Fig. 1: Left: ddm logo. Center: Original drawing of Schwarz from 1870 [12]. Right: Geometric parametrization of the ddm logo.

G. Ciaramella
Universität Konstanz, Germany, e-mail: gabriele.ciaramella@uni-konstanz.de

M. J. Gander
Université de Genève, Switzerland, e-mail: Martin.Gander@unige.ch

[1] "You have really nothing to do"!

[2] This logo was created by Benjamin Stocker, a friend for over 30 years of the second author and a computer scientist and web designer for SolNet.

R. Haynes et al. (eds.), *Domain Decomposition Methods in Science and Engineering XXV*,
Lecture Notes in Computational Science and Engineering 138,
https://doi.org/10.1007/978-3-030-56750-7_9

Here g is a sufficiently regular function and Ω is the ddm logo, obtained from the union of a disc Ω_1 and a rectangle Ω_2, as historically considered by Schwarz [12]; see Figure 1 (center). In this paper, we assume that Ω_1 is a unit disc, and Ω_2 has length $\delta + L$ and height $2\cos\alpha$. Here, δ, L and α are used to parametrize Ω; see Figure 1 (right). In particular, δ and L measure the overlapping and non-overlapping parts of Ω_2, and α is the angle that parametrizes the interface $\Gamma_1 := \overline{\partial\Omega_1 \cap \Omega_2}$. The other interface $\Gamma_2 := \overline{\partial\Omega_2 \cap \Omega_1}$ is clearly parametrized by δ and α, and it is composed by three segments whose vertices are $(\delta, 0)$, $(0, 0)$, $(0, 2\sin\alpha)$, and $(\delta, 2\sin\alpha)$. To avoid meaningless geometries (e.g., $\Omega_2 \setminus \Omega_1$ becomes a disjoint set), we assume that δ and α are non-negative and satisfy $\delta < 2\cos\alpha$.

In error form, the classical alternating Schwarz method for the solution to (1) is

$$\begin{aligned} \Delta e_1^n &= 0 \quad \text{in } \Omega_1, & \Delta e_2^n &= 0 \quad \text{in } \Omega_2, \\ e_1^n &= 0 \quad \text{on } \partial\Omega \cap \overline{\Omega}_1, & e_2^n &= 0 \quad \text{on } \partial\Omega \cap \overline{\Omega}_2, \\ e_1^n &= e_2^{n-1} \text{ on } \Gamma_1, & e_2^n &= e_1^n \text{ on } \Gamma_2, \end{aligned} \tag{2}$$

where the left subproblem is a Laplace problem on the disc and the right one on the rectangle. Assuming that one begins with a sufficiently regular initial guess e^0, then solving iteratively (2) one obtains the sequence $(e_1^n)_{n\in\mathbb{N}^+}$ of errors on the disc Ω_1 and the sequence $(e_2^n)_{n\in\mathbb{N}^+}$ of errors on the rectangle Ω_2. The functions e_1^n and e_2^n are continuous in their (open) domain, but can have jumps at the two points where $\partial\Omega_1$ and $\partial\Omega_2$ intersect, except if the initial guess satisfies the boundary conditions. How fast do these two sequences converge to zero? The estimate of the convergence rate of the Schwarz method for this particular geometry is not easy. Over the course of time, different analysis techniques have been proposed to study the classical Schwarz method: maximum principle analysis, see, e.g., [12, 10, 3], Fourier analysis, see, e.g., [6, 2], variational analysis, see, e.g., [9, 4], and stochastic analysis [10]. In the spirit of this historical manuscript, we estimate the convergence rate by using tools that are considered "classical" in domain decomposition methods: maximum principle, the Riemann mapping theorem, the Poisson kernel, and the Schwarz-Christoffel mapping[3]. However, we wish to remark that, to the best of our knowledge the results presented in this work are new, and that the techniques used to prove them can be in principle used to study other domains with complicated geometries, whose subdomains can be mapped into circles and (semi-)infinite rectangles.

2 Convergence analysis

We begin our analysis noticing that maximum principle arguments, as done in [3, Theorem 7], allow us to obtain the following convergence result; see also [8].

[3] The Schwarz-Christoffel mapping was discovered independently by Christoffel in 1867 [1] and Schwarz in 1869 [11]; see [5] for a review.

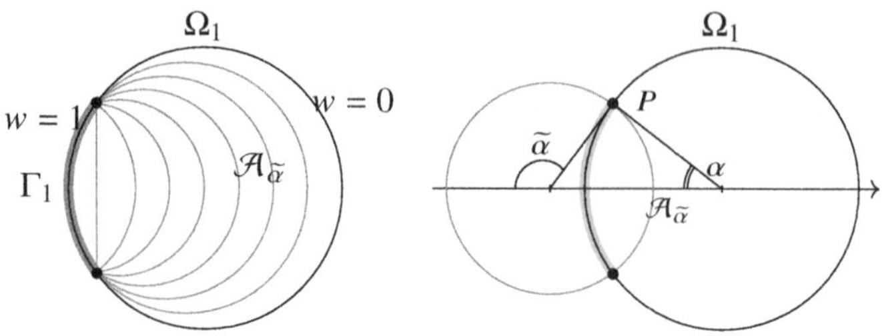

Fig. 2: Left: Level sets of w. Right: Geometric parametrization of a level set of w.

Theorem 1 (Convergence of the Schwarz method) *The Schwarz method* (2) *converges geometrically to the solution of* (1) *in the sense that there exists a convergence factor* $\rho < 1$ *such that*[4]

$$\max_{j=1,2} \|e_j^n\|_{\infty,\overline{\Omega}_j} \leq \rho^n \max_{j=1,2} \|e_j^0\|_{\infty,\overline{\Omega}_j}, \tag{3}$$

where $\rho = \big(\sup_{\Gamma_1} v_2\big)\big(\sup_{\Gamma_2} v_1\big)$, *with* v_j *solving for* $j = 1, 2$ *the problem*

$$\Delta v_j = 0 \text{ in } \Omega_j, \quad v_j = 1 \text{ on } \Gamma_j, \quad v_j = 0 \text{ on } \partial\Omega \cap \overline{\Omega}_j. \tag{4}$$

We thus have to study the two functions v_1 and v_2. Notice also the two sup in the definition of ρ could be replaced by max, as we see in what follows. We begin by studying v_1 and recalling the following result, which is proved in [3] by the Riemann mapping theorem and the Poisson kernel formula.

Lemma 1 *Problem* (4) *for* $j = 1$ *has a unique solution* w *which is harmonic in* Ω_1 *and constant on arcs of circles* $\mathcal{A}_{\widetilde{\alpha}}$ *passing through the two extrema of* Γ_1 *(see Fig. 2, left) and parametrized by angles* $\widetilde{\alpha}$ *between the horizontal line and the line that connects the center of the arc* $\mathcal{A}_{\widetilde{\alpha}}$ *to the point P (see Fig. 2, right), i.e.*

$$w(x, y) = \frac{\widetilde{\alpha} - \alpha}{\pi} \quad \forall (x, y) \in \mathcal{A}_{\widetilde{\alpha}}, \tag{5}$$

with $0 < w(x, y) < 1$ *for any* $(x, y) \in \Omega_1$ *and* $\alpha \leq \widetilde{\alpha} < \pi$. *Moreover, it holds that* $w(x, y) = \vartheta/\pi$ *for all* $(x, y) \in \mathcal{A}_{\widetilde{\alpha}}$, *where* ϑ *is the angle between the tangent to* $\mathcal{A}_{\widetilde{\alpha}}$ *in P and the tangent of* $\partial\Omega_2$ *in P; see Fig. 3 (left).*

Lemma 1 allows us to identify the sup of v_1 on Γ_2 with the max, that we estimate:

Lemma 2 (Estimated convergence factor on the disc) *Consider the function* v_1 *solving* (4) *for* $j = 1$. *It holds that*

$$\max_{\Gamma_2} v_1 = \begin{cases} \frac{1}{2} - \frac{\alpha}{\pi} & \text{if } \delta \geq \sin\alpha, \\ 1 - \frac{1}{\pi}\Big[\alpha + \arcsin\big(\frac{2\delta\sin\alpha}{\delta^2 + \sin^2\alpha}\big)\Big] & \text{if } \delta < \sin\alpha. \end{cases} \tag{6}$$

[4] The convergence rate is $-\log\rho$, see [7, Section 11.2.5].

Proof Lemma 1 implies that v_1 decays monotonically in Ω_1, in the sense that, according to formula (5) as $\widetilde{\alpha}$ decreases, the arc $\mathcal{A}_{\widetilde{\alpha}}$ is closer to $\partial\Omega_1 \setminus \Gamma_1$, and $v_1|_{\mathcal{A}_{\widetilde{\alpha}}}$ decreases monotonically. Therefore, to estimate the maximum of v_1 on Γ_2 we must find the arc that intersects Γ_2 on which v_1 has the highest value. To do so, we distinguish two cases: $\delta \geq \sin\alpha$ and $\delta < \sin\alpha$.

If $\delta \geq \sin\alpha$, then there exists an arc $\mathcal{A}$ (a semi-circle) that lies in the closure of the overlapping domain $\Omega_1 \cap \Omega_2$ and that is tangent to Γ_2 in the two points $\partial\Omega_1 \cap \partial\Omega_2$. Notice that if $\delta = \sin\alpha$, then $\mathcal{A}$ intersects Γ_2 also in the midpoint of its vertical segment. By the monotonicity of v_1, $\mathcal{A}$ is the arc intersecting Γ_2 on which v_1 attains the highest value. Since $\mathcal{A}$ is tangent to Γ_2 in both the points in $\partial\Omega_1 \cap \partial\Omega_2$, a simple geometric argument and the formula $v_1(x, y) = \vartheta/\pi$ allow us to obtain that $\max_{\Gamma_2} v_1 = \max_{\partial\Omega_1\cap\partial\Omega_2} v_1 = \frac{1}{2} - \frac{\alpha}{\pi}$.

Consider now that $\delta < \sin\alpha$. In this case, the monotonicity of v_1 implies that the arc that intersects Γ_2 on which v_1 attains the highest value is the one that passes through the two points in $\partial\Omega_1 \cap \partial\Omega_2$ and the midpoint of the vertical segment of Γ_2. Once this arc is found, direct calculations using simple geometric arguments and the formula $v_1(x, y) = \vartheta/\pi$ allow us to obtain the claim. □

Next, we focus on the function v_2 defined on the rectangle Ω_2. We begin recalling the following result proved in [3].

Lemma 3 *Let m denote the Möbius transformation that maps the half-plane $\mathcal{P} := \mathbb{R} \times \mathbb{R}^+$ onto the unit disc Ω_1. Recall the function w defined in Lemma 1. Then the function $\widehat{w}(\xi, \eta) := w(m(\xi, \eta))$ for all $(\xi, \eta) \in \mathcal{P}$ is harmonic in $\mathcal{P}$, it satisfies the boundary conditions $\widehat{w}(\xi, \eta) = 1$, for all (ξ, η) on the segment $m^{-1}(\Gamma_1)$ that lies on the horizontal line, and $\widehat{w}(\xi, \eta) = 0$, for all $(\xi, \eta) \in (\mathbb{R}\times\{0\})\setminus m^{-1}(\Gamma_1)$. Moreover $\widehat{w}$ is constant on arcs of circles passing through the extrema of $m^{-1}(\Gamma_1)$. Let Q be one of the two extrema of $m^{-1}(\Gamma_1)$ and let ϑ be the external angle between the tangent to one of these arcs, denoted by $\mathcal{A}_\vartheta$, in Q and the horizontal axis, then $\widehat{w}|_{\mathcal{A}_\vartheta} = \vartheta/\pi$.*

Notice that Lemma 3 allows us to identify the sup of v_2 on Γ_1 with the max. We can then prove the following lemmas.

Lemma 4 *Consider a semi-infinite strip Ω_2^∞ obtained by extending Ω_2 from the right to infinity and recall the half-plane $\mathcal{P}$ from Lemma 3.*

(a) *The Schwarz-Christoffel function that maps the semi-infinite strip onto the half-plane, denoted as $g : \Omega_2^\infty \to \mathcal{P}$, is given by*

$$g(x, y) = \begin{bmatrix} \cosh(x\frac{\pi}{2\sin\alpha})\cos(y\frac{\pi}{2\sin\alpha}) \\ \sinh(x\frac{\pi}{2\sin\alpha})\sin(y\frac{\pi}{2\sin\alpha}) \end{bmatrix}.$$

Moreover, g maps the interface Γ_2 onto the set $[g(\delta, 0), g(\delta, 2\sin\alpha)] \times \{0\}$.

(b) *Let v_2^∞ be a harmonic function in Ω_2^∞ such that $v_2^\infty = 1$ on Γ_2, $v_2^\infty = 0$ on $\partial\Omega_2^\infty \setminus \Gamma_2$ and $v_2^\infty(x, y) \to 0$ as $x \to \infty$. Let v_2 be the solution of (4) for $j = 2$. Then $v_2(x, y) < v_2^\infty(x, y)$ for all $(x, y) \in \Omega_2$.*

Proof Part (a): Recall the Schwarz-Christoffel function $f(\zeta) = C + K\,\mathrm{arcosh}(\zeta)$ for $\zeta \in \mathbb{C}$, where C and K are two constants in $\mathbb{C}$. It is well known that f maps the half-plane into any semi-infinite strip. Therefore, it is sufficient to determine the constants C and K by requiring that $f(1) = 0$ and $f(-1) = i2\sin\alpha$, where i is the imaginary unit. These conditions imply that the two corners of Γ_2 are mapped onto the points $\{-1, 1\}$ that lie on the real line in $\mathbb{C}$. We get $C = 0$ and $K = 2(\sin\alpha)/\pi$. Hence, $f(\zeta) = (2(\sin\alpha)/\pi)\,\mathrm{arcosh}(\zeta)$. Now, for any $z = x+iy = f(\zeta)$, we have that $\zeta = \cosh((x+iy)\pi/(2\sin\alpha))$. The function g is then obtained by using the formula $\cosh(a(x+iy)) = \cosh(ax)\cos(ay) + i\sinh(ax)\sin(ay)$, with $a = \pi/(2\sin\alpha)$. The last claim follows by the fact that $(g(x,0))_2 = (g(x, 2\sin\alpha))_2 = 0$ for any x and $(g(0,y))_2 = 0$ for any y and the properties of cosh and cos.

Part (b): Consider the function $p := v_2^\infty|_{\overline{\Omega}_2}$. Clearly p is harmonic in Ω_2 and it satisfies $p = v_2$ on Γ_2, $p = 0$ on $\partial\Omega_2 \cap \Omega_2^\infty$. However, by the maximum principle $p(x,y) > 0$ for all $(x,y) \in \partial\Omega_2 \setminus \Omega_2^\infty$. We can then decompose p as $p = v_2 + \widetilde{p}$, where $\widetilde{p}$ is harmonic in Ω_2, $\widetilde{p} = 0$ on $\partial\Omega_2 \setminus \Omega_2^\infty$ and $\widetilde{p} = p$ on $\partial\Omega_2 \cap \Omega_2^\infty$. By the maximum principle $\widetilde{p}(x,y) > 0$ for all $(x,y) \in \Omega_2$. Hence, $v_2^\infty|_{\overline{\Omega}_2}(x,y) = p(x,y) = v_2(x,y) + \widetilde{p}(x,y) > v_2(x,y)$ for all $(x,y) \in \Omega_2$ and the claim follows. □

Next, we parametrize the arc Γ_1 by an angle $\varphi \in [0,\pi]$ such that every point P on Γ_1 can be obtained as

$$P(\varphi) = \begin{bmatrix} x_P(\varphi) \\ y_P(\varphi) \end{bmatrix} := \begin{bmatrix} \delta + r(\varphi)\sin\varphi \\ \sin\alpha - r(\varphi)\cos\varphi \end{bmatrix},$$

where $r(\varphi) = -\cos\alpha\sin\varphi + \sqrt{\sin^2\alpha + \cos^2\alpha\sin^2\varphi}$. Using the function g in Lemma 4, we can map the arc Γ_1 into the half-plane and define $\widehat{\Gamma}_1 := g(\Gamma_1) = \{(\xi,\eta) \in \mathcal{P} : (\xi,\eta) = g(x_P(\varphi), y_P(\varphi)) \text{ for } \varphi \in [0,\pi]\}$. Notice that $\widehat{\Gamma}_1$ is a curve in the half-plane $\mathcal{P}$ and intersects the horizontal axis in the two points $g(\delta,0)$ and $g(\delta, 2\sin\alpha)$. We consider the following conjecture.

Conjecture Consider the arc Γ of the circle passing through the points $g(\delta,0)$ and $g(\delta,2\sin\alpha)$ and that intersects $\widehat{\Gamma}_1$ in $g(x_P(\pi/2), y_P(\pi/2))$. Then for any $\delta \geq 0$ and $\alpha \geq 0$ such that $\delta < 2\cos\alpha$, Γ is contained in the closure of the domain whose boundary is $\widehat{\Gamma}_1 \cup ([g(\delta,0), g(\delta,2\sin\alpha)] \times \{0\})$. □

A pictorial representation of Conjecture 1 is given in Fig. 3 (right). Notice that we have observed by direct numerical evaluation that Conjecture 1 always holds. We can then prove the following result.

Lemma 5 (Estimated convergence factor on the rectangle) *Let Conjecture 1 hold and recall the function v_2^∞ in Lemma 4. Then*

$$\max_{\Gamma_1} v_2 \leq v_2^\infty(x_P(\pi/2), y_P(\pi/2)) = \frac{1}{2} - \frac{1}{\pi}\arcsin(\kappa(\delta,\alpha)), \tag{7}$$

where

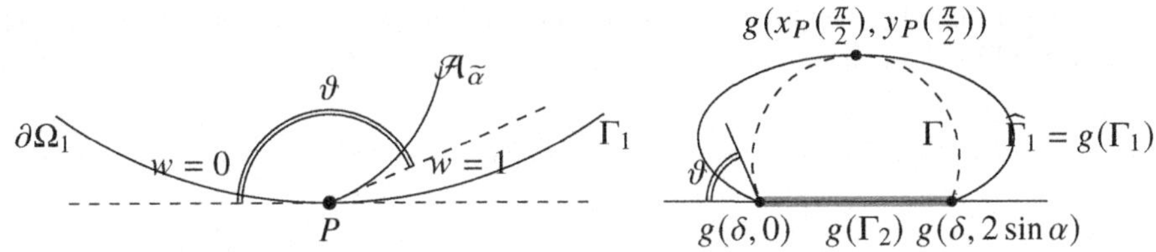

Fig. 3: Left: Geometry used in Lemma 1. Right: Geometric representation of Conjecture 1: the black solid curve represents $\widehat{\Gamma}_1 = g(\Gamma_1)$ and the black dashed arc of circle is Γ that passes through the points $g(\delta, 0), g(x_P(\frac{\pi}{2}), y_P(\frac{\pi}{2}))$ and $g(\delta, 2\sin\alpha)$. The set $g(\Gamma_2)$ is $[g(\delta, 0), g(\delta, 2\sin\alpha)] \times \{0\}$ and is marked in grey.

$$\kappa(\delta,\alpha) = \frac{\sinh^2\left(\frac{\pi(1+\delta-\cos\alpha)}{2\sin\alpha}\right) - \cosh^2\left(\frac{\pi\delta}{2\sin\alpha}\right)}{\sinh^2\left(\frac{\pi(1+\delta-\cos\alpha)}{2\sin\alpha}\right) + \cosh^2\left(\frac{\pi\delta}{2\sin\alpha}\right)}. \tag{8}$$

Proof Lemma 4 (b) implies that $v_2(x,y) \leq v_2^\infty(x,y)$ for all $(x,y) \in \Gamma_1$. Using the function g in Lemma 4 (a), we define $w^\infty(\xi,\eta) := v_2^\infty(x,y)$ for all $(x,y) \in \Omega_2^\infty$ and $(\xi,\eta) = g(x,y)$. Notice that $\max_{\Gamma_1} v_2^\infty = \max_{\widehat{\Gamma}_1} w^\infty$. The function w^∞ is harmonic in $\mathcal{P}$ and satisfies the conditions $w^\infty(\xi,0) = 1$ for $\xi \in [g(\delta,0), g(\delta, 2\sin\alpha)]$ and $w^\infty(\xi,0) = 0$ for $\xi \in \mathcal{R} \setminus [g(\delta,0), g(\delta,2\sin\alpha)]$. Hence, by using Lemma 3 we obtain that the function w^∞ is constant on arcs of circles $\mathcal{A}_\vartheta$ passing through the two points $g(\delta,0)$ and $g(\delta,2\sin\alpha)$. Moreover, the value of w^∞ on these arcs is given by $\widehat{w}|_{\mathcal{A}_\vartheta} = \vartheta/\pi$, where ϑ is defined in Lemma 3. This means that as ϑ decreases, the arc $\mathcal{A}_\vartheta$ becomes larger and the value $\widehat{w}|_{\mathcal{A}_\vartheta}$ decreases monotonically. Therefore, the value $\max_{\widehat{\Gamma}_1} w^\infty$ is given by the value of w^∞ on the arc Γ of the circle that passes through the two points $g(\delta,0)$ and $g(\delta,2\sin\alpha)$, that intersects $\widehat{\Gamma}_1$, and that corresponds to the largest ϑ; see Fig. 3 (right). By Conjecture 1, Γ is the arc of the circle that passes through the point $g(x_P(\frac{\pi}{2}), y_P(\frac{\pi}{2}))$. Notice that Γ is represented by a dashed line in Fig. 3 (right). Hence, $\max_{\widehat{\Gamma}_1} w^\infty = w^\infty(g(x_P(\pi/2), y_P(\pi/2)))$. The result follows by the formula $\widehat{w}|_{\mathcal{A}_\vartheta} = \vartheta/\pi$ and a direct calculation based on geometric arguments to obtain the angle ϑ characterizing Γ (see Fig. 3, right). □

We are now ready to prove our estimate of the convergence rate of the Schwarz method for the ddm logo.

Theorem 2 (Estimated convergence factor on the ddm logo) *The Schwarz method* (2) *converges in the sense of* (3)*, where*

$$\rho \leq \begin{cases} \left(\frac{1}{2} - \frac{1}{\pi}\arcsin(\kappa(\delta,\alpha))\right)\left(\frac{1}{2} - \frac{\alpha}{\pi}\right) & \text{if } \delta \geq \sin\alpha, \\ \left(\frac{1}{2} - \frac{1}{\pi}\arcsin(\kappa(\delta,\alpha))\right)\left[1 - \frac{1}{\pi}\left(\alpha + \arcsin\left(\frac{2\delta\sin\alpha}{\delta^2+\sin^2\alpha}\right)\right)\right] & \text{if } \delta < \sin\alpha, \end{cases} \tag{9}$$

with $\kappa(\delta,\alpha)$ *given in* (8).

Proof Recalling Theorem 1 and the formula $\rho = \left(\max_{\Gamma_1} v_2\right)\left(\max_{\Gamma_2} v_1\right)$, the estimate (9) follows using Lemmas 2 and 5. □

The estimated convergence factors obtained in Lemmas 2 and 5 and Theorem 2 are shown in Fig. 4. In particular, in Fig. 4 (left) the function (6) is shown. Fig. 4 (center)

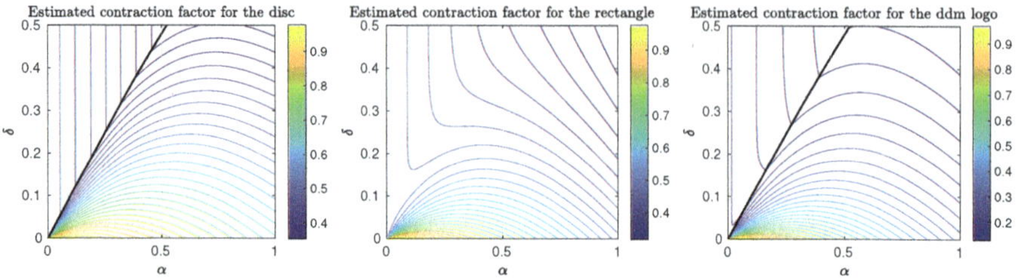

Fig. 4: Left: Values of $\max_{\Gamma_2} \nu_1$ as function of α and δ given in (6). Center: Estimate of $\max_{\Gamma_1} \nu_2$ given in (7). Right: Estimated convergence factor for the ddm logo given in (9).

represents the upper bound (7). Fig. 4 (right) shows the estimated convergence factor (9) for the ddm logo. The black curves in Fig. 4 (left and right) represent the function $\sin\alpha$ separating two regions according to (6) and (9).

3 Numerical experiments

We now compare our theoretical estimates with the numerical convergence behavior. We discretize the ddm logo by linear finite elements using Freefem[5]. Two finite element discretizations of the ddm logo are shown in Fig. 5. In order to accurately describe the behavior of the (continuous) Schwarz method, we used however in our experiments much finer meshes than the ones shown in Fig. 5. We solve problem (1) for a fixed $L = 2$ and different values of the parameters α and δ. Our results are shown in Fig. 6, where the decay of the error with respect to the number of iterations is represented. In particular, our theoretical estimates (solid lines) are compared with

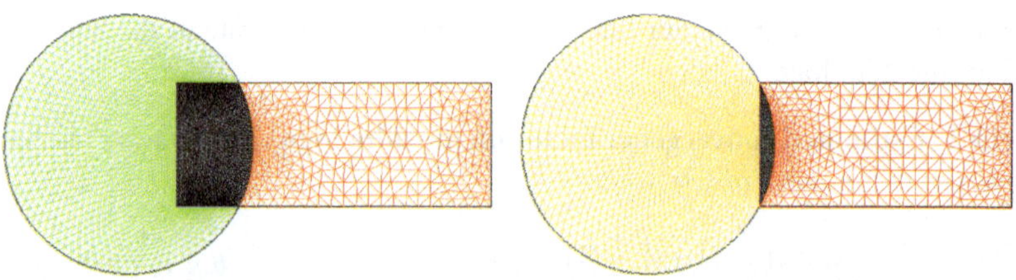

Fig. 5: Examples of finite element discretizations of the ddm logo obtained by Freefem. Left: $\alpha = 0.5$, $\delta = 0.5$ and $L = 2$. Right: $\alpha = 0.5$, $\delta = 0$ and $L = 2$.

[5] This finite-element code was designed by the first author and Felix Kwok for the DD Summer school organized by the second author at the University of Nice, June 19-21, 2018, and it was also used by the second author in his plenary lecture at the 25th domain decomposition conference.

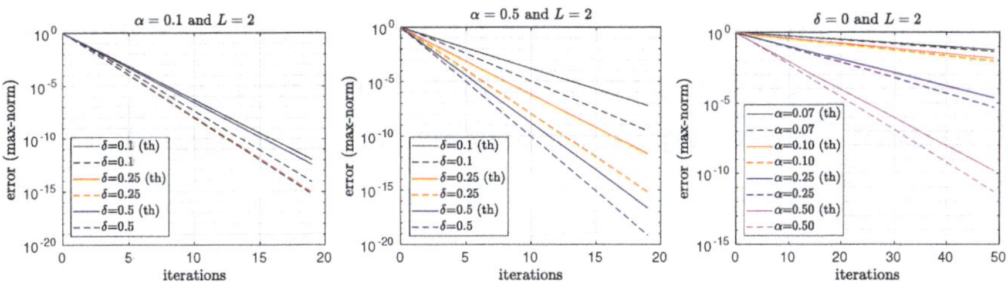

Fig. 6: Theoretical (solid line) and numerical (dashed line) convergences.

the numerical errors (dashed lines). The first two pictures in Fig. 6 (left and center) correspond to $\alpha = 0.1$ and $\alpha = 0.5$ and different values of $\delta > 0$. Notice that, even though our theoretical estimate is an upper bound for the true convergence rate, it describes very well the behavior of the method for different parameters. To study the sharpness of our results, we consider also the case with $\delta = 0$ and different values of α. The results of these experiments are shown in Fig. 6 (right), where one can clearly see that our results are very sharp for $\delta = 0$ and small values of α. The reason for this behavior is that our results are based on Theorem 1, where few estimates are present in the proof; see [3]. These are sharper when the dominating error is localized near the two points in $\partial\Omega_1 \cap \partial\Omega_2$ and the overlap is small.

References

1. Christoffel, E.B.: Sopra un problema proposto da Dirichlet. Ann. Mat. Pura Appl. Serie II **4**, 1–9 (1870)
2. Ciaramella, G., Gander, M.J.: Analysis of the parallel Schwarz method for growing chains of fixed-sized subdomains: Part I. SIAM J. Numer. Anal. **55**(3), 1330–1356 (2017)
3. Ciaramella, G., Gander, M.J.: Analysis of the parallel Schwarz method for growing chains of fixed-sized subdomains: Part II. SIAM J. Numer. Anal. **56**(3), 1498–1524 (2018)
4. Ciaramella, G., Gander, M.J.: Analysis of the parallel Schwarz method for growing chains of fixed-sized subdomains: Part III. to appear in ETNA (2018)
5. Driscoll, T.A., Trefethen, N.L.: Schwarz-Christoffel Mapping. Cambridge University Press (2002)
6. Gander, M.J.: Optimized Schwarz methods. SIAM J. Numer. Anal. **44**(2), 699–731 (2006)
7. Gander, W., Gander, M.J., Kwok, F.: Scientific computing-An introduction using Maple and MATLAB, vol. 11. Springer Science & Business (2014)
8. Kantorovich, L.V., Krylov, V.I.: Approximate Methods of Higher Analysis. Translated from the third Russian edition by Curtis D. Benster. Interscience Publishers Inc., New York (1958)
9. Lions, P.L.: On the Schwarz alternating method. I. First international symposium on domain decomposition methods for partial differential equations pp. 1–42 (1988)
10. Lions, P.L.: On the Schwarz alternating method. II. In: T. Chan, R. Glowinski, J. Périaux, O. Widlund (eds.) Second International Symposium on Domain Decomposition Methods for Partial Differential Equations, pp. 47–70. SIAM (1989)
11. Schwarz, H.A.: Conforme Abbildung der Oberflache eines Tetraeders auf die Oberflache einer Kugel. J. Reine Ange. Math. **20**, 70–105 (1869)
12. Schwarz, H.A.: Über einen Grenzübergang durch alternierendes Verfahren. Vierteljahresschrift Naturf. Ges. Zürich **50**, 272–286 (1870)

Additive Schwarz Preconditioners for a State Constrained Elliptic Distributed Optimal Control Problem Discretized by a Partition of Unity Method

Susanne C. Brenner, Christopher B. Davis, and Li-yeng Sung

1 Introduction

In this work, we are interested in solving a model elliptic optimal control problem of the following form: Find $(y, u) \in H_0^1(\Omega) \times L_2(\Omega)$ that minimize the functional

$$J(y, u) = \frac{1}{2}\int_\Omega (y - f)^2 dx + \frac{\beta}{2}\int_\Omega u^2 dx$$

subject to

$$-\Delta y = u \text{ in } \Omega, \quad y = 0 \text{ in } \partial\Omega, \tag{1}$$

and $y \leq \psi$ in Ω, where Ω is a convex polygon in $\mathbb{R}^2$ and $f \in L_2(\Omega)$. We also assume $\psi \in C^2(\Omega) \cap H^3(\Omega)$ and $\psi > 0$ on $\partial\Omega$.

Using elliptic regularity (cf. [7]) for (1), we can reformulate the model problem as follows: Find $y \in K$ such that

$$y = \underset{v\in K}{\operatorname{argmin}} \left[\frac{1}{2}a(v, v) - (f, v)\right], \tag{2}$$

where $K = \{v \in H^2(\Omega) \cap H_0^1(\Omega) : v \leq \psi \text{ in } \Omega\}$,

$$a(w, v) = \beta \int_\Omega \Delta w \Delta v dx + \int_\Omega wv dx \quad \text{and} \quad (f, v) = \int_\Omega fv dx.$$

Once y is calculated, then u can be determined by $u = -\Delta y$.

Susanne C. Brenner, Li-Yeng Sung
Department of Mathematics and Center for Computation and Technology, Louisiana State University, Baton Rouge, LA 70803, USA, e-mail: `brenner@math.lsu.edu`, `sung@math.lsu.edu`

Christopher B. Davis,
Foundation Hall 250, Department of Mathematics, Tennessee Tech University, Cookeville, TN 38505, e-mail: `CBDavis@tntech.edu`

R. Haynes et al. (eds.), *Domain Decomposition Methods in Science and Engineering XXV*,
Lecture Notes in Computational Science and Engineering 138,
https://doi.org/10.1007/978-3-030-56750-7_10

The minimization problem (2) is discretized in [4] by a partition of unity method (PUM). The goal of this paper is to use the ideas in [5] for an obstacle problem of clamped Kirchhoff plates to develop preconditioners for the discrete problems in [4]. We refer to these references for technical details and only present the important results here.

2 The Discrete Problem

We will use a variant of the PUM (cf. [11, 8, 1, 12]) to construct a conforming approximation space $V_h \subset H^2(\Omega) \cap H_0^1(\Omega)$. Below we present an overview of the construction of V_h.

Let $\{\Omega_i\}_{i=1}^n$ be an open cover of $\bar{\Omega}$ such that there exists a collection of nonnegative functions $\{\phi_i\}_{i=1}^n \in W_\infty^2(\mathbb{R}^2)$ with the following properties:

$$\begin{aligned} &\operatorname{supp}\phi_i \subset \Omega_i && \text{for } 1 \le i \le n, \\ &\sum_{i=1}^n \phi_i = 1 && \text{on } \Omega, \\ &|\phi_i|_{W_\infty^m(\mathbb{R}^2)} \le \frac{C}{(\operatorname{diam}\Omega_i)^m} && \text{for } 0 \le m \le 2,\ 1 \le i \le n. \end{aligned}$$

For $1 \le i \le n$, the local approximation space V_i consists of biquadratic polynomials satisfying the Dirichlet boundary conditions of (1), i.e. $v = 0$ on $\partial\Omega$ for all $v \in V_i$. Basis functions for V_i are tensor product Lagrange polynomials. Figure 1 (b) shows an illustration that depicts the interpolation nodes corresponding to the interior degrees of freedom for a given discretization.

In this work the patches $\{\Omega_i\}_{i=1}^n$ are open rectangles and $\{\phi_i\}_{i=1}^n$ are C^1 piecewise polynomial tensor product flat-top partition of unity functions. $\Omega_i^{\text{flat}} = \{x \in \Omega_i : \phi_i(x) = 1\}$. The interpolation nodes associated with V_i are distributed uniformly throughout Ω_i^{flat}, this is the reason the global basis functions have the Kronecker delta property. We will assume that the diameters of the patches are comparable to a mesh size h. We now define

$$V_h = \sum_{i=1}^n \phi_i V_i.$$

Let $\mathcal{N}_h$ be the set of all interior interpolation nodes used in the construction of V_h. The discrete problem is to find $y_h \in K_h$ such that

$$y_h = \underset{v \in K_h}{\operatorname{argmin}} \left[\frac{1}{2} a(v, v) - (f, v) \right], \tag{3}$$

where $K_h = \{v \in V_h : v(p) \le \psi(p)\ \forall p \in \mathcal{N}_h\}$.

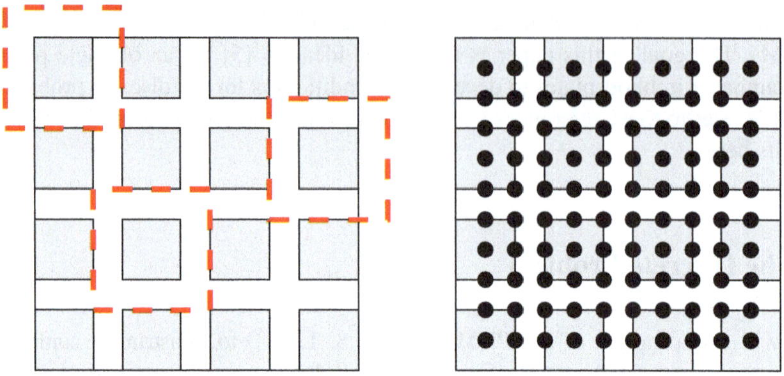

(a) (b)

Fig. 1: (a) Ω_i (bounded by dotted lines) and Ω_i^{flat} (shaded in grey) (b) nodes for the interior DOFs

By introducing a Lagrange multiplier $\lambda_h : \mathcal{N}_h \to \mathbb{R}$, the minimization problem (3) can be rewritten in the following form: Find $y_h \in K_h$ such that

$$\begin{aligned} a(y_h, v) - (f, v) &= -\sum_{p \in \mathcal{N}_h} \lambda_h(p) v(p) && \forall\, v \in V_h, \\ \lambda_h(p) &= \max(0, \lambda_h(p) + c(y_h(p) - \psi(p))) && \forall\, p \in \mathcal{N}_h, \end{aligned}$$

where c is a (large) positive number ($c = 10^8$ in our numerical experiments). This system can then be solved by a primal-dual active set (PDAS) algorithm (cf. [2, 3, 9, 10]). Given the k-th approximation (y_k, λ_k), the $(k+1)$-st iteration of the PDAS algorithm is to find (y_{k+1}, λ_{k+1}) such that

$$\begin{aligned} a(y_{k+1}, v) - (f, v) &= -\sum_{p \in \mathcal{N}_h} \lambda_{k+1}(p) v(p) && \forall v \in V_h, \\ y_{k+1}(p) &= \psi(p) && \forall p \in \mathfrak{A}_k, \\ \lambda_{k+1}(p) &= 0 && \forall p \in \mathcal{N}_h \backslash \mathfrak{A}_k, \end{aligned} \tag{4}$$

where $\mathfrak{A}_k = \{p \in \mathcal{N}_h : \lambda_k(p) + c(y_k(p) - \psi(p)) > 0\}$ is the set of active nodes determined from the approximations (y_k, λ_k). Below we present preconditioners for the linear systems encountered in (4).

3 The Preconditioners

The additive Schwarz preconditioners (cf. [6]) will be applied to a system associated with a subset $\tilde{\mathcal{N}}_h$ of $\mathcal{N}_h$. Let $\tilde{T}_h : V_h \to V_h$ be defined by

$$(\tilde{T}_h v)(p) = \begin{cases} v(p) & \text{if } p \in \tilde{\mathcal{N}}_h \\ 0 & \text{if } p \notin \tilde{\mathcal{N}}_h \end{cases}.$$

The approximation space for the subproblem is $\tilde{V}_h = \tilde{T}_h V_h$. The associated stiffness matrix is a symmetric positive definite operator $\tilde{A}_h : \tilde{V}_h \to \tilde{V}_h'$ defined by

$$\langle \tilde{A}_h v, w \rangle = a(v, w) \qquad \forall\, v, w \in \tilde{V}_h,$$

where $\langle \cdot, \cdot \rangle$ is the canonical bilinear form on $\tilde{V}_h' \times \tilde{V}_h$.

A One-Level Method Here we introduce a collection of shape regular subdomains $\{D_j\}_{j=1}^J$ with diam $D_j \approx H$ that overlap with each other by at most δ. Associated with each subdomain is a function space $V_j \subset \tilde{V}_h$ whose members vanish at the nodes outside D_j. Let $A_j : V_j \to V_j'$ be defined by

$$\langle A_j v, w \rangle = a(v, w) \qquad \forall v, w \in V_j.$$

The one-level additive Schwarz preconditioner $B_{\mathrm{OL}} : V_h' \to V_h$ is defined by

$$B_{\mathrm{OL}} = \sum_{j=1}^{J} I_j A_j^{-1} I_j^t,$$

where $I_j : V_j \to \tilde{V}_h$ is the natural injection.

Following the arguments in [5], we can obtain the following theorem.

Theorem 1 *There exists a positive constant C_{OL} independent of H, h, J, δ and $\tilde{\mathcal{N}}_h$ such that*

$$\kappa(B_{OL}\tilde{A}_h) \le C_{OL}\delta^{-3}H^{-1}.$$

Remark 1 The estimate given in Theorem 1 is identical to the one for the plate bending problem without an obstacle, i.e., the obstacle is invisible to the one-level additive Schwarz preconditioner.

A Two-Level Method Let $V_H \subset H^2(\Omega) \cap H_0^1(\Omega)$ be a coarse approximation space based on the construction in Section 2 where $H > h$. We assume the patches of V_H are of comparable size to the subdomains $\{D_j\}_{j=1}^J$. Let $\Pi_h : H^2(\Omega) \cap H_0^1(\Omega) \to V_h$ be the nodal interpolation operator. We define $V_0 \subset \tilde{V}_h$ by $V_0 = T_h \Pi_h V_H$, and $A_0 : V_0 \to V_0'$ by

$$\langle A_0 v, w \rangle = a(v, w) \qquad \forall\, v, w \in V_0.$$

The two-level additive Schwarz preconditioner $B_{\mathrm{TL}} : V_h' \to V_h$ is given by

$$B_{\mathrm{TL}} = \sum_{j=0}^{J} I_j A_j^{-1} I_j^t,$$

where $I_0 : V_0 \to \tilde{V}_h$ is the natural injection. Using the arguments in [5], we can obtain the following theorem.

Theorem 2 *There exists a positive constant C_{TL} independent of H, h, J, δ and $\tilde{N}_h$ such that*

$$\kappa(B_{TL}A_h) \leq C_{TL} \min\left((H/h)^4, \delta^{-3}H^{-1}\right).$$

Remark 2 The two-level method is scalable as long as H/h remains bounded.

Remark 3 The estimate given in Theorem 2 is different from the estimate for the plate bending problem without obstacles that reads

$$\kappa\left(B_{TL}A_h\right) \leq C\left(\frac{H}{\delta}\right)^3.$$

This difference is caused by the necessity of truncation in the construction of $\tilde{V}_0$ when the obstacle is present.

4 A Numerical Example

We consider Example 4.2 in [4], where $\Omega = (-0.5, 0.5)^2$, $\beta = 0.1$, $\psi = 0.01$, and $f = 10(\sin(2\pi(x_1 + 0.5)) + (x_2 + 0.5))$. We discretize (3) by the PUM with uniform rectangular patches so that $h \approx 2^{-\ell}$, where ℓ is the refinement level. As ℓ increases from 1 to 8, the number of degrees of freedom increases from 16 to 586756. The discrete variational inequalities are solved by the PDAS algorithm presented in Section 2, with $c = 10^8$.

For the purpose of comparison, we first solve the auxiliary systems in each iteration of the PDAS algorithm by the conjugate gradient (CG) method without a preconditioner. The average condition number during the PDAS iteration and the time to solve the variational inequality are presented in Table 1. The PDAS iterations fail to stop (DNC) within 48 hours beyond level 6.

Table 1: Average condition number (κ) and time to solve (t_{solve}) in seconds by the CG algorithm

ℓ	κ	t_{solve}
1	$3.1305\times10^{+2}$	2.6111×10^{-2}
2	$9.1118\times10^{+3}$	1.0793×10^{-1}
3	$2.0215\times10^{+5}$	9.7842×10^{-1}
4	$3.3705\times10^{+6}$	$3.3911\times10^{+1}$
5	$6.4346\times10^{+7}$	$6.2173\times10^{+2}$
6	$1.0537\times10^{+9}$	$8.8975\times10^{+3}$
7	DNC	DNC
8	DNC	DNC

We then solve the auxiliary systems by the preconditioned conjugate gradient (PCG) method, using the additive Schwarz preconditioners associated with J subdomains. The mesh size H for the coarse space V_H is $\approx 1/\sqrt{J}$. We say the PCG method has converged if $\|Br\|_2 \le 10^{-15}\|b\|_2$, where B is the preconditioner, r is the residual, and b is the load vector. The initial guess for the PDAS algorithm is taken to be the solution at the previous level, or 0 if $2^{2\ell} = J$. To obtain a good initial guess for the two-level method, the one-level method is used when $2^{2\ell} = J$. The subdomain problems and the coarse problem are solved by a direct method based on the Cholesky factorization on independent processors.

Small Overlap Here we apply the preconditioners in such a way that $\delta \approx h$. The average condition numbers of the linear systems over the PDAS iterations are presented in Table 2. We can see that these condition numbers are significantly smaller than those for the unpreconditioned case and the condition numbers for the two-level method are smaller than those for the one-level method. For each ℓ, as J increases the condition numbers for the two-level method are decreasing, which demonstrates the scalability of the two-level method (cf. Remark 2).

Table 2: Average condition number for small overlap: one-level (left) and two-level (right)

ℓ	$J=4$	$J=16$	$J=64$	$J=256$	$J=4$	$J=16$	$J=64$	$J=256$
1	$1.00\times10^{+0}$	-	-	-	$1.00\times10^{+0}$	-	-	-
2	$4.94\times10^{+0}$	$7.40\times10^{+0}$	-	-	$5.46\times10^{+0}$	$7.40\times10^{+0}$	-	-
3	$1.51\times10^{+1}$	$4.41\times10^{+1}$	$6.61\times10^{+1}$	-	$1.22\times10^{+1}$	$1.14\times10^{+1}$	$6.61\times10^{+1}$	-
4	$7.82\times10^{+1}$	$1.90\times10^{+2}$	$5.35\times10^{+2}$	$8.19\times10^{+2}$	$2.85\times10^{+1}$	$2.79\times10^{+1}$	$1.26\times10^{+1}$	$8.19\times10^{+2}$
5	$6.47\times10^{+2}$	$1.64\times10^{+3}$	$3.17\times10^{+3}$	$9.50\times10^{+3}$	$6.29\times10^{+1}$	$9.19\times10^{+1}$	$4.61\times10^{+1}$	$1.98\times10^{+1}$
6	$5.07\times10^{+3}$	$1.31\times10^{+4}$	$2.58\times10^{+4}$	$5.04\times10^{+4}$	$3.67\times10^{+2}$	$3.48\times10^{+2}$	$1.31\times10^{+2}$	$5.77\times10^{+1}$
7	$4.07\times10^{+4}$	$1.06\times10^{+5}$	$2.10\times10^{+5}$	$4.15\times10^{+5}$	$2.74\times10^{+3}$	$2.11\times10^{+3}$	$1.03\times10^{+3}$	$2.86\times10^{+2}$
8	$3.26\times10^{+5}$	$8.55\times10^{+5}$	$1.70\times10^{+6}$	$3.38\times10^{+6}$	$2.16\times10^{+4}$	$1.48\times10^{+4}$	$9.19\times10^{+3}$	$1.87\times10^{+3}$

The times to solve the problem for each method are presented in Table 3. By comparing them with the results in Table 1, we can see that both of the two methods represents progress. For comparison purposes, the faster time of the two methods is highlighted in red for each ℓ and J. As h decreases and J increases, the two-level method performs better than the one-level method. These results are consistent with Theorems 1 and 2.

Generous Overlap Here we apply the preconditioners in such a way that $\delta \approx H$. When $J = 4$ and $J = 16$ both methods fail to converge at $\ell = 8$ within 48 hours due to the large size of the local problems. The average condition numbers of the linear systems over the PDAS iterations are presented in Table 4. They agree with Theorems 1 and 2. We can also see that these condition numbers are smaller than those in the case of small overlap.

The times to solve the problem for each method are presented in Table 5. Again both methods are superior to the unpreconditioned method and the scalability of the two-level method is observed.

Table 3: Time to solve in seconds for small overlap: one-level (left) and two-level (right). Times highlighted in red are the fastest between the two methods.

ℓ	$J=4$	$J=16$	$J=64$	$J=256$	$J=4$	$J=16$	$J=64$	$J=256$
1	1.78×10^{+0}	-	-	-	1.78×10^{+0}	-	-	-
2	3.04×10^{-1}	1.55×10^{+1}	-	-	1.06×10^{+0}	1.55×10^{+1}	-	-
3	3.84×10^{-1}	1.07×10^{+1}	6.08×10^{+1}	-	1.08×10^{+0}	1.42×10^{+1}	6.08×10^{+1}	-
4	2.60×10^{+0}	4.18×10^{+1}	9.18×10^{+1}	3.55×10^{+2}	5.51×10^{+0}	5.83×10^{+1}	7.09×10^{+1}	3.55×10^{+2}
5	2.57×10^{+1}	1.11×10^{+2}	1.53×10^{+2}	3.54×10^{+2}	3.09×10^{+1}	1.14×10^{+2}	1.42×10^{+2}	1.46×10^{+2}
6	2.82×10^{+2}	2.69×10^{+2}	4.00×10^{+2}	4.63×10^{+2}	2.81×10^{+2}	2.06×10^{+2}	1.63×10^{+2}	1.50×10^{+2}
7	5.25×10^{+3}	1.91×10^{+3}	1.48×10^{+3}	1.58×10^{+3}	4.43×10^{+3}	1.18×10^{+3}	4.68×10^{+2}	2.98×10^{+2}
8	1.09×10^{+5}	2.90×10^{+4}	1.16×10^{+4}	6.85×10^{+3}	9.05×10^{+4}	2.04×10^{+4}	3.12×10^{+3}	8.80×10^{+2}

Table 4: Average condition number for generous overlap: one-level (left) and two-level (right)

ℓ	$J=4$	$J=16$	$J=64$	$J=256$	$J=4$	$J=16$	$J=64$	$J=256$
1	1.00×10^{+0}	-	-	-	1.00×10^{+0}	-	-	-
2	1.00×10^{+0}	7.40×10^{+0}	-	-	1.25×10^{+0}	7.40×10^{+0}	-	-
3	1.00×10^{+0}	7.84×10^{+0}	6.61×10^{+1}	-	1.25×10^{+0}	6.27×10^{+0}	6.61×10^{+1}	-
4	1.00×10^{+0}	7.56×10^{+0}	8.47×10^{+1}	8.19×10^{+2}	1.25×10^{+0}	6.47×10^{+0}	1.32×10^{+1}	8.19×10^{+2}
5	1.00×10^{+0}	8.29×10^{+0}	9.67×10^{+1}	1.48×10^{+3}	1.25×10^{+0}	7.15×10^{+0}	1.75×10^{+1}	1.73×10^{+1}
6	1.00×10^{+0}	8.36×10^{+0}	9.86×10^{+1}	1.47×10^{+3}	1.25×10^{+0}	7.45×10^{+0}	2.06×10^{+1}	2.03×10^{+1}
7	1.00×10^{+0}	8.43×10^{+0}	1.00×10^{+2}	1.49×10^{+3}	1.25×10^{+0}	7.63×10^{+0}	2.22×10^{+1}	2.59×10^{+1}
8	DNC	DNC	1.01×10^{+2}	1.51×10^{+3}	DNC	DNC	2.44×10^{+1}	2.82×10^{+1}

We now compare the generous overlap methods with the small overlap methods. In Table 5, the times in red are the ones where the method with generous overlap outperforms the method with small overlap. It is evident from Table 5 that the performance of the two-level method with generous overlap suffers from a high communication cost for small h and large J.

Table 5: Time to solve in seconds for generous overlap: one-level (left) and two-level (right). Times highlighted in red are faster than the corresponding method with small overlap.

ℓ	$J=4$	$J=16$	$J=64$	$J=256$	$J=4$	$J=16$	$J=64$	$J=256$
1	1.33×10^{-1}	-	-	-	1.33×10^{-1}	-	-	-
2	1.90×10^{-1}	1.66×10^{+1}	-	-	4.71×10^{-1}	1.66×10^{+1}	-	-
3	2.88×10^{-1}	7.17×10^{+0}	6.14×10^{+1}	-	6.47×10^{-1}	1.03×10^{+1}	6.14×10^{+1}	-
4	5.86×10^{+0}	2.54×10^{+1}	4.57×10^{+1}	3.55×10^{+2}	6.73×10^{+0}	3.45×10^{+1}	6.33×10^{+1}	3.55×10^{+2}
5	1.02×10^{+2}	7.34×10^{+1}	6.88×10^{+1}	1.57×10^{+2}	1.06×10^{+2}	8.17×10^{+1}	8.70×10^{+1}	1.48×10^{+2}
6	1.32×10^{+3}	5.21×10^{+2}	1.09×10^{+2}	1.50×10^{+2}	1.32×10^{+3}	5.46×10^{+2}	1.15×10^{+2}	1.12×10^{+2}
7	2.41×10^{+4}	8.12×10^{+3}	7.74×10^{+2}	3.00×10^{+2}	2.31×10^{+4}	8.41×10^{+3}	7.51×10^{+2}	1.97×10^{+2}
8	DNC	DNC	1.16×10^{+4}	1.64×10^{+3}	DNC	DNC	1.19×10^{+4}	1.13×10^{+3}

5 Conclusion

In this paper we present additive Schwarz preconditioners for the linear systems that arise from the PDAS algorithm applied to an elliptic distributed optimal control problem with pointwise state constraints discretized by a PUM. Based on the condition number estimates and the numerical results, the two-level method with small overlap appears to be the best choice for small h and large J.

Acknowledgements The work of the first and third authors was supported in part by the National Science Foundation under Grant No. DMS-16-20273. Portions of this research were conducted with high performance computing resources provided by Louisiana State University (http://www.hpc.lsu.edu).

References

1. I. Babuška, U. Banerjee, and J.E. Osborn. Survey of meshless and generalized finite element methods: a unified approach. *Acta Numer.*, 12:1—125, 2003.
2. M. Bergounioux, K. Ito, and K. Kunisch. Primal-dual strategy for constrained optimal control problems. *SIAM J. Control Optim.*, 37:1176—1194 (electronic), 1999.
3. M. Bergounioux and K. Kunisch. Primal-dual strategy for state-constrained optimal control problems. *Comput. Optim. Appl.*, 22:193—224, 2002.
4. S.C. Brenner, C.B. Davis, and L.-Y. Sung, A partition of unity method for a class of fourth order elliptic variational inequalities, *Comput. Methods Appl. Mech. Engrg.*, 276:612–626, 2014.
5. S.C. Brenner, C.B. Davis, and L.-Y. Sung, Additive Schwarz preconditioners for the obstacle problem of clamped Kirchhoff plates, arXiv:1809.06311 [math.NA]
6. M. Dryja and O.B. Widlund, An additive variant of the Schwarz alternating method in the case of many subregions. Techical Report 339, Department of Computer Science, Courant Institute, 1987.
7. P. Grisvard, *Elliptic Problems in Non Smooth Domains*. Pitman, Boston, 1985.
8. M. Griebel and M.A. Schweitzer. A particle-partition of unity method. II. Efficient cover construction and reliable integration. *SIAM J. Sci. Comput.*, 23:1655—1682, 2002.
9. M. Hintermüller, K. Ito, and K. Kunisch. The primal-dual active set strategy as a semismooth Newton method. *SIAM J. Optim.*, 13:865—888, 2003.
10. K. Ito and K. Kunisch. *Lagrange Multiplier Approach to Variational Problems and Applications*. Society for Industrial and Applied Mathematics, Philadelphia, PA, 2008.
11. J.M. Melenk and I. Babuška. The partition of unity finite element method: basic theory and applications. *Comput. Methods Appl. Mech. Engrg.*, 139:289—314, 1996.
12. H.-S. Oh, J.G. Kim, and W.-T. Hong. The piecewise polynomial partition of unity functions for the generalized finite element methods. *Comput. Methods Appl. Mech. Engrg.*, 197:3702—3711, 2008.

A Parallel Solver for a Preconditioned Space-Time Boundary Element Method for the Heat Equation

Stefan Dohr, Michal Merta, Günther Of, Olaf Steinbach, and Jan Zapletal

1 Introduction

In this note we describe a parallel solver for the discretized weakly singular space-time boundary integral equation of the spatially two-dimensional heat equation. The global space-time nature of the system matrices leads to improved parallel scalability in distributed memory systems in contrast to time-stepping methods where the parallelization is usually limited to spatial dimensions. We present a parallelization technique which is based on a decomposition of the input mesh into submeshes and a distribution of the corresponding blocks of the system matrices among processors. To ensure load balancing, the distribution is based on a cyclic decomposition of complete graphs [8, 9]. In addition, the solution of the global linear system requires an efficient preconditioner. We present a robust preconditioning strategy which is based on boundary integral operators of opposite order [6, 14].

The parallelization of the discretized space–time integral equation in distributed and shared memory is discussed in [5]. Here, we extend the parallel solver to the preconditioned system. We demonstrate the method for the spatially two-dimensional case. However, the presented results, particularly the parallelization in distributed memory and the stability results for the preconditioner, can be used to extend the method to the three-dimensional problem.

Let $\Omega \subset \mathbb{R}^2$ be a bounded domain with a Lipschitz boundary $\Gamma := \partial\Omega$ and $T > 0$. As a model problem we consider the initial Dirichlet boundary value problem for the heat equation

Stefan Dohr, Olaf Steinbach, Günther Of
Institute of Applied Mathematics, Graz University of Technology,
Steyrergasse 30, 8010 Graz, Austria
e-mail: dohr@math.tugraz.at, o.steinbach@tugraz.at, of@tugraz.at

Michal Merta, Jan Zapletal
IT4Innovations and Department of Applied Mathematics, VSB – Technical University of Ostrava,
17. listopadu 2172/15, 708 00 Ostrava-Poruba, Czech Republic
e-mail: michal.merta@vsb.cz, jan.zapletal@vsb.cz

R. Haynes et al. (eds.), *Domain Decomposition Methods in Science and Engineering XXV*,
Lecture Notes in Computational Science and Engineering 138,
https://doi.org/10.1007/978-3-030-56750-7_11

$$\alpha\partial_t u - \Delta_x u = 0 \text{ in } Q := \Omega\times(0,T),\ u = g \text{ on } \Sigma := \Gamma\times(0,T),\ u = u_0 \text{ in } \Omega \quad (1)$$

with a heat capacity constant $\alpha > 0$, the given initial datum $u_0 \in L^2(\Omega)$, and the boundary datum $g \in H^{1/2,1/4}(\Sigma)$. An explicit formula for the solution of (1) is given by the representation formula for the heat equation [1], i.e. for $(x,t) \in Q$ we have

$$\begin{aligned} u(x,t) &= (\widetilde{M}_0 u_0)(x,t) + (\widetilde{V} w)(x,t) - (Wg)(x,t) \\ &= \int_\Omega U^\star(x-y,t)u_0(y)\,\mathrm{d}y + \frac{1}{\alpha}\int_\Sigma U^\star(x-y,t-\tau)w(y,\tau)\,\mathrm{d}s_y\,\mathrm{d}\tau \\ &\quad - \frac{1}{\alpha}\int_\Sigma \frac{\partial}{\partial n_y}U^\star(x-y,t-\tau)g(y,\tau)\,\mathrm{d}s_y\,\mathrm{d}\tau, \end{aligned} \quad (2)$$

with $w := \partial_n u$ and $U^\star$ denoting the fundamental solution of the two-dimensional heat equation given by

$$U^\star(x-y,t-\tau) = \begin{cases} \dfrac{\alpha}{4\pi(t-\tau)}\exp\left(\dfrac{-\alpha|x-y|^2}{4(t-\tau)}\right) & \text{for } \tau < t, \\ 0 & \text{otherwise.} \end{cases}$$

The yet unknown Neumann datum $w \in H^{-1/2,-1/4}(\Sigma)$ can be found by applying the interior Dirichlet trace operator to (2) and solving the resulting weakly singular boundary integral equation

$$g(x,t) = (M_0 u_0)(x,t) + (Vw)(x,t) + ((\tfrac{1}{2}I - K)g)(x,t) \quad \text{for } (x,t) \in \Sigma. \quad (3)$$

The operators in (3) are obtained by composition of the heat potentials in (2) with the Dirichlet trace operator. The ellipticity [2] and boundedness of the single-layer operator $V\colon H^{-1/2,-1/4}(\Sigma) \to H^{1/2,1/4}(\Sigma)$ together with the boundedness of the double-layer operator $K\colon H^{1/2,1/4}(\Sigma) \to H^{1/2,1/4}(\Sigma)$ and the initial Dirichlet operator $M_0\colon L^2(\Omega) \to H^{1/2,1/4}(\Sigma)$ ensure unique solvability of (3).

We consider a space-time tensor product decomposition of Σ [2, 10, 11] and use the Galerkin method for the discretization of (3). For a triangulation $\Gamma_h = \{\gamma_i\}_{i=1}^{N_\Gamma}$ of the boundary Γ and a decomposition $I_h = \{\tau_k\}_{k=1}^{N_I}$ of the time interval $I := (0,T)$ we define $\Sigma_h := \{\sigma = \gamma_i \times \tau_k : i = 1,\dots,N_\Gamma;\ k = 1,\dots,N_I\}$, i.e. $\Sigma_h = \{\sigma_\ell\}_{\ell=1}^{N}$ with $N = N_\Gamma N_I$. In the two-dimensional case the space-time boundary elements σ are rectangular. A sample decomposition of the space-time boundary of $Q = (0,1)^3$ is shown in Fig. 1a.

We use the space $X_h^{0,0}(\Sigma_h) := \operatorname{span}\{\varphi_\ell^0\}_{\ell=1}^{N}$ of piecewise constant basis functions and the space $X_h^{1,0}(\Sigma_h) := \operatorname{span}\{\varphi_i^{10}\}_{i=1}^{N}$ of functions that are piecewise linear and globally continuous in space and piecewise constant in time for the approximations of the Cauchy data w and g, respectively. The initial datum u_0 is discretized by using the space of piecewise linear and globally continuous functions $S_h^1(\Omega_h) := \operatorname{span}\{\varphi_j^1\}_{j=1}^{M_\Omega}$, which is defined with respect to a given triangulation $\Omega_h := \{\omega_i\}_{i=1}^{N_\Omega}$ of the domain Ω.

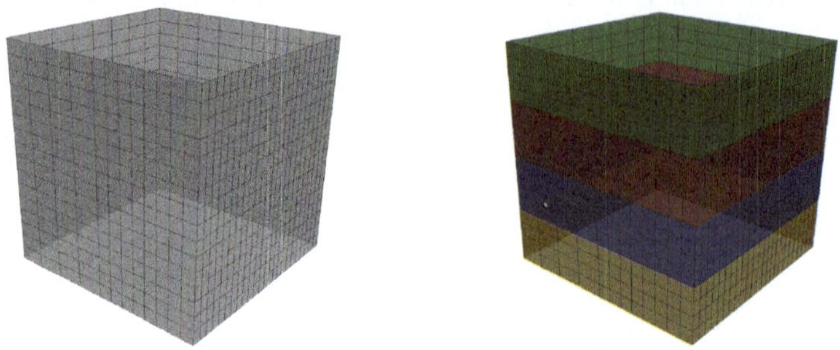

(a) Tensor product decomposition. (b) Submeshes.

Fig. 1: Sample space-time boundary decompositions for $Q = (0, 1)^3$ [5].

This leads to the system of linear equations

$$V_h \mathbf{w} = (\frac{1}{2} M_h + K_h)\mathbf{g} - M_h^0 \mathbf{u_0} \tag{4}$$

with

$$V_h[\ell, k] := \langle V\varphi_k^0, \varphi_\ell^0 \rangle_{L^2(\Sigma)}, \qquad K_h[\ell, i] := \langle K\varphi_i^{10}, \varphi_\ell^0 \rangle_{L^2(\Sigma)},$$
$$M_h^0[\ell, j] := \langle M_0\varphi_j^1, \varphi_\ell^0 \rangle_{L^2(\Sigma)}, \quad M_h[\ell, i] := \langle \varphi_i^{10}, \varphi_\ell^0 \rangle_{L^2(\Sigma)},$$

for $i, k, \ell = 1, \ldots, N$ and $j = 1, \ldots, M_\Omega$. Due to the ellipticity of the single-layer operator V the matrix V_h is positive definite and therefore (4) is uniquely solvable.

2 Operator Preconditioning

The boundary element discretization is done with respect to the whole space-time boundary Σ and since we want to solve (4) without an application of time-stepping schemes to make use of parallelization in time, we need to develop an efficient iterative solution technique. The linear system (4) with the positive definite but non-symmetric matrix V_h can be solved by a preconditioned GMRES method. Here we will apply a preconditioning technique based on boundary integral operators of opposite order [14], also known as operator or Calderon preconditioning [6].

First, we introduce the hypersingular operator D, which is defined as the negative Neumann trace of the double layer potential W in (2), i.e. $(Dv)(x,t) = -\partial_n(Wv)(x,t)$ for $(x,t) \in \Sigma$. The single-layer operator $V\colon H^{-1/2,-1/4}(\Sigma) \to H^{1/2,1/4}(\Sigma)$ and the hypersingular operator $D\colon H^{1/2,1/4}(\Sigma) \to H^{-1/2,-1/4}(\Sigma)$ are both elliptic [2] and the composition $DV\colon H^{-1/2,-1/4}(\Sigma) \to H^{-1/2,-1/4}(\Sigma)$ defines an operator of order zero. Thus, following [6], the Galerkin discretization of D allows the construction of a suitable preconditioner for V_h. While the discretization of the single-layer operator V is done with respect to $X_h^{0,0}(\Sigma_h)$, for the Galerkin discretization of the hypersingular operator D we need to use a conforming trial space $Y_h = \text{span}\,\{\psi_i\}_{i=1}^N \subset H^{1/2,1/4}(\Sigma)$, see also [4] for the spatially one-dimensional problem.

Theorem 1 ([6, 14]) *Assume that the discrete stability condition*

$$\sup_{0\neq v_h\in Y_h} \frac{\langle \tau_h, v_h\rangle_{L^2(\Sigma)}}{\|v_h\|_{H^{1/2,1/4}(\Sigma)}} \geq c_1^M \,\|\tau_h\|_{H^{-1/2,-1/4}(\Sigma)} \quad \textit{for all } \tau_h \in X_h^{0,0}(\Sigma_h) \tag{5}$$

holds. Then there exists a constant $c_\kappa > 1$ such that $\kappa\left(M_h^{-1}D_hM_h^{-\top}V_h\right) \leq c_\kappa$ where, for $k, \ell = 1, \ldots, N$,

$$D_h[\ell, k] = \langle D\psi_k, \psi_\ell\rangle_\Sigma\,, \quad M_h[\ell, k] = \langle \varphi_k^0, \psi_\ell\rangle_{L^2(\Sigma)}\,.$$

Thus we can use $C_V^{-1} = M_h^{-1}D_hM_h^{-\top}$ as a preconditioner for the matrix V_h. For the computation of the matrix D_h we use an alternative representation of the associated bilinear form which is attained by applying integration by parts, see [2, Theorem 6.1]. Note that the boundary element space Y_h is chosen to have the same dimension as $X_h^{0,0}(\Sigma_h)$ and thus, M_h is a square matrix. It remains to define a suitable boundary element space Y_h such that the mass matrix M_h is invertible and that the stability condition (5) is satisfied. In what follows we will discuss a possible choice.

We assume that the decompositions Γ_h and I_h are locally quasi-uniform. For the given boundary element mesh Γ_h we construct a dual mesh $\widetilde{\Gamma}_h := \{\tilde{\gamma}_\ell\}_\ell^{N_\Gamma}$ according to [7, 13] and assume, that $\widetilde{\Gamma}_h$ is locally quasi-uniform as well. For the discretization of the operator D we choose $Y_h = X_h^{1,0}(\widetilde{\Sigma}_h) \subset H^{1/2,1/4}(\Sigma)$, which denotes the space of functions that are piecewise linear and globally continuous in space and piecewise constant in time, defined with respect to the decomposition $\widetilde{\Gamma}_h$ and I_h, respectively. In order to prove the stability condition (5) we establish the $H^{1/2,1/4}(\Sigma)$-stability of the $L^2(\Sigma)$-projection $\widetilde{Q}_h^{1,0}\colon L^2(\Sigma) \to Y_h \subset L^2(\Sigma)$ defined by

$$\langle \widetilde{Q}_h^{1,0}v, \tau_h\rangle_{L^2(\Sigma)} = \langle v, \tau_h\rangle_{L^2(\Sigma)} \quad \text{for all } \tau_h \in X_h^{0,0}(\Sigma). \tag{6}$$

The Galerkin-Petrov variational problem (6) is uniquely solvable since the trial and test spaces satisfy a related stability condition [3]. When assuming appropriate local mesh conditions of Γ_h and $\widetilde{\Gamma}_h$, see [12, 13], we are able to establish the stability of $\widetilde{Q}_h^{1,0}\colon H^{1/2,1/4}(\Sigma) \to H^{1/2,1/4}(\Sigma)$, see [3] for a detailed discussion. Hence, there exists a constant $c_S > 0$ such that

$$\left\|\widetilde{Q}_h^{1,0}v\right\|_{H^{1/2,1/4}(\Sigma)} \leq c_S\,\|v\|_{H^{1/2,1/4}(\Sigma)} \quad \text{for all } v \in H^{1/2,1/4}(\Sigma). \tag{7}$$

The stability estimate (7) immediately implies the stability condition (5). Hence the condition number $\kappa(C_V^{-1}V_h)$ with $C_V^{-1} = M_h^{-1}D_hM_h^{-\top}$ is bounded.

3 Distributed Memory Parallelization

Distributed memory parallelization of the solver is based on the scheme presented in [8, 9] for spatial problems. In [5] we have extended the approach to support time-dependent problems for the heat equation. Let us briefly describe the method and refer the more interested readers to the above-mentioned papers.

To distribute the system among P processes the space-time mesh Σ is decomposed into P slices in the temporal dimension (see Fig. 1b) which splits the matrices $A \in \{V_h, K_h, D_h\}$ into $P \times P$ blocks

$$A = \begin{bmatrix} A_{0,0} & 0 & \cdots & 0 \\ A_{1,0} & A_{1,1} & \cdots & 0 \\ \vdots & \vdots & \ddots & \vdots \\ A_{P-1,0} & A_{P-1,1} & \cdots & A_{P-1,P-1} \end{bmatrix}.$$

The matrices are block lower triangular with lower triangular blocks on the main diagonal due to the properties of the fundamental solution and the selected discrete spaces. We aim to distribute the blocks among processes such that the number of shared mesh parts is minimal and each process owns a single diagonal block. For this purpose we consider each block $A_{i,j}$ as an edge (i, j) of a complete graph K_P on P vertices. The distribution problem corresponds to finding a suitable decomposition of K_P into P subgraphs $G_0, G_1, \ldots, G_{P-1}$. In [5, 8] we employ a cyclic decomposition algorithm – first, a generator graph G_0 on a minimal number of vertices (corresponding to blocks to be assembled by the process 0) is constructed; the remaining graphs $G_1, \ldots, G_{P-1}$ are obtained by a clock-wise rotation of G_0 along vertices of K_P placed on a circle. An example of the generating graph and the corresponding matrix decomposition for four processes is depicted in Fig. 2. In the case of the initial matrix M_h^0 we distribute block-rows of the matrix among processes. Similarly, since the matrix M_h is block diagonal, each process owns exactly one block of the matrix.

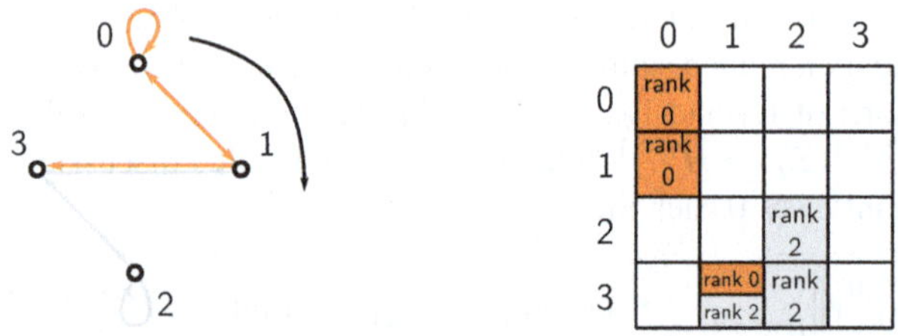

(a) Generating graph. (b) Block distribution.

Fig. 2: Distribution of the system matrix blocks among four processes [5].

In addition to the distributed memory parallelization by MPI, the assembly of the matrices is parallelized and vectorized in shared memory using OpenMP [5, 15]. Therefore, in our numerical experiments we usually employ hybrid parallelization using one MPI process per CPU socket and an appropriate number of OpenMP threads per process.

4 Numerical Experiments

The presented examples refer to the initial Dirichlet boundary value problem (1) in the space-time domain $Q := (0,1)^3$. The heat capacity constant is set to $\alpha = 10$. We consider the exact solution

$$u(x,t) := \exp\left(-\frac{t}{\alpha}\right) \sin\left(x_1 \cos\frac{\pi}{8} + x_2 \sin\frac{\pi}{8}\right) \quad \text{for } (x,t) = (x_1, x_2, t) \in Q$$

and determine the Dirichlet datum g and the initial datum u_0 accordingly. The linear system (4) is solved by the GMRES method with a relative precision of 10^{-8}.

Operator Preconditioning

As a preconditioner we use the discretization $C_V^{-1} = M_h^{-1} D_h M_h^{-\top}$ of the hypersingular operator D in the space $X_h^{1,0}(\widetilde{\Sigma}_h)$, while the Galerkin discretization of the integral equation (3) is done with respect to $X_h^{0,0}(\Sigma_h)$. Instead of using M_h in the preconditioner we computed a lumped mass matrix. Thus, the matrix becomes diagonal and the inverse can be applied efficiently.

The example corresponds to a globally uniform boundary element mesh with the mesh size $h = \mathcal{O}(2^{-L})$. Table 1 shows the iteration numbers of the non-preconditioned and preconditioned GMRES method. As expected, the iteration numbers of the preconditioned version are bounded due to the boundedness of $\kappa(C_V^{-1} V_h)$. For numerical results in the case of an adaptive refinement we refer to [3].

Table 1: Iteration numbers of the non-preconditioned GMRES method (It.) and the preconditioned GMRES method (It. prec.) in the case of uniform refinement. N denotes the number of boundary elements on level L.

L	N	It.	It. prec.
2	64	14	17
3	256	19	18
4	1 024	24	20
5	4 096	35	20
6	16 384	50	20
7	65 536	67	20
8	262 144	91	19
9	1 048 576	122	19

Table 2: Assembly of D_h for 65 536, 262 144, and 1 048 576 space-time elements.

nodes ↓	D_h assembly [s]			D_h speedup			D_h efficiency [%]		
mesh →	65k	262k	1M	65k	262k	1M	65k	262k	1M
1	184.1	—	—	1.0	—	—	100.0	—	—
2	92.0	—	—	2.0	—	—	100.1	—	—
4	46.8	—	—	3.9	—	—	98.4	—	—
8	23.8	373.6	—	7.7	1.0	—	96.7	100.0	—
16	11.8	186.1	—	15.6	2.0	—	97.3	100.4	—
32	5.9	91.9	—	31.0	4.1	—	96.7	101.7	—
64	3.0	47.0	747.0	60.5	7.9	1.0	94.6	99.3	100.0
128	—	24.0	376.9	—	15.6	2.0	—	97.3	99.1
256	—	—	193.5	—	—	3.9	—	—	96.5

Scalability in Distributed Memory

The numerical experiments for the scalability were executed on the Salomon cluster at IT4Innovations National Supercomputing Center in Ostrava, Czech Republic. The cluster is equipped with 1008 nodes with two 12-core Intel Xeon E5-2680v3 Haswell processors and 128 GB of RAM. Nodes of the cluster are interconnected by the InfiniBand 7D enhanced hypercube network.

We tested the assembly of the BEM matrix D_h. Computation times for the assembly of the matrices V_h, K_h, M_h^0, the related matrix-vector multiplication, and the evaluation of the solution in Q can be found in [5]. Strong scaling of the parallel solver was tested using a tensor product decomposition of Σ into 65 536, 262 144 and 1 048 576 space-time surface elements. We used up to 256 nodes (6 144 cores) of the Salomon cluster for the computations and executed two MPI processes per node. Each MPI process used 12 cores for the assembly of the matrix blocks.

In Table 2 the assembly times for D_h including the speedup and efficiency are listed. We obtain almost optimal parallel scalability. Note that the number of nodes is restricted by the number of elements of the temporal decomposition I_h. Conversely, for fine meshes we need a certain number of nodes to store the matrices.

5 Conclusion

In this note we have described a parallel space-time boundary element solver for the two-dimensional heat equation. The solver is parallelized using MPI in the distributed memory. The distribution of the system matrices is based on [5, 8, 9]. The space-time boundary mesh is decomposed into time slices which define blocks in the system matrices. These blocks are distributed among MPI processes using the graph decomposition based scheme. For a detailed discussion on shared memory parallelization see [5]. Moreover, we have introduced an efficient preconditioning strategy for the space-time system which is based on the use of boundary integral operators of opposite order. The preconditioner was then distributed with the presented parallelization technique.

The numerical experiments for the proposed preconditioning strategy confirm the theoretical findings, i.e. the boundedness of the iteration number of the iterative solver. We also tested the efficiency of the parallelization scheme for the preconditioner. The results show almost optimal scalability.

Acknowledgements The research was supported by the project 'Efficient parallel implementation of boundary element methods' provided jointly by the Ministry of Education, Youth and Sports (7AMB17AT028) and OeAD (CZ 16/2017). SD acknowledges the support provided by the International Research Training Group 1754, funded by the German Research Foundation (DFG) and the Austrian Science Fund (FWF). JZ and MM further acknowledge the support provided by The Ministry of Education, Youth and Sports from the National Programme of Sustainability (NPS II) project 'IT4Innovations excellence in science – LQ1602' and the Large Infrastructures for Research, Experimental Development and Innovations project 'IT4Innovations National Supercomputing Center – LM2015070'.

References

1. Arnold, D.N., Noon, P.J.: Boundary integral equations of the first kind for the heat equation. In: Boundary elements IX, Vol. 3 (Stuttgart, 1987), pp. 213–229. Comput. Mech., Southampton (1987)
2. Costabel, M.: Boundary integral operators for the heat equation. Integral Equations and Operator Theory **13**, 498–552 (1990). DOI:10.1007/BF01210400
3. Dohr, S., Niino, K., Steinbach, O.: Preconditioned space–time boundary element methods for the heat equation. In preparation
4. Dohr, S., Steinbach, O.: Preconditioned space-time boundary element methods for the one-dimensional heat equation. In: Domain Decomposition Methods in Science and Engineering XXIV., *Lect. Notes Comput. Sci. Eng.*, vol. 125, pp. 243–251. Springer, Cham (2018). DOI:10.1007/978-3-319-93873-8
5. Dohr, S., Zapletal, J., Of, G., Merta, M., Kravcenko, M.: A parallel space–time boundary element method for the heat equation. Comput. Math. Appl. **78**(9), 2852–2866 (2019). DOI:10.1016/j.camwa.2018.12.031
6. Hiptmair, R.: Operator Preconditioning. Comput. Math. Appl. **52**, 699–706 (2006)
7. Jerez-Hanckes, C., Hiptmair, R., Urzua, C.: Mesh-Independent Operator Preconditioning for Boundary Elements on Open Curves. SIAM J. Numer. Anal. **52**, 2295–2314 (2014)
8. Kravcenko, M., Merta, M., Zapletal, J.: Distributed fast boundary element methods for Helmholtz problems. Appl. Math. Comput. **362**, 1–15 (2019). DOI:10.1016/j.amc.2019.06.017
9. Lukas, D., Kovar, P., Kovarova, T., Merta, M.: A parallel fast boundary element method using cyclic graph decompositions. Numer. Algorithms **70**(4), 807–824 (2015). DOI:10.1007/s11075-015-9974-9
10. Messner, M., Schanz, M., Tausch, J.: A fast Galerkin method for parabolic space–time boundary integral equations. J. Comput. Phys. **258**, 15–30 (2014). DOI:10.1016/j.jcp.2013.10.029
11. Noon, P.: The Single Layer Heat Potential and Galerkin Boundary Element Methods for the Heat Equation. Thesis, University of Maryland (1988)
12. Steinbach, O.: On the stability of the L_2 projection in fractional Sobolev spaces. Numer. Math. **88**, 367–379 (2001)

13. Steinbach, O.: On a generalized L_2 projection and some related stability estimates in Sobolev spaces. Numer. Math. **90**, 775–786 (2002)
14. Steinbach, O., Wendland, W.: The construction of some efficient preconditioners in the boundary element method. Adv. Comput. Math. **9**, 191–216 (1998)
15. Zapletal, J., Merta, M., Maly, L.: Boundary element quadrature schemes for multi- and many-core architectures. Comput. Math. Appl. **74**(1), 157–173 (2017). DOI:10.1016/j.camwa.2017.01.018

On Inexact Solvers for the Coarse Problem of BDDC

Clark R. Dohrmann, Kendall H. Pierson, and Olof B. Widlund

1 Introduction

In this study, we present Balancing Domain Decomposition by Constraints (BDDC) preconditioners for three-dimensional scalar elliptic and linear elasticity problems in which the direct solution of the coarse problem is replaced by a preconditioner based on a smaller vertex-based coarse space. By doing so, the computational and memory requirements can be reduced significantly. Although the use of standard coarse spaces based on subdomain vertices (corners) alone has similar memory benefits, the associated rate of convergence is not attractive as the number of elements per subdomain grows [10]. This point is illustrated by a simple motivating example in the next section.

There exists a rich theory for Finite Element Tearing and Interconnecting Dual Primal (FETI-DP) and BDDC algorithms for scalar elliptic and linear elasticity problems in three dimensions (see, e.g., [10], [9] or §6.4.2 of [15]). In many cases, theoretical results for either FETI-DP or BDDC apply directly to the other because of the equivalence of eigenvalues of the preconditioned operators [13, 11, 1]. This equivalence does not hold in the present study because the basic FETI-DP algorithm [6] is not easily adapted to use a preconditioner instead of a direct solver for the symmetric and positive definite coarse problem. In contrast, such a change is accommodated easily by BDDC in both theory and practice [4]. Nevertheless, we expect that our approach could find use in the irFETI-DP algorithm described in [8].

The approach to preconditioning the BDDC coarse problem is motivated in part by more recent developments of small coarse spaces for domain decomposition

Clark R. Dohrmann
Sandia National Laboratories, Albuquerque, New Mexico, U.S.A., e-mail: crdohrm@sandia.gov

Kendall H. Pierson
Sandia National Laboratories, Albuquerque, New Mexico, U.S.A., e-mail: khpiers@sandia.gov

Olof B. Widlund
Courant Institute, New York, New York, U.S.A., e-mail: widlund@cims.nyu.edu

R. Haynes et al. (eds.), *Domain Decomposition Methods in Science and Engineering XXV*,
Lecture Notes in Computational Science and Engineering 138,
https://doi.org/10.1007/978-3-030-56750-7_12

algorithms [5]. Although that study was focused on overlapping Schwarz methods, similar ideas can be used to construct coarse spaces for preconditioning the BDDC coarse problem. Compared with larger edge-based or face-based coarse spaces, we find that similar condition number bounds can be achieved at much lower cost under certain assumptions on material property jumps between adjacent subdomains.

We note that three-level and multi-level BDDC algorithms [17, 16, 14] can also be viewed as using an inexact solver for the coarse problem, but such approaches are fundamentally different from ours. Namely, these algorithms construct and apply (recursively for multi-level approaches) a BDDC preconditioner for the original two-level coarse problem. In contrast, we do not introduce additional coarse levels and make use of standard two-level additive Schwarz concepts for preconditioning the coarse problem. One important result of using smaller coarse spaces is that larger numbers of subdomains are feasible before needing to use a three- or multi-level approach. Consequently, the number of coarse levels can potentially be reduced and result in fewer synchronization points for parallel implementations. We also note that approximate solvers of the coarse problem were introduced in [8] as in the context of a saddle-point formulation for FETI-DP.

Reducing the size of the coarse problem while retaining favorable convergence rates was also the subject of Algorithm D in [10]. The basic idea there was to use a coarse space based on a subset of subdomain edges and corners (vertices) rather than all of them. The authors note that their recipe for selecting such edges and corners is relatively complicated, but it can effectively reduce the coarse problem dimension. In contrast to their approach, the present one uses all subdomain edges, but replaces the direct solver for the coarse problem with a preconditioner.

A motivating example is presented in the next section for the proposed approach which is summarized in §3. Theoretical results for scalar elliptic and linear elasticity problems are presented in §4. Complete proofs are provided in the article, [2], that has appeared since this paper was submitted; it also contains implementation details, extensions to face-based coarse spaces, and additional numerical examples. The final section of this paper contains numerical results, which confirm the theory and demonstrate the computational advantages of our approach.

2 Motivation

To help motivate the proposed approach, consider a unit cube domain partitioned into 27 smaller cubic subdomains. Each of these subdomains is discretized using H/h lowest order hexahedral elements in each coordinate direction for the Poisson equation with constant material properties. Homogeneous essential boundary conditions are applied to one side of the domain and a random load vector b is used for the right-hand side of the linear system $Ax = b$. We note that our algorithm iterates on the interface problem $Su = g$ after eliminating residuals in subdomain interiors (initial static condensation step). Here, S is the Schur complement matrix for the interface problem.

We first consider coarse spaces based on subdomain vertices alone or edges alone. Table 1 shows the condition number estimates for the preconditioned operator along with the number of iterations needed to achieve a relative residual tolerance of 10^{-8} using the conjugate gradient algorithm preconditioned using BDDC. The fast growth of condition numbers in the third column is consistent with a condition number bound proportional to $(H/h)(1+\log(H/h))^2$ as given in Remark 2 of [10]. The shortcomings of using coarse spaces based on vertices alone were recognized early in the history of FETI-DP [7]. Notice the results for the proposed approach show significant improvements in comparison to the standard vertex (corner) based coarse space.

Table 1: Poisson equation results. Number of iterations (iter) and condition number estimates (cond) are shown for a unit cube domain constrained on one side and decomposed into 27 smaller cubic subdomains. In this table and others, H/h denotes the number of elements in each coordinate direction for each subdomain. More generally, H/h refers to the maximum ratio of subdomain diameter H_i to smallest element diameter h_i for any subdomain Ω_i.

	standard approach				proposed approach	
	vertices		edges			
H/h	iter	cond	iter	cond	iter	cond
4	28	27.1	12	2.36	14	2.50
8	38	75.2	14	2.93	16	3.13
12	45	132	16	3.37	18	3.59
16	47	195	17	3.73	19	3.97

Results are shown in Table 2 for increasing numbers of subdomains N and fixed $H/h = 8$. Notice the dimensions n_c of the coarse space for edge-based coarse spaces are significantly larger than those for the proposed approach. Again, the advantages of the new approach are evident in the final three columns of the table where the number of iterations and condition numbers are much smaller than those for the standard vertex-based coarse space.

Table 2: Poisson equation results. Coarse space dimension n_c and convergence results are shown for increasing numbers of subdomains N and fixed $H/h = 8$.

	standard approach						proposed approach		
	vertices			edges					
N	n_c	iter	cond	n_c	iter	cond	n_c	iter	cond
64	27	55	74.5	108	15	2.98	27	17	3.25
216	125	70	73.7	450	15	2.94	125	17	3.26
512	343	74	73.6	1176	15	2.95	343	17	3.30
1000	729	75	73.6	2430	15	2.95	729	17	3.32

A primary goal of this study is to present an approach that combines the best of both worlds. That is, an approach that has the attractive convergence rates of edge-based coarse spaces and the more streamlined computational requirements of a smaller vertex-based coarse space.

3 Overview of BDDC and Our Inexact Approach

The domain Ω for the problem is assumed to be partitioned into nonoverlapping subdomains $\Omega_1, \ldots, \Omega_N$. The set of interface points that are common to two or more subdomain boundaries is denoted by Γ, and the set of interface points for Ω_i is denoted by $\Gamma_i := \Gamma \cap \partial\Omega_i$. Finite element nodes on Γ_i are partitioned into different equivalence classes such as those of subdomain vertices, edges, or faces depending on which subdomain boundaries contain them (see, e.g., [3] or [5] for more details).

A two-level BDDC preconditioner (see, e.g., [3] can be expressed concisely in additive form as

$$M^{-1} = M_{local}^{-1} + \Phi_D K_c^{-1} \Phi_D^T, \tag{1}$$

where K_c is the coarse matrix and Φ_D is a weighted interpolation matrix. We note that the application of the local component M_{local}^{-1} requires solutions of problems local to each subdomain, which can be done in parallel.

The coarse matrix is obtained from the assembly of coarse subdomain matrices and given by

$$K_c = \sum_{i=1}^{N} R_{ic}^T K_{ic} R_{ic},$$

where K_{ic} is the coarse matrix for Ω_i and $u_{ic} = R_{ic} u_c$ is the restriction of a coarse vector u_c to Ω_i. Let M_c^{-1} denote a preconditioner for K_c which satisfies the bounds

$$\beta_1 u_c^T K_c^{-1} u_c \le u_c^T M_c^{-1} u_c \le \beta_2 u_c^T K_c^{-1} u_c \quad \forall u_c, \tag{2}$$

where $0 < \beta_1 \le \beta_2$. Defining the approximate BDDC preconditioner M_a^{-1} as

$$M_a^{-1} := M_{local}^{-1} + \Phi_D M_c^{-1} \Phi_D^T,$$

we find from (1) and (2) that

$$\begin{aligned} p^T M_a^{-1} p &= p^T (M^{-1} + \Phi_D (M_c^{-1} - K_c^{-1}) \Phi_D^T) p \\ &\le p^T (M^{-1} + (\beta_2 - 1) \Phi_D K_c^{-1} \Phi_D^T) p \\ &\le \max(1, \beta_2) p^T M^{-1} p. \end{aligned} \tag{3}$$

Similarly,

$$p^T M_a^{-1} p \ge \min(1, \beta_1) p^T M^{-1} p. \tag{4}$$

Let κ denote the condition number of the original BDDC preconditioned operator. It then follows from (3) and (4) that

$$\kappa_a \leq \frac{\max(1, \beta_2)}{\min(1, \beta_1)} \kappa, \tag{5}$$

where κ_a is the condition number of the approximate BDDC preconditioned operator. Here we only consider preconditioners for the coarse matrix K_c, but approximations for other components of the BDDC preconditioner have also been studied [12, 4].

The construction of the preconditioner M_c^{-1} for K_c was inspired in part by our recent work on small coarse spaces [5]. What we have called vertices thus far here are generalized there and called coarse nodes. We recall that the coarse degrees of freedom for BDDC or FETI-DP are often associated with average values over the different equivalence classes. The basic idea of the coarse component of the preconditioner M_c^{-1} is to approximate these averages using adjacent vertex values.

Using the notation of [5], let $C_{\mathcal{N}}$ denote the set of ancestor vertices for a nodal equivalence class $\mathcal{N}$ (e.g. $\mathcal{N}$ may be the nodes of a subdomain edge or face). Let u_Ψ denote a vector of vertex values. We introduce the coarse interpolation $u_{c0} = \Psi u_\Psi$ between vertex values and nodal equivalence class averages such that each of these averages equals the average of its ancestor vertex values. Thus, a row of Ψ associated with an edge of the center subdomain in the motivating example has two entries of 1/2 (one entry for each vertex at its ends), while all other entries are 0. Notice that the number of rows in Ψ is the number of active coarse degrees of freedom for the original BDDC preconditioner. For instance, if only edges are used this number equals the total number of subdomain edges.

The reduced coarse matrix is defined as $K_{cr} := \Psi^T K_c \Psi$. The number of rows and columns in K_{cr} is the number of vertices for scalar problems. We consider the following preconditioner for K_c.

$$M_c^{-1} = \Psi K_{cr}^{-1} \Psi^T + \operatorname{diag}(K_c)^{-1}, \tag{6}$$

where diag denotes the diagonal of the matrix (for elasticity problems the second term on the right hand side of (6) is block diagonal). Notice M_c^{-1} is simply a Jacobi preconditioner with an additive coarse correction. Thus, since the number of subdomains incident to an edge is bounded, a uniform upper bound on β_2 for M_c^{-1} can be obtained using a standard coloring argument. Therefore, the analysis focuses on obtaining lower bound estimates for β_1. We comment that higher quality local preconditioning can be used (e.g., replacing Jacobi smoothing by symmetric Gauss-Seidel). Indeed, the numerical results in §5 were obtained using such an approach.

4 Main Results

We presently restrict our attention to edge-based BDDC coarse spaces for both scalar elliptic and linear elasticity problems (cf. §4 and §5 of [5] for problem specifications). For the scalar case, we assume quasi-monotone edge-connected paths as in Assumptions 4.5 of [5]. For elasticity problems, we must make the stronger assumption of quasi-monotone face-connected paths as in Assumption 4.4 of [5]. We also assume that material properties are constant within each subdomain and that the ratio $(H_j/h_j)/(H_k/h_k)$ is uniformly bounded for any two subdomains Ω_j and Ω_k sharing any subdomain vertex.

Theorem 1 *For edge-based BDDC coarse spaces and with quasi-monotone edge-connected paths, the condition number of the preconditioned operator that is obtained by replacing the direct solver for the coarse problem by the preconditioner M_c^{-1} defined in (6) is bounded by*

$$\kappa_a \leq C(1 + \log(H/h))^2$$

for scalar elliptic problems.

Theorem 2 *For edge-based BDDC coarse spaces and with quasi-monotone face-connected paths, the condition number of the preconditioned operator that is obtained by replacing the direct solver for the coarse problem by the preconditioner M_c^{-1} defined in (6) is bounded by*

$$\kappa_a \leq C(1 + \log(H/h))^2$$

for compressible linear elasticity problems.

The proofs of these theorems use classical additive Schwarz theory, an estimate in Lemma 4.2 of [17], and a variety of standard domain decomposition estimates. Further, the analysis for linear elasticity relies on Korn inequalities and on rigid body fits of subdomain face deformations (cf. [9] for a related approach).

5 Numerical Results

The results in Tables 1 and 2 are in good agreement with the theory for the scalar case, and demonstrate that comparable performance to the standard edge-based BDDC preconditioner can be obtained more efficiently. Notice in Table 2 that the coarse space dimension n_c is approximately 3 times smaller for the proposed approach than that of the standard edge-based approach for larger numbers of subdomains. Similar results were obtained for linear elasticity (not shown), but the reductions in coarse space dimension were more modest.

The next example deals with a cubic domain decomposed into 64 smaller cubic subdomains and constrained on its left side. Three different distributions of material properties are considered as shown in Figure 1. The leftmost one has quasi-monotone face-connected paths, the middle one has quasi-monotone edge-connected paths, and the rightmost one has a checkerboard arrangement which is not covered by our theory.

The material properties in the lighter colored regions are given by $\rho = 1$ for the scalar case and $E = 1, \nu = 0.3$ for elasticity. Likewise, the other regions have $\rho = 10^3$,

$E = 10^3$, and $\nu = 0.3$. Results for the scalar case and elasticity are shown in Table 3. Consistent with the theory, condition numbers for the scalar case grow sublinearly with respect to H/h for both face-connected and edge-connected paths. As expected, similar growth in condition numbers is observed for linear elasticity in the case of face-connected paths. Recall that the case of edge-connected paths is not covered by our theory for elasticity, and much larger condition numbers are apparent in the table. Remarkably, very good results are obtained for the checkerboard arrangement of material properties for both the scalar case and linear elasticity.

Fig. 1: Material property distributions for a cube decomposed into 64 smaller cubic subdomains. The leftmost figure has quasi-monotone face-connected paths while the middle one only has quasi-monotone edge-connected paths. The rightmost figure shows a checkerboard arrangement of material properties.

Table 3: Results for the models in Figure 1.

scalar case						
	face-connected		edge-connected		checkerboard	
H/h	iter	cond	iter	cond	iter	cond
4	14	2.41	16	3.58	9	1.45
8	16	2.95	20	4.81	11	1.71
12	18	3.40	22	5.65	12	1.99
16	19	3.75	24	6.32	13	2.19
linear elasticity						
H/h	face-connected		edge-connected		checkerboard	
	iter	cond	iter	cond	iter	cond
4	25	6.10	40	72.9	24	6.55
8	33	11.1	53	113	31	11.1
12	38	14.8	61	137	35	14.4
16	42	17.8	68	154	38	16.9

Additional numerical results have been generated for face-based rather than edge-based coarse spaces, for unstructured meshes, and performance tests are given which show reduced compute times. They are reported in the article, [2], which has appeared since this conference paper was submitted.

In closing, we expect that the approach presented here could be combined with an adaptive coarse space to handle problems where material properties vary greatly within a subdomain. The basic idea would be to use existing adaptive approaches for challenging subdomains, while using the present approach for less problematic ones.

References

1. Brenner, S.C., Sung, L.Y.: BDDC and FETI-DP without matrices or vectors. Comput. Methods Appl. Mech. Engrg. **196**(8), 1429–1435 (2007). DOI:10.1016/j.cma.2006.03.012
2. Dohrmann, C., Pierson, K., Widlund, O.: Vertex-based preconditioners for the coarse problem of BDDC. SIAM J. Sci. Comput. **41**(5), A3021–A3044 (2019). DOI:10.1137/19M1237557
3. Dohrmann, C.R.: A preconditioner for substructuring based on constrained energy minimization. SIAM J. Sci. Comput. **25**(1), 246–258 (2003). DOI:10.1137/S1064827502412887
4. Dohrmann, C.R.: An approximate BDDC preconditioner. Numer. Linear Algebra Appl. **14**(2), 149–168 (2007). DOI:10.1002/nla.514
5. Dohrmann, C.R., Widlund, O.B.: On the design of small coarse spaces for domain decomposition algorithms. SIAM J. Sci. Comput. **39**(4), A1466–A1488 (2017). DOI:10.1137/17M1114272
6. Farhat, C., Lesoinne, M., LeTallec, P., Pierson, K., Rixen, D.: FETI-DP: a dual-primal unified FETI method. I. A faster alternative to the two-level FETI method. Internat. J. Numer. Methods Engrg. **50**(7), 1523–1544 (2001). DOI:10.1002/nme.76
7. Farhat, C., Lesoinne, M., Pierson, K.: A scalable dual-primal domain decomposition method. Numer. Linear Algebra Appl. **7**(7-8), 687–714 (2000). DOI:10.1002/1099-1506(200010/12)7:7/8<687::AID-NLA219>3.0.CO;2-S. Preconditioning techniques for large sparse matrix problems in industrial applications (Minneapolis, MN, 1999)
8. Klawonn, A., Rheinbach, O.: Inexact FETI-DP methods. Internat. J. Numer. Methods Engrg. **69**(2), 284–307 (2007). DOI:10.1002/nme.1758
9. Klawonn, A., Widlund, O.B.: Dual-primal FETI methods for linear elasticity. Comm. Pure Appl. Math. **59**(11), 1523–1572 (2006). DOI:10.1002/cpa.20156
10. Klawonn, A., Widlund, O.B., Dryja, M.: Dual-primal FETI methods for three-dimensional elliptic problems with heterogeneous coefficients. SIAM J. Numer. Anal. **40**(1), 159–179 (2002). DOI:10.1137/S0036142901388081
11. Li, J., Widlund, O.B.: FETI-DP, BDDC, and block Cholesky methods. Internat. J. Numer. Methods Engrg. **66**(2), 250–271 (2006). DOI:10.1002/nme.1553
12. Li, J., Widlund, O.B.: On the use of inexact subdomain solvers for BDDC algorithms. Comput. Methods Appl. Mech. Engrg. **196**(8), 1415–1428 (2007). DOI:10.1016/j.cma.2006.03.011
13. Mandel, J., Dohrmann, C.R., Tezaur, R.: An algebraic theory for primal and dual substructuring methods by constraints. Appl. Numer. Math. **54**(2), 167–193 (2005). DOI:10.1016/j.apnum.2004.09.022
14. Mandel, J., Sousedík, B., Dohrmann, C.R.: Multispace and multilevel BDDC. Computing **83**(2-3), 55–85 (2008). DOI:10.1007/s00607-008-0014-7
15. Toselli, A., Widlund, O.: Domain decomposition methods—algorithms and theory, *Springer Series in Computational Mathematics*, vol. 34. Springer-Verlag, Berlin (2005). DOI:10.1007/b137868
16. Tu, X.: Three-level BDDC in three dimensions. SIAM J. Sci. Comput. **29**(4), 1759–1780 (2007). DOI:10.1137/050629902
17. Tu, X.: Three-level BDDC in two dimensions. Internat. J. Numer. Methods Engrg. **69**(1), 33–59 (2007). DOI:10.1002/nme.1753

Simultaneous Approximation Terms for Elastic Wave Equations on Nonuniform Grids

Longfei Gao and David Keyes

1 Introduction

Numerical simulation of wave phenomena is routinely used in seismic studies, where simulated wave signals are compared against experimental ones to infer subterranean information. Various wave systems can be used to model wave propagation in earth media. Here, we consider the system of isotropic elastic wave equations described in Section 2. Various numerical methods can be applied to discretize such a system, among which the finite difference methods (FDMs) are still very popular, particularly for seismic exploration applications, due to their simplicity and efficiency.

However, when discretized on uniform grids, heterogeneity of the earth media will lead to oversampling in both space and time, undermining the efficiency of FDMs. Specifically, since spatial grid spacing is usually decided on a point-per-wavelength basis for wave simulations, uniform grid discretization will lead to oversampling in space for regions with higher wave-speeds. On the other hand, temporal step length is usually restricted by the Courant-Friedrichs-Lewy (CFL) stability condition for wave simulations using explicit time stepping methods, which will lead to oversampling in time for regions with lower wave-speeds.

For earth media, the wave-speeds tend to increase with depth due to sedimentation and consolidation. Contrast between the smallest and largest wave-speeds in earth media can be as high as fifty, cf. [1, p. 240], which entails significant oversampling for discretizations on uniform grids. These observations motivate us to consider the grid configuration illustrated in Figure 1, where two uniform grid regions are separated by a horizontal interface. The staggered grid discretization approach, which dates back to [12], is considered here, where different solution variables are discretized on different subgrids. In Figure 1, ratio of the grid spacings of the two regions is two. However, other ratios, not necessarily integers, can also be addressed with the

Longfei Gao & David Keyes
Division of Computer, Electrical and Mathematical Sciences and Engineering,
King Abdullah University of Science and Technology, Thuwal 23955-6900, Saudi Arabia
e-mail: longfei.gao@kaust.edu.sa & david.keyes@kaust.edu.sa

R. Haynes et al. (eds.), *Domain Decomposition Methods in Science and Engineering XXV*,
Lecture Notes in Computational Science and Engineering 138,
https://doi.org/10.1007/978-3-030-56750-7_13

methodology presented here. Furthermore, multiple grid layers can be combined together in a cascading manner to account for larger wave-speed contrasts.

In this work, we recap one of the earliest motivations of domain decomposition methods by demonstrating how to combine the two regions illustrated in Figure 1 without numerical artifacts. Specifically, we adopt the summation by parts (SBP) - simultaneous approximation terms (SATs) approach, which utilizes discrete energy analysis to guide the discretization. The overall semi-discretization is shown to be discretely energy conserving, preserving the analogous property in the continuous elastic wave system. The concept of SBP operators dates back to [7] while the technique of SATs was introduced in [2]. The two review papers [11, 3] provide comprehensive coverage of their developments. While the 2D elastic wave system is considered here to demonstrate the methodology, we expect the presented procedure to extend straightforwardly to the 3D case.

In the following, we describe the abstracted mathematical problem in Section 2, present the interface treatment in Section 3, provide numerical examples in Section 4, and summarize in Section 5.

2 Problem Description

We consider the 2D isotropic elastic wave equations posed as the following first-order dynamical system written in terms of velocity and stress:

$$\left\{\begin{aligned}
\frac{\partial v_x}{\partial t} &= \frac{1}{\rho}\left(\frac{\partial \sigma_{xx}}{\partial x} + \frac{\partial \sigma_{xy}}{\partial y}\right);\\
\frac{\partial v_y}{\partial t} &= \frac{1}{\rho}\left(\frac{\partial \sigma_{xy}}{\partial x} + \frac{\partial \sigma_{yy}}{\partial y}\right);\\
\frac{\partial \sigma_{xx}}{\partial t} &= (\lambda + 2\mu)\frac{\partial v_x}{\partial x} + \lambda\frac{\partial v_y}{\partial y} + S;\\
\frac{\partial \sigma_{xy}}{\partial t} &= \mu\frac{\partial v_y}{\partial x} + \mu\frac{\partial v_x}{\partial y};\\
\frac{\partial \sigma_{yy}}{\partial t} &= \lambda\frac{\partial v_x}{\partial x} + (\lambda + 2\mu)\frac{\partial v_y}{\partial y} + S,
\end{aligned}\right. \tag{1}$$

where v_x and v_y are particle velocities; σ_{xx}, σ_{xy} and σ_{yy} are stress components; ρ, λ and μ are density, first and second Lamé parameters that characterize the medium; S is the source term that drives the wave propagation. Lamé parameters λ and μ are related with the compressional and shear wave-speeds c_p and c_s through $\lambda = \rho(c_p^2 - 2c_s^2)$ and $\mu = \rho c_s^2$. For simplicity, the source term S is omitted in the upcoming discussion. All solution variables and their derivatives are assumed to be zero at the initial time. We consider periodic boundary condition for left and right boundaries and free-surface boundary condition for top and bottom boundaries.

The above system is equivalent to system (2), which is more natural for energy analysis and derivation of the interface treatment. In (2), the Einstein summation convention applies to subscript indices k and l. Coefficients s_{xxkl}, s_{xykl} and s_{yykl} are components of the compliance tensor, which can be expressed in terms of λ and μ. However, their exact expressions are not needed for the upcoming discussion. As explained later in Section 3, system (1) is still the one used for implementation.

$$\begin{cases} \rho \dfrac{\partial v_x}{\partial t} = \dfrac{\partial \sigma_{xx}}{\partial x} + \dfrac{\partial \sigma_{xy}}{\partial y}; \\ \rho \dfrac{\partial v_y}{\partial t} = \dfrac{\partial \sigma_{xy}}{\partial x} + \dfrac{\partial \sigma_{yy}}{\partial y}; \\ s_{xxkl} \dfrac{\partial \sigma_{kl}}{\partial t} = \dfrac{\partial v_x}{\partial x}; \\ s_{xykl} \dfrac{\partial \sigma_{kl}}{\partial t} = \dfrac{1}{2}\left(\dfrac{\partial v_y}{\partial x} + \dfrac{\partial v_x}{\partial y} \right); \\ s_{yykl} \dfrac{\partial \sigma_{kl}}{\partial t} = \dfrac{\partial v_y}{\partial y}. \end{cases} \tag{2}$$

The staggered grids illustrated in Figure 1 are used to discretize the above systems, where two uniform grid regions are separated by a horizontal interface, with a contrast ratio 1 : 2 in grid spacing. Both regions include the interface in discretization.

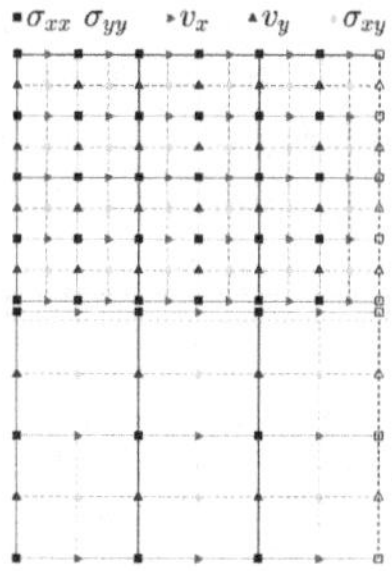

Fig. 1: Illustration of the grid configuration.

3 Methodology

In this section, we demonstrate how to couple the discretizations of system (1) on the two uniform grid regions illustrated in Figure 1 using the SBP-SAT approach. A similar work has been presented in [4] for acoustic wave equations. We will follow the methodology and terminology developed therein.

The continuous energy associated with system (2), and system (1) by equivalence, can be expressed as:

$$e = \int_{\Omega} \tfrac{1}{2} \rho v_i v_i d_{\Omega} + \int_{\Omega} \tfrac{1}{2} \sigma_{ij} s_{ijkl} \sigma_{kl} d_{\Omega} \,, \tag{3}$$

where Ω denotes a simply connected domain; the Einstein summation convention applies to subscript indices i, j, k and l. The two integrals of (3) correspond to the kinetic and potential parts of the continuous energy, respectively. Differentiating e with respect to time t and substituting the equations from (2), it can be shown that

$$\frac{de}{dt} = \int_{\partial\Omega} v_i \sigma_{ij} n_j d_{\partial\Omega}, \tag{4}$$

where $\partial\Omega$ denotes the boundary of Ω. For the free-surface boundary condition, i.e., $\sigma_{ij} n_j = 0$, and periodic boundary condition considered in this work, we have $de/dt = 0$, i.e., system (2) conserves energy e.

Spatially discretizing (2) with finite difference methods on a uniform grid leads to the following semi-discretized system:

$$\begin{cases} \mathcal{A}^{V_x}\rho^{V_x}\dfrac{dV_x}{dt} = \mathcal{A}^{V_x}\mathcal{D}_x^{\Sigma_{xx}}\Sigma_{xx} + \mathcal{A}^{V_x}\mathcal{D}_y^{\Sigma_{xy}}\Sigma_{xy}; \\ \mathcal{A}^{V_y}\rho^{V_y}\dfrac{dV_y}{dt} = \mathcal{A}^{V_y}\mathcal{D}_x^{\Sigma_{xy}}\Sigma_{xy} + \mathcal{A}^{V_y}\mathcal{D}_y^{\Sigma_{yy}}\Sigma_{yy}; \\ \mathcal{A}^{\Sigma_{xx}}S_{xxkl}^{\Sigma_{kl}}\dfrac{d\Sigma_{kl}}{dt} = \mathcal{A}^{\Sigma_{xx}}\mathcal{D}_x^{V_x}V_x; \\ \mathcal{A}^{\Sigma_{xy}}S_{xykl}^{\Sigma_{kl}}\dfrac{d\Sigma_{kl}}{dt} = \frac{1}{2}\mathcal{A}^{\Sigma_{xy}}\left(\mathcal{D}_x^{V_y}V_y + \mathcal{D}_y^{V_x}V_x\right); \\ \mathcal{A}^{\Sigma_{yy}}S_{yykl}^{\Sigma_{kl}}\dfrac{d\Sigma_{kl}}{dt} = \mathcal{A}^{\Sigma_{yy}}\mathcal{D}_y^{V_y}V_y, \end{cases} \tag{5}$$

where the Einstein summation convention applies only to k and l in the subscripts, but not to those appearing in the superscripts. Superscript such as ^{V_x} indicates the grid with which the underlying quantity or operator is associated. In (5), $\mathcal{D}$ symbolizes a finite difference matrix, while $\mathcal{A}$ symbolizes a diagonal norm matrix with its diagonal component loosely representing the area that the corresponding grid point occupies. From the implementation perspective, the norm matrices in (5) are redundant, but they will play an important role in deriving the proper interface treatment. These 2D finite difference matrices and norm matrices are constructed from their 1D counterparts via tensor product. Specifically,

$$\begin{aligned} &\mathcal{A}^{V_x} = \mathcal{A}_x^M \otimes \mathcal{A}_y^N, \quad \mathcal{A}^{V_y} = \mathcal{A}_x^N \otimes \mathcal{A}_y^M, \\ &\mathcal{A}^{\Sigma_{xy}} = \mathcal{A}_x^M \otimes \mathcal{A}_y^M, \quad \mathcal{A}^{\Sigma_{xx}} = \mathcal{A}^{\Sigma_{yy}} = \mathcal{A}_x^N \otimes \mathcal{A}_y^N, \end{aligned} \tag{6}$$

and

$$\begin{aligned} &\mathcal{D}_x^{V_x} = \mathcal{D}_x^M \otimes \mathcal{I}_y^N, \quad \mathcal{D}_x^{V_y} = \mathcal{D}_x^N \otimes \mathcal{I}_y^M, \quad \mathcal{D}_x^{\Sigma_{xy}} = \mathcal{D}_x^M \otimes \mathcal{I}_y^M, \quad \mathcal{D}_x^{\Sigma_{xx}} = \mathcal{D}_x^N \otimes \mathcal{I}_y^N, \\ &\mathcal{D}_y^{V_x} = \mathcal{I}_x^M \otimes \mathcal{D}_y^N, \quad \mathcal{D}_y^{V_y} = \mathcal{I}_x^N \otimes \mathcal{D}_y^M, \quad \mathcal{D}_y^{\Sigma_{xy}} = \mathcal{I}_x^M \otimes \mathcal{D}_y^M, \quad \mathcal{D}_y^{\Sigma_{yy}} = \mathcal{I}_x^N \otimes \mathcal{D}_y^N, \end{aligned} \tag{7}$$

where $\mathcal{I}$ symbolizes a 1D identity matrix. Superscript ^{N} indicates the '*normal*' grid that aligns with the boundaries while ^{M} indicates the '*modified*' grid that is staggered with respect to the '*normal*' grid. The specific forms of the 1D norm matrices and 1D finite difference matrices in (6) and (7) have been described in [5, p. 672][1]. By construction, they satisfy the following relations:

$$\mathcal{A}_x^N\mathcal{D}_x^M + \left(\mathcal{A}_x^M\mathcal{D}_x^N\right)^T = \mathbf{0}; \tag{8a}$$

$$\mathcal{A}_y^N\mathcal{D}_y^M + \left(\mathcal{A}_y^M\mathcal{D}_y^N\right)^T = \mathcal{E}_y^R\left(\mathcal{P}_y^R\right)^T - \mathcal{E}_y^L\left(\mathcal{P}_y^L\right)^T, \tag{8b}$$

where $\mathcal{E}_y^R$ and $\mathcal{E}_y^L$ are canonical basis vectors that select values of the solution variables defined on the N grid at the top and bottom boundaries, respectively, while $\mathcal{P}_y^R$ and $\mathcal{P}_y^L$ are projection vectors that extrapolate values of the solution variables defined on the M grid to the top and bottom boundaries, respectively.

The discrete energy associated with semi-discretized system (5) is defined as:

$$E = \frac{1}{2}V_i^T\left(\mathcal{A}^{V_i}\rho^{V_i}\right)V_i + \frac{1}{2}\Sigma_{ij}^T\left(\mathcal{A}^{\Sigma_{ij}}S_{ijkl}^{\Sigma_{kl}}\right)\Sigma_{kl}, \tag{9}$$

where, as in (5), the Einstein summation convention applies only to i, j, k, and l in the subscripts. Differentiating E with respect to time t and substituting the equations

[1] We use this specific set of operators here to demonstrate the methodology, while alternative choices exist (e.g., [9]), for which the presented methodology can still be applied with minor modifications.

from (5), it can be shown that

$$\frac{dE}{dt} = V_x^T \left[\mathcal{I}_x^M \otimes \mathcal{E}_y^R\right] \mathcal{A}_x^M \left[\mathcal{I}_x^M \otimes (\mathcal{P}_y^R)^T\right] \Sigma_{xy} + \Sigma_{yy}^T \left[\mathcal{I}_x^N \otimes \mathcal{E}_y^R\right] \mathcal{A}_x^N \left[\mathcal{I}_x^N \otimes (\mathcal{P}_y^R)^T\right] V_y$$
$$- V_x^T \left[\mathcal{I}_x^M \otimes \mathcal{E}_y^L\right] \mathcal{A}_x^M \left[\mathcal{I}_x^M \otimes (\mathcal{P}_y^L)^T\right] \Sigma_{xy} - \Sigma_{yy}^T \left[\mathcal{I}_x^N \otimes \mathcal{E}_y^L\right] \mathcal{A}_x^N \left[\mathcal{I}_x^N \otimes (\mathcal{P}_y^L)^T\right] V_y , \tag{10}$$

where the first two terms are associated with the top boundary while the last two terms are associated with the bottom boundary, as indicated by the respective selection and projection operators appearing in these terms. With the above discrete energy analysis result, we can now modify system (5) accordingly to account for boundary and interface conditions.

In the following, we use superscripts $^+$ and $^-$ to distinguish systems or terms from the upper and lower regions of Figure 1, respectively. To account for the free-surface boundary condition on the top boundary, i.e., $\sigma_{xy} = \sigma_{yy} = 0$, the first two equations in the upper region system are appended with penalty terms, i.e., SATs, as follows:

$$(+)\begin{cases} \mathcal{A}^{V_x^+} \rho^{V_x^+} \dfrac{dV_x^+}{dt} = \mathcal{A}^{V_x^+} \mathcal{D}_x^{\Sigma_{xx}^+} \Sigma_{xx}^+ + \mathcal{A}^{V_x^+} \mathcal{D}_y^{\Sigma_{xy}^+} \Sigma_{xy}^+ \\ \qquad + \eta_T^{V_x^+} \left[\mathcal{I}_x^{M^+} \otimes \mathcal{E}_y^{R^+}\right] \mathcal{A}_x^{M^+} \left\{\left[\mathcal{I}_x^{M^+} \otimes (\mathcal{P}_y^{R^+})^T\right] \Sigma_{xy}^+ - \mathbf{0}\right\}; \\ \mathcal{A}^{V_y^+} \rho^{V_y^+} \dfrac{dV_y^+}{dt} = \mathcal{A}^{V_y^+} \mathcal{D}_x^{\Sigma_{xy}^+} \Sigma_{xy}^+ + \mathcal{A}^{V_y^+} \mathcal{D}_y^{\Sigma_{yy}^+} \Sigma_{yy}^+ \\ \qquad + \eta_T^{V_y^+} \left[\mathcal{I}_x^{N^+} \otimes \mathcal{P}_y^{R^+}\right] \mathcal{A}_x^{N^+} \left\{\left[\mathcal{I}_x^{N^+} \otimes (\mathcal{E}_y^{R^+})^T\right] \Sigma_{yy}^+ - \mathbf{0}\right\}, \end{cases} \tag{11}$$

where $\eta_T^{V_x^+} = \eta_T^{V_y^+} = -1$ are penalty parameters. The forms of the penalty terms and values of the penalty parameters are chosen so that the energy-conserving property from the continuous system is preserved. To see this, we differentiate E from (9) with respect to time t as before. After substitution, the penalty terms in (11) bring two extra terms into dE/dt, i.e.,

$$\eta_T^{V_x^+} (V_x^+)^T \left[\mathcal{I}_x^{M^+} \otimes \mathcal{E}_y^{R^+}\right] \mathcal{A}_x^{M^+} \left[\mathcal{I}_x^{M^+} \otimes (\mathcal{P}_y^{R^+})^T\right] \Sigma_{xy}^+$$

and

$$\eta_T^{V_y^+} (V_y^+)^T \left[\mathcal{I}_x^{N^+} \otimes \mathcal{P}_y^{R^+}\right] \mathcal{A}_x^{N^+} \left[\mathcal{I}_x^{N^+} \otimes (\mathcal{E}_y^{R^+})^T\right] \Sigma_{yy}^+ .$$

By setting $\eta_T^{V_x^+} = \eta_T^{V_y^+} = -1$, these extra terms cancel out the first two terms in (10), which are associated with the top boundary. Similar modifications presented later in (12)-(14) are obtained by following the same procedure and rationale. It is worth mentioning that such procedure and rationale for deriving the proper boundary and interface treatment, particularly the usage of the energy method, is very similar to that for flux specification in discontinuous Galerkin methods, cf., for example, [6].

Similarly, to account for the free-surface boundary condition on the bottom boundary, the first two equations of the lower region system are modified as follows:

$$(-)\begin{cases} \mathcal{A}^{V_x^-} \rho^{V_x^-} \dfrac{dV_x^-}{dt} = \mathcal{A}^{V_x^-} \mathcal{D}_x^{\Sigma_{xx}^-} \Sigma_{xx}^- + \mathcal{A}^{V_x^-} \mathcal{D}_y^{\Sigma_{xy}^-} \Sigma_{xy}^- \\ \qquad + \eta_B^{V_x^-} \left[\mathcal{I}_x^{M^-} \otimes \mathcal{E}_y^{L^-}\right] \mathcal{A}_x^{M^-} \left\{\left[\mathcal{I}_x^{M^-} \otimes (\mathcal{P}_y^{L^-})^T\right] \Sigma_{xy}^- - \mathbf{0}\right\}; \\ \mathcal{A}^{V_y^-} \rho^{V_y^-} \dfrac{dV_y^-}{dt} = \mathcal{A}^{V_y^-} \mathcal{D}_x^{\Sigma_{xy}^-} \Sigma_{xy}^- + \mathcal{A}^{V_y^-} \mathcal{D}_y^{\Sigma_{yy}^-} \Sigma_{yy}^- \\ \qquad + \eta_B^{V_y^-} \left[\mathcal{I}_x^{N^-} \otimes \mathcal{P}_y^{L^-}\right] \mathcal{A}_x^{N^-} \left\{\left[\mathcal{I}_x^{N^-} \otimes (\mathcal{E}_y^{L^-})^T\right] \Sigma_{yy}^- - \mathbf{0}\right\}, \end{cases} \tag{12}$$

where $\eta_B^{V_x^-} = \eta_B^{V_y^-} = 1$ are the chosen penalty parameters.

To account for the interface conditions (cf. [10, p. 52]), i.e., $\sigma_{xy}^+ = \sigma_{xy}^-$, $\sigma_{yy}^+ = \sigma_{yy}^-$, $v_x^+ = v_x^-$, and $v_y^+ = v_y^-$, the upper and lower region systems are further modified by appending additional SATs as follows:

$$
(+)\left\{
\begin{aligned}
&\mathcal{A}^{V_x^+} \rho^{V_x^+} \frac{dV_x^+}{dt} = \mathcal{A}^{V_x^+} \mathcal{D}_x^{\Sigma_{xx}^+} \Sigma_{xx}^+ + \mathcal{A}^{V_x^+} \mathcal{D}_y^{\Sigma_{xy}^+} \Sigma_{xy}^+ \\
&\quad + \eta_I^{V_x^+} \left[\mathcal{I}_x^{M^+} \otimes \mathcal{E}_y^{L^+}\right] \mathcal{A}_x^{M^+} \left\{\left[\mathcal{I}_x^{M^+} \otimes (\mathcal{P}_y^{L^+})^T\right] \Sigma_{xy}^+ - \mathcal{T}^{M_-^+} \left(\left[\mathcal{I}_x^{M^-} \otimes (\mathcal{P}_y^{R^-})^T\right] \Sigma_{xy}^-\right)\right\}; \\
&\mathcal{A}^{V_y^+} \rho^{V_y^+} \frac{dV_y^+}{dt} = \mathcal{A}^{V_y^+} \mathcal{D}_x^{\Sigma_{xy}^+} \Sigma_{xy}^+ + \mathcal{A}^{V_y^+} \mathcal{D}_y^{\Sigma_{yy}^+} \Sigma_{yy}^+ \\
&\quad + \eta_I^{V_y^+} \left[\mathcal{I}_x^{N^+} \otimes \mathcal{P}_y^{L^+}\right] \mathcal{A}_x^{N^+} \left\{\left[\mathcal{I}_x^{N^+} \otimes (\mathcal{E}_y^{L^+})^T\right] \Sigma_{yy}^+ - \mathcal{T}^{N_-^+} \left(\left[\mathcal{I}_x^{N^-} \otimes (\mathcal{E}_y^{R^-})^T\right] \Sigma_{yy}^-\right)\right\}; \\
&\mathcal{A}^{\Sigma_{xx}^+} S_{xxkl}^{\Sigma_{kl}^+} \frac{d\Sigma_{kl}^+}{dt} = \mathcal{A}^{\Sigma_{xx}^+} \mathcal{D}_x^{V_x^+} V_x^+; \\
&\mathcal{A}^{\Sigma_{xy}^+} S_{xykl}^{\Sigma_{kl}^+} \frac{d\Sigma_{kl}^+}{dt} = \tfrac{1}{2} \mathcal{A}^{\Sigma_{xy}^+} \left(\mathcal{D}_x^{V_y^+} V_y^+ + \mathcal{D}_y^{V_x^+} V_x^+\right) \\
&\quad + \tfrac{1}{2} \eta_I^{\Sigma_{xy}^+} \left[\mathcal{I}_x^{M^+} \otimes \mathcal{P}_y^{L^+}\right] \mathcal{A}_x^{M^+} \left\{\left[\mathcal{I}_x^{M^+} \otimes (\mathcal{E}_y^{L^+})^T\right] V_x^+ - \mathcal{T}^{M_-^+} \left(\left[\mathcal{I}_x^{M^-} \otimes (\mathcal{E}_y^{R^-})^T\right] V_x^-\right)\right\}; \\
&\mathcal{A}^{\Sigma_{yy}^+} S_{yykl}^{\Sigma_{kl}^+} \frac{d\Sigma_{kl}^+}{dt} = \mathcal{A}^{\Sigma_{yy}^+} \mathcal{D}_y^{V_y^+} V_y^+ \\
&\quad + \eta_I^{\Sigma_{yy}^+} \left[\mathcal{I}_x^{N^+} \otimes \mathcal{E}_y^{L^+}\right] \mathcal{A}_x^{N^+} \left\{\left[\mathcal{I}_x^{N^+} \otimes (\mathcal{P}_y^{L^+})^T\right] V_y^+ - \mathcal{T}^{N_-^+} \left(\left[\mathcal{I}_x^{N^-} \otimes (\mathcal{P}_y^{R^-})^T\right] V_y^-\right)\right\},
\end{aligned}
\right.
\tag{13}
$$

$$
(-)\left\{
\begin{aligned}
&\mathcal{A}^{V_x^-} \rho^{V_x^-} \frac{dV_x^-}{dt} = \mathcal{A}^{V_x^-} \mathcal{D}_x^{\Sigma_{xx}^-} \Sigma_{xx}^- + \mathcal{A}^{V_x^-} \mathcal{D}_y^{\Sigma_{xy}^-} \Sigma_{xy}^- \\
&\quad + \eta_I^{V_x^-} \left[\mathcal{I}_x^{M^-} \otimes \mathcal{E}_y^{R^-}\right] \mathcal{A}_x^{M^-} \left\{\left[\mathcal{I}_x^{M^-} \otimes (\mathcal{P}_y^{R^-})^T\right] \Sigma_{xy}^- - \mathcal{T}^{M_+^-} \left(\left[\mathcal{I}_x^{M^+} \otimes (\mathcal{P}_y^{L^+})^T\right] \Sigma_{xy}^+\right)\right\}; \\
&\mathcal{A}^{V_y^-} \rho^{V_y^-} \frac{dV_y^-}{dt} = \mathcal{A}^{V_y^-} \mathcal{D}_x^{\Sigma_{xy}^-} \Sigma_{xy}^- + \mathcal{A}^{V_y^-} \mathcal{D}_y^{\Sigma_{yy}^-} \Sigma_{yy}^- \\
&\quad + \eta_I^{V_y^-} \left[\mathcal{I}_x^{N^-} \otimes \mathcal{P}_y^{R^-}\right] \mathcal{A}_x^{N^-} \left\{\left[\mathcal{I}_x^{N^-} \otimes (\mathcal{E}_y^{R^-})^T\right] \Sigma_{yy}^- - \mathcal{T}^{N_+^-} \left(\left[\mathcal{I}_x^{N^+} \otimes (\mathcal{E}_y^{L^+})^T\right] \Sigma_{yy}^+\right)\right\}; \\
&\mathcal{A}^{\Sigma_{xx}^-} S_{xxkl}^{\Sigma_{kl}^-} \frac{d\Sigma_{kl}^-}{dt} = \mathcal{A}^{\Sigma_{xx}^-} \mathcal{D}_x^{V_x^-} V_x^-; \\
&\mathcal{A}^{\Sigma_{xy}^-} S_{xykl}^{\Sigma_{kl}^-} \frac{d\Sigma_{kl}^-}{dt} = \tfrac{1}{2} \mathcal{A}^{\Sigma_{xy}^-} \left(\mathcal{D}_x^{V_y^-} V_y^- + \mathcal{D}_y^{V_x^-} V_x^-\right) \\
&\quad + \tfrac{1}{2} \eta_I^{\Sigma_{xy}^-} \left[\mathcal{I}_x^{M^-} \otimes \mathcal{P}_y^{R^-}\right] \mathcal{A}_x^{M^-} \left\{\left[\mathcal{I}_x^{M^-} \otimes (\mathcal{E}_y^{R^-})^T\right] V_x^- - \mathcal{T}^{M_+^-} \left(\left[\mathcal{I}_x^{M^+} \otimes (\mathcal{E}_y^{L^+})^T\right] V_x^+\right)\right\}; \\
&\mathcal{A}^{\Sigma_{yy}^-} S_{yykl}^{\Sigma_{kl}^-} \frac{d\Sigma_{kl}^-}{dt} = \mathcal{A}^{\Sigma_{yy}^-} \mathcal{D}_y^{V_y^-} V_y^- \\
&\quad + \eta_I^{\Sigma_{yy}^-} \left[\mathcal{I}_x^{N^-} \otimes \mathcal{E}_y^{R^-}\right] \mathcal{A}_x^{N^-} \left\{\left[\mathcal{I}_x^{N^-} \otimes (\mathcal{P}_y^{R^-})^T\right] V_y^- - \mathcal{T}^{N_+^-} \left(\left[\mathcal{I}_x^{N^+} \otimes (\mathcal{P}_y^{L^+})^T\right] V_y^+\right)\right\},
\end{aligned}
\right.
\tag{14}
$$

where $\eta_I^{V_x^+} = \eta_I^{\Sigma_{xy}^+} = \eta_I^{V_y^+} = \eta_I^{\Sigma_{yy}^+} = \frac{1}{2}$ and $\eta_I^{V_x^-} = \eta_I^{\Sigma_{xy}^-} = \eta_I^{V_y^-} = \eta_I^{\Sigma_{yy}^-} = -\frac{1}{2}$ are the chosen penalty parameters. Moreover, $\mathcal{T}^{M_-^+}$, $\mathcal{T}^{N_-^+}$ in (13) and $\mathcal{T}^{M_+^-}$, $\mathcal{T}^{N_+^-}$ in (14) are interpolation operators that satisfy the following relations:

$$
\mathcal{A}_x^{N^+} \mathcal{T}^{N_-^+} = \left(\mathcal{A}_x^{N^-} \mathcal{T}^{N_+^-}\right)^T \quad \text{and} \quad \mathcal{A}_x^{M^+} \mathcal{T}^{M_-^+} = \left(\mathcal{A}_x^{M^-} \mathcal{T}^{M_+^-}\right)^T . \tag{15}
$$

They operate on the interface only, e.g., $\mathcal{T}^{N_-^+}$ interpolates from lower region N grid points on the interface to upper region N grid points on the interface. Their derivations are usually assisted by symbolic computing software. For the interface illustrated in Figure 1, which has a 1 : 2 contrast in grid spacing, the operators $\mathcal{T}^{N_-^+}$ and $\mathcal{T}^{M_-^+}$ that we use here are characterized by the formulas in (16) and (17), respectively, for the collections of grid points illustrated in Figure 2; moreover, $\mathcal{T}^{N_+^-}$ and $\mathcal{T}^{M_+^-}$ can be derived from $\mathcal{T}^{N_-^+}$ and $\mathcal{T}^{M_-^+}$, respectively, via the relations in (15).

As in the case of SBP operators, these interpolation operators are not unique, either.

Fig. 2: Illustration of grid points involved for interpolation operators $\mathcal{T}^{N_-^+}$ (left) and $\mathcal{T}^{M_-^+}$ (right).

$$f(x_0^+) \leftarrow f(x_0^-);\; f(x_1^+) \leftarrow -\tfrac{1}{16}f(x_{-1}^-)+\tfrac{9}{16}f(x_0^-)+\tfrac{9}{16}f(x_1^-)-\tfrac{1}{16}f(x_2^-);\; f(x_2^+) \leftarrow f(x_1^-). \tag{16}$$

$$f(x_0^+) \leftarrow \tfrac{5}{32}f(x_0^-)+\tfrac{15}{16}f(x_1^-)-\tfrac{3}{32}f(x_2^-); \qquad f(x_1^+) \leftarrow -\tfrac{3}{32}f(x_0^-)+\tfrac{15}{16}f(x_1^-)+\tfrac{5}{32}f(x_2^-). \tag{17}$$

With the above choices on the SATs, it can be verified that the overall semi-discretization conserves the discrete energy E from (9). Now that the proper SATs have been derived, we can remove the norm matrices by dividing them from both sides of the equations in (11-14). From the implementation perspective, the appended SATs amount to modifying the corresponding derivative approximations, e.g., the SAT in the first equation of (14) modifies $\mathcal{D}_y^{\Sigma_{xy}^-}\Sigma_{xy}^-$. Written in terms of these modified derivative approximations, the above discretizations for system (2) can be easily reverted to forms that conform to system (1).

4 Numerical examples

The first example concerns a homogeneous medium characterized by parameters $\rho = 1\,\mathrm{kg/m^3}$, $c_p = 2\,\mathrm{m/s}$ and $c_s = 1\,\mathrm{m/s}$. The grid spacings of the upper and lower regions are chosen as 0.004 m and 0.008 m, respectively, while the time step length is chosen as 0.001 s, which is ~ 0.707 of the CFL limit associated with an infinite uniform grid with 0.004 m grid spacing. The rest of the numerical setup is the same as that for the first example of [4, p. 435], including sizes of the grids, source and receiver locations, and source profile.

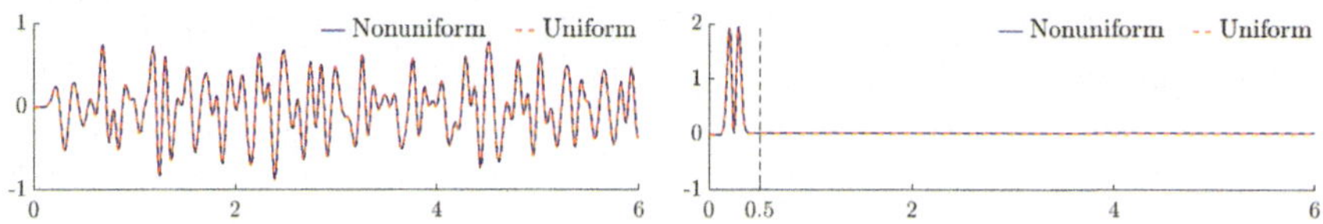

Fig. 3: Seismogram (left) and evolvement of discrete energy (right); Homogeneous media.

The recorded seismogram and evolvement of discrete energy for the first 6 s are displayed in Figure 3, where we observe good agreement between the uniform grid simulation result and the nonuniform grid simulation result using the presented SBP-SAT approach. The source term $\mathcal{S}$, cf. (1), which is omitted from the analysis, is responsible for the initial '*bumps*' in the evolvement of discrete energy. After the source effect tapers off at around 0.5 s (cf. [5, p. 684]), the discrete energy remains constant as expected (at a value ~0.0318).

The second example concerns a heterogeneous medium downsampled from the Marmousi2 model, cf. [8]. Wave-speeds c_p and c_s are illustrated in Figure 4.[2] Grid spacing is chosen as 2 m and 4 m for upper and lower regions (separated by the

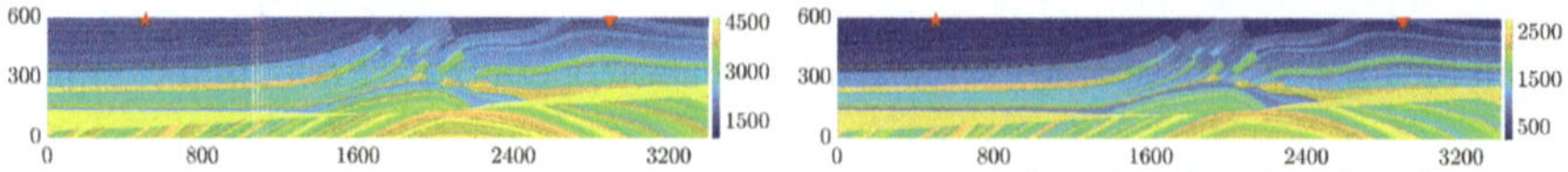

Fig. 4: Media parameters c_p (left) and c_s (right); Colorbar reflects wave-speed with unit m/s.

green dashed line), respectively. Time step length is chosen as 2e-4 s and 3e-4 s for uniform and nonuniform grid simulations, respectively. Same plots as in the previous example are displayed in Figure 5, from where similar observations can be made.

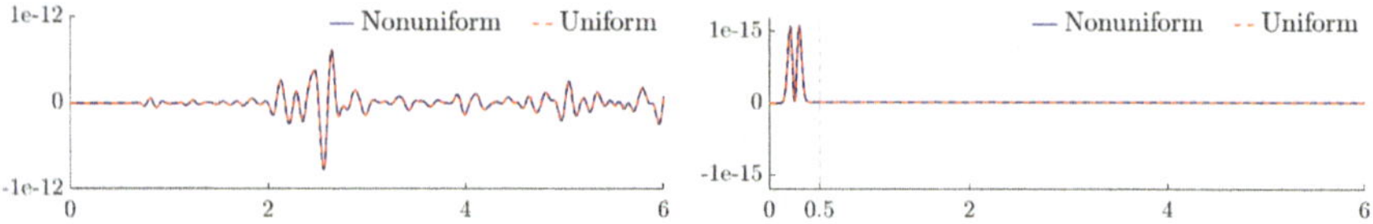

Fig. 5: Seismogram (left) and evolvement of discrete energy (right); Heterogeneous media.

In this example, the ratio between the numbers of spatial grid points in uniform and nonuniform grid simulations is $\sim$1.813. As a rough estimation, the amount of arithmetic operations per time step is assumed to be linearly proportional to the number of spatial grid points. We therefore expect the ratio between runtimes to be ~2.719, with an extra factor of 1.5 coming from the difference in total time steps. A test with our Matlab code reveals a ratio of ~2.681 in runtimes (average of 5 runs), which agrees well with the above complexity analysis result.

5 Summary

Finite difference discretization of the isotropic elastic wave system is considered. An interface treatment procedure is presented to connect two uniformly discretized regions with different grid spacings. The interface conditions are weakly imposed through carefully designed simultaneous approximation terms. The overall semi-discretization conserves a discrete energy that resembles the continuous physical energy, which is demonstrated on both homogeneous and heterogeneous media.

Acknowledgements Gao and Keyes gratefully acknowledge the support of KAUST's OSR under CCF-CAF/URF/1-2596. The authors would also like to thank KSL for computing resources. Part of this work was conducted while the first author was visiting IPAM (Sep - Dec 2018).

---

[2] To simplify the discussion, the distance between neighboring parameter grid points is assigned to 2 m, which is the same as the grid spacing used in uniform grid simulation. Bilinear interpolation is used when discretization grid points do not match parameter grid points due to grid staggering. We note here that media parameters for uniform grid and nonuniform grid simulations are sampled differently; thus, small discrepancies in simulation results should be allowed.

References

1. Bourbié, T., Coussy, O., Zinszner, B.: Acoustics of Porous Media. Institut Français du Pétrole Publications. Editions Technip (1987)
2. Carpenter, M.H., Gottlieb, D., Abarbanel, S.: Time-stable boundary conditions for finite-difference schemes solving hyperbolic systems: methodology and application to high-order compact schemes. J. Comput. Phys. **111**(2), 220–236 (1994). DOI:10.1006/jcph.1994.1057
3. Del Rey Fernández, D.C., Hicken, J.E., Zingg, D.W.: Review of summation-by-parts operators with simultaneous approximation terms for the numerical solution of partial differential equations. Comput. & Fluids **95**, 171–196 (2014). DOI:10.1016/j.compfluid.2014.02.016
4. Gao, L., Del Rey Fernández, D.C., Carpenter, M., Keyes, D.: SBP-SAT finite difference discretization of acoustic wave equations on staggered block-wise uniform grids. J. Comput. Appl. Math. **348**, 421–444 (2019). DOI:10.1016/j.cam.2018.08.040
5. Gao, L., Keyes, D.: Combining finite element and finite difference methods for isotropic elastic wave simulations in an energy-conserving manner. J. Comput. Phys. **378**, 665–685 (2019). DOI:10.1016/j.jcp.2018.11.031
6. Hesthaven, J.S., Warburton, T.: Nodal discontinuous Galerkin methods, *Texts in Applied Mathematics*, vol. 54. Springer, New York (2008). DOI:10.1007/978-0-387-72067-8. Algorithms, analysis, and applications
7. Kreiss, H.O., Scherer, G.: Finite element and finite difference methods for hyperbolic partial differential equations. In: Mathematical aspects of finite elements in partial differential equations. Academic Press (1974)
8. Martin, G.S., Wiley, R., Marfurt, K.J.: Marmousi2: An elastic upgrade for marmousi. The Leading Edge **25**(2), 156–166 (2006)
9. O'Reilly, O., Lundquist, T., Dunham, E.M., Nordström, J.: Energy stable and high-order-accurate finite difference methods on staggered grids. J. Comput. Phys. **346**, 572–589 (2017). DOI:10.1016/j.jcp.2017.06.030
10. Stein, S., Wysession, M.: An introduction to seismology, earthquakes, and earth structure. John Wiley & Sons (2009)
11. Svärd, M., Nordström, J.: Review of summation-by-parts schemes for initial-boundary-value problems. J. Comput. Phys. **268**, 17–38 (2014). DOI:10.1016/j.jcp.2014.02.031
12. Yee, K.: Numerical solution of initial boundary value problems involving maxwell's equations in isotropic media. IEEE Transactions on antennas and propagation **14**(3), 302–307 (1966)

Asynchronous One-Level and Two-Level Domain Decomposition Solvers

Christian Glusa, Erik G. Boman, Edmond Chow, Sivasankaran Rajamanickam, and Paritosh Ramanan

1 Introduction

Multilevel methods such as multigrid and domain decomposition are among the most efficient and scalable solvers developed to date. Adapting them to the next generation of supercomputers and improving their performance and scalability is crucial for exascale computing and beyond. Domain decomposition methods subdivide the global problem into subdomains, and then alternate between local solves and boundary data exchange. This puts significant stress on the network interconnect, since all processes try to communicate at once. On the other hand, during the solve phase, the network is under-utilized. The use of non-blocking communication can only alleviate this issue, but not solve it. In asynchronous methods, on the other hand, computation and communication occur concurrently, with some processes performing computation while others communicate, so that the network is consistently in use.

Unfortunately, the term "asynchronous" can have several different meanings in the literature. In computer science, it is sometimes used to describe communication patterns that are non-blocking, such that computation and communication can be overlapped. Iterative algorithms that use such "asynchronous" communication typically still yield the same iterates (results), just more efficiently. In applied mathematics, on the other hand, "asynchronous" denotes parallel algorithms where each process (processor) proceeds at its own speed without synchronization. Thus, asynchronous algorithms go beyond the widely used bulk-synchronous parallel (BSP) model. More importantly, they are mathematically different than synchronous methods and generate different iterates. The earliest work in this area was called "chaotic

Christian Glusa, Erik G. Boman, Sivasankaran Rajamanickam
Center for Computing Research, Sandia National Laboratories, Albuquerque, New Mexico 87185, USA, e-mail: \{caglusa, egboman, srajama\}@sandia.gov

Edmond Chow, Paritosh Ramanan
School of Computational Science and Engineering, College of Computing, Georgia Institute of Technology, Atlanta, Georgia, USA e-mail: paritoshpr@gatech.edu, echow@cc.gatech.edu

R. Haynes et al. (eds.), *Domain Decomposition Methods in Science and Engineering XXV*, Lecture Notes in Computational Science and Engineering 138,
https://doi.org/10.1007/978-3-030-56750-7_14

relaxation" [6]. Both approaches are expected to play an important role on future supercomputers. In this paper, we focus on the mathematically asynchronous methods.

Domain decomposition solvers [8, 16, 15] are often used as preconditioners in Krylov subspace iterations. Unfortunately, the computation of inner products and norms widely used in Krylov methods requires global communication. Global communication primitives, such as `MPI_Reduce`, asymptotically scale as the logarithm of the number of processes involved. This can become a limiting factor when very large process counts are used. The underlying domain decomposition method, however, can do away with globally synchronous communication, assuming the coarse problem in multilevel methods can be solved in a parallel way. Therefore, we will focus on using domain decomposition methods purely as iterative methods. We note, however, that the discussed algorithms could be coupled with existing pipelined methods [10] which alleviate the global synchronization requirement of Krylov solvers.

Another issue that is crucial to good scaling behavior is load imbalance. Load imbalance might occur due to heterogeneous hardware in the system, or due to local, problem specific causes, such as iteration counts for local sub-solves that vary from region to region. Especially the latter are difficult to predict, so that load balancing cannot occur before the actual solve. Therefore, a synchronous parallel application has to be idle until its slowest process has finished. In an asynchronous method, local computation can continue, and improve the quality of the global solution. An added benefit of asynchronous methods is that, since the interdependence of one subdomain on the others has been weakened, fault tolerance [4, 5] can be more easily achieved.

The main drawback of asynchronous iterations is the fact that deterministic behavior is sacrificed. Consecutive runs do not produce the same result. (The results do match up to a factor proportional to the convergence tolerance.) This also makes the mathematical analysis of asynchronous methods significantly more difficult than the analysis of their synchronous counterparts. Analytical frameworks for asynchronous linear iterations have long been available [6, 2, 3, 9], but generally cannot produce sharp convergence bounds except for in the simplest of cases.

2 Domain decomposition methods

We want to solve the system $\boldsymbol{A}\mathbf{u} = \mathbf{f}$, where $\boldsymbol{A} \in \mathbb{R}^{N\times N}$. Informally speaking, one-level domain decomposition solvers break up the global system into overlapping subproblems that cover the global system. The iteration alternates between computation of the global residual, involving communication, and local solves for corrections. Special attention is paid to unknowns in the overlap to avoid over-correction.

We use the notation of [8] and denote subdomain matrices by $\boldsymbol{A}_p$, restrictions by $\boldsymbol{R}_p$, and the discrete partition of unity by $\boldsymbol{D}_p$. The local form of the restricted additive Schwarz iteration (RAS) is given in Figure 1. A detailed derivation of the algorithm can be found in [11]. In fact, Figure 1 describes both the synchronous *and* the asynchronous version of RAS. In the synchronous version Line 4 is executed in lock step by all subdomains using non-blocking two-sided communication primitives. In

1: $\mathbf{w}_p \leftarrow \mathbf{0}$
2: **while** not converged **do**
3: Local residual: $\mathbf{t}_p \leftarrow \boldsymbol{D}_p \boldsymbol{R}_p \mathbf{f} - \boldsymbol{A}_p \boldsymbol{D}_p \mathbf{w}_p$
4: Accumulate: $\mathbf{r}_p \leftarrow \sum_{q=1}^{P} \boldsymbol{R}_p \boldsymbol{R}_q^T \mathbf{t}_q$
5: Solve: $\boldsymbol{A}_p \mathbf{v}_p = \mathbf{r}_p$
6: Update: $\mathbf{w}_p \leftarrow \mathbf{w}_p + \mathbf{v}_p$
7: **end while**
8: Post-process: $\mathbf{u}_p \leftarrow \sum_{q=1}^{P} \boldsymbol{R}_p \boldsymbol{R}_q^T \boldsymbol{D}_q \mathbf{w}_q$

Fig. 1: Restricted additive Schwarz (RAS) in local form. $\boldsymbol{A}_p$ are subdomain matrices, $\boldsymbol{R}_p$ are subdomain restrictions, $\boldsymbol{D}_p$ are the discrete partition of unity.

the asynchronous variant, each subdomain exposes a memory region to remote access via MPI one-sided primitives. On execution of Line 4, the relevant components of current local residual $\mathbf{t}_p$ are written to the neighboring subdomains, and the latest locally available data $\mathbf{t}_q$ from neighbors q is used.

In order to improve the scalability of the solver, a mechanism of global information exchange is required [15, 16]. Let $\boldsymbol{R}_0 \in \mathbb{R}^{n \times n_0}$ be the restriction from the fine grid problem to a coarser mesh, and let the coarse grid matrix $\boldsymbol{A}_0$ be given by the Galerkin relation $\boldsymbol{A}_0 = \boldsymbol{R}_0 \boldsymbol{A} \boldsymbol{R}_0^T$. The coarse grid solve can be incorporated in the RAS iteration in additive fashion $\mathbf{u}^{n+1} = \mathbf{u}^n + \left(\frac{1}{2}\boldsymbol{M}_{RAS}^{-1} + \frac{1}{2}\boldsymbol{R}_0^T \boldsymbol{A}_0^{-1} \boldsymbol{R}_0\right)(\mathbf{f} - \boldsymbol{A}\mathbf{u}^n)$, where $\boldsymbol{M}_{RAS}^{-1}$ denotes the preconditioner associated with the RAS iteration described above. We focus on the additive version, since it lends itself to asynchronous iterations: subdomain solves and coarse-grid solves are independent of each other. From the mathematical description of two-level additive RAS, one might be tempted to see the coarse-grid problem simply as an additional subdomain. However, subdomains determine the right-hand side for their local solve and correct it by transmitting boundary data to their neighbors. The coarse-grid, on the other hand, receives its entire right-hand side from the subdomains, and hence has to communicate with every single one of them.

In order to perform asynchronous coarse-grid solves, we therefore need to make sure that all the right-hand side data necessary for the solve has been received on the coarse grid. Moreover, corrections sent by the coarse grid should be used exactly once by the subdomains. This is achieved by not only allocating memory regions to hold the coarse grid right-hand side on the coarse grid rank and the coarse grid correction on the subdomains, but also Boolean variables that are polled to determine whether writing or reading right-hand side or solution is permitted. More precisely, writing of the local subdomain residuals to the coarse grid memory region of $\mathbf{r}_0$ is contingent upon the state of the Boolean variable `canWriteRHS`$_p$. (See Figure 2.) When `canWriteRHS`$_p$ is `True`, right-hand side data is written to the coarse grid, otherwise this operation is omitted. Here, the subscripts are used to signify the MPI rank owning the accessed memory region. As before, index 0 corresponds to the coarse grid and indices $1, \ldots, P$ correspond to the subdomains. To improve readability, we show access to a memory region on the calling process in light gray, while remote access is printed in dark gray. In a similar fashion, the coarse grid checks whether every subdomain has written a right-hand side to $\mathbf{r}_0$ by polling

while not converged **do**
  On subdomains
    Local residual: $\mathbf{t}_p \leftarrow \boldsymbol{D}_p \boldsymbol{R}_p \mathbf{f} - \boldsymbol{A}_p \boldsymbol{D}_p \mathbf{w}_p$
    if `canWriteRHS`$_p$ **then**
      $\mathbf{r}_0 \leftarrow \mathbf{r}_0 + \boldsymbol{R}_0 \boldsymbol{R}_p^T \mathbf{t}_p$
      `canWriteRHS`$_p \leftarrow$ `False`
      `RHSisReady`$_0[p] \leftarrow$ `True`
    end if
    Accumulate asynchronously:
    $\mathbf{r}_p \leftarrow \sum_{q=1}^{P} \boldsymbol{R}_p \boldsymbol{R}_q^T \mathbf{t}_q$
    Solve: $\boldsymbol{A}_p \mathbf{v}_p = \mathbf{r}_p$
    Update: $\mathbf{w}_p \leftarrow \mathbf{w}_p + \frac{1}{2} \mathbf{v}_p$
    if `solutionIsReady`$_p$ **then**
      Update: $\mathbf{w}_p \leftarrow \mathbf{w}_p + \frac{1}{2} \mathbf{c}_p$
      `solutionIsReady`$_p \leftarrow$ `False`
    end if
  On coarse grid
    if `RHSisReady`$_0[p]$ $\forall p = 1, \ldots, P$ **then**
      Solve $\boldsymbol{A}_0 \mathbf{v}_0 = \mathbf{r}_0$
      for $p = 1, \ldots, P$ **do**
        `RHSisReady`$_0[p] \leftarrow$ `False`
        `canWriteRHS`$_p \leftarrow$ `True`
        $\mathbf{c}_p \leftarrow \boldsymbol{R}_p \boldsymbol{R}_0^T \mathbf{v}_0$
        `solutionIsReady`$_p \leftarrow$ `True`
      end for
    else
      Sleep
    end if
end while
On subdomains
  Post-process synchronously
  $\mathbf{u}_p \leftarrow \sum_{q=1}^{P} \boldsymbol{R}_p \boldsymbol{R}_q^T \boldsymbol{D}_q \mathbf{w}_q$

Fig. 2: Asynchronous RAS with additive coarse grid in local form. Variables printed in light gray are exposed memory regions that are local to the calling process. Dark gray variables are remote memory regions.

the state of the local Boolean array `RHSisReady`$_0$. We notice that the algorithm is asynchronous despite the data dependencies. Coarse grid and subdomain solves do not wait for each other.

Since we determined by experiments that performance is adversely affected if the coarse grid constantly polls the status variable `RHSisReady`$_0$, we added a sleep statement into its work loop. The sleep interval should not be too large, since this results in under-usage of the coarse grid. Keeping the ratio of attempted coarse grid solves to actual performed coarse grid solves at around $1/20$ has been proven effective to us. This can easily be achieved by an adaptive procedure that counts solves and solve attempts and either increases or decreases the sleep interval accordingly.

We conclude this section with a note on convergence theory for the asynchronous case. Contrary to the synchronous case, where the condition $\rho(\boldsymbol{E}) < 1$ on the iteration matrix $\boldsymbol{E}$ of the method is necessary and sufficient for convergence, asynchronous convergence is guaranteed if $\boldsymbol{E}$ is a block H-matrix which is a P-contraction [9]. Obtaining a prediction for a rate at which the asynchronous method converges appears to be more elusive which is why we limit ourselves to experimental comparisons.

3 Numerical Experiments

The performance of linear iterative methods is typically measured by the average contraction factor per iteration $\widetilde{\rho} = (r_{\text{final}}/r_0)^{\frac{1}{K}}$, where r_0 is the norm of the initial residual vector, r_{final} the norm of the final residual vector, and K the number of iterations that were taken to decrease the residual from r_0 to r_{final}. For an asynchronous method, the number of iterations varies from subdomain to subdomain, and hence $\widetilde{\rho}$

is not well-defined. The following generalization permits us to compare synchronous methods with their asynchronous counterpart: $\widehat{\rho} = (r_{\text{final}}/r_0)^{\frac{\tau_{\text{sync}}}{T}}$. Here, T is the total iteration time, and τ_{sync} is the average time for a single iteration in the synchronous case. In the synchronous case, since $T = \tau_{\text{sync}} K$, $\widehat{\rho}$ recovers $\widetilde{\rho}$. The approximate contraction factor $\widehat{\rho}$ can be interpreted as the average contraction of the residual norm in the time of a single synchronous iteration.

We expect the performance of the asynchronous method relative to its synchronous counterpart to be essentially dependent on the communication stencil. Here, we limit ourselves to a simple 2D problem. Further experiments for more complicated PDEs and as well as in 3D are part of future work.

As a test problem, we solve $-\Delta u = f$ in $\Omega = [0, 1]^2$ subject to the boundary condition $u = 0$ on $\partial\Omega$, where the right-hand side is $f = 2\pi^2 \sin(\pi x) \sin(\pi y)$ and the corresponding solution is $u = \sin(\pi x) \sin(\pi y)$. We discretize Ω using a uniform triangular mesh and approximate the solution using piece-wise linear finite elements.

In classical synchronous iterative methods, a stopping criterion of the form $r < \varepsilon$ is evaluated at every iteration. Here, r is the norm of the residual vector and ε is a prescribed tolerance. The global quantity r needs to be computed as the sum of local contributions from all the subdomains. This implies that convergence detection in asynchronous methods is not straightforward, since collective communication primitives require synchronization. In the numerical examples below, we terminate the iteration using a simplistic convergence criterion where each process writes its local contribution to the residual norm to a master rank, say rank 0. The master rank sums the contributions, and exposes the result through another MPI window. Each subdomain can retrieve this estimate of the global residual norm, and terminates if it is smaller than the prescribed tolerance. This simplistic convergence detection mechanism has several drawbacks. For one, the global residual is updated by the master rank, which might not happen frequently enough. Hence it is possible that the iteration continues despite the true global residual norm already being smaller than the tolerance. Moreover, the mechanism puts an increased load on the network connection to the master rank, since every subdomain writes to its memory region. Finally, since the local contributions to the residual norm are not necessarily monotonically decreasing, the criterion might actually detect convergence when the true global residual is not yet smaller than the tolerance. The delicate topic of asynchronous convergence detection has been treated in much detail in the literature, and we refer to [1, 14] for an overview of more elaborate approaches.

All runs are performed on the Haswell partition of Cori at NERSC. While the code was written from scratch, the differences between the synchronous and the asynchronous code paths are limited, since only the communication layer and stopping criterion need to be changed. One MPI rank is used per core, i.e. 32 ranks per Haswell node. For the two-level method, the coarse grid solve is performed on a single rank. The underlying mesh is partitioned using METIS [12]. Both subdomain and coarse grid problems are factored and solved using the SuperLU [13, 7] direct solver. This choice is guided by the desire to eliminate the impact that inexact solves such as preconditioned iterative sub-solves might have on the overall convergence.

One-level RAS We compare synchronous and asynchronous one-level RAS in a strong scaling experiment, where we fix the global problem size to about 261,000 unknowns, and vary the number of subdomains between 4 and 256. We obviously cannot expect good scaling for this one-level method, since increasing the number of subdomains adversely affects the rate of convergence. In Figure 3a we display solve time, final residual norm and approximate rate of convergence. It can be observed that the synchronous method is faster for smaller subdomain count, yet comparatively slower for larger number of subdomains. The crossover point is at 64 subdomains.

An important question is whether the asynchronous method converges because every subdomain performs the same number of local iterations, and hence the asynchronous method just mirrors the synchronous one, merely with the communication method replaced. The histogram in Figure 3c shows that this is not the case. The number of local iterations varies significantly between 11,000 and 16,000 iterations. The problem was load balanced by the number of unknowns, thus the local solves are also approximately balanced but the communication is likely slightly imbalanced.

The advantage of asynchronous RAS becomes even clearer when the experiment is repeated with one of the subdomains being 50% larger, thereby artificially creating load imbalance. In Figure 3b we observe that the asynchronous method outperforms the synchronous one in all but the smallest run.

Two-level RAS In order to gauge the performance and scalability of the synchronous and asynchronous two-level RAS solvers, we perform a weak scaling experiments. We use 16, 64, 256 and 1024 subdomains. The local number of unknowns on each subdomain is kept constant at almost 20,000. The coarse grid problem increases in size proportionally to the number of subdomains, with approximately 16 unknowns per subdomain. In Figure 4a we plot the solution time, the achieved residual norm and the average contraction factor $\widehat{\rho}$. Both the synchronous and the asynchronous method reach the prescribed tolerance of 10^{-8}. Due to the lack of an efficient mechanism of convergence detection, the asynchronous method ends up iterating longer than necessary, so that the final residual often is smaller than 10^{-9}. The number of iterations in the synchronous case is about 110, whereas the number of local iterations in the asynchronous case varies between 110 and 150. (See Figure 4c.) One can observe that for 16, 64 and 256 subdomains, asynchronous and synchronous method take almost the same time. For 1024 subdomains, however, the synchronous method is seen to take drastically more time. For this case the size of the coarse grid is comparable to the size of the subdomains, and hence the coarse grid solve which exchanges information with all the subdomains slows down the overall progress. For the asynchronous case this is not observed, since the subdomains do not have to wait for information from the coarse grid. The third subplot of Figure 4a shows that the asynchronous method outperforms its synchronous equivalent in all but the smallest problem.

To further illustrate the effect of load imbalance, we repeat the previous experiment with one subdomain being 50% larger. The results are shown in Figure 4b. The results are consistent with the previous case, and the performance advantage of the asynchronous method over the synchronous one has increased. Even when the size

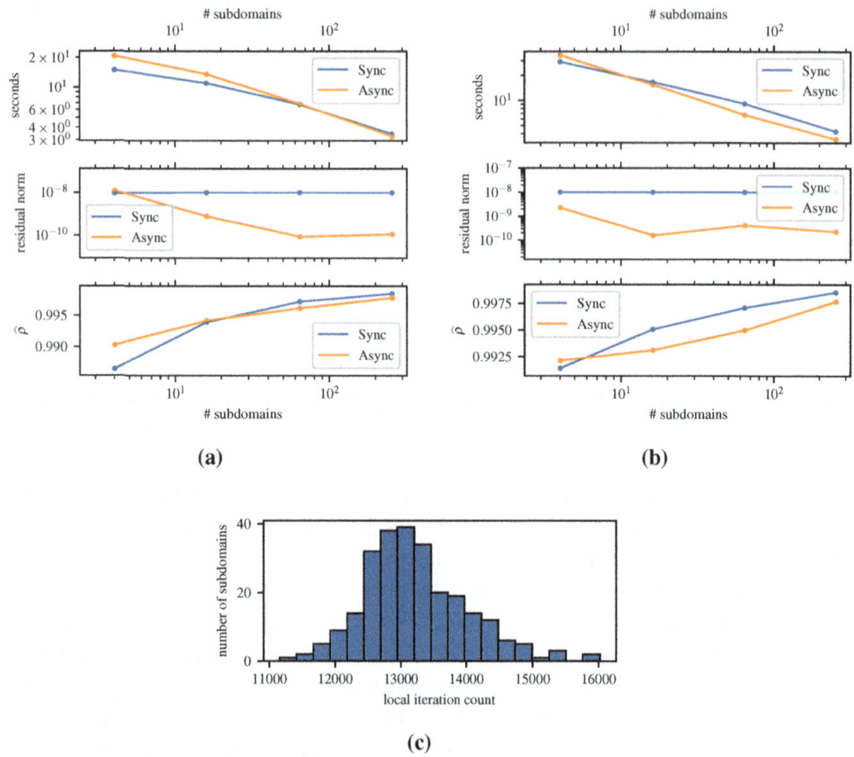

Fig. 3: *(a)* Performance of synchronous and asynchronous one-level RAS for a system size of approximately 261,000 unknowns. The subdomains are load balanced. From top to bottom: Solution time, final residual norm, and the resulting approximate contraction factor. *(b)* Performance of synchronous and asynchronous one-level RAS for a system size of approximately 261,000 unknowns under load imbalance: one subdomain is 50% larger than the rest. *(c)* Histogram of local iteration counts asynchronous one-level RAS with 256 subdomains in the balanced case.

of the coarse grid system is smaller than the size of the typical subdomain problem, the asynchronous method outperforms its synchronous counterpart.

Acknowledgements We thank Daniel Szyld for helpful discussions on asynchronous methods. This material is based upon work supported by the U.S. Department of Energy, Office of Science, Office of Advanced Scientific Computing Research, Applied Mathematics program under Award Numbers DE-SC-0016564. This research used resources of the National Energy Research Scientific Computing Center, a DOE Office of Science User Facility supported by the Office of Science of the U.S. Department of Energy under Contract No. DE-AC02-05CH11231. Sandia National Laboratories is a multimission laboratory managed and operated by National Technology and Engineering Solutions of Sandia, LLC, a wholly owned subsidiary of Honeywell International, Inc., for the U.S. Department of Energy's National Nuclear Security Administration under contract DE-NA0003525. This paper describes objective technical results and analysis. Any subjective views

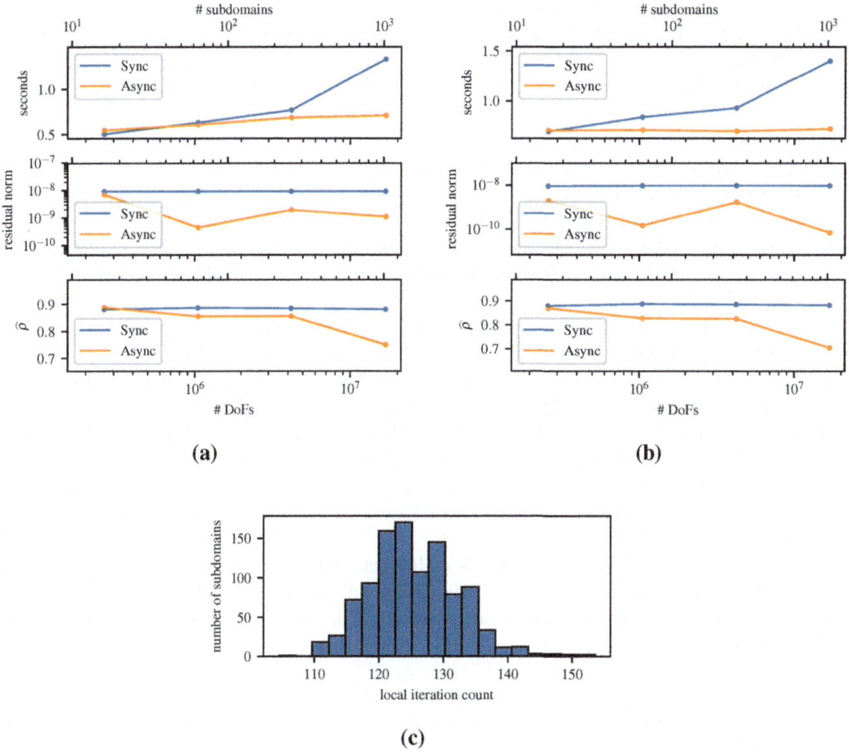

Fig. 4: *(a)* Weak scaling of synchronous and asynchronous two-level additive RAS, load balanced case. From top to bottom: Total solution time, final residual norm, and approximate contraction factor. *(b)* Weak scaling of synchronous and asynchronous two-level additive RAS under load imbalance: one subdomain is 50% larger than all the other ones. *(c)* Histogram of local iteration counts asynchronous two-level additive RAS with 1024 subdomains in the balanced case.

or opinions that might be expressed in the paper do not necessarily represent the views of the U.S. Department of Energy or the United States Government.

References

1. Bahi, J.M., Contassot-Vivier, S., Couturier, R., Vernier, F.: A decentralized convergence detection algorithm for asynchronous parallel iterative algorithms. IEEE Transactions on Parallel and Distributed Systems **16**(1), 4–13 (2005)
2. Baudet, G.M.: Asynchronous iterative methods for multiprocessors. Journal of the ACM (JACM) **25**(2), 226–244 (1978)
3. Bertsekas, D.P.: Distributed asynchronous computation of fixed points. Mathematical Programming **27**(1), 107–120 (1983)
4. Cappello, F., Geist, A., Gropp, B., Kale, L., Kramer, B., Snir, M.: Toward exascale resilience. International Journal of High Performance Computing Applications **23**(4), 374–388 (2009)

5. Cappello, F., Geist, A., Gropp, W., Kale, S., Kramer, B., Snir, M.: Toward exascale resilience: 2014 update. Supercomputing frontiers and innovations **1**(1), 5–28 (2014)
6. Chazan, D., Miranker, W.: Chaotic relaxation. Linear algebra and its applications **2**(2), 199–222 (1969)
7. Demmel, J.W., Eisenstat, S.C., Gilbert, J.R., Li, X.S., Liu, J.W.H.: A supernodal approach to sparse partial pivoting. SIAM J. Matrix Analysis and Applications **20**(3), 720–755 (1999)
8. Dolean, V., Jolivet, P., Nataf, F.: An introduction to domain decomposition methods: algorithms, theory, and parallel implementation, vol. 144. SIAM (2015)
9. Frommer, A., Szyld, D.B.: On asynchronous iterations. Journal of Computational and Applied Mathematics **123**(1), 201–216 (2000)
10. Ghysels, P., Ashby, T.J., Meerbergen, K., Vanroose, W.: Hiding global communication latency in the GMRES algorithm on massively parallel machines. SIAM Journal on Scientific Computing **35**(1), C48–C71 (2013)
11. Glusa, C., Ramanan, P., Boman, E.G., Chow, E., Rajamanickam, S.: Asynchronous One-Level and Two-Level Domain Decomposition Solvers. ArXiv e-prints (2018)
12. Karypis, G., Kumar, V.: A Fast and High Quality Multilevel Scheme for Partitioning Irregular Graphs. SIAM Journal on Scientific Computing **20**(1), 359–392 (1998)
13. Li, X., Demmel, J., Gilbert, J., iL. Grigori, Shao, M., Yamazaki, I.: SuperLU Users' Guide. Tech. Rep. LBNL-44289, Lawrence Berkeley National Laboratory (1999)
14. Magoulès, F., Gbikpi-Benissan, G.: Distributed convergence detection based on global residual error under asynchronous iterations. IEEE Transactions on Parallel and Distributed Systems (2017)
15. Smith, B., Bjorstad, P., Gropp, W.: Domain decomposition: parallel multilevel methods for elliptic partial differential equations. Cambridge university press (2004)
16. Toselli, A., Widlund, O.: Domain decomposition methods: algorithms and theory, vol. 3. Springer (2005)

Comparison of Continuous and Discrete Techniques to Apply Coarse Corrections

Martin J. Gander, Laurence Halpern, and Kévin Santugini-Repiquet

1 Introduction

There has been substantial attention on coarse correction in the domain decomposition community over the last decade, sparked by the interest of solving high contrast and multiscale problems, since in this case, the convergence of two-level domain decomposition methods is deteriorating when the contrast becomes large, see [1, 10, 16, 17, 11, 9, 8] and references therein. Our main interest here is not the content of the coarse spaces, but the way they are applied to correct the subdomain iterates. A classical way at the discrete level to apply coarse corrections, which led to the two level additive Schwarz method introduced in [2], is based on the residual like in multigrid: one computes the residual, projects it onto the coarse space, then solves a coarse problem which is for example obtained by a Galerkin projection of the fine system matrix on the coarse space, and then prolongates the correction by interpolation to the fine grid to add the correction to the current subdomain approximation. A complete analysis of this two level additive Schwarz preconditioner at the continuous level is given in [5], and for better coarse spaces, see [4, 6]. Another technique, also at the discrete level, is to use deflation, going back to the first coarse correction technique [15], where the functions spanning the coarse space are deflated, and then a deflated system is solved, see [14]. A further important class of coarse space correction techniques at the discrete level are the Balancing Domain Decomposition (BDD) methods [12, 13]. A more recent and very general approach at the continuous level for coarse correction is to approximately solve a

Martin J. Gander
Université de Genève, Switzerland e-mail: martin.gander@unige.ch

Laurence Halpern
Université Paris 13, France e-mail: halpern@math.univ-paris13.fr

Kévin Santugini-Repiquet
Bordeaux INP, IMB, UMR 5251, F-33400,Talence, France, e-mail: Kevin.Santugini-Repiquet@bordeaux-inp.fr

R. Haynes et al. (eds.), *Domain Decomposition Methods in Science and Engineering XXV*,
Lecture Notes in Computational Science and Engineering 138,
https://doi.org/10.1007/978-3-030-56750-7_15

transmission problem for the error, as described in [7], which also shows that for domain decomposition methods discontinuous coarse spaces are of interest, since subdomain solutions are in general discontinuous in their traces and/or fluxes at the interfaces. This observation led to the DCS-DMNV algorithm (Discontinuous Coarse Space - Dirichlet Minimization and Neumann Variational) at the continuous level, a two-level iterative domain decomposition algorithm introduced in [3].

We are interested here in understanding if there is a relation between the coarse corrections formulated at the discrete level by a residual correction, like in Additive Schwarz, and the coarse correction obtained at the continuous level solving a transmission problem. These two approaches seem at first to be very different, and to be able to compare them, we will precisely compute the coarse correction one obtains with these two approaches for the very simple model problem

$$\mathcal{L}u := \partial_{xx}u = f \quad \text{in } \Omega := (0,1), \quad u(0) = u(1) = 0, \tag{1}$$

and two subdomain iterates u_j, $j = 1,2$ on the subdomains $\Omega_1 := (0, \frac{1}{2} + L)$ and $\Omega_2 := (\frac{1}{2} - L, 1)$ which were obtained by an arbitrary domain decomposition method, i.e. the subdomain iterates simply satisfy the equation in (1) and the outer homogeneous boundary conditions, but no other interface condition at $\frac{1}{2} - L$ and $\frac{1}{2} + L$. They can thus come from a Schwarz method if $L > 0$, optimized Schwarz method for both $L > 0$ and $L = 0$, or a FETI or Neumann-Neumann method if $L = 0$. To compare continuous and discrete techniques, we also assume that we have a discretization of (1) leading to a linear system of equations

$$A\mathbf{u} = \mathbf{f}, \tag{2}$$

and two discrete subdomain iterates $\mathbf{u}_j$, $j = 1,2$.

2 Discrete Coarse Correction Based on the Residual

Suppose our coarse space is spanned by two continuous functions q_1 and q_2, see for example the hat functions (thick solid blue lines) in Fig. 1. Evaluating them on the grid used for the discretization leads to two vectors $\mathbf{q}_1$ and $\mathbf{q}_2$. To formulate the classical residual based coarse correction like in multigrid and used in Additive Schwarz, one puts the two row vectors $\mathbf{q}_1$ and $\mathbf{q}_2$ into the coarse restriction matrix R_0, and forms the coarse matrix $A_0 := R_0 A R_0^T$, like in a classical Galerkin approach. Having two approximate discrete subdomain solutions $\mathbf{u}_1$ and $\mathbf{u}_2$, one forms a global approximation using a partition of unity χ_j (diagonal matrices summing to the identity in this discrete setting with ones on the diagonal outside the overlap),

$$\tilde{\mathbf{u}} := \chi_1 \mathbf{u}_1 + \chi_2 \mathbf{u}_2, \tag{3}$$

and then corrects this approximation by the residual correction formula

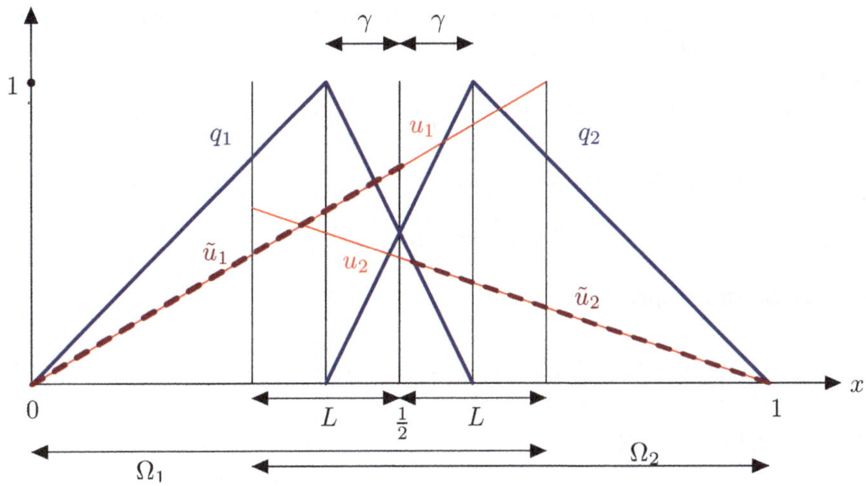

Fig. 1: Geometry with two subdomains Ω_j, coarse functions q_j and subdomain solutions u_j, $j = 1, 2$, which could be restricted to a non-overlapping decomposition to become $\tilde{u}_j$

$$\tilde{\mathbf{u}}^{new} := \tilde{\mathbf{u}} + R_0^T A_0^{-1} R_0(\mathbf{f} - A\tilde{\mathbf{u}}). \tag{4}$$

3 Continuous Coarse Correction Using a Transmission Problem

At the continuous level, a coarse correction can be computed by solving a transmission problem between the subdomains: for two approximate subdomain solution functions u_1 and u_2 shown as thin solid red lines in Figure 1, we restrict them first to a non-overlapping decomposition if the DD method used overlap,

$$\tilde{u}_1 := u_1|_{(0,\frac{1}{2})} \quad \text{and} \quad \tilde{u}_2 := u_2|_{(\frac{1}{2},1)}, \tag{5}$$

as shown with thick dashed dark red lines in Figure 1. If the DD method did not use overlap, we just denote by the tilde quantities $\tilde{u}_j$ the original iterates u_j, $j = 1, 2$. We then form the global approximation $\tilde{u}$ by gluing $\tilde{u}_1$ and $\tilde{u}_2$ together,

$$\tilde{u}(x) := \begin{cases} \tilde{u}_1(x) \text{ if } x \leq \frac{1}{2}, \\ \tilde{u}_2(x) \text{ if } x > \frac{1}{2}. \end{cases} \tag{6}$$

To compute the coarse correction, one can then for example use the DCS-DMNV technique, which we describe now using the coarse basis functions q_j shown with thick solid blue lines in Fig. 1 for the specific case when $\gamma = 0$: we define a continuous coarse space X_c and a discontinuous coarse space X_d by

$$X_c := \text{span}\{q_1 + q_2\}, \quad X_d := \text{span}\{q_1, q_2\}, \quad \gamma = 0. \tag{7}$$

Note that the glued solution $\tilde{u}$ lies in X_d. We then introduce a functional for measuring the jump in the approximate solution $\tilde{u}$ at the interfaces, which in our example would be at $x = \frac{1}{2}$,

$$q(v) := [v]^2(\frac{1}{2}), \quad [v](\frac{1}{2}) := v^+(\frac{1}{2}) - v^-(\frac{1}{2}). \tag{8}$$

To correct the approximation $\tilde{u}$, DCS-DMNV solves the minimization problem

$$\tilde{u}^{new} = \tilde{u} + \text{argmin}_{v \in V} q(\tilde{u} + v) \tag{9}$$

over the constraint space

$$V := \{v \in X_d : \int_\Omega v'(x)w'(x)\,dx = [\tilde{u}'](\frac{1}{2})w(\frac{1}{2}), \ \forall w \in X_c\}. \tag{10}$$

The underlying vector space is

$$V_0 := \text{span}(q_1 - q_2) \quad \text{and} \quad X_d = V_0 \oplus X_c. \tag{11}$$

4 Comparison of the Discrete and Continuous Techniques

In order to compare the discrete residual based coarse grid correction (4) to the continuous coarse correction (9) obtained by solving approximately a transmission problem using DCS-DMNV, we need to first formulate (4) at the continuous level for the specific case where the partition of unity (3) used in (4) glues the approximate subdomain solutions the same way as in DCS-DMNV which uses (6). Note that the glued function $\tilde{u}$ is a piece-wise C^∞ distribution, supported at $\frac{1}{2}$. This leads to

Theorem 1 *Let q_j for $j = 1, 2$ be the two hat functions in Figure 1,*

$$q_1 = \begin{cases} \frac{1}{\frac{1}{2}-\gamma}x & on\ (0, \frac{1}{2}-\gamma), \\ \frac{1}{2\gamma}(\frac{1}{2}+\gamma-x) & on\ (\frac{1}{2}-\gamma, \frac{1}{2}+\gamma), \\ 0 & on\ (\frac{1}{2}+\gamma, 1), \end{cases} \quad q_2 = \begin{cases} 0 & on\ (0, \frac{1}{2}-\gamma), \\ \frac{1}{2\gamma}(x-\frac{1}{2}+\gamma) & on\ (\frac{1}{2}-\gamma, \frac{1}{2}+\gamma), \\ \frac{1}{\frac{1}{2}-\gamma}(1-x) & on\ (\frac{1}{2}+\gamma, 1), \end{cases}$$

for $0 \le \gamma \le L$, and let the partition of unity (3) *be defined as in* (6). *Then the continuous equivalent to the discrete residual based coarse correction* (4) *is*

$$\tilde{u}^{new} = \tilde{u} + (\frac{1}{2} - \gamma)\left((\frac{1}{2}[\tilde{u}'](\frac{1}{2}) + [\tilde{u}](\frac{1}{2}))q_1 + (\frac{1}{2}[\tilde{u}'](\frac{1}{2}) - [\tilde{u}](\frac{1}{2}))q_2\right). \tag{12}$$

Proof If u is a piece-wise C^2 function with a finite number of jumps at $a_1, \ldots, a_N$, and T_u denotes the distribution corresponding to u, we obtain for the derivatives using the jumps formula

$$T'_u = T_{u'} + \sum_{i=1}^{N} [u](a_i)\delta_{a_i}, \quad T''_u = T_{u''} + \sum_{i=1}^{N} ([u'](a_i)\delta_{a_i} + [u](a_i)\delta'_{a_i}).$$

Recall that the derivative of the Dirac distribution δ_a is defined by $\delta'_a(\phi) = -\phi'(a)$ for ϕ a C^1 function in the neighborhood of a. If we now apply the jump formula to $\tilde{u}$ we constructed by gluing the subdomain solutions together in (6), we obtain for the residual we need for the computation of the coarse correction

$$r := f - T''_{\tilde{u}} = -[\tilde{u}](\frac{1}{2})\delta'_{\frac{1}{2}} - [\tilde{u}'](\frac{1}{2})\delta_{\frac{1}{2}}.$$

The continuous equivalent to the discrete coarse correction (4) is to search for a coarse correction function $U = \alpha q_1 + \beta q_2$ such that

$$(T''_U, q_1) = (r, q_1), \quad (T''_U, q_2) = (r, q_2), \tag{13}$$

and to add it to $\tilde{u}$ to obtain $\tilde{u}^{new}$. We will work equivalently for this proof instead with the basis $(q_1 + q_2, q_1 - q_2)$ to solve system (13), which will naturally reveal the role played by the sum (continuous) and difference (discontinuous in the limit when γ goes to zero) and prepare for the relation with the DCS-DMNV approach. Working with the sum and difference also simplifies the solution of the system. We thus project now the residual r onto V_0 defined in (11) and X_c defined in (7), for which we need the functions $q_1 + q_2$ and $q_1 - q_2$,

$$q_1+q_2 = \begin{cases} \frac{1}{\frac{1}{2}-\gamma}x & \text{on } (0, \frac{1}{2}-\gamma), \\ 1 & \text{on } (\frac{1}{2}-\gamma, \frac{1}{2}+\gamma), \\ \frac{1}{\frac{1}{2}-\gamma}(1-x) & \text{on } (\frac{1}{2}+\gamma, 1), \end{cases} \quad q_1-q_2 = \begin{cases} \frac{1}{\frac{1}{2}-\gamma}x & \text{on } (0, \frac{1}{2}-\gamma), \\ \frac{1}{2\gamma}(1-2x) & \text{on } (\frac{1}{2}-\gamma, \frac{1}{2}+\gamma), \\ -\frac{1}{\frac{1}{2}-\gamma}(1-x) & \text{on } (\frac{1}{2}+\gamma, 1). \end{cases}$$

Since $q_1 + q_2$ is constant equal to 1 in $(\frac{1}{2} - \gamma, \frac{1}{2} + \gamma)$, we obtain

$$(r, q_1 + q_2) = -[\tilde{u}'](\frac{1}{2}), \quad (r, q_1 - q_2) = -\frac{1}{\gamma}[\tilde{u}](\frac{1}{2}).$$

We search now for a coarse correction $U = \alpha'(q_1 + q_2) + \beta'(q_1 - q_2)$ such that

$$(T''_U, q_1 + q_2) = (r, q_1 + q_2), \quad (T''_U, q_1 - q_2) = (r, q_1 - q_2). \tag{14}$$

From the jumps formula, we find

$$T''_{q_1 \pm q_2} = [q'_1 \pm q'_2](\frac{1}{2} - \gamma)\delta_{\frac{1}{2}-\gamma} + [q'_1 \pm q'_2](\frac{1}{2} + \gamma)\delta_{\frac{1}{2}+\gamma},$$

which leads to

$$T''_{q_1+q_2} = \frac{-1}{\frac{1}{2} - \gamma}(\delta_{\frac{1}{2}-\gamma} + \delta_{\frac{1}{2}+\gamma}), \quad T''_{q_1-q_2} = \frac{-1}{2\gamma(\frac{1}{2} - \gamma)}(\delta_{\frac{1}{2}-\gamma} - \delta_{\frac{1}{2}+\gamma}).$$

Since $(q_1+q_2)(\frac{1}{2}-\gamma) = (q_1+q_2)(\frac{1}{2}+\gamma) = 1$ and $(q_1-q_2)(\frac{1}{2}-\gamma) = -(q_1-q_2)(\frac{1}{2}+\gamma) = 1$, we find that

$$T''_{q_1+q_2}(q_1+q_2) = 2\gamma T''_{q_1-q_2}(q_1-q_2) = \frac{-2}{\frac{1}{2}-\gamma}, \quad T''_{q_1\pm q_2}(q_1 \mp q_2) = 0.$$

Inserting this into (14) gives a simple diagonal system for α' and β', namely

$$\frac{-2}{\frac{1}{2}-\gamma}\alpha' = -[\tilde{u}'](\frac{1}{2}), \quad \frac{-1}{\gamma(\frac{1}{2}-\gamma)}\beta' = -\frac{1}{\gamma}[\tilde{u}](\frac{1}{2}),$$

and thus for the coarse correction

$$U = (\frac{1}{2}-\gamma)(\frac{1}{2}[\tilde{u}'](\frac{1}{2})(q_1+q_2)+[\tilde{u}](\frac{1}{2})(q_1-q_2)),$$

which concludes the proof. □

For the DCS-DMNV algorithm for computing the coarse correction described in Section 3, we obtain the following theorem:

Theorem 2 *The coarse correction computed by the DCS-DMNV algorithm* (9)-(10) *is given by*

$$\tilde{u}^{new} = \tilde{u} + (\frac{1}{2}[\tilde{u}](\frac{1}{2}) + \frac{1}{4}[\tilde{u}'](\frac{1}{2}))q_1 + (-\frac{1}{2}[\tilde{u}](\frac{1}{2}) + \frac{1}{4}[\tilde{u}'](\frac{1}{2}))q_2,$$

which is equal to the limit of the coarse correction computed by the residual correction approach given in (12) *when γ goes to zero.*

Proof The DCS-DMNV algorithm uses the spaces V_0 and X_c, which we defined in (7) and (11) using the hat functions q_1 and q_2 for the specific case where $\gamma = 0$, in which q_1 and q_2 are discontinuous at $x = \frac{1}{2}$, and we have

$$q_1+q_2 = \begin{cases} 2x & \text{on } [0,\frac{1}{2}], \\ 2(1-x) & \text{on } [\frac{1}{2},1], \end{cases} \qquad q_1-q_2 = \begin{cases} 2x & \text{on } [0,\frac{1}{2}), \\ -2(1-x) & \text{on } (\frac{1}{2},1]. \end{cases}$$

We first note that X_c and V_0 are orthogonal subspaces of L^2, and the same holds for their derivatives, since $\|q_1\| = \|q_2\|$ and $\|q_1'\| = \|q_2'\|$. We next identify the constraint space V from (10): the function

$$v := \alpha'(q_1+q_2) + \beta'(q_1-q_2)$$

belongs to V if and only if

$$\int_\Omega (\alpha'(q_1'+q_2') + \beta'(q_1'-q_2'))(x)(q_1'+q_2')(x)\,dx = [\tilde{u}'](\frac{1}{2})(q_1+q_2)(\frac{1}{2}),$$

which gives

$$4\alpha' = [\tilde{u}'](\frac{1}{2}).$$

This defines V as the affine line

$$V = V_0 + \frac{1}{4}[\tilde{u}'](\frac{1}{2})(q_1 + q_2).$$

Therefore $U = \frac{1}{4}[\tilde{u}'](\frac{1}{2})(q_1 + q_2) + \beta'(q_1 - q_2)$. Now the Euler equation for (9) is

$$q'(\tilde{u} + U) \cdot v := 2[\tilde{u} + U](\frac{1}{2})[v](\frac{1}{2}) = 0 \; \forall v \in V_0. \tag{15}$$

Since $[q_1 - q_2](\frac{1}{2}) = 2$, (15) yields $[\tilde{u} + U](\frac{1}{2}) = 0$, and since $q_1 + q_2$ is continuous at $x = \frac{1}{2}$,

$$[\tilde{u}](\frac{1}{2}) + \beta'[q_1 - q_2](\frac{1}{2}) = [\tilde{u}](\frac{1}{2}) - 2\beta' = 0.$$

Therefore

$$U = \frac{1}{2}[\tilde{u}](\frac{1}{2})(q_1 - q_2) + \frac{1}{4}[\tilde{u}'](\frac{1}{2})(q_1 + q_2),$$

and we see that this is indeed the limit as $\gamma \to 0$ of the system (12). □

5 Conclusions

We have shown that two apparently quite different approaches for computing a coarse correction in domain decomposition, namely the residual based approach at the discrete level, and the approximate solution of a transmission problem at the continuous level using DCS-DMNV, lead to the same coarse correction in the limit when the discretized approach is computed at the continuous level, provided that one uses a discontinuous partition of unity. It therefore does not matter in this case which approach is used for computing the coarse correction, they are equivalent.

We showed our result for a simplified setting of Laplace's equation in 1D and for two subdomains only, but the generalization to many subdomains in 1D does not pose any difficulties, one just has to use the jumps formula several times. The generalization to higher spatial dimensions is also possible and not difficult in the case of strip decompositions. The case when cross points are present would however require more care and does not follow trivially. For a more general operator than the Laplacian, the generalization is in principle also possible, but one essential ingredient is that the coarse space functions q_j must satisfy the homogeneous equation, which is in general a desirable property for coarse space functions, see [7] and references therein.

A further open question is how the coarse correction computation based on deflation, and the BDD technique, are related to the two methods we compared here. We are currently studying these two techniques for the same simple model problem presented here, and also the higher dimensional case.

References

1. Aarnes, J., Hou, T.Y.: Multiscale domain decomposition methods for elliptic problems with high aspect ratios. Acta Math. Appl. Sin. Engl. Ser. **18**(1), 63–76 (2002)
2. Dryja, M., Widlund, O.B.: An additive variant of the Schwarz alternating method for the case of many subregions. Tech. Rep. 339, also Ultracomputer Note 131, Department of Computer Science, Courant Institute (1987)
3. Gander, M.J., Halpern, L., Santugini-Repiquet, K.: Discontinuous coarse spaces for DD-methods with discontinuous iterates. In: Domain Decomposition Methods in Science and Engineering XXI. Springer LNCSE (2013)
4. Gander, M.J., Halpern, L., Santugini-Repiquet, K.: A new coarse grid correction for RAS. In: Domain Decomposition Methods in Science and Engineering XXI. Springer LNCSE (2013)
5. Gander, M.J., Halpern, L., Santugini-Repiquet, K.: Continuous analysis of the additive Schwarz method: A stable decomposition in H1 with explicit constants. ESAIM: Mathematical Modelling and Numerical Analysis **49**(3), 713–740 (2015)
6. Gander, M.J., Halpern, L., Santugini-Repiquet, K.: New coarse corrections for Optimized Restricted Additive Schwarz using PETSc. In: Domain Decomposition Methods in Science and Engineering XXV. submitted (2018)
7. Gander, M.J., Halpern, L., Santugini-Repiquet, K.: On optimal coarse spaces for domain decomposition and their approximation. In: Domain Decomposition Methods in Science and Engineering XXIV. Springer LNCSE (2018)
8. Gander, M.J., Loneland, A.: SHEM: An optimal coarse space for RAS and its multiscale approximation. In: Domain Decomposition Methods in Science and Engineering XXIII. Springer (2016)
9. Gander, M.J., Loneland, A., Rahman, T.: Analysis of a new harmonically enriched multiscale coarse space for domain decomposition methods. arXiv preprint arXiv:1512.05285 (2015)
10. Graham, I.G., Lechner, P.O., Scheichl, R.: Domain decomposition for multiscale PDEs. Numer. Math. **106**(4), 589–626 (2007)
11. Klawonn, A., Radtke, P., Rheinbach, O.: FETI-DP methods with an adaptive coarse space. SIAM Journal on Numerical Analysis **53**(1), 297–320 (2015)
12. Mandel, J.: Balancing domain decomposition. Communications in Numerical Methods in Engineering **9**(3), 233–241 (1993). `DOI:10.1002/cnm.1640090307`
13. Mandel, J., Brezina, M.: Balancing domain decomposition for problems with large jumps in coefficients. Math. Comp. **65**, 1387–1401 (1996)
14. Nabben, R., Vuik, C.: A comparison of deflation and coarse grid correction applied to porous media flow. SIAM Journal on Numerical Analysis **42**(4), 1631–1647 (2004)
15. Nicolaides, R.A.: Deflation conjugate gradients with application to boundary value problems. SIAM J. Num. An, **24**(2), 355–365 (1987). `DOI:doi:10.1137/0724027`
16. Pechstein, C., Scheichl, R.: Analysis of FETI methods for multiscale PDEs. Numer. Math. **111**(2), 293–333 (2008)
17. Spillane, N., Dolean, V., Hauret, P., Nataf, F., Pechstein, C., Scheichl, R.: Abstract robust coarse spaces for systems of PDEs via generalized eigenproblems in the overlaps. Numerische Mathematik **126**(4), 741–770 (2014)

On the Scalability of the Parallel Schwarz Method in One-Dimension

Gabriele Ciaramella, Muhammad Hassan, and Benjamin Stamm

1 Introduction and main results

An algorithm is said to be weakly scalable if it can solve progressively larger problems with an increasing number of processors in a fixed amount of time. According to classical Schwarz theory, the parallel Schwarz method (PSM) is not scalable (see, e.g., [2, 7]). Recent results in computational chemistry, however, have shed more light on the scalability of the PSM: surprisingly, in contrast with classical Schwarz theory, the authors in [1] provide numerical evidence that in some cases the one-level PSM converges to a given tolerance within the same number of iterations independently of the number N of subdomains. This behaviour is observed if fixed-sized subdomains form a "chain-like" domain such that the intersection of the boundary of each subdomain with the boundary of the global domain is non-empty. This result was subsequently rigorously proved in [3, 4, 5] for the PSM and in [2] for other one-level methods. On the other hand, this weak scalability is lost if the fixed-sized subdomains form a "globular-type" domain Ω, where the boundaries of many subdomains lie in the interior of Ω. The following question therefore arises: is it possible to quantify the lack of scalability of the PSM for cases where individual subdomains are entirely embedded inside the global domain? To do so, for increasing N one would need to estimate the number of iterations necessary to achieve a given tolerance.

Some isolated results in this direction do exist in the literature. For instance, in [2] a heuristic argument is used to explain why in the case of the PSM for the solution of a 1D Laplace problem an unfortunate initialisation leads to a contraction in the infinity norm being observed only after a number of iterations proportional to N.

Gabriele Ciaramella
Universität Konstanz e-mail: gabriele.ciaramella@uni-konstanz.de

Muhammad Hassan
RWTH Aachen University e-mail: hassan@mathcces.rwth-aachen.de

Benjamin Stamm
RWTH Aachen University e-mail: stamm@mathcces.rwth-aachen.de

R. Haynes et al. (eds.), *Domain Decomposition Methods in Science and Engineering XXV*,
Lecture Notes in Computational Science and Engineering 138,
https://doi.org/10.1007/978-3-030-56750-7_16

Similarly, for a special choice of overlapping subdomains, an elegant result can be found in [6] where the authors prove that the Schwarz waveform-relaxation, for the solution of the heat equation, contracts at most every $m+2$ iterations with m being an integer representing the maximum distance of the subdomains from the boundary. Nevertheless, the literature does not contain a comprehensive study of this problem for a general decomposition. Furthermore, existing results unfortunately do not provide a systematic approach to build on and extend in order to cover more general settings. Our goal therefore has been to develop a *framework* that can be applied to a broad class of overlapping subdomains, in multiple dimensions and containing sub-domains with an arbitrary number and type (double, triple, quadruple and so on) of intersections, such as for molecular domains in computational chemistry [1].

Of course, tackling this problem in a completely general setting is a daunting task. The purpose of the current article is to develop such a new framework and apply it to the PSM for the solution of a 1D Laplace problem as a 'toy' problem. The key elements of our proposed framework are

- the identification of an adequate norm for studying the properties of the Schwarz operator,
- the maximum principle,
- and the idea of tracking the propagation of the contraction towards the interior of the global domain Ω.

Our expectation is that a framework based on these ingredients can then be systematically extended to more general decompositions of the domain which can be quite complex in two and three dimensions. We emphasise that most (but not all) of the results we prove are either known or intuitively clear. Our true contribution is the new *analysis technique* that we introduce. On the one hand, this technique results in a deeper understanding of the method and leads to a sharper description of the contraction behaviour. On the other hand, the tools developed in this article also suggest a systematic roadmap to extend our results to more realistic problems in higher dimensions. In principle, this can be done by carefully tracking the propagation of the contraction through the subdomains as attempted in this manuscript. Note however that the contraction behaviour for a 1D reaction-diffusion equation is completely different from that of the 1D Laplace equation. This can be proved as shown in [2, 3] .

We first state our main results. We consider the Laplace equation in one-dimension. Let $L > 0$, we must find a function $e\colon [0, L] \to \mathbb{R}$ that solves

$$(e)''(x) = 0 \quad \forall\, x \in (0, L), \qquad e(0) = 0, \qquad e(L) = 0. \tag{1}$$

Clearly (1) represents an error equation whose solution is trivially $e = 0$. In order to apply the PSM to solve (1), we consider a decomposition $\Omega = \cup_{j=1}^{N} \Omega_j$, where $\Omega_j := (a_j, b_j)$ with $a_1 = 0$, $b_1 = a_1 + \ell$ and $a_{j+1} = j(\ell - \delta)$, $b_{j+1} = a_{j+1} + \ell$ for $j = 1, \ldots, N-1$. Here, $\ell > 0$ is the length of each subdomain, $\delta > 0$ the overlap, and it holds that $L = N\ell - (N-1)\delta$. Now, let $e_0\colon \Omega \to \mathbb{R}$ be some initialization. The PSM defines the sequences $\{e_j^n\}_{n\in\mathbb{N}}$ by solving for each $n \in \mathbb{N}$ the sub-problems

$$(e_j^n)''(x) = 0 \quad \forall\, x \in (a_j, b_j), \quad e_j^n(a_j) = e_{j-1}^{n-1}(a_j), \quad e_j^n(b_j) = e_{j+1}^{n-1}(b_j), \quad (2)$$

for each $j = 2, \ldots, N-1$ and

$$\begin{aligned} &(e_1^n)''(x) = 0 \quad \forall\, x \in (a_1, b_1), && e_1^n(a_1) = 0, && e_1^n(b_1) = e_2^{n-1}(b_1), \\ &(e_N^n)''(x) = 0 \;\forall\, x \in (a_N, b_N), && e_N^n(a_N) = e_{N-1}^{n-1}(a_N), && e_N^n(b_N) = 0. \end{aligned}$$

Solving (1) and (2) and defining for each $n \in \mathbb{N}$ the vector $\boldsymbol{e}^n \in \mathbb{R}^{2N}$ as

$$\boldsymbol{e}^n := \left[e_1^n(a_1)\; e_1^n(a_2)\; e_2^n(b_1)\; e_2^n(a_3)\; \cdots\; e_j^n(b_{j-1})\; e_j^n(a_{j+1})\; \cdots\; e_N^n(b_{N-1})\; e_N^n(b_N)\right]^\top,$$

it is possible to write the PSM iterations as $\boldsymbol{e}^{n+1} = T\boldsymbol{e}^n$. Here $T \in \mathbb{R}^{2N \times 2N}$ is a non-negative ($T_{j,k} \geq 0$), non-symmetric block tridiagonal matrix:

$$T = \begin{bmatrix} 0 & \widetilde{T}_2 & 0 & \cdots & 0 & 0 & 0 \\ T_1 & 0 & T_2 & \cdots & 0 & 0 & 0 \\ \ddots & \ddots & \ddots & \ddots & \ddots & \ddots & \ddots \\ 0 & 0 & 0 & \cdots & T_1 & 0 & T_2 \\ 0 & 0 & 0 & \cdots & 0 & \widetilde{T}_1 & 0 \end{bmatrix}, \quad \begin{aligned} T_1 &= \begin{bmatrix} 0 & 1-\frac{\delta}{\ell} \\ 0 & \frac{\delta}{\ell} \end{bmatrix}, \; \widetilde{T}_1 = \begin{bmatrix} 0 & 1-\frac{\delta}{\ell} \\ 0 & 0 \end{bmatrix}, \\ T_2 &= \begin{bmatrix} \frac{\delta}{\ell} & 0 \\ 1-\frac{\delta}{\ell} & 0 \end{bmatrix}, \; \widetilde{T}_2 = \begin{bmatrix} 0 & 0 \\ 1-\frac{\delta}{\ell} & 0 \end{bmatrix}. \end{aligned}$$

We denote by $\|\cdot\|$ the usual infinity norm and the corresponding induced matrix norm. Our goal is to analyze the convergence properties of the PSM sequence $\{e_j^n\}_{n\in\mathbb{N}}$ with respect to $\|\cdot\|$. Hence, we must study the properties of the matrix T. Our main results are summarized in the following theorem.

Theorem 1 *Let $N \in \mathbb{N}$ be the number of subdomains in Ω and $\mathbf{1}_N \in \mathbb{R}^{2N}$ the vector whose elements are all equal to* 1. *Then $\|T^n\| = \|T^n\mathbf{1}_N\| \leq 1$ for any $n \in \mathbb{N}$ and*

(a) $\|T^{\lceil \frac{N}{2} \rceil}\| < 1$ *and hence $\rho(T) < 1$, where $\rho(T)$ is the spectral radius of T.*
(b) $\|T^{n+1}\| < \|T^n\|$ *if N is even, and $\|T^{n+2}\| < \|T^n\|$ if N is odd, for $n \geq \lceil \frac{N}{2} \rceil$.*

Theorem 1 states clearly that the PSM converges. Moreover, it identifies $\mathbf{1}_N$ as the unit vector that maximises the ℓ_∞ operator norm of the iteration matrices $T^n, n \in \mathbb{N}$. This fact is then used to prove Theorem 1 (b): if initialized with $\mathbf{1}_N$, after $\lceil \frac{N}{2} \rceil$ iterations the PSM sequence contracts in the infinity norm at *every iteration* if N is even, or *every second iteration* if N is odd. Although proven for a 1-D problem, this result is much sharper than the one found in [6], which states that the PSM sequence contracts in the infinity norm at most every $\lceil \frac{N}{2} \rceil$ iterations. To prove Theorem 1 (b) we use Lemmas 4 and 5. These two technical results characterize precisely the shape of the vector $\boldsymbol{e}^n = T^n\mathbf{1}_N$ at every iteration n and clearly show how the contraction propagates from the two points of $\partial\Omega$ towards the subdomains in the middle of Ω.

We prove Theorem 1 in the following sections. In particular, in Section 2 we prove first that $\|T^n\| = \|T^n\mathbf{1}_N\|$ and then Theorem 1 (a). In Section 3 we prove Theorem 1 (b). Notice that using Theorem 1, one could also estimate the spectral radius of T as

$$\rho(T) \leq \|T^{\lceil \frac{N}{2} \rceil}\|^{1/\lceil \frac{N}{2} \rceil} = \left[1 - \left(\frac{\delta}{L}\right)^{\lceil \frac{N}{2} \rceil}\right]^{1/\lceil \frac{N}{2} \rceil}.$$

This can be proved by a direct calculation involving geometric arguments. However, since this is a quite conservative bound, we will not prove this result in this short article.

2 Proof of Theorem 1 (a)

In what follows, we use $P = \begin{bmatrix} 0 & 1 \\ 1 & 0 \end{bmatrix}$. Moreover, let $m \in \mathbb{N}$ and $\boldsymbol{v}, \boldsymbol{w} \in \mathbb{R}^m$, then $\boldsymbol{v} < \boldsymbol{w}$ (resp. $\boldsymbol{v} > \boldsymbol{w}$) means that $v_i < w_i$ (resp. $v_i > w_i$) for each $i \in \{1, \ldots, m\}$.

Lemma 1 *For all $n \in \mathbb{N}$ it holds that $\|T^n\| = \|T^n \mathbf{1}_N\|$.*

Proof Let $\boldsymbol{w} = T^n \mathbf{1}_N$. Then for each $i \in \{1, \ldots, 2N\}$ it holds that $w_i = \sum_{j=1}^{2N} (T^n)_{ij}$. Since T is non-negative, T^n is also non-negative for any $n \in \mathbb{N}$ and it holds that

$$\|T^n\| = \max_{i=1,\ldots,2N} \sum_{j=1}^{2N} (T^n)_{ij} = \max_{i=1,\ldots,2N} \left| \sum_{j=1}^{2N} (T^n)_{ij} \right| = \max_{i=1,\ldots,2N} |w_i| = \|T^n \mathbf{1}_N\|.$$

Next, let $a, b, c, d \in [0, 1)$ be real numbers such that $a < b \leq c < d$. Direct calculations show that the matrices T_1 and T_2 satisfy the following relations:

$$b\,\mathbf{1}_1 < T_1 \begin{bmatrix} a \\ b \end{bmatrix} + T_2 \mathbf{1}_1 = \begin{bmatrix} (1 - \frac{\delta}{\ell})b + \frac{\delta}{\ell} \\ \frac{\delta}{\ell} b + 1 - \frac{\delta}{\ell} \end{bmatrix} < \mathbf{1}_1, \tag{3}$$

$$b\,\mathbf{1}_1 \leq T_1 \begin{bmatrix} a \\ b \end{bmatrix} + T_2 \begin{bmatrix} c \\ d \end{bmatrix} = \begin{bmatrix} (1 - \frac{\delta}{\ell})b + \frac{\delta}{\ell} c \\ \frac{\delta}{\ell} b + (1 - \frac{\delta}{\ell})c \end{bmatrix} \leq c\,\mathbf{1}_1, \tag{4}$$

$$T_1 \mathbf{1}_1 + T_2 \mathbf{1}_1 = \mathbf{1}_1, \qquad T_1 \begin{bmatrix} a \\ b \end{bmatrix} + T_2 \begin{bmatrix} c \\ d \end{bmatrix} = P \left(T_1 \begin{bmatrix} d \\ c \end{bmatrix} + T_2 \begin{bmatrix} b \\ a \end{bmatrix} \right), \tag{5}$$

where the equality in (4) holds if and only if $b = c$.

Definition 1 Let $n \in \left\{1, \ldots, \left\lceil \frac{N}{2} \right\rceil\right\}$ be a natural number, we define $V^n \subset \mathbb{R}^{2N}$ as

$$V^n := \left\{ \boldsymbol{v} := (\boldsymbol{v}_1, \ldots, \boldsymbol{v}_N) : \begin{cases} \boldsymbol{v}_j < \mathbf{1}_1 & \text{if } j \in \{1, \ldots, n\} \cup \{N + 1 - n, \ldots, N\} \\ \boldsymbol{v}_j = \mathbf{1}_1 & \text{otherwise} \end{cases} \right\}.$$

We now state and prove the main result that will lead directly to Theorem 1 (a).

Lemma 2 *Let $n \in \left\{1, \ldots, \left\lceil \frac{N}{2} \right\rceil\right\}$ be a natural number and let $\boldsymbol{w} = T^n \mathbf{1}_N$. Then it holds that $T^n \mathbf{1}_N \in V^n$, and for all $j \in \{1, \ldots, N\}$ it holds that $\boldsymbol{w}_j = P\, \boldsymbol{w}_{N+1-j}$.*

Proof We proceed by induction on the iteration index n. Let $n = 1$ and $\boldsymbol{w} = T\mathbf{1}_N$. Then by definition of the iteration matrix T and the matrices $\widetilde{T}_1, \widetilde{T}_2$ it holds that

$$\boldsymbol{w}_1 = \widetilde{T}_2 \mathbf{1}_1 = \begin{bmatrix} 0 \\ 1 - \frac{\delta}{\ell} \end{bmatrix} < \mathbf{1}_1, \qquad \boldsymbol{w}_N = \widetilde{T}_1 \mathbf{1}_1 = \begin{bmatrix} 1 - \frac{\delta}{\ell} \\ 0 \end{bmatrix} < \mathbf{1}_1,$$

so that $\boldsymbol{w}_1 = P\boldsymbol{w}_N$. Furthermore, by (5) it holds that $\boldsymbol{w}_j = T_1\mathbf{1}_1 + T_2\mathbf{1}_1 = \mathbf{1}_1$ for all $j \in \{2, \ldots, N-1\}$, and thus $\boldsymbol{w}_j = P\boldsymbol{w}_{N+1-j}$. Hence, Lemma 2 holds for $n = 1$.

Assume now that Lemma 2 holds for some $n \in \left\{1, \ldots, \left\lceil \frac{N}{2} \right\rceil - 1\right\}$. We must show that Lemma 2 also holds for $n+1$. Let $\boldsymbol{u} = T^n\mathbf{1}_N$ and let $\boldsymbol{w} = T^{n+1}\mathbf{1}_N$. We proceed in three parts. First, we prove that the result holds in the case $n \geq 2$ for indices $j \in \{1, \ldots, n-1\}$, then we prove it for the index $j = n$ and finally for the index $j = n+1$. Note that it is necessary to proceed in these three steps since in each of these cases $\boldsymbol{w}_j$ depends on $\boldsymbol{u}_{j-1}$ and $\boldsymbol{u}_{j+1}$ which take different values depending on the index j.

1. $n \geq 2$ and $j \in \{1, \ldots, n-1\}$: Assume first that $j = 1$. It follows from the induction hypothesis that $\boldsymbol{u}_2 = P\boldsymbol{u}_{N-1} < \mathbf{1}_1$. A direct calculation similar to the one for the base case $n = 1$ reveals that $\boldsymbol{w}_1 = P\boldsymbol{w}_N < \mathbf{1}_1$. Now assume $j \neq 1$. It follows by the induction hypothesis that $\boldsymbol{u}_{j-1} = P\boldsymbol{u}_{N+2-j}$, and $P\boldsymbol{u}_{j+1} = \boldsymbol{u}_{N-j}$, and $\boldsymbol{u}_{j-1}, \boldsymbol{u}_{j+1} < \mathbf{1}_1$. We therefore obtain from (5) that
$$\boldsymbol{w}_j = T_1\boldsymbol{u}_{j-1} + T_2\boldsymbol{u}_{j+1} = P\big(T_1 P\boldsymbol{u}_{j+1} + T_2 P\boldsymbol{u}_{j-1}\big) = P\big(T_1\boldsymbol{u}_{N-j} + T_2\boldsymbol{u}_{N-j+2}\big) = P\boldsymbol{w}_{N-j+1} < \mathbf{1}_1.$$
2. $j = n$: The induction hypothesis implies that $\boldsymbol{u}_{n-1} = P\boldsymbol{u}_{N+2-n}$, $\boldsymbol{u}_{n-1} < \mathbf{1}_1$ and $\boldsymbol{u}_{n+1} = \mathbf{1}_1$. Hence (5) implies that
$$\boldsymbol{w}_n = T_1\boldsymbol{u}_{n-1} + T_2\mathbf{1}_1 = P(T_1\mathbf{1}_1 + T_2 P\boldsymbol{u}_{n-1}) = P(T_1\mathbf{1}_1 + T_2\boldsymbol{u}_{N+2-n}) = P\boldsymbol{w}_{N+1-n} < \mathbf{1}_1.$$
3. Let $j = n+1$: By the induction hypothesis we have that $\boldsymbol{u}_n = P\boldsymbol{u}_{N+1-n}$, $\boldsymbol{u}_{n+2} = P\boldsymbol{u}_{N-1-n}$, $\boldsymbol{u}_n < \mathbf{1}_1$ and $\boldsymbol{u}_{n+2} \leq \mathbf{1}_1$. Using (4) and (5) we get
$$\boldsymbol{w}_{n+1} = T_1\boldsymbol{u}_n + T_2\boldsymbol{u}_{n+2} = P(T_1 P\boldsymbol{u}_{n+2} + T_2 P\boldsymbol{u}_n) = P(T_1\boldsymbol{u}_{N-1-n} + T_2\boldsymbol{u}_{N+1-n}) = P\boldsymbol{w}_{N-n} < \mathbf{1}_1.$$
It remains to show that $\boldsymbol{w}_k = \mathbf{1}_1$ for all $k \in \{n+2, \ldots, \lceil \frac{N}{2} \rceil\} \cup \{\lceil \frac{N}{2} \rceil, \ldots, N-1-n\}$. The induction hypothesis yields that $\boldsymbol{u} = T^n\mathbf{1}_N \in V^n$. Hence, $\boldsymbol{u}_k = \mathbf{1}_1$ for all $k \in \{n+1, \ldots, \lceil \frac{N}{2} \rceil\} \cup \{\lceil \frac{N}{2} \rceil, \ldots, N-n\}$. The result now follows by applying Equation (5).

Lemma 2 implies that $T^{\lceil \frac{N}{2} \rceil}\mathbf{1}_N \in V^{\lceil \frac{N}{2} \rceil}$ so that $\|T^{\lceil \frac{N}{2} \rceil}\| = \|T^{\lceil \frac{N}{2} \rceil}\mathbf{1}_N\| < 1$, which is precisely Theorem 1 (a).

3 Proof of Theorem 1 (b)

We first prove an intermediate lemma.

Lemma 3 *Let $n \in \left\{2, \ldots, \left\lfloor \frac{N}{2} \right\rfloor - 2\right\}$, let $\boldsymbol{u} = T^n\mathbf{1}_N$ and $\boldsymbol{w} = T\boldsymbol{u}$. If for all $j \in \{1, \ldots, n\}$ it holds that*
$$(\boldsymbol{u}_j)_1 \leq (\boldsymbol{u}_j)_2, \quad \textit{and} \quad \boldsymbol{u}_j < \boldsymbol{u}_{j+1}, \tag{6}$$
then for all $j \in \{1, \ldots, n+1\}$ it holds that

$$(\boldsymbol{w}_j)_1 \leq (\boldsymbol{w}_j)_2, \quad \textit{and} \quad \boldsymbol{w}_j < \boldsymbol{w}_{j+1}. \tag{7}$$

Proof We prove the result by induction over the subdomain index j. The definition of the matrices $\widetilde{T}_1$ and $\widetilde{T}_2$ implies that $0 = (\boldsymbol{w}_1)_1 \leq (\boldsymbol{w}_1)_2$, and Equation (4) yields that

$$\boldsymbol{w}_1 = \begin{bmatrix} 0 \\ (1-\frac{\delta}{L})(\boldsymbol{u}_2)_1 \end{bmatrix} < \begin{bmatrix} (1-\frac{\delta}{L})(\boldsymbol{u}_1)_2 \\ (1-\frac{\delta}{L})(\boldsymbol{u}_3)_1 \end{bmatrix} \leq \begin{bmatrix} (1-\frac{\delta}{L})(\boldsymbol{u}_1)_2 + \frac{\delta}{L}(\boldsymbol{u}_3)_1 \\ \frac{\delta}{L}(\boldsymbol{u}_1)_2 + (1-\frac{\delta}{L})(\boldsymbol{u}_3)_1 \end{bmatrix} = \boldsymbol{w}_2.$$

We now proceed to the induction step. Assume that (7) holds for some $j \in \{1, \ldots, n\}$. We first show that $(\boldsymbol{w}_{j+1})_1 \leq (\boldsymbol{w}_{j+1})_2$. Equation (4) implies that it is sufficient to show that $\boldsymbol{u}_j \leq \boldsymbol{u}_{j+2}$. There are two cases: $j < n-1$ and $j \in \{n-1, n\}$. If $j < n-1$, then (6) yields the required result. If $j \in \{n-1, n\}$, then (6) and the fact that $\boldsymbol{u} \in V^n$ gives that $\boldsymbol{u}_j < \boldsymbol{u}_{j+2} = \mathbf{1}_1$.

Next, we show that $\boldsymbol{w}_{j+1} < \boldsymbol{w}_{j+2}$. Equation (4) implies that it is sufficient to show that $\boldsymbol{u}_j < \boldsymbol{u}_{j+1}$ and $\boldsymbol{u}_{j+2} \leq \boldsymbol{u}_{j+3}$. There are three cases: $j < n-1$, $j = n-1$ and $j = n$. If $j < n-1$ then (6) yields the required result. If $j = n-1$ then (6) yields that $\boldsymbol{u}_j < \boldsymbol{u}_{j+1}$ and the fact that $\boldsymbol{u} \in V^n$ gives that $\boldsymbol{u}_{j+2} = \boldsymbol{u}_{j+3} = \mathbf{1}_1$. If $j = n$ then (6) yields that $\boldsymbol{u}_j < \boldsymbol{u}_{j+1}$ and it remains to show that $\boldsymbol{u}_{j+2} \leq \boldsymbol{u}_{j+3}$. To this end, we recall that $n \leq \lfloor \frac{N}{2} \rfloor - 2$. Therefore, there are three sub-cases: $n < \lfloor \frac{N}{2} \rfloor - 2$ in which case $n+1 < j+2, j+3 \leq \lfloor \frac{N}{2} \rfloor$; $n = \lfloor \frac{N}{2} \rfloor - 2$ and N is even in which case $\boldsymbol{u}_{j+2} = P\boldsymbol{u}_{j+3}$; $n = \lfloor \frac{N}{2} \rfloor - 2$ and N is odd in which case $\boldsymbol{u}_{j+3} = P\boldsymbol{u}_{j+1}$. In all three sub-cases, we obtain that $\boldsymbol{u}_{j+2} = \boldsymbol{u}_{j+3} = \mathbf{1}_1$. □

Lemma 4 below describes the 'shape' of the vector $T^n\mathbf{1}_N$ for natural numbers $n < \lfloor \frac{N}{2} \rfloor$.

Lemma 4 *Let* $n \in \{1, \ldots, \lfloor \frac{N}{2} \rfloor - 1\}$ *be a natural number and let* $\boldsymbol{w} = T^n\mathbf{1}_N$. *Then for all* $j \in \{1, \ldots, n\}$ *it holds that*

$$(\boldsymbol{w}_j)_1 \leq (\boldsymbol{w}_j)_2, \quad (\boldsymbol{w}_{N+1-j})_2 \leq (\boldsymbol{w}_{N+1-j})_1, \quad \textit{and} \quad \boldsymbol{w}_j < \boldsymbol{w}_{j+1}, \quad \boldsymbol{w}_{N+1-j} < \boldsymbol{w}_{N-j}.$$

Proof By Lemma 2 it holds that $\boldsymbol{w}_j = P\boldsymbol{w}_{N+1-j}$ for each $j \in \{1, \ldots, n\}$ so it suffices to show that for each $n \in \{1, \ldots, \lfloor \frac{N}{2} \rfloor - 1\}$ and all $j \in \{1, \ldots, n\}$ it holds that

$$(\boldsymbol{w}_j)_1 \leq (\boldsymbol{w}_j)_2 \text{ and } \boldsymbol{w}_j < \boldsymbol{w}_{j+1}. \tag{8}$$

We prove the result by induction over the iteration number n. Let $n = 1$. The definition of the matrix $\widetilde{T}_2$ and (5) yield that $\boldsymbol{w}_1 = \begin{bmatrix} 0 \\ 1-\frac{\delta}{L} \end{bmatrix}$ and $\boldsymbol{w}_2 = \mathbf{1}_1$. Thus, (8) holds for $n = 1$. Next, let $n = 2$ and let $\boldsymbol{u} = T^1_N$. The definition of the matrix $\widetilde{T}_2$ together with Equations (3) and (5) yields $\boldsymbol{w}_1 = \begin{bmatrix} 0 \\ 1-\frac{\delta}{L} \end{bmatrix}$, $\boldsymbol{w}_1 < \boldsymbol{w}_2 < \mathbf{1}_1$ and $\boldsymbol{w}_3 = \mathbf{1}_1$. Thus, (8) holds for $n = 2$. Finally, assume that (8) holds for some $n \in \{2, \ldots, \lfloor \frac{N}{2} \rfloor - 2\}$. It follows from Lemma 3 that (8) also holds for $n + 1$. □

Next, Lemma 5 describes the ‘shape’ of the vector $T^n\mathbf{1}_N$ for natural numbers $n \geq \lfloor \frac{N}{2} \rfloor$. Together, Lemmas 4 and 5 establish that the vector $T^n\mathbf{1}_N$ is monotonically increasing as one moves from the extrema of Ω towards its centre.

Lemma 5 *Let* $n \geq \lfloor \frac{N}{2} \rfloor$ *be a natural number, and let* $\boldsymbol{w} = T^n\mathbf{1}_N$. *Then for all* $j \in \{1, \ldots, \lfloor \frac{N}{2} \rfloor - 1\}$ *it holds that*

$$(\boldsymbol{w}_j)_1 \leq (\boldsymbol{w}_j)_2, \quad (\boldsymbol{w}_{N+1-j})_2 \leq (\boldsymbol{w}_{N+1-j})_1, \quad \textit{and} \quad \boldsymbol{w}_j < \boldsymbol{w}_{j+1}, \quad \boldsymbol{w}_{N+1-j} < \boldsymbol{w}_{N-j}.$$

In addition, if N is an odd number, then $\boldsymbol{w}_{\lfloor \frac{N}{2} \rfloor} \leq \boldsymbol{w}_{\lceil \frac{N}{2} \rceil}$ and $\boldsymbol{w}_{\lfloor \frac{N}{2} \rfloor + 2} < \boldsymbol{w}_{\lceil \frac{N}{2} \rceil}$.

Proof Lemma 5 can be proven in a similar manner to Lemma 4 using a proof-by-induction on the iteration number n. We omit it here for brevity. □

We are now ready to prove our second main result.

Proof (Theorem 1* (b)*) Assume that $N \in \mathbb{N}$ is even. In view of Lemma 1, we must prove that $\|T^{n+1}\mathbf{1}_N\| < \|T^n\mathbf{1}_N\|$. Let $\boldsymbol{w} = T^{n+1}\mathbf{1}_N$ and $\boldsymbol{u} = T^n\mathbf{1}_N$. By Lemma 5, we know that $\|T^{n+1}\mathbf{1}_N\| = \|\boldsymbol{w}_{\lceil \frac{N}{2} \rceil}\|$ and $\|T^n\mathbf{1}_N\| = \|\boldsymbol{u}_{\lceil \frac{N}{2} \rceil}\|$.

Since $\boldsymbol{w} = T\boldsymbol{u}$, we have that $\boldsymbol{w}_{\lceil \frac{N}{2} \rceil} = T_1\boldsymbol{u}_{\lceil \frac{N}{2} \rceil - 1} + T_2\boldsymbol{u}_{\lceil \frac{N}{2} \rceil + 1}$. Since N is even, we obtain that $\lceil \frac{N}{2} \rceil + 1 = \frac{N}{2} + 1$ and thus, Lemma 2 yields that $\boldsymbol{u}_{\lceil \frac{N}{2} \rceil + 1} = P\boldsymbol{u}_{\lceil \frac{N}{2} \rceil}$. It follows that $\boldsymbol{w}_{\lceil \frac{N}{2} \rceil} = T_1\boldsymbol{u}_{\lceil \frac{N}{2} \rceil - 1} + T_2P\boldsymbol{u}_{\lceil \frac{N}{2} \rceil}$. From (4) we also obtain that

$$\boldsymbol{u}_{\lceil \frac{N}{2} \rceil - 1} \leq \boldsymbol{w}_{\lceil \frac{N}{2} \rceil} \leq P\boldsymbol{u}_{\lceil \frac{N}{2} \rceil}, \tag{9}$$

where the equality holds if and only if $(\boldsymbol{u}_{\lceil \frac{N}{2} \rceil - 1})_2 = (P\boldsymbol{u}_{\lceil \frac{N}{2} \rceil})_1 = (\boldsymbol{u}_{\lceil \frac{N}{2} \rceil})_2$. We know from Lemma 5 that $\boldsymbol{u}_{\lceil \frac{N}{2} \rceil - 1} < \boldsymbol{u}_{\lceil \frac{N}{2} \rceil}$, which yields that $(\boldsymbol{u}_{\lceil \frac{N}{2} \rceil - 1})_2 < (\boldsymbol{u}_{\lceil \frac{N}{2} \rceil})_2$. Hence the inequalities in (9) are strict: $\boldsymbol{u}_{\lceil \frac{N}{2} \rceil - 1} < \boldsymbol{w}_{\lceil \frac{N}{2} \rceil} < P\boldsymbol{u}_{\lceil \frac{N}{2} \rceil}$. Hence, we obtain that $\|T^{n+1}\mathbf{1}_N\| = \|\boldsymbol{w}_{\lceil \frac{N}{2} \rceil}\| < \|P\boldsymbol{u}_{\lceil \frac{N}{2} \rceil}\| = \|\boldsymbol{u}_{\lceil \frac{N}{2} \rceil}\| = \|T^n\mathbf{1}_N\|$. This completes the proof of the first assertion.

Assume now that $N \in \mathbb{N}$ is odd. Let $\boldsymbol{u} = T^n\mathbf{1}_N$, $\boldsymbol{w} = T^{n+1}\mathbf{1}_N$ and $\boldsymbol{y} = T^{n+2}\mathbf{1}_N$. Lemma 5 implies that

$$\|T^{n+2}\mathbf{1}_N\| = \|\boldsymbol{y}_{\lceil \frac{N}{2} \rceil}\|, \qquad \|T^{n+1}\mathbf{1}_N\| = \|\boldsymbol{w}_{\lceil \frac{N}{2} \rceil}\|, \qquad \|T^n\mathbf{1}_N\| = \|\boldsymbol{u}_{\lceil \frac{N}{2} \rceil}\|.$$

Since $\|T\| = 1$, we have

$$\|\boldsymbol{y}_{\lceil \frac{N}{2} \rceil}\| \leq \|\boldsymbol{w}_{\lceil \frac{N}{2} \rceil}\| \leq \|\boldsymbol{u}_{\lceil \frac{N}{2} \rceil}\|. \tag{10}$$

Clearly if $\|\boldsymbol{w}_{\lceil \frac{N}{2} \rceil}\| < \|\boldsymbol{u}_{\lceil \frac{N}{2} \rceil}\|$ then (10) yields that

$$\|T^{n+2}\mathbf{1}_N\| \leq \|T^{n+1}\mathbf{1}_N\| = \|\boldsymbol{w}_{\lceil \frac{N}{2} \rceil}\| < \|\boldsymbol{u}_{\lceil \frac{N}{2} \rceil}\| = \|T^n\mathbf{1}_N\|,$$

which is our claim. Suppose that $\|\boldsymbol{w}_{\lceil \frac{N}{2} \rceil}\| = \|\boldsymbol{u}_{\lceil \frac{N}{2} \rceil}\|$. We show that $\|\boldsymbol{y}_{\lceil \frac{N}{2} \rceil}\| < \|\boldsymbol{w}_{\lceil \frac{N}{2} \rceil}\|$ which implies our claim. To do so, since N is odd, we use Lemma 2 together with the facts that $\boldsymbol{y} = T\boldsymbol{w}$ and $\boldsymbol{w} = T\boldsymbol{u}$ to obtain that $\boldsymbol{y}_{\lceil \frac{N}{2} \rceil} = T_1\boldsymbol{w}_{\lceil \frac{N}{2} \rceil - 1} +$

$T_2 P\, \boldsymbol{w}_{\lceil \frac{N}{2} \rceil - 1}$ and $\boldsymbol{w}_{\lceil \frac{N}{2} \rceil} = T_1 \boldsymbol{u}_{\lceil \frac{N}{2} \rceil - 1} + T_2 P\, \boldsymbol{u}_{\lceil \frac{N}{2} \rceil - 1}$ which implies

$$\boldsymbol{y}_{\lceil \frac{N}{2} \rceil} = (\boldsymbol{w}_{\lceil \frac{N}{2} \rceil - 1})_2 \mathbf{1}_1, \qquad \boldsymbol{w}_{\lceil \frac{N}{2} \rceil} = (\boldsymbol{u}_{\lceil \frac{N}{2} \rceil - 1})_2 \mathbf{1}_1. \tag{11}$$

From Lemma 2, we know that

$$\boldsymbol{u}_{\lceil \frac{N}{2} \rceil} \overset{\text{Lemma 2}}{=} \boldsymbol{w}_{\lceil \frac{N}{2} \rceil} \overset{(11)}{=} (\boldsymbol{u}_{\lceil \frac{N}{2} \rceil - 1})_2 \mathbf{1}_1. \tag{12}$$

Using the fact that $\boldsymbol{w} = T\boldsymbol{u}$ and Equation (12) we have that

$$\boldsymbol{w}_{\lceil \frac{N}{2} \rceil - 1} \overset{(4)}{=} T_1 \boldsymbol{u}_{\lceil \frac{N}{2} \rceil - 2} + T_2\, \boldsymbol{u}_{\lceil \frac{N}{2} \rceil} \overset{(12)}{=} T_1 \boldsymbol{u}_{\lceil \frac{N}{2} \rceil - 2} + T_2 (\boldsymbol{u}_{\lceil \frac{N}{2} \rceil - 1})_2 \mathbf{1}_1.$$

Using (4) we obtain that

$$\boldsymbol{u}_{\lceil \frac{N}{2} \rceil - 2} \leq \boldsymbol{w}_{\lceil \frac{N}{2} \rceil - 1} \leq (\boldsymbol{u}_{\lceil \frac{N}{2} \rceil - 1})_2 \mathbf{1}_1. \tag{13}$$

where the equality holds if and only if $(\boldsymbol{u}_{\lceil \frac{N}{2} \rceil - 2})_2 = (\boldsymbol{u}_{\lceil \frac{N}{2} \rceil - 1})_2$. However, Lemma 5 implies that $\boldsymbol{u}_{\lceil \frac{N}{2} \rceil - 2} < \boldsymbol{u}_{\lceil \frac{N}{2} \rceil - 1}$ which immediately yields that $(\boldsymbol{u}_{\lceil \frac{N}{2} \rceil - 2})_2 < (\boldsymbol{u}_{\lceil \frac{N}{2} \rceil - 1})_2$. Hence the inequalities in (13) are strict and thus

$$\boldsymbol{w}_{\lceil \frac{N}{2} \rceil - 1} \overset{(13)}{<} (\boldsymbol{u}_{\lceil \frac{N}{2} \rceil - 1})_2 \mathbf{1}_1 \overset{(11)}{=} \boldsymbol{w}_{\lceil \frac{N}{2} \rceil}. \tag{14}$$

Recalling (11) we obtain that $\boldsymbol{y}_{\lceil \frac{N}{2} \rceil} \overset{(11)}{=} (\boldsymbol{w}_{\lceil \frac{N}{2} \rceil - 1})_2 \mathbf{1}_1 \overset{(14)}{<} \boldsymbol{w}_{\lceil \frac{N}{2} \rceil}$, which completes the proof. □

References

1. Cancès, E., Maday, Y., Stamm, B.: Domain decomposition for implicit solvation models. The Journal of Chemical Physics **139**, 054111 (2013)
2. Chaouqui, F., Ciaramella, G., Gander, M.J., Vanzan, T.: On the scalability of classical one-level domain-decomposition methods. Vietnam Journal of Mathematics **46**(4), 1053–1088 (2018)
3. Ciaramella, G., Gander, M.J.: Analysis of the parallel Schwarz method for growing chains of fixed-sized subdomains: Part I. SIAM J. Numer. Anal. **55**(3), 1330–1356 (2017)
4. Ciaramella, G., Gander, M.J.: Analysis of the parallel Schwarz method for growing chains of fixed-sized subdomains: Part II. SIAM J. Numer. Anal. **56 (3)**, 1498–1524 (2018)
5. Ciaramella, G., Gander, M.J.: Analysis of the parallel Schwarz method for growing chains of fixed-sized subdomains: Part III. Electronic Transactions on Numerical Analysis **49**, 210–243 (2018)
6. Gander, M.J., Zhao, H.: Overlapping Schwarz waveform relaxation for the heat equation in N dimensions. BIT Numerical Mathematics **42**(4), 779–795 (2002)
7. Toselli, A., Widlund, O.: Domain Decomposition Methods: Algorithms and Theory, vol. 34. Springer (2005)

Fully Discrete Schwarz Waveform Relaxation on Two Bounded Overlapping Subdomains

Ronald D. Haynes and Khaled Mohammad

1 Introduction

Overlapping Schwarz waveform relaxation (SWR) provides space–time parallelism by iteratively solving partial differential equations (PDEs) over a time window on overlapping spatial subdomains. SWR has been studied for many problems at the continuous and discrete levels. Gander and Stuart [5] and Giladi and Keller [6] have analyzed SWR for the heat equation on a finite spatial domain in the continuous and semi-discrete (in space) cases. Semi-discrete (in space) analysis for reaction diffusion equations on an infinite spatial domain can be found in [10]. Closely related work on applications of WR methods to RC type circuits can be found in [3, 2, 1] (continuous in time analysis), [11] (infinite circuit, discrete in time), [8, 9] (fractional order, infinite circuit, discrete and continuous resp. in time), and [12] (Volterra integro-PDEs, infinite spatial domain). Fully discrete analysis for Schrödinger's equation and the wave equation can be found in [7] and [4]. We provide an analysis of a full space–time discretization of SWR for the heat equation on two overlapping, bounded subdomains, which does not appear to be in the literature.

Consider the one dimensional heat equation $u_t = u_{xx} + f(x,t)$ for $-L < x < L$ and $0 < t \leq T$ subject to initial and boundary conditions $u(x,0) = u_0(x)$, $u(-L,t) = h_1(t)$, and $u(L,t) = h_2(t)$. Discretizing in space with central finite differences on $\Omega^h = \{x_m : x_{m+1} = x_m + \Delta x, m = -N, ..., N-1\}$, where $\Delta x = \frac{L}{N}$ and $x_{-N} = -L$, leads to the IVP

$$\frac{d\mathbf{u}(t)}{dt} = A\mathbf{u}(t) + \mathbf{f}(t),\ 0 < t \leq T,\ \mathbf{u}(0) = \mathbf{u}_0, \tag{1}$$

Ronald D. Haynes
Memorial University, St. John's, Newfoundland, Canada, e-mail: rhaynes@mun.ca

Khaled Mohammad
Memorial University, St. John's, Newfoundland, Canada, e-mail: km2605@mun.ca

R. Haynes et al. (eds.), *Domain Decomposition Methods in Science and Engineering XXV*, Lecture Notes in Computational Science and Engineering 138,
https://doi.org/10.1007/978-3-030-56750-7_17

where $\mathbf{u}(t)$ is the solution vector on the interior of Ω_h with components $u_m(t), m = -(N-1), \ldots, (N-1)$, which are the semi-discrete approximations of $u(x,t)$ at $x = x_m$. Here $A = \frac{1}{\Delta x^2}\text{tridiag}\{1, -2, 1\} \in \mathbb{R}^{(2N-1)\times(2N-1)}$,

$$\mathbf{f}(t) = (f(x_{-(N-1)}, t) + \frac{1}{\Delta x^2}h_1(t), f(x_{-(N-2)}, t), \ldots, f(x_{(N-2)}, t), f(x_{(N-1)}, t) + \frac{1}{\Delta x^2}h_2(t))^T$$

and $\mathbf{u}_0 = (u_0(x_{-(N-1)}), \ldots, u_0(x_{(N-1)}))^T$.

2 Semi–discretized SWR

To obtain the classical SWR solution of (1), we decompose Ω^h into two overlapping subdomains: $\Omega_1^h = \{x_{-N}, x_{-(N-1)}, \ldots, x_M\}$ and $\Omega_2^h = \{x_{-M}, x_{-(M-1)}, \ldots, x_N\}$ where the quantity $M \geq 1$ is an integer that determines the overlap size.

The classical semi-discrete SWR algorithm on the two subdomains, Ω_1^h and Ω_2^h, can be written as : for $k = 1, 2, \ldots$, for $j = 1, 2$ solve

$$\frac{d\mathbf{u}_j^k(t)}{dt} = A_j\mathbf{u}_j^k(t) + \mathbf{f}_j^k(t), \qquad 0 < t \leq T, \tag{2a}$$

where

$$\mathbf{u}_1^k(t) = (u_{1,-(N-1)}^k(t), u_{1,-(N-2)}^k(t), \ldots, u_{1,(M-1)}^k(t))^T, \tag{2b}$$

and

$$\mathbf{u}_2^k(t) = (u_{2,(-M+1)}^k(t), u_{2,(-M+2)}^k(t), \ldots, u_{2,(N-1)}^k(t))^T, \tag{2c}$$

are the subdomain iterates on the interior nodes of Ω_1^h and Ω_2^h. Here, for $j = 1, 2$, $A_j = \frac{1}{\Delta x^2}\text{tridiag}\{1, -2, 1\} \in \mathbb{R}^{N+M-1, N+M-1}$. The vectors $\mathbf{f}_j^k \in \mathbb{R}^{N+M-1}$, for $j = 1, 2$, are defined by

$$\mathbf{f}_1^k(t) = \bar{\mathbf{f}}_1(t) + \frac{1}{\Delta x^2}u_{1,M}^k(t)\delta_1 \quad \text{and} \quad \mathbf{f}_2^k(t) = \bar{\mathbf{f}}_2(t) + \frac{1}{\Delta x^2}u_{2,-M}^k(t)\delta_2, \tag{2d}$$

where $\delta_j \in \mathbb{R}^{N+M-1}$ for $j = 1, 2$, are the unit column vectors

$$\delta_1 = (0, \ldots, 0, 1)^T \quad \text{and} \quad \delta_2 = (1, 0, \ldots, 0)^T. \tag{2e}$$

The overbar notation indicates that $\bar{\mathbf{f}}_j$, for $j = 1, 2$, are the first and last $N + M - 1$ components of $\mathbf{f}$, respectively. This notation will be used throughout. The system (2a) is supplemented with an initial condition

$$\mathbf{u}_j^k(0) = \bar{\mathbf{u}}_j(0), \quad j = 1, 2, \tag{2f}$$

and boundary and transmission conditions

$$u^k_{1,-N}(t) = h_1(t), \qquad u^k_{1,M}(t) = u^{k-1}_{2,M}(t), \qquad 0 < t \leq T,$$
$$u^k_{2,-M}(t) = u^{k-1}_{1,-M}(t), \quad u^k_{2,N}(t) = h_2(t), \qquad 0 < t \leq T. \tag{2g}$$

Here $u^k_{j,m}(t)$ represents the numerical approximation of $u(x,t)$ at $x = x_m$ over Ω_j using the SWR algorithm at the kth iteration. To get the iteration started we must pick initial guesses for $u^0_{2,M}(t)$ and $u^0_{1,-M}(t)$.

3 Convergence Analysis

To analyze the fully discrete SWR we begin with a lemma which describes the single domain discrete solution of (1) using a backward Euler integrator.

Lemma 1 *The single domain solution at $t = t_n$, $\mathbf{u}(n)$, restricted to the interior of Ω^h_j, $\bar{\mathbf{u}}_j(n)$, for $j = 1,2$, using a backward Euler integrator for the semi–discrete heat equation (1), is the unique solution of the subsystems*

$$(I_1 - \Delta t A_1)\bar{\mathbf{u}}_1(n) - \mu u_M(n)\delta_1 = \bar{\mathbf{u}}_1(n-1) + \Delta t \bar{\mathbf{f}}_1(n),$$
$$(I_2 - \Delta t A_2)\bar{\mathbf{u}}_2(n) - \mu u_{-M}(n)\delta_2 = \bar{\mathbf{u}}_2(n-1) + \Delta t \bar{\mathbf{f}}_2(n),$$

for $n = 1,2,\ldots$. Here $\mu = \Delta t/\Delta x^2$, $\delta_\mathbf{j}$, for $j = 1,2$, are defined in (2e), $u_M(n)$ and $u_{-M}(n)$ are the single domain solutions at the interior interface nodes at time t_n, and $I_{1,2}$ are $(N+M-1)\times(N+M-1)$ identity matrices. Here $\bar{\mathbf{f}}_\mathbf{j}(\mathbf{n}) \equiv \bar{\mathbf{f}}_\mathbf{j}(\mathbf{t_n})$ for $j = 1,2$.

Similar expressions for the SWR approximations are given in the next lemma.

Lemma 2 *The solution of (2a)–(2g) using a backward Euler integrator at $t = t_n$, $\mathbf{u}^k_j(n)$, for $j = 1,2$, at the kth iteration, are the unique solutions of the subsystems*

$$(I_1 - \Delta t A_1)\mathbf{u}^\mathbf{k}_\mathbf{1}(n) = \mathbf{u}^\mathbf{k}_\mathbf{1}(n-1) + \Delta t \mathbf{f}^\mathbf{k}_\mathbf{1}(n),$$
$$(I_2 - \Delta t A_2)\mathbf{u}^\mathbf{k}_\mathbf{2}(n) = \mathbf{u}^\mathbf{k}_\mathbf{2}(n-1) + \Delta t \mathbf{f}^\mathbf{k}_\mathbf{2}(n),$$

for $n = 1,2,\ldots$. Here $\mathbf{f}^k_j(n) \equiv \mathbf{f}^k_j(t_n)$, for $j = 1,2$, where $\mathbf{f}^\mathbf{k}_\mathbf{j}(t)$ are defined in (2d).

We denote the error between the single domain and SWR solutions at time step n by $\mathbf{e}^k_j(n) = \mathbf{u}^k_j(n) - \bar{\mathbf{u}}_j(n)$ for $j = 1,2$. Simply subtracting the representations of the single domain and SWR solutions from the previous two lemmas gives the following result.

Lemma 3 *For $j = 1,2$, $k = 1,2,\ldots$ and $n = 1,2,\ldots$ the errors, $\mathbf{e}^k_j(n)$, satisfy*

$$(I_1 - \Delta t A_1)\mathbf{e}^\mathbf{k}_\mathbf{1}(n) = \mathbf{e}^\mathbf{k}_\mathbf{1}(n-1) + \mu \mathbf{e}^\mathbf{k}_\mathbf{1,M}(n)\delta_1,$$
$$(I_2 - \Delta t A_2)\mathbf{e}^\mathbf{k}_\mathbf{2}(n) = \mathbf{e}^\mathbf{k}_\mathbf{2}(n-1) + \mu \mathbf{e}^\mathbf{k}_\mathbf{2,-M}(n)\delta_2,$$

with initial condition

$$\mathbf{e}_j^k(0) = \bar{\mathbf{0}}_j, \quad \textit{for } j = 1, 2,$$

and boundary conditions

$$e_{1,M}^k(n) = e_{2,M}^{k-1}(n), \quad e_{1,-N}^k(n) = 0,$$
$$e_{2,-M}^k(n) = e_{1,-M}^{k-1}(n), \quad e_{2,N}^k(n) = 0.$$

Here $\bar{\mathbf{0}}_\mathbf{j} \in \mathbb{R}^{N+M-1}$, *for* $j = 1, 2$, *is the zero vector.*

Using the boundary values and the definition of $A_{1,2}$ and $\delta_{1,2}$ we obtain the following lemma.

Lemma 4 *Component-wise, for* $j = 1, 2$, $k = 1, 2, \ldots$ *and* $n = 1, 2, \ldots$ *the errors* $\mathbf{e}_{j,m}^k(n)$ *satisfy*

$$-\mu e_{1,m-1}^k(n) + (1 + 2\mu)e_{1,m}^k(n) - \mu e_{1,m+1}^k(n) = e_{1,m}^k(n-1), \textit{ for } m = -(N-1), \ldots, M-1,$$
$$-\mu e_{2,m-1}^k(n) + (1 + 2\mu)e_{2,m}^k(n) - \mu e_{2,m+1}^k(n) = e_{2,m}^k(n-1), \textit{ for } m = -(M-1), \ldots, N-1.$$

To analyze these recursions for the error we need the discrete Laplace transform. The discrete Laplace transform for a general vector $v = (v(0), v(1), \ldots)^T$, defined on a regular grids with time step Δt is

$$\hat{v}(s) = \frac{\Delta t}{\sqrt{2\pi}} \sum_{n=0}^{\infty} z^{-n} v(n),$$

where $z = e^{s\Delta t}$, $s = \sigma + i\omega$, $\sigma > 0$ and $-\pi/T \le \omega \le \pi/\Delta t$.

The recursions for the discrete Laplace transforms of the errors are recorded in the next lemma.

Lemma 5 *For* $j = 1, 2$, $k = 1, 2, \ldots$ *and* $n = 1, 2, \ldots$ *the discrete Laplace transform of errors* $\hat{\mathbf{e}}_{j,m}^k(n)$ *satisfy*

$$\mu \hat{e}_{1,m-1}^k(s) - (2\mu + \eta)\hat{e}_{1,m}^k(s) + \mu \hat{e}_{1,m+1}^k(s) = 0, \quad m = -(N-1), \ldots, (M-1)$$

and

$$\mu \hat{e}_{2,m-1}^k(s) - (2\mu + \eta)\hat{e}_{2,m}^k(s) + \mu \hat{e}_{2,m+1}^k(s) = 0, \quad m = -(M-1), \ldots, (N-1).$$

The Laplace transform of the initial error gives

$$\hat{\mathbf{e}}_j^k(0) = \mathbf{0}_j, \quad \textit{for } j = 1, 2$$

and the Laplace transforms of the boundary conditions are

$$\hat{e}_{1,M}^k(s) = \hat{e}_{2,M}^{k-1}(s), \quad \hat{e}_{1,-N}^k(s) = 0,$$
$$\hat{e}_{2,-M}^k(s) = \hat{e}_{1,-M}^{k-1}(s), \quad \hat{e}_{2,N}^k(s) = 0,$$

where $\mu = \frac{\Delta t}{\Delta x^2}$, $\eta = \frac{z-1}{z}$ *and* $z = e^{s\Delta t}$.

The general solutions of these recursion relations are given in the next two lemmas.

Lemma 6 *The general solutions of the recursions for the Laplace transforms of the error are given by*

$$\hat{e}_{j,m}^k(s) = a_j^k \lambda_+^m + b_j^k \lambda_+^{-m}, \quad \textit{for } j = 1, 2, \tag{3}$$

where λ_+ *solves* $\mu - (2\mu + \eta)\lambda + \mu\lambda^2 = 0$ *and is given explicitly by* $\lambda_+ = \frac{(2\mu+\eta)+\sqrt{(2\mu+\eta)^2-4\mu^2}}{2\mu}$, $\mu = \frac{\Delta t}{\Delta x^2}$, $\eta = \frac{z-1}{z}$ *and* $z = e^{s\Delta t}$ *where the coefficients* $(a_j^k, b_j^k)^T =: \mathbf{c}_j^k$ *are shown to satisfy a simple fixed point iteration in the next lemma.*

Note: in the expression above for λ_+, we have chosen the square root with positive real part.

Lemma 7 *The coefficients in the general solution for the Laplace transform of the error,* $\mathbf{c}_j^k = (a_j^k, b_j^k)^T$, *for* $j = 1, 2$, *satisfy*

$$\begin{pmatrix} \mathbf{c}_1^k \\ \mathbf{c}_2^k \end{pmatrix} = \Gamma \begin{pmatrix} \mathbf{c}_1^{k-2} \\ \mathbf{c}_2^{k-2} \end{pmatrix},$$

where the contraction matrix, Γ, *is the block diagonal matrix*

$$\Gamma = \begin{pmatrix} S_1 & \\ & S_2 \end{pmatrix},$$

where

$$S_1 = \Lambda_1^{-1}\Theta_1\Lambda_2^{-1}\Theta_2 \quad \textit{and} \quad S_2 = \Lambda_2^{-1}\Theta_2\Lambda_1^{-1}\Theta_1,$$

and

$$\Lambda_1 = \begin{pmatrix} \lambda_+^{-N} & \lambda_+^{N} \\ \lambda_+^{M} & \lambda_+^{-M} \end{pmatrix}, \Lambda_2 = \begin{pmatrix} \lambda_+^{-M} & \lambda_+^{M} \\ \lambda_+^{N} & \lambda_+^{-N} \end{pmatrix}, \Theta_1 = \begin{pmatrix} 0 & 0 \\ \lambda_+^{M} & \lambda_+^{-M} \end{pmatrix}, \Theta_2 = \begin{pmatrix} \lambda_+^{-M} & \lambda_+^{M} \\ 0 & 0 \end{pmatrix}.$$

Proof The boundary conditions, obtained from (3), can be written as

$$\Lambda_2 \mathbf{c}_2^{\mathbf{k}} = \Theta_2 \mathbf{c}_1^{\mathbf{k-1}} \qquad \text{and} \qquad \Lambda_1 \mathbf{c}_1^{\mathbf{k}} = \Theta_1 \mathbf{c}_2^{\mathbf{k-1}},$$

from which the result follows. □

To ultimately show convergence of the discrete SWR algorithm we show that for $j = 1, 2$, $\mathbf{c}_j^k$ tends to zero as k tends to infinity. A straightforward, but slightly tedious calculation, gives the following explicit representation of $\rho(\Gamma)$.

Lemma 8 *The spectral radius of the contraction matrix* Γ *above,* $\rho(\Gamma)$, *is*

$$\rho(\Gamma) = \left| \frac{\lambda_+^{(N-M)} - \lambda_+^{-(N-M)}}{\lambda_+^{(N+M)} - \lambda_+^{-(N+M)}} \right|^2,$$

where $\lambda_+ = \frac{(2\mu+\eta)+\sqrt{(2\mu+\eta)^2-4\mu^2}}{2\mu}$, $\mu = \frac{\Delta t}{\Delta x^2}$, $\eta = \frac{z-1}{z}$ *and* $z = e^{s\Delta t}$.

Proof Direct calculation gives

$$S_1 = \chi \begin{pmatrix} 1-\lambda_+^{2(N-M)} & \lambda_+^{2M}-\lambda_+^{2N} \\ \lambda_+^{-2M}-\lambda_+^{-2N} & 1-\lambda_+^{-2(N-M)} \end{pmatrix} \quad \text{and} \quad S_2 = \chi \begin{pmatrix} 1-\lambda_+^{-2(N-M)} & \lambda_+^{-2M}-\lambda_+^{-2N} \\ \lambda_+^{2M}-\lambda_+^{2N} & 1-\lambda_+^{2(N-M)} \end{pmatrix},$$

where $\chi = -\left(\lambda_+^{N+M} - \lambda_+^{-(N+M)}\right)^{-2}$. It is easy to see that $\det(S_1) = \det(S_2) = 0$, hence $\rho(S_1) = |tr(S_1)|$ and $\rho(S_2) = |tr(S_2)|$, where $tr(\cdot)$ denotes the trace of the matrix. We also note $tr(S_1) = tr(S_2)$. Computing the trace gives the required result. □

From the form of the contraction factor in the previous lemma it is not clear that the algorithm converges. We may rewrite the contraction factor as follows.

Lemma 9 *Using the mapping,* $\lambda_+ = e^v$, *where* $v = \zeta + i\varphi$, *the spectral radius of the contraction matrix,* $\rho(\Gamma)$, *can be written as*

$$\rho(\Gamma) = \left|\frac{\sinh((N-M)v)}{\sinh((N+M)v)}\right|^2. \tag{4}$$

or

$$\rho(\zeta,\varphi) = \frac{2p(\zeta,\varphi) - \sin(2N\varphi)\sin(2M\varphi) - \sinh(2N\zeta)\sinh(2M\zeta)}{2p(\zeta,\varphi) + \sin(2N\varphi)\sin(2M\varphi) + \sinh(2N\zeta)\sinh(2M\zeta)}, \tag{5}$$

where

$$\begin{aligned} p(\zeta,\varphi) = {} & \sinh^2(N\zeta)\cosh^2(M\zeta) + \sinh^2(M\zeta)\cosh^2(N\zeta) \\ & + \sin^2(N\varphi)\cos^2(M\varphi) + \sin^2(M\varphi)\cos^2(N\varphi). \end{aligned} \tag{6}$$

Proof Using the substitution $\lambda_+ = e^v$ and the definition of the hyperbolic sine function we arrive at (4). Now using $v = \zeta + i\varphi$ and hyperbolic trigonometric identities, the contraction rate (4) can be written as

$$\rho(\zeta,\varphi) = \left|\frac{\sinh((N-M)\zeta)\cos((N-M)\varphi) + i\cosh((N-M)\zeta)\sin((N-M)\varphi)}{\sinh((N+M)\zeta)\cos((N+M)\varphi) + i\cosh((N+M)\zeta)\sin((N+M)\varphi)}\right|^2. \tag{7}$$

Simplifying the modulus in (7) gives

$$\rho(\zeta,\varphi) = \frac{\sinh^2((N-M)\zeta)\cos^2((N-M)\varphi) + \cosh^2((N-M)\zeta)\sin^2((N-M)\varphi)}{\sinh^2((N+M)\zeta)\cos^2((N+M)\varphi) + \cosh^2((N+M)\zeta)\sin^2((N+M)\varphi)}. \tag{8}$$

Again using hyperbolic trigonometric identities we arrive at (5) where p is as defined in (6). □

To show the spectral radius is strictly less one a more detailed analysis of λ_+ is necessary.

Lemma 10 *The quantity* $\eta = (z-1)/z$ *in the expression for* λ_+ *satisfies* $Re(\eta) > 0$ *and hence* $Re(\lambda_+) > 1$.

Proof Consider $\eta = (z-1)/z$ where $z = e^{s\Delta t}$, $s = \sigma + i\omega$, $\sigma > 0$ and $\pi/T \leq |\omega| \leq \pi/\Delta t$. The real part of η is given by $\mathrm{Re}(\eta) = 1 - e^{-\sigma\Delta t}\cos(\omega\Delta t)$ which is easily seen to be positive for $\sigma > 0$ and $\pi/T \leq |\omega| \leq \pi/\Delta t$. The real part of λ_+ is given by $\mathrm{Re}(\lambda_+) = 1 + \frac{\mathrm{Re}(\eta)}{2\mu} + \frac{\mathrm{Re}(\sqrt{\eta^2+4\mu\eta})}{2\mu}$. The conclusion $\mathrm{Re}(\lambda_+) > 1$ then follows from the fact that $\mathrm{Re}(\eta) > 0$ and the choice of the square root in λ_+. □

The following inequality will finally lead us to the main result.

Lemma 11 *If* $\lambda_+ = e^{\zeta+i\varphi}$ *then*

$$\sinh(K\zeta) > |\sin(K\varphi)|, \tag{9}$$

for any integer $K \geq 1$.

Proof Recall that λ_+ satisfies $\mu - (2\mu + \eta)\lambda_+ + \mu\lambda_+^2 = 0$. Substituting $\lambda_+ = e^{\zeta+i\varphi}$, multiplying by $e^{-(\zeta+i\varphi)}$ and dividing by 2μ, we find $\frac{e^{\zeta+i\varphi}+e^{-(\zeta+i\varphi)}}{2} = 1+\frac{\eta}{2\mu}$. Using the definition of the hyperbolic cosine function and splitting the real and the imaginary parts of η we have

$$\cosh(\zeta + i\varphi) = \left(1 + \frac{\mathrm{Re}(\eta)}{2\mu}\right) + i\frac{\mathrm{Im}(\eta)}{2\mu}.$$

Since $\mathrm{Re}(\eta) > 0$ then clearly $|\cosh(\zeta + i\varphi)|^2 > 1$.

Induction is used to prove (9). Using Euler's formula, hyperbolic trigonometric identities and simplifying the square of the modulus, $|\cosh(\zeta + i\varphi)|^2 > 1$ becomes

$$\cosh^2(\zeta)\cos^2(\varphi) + \sinh^2(\zeta)\sin^2(\varphi) > 1,$$

which simplifies to $\sinh^2(\zeta) > \sin^2(\varphi)$. Since $Re(\lambda_+) = e^{\zeta}cos(\varphi) > 1$, then $\zeta > 0$ and hence $\sinh(\zeta) > 0$. Taking the square root of both sides of the inequality $\sinh^2(\zeta) > \sin^2(\varphi)$ then gives the base case in the induction argument.

The induction step then follows using the base inequality, hyperbolic trigonometric identities, properties of the hyperbolic and trigonometric functions and the triangle inequality. □

We now arrive at the final and main result.

Theorem 1 *The fully discrete SWR algorithm which results from applying the backward Euler time integrator to (2a)–(2g) converges to the single domain discrete solution on the interior of* Ω_j^h, *for* $j = 1, 2$.

Proof We are now in a position to prove that $\rho(\Gamma) < 1$. The spectral radius of the contraction matrix, $\rho(\Gamma)$, is given in (5) where p is given in (6). Since $p > 0$, then clearly $\rho(\zeta, \varphi) < 1$ if $\sin(2N\varphi)\sin(2M\varphi) + \sinh(2N\zeta)\sinh(2M\zeta) > 0$. This inequality follows from Lemma 11 for $K = 2N$ and $K = 2M$. To see this, we consider different cases for the sign of $\sin(2N\varphi)$ and $\sin(2M\varphi)$. Since $\zeta > 0$ we have $\sinh(2N\zeta) > 0$ and $\sinh(2M\zeta) > 0$. There are two cases to consider: if $\sin(2N\varphi)$ and $\sin(2M\varphi)$ have the same or opposite signs. If they have the same sign then the inequality above is obvious. If they have opposite signs then Lemma 11 gives the result.

4 Conclusions

In this paper we have obtained an explicit contraction rate for the discrete Laplace transform of the error for the fully discretized SWR algorithm applied to the heat equation on two overlapping bounded domains. Further analysis, with other families of time integrators and an arbitrary number of subdomains will appear elsewhere.

Acknowledgement We would like to thank Felix Kwok for several discussions related to this work.

References

1. Al-Khaleel, M., Gander, M.J., Ruehli, A.E.: A mathematical analysis of optimized waveform relaxation for a small RC circuit. Applied Numerical Mathematics **75**, 61–76 (2014)
2. Al-Khaleel, M.D., Gander, M.J., Ruehli, A.E.: Optimization of transmission conditions in waveform relaxation techniques for RC circuits. SIAM Journal on Numerical Analysis **52**(2), 1076–1101 (2014)
3. Gander, M., Ruehli, A.: Optimized waveform relaxation methods for RC type circuits. IEEE Transactions on Circuits and Systems I: Regular Papers **51**(4), 755–768 (2004)
4. Gander, M.J., Halpern, L., Nataf, F.: Optimal Schwarz waveform relaxation for the one dimensional wave equation. SIAM Journal on Numerical Analysis **41**(5), 1643–1681 (2003). DOI:10.1137/S003614290139559X. 00148
5. Gander, M.J., Stuart, A.M.: Space-time continuous analysis of waveform relaxation for the heat equation. SIAM Journal on Scientific Computing **19**(6), 2014–2031 (1998)
6. Giladi, E., Keller, H.B.: Space-time domain decomposition for parabolic problems. Numerische Mathematik **93**(2), 279–313 (2002). DOI:10.1007/s002110100345
7. Halpern, L., Szeftel, J.: Optimized and quasi-optimal Schwarz waveform relaxation for the one-dimensional Schrödinger equation. Mathematical Models and Methods in Applied Sciences **20**(12), 2167–2199 (2010). DOI:10.1142/S0218202510004891. 00029
8. Shu-Lin Wu, M., Al-Khaleel, M.: Parameter optimization in waveform relaxation for fractional-order RC circuits. Circuits and Systems I: Regular Papers, IEEE Transactions on **64**(7), 1781–1790 (2017)
9. Wu, S.L., Al-Khaleel, M.: Convergence analysis of the Neumann–Neumann waveform relaxation method for time-fractional RC circuits. Simulation Modelling Practice and Theory **64**, 43–56 (2016)
10. Wu, S.L., Al-Khaleel, M.D.: Semi-discrete Schwarz waveform relaxation algorithms for reaction diffusion equations. BIT **54**(3), 831–866 (2014)
11. Wu, S.L., Al-Khaleel, M.D.: Optimized waveform relaxation methods for RC circuits: discrete case. ESAIM: Mathematical Modelling and Numerical Analysis **51**(1), 209–223 (2017)
12. Wu, S.L., Xu, Y.: Convergence analysis of Schwarz waveform relaxation with convolution transmission conditions. SIAM Journal on Scientific Computing **39**(3), 890–921 (2017)

Local Spectra of Adaptive Domain Decomposition Methods

Alexander Heinlein, Axel Klawonn, and Martin J. Kühn

1 Introduction

For second order elliptic partial differential equations, such as diffusion or elasticity, with arbitrary and high coefficient jumps, the convergence rate of domain decomposition methods with classical coarse spaces typically deteriorates. One remedy is the use of adaptive coarse spaces, which use eigenfunctions computed from local generalized eigenvalue problems to enrich the standard coarse space; see, e.g., [19, 6, 5, 4, 22, 23, 3, 16, 17, 14, 7, 8, 24, 1, 20, 2, 13, 21, 10, 9, 11]. This typically results in a condition number estimate of the form

$$\kappa \leq C \,\mathrm{tol} \quad \text{or} \quad \kappa \leq C \frac{1}{\mathrm{tol}} \tag{1}$$

of the preconditioned system, where C is independent of the coefficient function and tol is a tolerance for the selection of the eigenfunctions.

Obviously, the robustness of the adaptive domain decomposition methods is therefore closely related to the choice of tol. Whereas for a pessimistic choice, i.e., tol ≈ 1, the adaptive coarse space can resort to a direct solver, a very optimistic choice can lead to bad convergence behavior of the method.

In this article, we will compare the spectra of the generalized eigenvalue problems of several adaptive coarse spaces for overlapping as well as nonoverlapping domain decomposition methods. The spectra are of interest because they provide information for choosing an adequate tolerance splitting bad and good eigenmodes as well as

Alexander Heinlein and Axel Klawonn
Department of Mathematics and Computer Science, University of Cologne, Weyertal 86-90, 50931 Köln, Germany. E-mail: {alexander.heinlein,axel.klawonn}@uni-koeln.de.
Center for Data and Simulation Science, University of Cologne, Germany, url: http://www.cds.uni-koeln.de

Martin J. Kühn
CERFACS (Centre Européen de Recherche et de Formation Avancée en Calcul Scientifique), 42 Avenue Gaspard Coriolis, 31057 Toulouse Cedex 01, France; e-mail: martin.kuehn@cerfacs.fr

R. Haynes et al. (eds.), *Domain Decomposition Methods in Science and Engineering XXV*, Lecture Notes in Computational Science and Engineering 138,
https://doi.org/10.1007/978-3-030-56750-7_18

about the resulting dimension of the adaptive coarse spaces. Therefore, we will consider certain representative examples of coefficient functions in two dimensions.

Note that we are not going to discuss other important properties of the adaptive coarse spaces considered here, such as

- condition number and iteration counts of the methods,
- costs for the computation of the eigenvalue problems and the coarse basis functions, respectively,
- necessary communication in a parallel implementation and the ratio of local and global work.

Thus, we do not claim to draw a general comparison of the different adaptive methods. We only want to discuss reasonable choices for the user-defined tolerance for different, exemplary coefficient distributions and the different types of eigenvalue problems. We hope that this gives some insight for further discussions.

Model problems and domain decomposition notation We consider the variational form of a second order elliptic partial differential equation, such as diffusion or elasticity, and denote the coefficient by $\rho \in \mathbb{R}^+$ which is assumed to be constant on each finite element. In matrix form, the problem reads $Ax = b$.

Now, let Ω be decomposed into nonoverlapping subdomains $\Omega_1, \ldots, \Omega_N$ and Γ be the interface of this domain decomposition. We define corresponding subdomain stiffness matrices $A^{(i)}$ with Neumann boundary conditions on $\partial\Omega_i$, $i = 1, \ldots, N$ and the block diagonal matrix $A_N := \text{blockdiag}_i(A^{(i)})$ which is not assembled in the interface degrees of freedom. For an edge $\mathcal{E}$ or its closure $\overline{\mathcal{E}}$ shared by the subdomains Ω_i and Ω_j, we obtain the matrix $A_a^{(i,j)}$ by assembly of the degrees of freedom on $\mathcal{E}$ or $\overline{\mathcal{E}}$, respectively, in the matrix $A_{na}^{(i,j)} := \text{blockdiag}(A^{(i)}, A^{(j)})$.

The Schur complements with respect to $\mathcal{Z} = \mathcal{E}$, $\mathcal{Z} = \overline{\mathcal{E}}$, or any other $\mathcal{Z} \subset \Gamma$ are obtained from $A_{na}^{(i,j)}$ or $A_a^{(i,j)}$ by elimination of all remaining local degrees of freedom $\mathcal{Z}^C$:

$$S_{*,\mathcal{Z}}^{(i,j)} := A_{*,\mathcal{Z}\mathcal{Z}}^{(i,j)} - A_{*,\mathcal{Z}\mathcal{Z}^C}^{(i,j)} \left(A_{*,\mathcal{Z}^C\mathcal{Z}^C}^{(i,j)}\right)^{-1} A_{*,\mathcal{Z}^C\mathcal{Z}}^{(i,j)}$$

with $* \in \{a, na\}$. We also need $S_{\mathcal{Z}}^{(i)} := A_{\mathcal{Z}\mathcal{Z}}^{(i)} - A_{\mathcal{Z}\mathcal{Z}^C}^{(i)} \left(A_{\mathcal{Z}^C\mathcal{Z}^C}^{(i)}\right)^{-1} A_{\mathcal{Z}^C\mathcal{Z}}^{(i)}$.

In addition to that, let the matrices $A_{\mathcal{E}}$ and $M_{\mathcal{E}}$ be matrix discretizations of the one-dimensional bilinear forms $a_{\mathcal{E}}(u, v) := \int_{\mathcal{E}} \rho_{\mathcal{E},\max}\, D_{x^t} u\, D_{x^t} v\, dx$ and $b_{\mathcal{E}}(u, v) := h^{-1} \sum_{x_k \in \mathcal{E}} \beta_k u(x_k)\, v(x_k)$ with $\beta_k := \sum_{\{t \in \tau_h : k \in \text{dof}(t)\}} \rho_t$. Here, ρ_t is the constant coefficient on the element $t \in \tau_h$ and $\rho_{\mathcal{E},\max}(x) := \max\left\{\lim_{y_i \in \Omega_i \to x} \rho(y_i), \lim_{y_j \in \Omega_j \to x} \rho(y_j)\right\}$. D_{x^t} denotes the tangent derivative with respect to the edge e_{ij}, and the x_k correspond to the finite element nodes on the edge. Consequently, $A_{\mathcal{E}}$ and $M_{\mathcal{E}}$ are the stiffness matrix and a scaled lumped mass matrix on the edge $\mathcal{E}$.

2 Various adaptive coarse spaces in domain decomposition

Overlapping Schwarz methods We extend the nonoverlapping subdomains to overlapping subdomains $\Omega_1', \ldots, \Omega_N'$ and consider two-level overlapping Schwarz methods of the form

$$M_{OS-2}^{-1} = \Phi A_0^{-1} \Phi^T + \sum_{i=1}^{N} R_i^T A_i^{-1} R_i,$$

with overlapping matrices $A_i = R_i A R_i^T$, $i = 1, ..., N$, where R_i is the restriction matrix to the overlapping subdomain Ω_i', and the coarse matrix $A_0 = \Phi^T A \Phi$. Here, the columns of Φ are the coarse basis functions. We consider three different adaptive coarse spaces for overlapping Schwarz methods, i.e., the Spectral Harmonically Enriched Multiscale (SHEM) [7], the Overlapping Schwarz Approximate Component Mode Synthesis (OS-ACMS) [10], and the Adaptive Generalized Dryja-Smith-Widlund (AGDSW) [9, 11] coarse spaces.

In all these approaches, the coarse space consists of vertex- and edge-based energy-minimizing basis functions, i.e., the interior values Φ_I are given by $\Phi_I := -A_{II}^{-1} A_{I\Gamma} \Phi_\Gamma$ for given interface values Φ_Γ. The vertex-based basis functions are nodal basis functions of Multiscale Finite Element Method (MsFEM) [12] type with different choices of edge values; cf. [7, 10, 9, 11]. The edge-based basis functions are energy-minimizing extensions of the solutions of generalized eigenvalue problems corresponding to the edges of the nonoverlapping domain decomposition.

For an edge $\mathcal{E}$ of the nonoverlapping domain decomposition, we consider the following edge eigenvalue problems.

(Ov1) SHEM coarse space [7]: find $(\tau_\mathcal{E}, \mu_\mathcal{E}) \in V_0^h(\mathcal{E}) \times \mathbb{R}$ s. t.

$$\theta^T A_\mathcal{E} \tau_\mathcal{E} = \mu_\mathcal{E}^{-1} \theta^T M_\mathcal{E} \tau_\mathcal{E} \quad \forall \theta \in V_0^h(\mathcal{E}).$$

(Ov2) OS-ACMS coarse space [10]: find $(\tau_\mathcal{E}, \mu_\mathcal{E}) \in V_0^h(\mathcal{E}) \times \mathbb{R}$ s. t.

$$\theta^T S_{\overline{\mathcal{E}}}^{(i,j)} \tau_\mathcal{E} = \mu_\mathcal{E}^{-1} \theta^T A_{\overline{\mathcal{E}}\overline{\mathcal{E}}} \tau_\mathcal{E} \quad \forall \theta \in V_0^h(\mathcal{E}).$$

(Ov3) AGDSW coarse space [9, 11]: find $(\tau_\mathcal{E}, \mu_\mathcal{E}) \in V_0^h(\mathcal{E}) \times \mathbb{R}$ s. t.

$$\theta^T S_{\mathcal{E}}^{(i,j)} \tau_\mathcal{E} = \mu_\mathcal{E}^{-1} \theta^T A_{\mathcal{E}\mathcal{E}} \tau_\mathcal{E} \quad \forall \theta \in V_0^h(\mathcal{E}).$$

Let the reciprocal eigenvalues $\mu_\mathcal{E}$ be ordered nondescendingly. Then, we select eigenpairs with $\mu_\mathcal{E} > \text{tol}$ to obtain a condition number estimate of the form $\kappa(M_{OS2}^{-1} A) \leq C\text{tol}$ that is independent of the coefficient function ρ. Note that we use the reciprocal eigenvalue only for comparison with the adaptive coarse spaces for nonoverlapping domain decomposition methods. For the AGDSW coarse space, the matrix on the left hand side is singular. Therefore, we obtain infinity reciprocal eigenvalues in our numerical results.

Nonoverlapping methods In the nonoverlapping domain decomposition methods FETI-1 and FETI-DP, we use the block diagonal matrix A_N and introduce a jump operator B for the interface with $B := (B_1, \ldots, B_N)$, $u = (u_1^T, \ldots, u_N^T)^T$, and $u_i : \Omega_i \to \mathbb{R}$, $i = 1, \ldots, N$ such that $Bu = 0$ if and only if u is continuous across the interface. The FETI master system is given by

$$\begin{bmatrix} A_N & B^T \\ B & 0 \end{bmatrix} \begin{bmatrix} u \\ \lambda \end{bmatrix} = \begin{bmatrix} f \\ 0 \end{bmatrix}.$$

In FETI-1, the null space of A_N is handled by a projection P such that we solve the following system reduced to the Lagrange multipliers and preconditioned by the nonadaptive, projected Dirichlet preconditioner $PM_D^{-1}P^T$

$$PM_D^{-1}P^T BA_N^+ B^T P^T \lambda = PM_D^{-1}P^T d$$

with corresponding right hand side $PM_D^{-1}P^T d$. We have $M_D^{-1} = B_D \text{blockdiag}(S_{\Gamma_i}^{(i)}) B_D^T$, where B_D is a scaled variant of B. In FETI-DP, we subassemble A_N in a selected number of degrees of freedom on the interface, e.g., all vertices, and denote the resulting nonsingular matrix by $\widetilde{A}_N$. In the nonadaptive case, we then solve the preconditioned system

$$M_D^{-1} B\widetilde{A}_N^{-1} B^T \lambda = M_D^{-1}\widetilde{d}.$$

Adaptive constraints can then be enforced by, e.g., a second projection P_0 (see [23] for FETI-1 or [17, 14] for FETI-DP) or via a generalized transformation-of-basis approach; see [15]. In FETI-1/-DP and BDD(C) methods, the operator $P_D = B_D^T B$ is used for proving condition number bounds and thus also appears in some generalized eigenvalue problems.

In this paper, we consider the the GenEO eigenvalue problems for FETI-1 (or BDD methods); see [23]; which were first introduced for overlapping Schwarz methods; see [22]. A P_D-based estimate based coarse space was motivated in [19]. There, P_D was localized to $P_{D,\mathcal{E}}$ by extracting from B and B_D the rows only considering the jumps on the corresponding edge (in 2D). A condition number bound for the 2D case was proven in [17]. The method was extended to a robust three dimensional version in [14]. We present results with ρ-scaling as (NOv2a) and deluxe-scaling as (NOv2b). Another P_D-based coarse space was proposed by [3] for BDDC with deluxe-scaling. In the eigenvalue problems, the matrix operator $A : B = A(A+B)^+B$ is used and the cutoff of the interface Schur complement at the edge $S_{\Gamma_i|\mathcal{E}}^{(i)}$ is used on the right hand side. The energy comparison was generalized to arbitrary scaling matrices $D^{(i)}$ in [17]. Extensions of this method to three dimensions were considered, e.g., in [24, 1, 20, 2, 13]. We present results for ρ-scaling as (NOv3a) and deluxe-scaling as (NOv3b).

(NOv1) GenEO coarse space (FETI-1/BDD) [23]: find $(\tau_{\Gamma_i}, \mu_{\Gamma_i}) \in V^h(\Gamma_i) \times \mathbb{R}$ s. t.

$$\theta^T S_{\Gamma_i}^{(i)} \tau_{\Gamma_i} = \mu_{\Gamma_i}^{-1} \theta^T \left(B_i^T M_D^{-1} B_i\right) \tau_{\Gamma_i} \quad \forall \theta \in V^h(\Gamma_i).$$

(NOv2) P_D-based coarse space no. 1 (FETI-DP/BDDC) [19]: find $(\tau_{\Gamma_i}, \mu_{\Gamma_i}) \in \left(\ker S_{na,\Gamma_{ij}}^{(i,j)}\right)^\perp \times \mathbb{R}$ s. t.

$$\theta^T P_{D,\mathcal{E}}^T S_{na,\Gamma_{ij}}^{(i,j)} P_{D,\mathcal{E}} \tau_{\Gamma_{ij}} = \mu_{\Gamma_{ij}} \theta^T S_{na,\Gamma_{ij}}^{(i,j)} \tau_{\Gamma_{ij}} \quad \forall \theta \in \left(\ker S_{na,\Gamma_{ij}}^{(i,j)}\right)^\perp.$$

(NOv3) P_D-based coarse space no. 2 (FETI-DP/BDDC) [3]: find $(\tau_\mathcal{E}, \mu_\mathcal{E}) \in V_0^h(\mathcal{E}) \times \mathbb{R}$ s. t.

$$\theta^T S_\mathcal{E}^{(i)} : S_\mathcal{E}^{(j)} \tau_\mathcal{E} = \mu_\mathcal{E} \theta^T \left(D_\mathcal{E}^{(j),T} S_{\Gamma|\mathcal{E}}^{(i)} D_\mathcal{E}^{(j)} + D_\mathcal{E}^{(i),T} S_{\Gamma|\mathcal{E}}^{(j)} D_\mathcal{E}^{(i)}\right) \tau_\mathcal{E} \quad \forall \theta \in V_0^h(\mathcal{E})$$

Let the (reciprocal) eigenvalues be ordered nondescendingly. Then, we select eigenpairs with $\mu_{\Gamma_i}^{-1}$, $\mu_{\Gamma_{ij}}$, or $\mu_{\mathcal{E}}$ greater than tol. For the (NOv1) and the (NOv3) eigenvalue problems, the matrix on the left hand side is singular, therefore, we obtain infinity (reciprocal) eigenvalues in our numerical results. For (NOv1), note that the authors of [23] do not incorporate the eigenvectors corresponding to zero eigenvalues into the coarse space. With all three eigenvalue problems (NOv1)-(NOv3), we then obtain adaptive methods with a condition number bound $\kappa \leq C\text{tol}$ that is independent of the coefficient function ρ.

3 Numerical results

In this section, we present results for a diffusion problem on $\Omega = (0,1)^2$ decomposed into nine subdomains. We used a rectangular domain decomposition and slightly curved edges for the subdomain in the center to prevent the appearance of symmetric effects. We set homogeneous Dirichlet boundary conditions for the edge with $x = 0$ and homogeneous Neumann boundary conditions elsewhere.

The local spectra of the different adaptive coarse spaces for eight different coefficient distributions are shown in Figures 1 and 2. The critical eigenvalues and reciprocal eigenvalues, respectively, are displayed above the spectral gap, which is hatched in gray. They are plotted side by side if they are close to each other. A wide spectral gap simplifies the choice of an appropriate tolerance tol. In addition to that, the number of critical eigenvalues is related to the dimension of the coarse space. Note that the condition number estimate (1) guarantees fast convergence of all different approaches for arbitrary coefficient distributions if a suitable tolerance is chosen. However, as can be observed from our results, there are significant differences in the width of the spectral gap and the number of critical eigenvalues for the depicted model problems.

The use of harmonic extensions in the eigenvalue problems of the OS-ACMS coarse space can reduce the number of bad eigenmodes compared to the cheaper one-dimensional integrals in the related SHEM coarse space. A similar behavior can be observed for the expensive deluxe-scaling compared to the cheaper ρ-scaling for the P_D-based approaches for FETI-DP/BDDC. For several coefficient distributions, the width of the spectral gap is larger than two orders of magnitude for all approaches, whereas it is quite small, e.g., for channel-type coefficient distributions.

Note that the plots in Figures 1 and 2 contain much more information, which we cannot discuss here due to lack of space. We hope that the results presented here give some insight for further investigations. Further investigations in three dimensions are also of high interest. This is however out of the scope of this paper. A small comparison between the 3D version algorithms of columns (Nov2a) and (Nov2b) can be found in [18, Sec. 6.5.3]. For overlapping Schwarz methods, a comparison between different 3D approaches including (Ov3) can be found in [11].

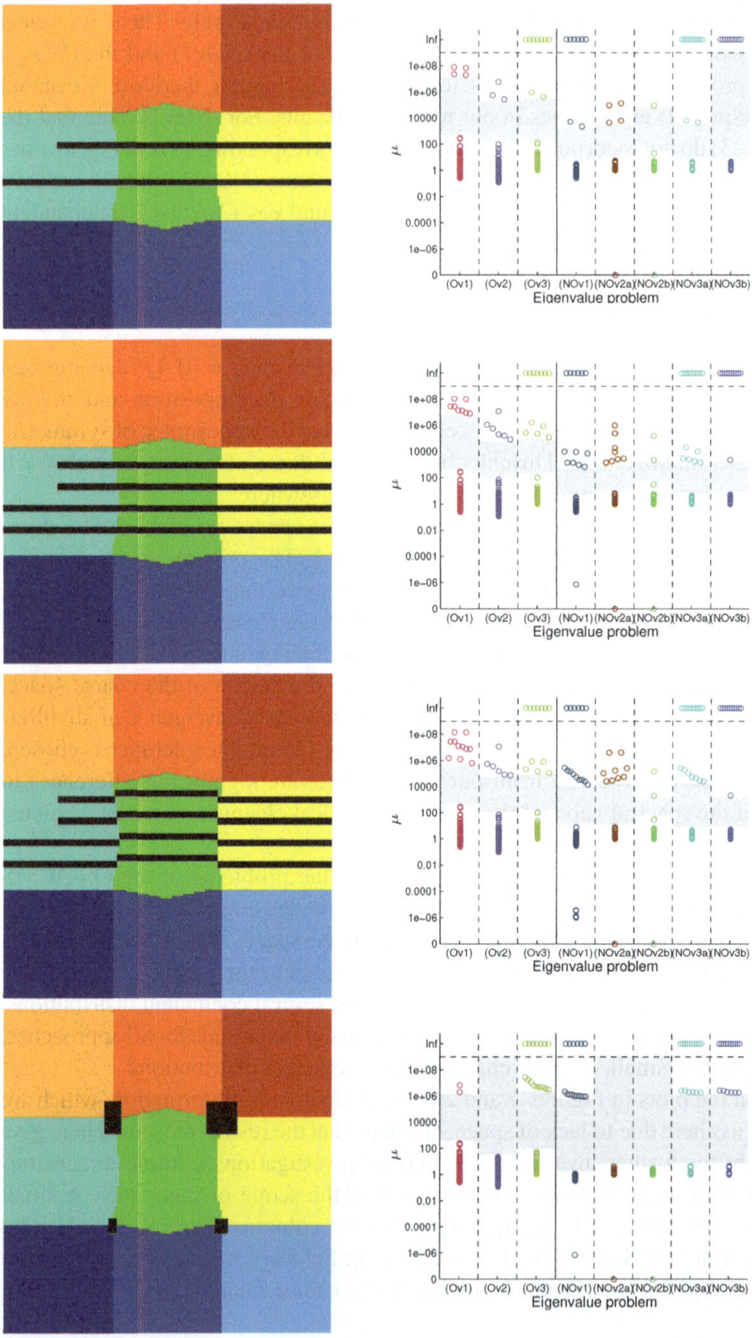

Fig. 1: Left: domain decompositions and high coefficient components ($\rho = 1e6$, black) for several exemplary coefficient distributions. Right: corresponding (reciprocal) eigenvalues μ. Large eigenvalues (> 500) are distributed horizontally to visualize their number. The gap between good and bad eigenmodes is shown in gray.

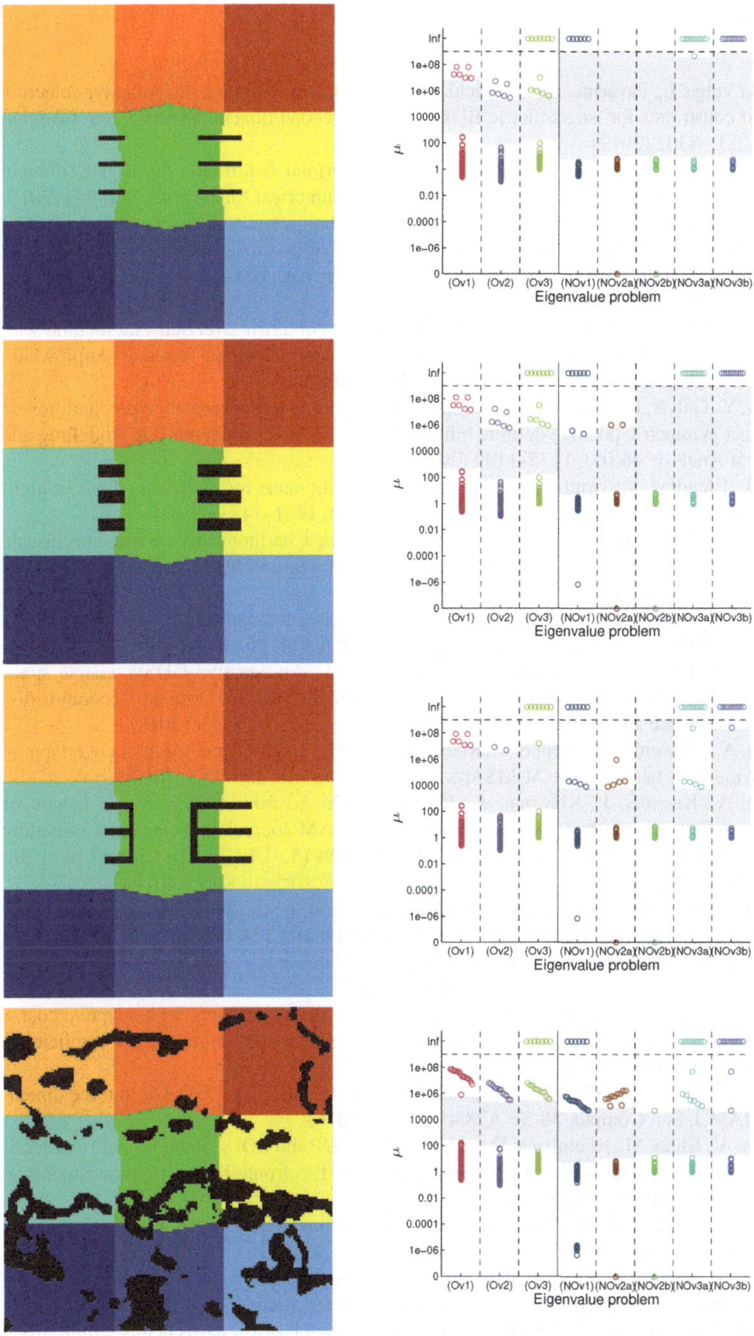

Fig. 2: Left: domain decompositions and high coefficient components ($\rho = 1e6$, black) for several exemplary coefficient distributions. Bottom coefficient function generated from the microstructure of a dual-phase steel; courtesy of Jörg Schröder, University of Duisburg-Essen, Germany, originating from a cooperation with ThyssenKruppSteel. Right: corresponding (reciprocal) eigenvalues μ. Large eigenvalues (> 500) are distributed horizontally to visualize their number. The gap between good and bad eigenmodes is shown in gray.

References

1. Beirão da Veiga, L., Pavarino, L.F., Scacchi, S., Widlund, O.B., Zampini, S.: Adaptive selection of primal constraints for isogeometric BDDC deluxe preconditioners. SIAM J. Sci. Comput. **39**(1), A281–A302 (2017)
2. Calvo, J.G., Widlund, O.B.: An adaptive choice of primal constraints for BDDC domain decomposition algorithms. Electronic Transactions on Numerical Analysis **45**, 524–544 (2016)
3. Dohrmann, C., Pechstein, C.: In C. Pechstein, Modern domain decomposition solvers - BDDC, deluxe scaling, and an algebraic approach. Slides to a talk at NuMa Seminar, JKU Linz, December 10th, 2013, `http://people.ricam.oeaw.ac.at/c.pechstein/pechstein-bddc2013.pdf` (2013)
4. Dolean, V., Nataf, F., Scheichl, R., Spillane, N.: Analysis of a two-level Schwarz method with coarse spaces based on local Dirichlet-to-Neumann maps. Comput. Methods Appl. Math. **12**(4), 391–414 (2012). `DOI:10.2478/cmam-2012-0027`
5. Efendiev, Y., Galvis, J., Lazarov, R., Willems, J.: Robust domain decomposition preconditioners for abstract symmetric positive definite bilinear forms. ESAIM: Mathematical Modelling and Numerical Analysis **46**(05), 1175–1199 (2012)
6. Galvis, J., Efendiev, Y.: Domain decomposition preconditioners for multiscale flows in high-contrast media. Multiscale Modeling & Simulation **8**(4), 1461–1483 (2010)
7. Gander, M.J., Loneland, A., Rahman, T.: Analysis of a new harmonically enriched multiscale coarse space for domain decomposition methods. Tech. rep., arxiv.org (2015). URL `http://arxiv.org/abs/1512.05285`
8. Haferssas, R., Jolivet, P., Nataf, F.: A robust coarse space for optimized Schwarz methods: SORAS-GenEO-2. C. R. Math. Acad. Sci. Paris **353**(10), 959–963 (2015)
9. Heinlein, A., Klawonn, A., Knepper, J., Rheinbach, O.: An adaptive GDSW coarse space for two-level overlapping Schwarz methods in two dimensions. In: Domain Decomposition Methods in Science and Engineering XXIV, *LNCSE*, vol. 125. Springer (2018)
10. Heinlein, A., Klawonn, A., Knepper, J., Rheinbach, O.: Multiscale coarse spaces for overlapping Schwarz methods based on the ACMS space in 2D. ETNA **48**, 156–182 (2018)
11. Heinlein, A., Knepper, J., Klawonn, A., Rheinbach, O.: Adaptive GDSW coarse spaces for overlapping Schwarz methods in three dimensions. SIAM Journal on Scientific Computing **41**(5), A3045–A3072 (2019). `DOI:10.1137/18M1220613`. URL `https://doi.org/10.1137/18M1220613`
12. Hou, T.Y., Wu, X.H.: A multiscale finite element method for elliptic problems in composite materials and porous media. Journal of Computational Physics **134**(1), 169 – 189 (1997). `DOI:http://dx.doi.org/10.1006/jcph.1997.5682`. URL `http://www.sciencedirect.com/science/article/pii/S0021999197956825`
13. Kim, H.H., Chung, E., Wang, J.: BDDC and FETI-DP preconditioners with adaptive coarse spaces for three-dimensional elliptic problems with oscillatory and high contrast coefficients. J. Comput. Phys. **349**, 191–214 (2017)
14. Klawonn, A., Kühn, M., Rheinbach, O.: Adaptive coarse spaces for FETI-DP in three dimensions. SIAM J. Sci. Comput. **38**(5), A2880–A2911 (2016)
15. Klawonn, A., Kühn, M., Rheinbach, O.: Adaptive FETI-DP and BDDC methods with a generalized transformation of basis for heterogeneous problems. Electronic Transactions on Numerical Analysis (ETNA) **49**, 1–27 (2018)
16. Klawonn, A., Radtke, P., Rheinbach, O.: FETI-DP methods with an adaptive coarse space. SIAM J. Numer. Anal. **53**(1), 297–320 (2015)
17. Klawonn, A., Radtke, P., Rheinbach, O.: A comparison of adaptive coarse spaces for iterative substructuring in two dimensions. Electron. Trans. Numer. Anal. **45**, 75–106 (2016)
18. Kühn, M.J.: Adaptive FETI-DP and BDDC methods for highly heterogeneous elliptic finite element problems in three dimensions. Ph.D. thesis, Universität zu Köln (2018)
19. Mandel, J., Sousedík, B.: Adaptive selection of face coarse degrees of freedom in the BDDC and the FETI-DP iterative substructuring methods. Comput. Methods Appl. Mech. Engrg. **196**(8), 1389–1399 (2007)

20. Oh, D., Widlund, O.B., Zampini, S., Dohrmann, C.R.: BDDC Algorithms with deluxe scaling and adaptive selection of primal constraints for Raviart-Thomas vector fields. Math. Comp. **87**(310), 659–692 (2018)
21. Pechstein, C., Dohrmann, C.R.: A unified framework for adaptive BDDC. Electron. Trans. Numer. Anal. **46**, 273–336 (2017)
22. Spillane, N., Dolean, V., Hauret, P., Nataf, F., Pechstein, C., Scheichl, R.: Abstract robust coarse spaces for systems of PDEs via generalized eigenproblems in the overlaps. Numer. Math. **126**(4), 741–770 (2014)
23. Spillane, N., Rixen, D.J.: Automatic spectral coarse spaces for robust finite element tearing and interconnecting and balanced domain decomposition algorithms. Internat. J. Numer. Methods Engrg. **95**(11), 953–990 (2013)
24. Zampini, S.: PCBDDC: a class of robust dual-primal methods in PETSc. SIAM J. Sci. Comput. **38**(5), S282–S306 (2016)

FROSch: A Fast And Robust Overlapping Schwarz Domain Decomposition Preconditioner Based on Xpetra in Trilinos

Alexander Heinlein, Axel Klawonn, Sivasankaran Rajamanickam, and Oliver Rheinbach

1 Introduction

This article describes a parallel implementation of a two-level overlapping Schwarz preconditioner with the GDSW (Generalized Dryja–Smith–Widlund) coarse space described in previous work [12, 10, 15] into the Trilinos framework; cf. [16]. The software is a significant improvement of a previous implementation [12]; see Sec. 4 for results on the improved performance.

In the software, now named `FROSch` (Fast and Robust Overlapping Schwarz), efforts were made for the seamless integration into the open-source Trilinos framework, and to allow the use of heterogeneous architectures, such as those with NVIDIA accelerators. These goals were achieved in the following way:

1. The GDSW preconditioner, i.e., the `FROSch` library, is now part of Trilinos as a subpackage of the package `ShyLU`. The `ShyLU` package provides distributed-memory parallel domain decomposition solvers, and node-level direct solvers for the subdomains. Currently, `ShyLU` has two other domain decomposition solvers, i.e., a Schur complement solver [19] and an implementation of the BDDC method by Clark Dohrmann, and node-level (in)complete `LU` factorizations (`basker` [2]), `fastilu` [18]) , Cholesky factorization (`tacho` [17]) and triangular solves (`hts` [3]).

Alexander Heinlein and Axel Klawonn
Department of Mathematics and Computer Science, University of Cologne, Weyertal 86-90, 50931 Köln, Germany. E-mail: `{alexander.heinlein,axel.klawonn}@uni-koeln.de`.
Center for Data and Simulation Science, University of Cologne, Germany, url: `http://www.cds.uni-koeln.de`

Sivasankaran Rajamanickam
Center for Computing Research, Scalable Algorithms Department, Sandia National Laboratories, Albuquerque, NM 87123. e-mail: `srajama@sandia.gov`

Oliver Rheinbach
INMO, Technische Universität Bergakademie Freiberg, Akademiestr. 6, 09599 Freiberg, Germany. E-mail: `oliver.rheinbach@math.tu-freiberg.de`.

R. Haynes et al. (eds.), *Domain Decomposition Methods in Science and Engineering XXV*, Lecture Notes in Computational Science and Engineering 138,
https://doi.org/10.1007/978-3-030-56750-7_19

2. `FROSch` now supports the `Kokkos` programming model through the use of the `Tpetra` stack in Trilinos. The `FROSch` library can therefore profit from the efforts of the `Kokkos` package to obtain performance portability by template meta-programming, on modern hybrid architectures with accelerators. During this process the GDSW code has been modified and improved significantly. The resulting `FROSch` library is now designed such that different types of Schwarz operators can be added and combined more easily. Consequently, various different Schwarz preconditioners can be constructed using the `FROSch` framework. Recently, `FROSch` has been used in a three-level GDSW implementation [13, 14] and for the solution of incompressible fluid flow problems [11].

2 The GDSW Preconditioner

We are concerned with finding the solution of a sparse linear system

$$Ax = b, \tag{1}$$

arising from a finite element discretization with finite element space $V = V^h(\Omega)$ of an elliptic problem, such as, a Laplace problem, on a domain $\Omega \subset \mathbb{R}^d$, $d = 2, 3$, with sufficient Dirichlet boundary conditions. The GDSW preconditioner [4, 5] is a two-level additive overlapping Schwarz preconditioner with exact local solvers (cf. [21]) using a coarse space constructed from energy-minimizing functions. It is meant to be used in combination with the Krylov methods from the packages `Belos` [1] or `AztecOO`. In particular, let Ω be decomposed into N nonoverlapping subdomains Ω_i, $i = 1, \ldots, N$, and overlapping sudomains Ω_i', $i = 1, \ldots, N$, respectively, and $V_i = V^h(\Omega_i')$, $i = 1, \ldots, N$, be the corresponding local finite element spaces. Further, we define standard restriction operators $R_i : V \to V_i$, $i = 1, \ldots, N$, from the global to the local finite element spaces. Then, the Schwarz operator of the GDSW method can be written in the form

$$P_{\text{GDSW}} = M_{\text{GDSW}}^{-1} A = \Phi A_0^{-1} \Phi^T A + \sum_{i=1}^{N} R_i^T A_i^{-1} R_i A, \tag{2}$$

where $A_0 = \Phi^T A \Phi$ is the coarse space matrix, and the matrices $A_i = R_i A R_i^T$, $i = 1, \ldots, N$, represent the overlapping local problems; cf. [5]. The matrix Φ is the essential ingredient of the GDSW preconditioner. It is composed of coarse space functions which are discrete harmonic extensions from the interface to the interior degrees of freedom of nonoverlapping subdomains. The values on the interface are typically chosen as restrictions of the elements of the null space of the operator $\hat{A}$ to the edges, vertices, and faces of the decomposition, where $\hat{A}$ is the global matrix corresponding to A but with homogeneous Neumann boundary condition. Therefore, for a scalar elliptic problem, the coarse basis functions form a partition of unity on all subdomains that do not touch the Dirichlet boundary. The condition number of the GDSW Schwarz operator is bounded as

$$\kappa(P_{\mathrm{GDSW}}) \leq C\left(1+\frac{H}{\delta}\right)\left(1+\log\left(\frac{H}{h}\right)\right)^2, \tag{3}$$

where h is the size of a finite element, H the size of a nonoverlapping subdomain, and δ the width of the overlap; see [4, 5, 6]. The exponent of the logarithmic term can be reduced to 1 for variants of the GDSW coarse space; see, e.g., [7, 8].

However, the dimension of the standard GDSW coarse space is in the order of $\dim(V_0) = O(\dim(\mathrm{null}(\hat{A}))(N_{\mathcal{V}}+N_{\mathcal{E}}+N_{\mathcal{F}}))$, where $N_{\mathcal{V}}$, $N_{\mathcal{E}}$, and $N_{\mathcal{F}}$ are the global numbers of vertices, edges, and faces of the nonoverlapping domain decomposition, respectively. The dimension of the coarse space is fairly high. Therefore, GDSW coarse spaces of reduced dimension have very recently been introduced in [8]; see also [15] for parallel results. For general problems, the dimension of the reduced GDSW coarse spaces is $\dim(V_0) = O(\dim(\mathrm{null}(\hat{A}))(N_{\mathcal{V}}))$, which is, especially for unstructured decompositions, significantly smaller. Both types of GDSW coarse spaces are implemented in `FROSch`, and in Sec. 4, we present performance results.

3 Software Design of the FROSch Library

During the integration of the `FROSch` library into Trilinos, the code was substantially restructured. In particular, in the transition from the Trilinos `Epetra` (used in [12]) to the newer `Xpetra` sparse matrix infrastructure, it was extended to a framework of Schwarz preconditioners. Additionally, parts of the code have been improved and functionality has been added. As opposed to [12], `FROSch` is completely based on `Xpetra`.

A Framework for Schwarz Preconditioners As described in Sec. 2, the GDSW preconditioner is a two-level overlapping Schwarz method using a specific coarse space. The GDSW Schwarz operator is of the form

$$P_{2\text{-Lvl}} = \underbrace{\Phi A_0^{-1}\Phi^T A}_{P_0} + \sum_{i=1}^{N} \underbrace{R_i^T A_i^{-1} R_i A}_{P_i};$$

cf. (2); and therefore, it is the sum of local overlapping Schwarz operators P_i, $i = 1, \ldots, N$, and a global coarse Schwarz operator P_0. There are different ways to compose Schwarz operators P_i, $i = 0, \ldots, N$, e.g.:

Additive:	P_{ad}	$= \sum_{i=0}^{N} P_i$
Multiplicative:	P_{mu}	$= I - (I-P_N)(I-P_{N-1})\cdots(I-P_0)$
	$P_{\mathrm{mu-sym}}$	$= I - \prod_{i=0}^{N}(I-P_i)\prod_{i=0}^{N-1}(I-P_{N-1-i})$
Hybrid:	$P_{\mathrm{hy-1}}$	$= I-(I-P_0)\left(I-\sum_{i=0}^{N} P_i\right)(I-P_0)$
	$P_{\mathrm{hy-2}}$	$= \alpha P_0 + I - (I-P_N)\cdots(I-P_1);$

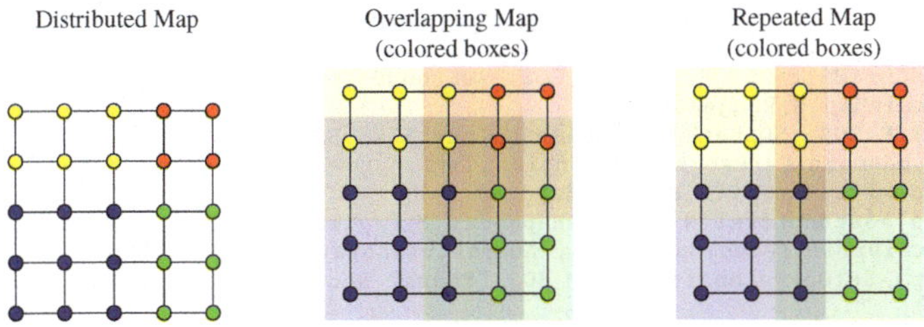

Fig. 1: Heuristic reconstruction of the domain decomposition interface: uniquely distributed map (left); extension of the uniquely distributed map by one layer of elements resulting in an overlapping map, where the overlap contains the interface (middle); by selection, using the lower subdomain ID, the interface is defined (right).

cf. [21]. Using the FROSch library, it is simple to construct the different variants once the ingredients are set up. Let us explain this based on the example of the class `GDSWPreconditioner` in FROSch, which is derived from the abstract class `SchwarzPreconditioner` and contains an implementation of the GDSW preconditioner: in FROSch, the `SumOperator` is used to combine Schwarz operators in an additive way. The additive first level is implemented in the class `AlgebraicOverlappingOperator` and the coarse level of the GDSW preconditioner in the class `GDSWCoarseOperator`. Therefore, the `GDSWPreconditioner` is basically just the following composition of Schwarz operators:

```
GDSWPreconditioner = SumOperator( AlgebraicOverlappingOperator,
                                  GDSWCoarseOperator )
```

By replacing the `SumOperator` by a `ProductOperator`, the levels can be coupled in a multiplicative way. The different classes for Schwarz operators are all derived from an abstract `SchwarzOperator`, and the classes `SchwarzOperator` and `SchwarzPreconditioner` are both derived from the abstract `Xpetra::Operator`.

Transition from Epetra to Xpetra To facilitate the use of FROSch on novel architectures, the code was ported completely from `Epetra` data structures to `Xpetra`. As `Xpetra` provides a lightweight interface to `Epetra` as well as `Tpetra`, FROSch can now profit from the computational kernels from `Kokkos`, while maintaining compatibility to older `Epetra`-based software such as LifeV [9].

Improvement of the Code & Additional Functionality The efficiency of the code was improved and new functionality was added as part of this redesign. In particular, the routines for the computation of local-to-global mappings and the identification of the interface components have been rewritten and therefore improved with respect to their performance; see Sec. 4 for the numerical results.

Two important features have been added. First, we have introduced the possibility to reconstruct a domain decomposition interface algebraically based on a unique distribution of the degrees of freedom into subdomains and the nonzero pattern of the matrix; cf. Fig. 1. This works particularly well for scalar elliptic problems and piecewise linear elements. In general, the best performance is obtained when a `RepeatedMap` is provided by the user; cf. Fig. 2. This map corresponds to the

Previous implementation from [12]:

```
Teuchos::RCP<SOS::SOS> M_SOS(new SOS::SOS(numVectors,numSubdomainsPerProcess,
     M_DomainMap,M_RangeMap));
Teuchos::RCP<SOS::SOSSetUp> M_SOSSetUp(new SOS::SOSSetUp(
     numSubdomainsPerProcess,dimension,dofs,M_rowMatrixTeuchos,M_DomainMap));
M_SOSSetUp->FirstLevel(M_ProcessMapOverlap);
M_SOSSetUp->SecondLevel(M_ProcessMapNodes,M_ProcessMap,SOS::LifeVOrdering,
     M_LocalDirichletBoundaryDofs,"Mumps",useRotations,M_LocalNodeList);
M_SOSSetUp->SetUpPreconditioner(M_SOS,"Mumps",secondLevelSolverParameterList,
     Type);
```

Current implementation Shylu/FROSch:

```
Teuchos::RCP<FROSch::GDSWPreconditioner<SC,LO,GO,NO> > FROSchGDSW(new FROSch::
     GDSWPreconditioner<SC,LO,GO,NO>(A,ParameterList);
FROSchGDSW->initialize(Dimension,Overlap,RepeatedMap);
FROSchGDSW->compute();
```

Fig. 2: Comparison of the user-interface for the previous implementation of the GDSW solver (top) and the current implementation in FROSch (bottom). The setup is split into the `initialize` and `compute` phases instead of the two levels.

nonoverlapping domain decomposition and is replicated in the interface degrees of freedom. Secondly, we have introduced a function that identifies Dirichlet boundary conditions based on matrix rows with only diagonal entries. This is important because the coarse basis functions are zero on the Dirichlet boundary.

User Interface The user-interface of the FROSch library has been completely re-designed. Compared to the previous implementation, where the setup of the preconditioner was split into the first and second level, it is now split into the phases `initialize` and `compute`, also reducing the number of required lines of code to construct the GDSW preconditioner; cf. Fig. 2. In the `initialize` phase, all data structures that correspond to the structure of the problem are built, i.e., the index sets of the overlapping subdomains and the interface are identified and the interface values of the coarse basis are computed. In the `compute` phase, all computations related to the values of the matrix A are performed, i.e., the overlapping problems are factorized, the interior values of the coarse basis are computed, and the coarse problem is assembled and factorized. Therefore, the `initialize` and `compute` phases can be seen as the symbolic and the numerical factorizations of a direct solver: if only the the values in the matrix A change, the preconditioner can be updated using `compute`, and if the structure of the problem is changed, `initialize` has to be called to update the preconditioner. Also, FROSch provides a `Stratimikos` interface for easier use in applications; `Stratimikos` provides a unified framework for solvers and preconditioners in Trilinos.

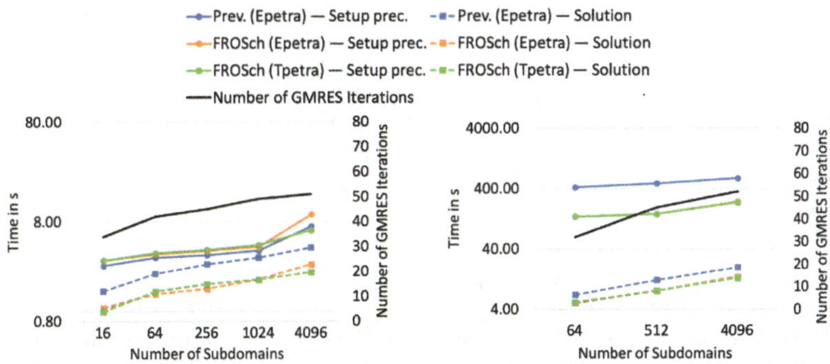

Fig. 3: Weak scalability of the two-level Schwarz preconditioner using the GDSW coarse space for the Poisson model problem: (Left) in two dimensions with overlap $\delta = 5h$ and $H/h = 100$ (approximately 50k degrees of freedom per sudomain); (Right) in two dimensions with overlap $\delta = 2h$ and $H/h = 14$ (approximately 50k degrees of freedom per sudomain). Comparison of the previous implementation (blue) and the current implementation in `FROSch`, i.e., the `Epetra` (orange) and the `Tpetra` (green) versions available through the `Xpetra` interface. The number of iterations (black) are identical for all versions.

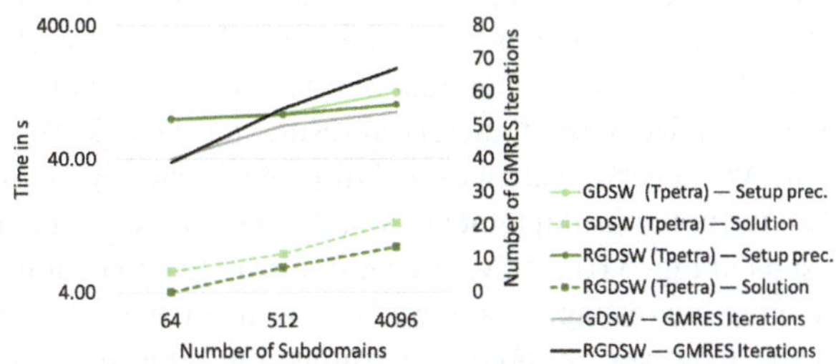

Fig. 4: Weak scalability of the two-level Schwarz preconditioner with overlap $\delta = 1h$ for the Poisson model problem in three dimensions with $H/h = 14$ (approximately 35k degrees of freedom per subdomain): comparison of the GDSW and the RGDSW coarse space using the `Tpetra` version of the `FROSch` implementation.

4 Performance of the New FROSch Software

Here, the performance of the new software is compared against the previous implementation. We consider a Poisson model problem on $\Omega \subset \mathbb{R}^d$, $d = 2, 3$, with full Dirichlet boundary condition, discretized by piecewise quadratic finite elements. We compare the performance of the previous implementation, which is based on `Epetra`, and the current implementation in `FROSch`. In particular, the `Epetra` and the `Tpetra` version of the current implementation, which are both available through the `Xpetra` interface, are compared. As a Krylov-solver GMRES from `Belos` [1] is used with a relative tolerance of 10^{-7} for the unpreconditioned residual. For the local and coarse problems, the native direct solver in Trilinos, `KLU`, is used; only in Fig. 5, `Mumps` is used as the direct solver. We always use one subdomain per processor core. The computations were performed on the magnitUDE supercomputer at Universität Duisburg-Essen, which has 15k cores (Intel Xeon E5-2650v4, 12C, 2.2GHz) and

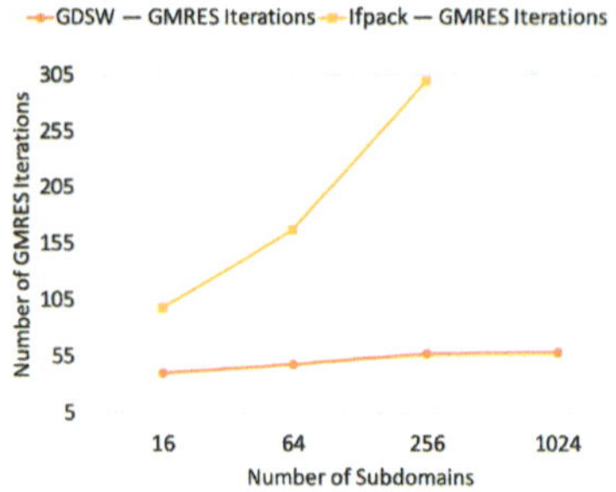

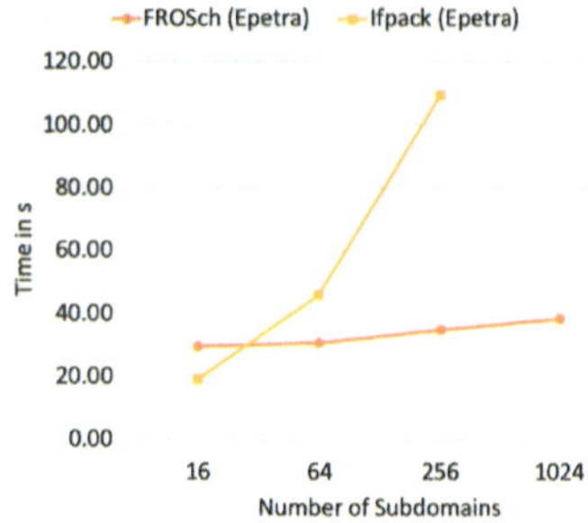

Fig. 5: Weak scalability for the Poisson model problem in two dimensions with $H/h = 200$ (approximately 195k degrees of freedom per sudomain): comparison of FROSch using the GDSW coarse space and the one-level overlapping Schwarz preconditioner Ifpack with overlap $\delta = 20h$; numbers of GMRES iterations (left) and total solver times (right). Using Mumps for all direct solves. For 1 024 subdomains, Ifpack did not converge within 500 GMRES iterations.

a total memory of 36 096 GB. Here, we do not exploit any node parallelism when using Tpetra. We consider the setup phase and the solution phase and include the identification of the interface components in the setup phase. This part does not scale very well and can takes a significant amount of time for a large number of processes; cf. [12]. In Fig. 3 (left), we present numerical results for the GDSW preconditioner in two dimensions. We observe that, in the solution phase, the new implementation is always faster than the previous implementation. The time for the setup phase is comparable. The results in Fig. 3 (right), where we compare the preconditioners in three dimensions, are more interesting. Again, we observe that the solution phase is faster by a similar factor. However, in three dimensions, the setup phase in the FROSch implementation is much faster compared to the previous implementation. We also observe that the Tpetra version is always slightly faster than the Epetra version of the new code. In Fig. 4, the GDSW and the RGDSW coarse spaces are compared for the Tpetra version of the FROSch implementation. We observe that, due to the increasing dimension of the coarse space, the computation time can be improved when using reduced dimension coarse spaces. This effect becomes stronger when the number of subdomains is increased; cf. [15]. Finally, we present a comparison of FROSch using the GDSW coarse space and Ifpack [20], i.e., a one-level overlapping Schwarz preconditioner, in Fig. 5. We observe that Ifpack does not scale as it lacks a second level. Already for 64 subdomains, FROSch converges much faster, and for 1 024 subdomains, Ifpack does not converge within a maximum number of 500 GMRES iterations.

Conclusion We presented the new Trilinos library FROSch that allows the flexible construction of different overlapping Schwarz methods. The FROSch implementation of the GDSW preconditioner is significantly faster than the previous one.

Acknowledgements The authors gratefully acknowledge the computing time granted by the Center for Computational Sciences and Simulation (CCSS) at Universität Duisburg-Essen and provided on the supercomputer magnitUDE (DFG grants INST 20876/209-1 FUGG, INST 20876/243-1 FUGG) at Zentrum für Informations- und Mediendienste (ZIM). Sandia National Laboratories is a multimission laboratory managed and operated by National Technology and Engineering Solutions of Sandia, LLC., a wholly owned subsidiary of Honeywell International, Inc., for the U.S. Department of Energy's National Nuclear Security Administration under contract DE-NA-0003525.

References

1. Bavier, E., Hoemmen, M., Rajamanickam, S., Thornquist, H.: Amesos2 and Belos: Direct and iterative solvers for large sparse linear systems. Scientific Programming **20**(3), 241–255 (2012)
2. Booth, J.D., Ellingwood, N.D., Thornquist, H.K., Rajamanickam, S.: Basker: Parallel sparse lu factorization utilizing hierarchical parallelism and data layouts. Parallel Computing **68**, 17 – 31 (2017). DOI:https://doi.org/10.1016/j.parco.2017.06.003. URL http://www.sciencedirect.com/science/article/pii/S0167819117300868
3. Bradley, A.M.: A hybrid multithreaded direct sparse triangular solver. In: Proceedings of CSC16, pp. 13–22. SIAM (2016)
4. Dohrmann, C.R., Klawonn, A., Widlund, O.B.: Domain decomposition for less regular subdomains: overlapping Schwarz in two dimensions. SIAM J. Numer. Anal. **46**(4), 2153–2168 (2008)
5. Dohrmann, C.R., Klawonn, A., Widlund, O.B.: A family of energy minimizing coarse spaces for overlapping Schwarz preconditioners. In: Domain decomposition methods in science and engineering XVII, *LNCSE*, vol. 60, pp. 247–254. Springer, Berlin (2008)
6. Dohrmann, C.R., Widlund, O.B.: Hybrid domain decomposition algorithms for compressible and almost incompressible elasticity. IJNME **82**(2), 157–183 (2010)
7. Dohrmann, C.R., Widlund, O.B.: An alternative coarse space for irregular subdomains and an overlapping Schwarz algorithm for scalar elliptic problems in the plane. SIAM J. Numer. Anal. **50**(5), 2522–2537 (2012). DOI:10.1137/110853959
8. Dohrmann, C.R., Widlund, O.B.: On the design of small coarse spaces for domain decomposition algorithms. SIAM J. Sci. Comput. **39**(4), A1466–A1488 (2017). DOI:10.1137/17M1114272
9. Formaggia, L., Fernandez, M., Gauthier, A., Gerbeau, J.F., Prud'homme, C., Veneziani, A.: The LifeV Project. Web. Http://www.lifev.org
10. Heinlein, A.: Parallel overlapping Schwarz preconditioners and multiscale discretizations with applications to fluid-structure interaction and highly heterogeneous problems. PhD thesis, Universität zu Köln, Germany (2016)
11. Heinlein, A., Hochmuth, C., Klawonn, A.: Monolithic overlapping Schwarz domain decomposition methods with GDSW coarse spaces for incompressible fluid flow problems. SIAM Journal on Scientific Computing **41**(4), C291–C316 (2019). DOI:10.1137/18M1184047. URL https://doi.org/10.1137/18M1184047
12. Heinlein, A., Klawonn, A., Rheinbach, O.: A parallel implementation of a two-level overlapping Schwarz method with energy-minimizing coarse space based on Trilinos. SIAM J. Sci. Comput. **38**(6), C713–C747 (2016). DOI:10.1137/16M1062843
13. Heinlein, A., Klawonn, A., Rheinbach, O., Röver, F.: A Three-Level Extension of the GDSW Overlapping Schwarz Preconditioner in Two Dimensions, pp. 187–204. Springer International Publishing, Cham (2019). DOI:10.1007/978-3-030-14244-5_10. URL https://doi.org/10.1007/978-3-030-14244-5_10
14. Heinlein, A., Klawonn, A., Rheinbach, O., Röver, F.: A three-level extension of the GDSW overlapping Schwarz preconditioner in three dimensions. Tech. rep. (March 2019). Accepted for publication to LNCSE.
15. Heinlein, A., Klawonn, A., Rheinbach, O., Widlund, O.B.: Improving the parallel performance of overlapping Schwarz methods by using a smaller energy minimizing coarse space. In: Domain Decomposition Methods in Science and Engineering XXIV, pp. 383–392. Springer International Publishing, Cham (2018)
16. Heroux, M.A., Bartlett, R.A., Howle, V.E., Hoekstra, R.J., Hu, J.J., Kolda, T.G., Lehoucq, R.B., Long, K.R., Pawlowski, R.P., Phipps, E.T., Salinger, A.G., Thornquist, H.K., Tuminaro,

R.S., Willenbring, J.M., Williams, A., Stanley, K.S.: An overview of the Trilinos project. ACM Trans. Math. Softw. **31**(3), 397–423 (2005). DOI:http://doi.acm.org/10.1145/1089014.1089021

17. Kim, K., Edwards, H.C., Rajamanickam, S.: Tacho: Memory-scalable task parallel sparse cholesky factorization. In: 2018 IEEE International Parallel and Distributed Processing Symposium Workshops (IPDPSW), pp. 550–559. IEEE (2018)
18. Patel, A., Boman, E., Rajamanickam, S., Chow, E.: Cross platform fine grained ILU and ILDL factorizations using Kokkos. In: Proceedings of the CCR (2015)
19. Rajamanickam, S., Boman, E.G., Heroux, M.A.: ShyLU: A hybrid-hybrid solver for multicore platforms. In: 2012 IEEE 26th International Parallel and Distributed Processing Symposium, pp. 631–643 (2012). DOI:10.1109/IPDPS.2012.64
20. Sala, M., Heroux, M.: Robust algebraic preconditioners with IFPACK 3.0. Tech. Rep. SAND-0662, Sandia National Laboratories (2005)
21. Toselli, A., Widlund, O.: Domain decomposition methods—algorithms and theory, *Springer Series in Computational Mathematics*, vol. 34. Springer-Verlag, Berlin (2005)

A Three-level Extension of the GDSW Overlapping Schwarz Preconditioner in Three Dimensions

Alexander Heinlein, Axel Klawonn, Oliver Rheinbach, and Friederike Röver

1 The Standard GDSW Preconditioner

The GDSW (Generalized Dryja–Smith–Widlund) preconditioner is a two-level overlapping Schwarz domain decomposition preconditioner [23] with exact local solvers [5, 4]. The GDSW preconditioner can be written in the form

$$M_{\mathrm{GDSW}}^{-1} = \underbrace{\Phi K_0^{-1} \Phi^T}_{\text{Coarse Level}} + \underbrace{\sum_{i=1}^{N} R_i^T K_i^{-1} R_i}_{\text{First Level}}, \tag{1}$$

where $K_0 = \Phi^T K \Phi$ is the coarse matrix and the $K_i = R_i K R_i^T$, $i = 1, ..., N$, correspond to the local overlapping subdomain problems. By $V_1, \ldots, V_N$, we denote the local subspaces corresponding to the overlapping subdomains, and V_0 denotes the corresponding coarse space. The restriction operators on the subdomain level are defined as $R_i : V^h(\Omega) \to V_i := V^h(\Omega_i')$ for $i = 1, \ldots, N$. The columns of the matrix Φ correspond to the coarse basis function which are chosen to to be discrete harmonic extensions from the interface of the nonoverlapping decomposition to the interior degrees of freedom. The interface values are restrictions of the elements of the null space of the operator to the edges, vertices, and faces. For linear elliptic problems, the condition number of the Schwarz operator is bounded by

Alexander Heinlein and Axel Klawonn
Department of Mathematics and Computer Science, University of Cologne, Weyertal 86-90, 50931 Köln, Germany. E-mail: {alexander.heinlein,axel.klawonn}@uni-koeln.de.
Center for Data and Simulation Science, University of Cologne, Germany, url: http://www.cds.uni-koeln.de

Oliver Rheinbach, Friederike Röver
Institut für Numerische Mathematik und Optimierung, Technische Universität Bergakademie Freiberg, Akademiestr. 6, 09599 Freiberg, Germany. E-mail: {oliver.rheinbach, friederike.roever}@math.tu-freiberg.de.

R. Haynes et al. (eds.), *Domain Decomposition Methods in Science and Engineering XXV*, Lecture Notes in Computational Science and Engineering 138,
https://doi.org/10.1007/978-3-030-56750-7_20

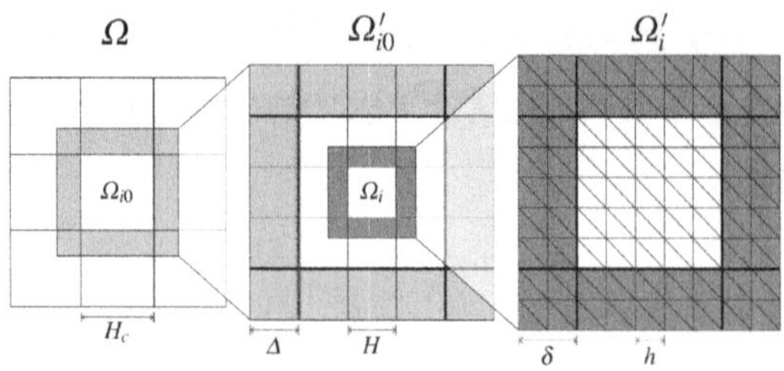

Fig. 1: Structured decomposition of an exemplary two-dimensional computational domain Ω into nonoverlapping subregions Ω_{i0}(left), a zoom into one overlapping subregion Ω'_{i0} consisting of subdomains Ω_i (middle), and a zoom into one overlapping subdomain Ω'_i (right). Each level of zoom corresponds to one level of the preconditioner; image taken from [13].

$$\kappa(M_{\mathrm{GDSW}}^{-1}K) \leq C\left(1+\frac{H}{\delta}\right)\left(1+\log\left(\frac{H}{h}\right)\right)^2, \tag{2}$$

where h is the size of a finite element, H the size of a nonoverlapping subdomain, and δ the width of the overlap; see [4, 5, 6]. Even better condition number estimates are available. For example, the power of the logarithmic term can be reduced to 1 using Option 1 in [7]. An important advantage of the GDSW preconditioner is that it can be constructed in an algebraic fashion from the fully assembled matrix K and without the need of an additional coarse triangulation. This will also facilitate the construction of the three-level GDSW preconditioner presented in the following section.

2 The Three-Level GDSW Preconditioner

If a direct solver is used for the solution of the coarse problem in (1), this can become a bottleneck for a large number of subdomains; cf. [11, 9]. As a remedy, in this paper, we apply the GDSW preconditioner recursively to the coarse problem, resulting in a three-level extension of the GDSW preconditioner; see [13] for the corresponding algorithm in two dimensions. Our three-level GDSW method is related to the three-level BDDC method [24]. A further recursive application of the preconditioner, resulting in a multilevel extension similar to multi-level BDDC methods [18, 2, 16], multilevel Schwarz methods [17, 21], or multigrid methods [8], is algorithmically straightforward but out of the scope of this paper. The scalability of the two-level method can also be improved by reducing the size of the GDSW coarse space; cf. [14, 7]. There, instead of using coarse basis functions corresponding to subdomain edges, vertices, and, faces, new basis functions are constructed, e.g., corresponding only to the vertices. In this paper, we will construct three-level GDSW methods using standard as well as reduced dimension coarse spaces.

To define the three-level GDSW preconditioner, we decompose the domain Ω into nonoverlapping subregions Ω_{i0} of diameter H_c; see [24] and Figure 1 for a graphical representation of the decomposition Ω in two dimensions. Each subregion is decomposed into nonoverlapping subdomains of diameter about H. Extending each subregion Ω_{i0} to Ω'_{i0} by recursively adding layers of subdomains, an overlapping decomposition into subregions is obtained. The overlap on subregion level is denoted by Δ; the overlap on the subdomain level is denoted by δ, consistent with the notation of the two-level method; see Figure 1.

The three-level GDSW preconditioner then is defined as

$$M_{3GDSW}^{-1} = \underbrace{\Phi\Big(\overbrace{\Phi_0 K_{00}^{-1} \Phi_0^T}^{\text{Third Level}} + \overbrace{\sum_{i=1}^{N_0} R_{i0}^T K_{i0}^{-1} R_{i0}}^{\text{Second Level}} \Big) \Phi^T}_{\text{Coarse Levels}} + \underbrace{\sum_{j=1}^{N} R_j^T K_j^{-1} R_j}_{\text{First Level}}, \tag{3}$$

where $K_{00} = \Phi_0^T K_0 \Phi_0$ and $K_{i0} = R_{i0} K_0 R_{i0}^T$. On the subregion level, we define the restriction operators to the overlapping subregions Ω'_{i0} as $R_{i0} : V^0 \rightarrow V_i^0 := V^0(\Omega'_{i0})$ for $i = 1, ..., N_0$. The respective coarse space is denoted as V_{00} and spanned by the coarse basis functions Φ_0.

3 Implementation and Software Libraries

The parallel three-level GDSW implementation discussed in this paper is based on [9, 11, 12] and uses the Trilinos Epetra linear algebra package. A recent Xpetra version (FROSch - Fast and Robust Overlapping Schwarz framework [10]) is now part of the Trilinos [15] package ShyLU [19].

To test our three-level GDSW implementation, we consider the Poisson problem on the unit cube $[0, 1]^3$ with homogenous Dirichlet boundary conditions on $\partial\Omega$. We use structured domain decompositions into subregions and subdomains; see Figure 1 for a representation of the two-dimensional case. Our model problem is discretized using piecewise linear finite elements. As a default Krylov method, we apply the GMRES method provided by the Trilinos package Belos [3]. Trilinos version 12.11 (Dev) is used; cf. [15].

All numerical experiments were carried out on the JUQUEEN supercomputer at JSC Julich. We use the IBM XL C/C++ compiler for Blue Gene V.12.1, and Trilinos is linked to the ESSL.

To solve the overlapping subdomain and subregion problems and the coarse problem, we always use MUMPS 4.10.0 [1] in symmetric, sequential mode, and interfaced through the Trilinos package Amesos [20]. For our experiments, we always have a one-to-one correspondence of subdomains and processor cores. We use the relative stopping criterion $\|r^k\|_2/\|r^0\|_2 \leq 10^{-6}$. Moreover, we assume that we have a fast and scalable method to identify interface degrees of freedom. That cost is therefore neglected in this paper.

3.1 Weak Parallel Scalability of the Three-Level GDSW Preconditioner

In this section, we focus on the weak scalability of our preconditioners. For the numerical scalability of the three-level GDSW preconditioner in two dimensions, more detailed numerical results can be found in [13]. We also compare results for the two and three-level methods using the standard coarse space (denoted by GDSW) and the reduced dimension coarse space (denoted by RGDSW). In particular, we use *Option 1*, which is the completely algebraic variant of the RGDSW coarse space; cf. [7] or [14], respectively.

The number of Krylov iterations is presented in Figure 2 and Table 1. Note that the standard two-level GDSW method fails for more than 13 824 cores since the coarse problem could not be factored any more due to memory limits. All other methods show numerical scalability for up to 64 000 cores. This includes the two-level RGDSW method, which is a remarkable result since RGDSW coarse space is smaller than the standard GDSW coarse space but the coarse basis function have, on average, a larger support (see also Table 2); cf. also [14].

Our results show, that the numerical scalability of both two-level methods is slightly better; cf. Figure 2 and Table 1. Moreover, the number of iterations is higher by almost a factor of two for both three-level methods; this is, however, not surprising since the direct coarse solver is replaced by a (two-level) preconditioner.

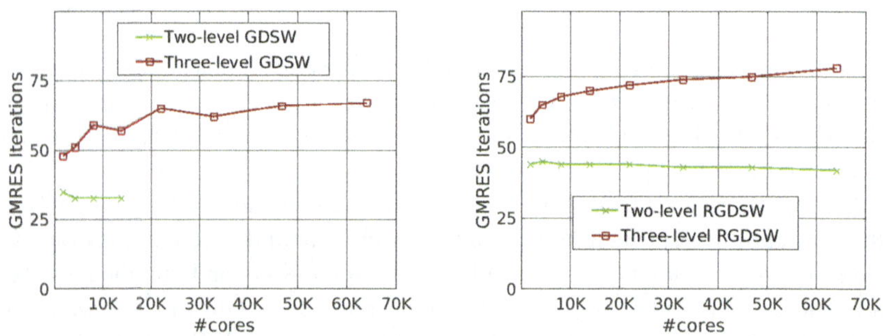

Fig. 2: Weak numerical scalability of the two- and three-level GDSW (left) and the RGDSW (right) preconditioner. All methods are numerically scalable; see Table 1 for the corresponding data.

Let us now consider the computing times, which are more favorable for the three-level methods; see Figure 3 and Table 1. By *Solver Time*, we denote the time to solution, which is the sum of the time for the setup of the preconditioner, denoted *Setup Time*, and the time for the Krylov iteration, which we denote *Krylov Time*. The *Setup Time* includes the factorizations of the matrices on the different levels using the MUMPS sparse direct solver. For RGDSW coarse space, the three-level method is faster than the two-level methods for 4 096 cores and more; see Figure 3 and Table 1. The two-level RGDSW method is consistently the fastest method from 1 728 to to 32 768 cores. However, for 46 656 and 64 000 cores, the three-level method is faster.

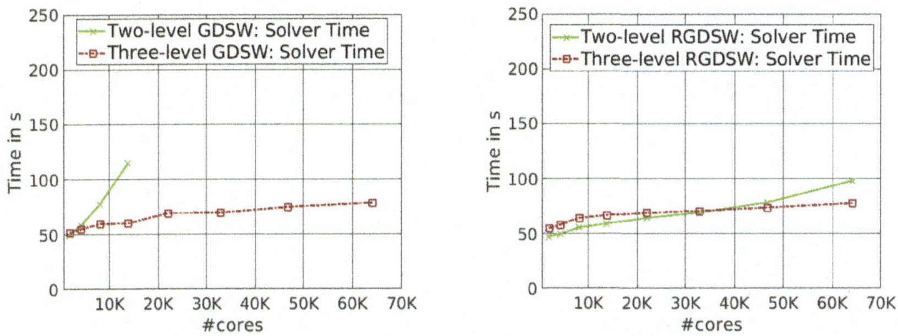

Fig. 3: Weak parallel scalability of the two- and three-level methods using the standard (left) and the reduced coarse space (right); see Table 1 for the data.

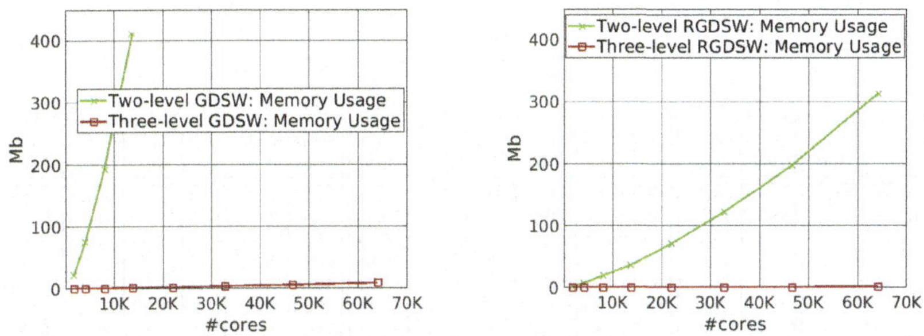

Fig. 4: Memory usage of the MUMPS direct solver for the factorization of the coarse matrix K_0 and K_{00} for the two-level and three-level GDSW method using the standard (left) and reduced coarse space (right); see Table 2 for the corresponding data.

For the largest problem with 1.72 billion degrees of freedom, the *Solver Time* for three-level RGDSW precondtioner (77.7s *Solver Time*) more than 20% faster than two-level RGDSW preconditioner (98.3s *Solver Time*) and also slightly faster than the three-level GDSW preconditioner (78.7s *Solver Time*). However, considering the size of K_0, we expect the two-level RGDSW to fail beyond 100 000 cores while both three-level methods will continue to scale; also cf. the memory usage for the factorazation of K_{00} in Figure 4 and Table 2.

Acknowledgements This work was supported in part by the German Research Foundation (DFG) through the Priority Programme 1648 "Software for Exascale Computing" (SPPEXA) under grants RH 122/3-2 and KL 2094/4-2. Also, this work is in part supported under grant RH 122/5-2 .

The authors gratefully acknowledge the computing the Gauss Centre for Supercomputing e. V. (*www.gauss-centre.eu*) for providing computing time on the GCS Supercomputer JUQUEEN BG/Q supercomputer [22] at JSC Jülich. GCS is the alliance of the three national supercomputing centres HLRS (Universität Stuttgart), JSC (Forschungszentrum Jülich), and LRZ (Bayrische Akademie der Wissenschaften).

#Sub-domains = #cores	Two-level GDSW				Three-level GDSW				Two-level RGDSW				Three-level RGDSW			
	Iter	Solver Time	Setup Time	Krylov Time	Iter	Solver Time	Setup Time	Krylov Time	Iter	Solver Time	Setup Time	Krylov Time	Iter	Solver Time	Setup Time	Krylov Time
1 728	35	50.2 s	30.9 s	19.4 s	48	51.8 s	28.3 s	23.4 s	44	**47.9 s**	26.9 s	21.1 s	60	55.2 s	26.2 s	28.9 s
4 096	33	58.7 s	35.5 s	23.2 s	51	55.1 s	30.1 s	25.0 s	45	**50.0 s**	27.6 s	22.4 s	65	58.3 s	26.7 s	31.6 s
8 000	33	77.7 s	46.3 s	31.4 s	59	60.0 s	30.2 s	29.8 s	44	**56.1 s**	32.3 s	23.8 s	68	64.4 s	30.8 s	33.7 s
13 824	33	115.2 s	69.1 s	46.0 s	57	60.4 s	31.3 s	29.1 s	44	**59.6 s**	33.3 s	26.3 s	70	67.0 s	31.9 s	35.1 s
21 952	—	—	—	—	65	69.5 s	35.0 s	34.6 s	44	**64.7 s**	34.6s	30.1 s	72	69.0 s	32.1 s	36.9 s
32 768	—	—	—	—	62	69.8 s	36.2 s	33.6 s	43	**69.4 s**	35.2 s	34.2 s	74	70.8 s	32.6 s	38.2 s
46 656	—	—	—	—	66	74.8 s	37.1 s	37.6 s	43	78.6 s	37.2 s	41.4 s	75	**73.8 s**	33.7 s	40.2 s
64 000	—	—	—	—	67	78.7 s	38.5 s	40.2 s	42	98.3 s	50.2 s	48.1 s	78	**77.7 s**	34.8 s	42.9 s

Table 1: By *Iter*, we denote number of Krylov iterations. The *Solver Time* is the sum of the *Setup Time* and *Krylov Time*. We have $H/h = 30$, $H/\delta = 15$, $H_c/H = 4$, and $H_c/\Delta = 4$. Also see Figure 2 and Figure 3. The fastest *Solver Time* is printed in bold.

References

1. Amestoy, P.R., Duff, I.S., L'Excellent, J.Y., Koster, J.: A fully asynchronous multifrontal solver using distributed dynamic scheduling. SIAM J. Matrix Anal. Appl. **23**(1), 15–41 (2001)
2. Badia, S., Martín, A.F., Principe, J.: Multilevel balancing domain decomposition at extreme scales. SIAM J. Sci. Comput. **38**(1), C22–C52 (2016). DOI:10.1137/15M1013511
3. Bavier, E., Hoemmen, M., Rajamanickam, S., Thornquist, H.: Amesos2 and Belos: Direct and iterative solvers for large sparse linear systems. Scientific Programming **20**(3), 241–255 (2012)
4. Dohrmann, C.R., Klawonn, A., Widlund, O.B.: Domain decomposition for less regular subdomains: overlapping Schwarz in two dimensions. SIAM J. Numer. Anal. **46**(4), 2153–2168 (2008)
5. Dohrmann, C.R., Klawonn, A., Widlund, O.B.: A family of energy minimizing coarse spaces for overlapping Schwarz preconditioners. In: Domain decomposition methods in science and engineering XVII, *Lect. Notes Comput. Sci. Eng.*, vol. 60, pp. 247–254. Springer, Berlin (2008)
6. Dohrmann, C.R., Widlund, O.B.: Hybrid domain decomposition algorithms for compressible and almost incompressible elasticity. Internat. J. Numer. Methods Engrg. **82**(2), 157–183 (2010)

#Subdomains = #Cores	Size of K_0	Factorization Time	Forward-Backward	Memory Usage	Size of K_{00}	Factorization Time	Forward-Backward	Memory Usage
	Two-level GDSW				Three-level GDSW			
1 728	10 439	1.28 s	2.08 s	23 Mb	98	<0.01 s	0.03 s	1 Mb
4 096	25 695	4.43 s	5.17 s	76 Mb	279	0.01 s	0.09 s	1 Mb
8 000	51 319	11.25 s	11.31 s	193 Mb	604	0.02 s	0.21 s	1 Mb
13 824	89 999	29.58 s	20.46 s	412 Mb	1 115	0.04 s	0.35 s	2 Mb
21 952		—	—	—	1 854	0.09 s	0.68 s	2 Mb
32 768		—	—	—	2 863	0.15 s	0.99 s	4 Mb
46 656		—	—	—	4 184	0.25 s	1.55 s	6 Mb
64 000		—	—	—	5 589	0.40 s	2.28 s	9 Mb
	Two-level RGDSW				Three-level RGDSW			
1 728	1 331	0.06 s	0.3 s	3 Mb	8	<0.01 s	0.01 s	1 Mb
4 096	3 375	0.25 s	0.87 s	8 Mb	27	<0.01 s	0.02 s	1 Mb
8 000	6 859	0.74 s	1.73 s	20 Mb	64	<0.01 s	0.03 s	1 Mb
13 824	12 167	1.81 s	3.02 s	37 Mb	125	<0.01 s	0.05 s	1 Mb
21 952	19 683	3.66 s	5.31 s	71 Mb	216	<0.01 s	0.08 s	1 Mb
32 768	29 791	6.15 s	8.25 s	122 Mb	343	0.01 s	0.13 s	1 Mb
46 656	42 875	10.39 s	12.53 s	198 Mb	512	0.02 s	0.19 s	1 Mb
64 000	59 319	16.80 s	16.96 s	313 Mb	729	0.03 s	0.27 s	2 Mb

Table 2: Costs for solving the problem on the coarsest level, i.e., using K_0 in the standard two-level GDSW and RGDSW preconditioner and using K_{00} in the three-level GDSW and RGDSW preconditioner. Here, *Factorization Time* is the time Amesos reports for the MUMPS sparse direct solver for the sum of symbolic and numerical factorization of K_0 and K_{00}, respectively; *Forward-Backward* is the sum of all times spent in forward-backward substitutions during the Krylov iteration; *Memory Usage* is the estimated amount of memory allocated by MUMPS during the factorization. See Table 1 for the corresponding *Solver Time*, *Setup Time* and *Krylov Time*. Also see Figures 5, 4.

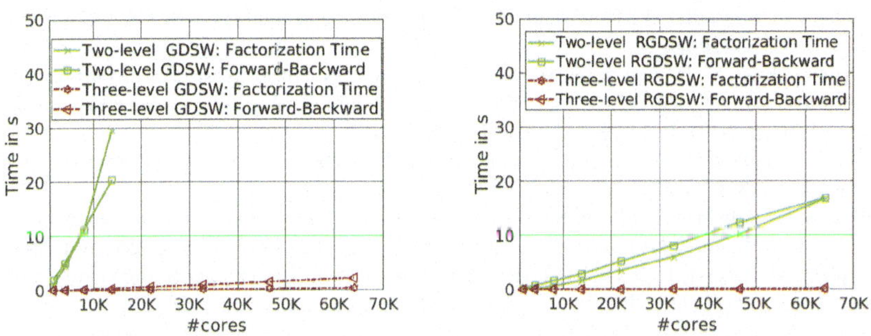

Fig. 5: Computing time for solving the problem on the coarsest level, i.e., using K_0 in the standard two-level method preconditioner ans using K_{00} for the three-level GDSW preconditioner and using the standard coarse space (left) and respectively the reduced coarse space (right). See Table 2 for the corresponding data.

7. Dohrmann, C.R., Widlund, O.B.: On the design of small coarse spaces for domain decomposition algorithms. SIAM J. Sci. Comput. **39**(4), A1466–A1488 (2017)
8. Hackbusch, W.: Multigrid methods and applications, *Springer Series in Computational Mathematics*, vol. 4. Springer-Verlag, Berlin (1985)
9. Heinlein, A.: Parallel overlapping Schwarz preconditioners and multiscale discretizations with applications to fluid-structure interaction and highly heterogeneous problems. Ph.D. thesis, Universität zu Köln (2016)
10. Heinlein, A., Klawonn, A., Rajamanickam, S., Rheinbach, O.: FROSch – a parallel implementation of the GDSW domain decomposition preconditioner in Trilinos (2020). In preparation

11. Heinlein, A., Klawonn, A., Rheinbach, O.: A parallel implementation of a two-level overlapping Schwarz method with energy-minimizing coarse space based on Trilinos. SIAM J. Sci. Comput. **38**(6), C713–C747 (2016)
12. Heinlein, A., Klawonn, A., Rheinbach, O.: Parallel Two-Level Overlapping Schwarz Methods in Fluid-Structure Interaction, pp. 521–530. LNCSE vol 112. Springer International Publishing, Cham (2016)
13. Heinlein, A., Klawonn, A., Rheinbach, O., Röver, F.: A Three-Level Extension of the GDSW Overlapping Schwarz Preconditioner in Two Dimensions, pp. 187–204. Springer International Publishing, Cham (2019). DOI:10.1007/978-3-030-14244-5_10. URL https://doi.org/10.1007/978-3-030-14244-5_10
14. Heinlein, A., Klawonn, A., Rheinbach, O., Widlund, O.B.: Improving the parallel performance of overlapping Schwarz methods by using a smaller energy minimizing coarse space. In: P.E. Bjørstad, S.C. Brenner, L. Halpern, H.H. Kim, R. Kornhuber, T. Rahman, O.B. Widlund (eds.) Domain Decomposition Methods in Science and Engineering XXIV, pp. 383–392. Springer International Publishing, Cham (2018)
15. Heroux, M.A., Bartlett, R.A., Howle, V.E., Hoekstra, R.J., Hu, J.J., Kolda, T.G., Lehoucq, R.B., Long, K.R., Pawlowski, R.P., Phipps, E.T., Salinger, A.G., Thornquist, H.K., Tuminaro, R.S., Willenbring, J.M., Williams, A., Stanley, K.S.: An overview of the Trilinos Project. ACM Trans. Math. Software **31**(3), 397–423 (2005)
16. J. Sístek B. Sousedík, J.M., Burda, P.: Parallel implementation of multilevel BDDC. In: Numerical mathematics and advanced applications 2011 (2011). 9th European conference on numerical mathematics and advanced applications, Leicester, UK, September 5–9, 2011
17. Kong, F., Cai, X.C.: A highly scalable multilevel Schwarz method with boundary geometry preserving coarse spaces for 3D elasticity problems on domains with complex geometry. SIAM J. Sci. Comput. **38**(2), C73–C95 (2016). DOI:10.1137/15M1010567
18. Mandel, J., Sousedík, B., Dohrmann, C.R.: Multispace and multilevel BDDC. Computing **83**(2-3), 55–85 (2008)
19. Rajamanickam, S., Boman, E.G., Heroux, M.A.: Shylu: A hybrid-hybrid solver for multicore platforms. In: 2012 IEEE 26th International Parallel and Distributed Processing Symposium, pp. 631–643 (2012). DOI:10.1109/IPDPS.2012.64
20. Sala, M., Stanley, K., Heroux, M.: On the design of interfaces to sparse direct solvers. ACM Trans. Math. Software (TOMS) **34**(2) (2008)
21. Scacchi, S.: A hybrid multilevel Schwarz method for the bidomain model. Comput. Methods Appl. Mech. Engrg. **197**(45-48), 4051–4061 (2008). DOI:10.1016/j.cma.2008.04.008
22. Stephan, M., Docter, J.: JUQUEEN: IBM Blue Gene/Q® Supercomputer System at the Julich Supercomputing Centre. Journal of large-scale research facilities **1**, A1 (2015)
23. Toselli, A., Widlund, O.: Domain decomposition methods—algorithms and theory, *Springer Series in Computational Mathematics*, vol. 34. Springer-Verlag, Berlin (2005)
24. Tu, X.: Three-level BDDC in two dimensions. Internat. J. Numer. Meth. Engrg **69**(1), 33–59 (2007). DOI:10.1002/nme.1753

Non-geometric Convergence of the Classical Alternating Schwarz Method

Gabriele Ciaramella and Richard M. Höfer

1 Introduction

Let Ω be a domain in $\mathbb{R}^n$ and $f \in L^2(\Omega)$ be a given function. Consider the Laplace problem

$$\Delta u = f \text{ in } \Omega, \quad u = 0 \text{ on } \partial\Omega. \tag{1}$$

In error form, the alternating Schwarz method for the solution to (1) is

$$\begin{aligned} \Delta e_1^n &= 0 && \text{in } \Omega_1, & \Delta e_2^n &= 0 && \text{in } \Omega_2,\\ e_1^n &= 0 && \text{on } \partial\Omega \cap \overline{\Omega}_1, & e_2^n &= 0 && \text{on } \partial\Omega \cap \overline{\Omega}_2,\\ e_1^n &= e_2^{n-1} && \text{on } \Gamma_1, & e_2^n &= e_1^n && \text{on } \Gamma_2. \end{aligned} \tag{2}$$

Given any initial guess $e_0 \in V := H_0^1(\Omega)$ and solving iteratively (2), one obtains the sequence $(e_1^n)_{n\in\mathbb{N}^+} \subset H^1(\Omega_1)$ of errors in Ω_1 and the sequence $(e_2^n)_{n\in\mathbb{N}^+} \subset H^1(\Omega_2)$ of errors in Ω_2. Let us define the sequence $(e_k)_{k\in\mathbb{N}^+} \subset V$ as

$$e_k := \begin{cases} e_1^k & \text{in } \overline{\Omega}_1 \\ e_2^{k-1} & \text{in } \overline{\Omega} \setminus \Omega_1 \end{cases} \text{ for } k \text{ odd, and } \quad e_k := \begin{cases} e_2^k & \text{in } \overline{\Omega}_2 \\ e_1^{k-1} & \text{in } \overline{\Omega} \setminus \Omega_2 \end{cases} \text{ for } k \text{ even.}$$

We denote by V_1 and V_2 the extensions by zero in Ω of $H_0^1(\Omega_1)$ and $H_0^1(\Omega_2)$. Their orthogonal complements $V_1^\perp$ and $V_2^\perp$ in V with respect to the inner product $\langle\cdot,\cdot\rangle := (\nabla\cdot,\nabla\cdot)_{L^2}$ are of the form

$$V_j^\perp = \{v \in H_0^1(\Omega) : \Delta v = 0 \text{ in } \Omega\backslash\overline{\Omega_j}\} \tag{3}$$

G. Ciaramella
Universität Konstanz, Germany, e-mail: gabriele.ciaramella@uni-konstanz.de

R. M. Höfer
University of Bonn, Germany, e-mail: hoefer@iam.uni-bonn.de

R. Haynes et al. (eds.), *Domain Decomposition Methods in Science and Engineering XXV*,
Lecture Notes in Computational Science and Engineering 138,
https://doi.org/10.1007/978-3-030-56750-7_21

for $j = 1, 2$. It is then possible to show that (2) is equivalent to the alternating projection method (APM), $e_k := P_{V_2^\perp} P_{V_1^\perp} e_{k-1}$, for $k \in \mathbb{N}^+$, where $P_{V_j^\perp}$ denote the orthogonal projections onto $V_j^\perp$, $j = 1, 2$; [11, 5].

For an arbitrary Hilbert space V and two closed subspaces V_1 and V_2, von Neumann [12] and Halperin [10] proved that $e_k \to 0$ whenever $\overline{V_1 + V_2} = V$. Moreover, if V_1+V_2 is closed, then the convergence is geometric, i.e. there exists $\theta < 1$ such that for all $e_0 \in V$ it holds that $\|e_k\| \leq \theta^k \|e_0\|$. In the particular case of only two subspaces V_1 and V_2, it is proven that the optimal θ is $\text{incl}(V_1, V_2)$, with $0 \leq \text{incl}(V_1, V_2) \leq 1$ the inclination between the subspaces V_1 and V_2, and that $\theta = \text{incl}(V_1, V_2) < 1$ if and only if $V_1 + V_2$ is closed; see, e.g., [6, 5].

In the context of Schwarz method, P.L. Lions proves in [11] that an overlapping decomposition $\Omega = \Omega_1 \cup \Omega_2$ guarantees that $\overline{V_1 + V_2} = V$, and gives sufficient conditions for $V_1 + V_2 = \overline{V_1 + V_2}$ to hold; see also [5, Lemma 2.16 and Theorem 2.17]. These conditions hold if the overlap $\Omega_1 \cap \Omega_2$ is a sufficiently regular domain. A natural question arises: what happens if $\Omega_1 \cap \Omega_2$ is not regular enough (e.g., non-Lipschitz)? Is the geometric convergence still guaranteed in this case?

We show in this paper that if $\Omega_1 \cap \Omega_2$ is non-Lipschitz, then V_1+V_2 is not necessarily closed. Classical abstract results state that in this case the APM converges 'arbitrarily slowly' [7, 8, 1]:

Definition 1 (Arbitrarily slow convergence (ASC)) The APM is said to converge arbitrarily slowly if for every sequence $(f_n)_{n\in\mathbb{N}} \subset \mathbb{R}_+$ with $f_n \to 0$ and for all $\varepsilon > 0$ there exists $e_0 \in V$ with $\|e_0\| < \sup_n f_n + \varepsilon$ and $\|e_k\| \geq f_n$ for all n.

An ASC is quite difficult to observe and characterize. Therefore, we introduce the notion of 'non-geometric' convergence:

Definition 2 (Non-geometric convergence (NGC)) The APM is said to converge non-geometrically if there is no $\theta < 1$ such that for all $e_0 \in V$ it holds that $\|e_k\| \leq \theta^k \|e_0\|$. Moreover, we say that a vector $e_0 \in V$ leads to NCG, if there exists no $\theta < 1$ such that $\|e_k\| \leq \theta^k \|e_0\|$.

To the best of our knowledge, the case of a non-closed sum $V_1 + V_2$ is not studied in the literature of classical Schwarz theory. Moreover, also the literature concerning the more general framework of the APM presents surprisingly few results for this problem. The aim of our work is to study ASC and NGC of the classical Schwarz method and hence to shed more light on the issue of 'slow convergence' of the APM. To do so, in Section 2 we present a domain decomposition example that leads to two subspaces V_1 and V_2 whose sum is not closed. Section 3 focuses on theoretical results about NGC and ASC of the APM. In Section 4, we consider again the example from Section 2 and discuss the dependence of the convergence rate on the initial function e_0. Moreover, we precisely characterize a dense subset of the set of all functions leading to geometric convergence. Finally, results of numerical experiments are presented in Section 5.

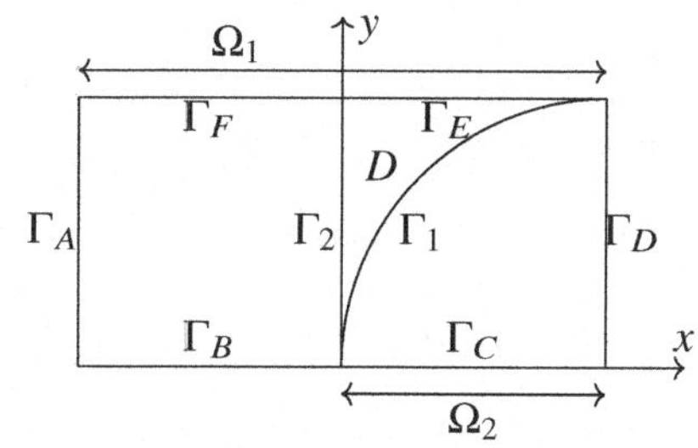

Fig. 1: Decomposition $\Omega = \Omega_1 \cup \Omega_2$ with $D = \{(x, y) \in \Omega \colon x > 0, y > x^\alpha\} = \Omega_1 \cap \Omega_2$ and $\alpha < 1$.

2 Domain decomposition with non-Lipschitz overlap

Consider a domain $\Omega = (-1, 1)\times(0, 1)$ and two subdomains $\Omega_1 = (-1, 0]\times(0, 1)\cup D$ and $\Omega_2 = (0, 1) \times (0, 1)$ with $D = \{(x, y) \in \Omega \colon x > 0, y > x^\alpha\}$ for some $\alpha > 0$. Clearly, the overlapping decomposition $\Omega_1 \cup \Omega_2 = \Omega$ holds, and D is the overlap; see Fig. 1. The following theorem shows that, if $\alpha < 1$ (hence D is a non-Lipschitz domain), then the decomposition $\Omega = \Omega_1 \cup \Omega_2$ leads to two subspaces V_1 and V_2 of V whose sum is not closed.

Theorem 1 (Non-closedness of $\mathbf{V_1 + V_2}$) *Let V_j denote the extension by zero in Ω of $H_0^1(\Omega_j)$ for $j = 1, 2$. Then $\overline{V_1 + V_2} = H_0^1(\Omega)$, but $V_1 + V_2 \neq V$ for any $\alpha < 1$.*

Proof Let $v \in \overline{V_1 + V_2}^\perp$. Then $v \in V_j^\perp$ (see (3)), for $j = 1, 2$. In particular $\Delta v = 0$ in Ω, thus $v = 0$. This proves that $\overline{V_1 + V_2}^\perp = \{0\}$ and the first claim follows.

To prove the second statement, we consider the function $v = (r^\beta \sin\phi)\psi$, where (r, ϕ) denote polar coordinates and $\psi \in C^1(\overline{\Omega})$ is a cut-off function with $\psi = 0$ on $\partial\Omega \setminus \{y = 0\}$ and $\psi = 1$ in $[-2^{-\alpha^{-1}}, 2^{-\alpha^{-1}}] \times [0, \frac{1}{2}]$. A direct calculation shows that $v \in H_0^1(\Omega)$ for $\beta > 0$, and we now prove that $v \notin V_1 + V_2$. To do so, assume for the sake of contradiction that there are $v_1 \in V_1$ and $v_2 \in V_2$ such that $v = v_1 + v_2$. Clearly, it must hold that $v_1 = v$ on $\{x = 0\}$ and $v_1 = 0$ on $\{(x, x^\alpha) \colon 0 \le x \le 1\}$. Let $\gamma(y) := y^{\alpha^{-1}}$. Then $v_1(\gamma(y), y) = 0$ and we get $-v_1(0, y) = \int_0^{\gamma(y)} \partial_x v_1(t, y)\, dt$.[1] Hence, we have

$$\|\nabla v_1\|_{L^2}^2 \ge \int_0^{\frac{1}{2}} \int_0^{\gamma(y)} |\partial_x v_1(t, y)|^2\, dt\, dy \ge \int_0^{\frac{1}{2}} \left[\int_0^{\gamma(y)} \partial_x v_1(t, y)\, dt \right]^2 \frac{1}{\gamma(y)}\, dy$$
$$= \int_0^{\frac{1}{2}} \frac{v_1^2(0, y)}{\gamma(y)} = \int_0^{\frac{1}{2}} \frac{y^{2\beta}}{y^{\alpha^{-1}}},$$

[1] Strictly speaking this is not necessarily meaningful due to possible lack of regularity of v_1. However, it is true for smooth functions and therefore one can argue by density.

which implies that $\|\nabla v_1\|_{L^2} = \infty$ if $2\beta - \alpha^{-1} \leq -1$, i.e., if $\alpha \leq \frac{1}{1+2\beta}$. Thus, for any $\alpha < 1$, this shows that $v_1 \notin V_1$ if we choose $\beta > 0$ sufficiently small, which leads to a contradiction. Hence the second claim follows. □

Consider for any $\varepsilon \in (0,1)$ and $\lambda > 0$ the sets

$$X_{\lambda,\varepsilon} := \{u \in H_0^1(\Omega) \colon u(0,y) \geq \lambda y^\beta \text{ for a.e. } y \in (0,\varepsilon)\}, \tag{4}$$

where the inequality has to be understood in the sense of traces. Notice that $\cup_{\lambda>0} X_{\lambda,\varepsilon}$ is dense in V for any $0 < \varepsilon < 1$. Moreover, if $\beta = (\alpha^{-1} - 1)/2$, then according to the proof of Theorem 1 it holds that $X_{\lambda,\varepsilon} \subset V \setminus (V_1 + V_2)$. Hence, Theorem 3 implies that any $e_0 \in X_{\lambda,\varepsilon}$ leads to a NGC.

In view of Theorem 1, the geometric convergence of the Schwarz method (as APM) does not hold. This is due to results that we discuss in Section 3.

3 'Slow' convergence in the abstract framework of the APM

Consider an arbitrary Hilbert space $(V, \langle\cdot,\cdot\rangle)$ and two closed subspaces V_1 and V_2 such that $V_1 + V_2 \neq \overline{V_1 + V_2}$. Denote by $\|\cdot\|$ the norm induced by $\langle\cdot,\cdot\rangle$.[2] Does the APM, corresponding to the iteration operator $P_{V_2^\perp} P_{V_1^\perp}$, converge geometrically? The answer is negative and given in Theorem 2.

Theorem 2 (On the geometric convergence of the APM) *Let $V_1, V_2 \subset V$ be closed subspaces of a Hilbert space with $\overline{V_1 + V_2} = V$. Let $\|\cdot\|'$ be the operator norm induced by $\|\cdot\|$. The following statements are equivalent.*

(i) $V_1 + V_2 = V$.
(ii) $\|P_{V_2^\perp} P_{V_1^\perp}\|' < 1$.
(iii) There exists $\theta \in [0,1)$ such that $\forall e_0 \in V$ and $\forall k \in \mathbb{N}$ $\|(P_{V_2^\perp} P_{V_1^\perp})^k e_0\| \leq \theta^k \|e_0\|$.
(iv) For all $e_0 \in V$ there is $\theta_{e_0} \in [0,1)$ such that $\forall k \in \mathbb{N}$ $\|(P_{V_2^\perp} P_{V_1^\perp})^k e_0\| \leq \theta_{e_0}^k \|e_0\|$.

Proof The implication from (i) to (ii) is well known; see, e.g., [11, 5]. Clearly (ii) implies (iii), and (iii) implies (iv). It remains to prove that (iv) implies (i). To do so, let $e_0 \in V$, and denote $e_k = (P_{V_2^\perp} P_{V_1^\perp})^k e_0$. We observe that (iv) implies that $\lim_{k\to\infty} e_k = 0$ and that the series $y := \sum_{k=0}^\infty e_k$ is absolutely convergent. Moreover, we have

$$P_{V_2^\perp} P_{V_1^\perp} e_0 = (1 - P_{V_2})(1 - P_{V_1}) e_0 = e_0 - P_{V_2} P_{V_1^\perp} e_0 - P_{V_1} e_0.$$

[2] Notice that we consider here the same notation (namely the symbols V, V_1, V_2, $\langle\cdot,\cdot\rangle$ and $\|\cdot\|$) used in the other sections to describe a more abstract setting. However, it is clear from the context whether the notation refers to an abstract Hilbert space setting or to the precise domain decomposition setting.

By induction and using that $\lim_{k\to\infty} e_k = 0$ and that the series $y := \sum_{k=0}^{\infty} e_k$ converges absolutely, we obtain

$$e_0 = (P_{V_1} + P_{V_2}P_{V_1^\perp})e_0 + e_1 = (P_{V_1} + P_{V_2}P_{V_1^\perp}) \sum_{k=0}^{n} e_k + e_{n+1}$$
$$= (P_{V_1} + P_{V_2}P_{V_1^\perp})y \in V_1 + V_2.$$

Since $e_0 \in V$ was arbitrary, the claim follows. □

Theorem 2 implies that, if $V_1 + V_2$ is not closed, then there exists an initial function e_0 such that the APM sequence $(e_k)_{k\in\mathbb{N}}$ does not converge geometrically. The issue of the rate of convergence of the APM when $V_1 + V_2$ is not closed has first been addressed by Franchetti and Light, who prove in [9] the following result.

Theorem 3 (NGC of the APM) *Let $V_1, V_2 \subset V$ be as in Theorem 2 and assume that $V_1 + V_2$ is not closed. Then, for all $e_0 \in V \setminus (V_1 + V_2)$ it holds that $\sum_{k=1}^{\infty} \frac{\|e_k\|}{\sqrt{k}} = \infty$. In particular, the convergence is NGC.*

Theorem 3 states that for any initial function $e_0 \in V \setminus (V_1 + V_2)$ the convergence of the APM is much slower than geometric. Moreover, in the same paper, the authors provide an example of a non-closed sum $V_1 + V_2$ leading to ASC. In 1997, Bauschke, Borwein and Lewis proved in [3] that ASC holds whenever $V_1 + V_2$ is not closed. However, Bauschke, Deutsch and Hundal pointed out later in [4] that the proof of this result given in [3] is erroneous, and they give a different approach to obtain the same result:

Theorem 4 (Dichotomy between ASC and non-closedness of $\mathbf{V_1+V_2}$) *Let $V_1, V_2 \subset V$ be as in Theorem 2. Then, exactly one of the following two statements holds:*

(1) $V_1 + V_2$ is closed. Then the convergence is geometric.
(2) $V_1 + V_2$ is not closed. Then the convergence is arbitrarily slow.

In 2010, Deutsch and Hundal studied ASC for a general class of operators on Banach spaces [7, 8]. Their results include Theorem 4, also in the case of more than two subspaces. Independently, the same results have been proved in 2011 by Badea, Grivaux and Müller [1]. In the same paper it is shown that, if $V_1 + V_2$ is not closed, then, for any positive sequence $(f_n)_{n\in\mathbb{N}}$, the set $\{e_0 \in V : \|e_n\| \geq f_n \text{ for a. e. } n \in \mathbb{N}\}$ is dense in V.

We have seen that if $V_1 + V_2$ is not closed, then the APM converges arbitrarily slow and the convergence is much slower than geometric at least for any initial vector $e_0 \in V \setminus (V_1 + V_2)$, and that the set of all e_0 leading to ASC is dense in V. However, what is the dependence of the convergence rate on the initial vector e_0? Can one characterize the set of all e_0 leading to geometric convergence? In the papers mentioned above, there are only a few sentences hinting on the dependence of the convergence rate on the starting point e_0. In [2], Badea and Seifert have shown that one can always find a dense subset $W \subset V$ for which 'super-polynomially fast convergence' holds. However, it seems difficult to characterize such a subset in a concrete example. In the following section, we discuss the dependence of the convergence rate on the starting point e_0 for the specific example from Section 2. In particular, we provide rigorous results on the regularity that is needed for an initial function e_0 to lead to geometric convergence, and show that the set of these initial functions is a dense subset of V if the overlap of the domains is not too rough ($\alpha > 1/3$).

4 The dependence of the convergence rate on the initial function

Consider the domain decomposition studied in Section 2 with a non-Lipschitz overlap D. Recall also $V = H_0^1(\Omega)$ and the two subspaces V_1 and V_2 whose orthogonal complements are given in (3). For which initial functions $e_0 \in V$ does the Schwarz method converge geometrically?

Probably the functions that come first to the mind of the reader are the ones in V that vanish on the interface $\Gamma_1 := \overline{\partial\Omega_1 \cap \Omega}$. For these functions, the Schwarz method (2) converges in only one step. Indeed, with $F := \{v \in V : v = 0 \text{ on } \partial\Omega_1 \cap \Omega\}$, we see that $\ker(P_{V_2^\perp} P_{V_1^\perp}) = V_1 \oplus (V_1^\perp \cap V_2) = V_1 \oplus \{v \in V : v = 0 \text{ in } \overline{\Omega_1}\} = F$, where we used (3). It is not difficult to see that, if $u \notin F$, then the iteration will not yield the exact result after any finite number of iterations. Moreover, F is not the maximal set of functions that lead to geometric convergence. This is clearly shown by Theorem 5 below. To prove it, we need the following lemma.

Lemma 1 *Let $V_1, V_2 \subset V$ be as in Theorem 2, and let $W \subset V_1 + V_2$ be a closed subspace which is invariant under $P_{V_2^\perp} P_{V_1^\perp}$. Then, there exists $\theta < 1$ such that $\|(P_{V_2^\perp} P_{V_1^\perp})^k e_0\| \leq \theta^k \|e_0\|$ for all $e_0 \in W$.*

Proof The result follows by the same arguments used in [11, Theorem I.1]. □

Theorem 5 (A set of initial functions leading to geometric convergence) *Recall the domain decomposition given in Theorem 1 and the corresponding parameter α. Consider for an arbitrary $\lambda > 0$ the set*

$$W_\lambda := \left\{v \in V : v(x,y) \leq \lambda y \text{ for almost all } (x,y) \in \Omega\right\}.$$

For all $1 > \alpha > 1/3$, the sets W_λ are closed subspaces of $V_1 + V_2$ and invariant under $P_{V_2^\perp} P_{V_1^\perp}$. Moreover, $\cup_{\lambda>0} W_\lambda$ is dense in V, and for any $\lambda > 0$ there exists $\theta < 1$ such that

$$\|(P_{V_2^\perp} P_{V_1^\perp})^k e_0\| \leq \theta^k \|e_0\| \quad \text{for all } e_0 \in W_\lambda.$$

Proof Notice that W_λ are closed subspaces of V and $C_c^\infty(\Omega) \subset \cup_{\lambda>0} W_\lambda$. Hence $\cup_{\lambda>0} W_\lambda$ is dense in V.

To show that $W_\lambda \subset V_1 + V_2$, we define the cut-off function $\eta \colon \Omega \to \mathbb{R}$ by

$$\eta(x,y) = \begin{cases} 0 & \text{in } \Omega_1 \setminus \Omega_2, \\ 1 & \text{in } \Omega_2 \setminus \Omega_1, \\ xy^{-\alpha^{-1}} & \text{in } D = \Omega_1 \cap \Omega_2. \end{cases}$$

Then, for $(x, y) \in D$ we have

$$|\nabla\eta(x,y)| = \left|(y^{-\alpha^{-1}}, -\alpha^{-1}xy^{-\alpha^{-1}-1})\right| \leq C(\alpha)y^{-\alpha^{-1}}. \tag{5}$$

Let now $\lambda > 0$ be fixed and let $w \in W_\lambda$. Then we claim $\eta w \in V_1$ and $(1-\eta)w \in V_2$. Using (5) and recalling that $\alpha > 1/3$, we get

$$\begin{aligned}\|(\nabla\eta)w\|^2_{L^2(\Omega)} &\leq \int_D C(\alpha)\lambda^2 \frac{y^2}{y^{2\alpha^{-1}}} = C(\alpha)\lambda^2 \int_0^1 \int_0^{y^{\alpha^{-1}}} \frac{y^2}{y^{2\alpha^{-1}}}\, dx\, dy \\ &= C(\alpha)\lambda^2 \int_0^1 \frac{y^2}{y^{\alpha^{-1}}}\, dy = C(\alpha)\lambda^2 \frac{1}{3-\alpha^{-1}}.\end{aligned}$$

Noticing $\eta \leq 1$ in Ω, the above estimate shows $\eta w \in V_1$ and $(1-\eta)w \in V_2$.

Next, we show that W_λ are invariant under $P_{V_i^\perp}$, $i = 1, 2$. Let $w \in W_\lambda$. Then $v := P_{V_1^\perp}w$ is the unique function such that $\Delta v = 0$ in Ω_1 with $v = w$ in $\Omega \setminus \Omega_1$. Therefore, since the function $\varphi(x, y) = \lambda y$ is harmonic, the maximum principle implies that $v \leq \varphi$ in Ω_1 and clearly also that $v = w \leq \varphi$ in $\Omega \setminus \Omega_1$. Hence $v \in W_\lambda$. The invariance under $P_{V_2^\perp}$ is analogous. Therefore, we obtain that W_λ is invariant under $P_{V_2^\perp}P_{V_1^\perp}$. Finally, the geometric convergence follows from Lemma 1. □

Theorem 5 says that for $\alpha > 1/3$ we have geometric convergence for all $e_0 \in \cup_{\lambda>0}W_\lambda$. The restriction $\alpha > 1/3$ is optimal. To see it, recall the sets $X_{\lambda,\varepsilon}$ defined in (4) and that $X_{\lambda,\varepsilon} \subset V \setminus (V_1 + V_2)$. Hence, Theorem 3 guarantees that any $e_0 \in X_{\lambda,\varepsilon}$ leads to a NGC. However, for $\alpha \leq 1/3$, $X_{\lambda,\varepsilon}$ and W_λ have non-trivial intersections. Therefore, if $\alpha \leq 1/3$, then there exists $e_0 \in W_\lambda$, in particular $e_0 \in W_\lambda \cap X_{\lambda,\varepsilon}$, that leads to NGC.

5 Numerical experiments

In this section, we present a numerical study of the NGC of the Schwarz method corresponding to the domain decomposition given in Fig. 1. The (monodomain) problem is discretized by linear finite elements using the software Freefem. The discrete meshes for $\Omega_1 \setminus \Omega_2$, D and $\Omega_2 \setminus \Omega_1$ are obtained by the mesh generator of Freefem where we discretized the boundary components Γ_A, Γ_D, Γ_E and Γ_F with 10 points and Γ_B, Γ_C, Γ_1 and Γ_2 with $10N$ points with a positive integer N. This choice is motivated by the higher accuracy needed close to the singularity point of ∂D. The results of our numerical experiments are shown in Fig. 2, where we plot the value $1 - \|e_n\|/\|e_{n-1}\|$ for the iteration count $n = 1, \ldots, 2000$. The numerical procedure is stopped only if $\|e_n\| < 10^{-16}$ or if the value $1 - \|e_n\|/\|e_{n-1}\|$ becomes too small (or negative). Clearly, if $1 - \|e_n\|/\|e_{n-1}\|$ becomes constant as n grows, then the method reached a geometric convergence regime. On the other hand, if $1 - \|e_n\|/\|e_{n-1}\| \to 0$ as n grows, then the method converges non-geometrically. Motivated by Theorems 1 and 5, we study the numerical behavior of the Schwarz method for an overlap characterized by $\alpha = \frac{\theta}{2} + \frac{1-\theta}{3}$ for different θ in $[0, 1]$, an initial guess $e_0 \in W_1$, and different N. In particular, according to Theorem 5, we expect

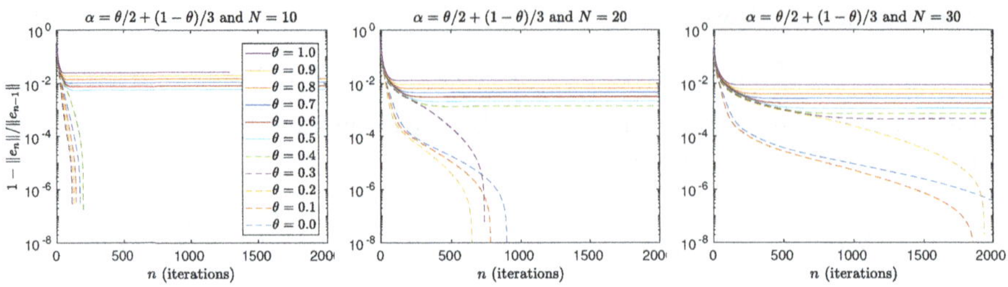

Fig. 2: Convergence behavior of the Schwarz method. The value $1 - \|e_n\|/\|e_{n-1}\|$ is shown for $N = 10$ (left), $N = 20$ (center), $N = 30$ (right).

geometric convergence for any $\theta \in (0, 1]$ and NGC for $\theta = 0$. In Fig. 2, we see that for $N = 10$ the Schwarz method is numerically geometric convergent for $\theta \in [1/2, 1]$ (solid lines), but not for $\theta < 1/2$ (dashed lines). However, if one refines the mesh with $N = 20$ and $N = 30$, then geometric convergence holds also for $\theta = 0.4$ and $\theta = 0.3$. Moreover, for bigger N also the curves for smaller θ are less steep and show a behavior closer to the proved geometric convergence. Finally, we wish to remark that, according to our experience, a more precise numerical description of the correct theoretical behavior for θ approaching zero is hard. This is mainly due to the non-Lipschitz overlap, where a correct numerical discretization is not trivial. Therefore, further studies would be needed. These are beyond the scope of this short manuscript, and we hope to consider them in future work.

References

1. Badea, C., Grivaux, S., Müller, V.: The rate of convergence in the method of alternating projections. Algebra i Analiz **23**(3), 1–30 (2011)
2. Badea, C., Seifert, D.: Ritt operators and convergence in the method of alternating projections. J. Approx. Theory **205**, 133–148 (2016). DOI:10.1016/j.jat.2016.02.001
3. Bauschke, H., Borwein, J., Lewis, A.: The method of cyclic projections for closed convex sets in Hilbert space. In: Recent developments in optimization theory and nonlinear analysis (Jerusalem, 1995), *Contemp. Math.*, vol. 204, pp. 1–38. Amer. Math. Soc., Providence, RI (1997)
4. Bauschke, H., Deutsch, F., Hundal, H.: Characterizing arbitrarily slow convergence in the method of alternating projections. Int. Trans. Oper. Res. **16**(4), 413–425 (2009)
5. Ciaramella, G., Gander, M.: Analysis of the parallel Schwarz method for growing chains of fixed-sized subdomains: Part III. ETNA **48**, 201–243 (2018)
6. Deutsch, F.: Best approximation in inner product spaces, *CMS Books in Mathematics/Ouvrages de Mathématiques de la SMC*, vol. 7. Springer-Verlag, New York (2001). DOI:10.1007/978-1-4684-9298-9
7. Deutsch, F., Hundal, H.: Slow convergence of sequences of linear operators I: almost arbitrarily slow convergence. J. Approx. Theory **162**(9), 1701–1716 (2010)
8. Deutsch, F., Hundal, H.: Slow convergence of sequences of linear operators II: arbitrarily slow convergence. J. Approx. Theory **162**(9), 1717–1738 (2010)
9. Franchetti, C., Light, W.: On the von Neumann alternating algorithm in Hilbert space. J. Math. Anal. Appl. **114**(2), 305–314 (1986)

10. Halperin, I.: The product of projection operators. Acta Sci. Math. (Szeged) **23**, 96–99 (1962)
11. Lions, P.: On the Schwarz alternating method. I. In: First International Symposium on Domain Decomposition Methods for Partial Differential Equations (Paris, 1987), pp. 1–42. SIAM, Philadelphia, PA (1988)
12. von Neumann, J.: On rings of operators. Reduction theory. Ann. of Math. (2) **50**, 401–485 (1949)

Global–in–Time Domain Decomposition for a Nonlinear Diffusion Problem

Elyes Ahmed, Caroline Japhet and Michel Kern

1 Introduction

We study a simplified model for two–phase flow in porous media, where the medium is made of two (or more) different *rock types*. Each rock type is a subdomain with a distinct capillary pressure function so that the saturation becomes discontinuous across the interface between the different regions. This leads to the phenomenon of capillary trapping (see [12] or [4]).

In this paper we develop a non-overlapping domain decomposition method that combines the Optimized Schwarz Waveform Relaxation method with Robin transmission conditions and the discontinuous Galerkin method in time. The domain decomposition method we present is global-in-time, which provides flexibility for using non-matching time grids so as to handle the very different time scales that occur in the different rocks of the porous medium. The method is a generalization of previous work on linear diffusion or diffusion–advection problems [8, 9].

We state briefly the physical model, referring to [1, 4] for further details. Let Ω be a bounded open subset of $\mathbb{R}^d$ ($d = 2$ or 3), assumed to be polygonal, with Lipschitz continuous boundary. We assume that the porous medium Ω is heterogeneous and made-up of two rock types, represented by polygonal subsets $(\Omega_i)_{i\in\{1,2\}}$ (the restriction to two subdomain is only to simplify the exposition, and indeed the example given in section 4 uses more than 2 subdomains). The subdomains share

Elyes Ahmed
Inria, 2 rue Simone iff, 75589 Paris, France,
current address Department of Mathematics, University of Bergen, P. O. Box 7800, N-5020 Bergen, Norway, e-mail: Elyes.Ahmed@uib.no

Caroline Japhet
Université Paris 13, Sorbonne Paris Cité, LAGA, CNRS(UMR 7539), 93430, Villetaneuse, France, e-mail: japhet@math.univ-paris13.fr

Michel Kern
Inria, 2 rue Simone iff, 75589 Paris, France, e-mail: michel.kern@inria.fr,
Université Paris-Est, CERMICS (ENPC), 77455 Marne-la-Vallée 2, France.

R. Haynes et al. (eds.), *Domain Decomposition Methods in Science and Engineering XXV*,
Lecture Notes in Computational Science and Engineering 138,
https://doi.org/10.1007/978-3-030-56750-7_22

the interface $\Gamma = \overline{\Omega}_1 \cap \overline{\Omega}_2$. We suppose that each subdomain Ω_i is homogeneous, in that the physical properties depend on space only through the subdomain index.

We consider the following nonlinear diffusion problem (for some time $T > 0$)

$$\partial_t u_i - \nabla \cdot (\lambda_i(u_i)\nabla \pi_i(u_i)) = 0, \quad \text{in } \Omega_i \times (0,T),\ i = 1,2, \tag{1}$$

for scalar unknowns $u_i = u_{|\Omega_i} : \Omega_i \times (0,T) \to [0,1]$ representing the gas saturation. This model can be obtained from the complete two-phase flow model by neglecting the advection terms in the saturation equation, so that the saturation and pressure equations become completely decoupled (see [4] for details). In [4], this simplified model is shown to allow gas trapping in low capillary pressure regions. The functions $\pi_i : [0,1] \to \mathbb{R}$ (Lipschitz and strictly increasing) and $\lambda_i : [0,1] \to \mathbb{R}$ are respectively the capillary pressure and the global mobility of the gas in subdomain Ω_i. Initial data $u_0 \in L^2(\Omega)$ is given with $u_0(x) \in [0,1]$ for a.e. $x \in \Omega$, and for simplicity we assume homogeneous Neumann boundary conditions on $\partial\Omega$.

Transmission conditions across the interface $\Gamma \times [0,T]$ are needed to complement (1). In order to handle the three different cases where both phases can flow across the interface, or where only one phase flows, and the other phase is trapped in a subdomain, one introduces *truncated* capillary pressure curves (see [4] or [1] for details), defined by

$$\bar{\pi}_1(u) = \max(\pi_1(u), \pi_2(0)), \qquad \bar{\pi}_2(u) = \min(\pi_2(u), \pi_1(1)).$$

The transmission conditions are then given by

$$\left.\begin{aligned} \bar{\pi}_1(u_1) &= \bar{\pi}_2(u_2) \\ \lambda_1 \nabla \pi_1(u_1) \cdot \mathbf{n}_1 &= -\lambda_2 \nabla \pi_2(u_2) \cdot \mathbf{n}_2 \end{aligned}\right\} \text{ on } \Gamma \times (0,T), \tag{2}$$

where $\mathbf{n}_i$ is the unit, outward pointing, normal vector field on $\partial\Omega_i$.

In the next section, this physical problem is rewritten in a form better suited for mathematical and numerical analysis. In particular, the existence of a weak solution of the local Robin problems is addressed. A semi–discrete formulation based on discontinuous Galerkin in time is given in section 3 and numerical experiments using a finite volume method are described in section 4.

2 Space–time domain decomposition at the continuous level

The model stated above is well adapted to physical modeling, but is difficult to handle mathematically because of the low regularity of the solutions. To obtain mathematical results, it has been found useful to introduce the Kirchhoff transformation [4], so that λ_i and π_i are replaced by a single function φ_i. Following [3, 4], one also introduces new functions $(\Pi_i)_{i=1,2}$ that satisfy

$$\bar{\pi}_1(u_1) = \bar{\pi}_2(u_2) \Leftrightarrow \Pi_1(u_1) = \Pi_2(u_2),\ \forall (u_1, u_2) \in [0,1]^2.$$

Defining the global function $\Pi_g(x,t) = \Pi_i(u_i(x,t))$, for $x \in \Omega_i$, $t \in (0,T)$, it is shown in the above references that $\Pi_g(u) \in L^2(0,T;H^1(\Omega))$, which gives a meaning to the first transmission condition in (4) below.

In terms of the new functions, the problem becomes

$$\partial_t u_i - \Delta\varphi_i(u_i) = 0, \quad \text{in } \Omega_i \times (0,T), \qquad u_i(\cdot,0) = u_0, \quad \text{in } \Omega_i, \tag{3}$$

together with a Neumann boundary condition on $\partial\Omega_i \backslash \Gamma$ and the transmission conditions

$$\begin{aligned} \Pi_1(u_1) &= \Pi_2(u_2) \\ \nabla\varphi_1(u_1)\cdot\mathbf{n}_1 &= -\nabla\varphi_2(u_2)\cdot\mathbf{n}_2, \end{aligned} \quad \text{on } \Gamma \times (0,T). \tag{4}$$

An existence theorem is known for the transmission problem (3), (4), see [3, 4] where existence of a suitably defined weak solution is proved.

An equivalent formulation to the model problem (3)–(4) can be obtained by replacing (4) with equivalent Robin transmission conditions on $\Gamma \times (0,T)$:

$$\begin{aligned} \nabla\varphi_1(u_1)\cdot\mathbf{n}_1 + \alpha_1\Pi_1(u_1) &= -\nabla\varphi_2(u_2)\cdot\mathbf{n}_2 + \alpha_1\Pi_2(u_2), \\ \nabla\varphi_2(u_2)\cdot\mathbf{n}_2 + \alpha_2\Pi_2(u_2) &= -\nabla\varphi_1(u_1)\cdot\mathbf{n}_1 + \alpha_2\Pi_1(u_1), \end{aligned} \quad \text{on } \Gamma \times (0,T), \tag{5}$$

where α_1 and α_2 are free parameters that can be chosen to enhance the convergence of the method (see [7, 8] for linear problems and [2] for a reaction-diffusion problem with nonlinear source term). It is shown in [1] how the Robin transmission conditions can be extended to Ventcell transmission conditions, to further improve the convergence of the method.

The Optimized Schwarz Waveform Relaxation method with nonlinear Robin transmission conditions (NL–OSWR) is defined by the following iterations for $k \geq 0$, where Ψ_i^0 is a given initial Robin guess on $\Gamma \times (0,T)$ for $i = 1,2$:

$$\begin{aligned} \partial_t u_i^k - \Delta\varphi_i(u_i^k) &= 0, && \text{in } \Omega_i \times (0,T), \\ \nabla\varphi_i(u_i^k)\cdot\mathbf{n}_i + \alpha_i\Pi_i(u_i^k) &= \Psi_i^{k-1}, && \text{on } \Gamma \times (0,T), \end{aligned} \tag{6}$$

with suitable initial and boundary conditions, then set

$$\Psi_i^k := -\nabla\varphi_j(u_j^k)\cdot\mathbf{n}_j + \alpha_i\Pi_j(u_j^k), \quad j = (3-i),\ k \geq 1. \tag{7}$$

We give an existence result for the subdomain problem, namely problem (6) with the iteration k and the subdomain Ω_i fixed. Because of the non-linear Robin boundary condition, the result is not standard (references [3] and [4] both assume Neumann boundary conditions). For the rest of this section, we denote by $\mathcal{O} \subset \mathbb{R}^d$ a polygonal domain with Lipschitz boundary (that plays the role of one the Ω_i), and denote by Γ the part of the boundary of $\mathcal{O}$ along which the Robin boundary condition applies. First a notion of weak solution is defined:

Definition 1 (Weak solution for the local Robin problem)

A function u is said to be a weak solution of problem (6) (with initial condition u_0 and homogeneous Neumann boundary condition on $\partial\mathcal{O}\backslash\Gamma$ if it satisfies:

1. $u \in L^\infty(\mathcal{O} \times (0,T))$, $\quad 0 \le u \le 1$ for a.e. in $\mathcal{O} \times (0,T)$,
2. $\varphi(u) \in L^2(0,T;H^1(\mathcal{O}))$, and $\Pi(u) \in L^2(0,T;H^1(\mathcal{O}))$,
3. For all $\psi \in C_{\text{test}} = \left\{h \in H^1(\mathcal{O} \times (0,T)),\ h(.,T) = 0\right\}$,

$$
\begin{aligned}
&-\int_0^T \int_{\mathcal{O}} u(\mathbf{x},t)\partial_t \psi(\mathbf{x},t)\,\mathrm{d}\mathbf{x}\mathrm{d}t - \int_{\mathcal{O}} u_0(\mathbf{x})\psi(\mathbf{x},0)\,\mathrm{d}\mathbf{x} \\
&\quad + \int_0^T \int_{\mathcal{O}} \nabla\varphi(u(\mathbf{x},t)) \cdot \nabla\psi(\mathbf{x},t)\,\mathrm{d}\mathbf{x}\mathrm{d}t - \int_0^T \int_\Gamma \alpha\Pi(u(\mathbf{x},t))\psi\,\mathrm{d}\gamma(\mathbf{x})\mathrm{d}t \\
&\qquad = \int_0^T \int_\Gamma \Psi(\mathbf{x},t)\psi\,\mathrm{d}\gamma(\mathbf{x})\mathrm{d}t, \qquad (8)
\end{aligned}
$$

where $\mathrm{d}\gamma(\mathbf{x})$ is the $(d-1)$-dimensional Lebesgue measure on $\partial\mathcal{O}$.

We then have an existence theorem for the sub-domain problem

Theorem 1 *Assume that:*

1. *the initial condition u_0 is in $L^\infty(\mathcal{O})$ and satisfies $u_0(x) \in [0,1]$ for a.e. $x \in \mathcal{O}$;*
2. *the right-hand side $\Psi \in L^2(\mathcal{O} \times (0,T))$;*
3. *the function φ is Lipschitz continuous and strictly increasing on $(0,1)$;*
4. *the function Π is continuous and non–decreasing on $(0,1)$;*
5. *the Robin coefficient α is chosen such that:*

$$
0 < \Psi(x,t) < \alpha\Pi(1), \quad \forall (x,t) \in \mathcal{O} \times (0,T). \qquad (9)
$$

Then there exists a weak solution to Problem (6) *in the sense of Definition 1.*

The proof is beyond the scope of this article, and will be the topic of a future paper. It is an adaptation to nonlinear Robin boundary conditions of the proof in [3, 4], and is based on the convergence of a finite volume scheme.

Note that in the context of the NL–OSWR method assumption (9) will have to be checked iteratively to prove that the algorithm is well posed (see section 3).

3 Semi–discrete space–time domain decomposition with different time steps in the subdomains

We introduce a non–conforming time discretization, that is each subdomain Ω_i has its own time discretization, by using a (lowest order) Discontinuous Galerkin (DG) time discretization on each subdomain, together with a projection across the interface (see [7, 8] for an analysis in the linear case). More precisely, for integers M_i, define $\delta t_i = T/M_i$, and denote by $\mathcal{T}_i$ the partition of $[0,T]$ in sub-intervals J_i^n of size δt_i, where $J_i^n = (t_i^{n-1}, t_i^n]$, with $t_i^n = n\delta t_i$, for $n = 0, \ldots, M_i$.

For $i = 1, 2$, we introduce the space

$$\mathcal{P}^0_{\mathcal{T}_i} := \{u_i(\cdot,t) : (0,T) \to L^2(\Gamma);\ u_i(\cdot,t) \text{ is constant on } J_i^n,\ 1 \le n \le M_i\}.$$

A function in $\mathcal{P}^0_{\mathcal{T}_i}$ is thus defined by the M_i functions $\{u_i^n := u_i(\cdot,t)|_{J_i^n}\}_{1\le n\le M_i}$ in $L^2(\Gamma)$. In order to deal with the non–conformity in time, we introduce the L^2 projection operator $P_{i,j}$ from $\mathcal{P}^0_{\mathcal{T}_j}(L^2(\Gamma))$ onto $\mathcal{P}^0_{\mathcal{T}_i}(L^2(\Gamma))$, i.e., for $\phi \in \mathcal{P}^0_{\mathcal{T}_j}(L^2(\Gamma))$, $(P_{i,j}\phi)|_{J_i^n}$ is the average value of ϕ on J_i^n, for $n = 1, ..., M_i$:

$$(P_{i,j}\phi)|_{J_i^n} = \frac{1}{\delta t_i} \sum_{\ell=1}^{M_j} \int_{J_j^\ell \cap J_i^n} \phi.$$

The semi-discrete counterpart in time of the NL–OSWR method (6)–(7) with possibly different time grids in the subdomains can be written as follows:

For $i = 1, 2$, given initial iterates $\Psi_i^0 \in \mathcal{P}^0_{\mathcal{T}_i}$ and starting from the initial condition $u_i^{0,0} = u_0|_{\Omega_i}$, a semi–discrete solution $\left(u_i^{k,n}\right)_{1\le n\le M_i}$ at step k of the algorithm is computed by solving, for $n = 1, \ldots, M_i$,

$$\begin{aligned} \frac{u_i^{k,n} - u_i^{k,n-1}}{\delta t_i} - \Delta\varphi_i(u_i^{k,n}) &= 0 && \text{in } \Omega_i, \\ \nabla\varphi_i(u_i^{k,n}) \cdot \mathbf{n}_i + \alpha_i \Pi_i(u_i^{k,n}) &= \frac{1}{\delta t_i} \int_{J_i^n} \Psi_i^{k-1}(t)\, dt, && \text{on } \Gamma \times (0,T). \end{aligned} \tag{10}$$

Then we set

$$\Psi_i^k := P_{i,j}\Big(- \nabla\varphi_j(u_j^k(t))\cdot\mathbf{n}_j + \alpha_i \Pi_j(u_j^k(t))\Big), \quad j = (3-i),\ k \ge 1. \tag{11}$$

The projections in (11) between arbitrary grids are performed using the algorithm with linear complexity introduced in [5, 6].

Last, we check that the NL–OSWR algorithm is well posed. That is, we need to verify that assumption (9) holds for every iteration. The initial iterate and the Robin coefficients are chosen such that it holds for $k = 0$. We have been able to show that this remains true throughout the algorithm only in the case when the capillary pressure functions satisfy

$$\pi_1(0) = \pi_2(0) \text{ and } \pi_1(1) = \pi_2(1).$$

4 Numerical experiment

The domain Ω is the unit cube, decomposed into two subdomains with two rock types (see figure 1). The mobilities and capillary pressure functions are given by

$$\lambda_{o,i}(u) = u,\ i \in \{1,2\}, \quad \pi_1(u) = 5u^2, \quad \text{and} \quad \pi_2(u) = 5u^2 + 1.$$

The initial condition is that the domain contains some quantity of gas, situated only within Ω_1, more precisely, $u_0(\mathbf{x}) = 0.9$ for $x_1 < 0.4$ and 0 otherwise. The domain is discretized by a mesh of $20 \times 20 \times 20$ elements, the time discretization is non–conforming, with constant time steps in each subdomain $\delta t_1 = 10^{-3}$, and $\delta t_2 = \frac{1}{8}10^{-2}$.

The full discretization is carried out with a two–point finite volume scheme [4]. The method is implemented with the Matlab Reservoir Simulation Toolbox [11]. The nonlinear subdomain problem is solved with Newton's method. The only change required to the finite volume scheme to cope with a non–conforming time scheme is the projection of the right hand side of the transmission condition on the grid of the current subdomain, as shown on eq. (11). This is what makes the choice of a DG formalism important, together with a global in time DD method. The resulting scheme is non-conforming in time, and the equivalence with the physical transmission conditions no longer holds.

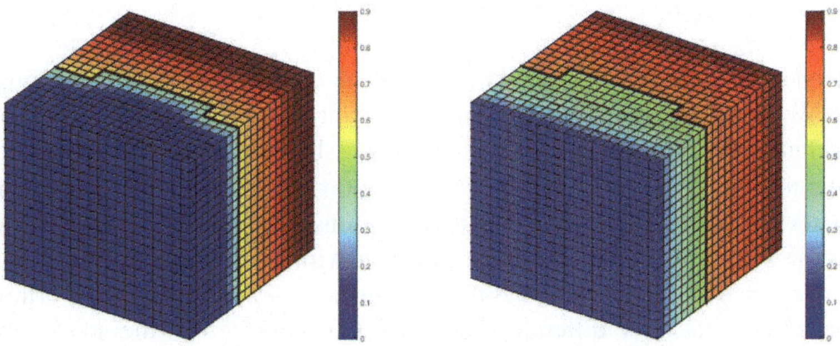

Fig. 1: Test case 1: Saturation $u(t)$ for $t = 0.3$ and $t = 3$

The evolution of the saturation at two time steps is shown in Fig. 1. We remark that at the beginning of the simulation, approximately until $t \approx 0.02$, the gas cannot penetrate to the domain Ω_2, since the capillary pressure is lower than the threshold value $\pi_2(0) = 1$, which is known as the entry pressure. The saturation of the trapped gas in Ω_1 as well as the capillary pressure increase until the capillary pressure reaches the entry pressure.

We study the convergence behavior of the NL–OSWR algorithm. The tolerance for Newton's method is fixed to 10^{-8}. The tolerance of the NL–OSWR algorithm is 10^{-6}. The Robin parameters are chosen for the two subdomains so as to minimize the convergence rate of a linearized version of the problem. Precisely, we take in the model problem the capillary pressure as unknown, then linearize the nonlinear terms, leading to determine the optimal Robin parameters for a linear diffusion problem with discontinuous coefficients similar to that in [8, 10]. We show in Fig. 2 (right) the relative residuals comparing the convergence history with the parameters calculated numerically by minimizing the convergence factor for the linearized problem and that of with the best parameters located in the zone giving the smaller errors after the same number of iterations (see Fig. 2 left).

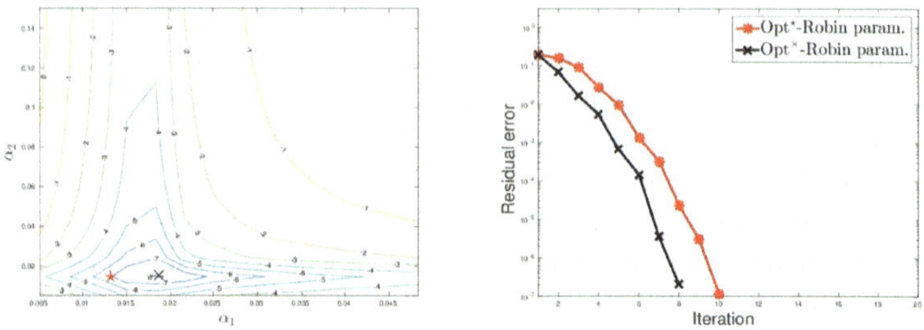

Fig. 2: Test case 1: Left: Level curves for the residual error obtained after 10 iterations for various values of the parameters α_1 and α_2. The star (in magenta) marked the parameters obtained with the minimization process of the convergence factor applied to the linearized problem which is close to the best one marked by times symbol (in black). Right: The convergence curves.

We now analyze the efficiency in time of the method with nonconforming time steps. We compute a reference solution as the converged multidomain solution with conforming fine time grids $\delta t_f = \frac{1}{4}10^{-3}$, and where the relative residual is taken smaller than 10^{-12}. We then compare the solution obtained with the nonconforming time steps, as described above with two solutions computed first with conforming fine time steps ($\delta t_1 = \delta t_2 = 10^{-3}$) and then with conforming coarse time steps (($\delta t_1 = \delta t_2 = \frac{1}{8}10^{-2}$)). Fig. 3 shows the error in the saturation along a line orthogonal to the interface at three different time steps. One can see that the nonconforming solution as well as the solution with conforming and fine steps are in close agreement with the reference solution, whereas the solution with coarse time steps has a larger error. This confirms that nonconforming time grids with respect to the rock type numerically preserve the accuracy in time of the multidomain solution.

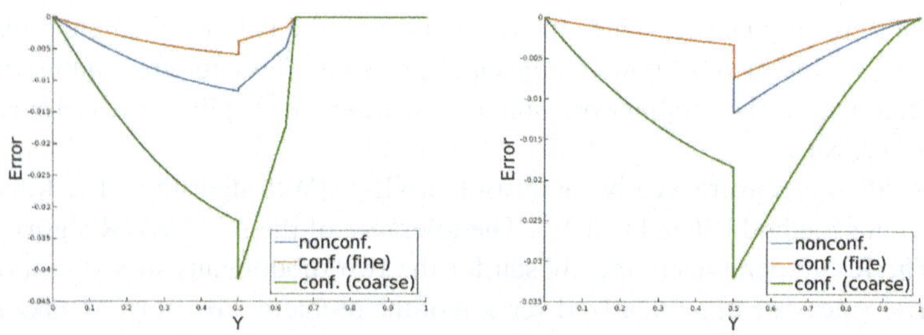

Fig. 3: Test case 1. Error in saturation along a line orthogonal to the interface, nonconforming and conforming (coarse and fine) time-steps. Left $T = T_f/20$, right, $T = T_f$.

Other examples with more physical content can be found in [1].

Acknowledgements This work was funded by ANR DEDALES under grant ANR-14-CE23-0005.

We thank the two referees whose careful comments helped improve the content of the paper.

References

1. Ahmed, E., Ali Hassan, S., Japhet, C., Kern, M., Vohralík, M.: A posteriori error estimates and stopping criteria for space-time domain decomposition for two-phase flow between different rock types. The SMAI journal of computational mathematics **5**, 195–227 (2019). DOI:10.5802/smai-jcm.47. URL https://smai-jcm.centre-mersenne.org/item/SMAI-JCM_2019__5__195_0
2. Caetano, F., Gander, M.J., Halpern, L., Szeftel, J.: Schwarz waveform relaxation algorithms for semilinear reaction-diffusion equations. Netw. Heterog. Media **5**(3), 487–505 (2010). DOI:10.3934/nhm.2010.5.487
3. Cancès, C.: Nonlinear parabolic equations with spatial discontinuities. NoDEA Nonlinear Differential Equations Appl. **15**(4-5), 427–456 (2008). DOI:10.1007/s00030-008-6030-7
4. Enchéry, G., Eymard, R., Michel, A.: Numerical approximation of a two-phase flow problem in a porous medium with discontinuous capillary forces. SIAM J. Numer. Anal. **43**(6), 2402–2422 (2006). DOI:10.1137/040602936
5. Gander, M.J., Japhet, C.: Algorithm 932: PANG: software for nonmatching grid projections in 2D and 3D with linear complexity. ACM Trans. Math. Software **40**(1), Art. 6, 25 (2013). DOI:10.1145/2513109.2513115
6. Gander, M.J., Japhet, C., Maday, Y., Nataf, F.: A new cement to glue nonconforming grids with Robin interface conditions: the finite element case. In: Domain decomposition methods in science and engineering, *Lect. Notes Comput. Sci. Eng.*, vol. 40, pp. 259–266. Springer, Berlin (2005). DOI:10.1007/3-540-26825-1_24
7. Halpern, L., Japhet, C., Szeftel, J.: Optimized Schwarz waveform relaxation and discontinuous Galerkin time stepping for heterogeneous problems. SIAM J. Numer. Anal. **50**(5), 2588–2611 (2012). DOI:10.1137/120865033
8. Hoang, T.T.P., Jaffré, J., Japhet, C., Kern, M., Roberts, J.E.: Space-time domain decomposition methods for diffusion problems in mixed formulations. SIAM J. Numer. Anal. **51**(6), 3532–3559 (2013). DOI:10.1137/130914401
9. Hoang, T.T.P., Japhet, C., Kern, M., Roberts, J.E.: Space-time domain decomposition for advection-diffusion problems in mixed formulations. Math. Comput. Simulation **137**, 366–389 (2017). DOI:10.1016/j.matcom.2016.11.002
10. Lemarié, F., Debreu, L., Blayo, E.: Toward an optimized global-in-time Schwarz algorithm for diffusion equations with discontinuous and spatially variable coefficients. Part 2: The variable coefficients case. Electron. Trans. Numer. Anal. **40**, 170–186 (2013)
11. Lie, K.A., Krogstad, S., Ligaarden, I.S., Natvig, J.R., Nilsen, H.M., Skaflestad, B.: Open source MATLAB implementation of consistent discretisations on complex grids. Comput. Geosci. **16**(2), 297–322 (2012). DOI:10.1007/s10596-011-9244-4
12. Van Duijn, C.J., Molenaar, J., De Neef, M.J.: The effect of capillary forces on immiscible two-phase flow in heterogeneous porous media. Transport in Porous Media **21**(1), 71–93 (1995). DOI:10.1007/BF00615335

A Two-level Overlapping Schwarz Method Using Energy Minimizing Multiscale Finite Element Functions

Hyea Hyun Kim, Eric T. Chung, and Junxian Wang

1 Introduction

In this paper, a two-level overlapping Schwarz algorithm is proposed for solving finite element discretization of the following model problem,

$$\int_{\Omega} \rho(x)\nabla u(x) \cdot \nabla v(x)\, dx = \int_{\Omega} f(x)v(x)\, dx, \quad \forall v(x) \in H_0^1(\Omega), \tag{1}$$

where $u(x)$ is in the Sobolev space $H_0^1(\Omega)$, the space of integrable functions with their weak derivatives of the first order being square integrable. The coefficient $\rho(x)$ can be highly varying and random with high contrast inside Ω. For such model problems, the standard coarse problem in the two-level overlapping Schwarz algorithm often fails and a more robust coarse problem is required.

A new idea here is that we will form the coarse problem by utilizing multiscale finite element functions proposed in [2]. The multiscale finite element functions are obtained by solving certain constrained energy minimizing problems where the constraints are formed by using a set of selected eigenvectors from a generalized eigenvalue problem in each overlapping subdomain. The generalized eigenvalue problem is similar to that considered in [4]. In their work, the eigenvectors are directly used to form the coarse basis functions and the resulting preconditioner is

Hyea Hyun Kim
Department of Applied Mathematics and Institute of Natural Sciences, Kyung Hee University, Korea hhkim@khu.ac.kr

Eric T. Chung
Department of Mathematics, The Chinese University of Hong Kong, Hong Kong SAR tschung@math.cuhk.edu.hk

Junxian Wang
Hunan Key Laboratory for Computation and Simulation in Science and Engineering, School of Mathematics and Computational Science, Xiangtan University, P.R.China wangjunxian@xtu.edu.cn

R. Haynes et al. (eds.), *Domain Decomposition Methods in Science and Engineering XXV*, Lecture Notes in Computational Science and Engineering 138,
https://doi.org/10.1007/978-3-030-56750-7_23

shown to have a condition number robust with respect to the contrast of the coefficient but dependent on the overlapping width in the subdomain partition.

The advantage in our new method is that the resulting coarse basis functions provide a more robust coarse problem and thus the condition number of the resulting preconditioner becomes robust to the overlapping width as well as the contrast in the model coefficient. The idea was originated from [2] where it was shown that such constrained energy minimizing finite element functions can approximate the solution of the model problem with the errors dependent on the coarse mesh size but independent of the contrast in the model coefficient. One disadvantage of our approach is that the constrained minimization problem needs to be solved in the whole domain. To overcome this heavy cost, we can localize the minimization problem on each subregion and use the solution to form the coarse basis functions. In [2], it was shown that the error between the full minimization solution and the localized one decays exponentially as a function of the subregion size. From that result, we can expect that the proposed preconditioner with these localized coarse basis functions also share the same good quality, i.e., is robust with respect to the overlapping width as well as the contrast in the coefficient. More detailed analysis and extensive numerical tests will be given later in a full version of this short proceeding paper [14].

We note that the similar idea, enriching the coarse problem by using adaptively chosen eigenvectors from generalized eigenvalue problems on each subdomain or on each subdomain interface, has been also extensively developed for other types of domain decomposition algorithms, such as, FETI(-DP), BDD(C), and additive-Schwarz algorithms, see [10, 3, 9, 1, 8, 11, 13, 5, 6, 7].

2 Multiscale finite element basis functions

For finite element approximation of the solution of the model problem (1), we introduce a piecewise linear conforming finite element space V_h $(\subset H_0^1(\Omega))$ defined for a triangulation $\mathcal{T}_h$ of Ω. We assume that the triangulation is fine enough to resolve the variation in the coefficient $\rho(x)$ in the following sense,

$$\max_{\tau\in\mathcal{T}_h} \frac{\max_{x\in\tau}\rho(x)}{\min_{x\in\tau}\rho(x)} \leq C, \tag{2}$$

for a given constant C.

We partition the domain Ω into overlapping subdomains $\{\Omega_i\}_{i=1}^N$ where each Ω_i is a connected union of triangles in $\mathcal{T}_h$. For a given overlapping subdomain partition, we introduce a partition of unity $\{\theta_i(x)\}_{i=1}^N$, where $\sum_{i=1}^N \theta_i(x) = 1$ and each $\theta_i(x)$ is supported in Ω_i.

We consider the following generalized eigenvalue problem in each subdomain Ω_i:

$$a_i(\phi_j^{(i)}, w) = \lambda_j^{(i)} s_i(\phi_j^{(i)}, w), \quad \forall w \in V(\Omega_i),$$

where $V(\Omega_i)$ is the restriction of the functions in V_h to the subdomain Ω_i and the local bilinear forms are defined as

$$a_i(v,w) := \int_{\Omega_i} \rho(x)\nabla v \cdot \nabla w \, dx, \quad s_i(v,w) := \int_{\Omega_i} \rho(x)|\nabla\theta_i(x)|^2 v\, w \, dx.$$

We let the eigenvalues $\lambda_j^{(i)}$ be arranged in ascending order and choose the eigenvectors $\phi_j^{(i)}$ with their associate eigenvalues $\lambda_j^{(i)}$ smaller than a given tolerance value Λ, i.e., $\lambda_j^{(i)} < \Lambda$. We use the notation l_i for the number of such eigenvectors.

We first form an auxiliary multiscale finite element space by collecting all the selected eigenvectors

$$V_{aux} := \left\{ \phi_j^{(i)} \mid i = 1, \cdots, N, \ j = 1, \cdots, l_i \right\}.$$

We introduce the following definition for a function v in V_h: v is $\phi_j^{(i)}$-orthogonal if $s_i(v, \phi_j^{(i)}) = 1$ and $s_k(v, \phi_l^{(k)}) = 0$ for $k \neq i, \ l = 1, \cdots, l_k, \ k = i, \ l = 1, \cdots, j-1, j+1, \cdots, l_i$. We obtain a set of coarse basis functions $\psi_j^{(i)}$ as the solution of the following constrained minimization problem:

$$\psi_j^{(i)} = \operatorname{argmin}\{a(\psi,\psi) \mid \psi \in V_h, \ \psi \text{ is } \phi_j^{(i)}\text{-orthogonal.}\}, \tag{3}$$

where

$$a(u,v) := \int_{\Omega} \rho(x)\nabla u \cdot \nabla v \, dx.$$

The coarse space V_{glb} defined as a span of these functions $\psi_j^{(i)}$ can be shown to have the following property: V_{glb} is the orthogonal complement of $\widetilde{V}$ with respect to the bilinear form $a(\cdot,\cdot)$, where the space $\widetilde{V}$ is defined by

$$\widetilde{V} := \{v \in V_h \mid s_i(v, \phi_j^{(i)}) = 0, \ i = 1, \cdots, N, j = 1, \cdots, l_i\}. \tag{4}$$

As proposed in [2], we can consider a more practical relaxed constrained energy minimizing problem:

$$\psi_{r,j}^{(i)} = \operatorname{argmin} \left\{ a(\psi,\psi) + s(\pi\psi - \phi_j^{(i)}, \pi\psi - \phi_j^{(i)}) \mid \forall \psi \in V_h \right\}, \tag{5}$$

where

$$\pi\psi := \sum_{i=1}^{N} \sum_{j=1}^{l_i} s_i(\psi, \phi_j^{(i)})\phi_j^{(i)}, \quad s(v,w) = \sum_{i=1}^{N} s_i(v,w).$$

We note that the function $\psi_{r,j}^{(i)}$ in (5) will satisfy the same orthogonal property with respect to the resulting coarse space as that from (3) and it can be found by solving the following problem: find $\psi_{r,j}^{(i)}$ in V_h such that

$$a(\psi_{r,j}^{(i)}, v) + s(\pi\psi_{r,j}^{(i)}, \pi v) = s(\phi_j^{(i)}, \pi v), \quad \forall v \in V_h. \tag{6}$$

Let V_{glb} be the coarse space obtained from the $\psi_{r,j}^{(i)}$ of the above relaxed constrained problem. From (6), the following orthogonal property holds

$$a(\psi_{r,j}^{(i)}, v) = 0, \quad \forall v \in \widetilde{V}$$

and we thus obtain

$$V_h = \widetilde{V} \oplus V_{glb}.$$

We note that $V_h = V_{glb}^{\perp} \oplus V_{glb}$ and that $\widetilde{V}$ is contained in $V_{glb}^{\perp}$. Since the dimension of $V_{glb}^{\perp}$ is equal to the dimension of $\widetilde{V}$, see (4), we have $\widetilde{V} = V_{glb}^{\perp}$. In the following, we will use the space V_{glb} defined by the $\psi_{r,j}^{(i)}$ in (6) as the coarse space of the two-level overlapping Schwarz algorithm.

3 Two-level overlapping Schwarz algorithm

In this section, we propose a two-level overlapping Schwarz preconditioner for the finite element discretization of the model problem in (1), i.e.,

$$Au = b.$$

We introduce the local finite element space $V_0(\Omega_i)$, which is the restriction of functions in V_h to Ω_i and vanishing on $\partial\Omega_i$. We define the local problem matrix by

$$\langle A_i v, w\rangle := \int_{\Omega_i} \rho(x)\nabla v \cdot \nabla w \, dx, \forall v, \; w \in V_0(\Omega_i).$$

We introduce the restriction R_i from V_h to $V_0(\Omega_i)$ and denote by R_i^T the extension from $V_0(\Omega_i)$ by zero to V_h . We define the coarse problem matrix by

$$A_0 = a(\psi_{r,j}^{(i)}, \psi_{r,q}^{(k)}), \; i, k = 1, \cdots, N, \text{ and } j = 1, \cdots, l_i, \; q = 1, \cdots, l_k.$$

We note that the size of the matrix A_0 is identical to the dimension of V_{glb}. We introduce R_0 as the matrix with rows consisting of the nodal values of $\psi_{r,j}^{(i)}$ in V_{glb} and define the two-level overlapping Schwarz preconditioner as

$$R_0^T A_0^{-1} R_0 + \sum_{i=1}^{N} R_i^T A_i^{-1} R_i. \tag{7}$$

For the overlapping Schwarz method, the upper bound estimate can be obtained from a coloring argument. We will only need to work on the following lower bound estimate, see [12]:

Lemma 1 *Let the triangulation $\mathcal{T}_h$ satisfy the assumption in (2). For any given u in V_h, there exists $\{u_i\}_{i=0}^N$, $u_i \in V_0(\Omega_i)$, $i \geq 1$ and $u_0 \in V_{glb}$, such that*

$$u = u_0 + \sum_{i=1}^N u_i$$

and

$$a(u_0, u_0) + \sum_{i=1}^N a(u_i, u_i) \leq C_0^2 a(u, u)$$

with the constant C_0 independent of $\rho(x)$ and the overlapping width in the subdomain partition.

Proof For the proof we will choose u_0 as the solution of

$$a(u_0, v) = a(u, v), \quad \forall v \in V_{glb}$$

and choose u_i as

$$u_i = I^h(\theta_i(u - u_0)),$$

where $I^h(v)$ denotes the nodal interpolant of v to the space V_h. We note that $u - u_0$ is in $V_{glb}^{\perp}$ and also in $\widetilde{V}$ since $V_{glb}^{\perp} = \widetilde{V}$.

We can see that u_i is supported in Ω_i by construction and then obtain

$$\begin{aligned}
\sum_{i=1}^N a(u_i, u_i) &= \sum_{i=1}^N \int_{\Omega_i} \rho |\nabla I^h(\theta_i(u - u_0))|^2 \, dx \\
&\leq C_I \sum_{i=1}^N \int_{\Omega_i} \rho |\nabla(\theta_i(u - u_0))|^2 \, dx \\
&\leq 2C_I \sum_{i=1}^N \left(\int_{\Omega_i} \rho |\nabla(u - u_0)|^2 \, dx + \int_{\Omega_i} \rho |\nabla \theta_i|^2 (u - u_0)^2 \, dx \right) \\
&\leq 2C_I \sum_{i=1}^N (1 + \Lambda^{-1}) \int_{\Omega_i} \rho |\nabla(u - u_0)|^2 \, dx
\end{aligned}$$

where the constant C_I depends on the stability of the interpolation I^h and the constant C depends on the number of overlapping subdomains intersecting with Ω_i. In the above, we used the assumption (2) on $\mathcal{T}_h$ in the first inequality, and also that $u - u_0$ in $V_{glb}^{\perp}(= \widetilde{V})$ and thus get the third inequality with Λ^{-1}. Using that $a(u - u_0, u - u_0) + a(u_0, u_0) = a(u, u)$, we obtain the resulting bound. □

Theorem 1 *For the proposed preconditioner, the condition number bound is obtained as*

$$\kappa((R_0^T A_0^{-1} R_0 + \sum_{i=1}^N R_i^T A_i^{-1} R_i) A) \leq C_1 C_0^{-2},$$

where C_1 is the constant in the coloring argument, and C_0 is the constant in Lemma 1.

We note that the constant $C_0^2 = 2CC_I(1+\Lambda^{-1})$ is independent of $\rho(x)$ as well as the overlapping width, which is improvement over the previous work in [4].

On the other hand, the computation of $\psi_{r,j}^{(i)}$ requires us to solve the relaxed constrained minimization problem in the global space V_h. In practice, we can solve the same problem in a subspace of V_h, where the functions are restricted to a subregion $\widetilde{\Omega}_i$ containing Ω_i. In more detail, we solve

$$\psi_{j,ms}^{(i)} = \operatorname{argmin}\left\{a(\psi,\psi) + s(\pi(\psi)-\phi_j^{(i)}, \pi(\psi)-\phi_j^{(i)}) \mid \forall\psi \in V_h \bigcap H_0^1(\widetilde{\Omega}_i)\right\}.$$

From the above minimization problem, we obtain $\psi_{j,ms}^{(i)}$ and denote by $\Psi_{j,ms}^{(i)}$, the extension of $\psi_{j,ms}^{(i)}$ by zero to a function in V_h. We then define V_{ms} by

$$V_{ms} := \operatorname{span}\{\Psi_{ms,j}^{(i)} \mid i = 1,\cdots,N,\ j = 1,\cdots,l_i\}.$$

We can propose the following more practical preconditioner

$$M_{ms}^{-1} = \sum_{i=1}^{N} R_i^T A_i^{-1} R_i + R_{0,ms}^T A_{0,ms}^{-1} R_{0,ms}, \tag{8}$$

where $A_{0,ms}$ and $R_{0,ms}$ are defined similarly as before by replacing V_{glb} with V_{ms}.

4 Numerical results

In Table 1, we present some numerical results for a 2D model problem. We use the coarse problem obtained from the more practical space V_{ms}. Though we do not have an estimate of the condition numbers for this case, we can expect a similar performance to that with V_{glb}. The domain Ω is a unit square partitioned into $N \times N$ uniform squares. Each square is partitioned into uniform triangles with m elements on each edge of a square where the triangles form a mesh, $\mathcal{T}_h$. Each square is extended by d layers of fine triangles and the extended squares form the overlapping subdomain partition. In our experiment, we consider a random coefficient with its value varying between 10^{-3} to 10^3 inside the domain, and show the number of iterations and the number of primal unknowns for various subdomain partitions and for various overlapping width d. The minimization problem is solved in a smaller region $\widetilde{\Omega}_i$, which is obtained by extending each square by only one layer of neighboring squares. We can observe that the proposed method is robust with respect to the overlapping width d as well as the variation in $\rho(x)$.

Acknowledgements The first author was supported by the National Research Foundation of Korea(NRF) grants funded by NRF-2015R1A5A1009350 and by NRF-2019R1A2C1010090, the second author was supported by the Hong Kong RGC General Research Fund (Project 14317516)

Table 1: Performance of the proposed method with $\Lambda = (1 + \log m)$: d (number of elements in the overlap) $iter$ (number of iterations), $\lambda_{\min}$ (minimum eigenvalues), $\lambda_{\max}$ (maximum eigenvalues), and pD (number of coarse basis functions per subdomain).

$N(m)$	d	$iter$	λ_{min}	λ_{max}	pD
3(10)	1	22	0.60	4.33	4.67
	2	23	1.00	4.78	6.67
	3	23	1.00	4.99	7.78
	4	23	1.00	4.99	11.11
	5	23	1.00	4.99	13.22
4(10)	1	24	0.65	4.30	4.63
	2	24	0.99	4.86	7.38
	3	24	1.00	4.98	9.94
	4	24	1.00	4.99	12.13
	5	24	1.00	4.99	13.94
5(10)	1	31	0.47	4.56	4.40
	2	25	0.87	4.96	6.12
	3	25	1.00	4.99	8.36
	4	26	0.87	4.99	10.52
	5	24	1.00	4.99	12.40

and the CUHK Direct Grant for Research 2016-17, and the third author was supported by the National Natural Science Foundation of China (Project number 11201398, 11971414), Scientific Research Fund of Hunan Provincial Education Department (Project number 18B082) and Hunan Provincial Civil Military Integration Industrial Development Project "Adaptive Multilevel Solver and Its Application in ICF Numerical Simulation".

References

1. Calvo, J.G., Widlund, O.B.: An adaptive choice of primal constraints for BDDC domain decomposition algorithms. Electron. Trans. Numer. Anal. **45**, 524–544 (2016)
2. Chung, E.T., Efendiev, Y., Leung, W.T.: Constraint energy minimizing generalized multiscale finite element method. Comput. Methods Appl. Mech. Engrg. **339**, 298–319 (2018)
3. Dolean, V., Nataf, F., Scheichl, R., Spillane, N.: Analysis of a two-level Schwarz method with coarse spaces based on local Dirichlet-to-Neumann maps. Comput. Methods Appl. Math. **12**(4), 391–414 (2012)
4. Galvis, J., Efendiev, Y.: Domain decomposition preconditioners for multiscale flows in high-contrast media. Multiscale Model. Simul. **8**(4), 1461–1483 (2010)
5. Heinlein, A., Klawonn, A., Knepper, J., Rheinbach, O.: Adaptive GDSW coarse spaces for over- lapping Schwarz methods in three dimensions. Technical report. University of Cologne, October 2018. URL `https://kups.ub.uni-koeln.de/8756/`
6. Heinlein, A., Klawonn, A., Knepper, J., Rheinbach, O.: An adaptive GDSW coarse space for two-level overlapping Schwarz methods in two dimensions. In: Domain decomposition methods in science and engineering XXIV, *Lect. Notes Comput. Sci. Eng.*, vol. 125, pp. 373–382. Springer, Heidelberg (2018)
7. Heinlein, A., Klawonn, A., Knepper, J., Rheinbach, O.: Multiscale coarse spaces for overlapping Schwarz methods based on the ACMS space in 2D. Electron. Trans. Numer. Anal. **48**, 156–182 (2018)

8. Kim, H.H., Chung, E., Wang, J.: BDDC and FETI-DP preconditioners with adaptive coarse spaces for three-dimensional elliptic problems with oscillatory and high contrast coefficients. J. Comput. Phys. **349**, 191–214 (2017)
9. Klawonn, A., Kühn, M., Rheinbach, O.: Adaptive coarse spaces for FETI-DP in three dimensions. SIAM J. Sci. Comput. **38**(5), A2880–A2911 (2016)
10. Mandel, J., Sousedík, B., Šístek, J.: Adaptive BDDC in three dimensions. Math. Comput. Simulation **82**(10), 1812–1831 (2012)
11. Pechstein, C., Dohrmann, C.R.: A unified framework for adaptive BDDC. Electron. Trans. Numer. Anal. **46**, 273–336 (2017)
12. Toselli, A., Widlund, O.: Domain decomposition methods—algorithms and theory, *Springer Series in Computational Mathematics*, vol. 34. Springer-Verlag, Berlin (2005)
13. Beirão da Veiga, L., Pavarino, L.F., Scacchi, S., Widlund, O.B., Zampini, S.: Adaptive selection of primal constraints for isogeometric BDDC deluxe preconditioners. SIAM J. Sci. Comput. **39**(1), A281–A302 (2017)
14. Wang, J., Chung, E., Kim, H.H.: A two-level overlapping Schwarz method with energy-minimizing multiscale coarse basis functions. arXiv:1901.00112 [math.NA] (2019)

Machine Learning in Adaptive FETI-DP – A Comparison of Smart and Random Training Data

Alexander Heinlein, Axel Klawonn, Martin Lanser, and Janine Weber

1 Introduction

The convergence rate of classical domain decomposition methods for diffusion or elasticity problems usually deteriorates when large coefficient jumps occur along or across the interface between subdomains. In fact, the constant in the classical condition number bounds [11, 12] will depend on the coefficient jump. Therefore, several adaptive approaches to enrich the coarse space with additional coarse modes or primal constraints, which are constructed from the solutions of localized eigenvalue problems, have been developed to overcome this limitation, e.g., [7, 6, 13, 14, 2, 3]. For many realistic coefficient distributions, however, only a few adaptive constraints on a few edges or faces are necessary for robustness. Although some heuristic approaches [6, 7] exist to reduce the number of eigenvalue problems that have to be solved, in general, we do not know in advance on which edges or faces additional adaptive constraints are needed to obtain a robust algorithm.

To overcome this issue, we consider an approach to train a neural network to predict the geometric location of adaptive constraints in a preprocessing step, i.e., to make the decision whether or not we have to solve a certain eigenvalue problem. First results using this machine learning based strategy in the context of adaptive domain decomposition methods for a concrete and carefully designed set of training data can be found in [5]. Here, in addition to [5], we test the feasibility of using randomly

Alexander Heinlein, Axel Klawonn, and Martin Lanser
Department of Mathematics and Computer Science, University of Cologne, Weyertal 86-90, 50931 Köln, Germany, e-mail: `alexander.heinlein@uni-koeln.de`, `axel.klawonn@uni-koeln.de`, `martin.lanser@uni-koeln.de`, url: `http://www.numerik.uni-koeln.de`
Center for Data and Simulation Science, University of Cologne, url: `http://www.cds.uni-koeln.de`

Janine Weber
Department of Mathematics and Computer Science, University of Cologne, Weyertal 86-90, 50931 Köln, Germany, e-mail: `janine.weber@uni-koeln.de`, url: `http://www.numerik.uni-koeln.de`

R. Haynes et al. (eds.), *Domain Decomposition Methods in Science and Engineering XXV*,
Lecture Notes in Computational Science and Engineering 138,
https://doi.org/10.1007/978-3-030-56750-7_24

generated training data for the neural network. Random training data can be easily generated without any knowledge of the considered model problem, and therefore, the approach discussed here is more general compared to [5]; however, a larger set of training data may be required; cf. section 4. We provide numerical results for four different sets of training data to train the neural network and compare the robustness of the resulting algorithms both with respect to the training and validation data as well as for a concrete test problem.

We focus on a stationary diffusion problem in two dimensions and a certain adaptive coarse space technique for the FETI-DP (Finite Element Tearing and Interconnecting - Dual-Primal) algorithm [13, 14]. The adaptive coarse space is implemented using a balancing preconditioner. Let us remark that all strategies introduced here and in [5] can be generalized for arbitrary adaptive domain decomposition methods in two dimensions.

2 Model Problem and Adaptive FETI-DP

As a model problem, we use a stationary diffusion problem in two dimensions with various highly heterogenous coefficient functions $\rho : \Omega := [0, 1] \times [0, 1] \rightarrow \mathbb{R}$, i.e., the weak formulation of

$$\begin{aligned} \operatorname{div}(\rho \nabla u) &= -1 \text{ in } \Omega \\ u &= \ \ 0 \text{ on } \partial\Omega. \end{aligned} \tag{1}$$

In this paper, we apply the proposed machine learning based strategy to a certain adaptive FETI-DP method. We thus decompose the domain Ω into $N \in \mathbb{N}$ nonoverlapping subdomains $\Omega_i, i = 1, \ldots, N$. Due to space limitations, we do not explain the standard FETI-DP algorithm in detail. For a detailed description of the FETI-DP algorithm, see, e.g., [12]. Note that we choose all vertices as primal variables.

As already mentioned, for arbitrary and complex coefficient functions ρ, using solely primal vertex constraints is not sufficient to obtain a robust algorithm or condition number bound, respectively. Additional adaptive constraints, resulting from the solution of localized eigenvalue problems, are needed to enrich the coarse space and guarantee robustness. In our case, the adaptive constraints are implemented in FETI-DP by using a balancing preconditioner. For a detailed description of projector or balancing preconditioning, see [10]. In all our computations, we exclusively use ρ-scaling. Please note that also other approaches are possible to enforce coarse constraints, e.g., a transformation of basis approach [12].

The main idea of the concrete adaptive FETI-DP algorithm [13, 14] is to solve a local generalized eigenvalue problem for each edge between two neighboring subdomains. For a detailed description of the specific local edge eigenvalue problem as well as the resulting enforced coarse constraints, see [13, 14]. Usually, it is not known in advance on which edges additional coarse components are necessary. Although the solution of the different eigenvalue problems and thus the computation of the adaptive constraints can be parallelized, building the adaptive coarse space can make up the larger part of the overall time to solution. As already mentioned,

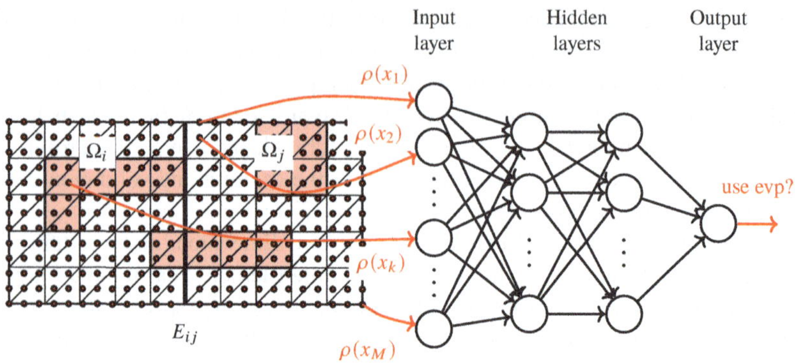

Fig. 1: Sampling of the coefficient function ρ; white color corresponds to a low coefficient and red color to a high coefficient. In this representation, the samples are used as input data for a neural network with two hidden layers. Figure from [5, Fig. 2].

we suggest a machine learning based approach to avoid the solution of unnecessary eigenvalue problems in order to save compute time.

3 Machine Learning for Adaptive FETI-DP

Our approach is to train a neural network to automatically make the decision whether an adaptive constraint needs to be enforced on a specific edge, or not, to retain the robustness of the algorithm, depending on a user-given tolerance TOL.

Supervised machine learning in general approximates nonlinear functions, which associate input and output data. The training of a neural network in supervised machine learning corresponds to the solution of a high-dimensional nonlinear optimization problem. In this paper, we use a dense feedfoward neural network, or more precisely, a multilayer perceptron. For more details on multilayer perceptrons, see, e.g., [15, 1, 16]. As input data for our neural network, we use samples of the coefficient functions within the two subdomains adjacent to an edge; cf. Figure 1. We use a sampling approach which is independent of the finite element discretization. In particular, we use a fixed number of sampling points for all mesh resolutions but assume the sampling grid to resolve all geometric details of the coefficient function. Our sampling grid is oriented to the tangential and orthogonal direction of an edge. Therefore, our approach is also valid for more general subdomain geometries than square subdomains; see also [5]. As output of the neural network, we save the information whether an adaptive coarse constraint has to be computed for the considered edge or not. Our neural network consists of three hidden layers with 30 neurons each. For all hidden layers, we use the ReLU function as an activation function and a dropout rate of 20%. For the training of the neural network, we use the stochastic gradient descent algorithm with an adaptive scaling of the learning rate and a batch size of 100. As an optimizer for the stochastic gradient descent method, we use

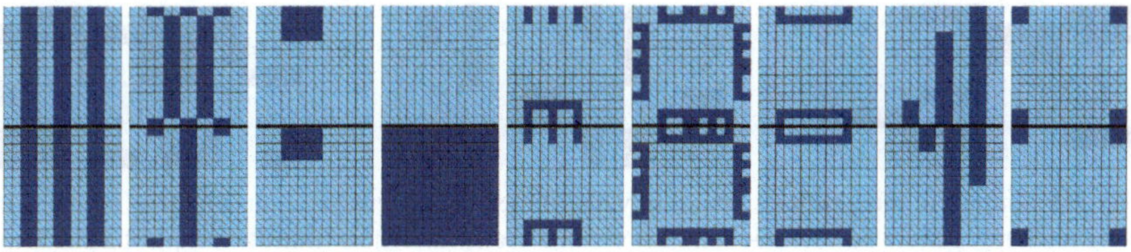

Fig. 2: Nine different types of coefficient functions used for training and validation of the neural network. The inclusions, channels, boxes, and combs with high coefficient are displaced, modified in sized, and mirrored with respect to the edge in order to generate the complete training data set.

the Adam (Adaptive moments) optimizer. All the aforementioned parameters result from applying a grid-search algorithm with cross-validation over a discrete space of hyper-parameters for the neural network; see [5] for more details on the parameters.

For the numerical results presented in this paper, we only train on two regular subdomains sharing a straight edge. Regarding the coefficient functions, we use different sets of coefficient distributions to generate different sets of training data. For the first set of training data, we use a total of 4,500 configurations varying the coefficient distributions as shown in Figure 2. These coefficient distributions are inspired by those used in [8, 2], and all coefficient distributions shown in Figure 2 are varied in size, location, and orientation to obtain the full set of training data. We refer to this set of training data, which already has been used in [5], as *smart data*.

Next, we consider random data to train the neural network. Let us note that a completely random coefficient distribution is not appropriate since in this case coefficient jumps appear at almost all edges. Thus, for almost every edge an eigenvalue problem has to be solved. This yields a neural network which overestimates the number of eigenvalue problems needed and thus leads to a large number of false positive edges in the test data which here is given by the microsection problem.

Thus, as a second set of training data, we use a slightly more structured set of randomly generated coefficients with a varying ratio of high and low coefficient values. For the first part of this training set, we randomly generate the coefficient for each pixel, consisting of two triangular finite elements, independently and only control the ratio of high and low coefficient values. Here, we use 30%, 20%, 10%, and 5% of high coefficient values. For the second part, we also control the distribution of the coefficients to a certain degree by randomly generating either horizontal or vertical stripes of a maximum length of four or eight pixels, respectively; see Figure 3. Additionally, we generate new coefficient distibutions by superimposing pairs of horizontal and vertical coefficient distributions. We refer to this second set of training data as *random data*.

To generate the output data that is necessary to train the neural network, we solve the eigenvalue problems as described in [13, 14] for all the aforementioned training and validation configurations. Here, we basically propose two different classification approaches as already considered in [5]. The first approach is referred to as 'two-class classification' and classifies an edge to belong to class 0 if no adaptive constraint needs to be added to the coarse space for the respective edge, depending on the

Fig. 3: Examples of three different randomly distributed coefficient functions obtained by using the same randomly generated coefficient for a horizontal (**left**) or vertical (**middle**) stripe of a maximum length of four finite element pixels, as well as by pairwise superimposing (**right**).

user-based tolerance TOL. It is classified to belong to class 1 if at least one adaptive constraint needs to be added. We further provide a second approach, which is referred to as 'three-class classification'. Here, besides class 0, we further distinguish between class 1, for edges where exactly one adaptive constraint needs to be added to the coarse space, and class 2, for edges where more than one constraint is necessary. For class 1, we replace the eigenvalue problem and the resulting eigenvector by a single edge constraint designed using ρ, which therefore also avoids the solution of some eigenvalue problems. These edge constraints can be interpreted as a generalization of the weighted edge averages suggested in [9] and are robust for a broader range of heterogeneities; see [4] for a detailed discussion. For all our training and validation data, we use a tolerance of $TOL = 100$ to generate the output for each edge.

4 Numerical Results

In this section, we compare the performance of the proposed machine learning based adaptive strategy for the FETI-DP algorithm using different sets of training and validation data to train our neural network. In particular, we use a set of 4,500 *smart data* configurations (denoted by *'S'*) and sets of 4,500 and 9,000 *random data* configurations (denoted by *'R1'* and *'R2'*, respectively) each individually as well as a combination of 4,500 *smart* and 4,500 *random data* configurations, which will be denoted by *'SR'*. Note that we did not observe a significant improvement for the larger number of 18,000 random data configurations.

First, we will present results for the whole set of training data using cross-validation and a fixed ratio of 20% as validation data to test the generalization properties of our neural network. Please note that due to different heterogeneity of the various training data, the accuracies in Table 1 are not directly comparable with each other. However, the results in Table 1 serve as a sanity check to prove that the trained model is able to generate appropriate predictions. We will then use 10 different randomly chosen subsections of a microsection of a dual-phase steel as shown in Figure 4 (right) as a test problem for the trained neural network. In all the computations, we consider $\rho = 1e6$ in the black part of the microsection and $\rho = 1$ elsewhere. Here, we use a regular decomposition of the domain Ω into 64 square subdomains, a subdomain size of $H/h = 64$, and a tolerance of $TOL = 100$. Please note, that also other mesh resolutions of the finite element mesh can be

Table 1: Results on the complete training data set; the numbers are averages over all training configurations. We show the ML-threshold (τ), the number of false positives (**fp**), the number of false negatives (**fn**), and the accuracy in the classification (**acc**). We define the accuracy as the number of true positives and true negatives divided by the total number of training configurations.

training configuration	τ	fp	fn	acc	class 0	class 1	class 2
4,500 smart data S, two-class	0.45	8.8%	1.9%	89.2%	67%	33%	-
	0.5	5.4%	5.1%	89.5%			
4,500 smart data S, three-class	0.4	5.1%	1.0%	93.9%	67%	20%	13%
	0.5	3.2%	2.3%	94.5%			
4,500 random data R1, two-class	0.45	11.4%	6.7%	81.9%	49%	51%	-
	0.5	8.8%	9.0%	82.2%			
4,500 random data R1, three-class	0.4	9.1%	7.1%	83.8%	49%	39%	12%
	0.5	8.9%	7.0%	84.1%			
9,000 random data R2, two-class	0.45	9.6%	5.3%	85.1%	53%	47%	-
	0.5	7.2%	7.5%	85.3%			
9,000 random data R2, three-class	0.4	10.7%	4.4%	84.9%	53%	28%	19%
	0.5	7.4%	6.9%	85.7%			
4,500 smart + 4,500 random data SR, two-class	0.45	5.1%	2.1%	92.8%	58%	42%	-
	0.5	3.4%	3.5%	93.1%			
4,500 smart + 4,500 random data SR, three-class	0.4	5.2%	2.0%	92.8%	58%	29.5%	12.5%
	0.5	4.3%	2.2%	93.5%			

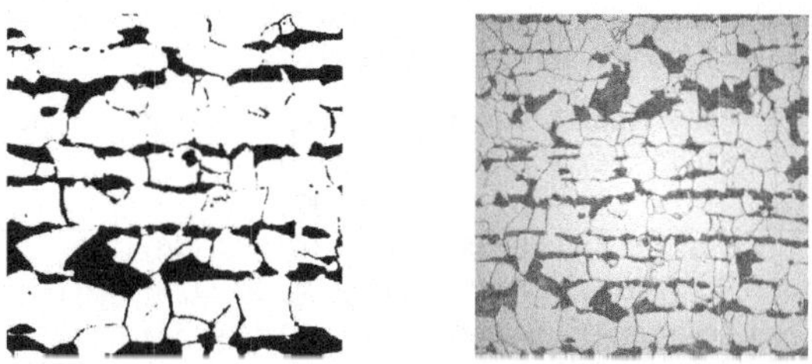

Fig. 4: Left: Subsection of a microsection of a dual-phase steel obtained from the image on the right. We consider $\rho = 1e6$ in the black part and $\rho = 1$ elsewhere. **Right:** Complete microsection of a dual-phase steel. Right image: Courtesy of Jörg Schröder, University of Duisburg-Essen, Germany, orginating from a cooperation with ThyssenKruppSteel.

used without affecting the accuracy of our classification algorithm as long as the coefficient function is constant on each finite element; see also [5]. For the test data, we will only compute the local eigenvalue problems on edges which are classified as critical (class 1 or 2) by the neural network. On all uncritical edges (class 0), we do not enforce any constraints. We use an ML (Machine Learning) threshold τ of 0.5 and 0.45 for the two-class classification as well as 0.5 and 0.4 for the three-class classification, respectively, for the decision boundary between critical and uncritical edges. A lower threshold decreases the false negative rate of the predictions and thus increases the robustness of our algorithm. All computations are performed using the machine learning implementations in TensorFlow and Scikit-learn as well as our Matlab implementation of the adaptive FETI-DP method.

Table 2: Comparison of standard FETI-DP, adaptive FETI-DP, and ML-FETI-DP for **regular domain decompositions** for the **two-class model**, for 10 different subsections of the microsection in Figure 4 (right). Here, training data is denoted as **T-Data**. We show the ML-threshold (τ), the condition number (**cond**), the number of CG iterations (**it**), the number of solved eigenvalue problems (**evp**), the number of false positives (**fp**), the number of false negatives (**fn**), and the accuracy in the classification (**acc**). We show the average values as well as the maximum values (in brackets).

Alg.	**T-Data**	τ	**cond**	**it**	**evp**	**fp**	**fn**	**acc**
standard	-	-	1.5e6 (2.2e6)	>300	0	-	-	-
adaptive	-	-	11.0 (15.9)	34.6 (38)	112.0 (112)	-	-	-
ML	S	0.5	8.6e4 (9.7e4)	39.5 (52)	45.0 (57)	1.6 (2)	1.9 (3)	0.97 (0.96)
	S	0.45	11.0 (15.9)	34.6 (38)	46.9 (59)	**4.4 (6)**	**0 (0)**	0.96 (0.94)
	R1	0.5	1.3e5 (1.6e5)	49.8 (52)	43.2 (44)	7.4 (8)	3.8 (4)	0.88 (0.87)
	R1	0.45	11.0 (15.9)	34.6 (38)	53.8 (58)	14.6 (16)	0 (0)	0.86 (0.84)
	R2	0.5	1.5e5 (1.6e5)	50.2 (51)	40.4 (41)	5.6 (6)	3.4 (4)	0.91 (0.89)
	R2	0.45	11.0 (15.9)	34.6 (38)	50.4 (52)	11.2 (12)	0 (0)	0.90 (0.87)
	SR	0.5	9.6e4 (9.8e4)	45.8 (48)	38.2 (39)	1.8 (2)	1.6 (2)	0.96 (0.95)
	SR	0.45	11.0 (15.9)	34.6 (38)	43.4 (44)	4.8 (5)	0 (0)	0.96 (0.94)

Table 3: Comparison of standard FETI-DP, adaptive FETI-DP, and ML-FETI-DP for **regular domain decompositions** for the **three-class model**, for 10 different subsections of the microsection in Figure 4 (right). Here, training data is denoted as **T-Data**. By **e-avg** we denote the generalized edge average described at the end of Section 3. See Table 2 for the column labeling. We show the average values as well as the maximum values (in brackets).

Alg.	**T-Data**	τ	**cond**	**it**	**evp**	**e-avg**	**fp**	**fn**	**acc**
standard	-	-	1.5e6 (2.2e6)	>300	0	-	-	-	-
adaptive	-	-	11.0 (15.9)	34.6 (38)	112.0 (112)	-	-	-	-
ML	S	0.5	147.4 (271.4)	48.8 (58)	4.2 (10)	43.6 (46)	1.8 (3)	1.4 (3)	0.97 (0.95)
	S	0.4	12.4 (16.4)	34.8 (39)	16.0 (24)	24.2 (28)	10.6 (16)	0 (0)	0.90 (0.85)
	R1	0.5	1.8e4 (1.8e4)	70.4 (72)	7.4 (8)	31.4 (33)	1.8 (2)	1.2 (2)	0.97 (0.94)
	R1	0.4	12.4 (16.4)	34.8 (39)	28.2 (30)	20.4 (26)	16.4 (21)	0 (0)	0.84 (0.83)
	R2	0.5	2.2e4 (2.5e4)	69.4 (72)	5.8 (7)	28.8 (31)	1.6 (2)	1.2 (2)	0.96 (0.95)
	R2	0.4	12.4 (16.4)	34.8 (39)	23.6 (24)	17.0 (18)	15.6 (17)	0 (0)	0.84 (0.83)
	SR	0.5	142.5 (286.3)	52.4 (66)	7.2 (9)	30.6 (32)	1.8 (2)	1.8 (2)	0.96 (0.94)
	SR	0.4	12.4 (16.4)	34.8 (39)	18.2 (20)	16.2 (18)	**10.0 (12)**	**0 (0)**	0.90 (0.89)

Results on the training data: With respect to the training data, the results in terms of accuracy in Table 1 show that, besides training the neural network with the set of smart data, also the training with randomly generated coefficient functions as well as with a combination of both training sets lead to an appropriate model. Thus, it is reasonable to apply all of the trained models to our test problem in form of microsection subsections.

Results on microsection subsections: For the mircosections and the two-class classification, see Table 2, all four different training data sets result in a robust algorithm when using an ML threshold $\tau = 0.45$. For all these approaches, we obtain no false negative edges, which are critical for the convergence of the algorithm. However, the usage of 4,500 and 9,000 random data (see R1 and R2) results in a higher number of false positive edges compared to the sole use of 4,500 smart data, resulting in a larger number of computed eigenvalue problems. Also for the three-class classification, see Table 3, the usage of all four aforementioned training data sets results in zero false negative edges when using the ML threshold $\tau = 0.4$. In this case, we further obtain a quantitatively smaller difference between the training data S and R1 or R2, respectively, in terms of false positive edges than for the two-class classification.

As a conclusion, we observe that we were able to achieve comparable results when using randomly generated coefficient distributions as training data compared to the manually selected smart data; this is beneficial since the random data can be generated without a priori knowledge. However, we need a higher number of random data and a slight structure in the random coefficient distributions to achieve the same accuracy as for the smart data. It also seems possible to slightly improve the performance of the neural network trained using a combination of smart data and random data for the training.

References

1. Goodfellow, I., Bengio, Y., Courville, A.: Deep learning, vol. 1. MIT press Cambridge (2016)
2. Heinlein, A., Klawonn, A., Knepper, J., Rheinbach, O.: Multiscale coarse spaces for overlapping Schwarz methods based on the ACMS space in 2d. Electron. Trans. Numer. Anal. **48**, 156–182 (2018)
3. Heinlein, A., Klawonn, A., Knepper, J., Rheinbach, O.: An adaptive GDSW coarse space for two-level overlapping Schwarz methods in two dimensions. In: Domain Decomposition Methods in Science and Engineering XXIV, *LNCSE*, vol. 125, pp. 373–382. Springer (2019). DOI:10.1007/978-3-319-93873-835
4. Heinlein, A., Klawonn, A., Lanser, M., Weber, J.: A Frugal FETI-DP and BDDC Coarse Space for Heterogeneous Problems (2019). TR series, Center for Data and Simulation Science, University of Cologne, Germany, Vol. 2019-18. https://kups.ub.uni-koeln.de/10363/. Submitted for publication
5. Heinlein, A., Klawonn, A., Lanser, M., Weber, J.: Machine Learning in Adaptive Domain Decomposition Methods - Predicting the Geometric Location of Constraints. SIAM J. Sci. Comput. **41**(6), A3887–A3912 (2019)
6. Klawonn, A., Kühn, M., Rheinbach, O.: Adaptive coarse spaces for FETI-DP in three dimensions. SIAM J. Sci. Comput. **38**(5), A2880–A2911 (2016)
7. Klawonn, A., Kühn, M., Rheinbach, O.: Adaptive FETI-DP and BDDC methods with a generalized transformation of basis for heterogeneous problems. Electron. Trans. Numer. Anal. **49**, 1–27 (2018)
8. Klawonn, A., Radtke, P., Rheinbach, O.: A comparison of adaptive coarse spaces for iterative substructuring in two dimensions. Electron. Trans. Numer. Anal. **45**, 75–106 (2016)
9. Klawonn, A., Rheinbach, O.: Robust FETI-DP methods for heterogeneous three dimensional elasticity problems. Comput. Methods Appl. Mech. Engrg. **196**(8), 1400–1414 (2007)
10. Klawonn, A., Rheinbach, O.: Deflation, projector preconditioning, and balancing in iterative substructuring methods: connections and new results. SIAM J. Sci. Comput. **34**(1), A459–A484 (2012)
11. Klawonn, A., Rheinbach, O., Widlund, O.B.: An analysis of a FETI-DP algorithm on irregular subdomains in the plane. SIAM J. Numer. Anal. **46**(5), 2484–2504 (2008)
12. Klawonn, A., Widlund, O.B.: Dual-primal FETI methods for linear elasticity. Comm. Pure Appl. Math. **59**(11), 1523–1572 (2006)

13. Mandel, J., Sousedík, B.: Adaptive selection of face coarse degrees of freedom in the BDDC and the FETI-DP iterative substructuring methods. Comput. Methods Appl. Mech. Engrg. **196**(8), 1389–1399 (2007)
14. Mandel, J., Sousedík, B., Sístek, J.: Adaptive BDDC in three dimensions. Math. Comput. Simulation **82**(10), 1812–1831 (2012)
15. Müller, A., Guido, S.: Introduction to Machine Learning with Python: A Guide for Data Scientists. O'Reilly Media (2016)
16. Shalev-Shwartz, S., Ben-David, S.: Understanding Machine Learning. Cambridge University Press (2014)

Nonoverlapping Additive Schwarz Method for hp-DGFEM with Higher-order Penalty Terms

Piotr Krzyżanowski and Marcus Sarkis

1 Introduction

Let us consider a second order elliptic equation

$$-\operatorname{div}(\varrho \nabla u) = f \ \text{ in } \Omega \quad \text{and} \quad u = 0 \ \text{ in } \partial\Omega. \tag{1}$$

The problem is discretized by an h-p symmetric interior higher-order [4] discontinuous Galerkin finite element method. In a K-th order multipenalty method, one penalizes the jumps of scaled normal higher-order derivatives up to order K across the interelement boundaries — so the standard interior penalty method corresponds to taking $K = 0$. The idea to penalize the discontinuity in the flux ($K = 1$) of the discrete solution was introduced by Douglas and Dupont [6]. It addresses the observation that the flux (which is an important quantity in many applications) of the accurate solution is continuous. Giving the user a possibility to control the inevitable violation of this principle makes the discretization method more robust and conservative. Recently, flux jump penalization has been used to improve stability properties of an unfitted Nitsche's method [5], the case $K > 1$ was also considered in [1] for the immersed finite element method to obtain higher-order discretizations.

A nonoverlapping additive Schwarz method [7], [3] is applied to precondition the discrete equations. For more flexibility and enhanced parallelism, we formulate our results addressing the case when the subdomains (where the local problems are solved in parallel) are potentially smaller than the coarse grid cells [8]. By allowing small subdomains of diameter $H \leq \mathcal{H}$, the local problems are cheaper to solve and the amount of concurrency of the method is substantially increased. A by-product of

Piotr Krzyżanowski
University of Warsaw, Poland, e-mail: p.krzyzanowski@mimuw.edu.pl

Marcus Sarkis
Worchester Polytechnic Institute, e-mail: msarkis@wpi.edu

R. Haynes et al. (eds.), *Domain Decomposition Methods in Science and Engineering XXV*,
Lecture Notes in Computational Science and Engineering 138,
https://doi.org/10.1007/978-3-030-56750-7_25

this approach is more flexibility in assigning subdomain problems to processors for load balancing in coarse grain parallel processing.

The paper is organized as follows. In Section 2, the differential problem and its discontinuous Galerkin multipenalty discretization are formulated. In Section 3, a nonoverlapping two-level, three-grid additive ASM for solving the discrete problem is designed and analyzed under the assumption that the coarse mesh resolves the discontinuities of the coefficient, that the variation of the mesh size and of the polynomial degree are locally bounded, and that the original problem satisfies some regularity assumption. Section 4 presents some numerical experiments.

For nonnegative scalars x, y, we shall write $x \lesssim y$ if there exists a positive constant C, such that $x \leq Cy$ with C independent of: x, y, the fine, subdomain and coarse mesh parameters $h, H, \mathcal{H}$, the orders of the finite element spaces p, q, the order of the multipenalty method (K, L), and of jumps of the diffusion coefficient ϱ as well. If both $x \lesssim y$ and $y \lesssim x$, we shall write $x \simeq y$.

The norm of a function f from the Sobolev space $H^k(S)$ will be denoted by $||f||_{k,S}$, while the seminorm of f will be denoted by $|f|_{k,S}$. For short, the L^2-norm of f will then be denoted by $|f|_{0,S}$.

2 High-order penalty *h-p* discontinuous Galerkin discretization

Let Ω be a bounded open convex polyhedral domain in R^d, $d \in \{2,3\}$, with a Lipschitz boundary $\partial\Omega$. We consider the following variational formulation of (1): Find $U^* \in H_0^1(\Omega)$ such that for a prescribed $f \in L^2(\Omega)$ and $\varrho \in L^\infty(\Omega)$

$$a(U^*, v) = (f, v)_\Omega, \qquad \forall v \in H_0^1(\Omega), \tag{2}$$

where

$$a(u, v) = \int_\Omega \varrho \nabla u \cdot \nabla v \, dx, \qquad (f, v)_\Omega = \int_\Omega f v \, dx.$$

We assume that there exists a constant α such that $1 \leq \varrho \leq \alpha$ a.e. in Ω so that (2) is well–posed. We also assume that ϱ is piecewise constant, i.e. Ω can be partitioned into nonoverlapping polyhedral subregions with the property that ϱ restricted to any of these subregions is some positive constant, see assumption (5) later on.

Let $\mathcal{T}_h = \{\tau_1, \ldots, \tau_{N_h}\}$ denote an affine nonconforming partition of Ω, where τ_i are either triangles in 2-D or tetrahedra in 3-D. For $\tau \in \mathcal{T}_h$ we set $h_\tau = \mathrm{diam}(\tau)$. By $\mathcal{E}_h^{\mathrm{in}}$ we denote the set of all common (internal) faces (edges in 2-D) of elements in $\mathcal{T}_h$, such that $e \in \mathcal{E}_h^{\mathrm{in}}$ iff $e = \partial\tau_i \cap \partial\tau_j$ is of positive measure. We will use the symbol $\mathcal{E}_h$ to denote the set of all faces (edges in 2-D) of the fine mesh $\mathcal{T}_h$, that is those either in $\mathcal{E}_h^{\mathrm{in}}$ or on the boundary $\partial\Omega$. For $e \in \mathcal{E}_h$ we set $h_e = \mathrm{diam}(e)$. We assume that $\mathcal{T}_h$ is shape- and contact–regular, that is, it admits a matching submesh $\mathcal{T}_{\hat{h}}$ which is shape–regular and such that for any $\tau \in \mathcal{T}_h$ the ratios of h_τ to diameters of simplices in $\mathcal{T}_{\hat{h}}$ covering τ are uniformly bounded by an absolute constant. As a consequence, if $e = \partial\tau_i \cap \partial\tau_j$ is of positive measure, then $h_e \simeq h_{\tau_i} \simeq h_{\tau_j}$. We shall refer to $\mathcal{T}_h$ as the "fine mesh". Throughout the paper we will assume that the fine mesh is chosen in such a way that $\varrho_{|\tau}$ is already constant for all $\tau \in \mathcal{T}_h$.

We define the finite element space V_h^p in which problem (2) is approximated,

$$V_h^p = \{v \in L^2(\Omega) : v_{|\tau} \in \mathbb{P}_{p_\tau} \text{ for } \tau \in \mathcal{T}_h\} \tag{3}$$

where $\mathbb{P}_{p_\tau}$ denotes the set of polynomials of degree not greater than p_τ. We shall assume that $1 \le p_\tau$ and that polynomial degrees have bounded local variation, that is, if $e = \partial\tau_i \cap \partial\tau_j \in \mathcal{E}_h^{\text{in}}$, then $p_{\tau_i} \simeq p_{\tau_j}$.

On $e \in \mathcal{E}_h^{\text{in}}$ such that $e = \partial\tau^+ \cap \partial\tau^-$, we define

$$\overline{\varrho} = \frac{\varrho^+ + \varrho^-}{2}, \qquad \omega^\pm = \frac{\varrho^\pm}{\varrho^+ + \varrho^-}, \qquad \underline{\varrho} = \frac{2\varrho^+\varrho^-}{\varrho^+ + \varrho^-}$$

with the standard notation $\rho^\pm = \rho_{|\tau^\pm}$, and then define weighted averages

$$\{\varrho\nabla u\} = \omega^- \varrho^+\nabla u^+ + \omega^+ \varrho^-\nabla u^- = \frac{\underline{\varrho}}{2}(\nabla u^+ + \nabla u^-)$$

and jumps

$$[u] = u^+n^+ + u^-n^-,$$

where $n^\pm$ denotes the outward unit normal vector to $\tau^\pm$. We note that when $\varrho^+ = \varrho^- = 1$, then $\overline{\varrho} = \underline{\varrho} = 1$ and the weighted average reduces to the usual arithmetic average. We set

$$\gamma_0 = \frac{\overline{p}^2}{\underline{h}}\delta_0, \qquad \gamma_k = \frac{\underline{h}^{2k-1}}{\overline{p}^{2k}}\delta_k, \qquad \tilde{\gamma}_k = \frac{\underline{h}^{2k-1}}{\overline{p}^{2k}}\tilde{\delta}_k,$$

with

$$\underline{h} = \min\{h_+, h_-\}, \qquad \overline{p} = \max\{p_+, p_-\}.$$

where for simplicity we write $h_\pm, p_\pm$ for $h_{\tau^\pm}$ (or $p_{\tau^\pm}$, respectively). The parameters $\delta_0 > 0$ and $\delta_k, \tilde{\delta}_k \ge 0$ where $k \ge 1$ are some prescribed constants. We collect all δ_k in a multi-parameter $\delta = (\delta_0, \delta_1, \tilde{\delta}_1, \ldots)$.

On e which lies on $\partial\Omega$ and belongs to the face of $\tau \in \mathcal{T}_h$, we prescribe $\underline{\varrho} = \varrho$ and

$$\{\varrho\nabla u\} = \varrho\nabla u, \quad [u] = un, \quad \gamma_0 = \frac{p_\tau^2}{h_\tau}\delta_0, \quad \gamma_k = \frac{h_\tau^{2k-1}}{p_\tau^{2k}}\delta_k, \quad \tilde{\gamma}_k = \frac{h_\tau^{2k-1}}{p_\tau^{2k}}\tilde{\delta}_k.$$

Inspired by [1], we discretize (2) by the symmetric weighted interior (K, L)-th order multipenalty discontinuous Galerkin method: Find $u^* \in V_h^p$ such that

$$\mathcal{A}_h^{p,KL}(u^*, v) = (f, v)_\Omega - \sum_{k=1}^{L}\sum_{e\in\mathcal{E}_h^{\text{in}}} \frac{\tilde{\gamma}_{k+2}}{\overline{\varrho}}\langle[\varrho\frac{\partial^k f}{\partial n^k}], [\varrho\frac{\partial^k \Delta v}{\partial n^k}]\rangle_{e}, \qquad \forall v \in V_h^p, \tag{4}$$

where

$$\mathcal{A}_h^{p,KL}(u,v) = A_h^{p,KL}(u,v) - F_h^p(u,v) - F_h^p(v,u)$$

and

$$A_h^{p,KL}(u,v) = \sum_{\tau\in\mathcal{T}_h} (\varrho\,\nabla u, \nabla v)_\tau + \sum_{e\in\mathcal{E}_h} \gamma_0 \underline{\varrho}\langle [u],[v]\rangle_e$$
$$+ \sum_{k=1}^{K} \sum_{e\in\mathcal{E}_h^{\text{in}}} \frac{\gamma_k}{\overline{\varrho}} \langle [\varrho \frac{\partial^k u}{\partial n^k}], [\varrho \frac{\partial^k v}{\partial n^k}]\rangle_e + \sum_{k=1}^{L} \sum_{e\in\mathcal{E}_h^{\text{in}}} \frac{\tilde{\gamma}_{k+2}}{\overline{\varrho}} \langle [\varrho \frac{\partial^k \Delta u}{\partial n^k}], [\varrho \frac{\partial^k \Delta v}{\partial n^k}]\rangle_e,$$
$$F_h^p(u,v) = \sum_{e\in\mathcal{E}_h} \langle \{\varrho \nabla u\}, [v]\rangle_e.$$

Here for $\tau \in \mathcal{T}_h$ and $e \in \mathcal{E}_h$ we use the standard notation: $(u,v)_\tau = \int_\tau u\, v\, dx$ and $\langle u,v\rangle_e = \int_e u\, v\, d\sigma$. This discretization generalizes the multipenalty method, introduced by Arnold in [4] with $L = 0$, to the case of discontinuous coefficient and takes into account the explicit dependence on the polynomial degree p. In particular, for $(K,L) = (0,0)$, a standard symmetric weighted interior penalty method is restored, with

$$A_h^{p,00}(u,v) = \sum_{\tau\in\mathcal{T}_h} (\varrho\,\nabla u, \nabla v)_\tau + \sum_{e\in\mathcal{E}_h} \gamma_0 \underline{\varrho}\langle [u],[v]\rangle_e.$$

Moreover, for $\varrho \equiv 1$ and $(K,L) = (1,0)$, problem (4) corresponds to the method by Douglas and Dupont [6]. The case of $L > 0$ has been considered e.g. in [1]. It is known [4] that for sufficiently large penalty constant δ_0 problem (4) is well–defined.

3 Nonoverlapping additive Schwarz method

Let us introduce the subdomain grid $\mathcal{T}_H$ as a partition of Ω into N_H disjoint open polygons (polyhedrons in 3-D) Ω_i, $i = 1,\ldots,N_H$, such that $\bar{\Omega} = \bigcup_{i=1,\ldots,N_H} \bar{\Omega}_i$ and that each Ω_i is a union of certain elements from the fine mesh $\mathcal{T}_h$. We shall retain the common notion of "subdomains" while referring to elements of $\mathcal{T}_H$. We set $H_i = \operatorname{diam}(\Omega_i)$ and $H = (H_1,\ldots,H_{N_H})$. We assume that there exists a reference simply-connected polygonal (polyhedral in 3-D) domain $\hat{\Omega} \subset R^d$ with Lipschitz boundary, such that every Ω_i is affinely homeomorphic to $\hat{\Omega}$ and that the aspect ratios of Ω_i are bounded independently of h and H. Moreover, we assume that the number of neighboring regions in $\mathcal{T}_H$ is uniformly bounded by an absolute constant $\mathcal{N}$.

Next, let $\mathcal{T}_\mathcal{H}$ be a shape-regular affine triangulation by triangles in 2-D or tetrahedra in 3-D, with diameter $\mathcal{H}$. We denote the elements of $\mathcal{T}_\mathcal{H}$ by D_n and we call this partition the "coarse grid" and assume that ϱ is piecewise constant on $\mathcal{T}_\mathcal{H}$:

$$\varrho_{|D_n} = \varrho_n \qquad \forall 1 \le n \le N_\mathcal{H}. \tag{5}$$

Let us define the standard decomposition of V_h^p, cf. [3], [8]:

$$V_h^p = V_0 + V_1 + \ldots + V_{N_H}, \tag{6}$$

where the coarse space consists of functions which are polynomials inside each element of the coarse grid:

$$V_0 = \{v \in V_h^p : v_{|D_n} \in \mathbb{P}_q \text{ for all } n = 1, \ldots, N_{\mathcal{H}}\} \tag{7}$$

where $1 \leq q \leq \min\{p_\tau : \tau \in \mathcal{T}_h\}$. Next, for $i = 1, \ldots, N_H$ we set

$$V_i = \{v \in V_h^p : v_{|\Omega_j} = 0 \text{ for all } j \neq i\}.$$

One can view V_0 as a rough approximation to V_h^p (using coarser grid and lower order polynomials), cf. condition (11). Note that V_h^p already is a direct sum of spaces $V_1, \ldots, V_{N_H}$ and when $\mathcal{T}_{\mathcal{H}} = \mathcal{T}_H$, this decomposition coincides with [3]. Next, with fixed $0 \leq r \leq K$ and $0 \leq s \leq L$, we define inexact solvers $T_i : V_h^p \to V_i$, by

$$A_h^{p,rs}(T_i u, v) = \mathcal{A}_h^{p,KL}(u, v) \qquad \forall v \in V_i, \qquad 0 \leq i \leq N_H, \tag{8}$$

so that for $1 \leq i \leq N_H$ one has to solve only a relatively small system of linear equations on subdomain Ω_i (a "local problem") for $u_i = T_i u|_{\Omega_i}$. These subdomain problems are independent of each another and can be solved in parallel. The preconditioned operator is

$$T = T_0 + T_1 + \ldots + T_{N_H}. \tag{9}$$

Obviously, T is symmetric with respect to $\mathcal{A}_h^{p,KL}(\cdot, \cdot)$. For D_n in $\mathcal{T}_{\mathcal{H}}$ let us define an auxiliary seminorm

$$|||u|||^2_{D_n,\text{in}} = \sum_{\tau \in \mathcal{T}_h(D_n)} \varrho_n |\nabla u|^2_{0,\tau} + \sum_{e \in \mathcal{E}_h^{\text{in}}(D_n)} \gamma_0 \varrho_n |[u]|^2_{0,e}, \tag{10}$$

where $\mathcal{E}_h^{\text{in}}(D_n) = \{e \in \mathcal{E}_h : e \subset \bar{D}_n \setminus \partial D_n\}$.

Theorem 1 *Let us set $r = s = 0$ in* (8) *and assume that for each $u \in V_h^p$ there exists $u^{(0)} \in V_0$ satisfying*

$$\sum_{n=1}^{N_{\mathcal{H}}} \left(\frac{\varrho_n q_n^2}{\mathcal{H}_n^2} |u - u^{(0)}|^2_{0,D_n} + |||u - u^{(0)}|||^2_{D_n,\text{in}} \right) \lesssim \mathcal{A}_h^{p,00}(u, u). \tag{11}$$

Then the operator T defined in (9) *satisfies*

$$\beta^{-1} \mathcal{A}_h^{p,KL}(u, u) \lesssim \mathcal{A}_h^{p,KL}(Tu, u) \lesssim (K+L+1) \mathcal{A}_h^{p,KL}(u, u) \qquad \forall u \in V_h^p, \tag{12}$$

where

$$\beta = \max_{n=1,\ldots,N_{\mathcal{H}}} \left\{ \frac{\mathcal{H}_n^2}{q_n} \max_{i:\Omega_i \subset D_n} \left\{ \frac{\overline{p}_i^2}{\underline{h}_i H_i} \right\} \right\}$$

with $\underline{h}_i = \min\{h_\tau : \tau \in \mathcal{T}_h(\Omega_i)\}$ *and* $\overline{p}_i = \max\{p_\tau : \tau \in \mathcal{T}_h(\Omega_i)\}$. *Therefore, the condition number of* T *is* $O(\beta \cdot (K + L + 1))$.

Proof According to the general theory of ASM [11], it suffices to check three conditions. The strengthened Cauchy–Schwarz inequality holds with a constant independent of the parameters, due to our assumption that the number of neighbouring subdomains is bounded by an absolute constant.

For the local stability condition, it suffices to prove that for any k

$$\sum_{e\in\mathcal{E}_h^{\text{in}}} \frac{\gamma_k}{\overline{\varrho}} \|[\varrho \frac{\partial^k u}{\partial n^k}]\|_{0,e}^2 \lesssim A_h^{p,00}(u,u) \text{ and } \sum_{e\in\mathcal{E}_h^{\text{in}}} \frac{\tilde{\gamma}_{k+2}}{\overline{\varrho}} \|[\varrho \frac{\partial^k \Delta u}{\partial n^k}]\|_{0,e}^2 \lesssim A_h^{p,00}(u,u). \tag{13}$$

We prove the first inequality, the other can be proved analogously. On $e = \partial\tau^+ \cap \partial\tau^-$, we have (denoting by n either n^+ or n^-)

$$\frac{1}{\overline{\varrho}} \|[\varrho \frac{\partial^k u}{\partial n^k}]\|_{0,e}^2 \lesssim \frac{(\varrho^+)^2}{\overline{\varrho}} |\frac{\partial^k u^+}{\partial n^k}|_{0,e}^2 + \frac{(\varrho^-)^2}{\overline{\varrho}} |\frac{\partial^k u^-}{\partial n^k}|_{0,e}^2 \lesssim \varrho^+ |\frac{\partial^k u^+}{\partial n^k}|_{0,e}^2 + \varrho^- |\frac{\partial^k u^-}{\partial n^k}|_{0,e}^2,$$

since $(\varrho^\pm)^2/\overline{\varrho} = \omega^\pm \varrho^\pm \leq \varrho^\pm$. Now, by the trace inequality [4], we have $|\frac{\partial^k u}{\partial n^k}|_{0,e}^2 \lesssim \frac{1}{h_\tau}|u|_{k,\tau}^2 + h_\tau |u|_{k+1,\tau}^2$, so applying k times the inverse inequality we arrive at

$$\frac{1}{\overline{\varrho}} \|[\varrho \frac{\partial^k u}{\partial n^k}]\|_{0,e}^2 \lesssim \varrho^+ \frac{p_+^{2k}}{h_+^{2k-1}} |u^+|_{1,\tau^+}^2 + \varrho^- \frac{p_-^{2k}}{h_-^{2k-1}} |u^-|_{1,\tau^-}^2,$$

which yields

$$\sum_{e\in\mathcal{E}_h^{\text{in}}} \gamma_k \frac{1}{\overline{\varrho}} \langle [\varrho \frac{\partial^k u}{\partial n^k}], [\varrho \frac{\partial^k u}{\partial n^k}] \rangle_e \lesssim \sum_{\tau\in\mathcal{T}_h} \varrho\, |u|_{1,\tau}^2 \lesssim A_h^{p,00}(u,u).$$

Summing (13) over k, we complete the stability estimate

$$\mathcal{A}_h^{p,KL}(u,u) \lesssim (K + L + 1)\, A_h^{p,00}(u,u) \qquad \forall u \in V_i, \quad \forall 0 \leq i \leq N_H,$$

from which the right inequality in (12) already follows.

Finally, to prove the existence of a stable decomposition, from [9] we have that there exists a decomposition of $u = \sum_{i=0}^{N_H} u^{(i)}$, with $u^{(i)} \in V_i$, such that

$$\sum_{i=0}^{N_H} A_h^{p,00}(u^{(i)}, u^{(i)}) \lesssim \beta\, \mathcal{A}_h^{p,00}(u,u) \qquad \forall u \in V_h^p.$$

Since $\mathcal{A}_h^{p,00}(u,u) \leq \mathcal{A}_h^{p,KL}(u,u)$, we conclude that $\sum_{i=0}^{N_H} A_h^{p,00}(u^{(i)}, u^{(i)}) \lesssim \beta \mathcal{A}_h^{p,KL}(u,u)$, which gives us the left inequality in (12). □

Remark 1 Analogous result holds if, instead of the simplified form $A_h^{p,00}(\cdot,\cdot)$, we choose $A_h^{p,KL}(\cdot,\cdot)$ while defining local and coarse solvers T_i, $i = 0, 1, \ldots, N_H$, as we do in the following section.

Remark 2 In [10], sufficient conditions are provided for (11) to hold.

4 Numerical experiments

Let us choose the unit square $[0,1]^2$ as the domain Ω and consider (2) with $\varrho = 1$ in Ω. We do not investigate the influence of the intermediate grid $\mathcal{T}_H$, referring the reader to [9] for these results. Instead, we set $\mathcal{T}_H = \mathcal{T}_{\mathcal{H}}$ and use two levels of nested grids on Ω. For a prescribed integer $\mathcal{M}$, we divide Ω into $N_{\mathcal{H}} = 2^{\mathcal{M}} \times 2^{\mathcal{M}}$ squares of equal size. This coarse grid $\mathcal{T}_{\mathcal{H}}$ is then refined into a uniform fine triangulation $\mathcal{T}_h$ based on a square $2^m \times 2^m$ grid ($m \geq \mathcal{M}$) with each square split into two triangles of identical shape. Hence, the grid parameters are $h = 2^{-m}$, $\mathcal{H} = H = 2^{-\mathcal{M}}$. We set $L = 0$ and discretize problem (2) on the fine mesh $\mathcal{T}_h$ using (4) with $\delta_0 = 8$, $\delta_1 = \ldots = \delta_K = 2$ (if not specified otherwise) and equal polynomial degree p across all elements in $\mathcal{T}_h$. For the coarse problem, we set $q = p$. We always take $(r,s) = (K,L) = (K,0)$ while defining the inexact solvers, which seems to give preferable constants in (12). Our implementation makes use of the FEniCS [2] and MATLAB software packages.

In the following tables we report the number of Preconditioned Conjugate Gradient iterations for the operator T required to reduce the initial norm of the preconditioned residual by a factor of 10^8 and (in parentheses) the condition number of T estimated from the PCG convergence history. We always choose a random vector for the solution and a zero as the initial guess.

$\mathcal{H}$	iter (cond)
1/2	120 (328)
1/4	90 (157)
1/8	64 (71)
1/16	60 (60)

Table 1: Dependence on the coarse mesh size $\mathcal{H}$. Fixed $h = 1/64$, $p = 3$, $K = 3$.

p	iter (cond)
1	26 (11)
2	34 (21)
3	42 (34)
4	50 (50)
5	59 (70)

Table 2: Dependence on the polynomial degree p. Fixed $h = 1/16$, $\mathcal{H} = 1/4$, $K = 1$.

While the results with respect to $\mathcal{H}$ and p smoothly follow the theory developed, cf. Tables 1 and 2, the dependence on K is less regular, initially with superlinear increase, as reported in Table 3. Moreover, from Table 4 we observe that higher values of the penalization parameters δ_k, $k \geq 1$, adversely influence the convergence rate which is a drawback of this, otherwise simple and efficient, domain decomposition method.

K	iter (cond)
0	61 (81)
1	59 (70)
2	81 (124)
3	142 (389)
4	214 (902)
5	222 (1016)

Table 3: Dependence on the number of penalty terms K. Fixed $h = 1/16$, $\mathcal{H} = 1/4$, $p = 5$.

δ_1	iter (cond)
$2 \cdot 10^0$	29 (15)
$2 \cdot 10^1$	41 (29)
$2 \cdot 10^2$	102 (217)
$2 \cdot 10^3$	305 (2096)

Table 4: Dependence on the flux penalty parameter γ_1. Fixed $p = 3$, $K = 1$.

Acknowledgements The authors wish to thank two anonymous referees whose comments and remarks helped to improve the paper substantially. The research of Piotr Krzyżanowski has been partially supported by the Polish National Science Centre grant 2016/21/B/ST1/00350; Marcus Sarkis was supported by NSF-MPS 1522663.

References

1. Adjerid, S., Guo, R., Lin, T.: High degree immersed finite element spaces by a least square method. Int. J. Numer. Anal. Mod. **14**(4-5) (2017)
2. Alnæs, M.S., Blechta, J., Hake, J., Johansson, A., Kehlet, B., Logg, A., Richardson, C., Ring, J., Rognes, M.E., Wells, G.N.: The FEniCS Project Version 1.5. Archive of Numerical Software **3**(100) (2015). DOI:10.11588/ans.2015.100.20553
3. Antonietti, P.F., Houston, P.: A class of domain decomposition preconditioners for hp-discontinuous Galerkin finite element methods. J. Sci. Comput. **46**(1), 124–149 (2011). DOI:10.1007/s10915-010-9390-1. URL http://dx.doi.org/10.1007/s10915-010-9390-1
4. Arnold, D.N.: An interior penalty finite element method with discontinuous elements. SIAM J. Numer. Anal. **19**(4), 742–760 (1982)
5. Burman, E., Guzmán, J., Sánchez, M.A., Sarkis, M.: Robust flux error estimation of an unfitted Nitsche method for high-contrast interface problems. IMA Journal of Numerical Analysis **38**(2), 646–668 (2018). DOI:10.1093/imanum/drx017. URL http://dx.doi.org/10.1093/imanum/drx017
6. Douglas Jr., J., Dupont, T.: Interior penalty procedures for elliptic and parabolic Galerkin methods. In: Computing methods in applied sciences (Second Internat. Sympos., Versailles, 1975), pp. 207–216. Lecture Notes in Phys., Vol. 58. Springer, Berlin (1976)
7. Dryja, M.: On discontinuous Galerkin methods for elliptic problems with discontinuous coefficients. Comput. Methods Appl. Math. **3**(1), 76–85 (electronic) (2003)
8. Dryja, M., Krzyżanowski, P.: A massively parallel nonoverlapping additive Schwarz method for discontinuous Galerkin discretization of elliptic problems. Num. Math. **132**(2), 347–367 (2015). DOI:10.1007/s00211-015-0718-5. URL http://dx.doi.org/10.1007/s00211-015-0718-5
9. Krzyżanowski, P.: On a nonoverlapping additive Schwarz method for h-p discontinuous Galerkin discretization of elliptic problems. Num. Meth. PDEs **32**(6), 1572–1590 (2016)
10. Krzyżanowski, P.: Nonoverlapping three grid Additive Schwarz for hp-DGFEM with discontinuous coefficients. In: Domain Decomposition Methods in Science and Engineering XXIV, *Lecture Notes in Computer Science and Engineering*, vol. 125, pp. 455–463. Springer (2018)
11. Toselli, A., Widlund, O.: Domain decomposition methods—algorithms and theory, *Springer Series in Computational Mathematics*, vol. 34. Springer-Verlag, Berlin (2005)

A Closer Look at Local Eigenvalue Solvers for Adaptive FETI-DP and BDDC

Axel Klawonn, Martin J. Kühn, and Oliver Rheinbach

1 Introduction

In order to obtain a scalable domain decomposition method (DDM) for elliptic problems, a coarse space is necessary and an associated coarse problem has to be solved in each iteration. In the presence of arbitrary, large coefficient jumps or in case of almost incompressible elastic materials, the convergence rate of standard DDM deteriorates. In recent years, many authors have proposed the use of different (local, generalized) eigenvalue problems to develop problem dependent, adaptive coarse spaces in order to ensure or accelerate the convergence of the method; see, e.g., [2, 17, 6, 18, 5, 4, 22, 23, 14, 9, 10, 1, 19, 3, 20]. These methods are very robust and in many cases, a condition number estimate of the form

$$\text{cond} \leq C\,\text{TOL}, \tag{1}$$

exists. Here, TOL is an a priori, user defined tolerance for the solution of the eigenvalue problems and $C > 0$ a constant that only depends on geometric parameters, e.g., maximum number of edges of a subdomain; cf., (3). However, in order to make their use feasible, many issues have to be considered in the parallel implementation.

Axel Klawonn
Department of Mathematics and Computer Science, University of Cologne, Weyertal 86-90, 50931 Köln, Germany. e-mail: axel.klawonn@uni-koeln.de.
Center for Data and Simulation Science, University of Cologne, Germany, url: http://www.cds.uni-koeln.de

Martin J. Kühn
CERFACS (Centre Européen de Recherche et de Formation Avancée en Calcul Scientifique), 42 Avenue Gaspard Coriolis, 31057 Toulouse Cedex 01, France; e-mail: martin.kuehn@cerfacs.fr

Oliver Rheinbach
Technische Universität Bergakademie Freiberg, Fakultät für Mathematik und Informatik, Institut für Numerische Mathematik und Optimierung, 09596 Freiberg, Germany; e-mail: oliver.rheinbach@math.tu-freiberg.de

R. Haynes et al. (eds.), *Domain Decomposition Methods in Science and Engineering XXV*,
Lecture Notes in Computational Science and Engineering 138,
https://doi.org/10.1007/978-3-030-56750-7_26

In [16, 13], we have seen that the computational overhead of the solution process of the local eigenvalue problems in adaptive FETI-DP is not negligible. Consequently, in this paper, we focus on some aspects of the local eigenvalue solution process that have not been studied or documented elsewhere. Certainly, load balancing of the eigenvalue problems is a very important task but this issue is out the scope of this paper and will be discussed in detail in [13].

2 Model problem, domain decomposition, and notation

As a model problem, we consider three-dimensional linear elasticity, discretized with piecewise quadratic conforming finite elements. The domain is decomposed into nonoverlapping subdomains. Due to page restrictions, for further details, we refer to [10, Section 2] or [11, Section 2].

As it is standard in FETI-DP, we assemble the local stiffness matrices $K^{(i)}$ and compute the local Schur complements $S^{(i)}$ on the interface, $i = 1, \ldots, N$. Starting with the block-diagonal matrix S built from the the local Schur complements, we get the global matrix $\widetilde{S}$ by finite element subassembly in only a few a priori chosen primal variables (i.e., all vertices). In order to enforce continuity on the remaining a priori dual degrees of freedom on the interface, we introduce a jump operator B as well as a scaled variant B_D. Different scaling choices are available in the literature. We then obtain the FETI-DP system, which is reduced to the Lagrange multipliers enforcing continuity in the a priori dual variables,

$$M_D^{-1}F := B_D S B_D^T B \widetilde{S}^{-1} B \lambda = B_D S B_D^T d = M_D^{-1} d$$

with corresponding right hand side d; see, e.g., [11, Section 3] for further details.

3 Adaptive FETI-DP

In adaptive FETI-DP, as proposed in two dimensions in [17], and in three dimensions in [10], local generalized eigenvalue problems are solved on each pair of subdomains Ω_i and Ω_j sharing either a face $\mathcal{Z} = \mathcal{F}$ or an edge $\mathcal{Z} = \mathcal{E}$. By extracting all the rows of B and B_D corresponding to dual degrees of freedom of Ω_i and Ω_j and belonging to the closure of $\mathcal{Z}$, we can define the localized variants $B_{\mathcal{Z}}$, $B_{D,\mathcal{Z}}$ and $P_{D,\mathcal{Z}} := B_{D,\mathcal{Z}}^T B_{\mathcal{Z}}$. The localized Schur complement is defined as $S_{ij} :=$ blockdiag$\big(S^{(i)}, S^{(j)}\big)$. The local generalized eigenvalue problem then writes: find $w_{ij} \in (\ker S_{ij})^{\perp}$ with $\mu_{ij} >$ TOL, such that

$$(P_{D,\mathcal{Z}^{ij}} v_{ij}, S_{ij} P_{D,\mathcal{Z}^{ij}} w_{ij}) = \mu_{ij} (v_{ij}, S_{ij} w_{ij}) \quad \forall v_{ij} \in (\ker S_{ij})^{\perp}, \tag{2}$$

cf. [17, Sections 3 and 4] and [10, Section 5] or [16, Section 5] for a more detailed description.

The constraints obtained on a local basis can then be enforced by different techniques. Here, we use the generalized transformation-of-basis approach proposed in [12] and obtain the preconditioner $\widehat{M}_T^{-1}$, the modified system matrix $\widehat{F}$, and the condition number bound

$$\kappa(\widehat{M}_T^{-1}\widehat{F}) \leq 4\max\{N_{\mathcal{F}}, N_{\mathcal{E}}M_{\mathcal{E}}\}^2\mathrm{TOL}, \tag{3}$$

where $N_{\mathcal{F}}$ denotes the maximum number of faces of a subdomain, $N_{\mathcal{E}}$ the maximum number of edges of a subdomain, $M_{\mathcal{E}}$ the maximum multiplicity of an edge; see [11].

4 Numerical results

Fig. 1: A composite material on the unit cube for 216 subdomains: 36 beams (left) and 64 beams (right) of a stiff material with $E_2 = 1e+6$, shown in dark purple, are surrounded by a soft matrix material with $E_1 = 1$. A part of the mesh with $1/h = 54$ (left), $1/h = 30$ (right) and the irregular decomposition using METIS is shown in different (half-transparent) colors.

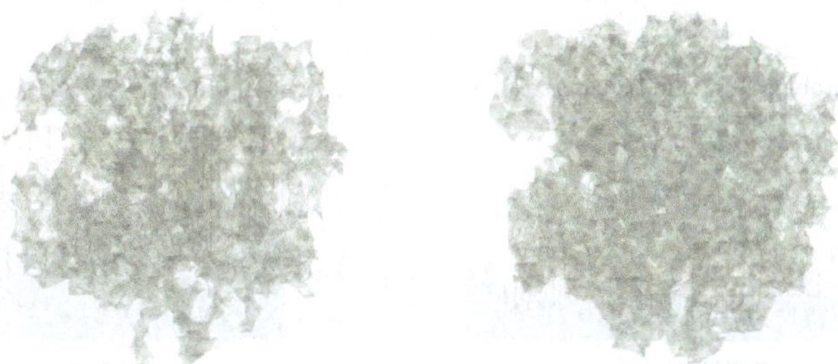

Fig. 2: Stiff material in a foam-like structure with ~15% (left) and ~26% (right) high coefficients with $E_2 = 1e+6$. The structure is surrounded by a soft matrix material with $E_1 = 1$. The stiff material is shown smoothed and half-transparent, the surrounding matrix material is not shown.

In the following, we consider the unit cube with zero Dirichlet boundary conditions on the face with $x_1 = 0$ and zero Neumann boundary conditions elsewhere. The domain decomposition is obtained from the METIS partitioner using the options `-ncommon=3` and `-contig`. We apply our adaptive method to four different materials.

The first material is considered for $1/h \in \{30, 54\}$ and 36 beams with a Young modulus of 1e+6 that run from the face with $x_1 = 0$ to the face with $x_1 = 1$; see Fig. 1 (left). The remaining part of the material has a Young modulus of one. In the second material, we have a larger number of 64 thinner beams; see Fig. 1 (right).

The third and fourth materials are stiff foam-like materials surrounded by a soft matrix material. They are obtained by using a pseudo-random number generator and adjacency structures of the tetrehedra and they merely differ by the amount of stiff material inside the unit cube; see Fig. 2.

We now focus on the solution process for the local eigenvalue problems. In recent works, [10, 16, 13], we have developed heuristic strategies to discard eigenvalue problems based on the coefficients, or in a more realistic setting, based on scaling information or the entries of the stiffness matrix which are more likely to be available. In following, we will show that our most recent heuristic strategy (see [13]) is successful in discarding unnecessary eigenvalue problems without, on the other hand, discarding necessary ones.

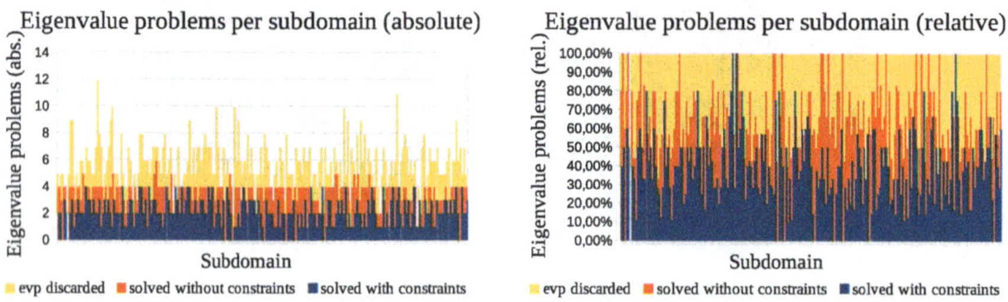

Fig. 3: Number of eigenvalue problems for each subdomain: discarded by our heuristic strategy [13] in yellow; solved and yielding constraints in blue; solved but not resulting in constraints in red; for the composite material with 36 beams; in absolute (left) and relative (right) numbers.

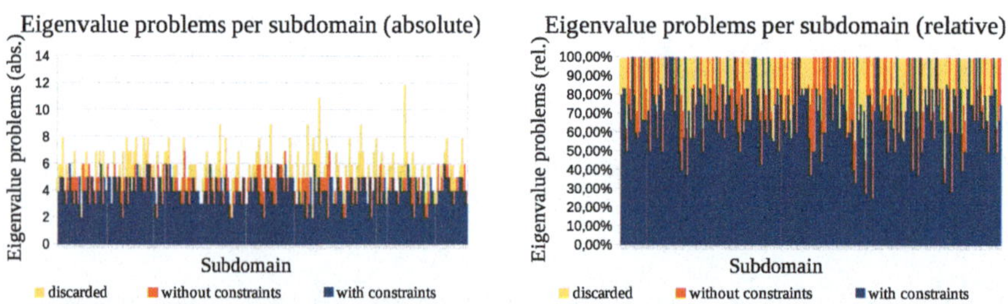

Fig. 4: Number of eigenvalue problems for each subdomain: discarded by our heuristic strategy [13] in yellow; solved and yielding constraints in blue; solved but not resulting in constraints in red; for the composite material with 64 beams; in absolute (left) and relative (right) numbers.

In the present implementation, for every face or edge, we consider the diagonal entries of the local subdomain stiffness matrices corresponding to interior nodes of these faces or edges. We have two criteria. First, if the ratio of the smallest and the largest entry is larger than a certain threshold or possibly second, if all entries are large, then the corresponding eigenvalue problem is solved. Otherwise it is discarded. For a more detailed description, see [13].

There are different situations in which eigenvalue problems become superfluous for the reduction of the condition number. One obvious reason is the nonexistence of jumps in the neighborhood of the face or edge. One could then apply slab techniques; see, e.g. [21, 7]. In our heuristics, we focus on these eigenvalue problems which we classify as *unnecessary*.

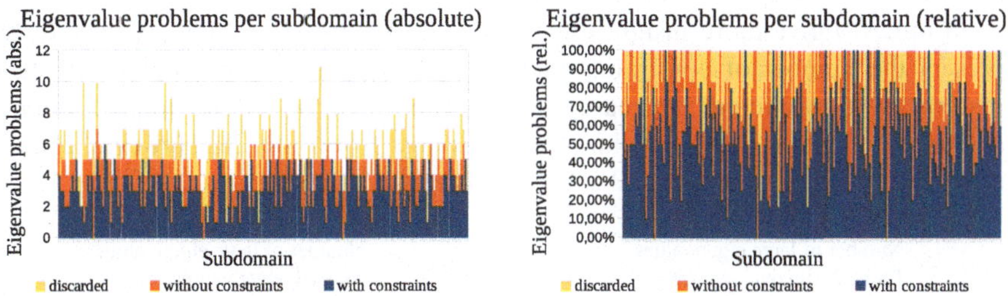

Fig. 5: Number of eigenvalue problems for each subdomain: discarded by our heuristic strategy [13] in yellow; solved and yielding constraints in blue; solved but not resulting in constraints in red; for the foam-like composite with ~15% high coefficients; in abs. (left) and rel. (right) numbers.

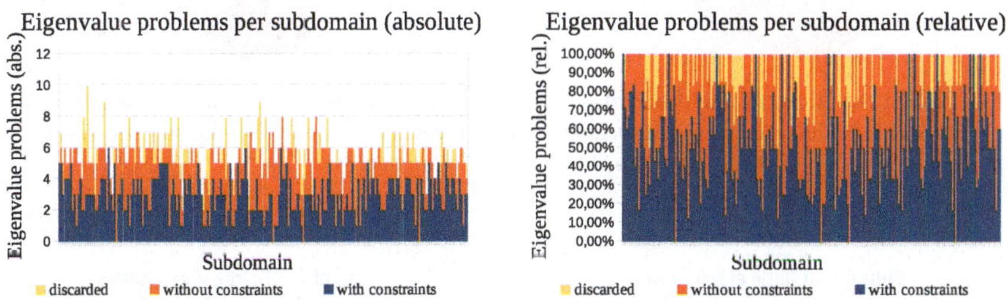

Fig. 6: Number of eigenvalue problems for each subdomain: discarded by our heuristic strategy [13] in yellow; solved and yielding constraints in blue; solved but not resulting in constraints in red; for the foam-like composite with ~26% high coefficients; in abs. (left) and rel. (right) numbers.

However, there might be eigenvalue problems on coefficient distributions which satisfy all assumptions for weighted Poincare inequalities. For arbitrary situations, the numerical necessity of certain eigenvalue problems becomes even more complex and is not yet fully understood. In [8], we have presented a short numerical study which gives a little more insight for typical situations but which might also raise new questions since not all configurations introduce as many bad modes as expected.

In the case of 36 beams, we see that we can discard a large number of eigenvalue problems (i.e., 37%) while 38% of eigenvalue problems yield large eigenvalues and thus adaptive constraints; see Fig. 3. In the case of 64 beams, we see that we only discard 16% of eigenvalue problems but, here, more than 70% of eigenvalue problems yield large eigenvalues and thus adaptive constraints; see Fig. 4. In both cases, the percentage of eigenvalue problems that were solved without yielding constraints is small, i.e., 25% and 14%. For both foam-like composites a little more than 50% of the eigenvalue problems have to be solved to reduce the condition number to about TOL (here $\mathrm{TOL} = 50 \log \left(N/n_i\right)^{1/3}$, where n_i is the number of local nodes). Between 26% and 39% of the eigenvalue problems were solved there without yielding constraints; see Fig. 5 and Fig. 6. Still 10% to 20% of the eigenvalue problems are detected as discardable, and thus the total algorithm is accelerated. We can summarize that our strategy can successfully identify and discard many unnecessary eigenvalue problems while keeping all necessary ones.

In order to give more insight into the eigenvalue problems that are solved, we present the number of constraints yielded for each eigenvalue problem for each material in four different pie charts; see Fig. 5 and Fig. 6. We see that the number of constraints for each eigenvalue problem range from 1 to 28. However, a large majority always gives between 2 and 12 constraints.

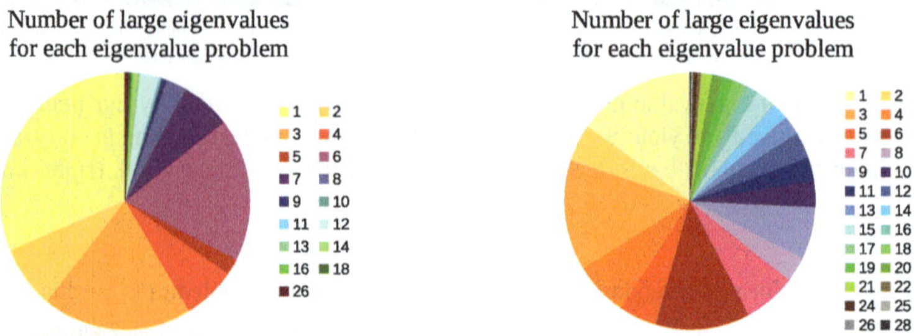

Fig. 7: Number of eigenvalue problems with given number of large eigenvalues for the composite material with 36 (left) and 64 (right) beams. In these presentations, only the blue marked eigenvalue problems of Fig. 3 and Fig. 4 are considered to give more details.

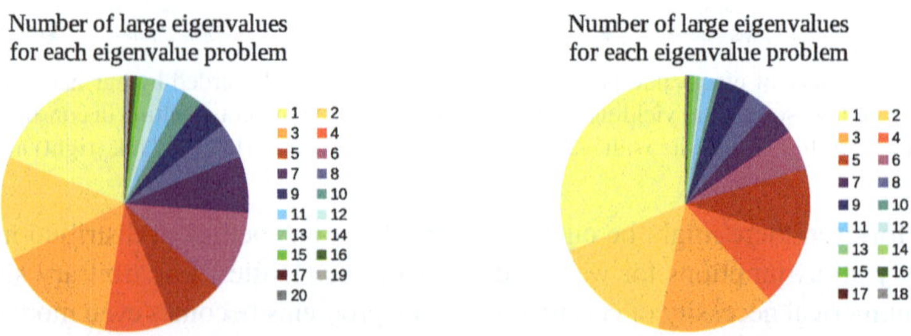

Fig. 8: Number of eigenvalue problems with given number of large eigenvalues for the foam-like composite material with 15% (left) and 26% (right) high coefficients. In these presentations, only the blue marked eigenvalue problems of Fig. 5 and Fig. 6 are considered to give more details.

Finally, we focus on the important topic of block sizes in the SLEPc Krylov-Schur solver. As motivated by [17] and our tests in [10, 11], we have already opted for an approximate solution of the eigenvalue problems by carrying out only a few steps of the iterative block scheme. Justified by the idea that the LOBPCG block solver of [15] could accelerate the convergence on extreme eigenvalues we have mostly used a block size of 10. Here, we study the timings of the global algorithms by varying the block size of the local Krylov-Schur algorithm for our four materials.

In Table 1, we have presented iteration counts and estimated condition numbers in order to show that the chosen block size does not effect the convergence of the global PCG scheme. That means that the constraints obtained with different block sizes do not differ in quality. In Fig. 9, we see that the use of smaller block sizes or even a single vector iteration might be favorable with respect to time to solution.

block size Krylov-Schur	36 beams $(1/h = 36)$		64 beams $(1/h = 30)$		~15% foam-like $(1/h = 30)$		~26% foam-like $(1/h = 30)$	
	κ	*its*	κ	*its*	κ	*its*	κ	*its*
1	5.48e+01	60	6.29e+1	62	7.21e+01	62	5.99e+01	63
3	5.48e+01	60	6.29e+1	62	7.21e+01	61	5.99e+01	62
6	5.48e+01	60	6.29e+1	62	7.21e+01	61	5.99e+01	61
9	5.48e+01	60	6.30e+1	63	7.21e+01	62	5.99e+01	64
12	5.48e+01	60	6.29e+1	62	7.21e+01	61	5.99e+01	61
15	5.48e+01	61	6.29e+1	62	7.21e+01	62	5.99e+01	61
18	5.49e+01	61	6.30e+1	63	7.21e+01	61	5.99e+01	61

Table 1: Condition number and iteration count of the global FETI-DP solver for different composite materials for 216 subdomains with different block sizes for the iterative Krylov-Schur eigenvalue solver and $\mathrm{TOL} = 50 \log \left(N/n_i\right)^{1/3}$, where n_i is the number of local nodes.

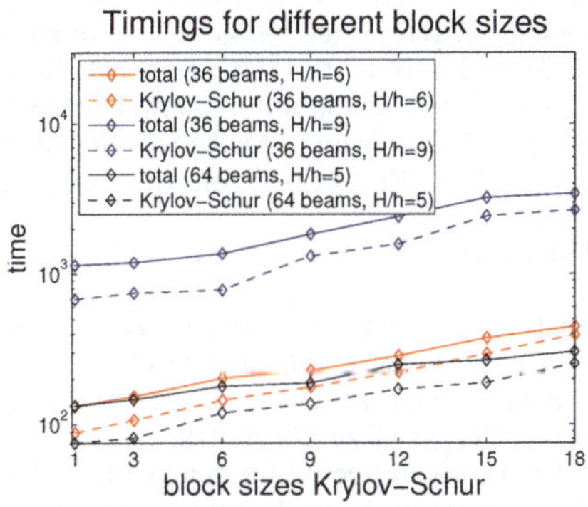

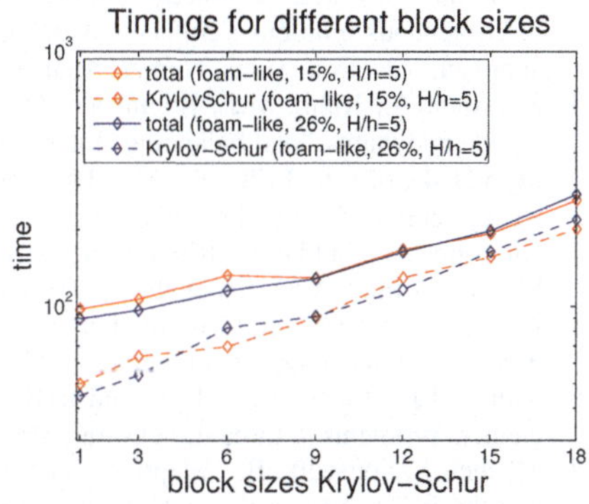

Fig. 9: Total global time and total local time needed by the Rayleigh-Ritz procedures in the Krylov-Schur scheme to approximately compute the largest eigenvectors of the generalized eigenvalue problems. Composite with 36 beams and 64 beams (left) and foam-like composite with ~15% and ~26% high coefficients (right).

References

1. Beirão da Veiga, L., Pavarino, L.F., Scacchi, S., Widlund, O.B., Zampini, S.: Adaptive selection of primal constraints for isogeometric BDDC deluxe preconditioners. SIAM J. Sci. Comput. **39**(1), A281–A302 (2017)
2. Bjørstad, P., Koster, J., Krzyżanowski, P.: Domain decomposition solvers for large scale industrial finite element problems. In: Applied Parallel Computing. New Paradigms for HPC in Industry and Academia, *Lecture Notes in Comput. Sci.*, vol. 1947, pp. 373–383. Springer, Berlin (2001)
3. Calvo, J.G., Widlund, O.B.: An adaptive choice of primal constraints for BDDC domain decomposition algorithms. Electron. Trans. Numer. Anal. **45**, 524–544 (2016)

4. Dohrmann, C., Pechstein, C.: In C. Pechstein, Modern domain decomposition solvers - BDDC, deluxe scaling, and an algebraic approach. NuMa Seminar, `http://people.ricam.oeaw.ac.at/c.pechstein/pechstein-bddc2013.pdf` (2013)
5. Dolean, V., Nataf, F., Scheichl, R., Spillane, N.: Analysis of a two-level Schwarz method with coarse spaces based on local Dirichlet-to-Neumann maps. Comput. Methods Appl. Math. **12**(4), 391–414 (2012)
6. Galvis, J., Efendiev, Y.: Domain decomposition preconditioners for multiscale flows in high-contrast media. Multiscale Model. Simul. **8**(4), 1461–1483 (2010)
7. Gippert, S., Klawonn, A., Rheinbach, O.: Analysis of FETI-DP and BDDC for linear elasticity in 3D with almost incompressible components and varying coefficients inside subdomains. SIAM J. Numer. Anal. **50**(5), 2208–2236 (2012)
8. Heinlein, A., Klawonn, A., Kühn, M.J.: Local spectra of adaptive domain decomposition methods. In: Domain Decomposition Methods in Science and Engineering XXV. Springer International Publishing, Cham (2020). Also Preprint at `https://kups.ub.uni-koeln.de/9019/`.
9. Kim, H.H., Chung, E.T.: A BDDC algorithm with enriched coarse spaces for two-dimensional elliptic problems with oscillatory and high contrast coefficients. Multiscale Model. Simul. **13**(2), 571–593 (2015)
10. Klawonn, A., Kühn, M., Rheinbach, O.: Adaptive coarse spaces for FETI-DP in three dimensions. SIAM J. Sci. Comput. **38**(5), A2880–A2911 (2016)
11. Klawonn, A., Kühn, M., Rheinbach, O.: Adaptive FETI-DP and BDDC methods with a generalized transformation of basis for heterogeneous problems. Electron. Trans. Numer. Anal. **49**, 1–27 (2018)
12. Klawonn, A., Kühn, M., Rheinbach, O.: Coarse spaces for FETI-DP and BDDC methods for heterogeneous problems: connections of deflation and a generalized transformation-of-basis approach. Electron. Trans. Numer. Anal. **52**, 43–76 (2020). `DOI:10.1553/etna_vol52s43`
13. Klawonn, A., Kühn, M.J., Rheinbach, O.: Parallel adaptive feti-dp using lightweight asynchronous dynamic load balancing. International Journal for Numerical Methods in Engineering **121**(4), 621–643 (2020). `DOI:10.1002/nme.6237`. URL `https://onlinelibrary.wiley.com/doi/abs/10.1002/nme.6237`
14. Klawonn, A., Radtke, P., Rheinbach, O.: FETI-DP methods with an adaptive coarse space. SIAM J. Numer. Anal. **53**(1), 297–320 (2015)
15. Knyazev, A.V.: Toward the optimal preconditioned eigensolver: Locally optimal block preconditioned conjugate gradient method. SIAM J. Sci. Comput. **23**(2), 517–541 (2001)
16. Kühn, M.J.: Adaptive FETI-DP and BDDC methods for highly heterogeneous elliptic finite element problems in three dimensions. Ph.D. thesis, Universität zu Köln (2018)
17. Mandel, J., Sousedík, B.: Adaptive selection of face coarse degrees of freedom in the BDDC and the FETI-DP iterative substructuring methods. Comput. Methods Appl. Mech. Engrg. **196**(8), 1389–1399 (2007)
18. Nataf, F., Xiang, H., Dolean, V.: A two level domain decomposition preconditioner based on local Dirichlet-to-Neumann maps. C. R. Math. Acad. Sci. Paris **348**(21-22), 1163–1167 (2010)
19. Oh, D.S., Widlund, O.B., Zampini, S., Dohrmann, C.R.: BDDC Algorithms with deluxe scaling and adaptive selection of primal constraints for Raviart-Thomas vector fields. Math. Comp. **87**(310), 659–692 (2018)
20. Pechstein, C., Dohrmann, C.R.: A unified framework for adaptive BDDC. Electron. Trans. Numer. Anal. **46**, 273–336 (2017)
21. Pechstein, C., Scheichl, R.: Analysis of FETI methods for multiscale PDEs. Numer. Math. **111**(2), 293–333 (2008)
22. Spillane, N., Dolean, V., Hauret, P., Nataf, F., Pechstein, C., Scheichl, R.: Abstract robust coarse spaces for systems of PDEs via generalized eigenproblems in the overlaps. Numer. Math. **126**(4), 741–770 (2014)
23. Spillane, N., Rixen, D.J.: Automatic spectral coarse spaces for robust finite element tearing and interconnecting and balanced domain decomposition algorithms. Internat. J. Numer. Methods Engrg. **95**(11), 953–990 (2013)

A New Parareal Algorithm for Time-Periodic Problems with Discontinuous Inputs

Martin J. Gander, Iryna Kulchytska-Ruchka, and Sebastian Schöps

1 Introduction

Time-periodic problems appear naturally in engineering applications. For instance, the time-periodic steady-state behavior of an electromagnetic device is often the main interest in electrical engineering, because devices are operated most of their life-time in this state. Depending on the size and complexity of the underlying system, the search for a time-periodic solution might however be prohibitively expensive. Special techniques were developed for the efficient computation of such solutions, like the *time-periodic explicit error correction method* [9], which accelerates calculations by correcting the solution after each half period, or the method presented in [1], which leads to faster computations of periodic solutions by determining suitable initial conditions.

The Parareal algorithm was invented in [10] for the parallelization of evolution problems in the time direction. A detailed convergence analysis when applied to linear ordinary and partial differential equations with smooth right-hand sides can be found in [6], for nonlinear problems, see [3]. In [5], a new Parareal algorithm was introduced and analyzed for problems with discontinuous sources. The main idea of the method is to use a smooth approximation of the original signal as the input for the coarse propagator. In [4], a Parareal algorithm for nonlinear time-periodic problems was presented and analyzed. Our interest here is in time-periodic steady-state solutions of problems with quickly-switching discontinuous excitation, for which we will introduce and study a new periodic Parareal algorithm.

Martin J. Gander
Section de Mathématiques, University of Geneva, 2-4 Rue du Lièvre, CH-1211 Geneva, Switzerland, e-mail: martin.gander@unige.ch

Iryna Kulchytska-Ruchka and Sebastian Schöps
Institut für Teilchenbeschleunigung und Elektromagnetische Felder, Technische Universität Darmstadt, Schlossgartenstrasse 8, D-64289 Darmstadt, Germany, e-mail: kulchytska@gsc.tu-darmstadt.de and e-mail: schoeps@temf.tu-darmstadt.de

R. Haynes et al. (eds.), *Domain Decomposition Methods in Science and Engineering XXV*,
Lecture Notes in Computational Science and Engineering 138,
https://doi.org/10.1007/978-3-030-56750-7_27

2 Parareal for time-periodic problems with discontinuous inputs

We consider a time-periodic problem given by a system of ordinary differential equations (ODEs) of the form

$$\boldsymbol{u}'(t) = \boldsymbol{f}(t, \boldsymbol{u}(t)), \quad t \in \mathcal{I}, \qquad \boldsymbol{u}(0) = \boldsymbol{u}(T), \tag{1}$$

with the right-hand side (RHS) $\boldsymbol{f} : \mathcal{I} \times \mathbb{R}^n \to \mathbb{R}^n$ and $\boldsymbol{f}(0, \boldsymbol{v}) = \boldsymbol{f}(T, \boldsymbol{v})$ for all n-dimensional vectors $\boldsymbol{v}$ and the solution $\boldsymbol{u} : \mathcal{I} \to \mathbb{R}^n$ on the time interval $\mathcal{I} := (0, T)$.

In power engineering, electrical devices are often excited with a pulse-width-modulated (PWM) signal [2], which is a discontinuous function with quickly-switching dynamics. For applications such as motors or transformers a couple of tens of kHz might be used as the switching frequency [11]. To solve time-periodic problems of the form (1) supplied with such inputs with our new periodic Parareal algorithm, we assume that the RHS can be split into a sufficiently smooth bounded function $\bar{\boldsymbol{f}}$ and the corresponding discontinuous remainder $\tilde{\boldsymbol{f}}$ as

$$\boldsymbol{f}(t, \boldsymbol{u}(t)) := \bar{\boldsymbol{f}}(t, \boldsymbol{u}(t)) + \tilde{\boldsymbol{f}}(t), \quad t \in \mathcal{I}. \tag{2}$$

Such a decomposition can be achieved by a Fourier series expansion of the time-dependent input, then $\bar{\boldsymbol{f}}$ may be excited by a few principal harmonics, see [5].

We decompose $[0, T]$ into N subintervals $[T_{n-1}, T_n]$, $n = 1, \ldots, N$ with $T_0 = 0$ and $T_N = T$, and introduce the fine propagator $\mathcal{F}(T_n, T_{n-1}, \mathbf{U}_{n-1}^{(k)})$ which computes an accurate solution at time T_n of the initial-value problem (IVP)

$$\boldsymbol{u}_n'(t) = \boldsymbol{f}(t, \boldsymbol{u}_n(t)), \quad t \in (T_{n-1}, T_n], \qquad \boldsymbol{u}_n(T_{n-1}) = \boldsymbol{U}_{n-1}^{(k)}. \tag{3}$$

The corresponding coarse propagator $\bar{\mathcal{G}}(T_n, T_{n-1}, \mathbf{U}_{n-1}^{(k)})$ computes an inexpensive approximation at time T_n of the corresponding IVP having the reduced RHS $\bar{\boldsymbol{f}}(t, \boldsymbol{u}(t))$,

$$\bar{\boldsymbol{u}}_n'(t) = \bar{\boldsymbol{f}}(t, \bar{\boldsymbol{u}}_n(t)), \quad t \in (T_{n-1}, T_n], \qquad \bar{\boldsymbol{u}}_n(T_{n-1}) = \boldsymbol{U}_{n-1}^{(k)}. \tag{4}$$

Our new periodic Parareal algorithm then computes for $k = 0, 1, \ldots$ and $n = 1, \ldots, N$

$$\mathbf{U}_0^{(k+1)} = \mathbf{U}_N^{(k)}, \tag{5}$$

$$\mathbf{U}_n^{(k+1)} = \mathcal{F}(T_n, T_{n-1}, \mathbf{U}_{n-1}^{(k)}) + \bar{\mathcal{G}}(T_n, T_{n-1}, \mathbf{U}_{n-1}^{(k+1)}) - \bar{\mathcal{G}}(T_n, T_{n-1}, \mathbf{U}_{n-1}^{(k)}), \tag{6}$$

until the jumps at the synchronization points T_n, $n = 1, \ldots, N-1$ as well as the periodicity error between $\mathbf{U}_0^{(k)}$ and $\mathbf{U}_N^{(k)}$ are reduced to a given tolerance. The initial guesses $\mathbf{U}_n^{(0)}$, $n = 0, \ldots, N$ for (5)-(6) can be computed by

$$\boldsymbol{U}_n^{(0)} := \bar{\mathcal{G}}(T_n, T_{n-1}, \boldsymbol{U}_{n-1}^{(0)}), \quad n = 1, \ldots, N. \tag{7}$$

We note that the correction (5) does not impose a strict periodicity, but a relaxed one, since the end value at the kth iteration $\mathbf{U}_N^{(k)}$ is used to update the initial approximation $\mathbf{U}_0^{(k+1)}$ at the next iteration. This approach was introduced for time-periodic problems in [4] and was named PP-IC (which stands for *periodic Parareal with initial-value coarse problem*). In contrast to this method, where both coarse and fine propagators solve the IVP (3), our iteration (5)-(6) uses a reduced dynamics on the coarse level, described by (4). Convergence of the PP-IC algorithm was analysed in [4]. We extend this analysis now to the new Parareal iteration (5)-(6), applied to a model problem.

3 Convergence of the new periodic Parareal iteration

We consider the linear time-periodic scalar ODE

$$u'(t) + \kappa u(t) = f(t), \quad t \in (0,T), \qquad u(0) = u(T), \tag{8}$$

with a T-periodic discontinuous RHS $f : [0,T] \to \mathbb{R}$, a constant $\kappa \in \mathbb{R} : \kappa > 0$, and the solution function $u : [0,T] \to \mathbb{R}$ we want to compute.

In order to investigate the convergence of the new periodic Parareal algorithm (5)-(6) applied to (8), we introduce several assumptions. Let the time interval $[0,T]$ be decomposed into subintervals of equal length $\Delta T = T/N$. We assume that the fine propagator $\mathcal{F}$ is exact, and we can thus write the solution of the IVP for the ODE in (8) at T_n, starting from the initial value $U_{n-1}^{(k)}$ at T_{n-1} as

$$\mathcal{F}\left(T_n, T_{n-1}, U_{n-1}^{(k)}\right) = e^{-\kappa \Delta T}\, U_{n-1}^{(k)} + \int_{T_{n-1}}^{T_n} e^{-\kappa(T_n - s)} f(s) ds. \tag{9}$$

Next, introducing a smooth and slowly-varying RHS $\bar{f}$ by $f = \bar{f} + \tilde{f}$, we let the coarse propagator $\bar{\mathcal{G}}$ be a one-step method, applied to

$$\bar{u}_n'(t) + \kappa \bar{u}_n(t) = \bar{f}(t), \quad t \in (T_{n-1}, T_n], \qquad \bar{u}_n(T_{n-1}) = U_{n-1}^{(k)}. \tag{10}$$

Using the stability function $\mathcal{R}\,(\kappa \Delta T)$ of the one-step method, one can then write

$$\bar{\mathcal{G}}\left(T_n, T_{n-1}, U_{n-1}^{(k)}\right) = \mathcal{R}\,(\kappa \Delta T)\, U_{n-1}^{(k)} + \xi_n\left(\bar{f}, \kappa \Delta T\right), \tag{11}$$

where function ξ_n corresponds to the RHS discretized on $[T_n, T_{n-1}]$ with the one-step method. We also assume that

$$|\mathcal{R}\,(\kappa \Delta T)| + \left|e^{-\kappa \Delta T} - \mathcal{R}\,(\kappa \Delta T)\right| < 1. \tag{12}$$

Using (9) and (11) and following [4], the errors $e_n^{(k+1)} := u(T_n) - U_n^{(k+1)}$ of the new periodic Parareal algorithm (5)-(6) applied to the model problem (8) satisfy for

$n = 1, 2, \ldots, N$ the relation

$$\begin{aligned} e_n^{(k+1)} &= u(T_n) - \mathcal{F}(T_n, T_{n-1}, U_{n-1}^{(k)}) - \bar{\mathcal{G}}(T_n, T_{n-1}, U_{n-1}^{(k+1)}) + \bar{\mathcal{G}}(T_n, T_{n-1}, U_{n-1}^{(k)}) \\ &= e^{-\kappa\Delta T} u(T_{n-1}) + \int_{T_{n-1}}^{T_n} e^{-\kappa(T_n - s)} f(s) ds - e^{-\kappa\Delta T} U_{n-1}^{(k)} - \int_{T_{n-1}}^{T_n} e^{-\kappa(T_n - s)} f(s) ds \\ &\quad - \left(\mathcal{R}(\kappa\Delta T)\, U_{n-1}^{(k+1)} + \xi_n\left(\bar{f}, \kappa\Delta T\right)\right) + \left(\mathcal{R}(\kappa\Delta T)\, U_{n-1}^{(k)} + \xi_n\left(\bar{f}, \kappa\Delta T\right)\right) \\ &= \mathcal{R}(\kappa\Delta T)\, e_{n-1}^{(k+1)} + \left(e^{-\kappa\Delta T} - \mathcal{R}(\kappa\Delta T)\right) e_{n-1}^{(k)}. \end{aligned} \tag{13}$$

Similarly, the initial error satisfies $e_0^{(k+1)} = e_N^{(k)}$. A key observation here is that there is no explicit reference to the right-hand sides f or $\bar{f}$ in (13): the corresponding terms cancel both between the exact solution and the (exact) fine solver, and also between the two coarse solvers! Collecting the errors in the error vector $\boldsymbol{e}^{(k)} := \left(e_0^{(k)}, e_1^{(k)}, \ldots, e_N^{(k)}\right)^T$, we obtain from (13) the same fixed-point iteration as in [4]

$$\boldsymbol{e}^{(k+1)} = S\boldsymbol{e}^{(k)}, \tag{14}$$

where the matrix S is given by

$$S = \begin{bmatrix} 1 & & & 0 \\ -\mathcal{R}(\kappa\Delta T) & 1 & & \\ & \ddots & \ddots & \\ & & -\mathcal{R}(\kappa\Delta T) & 1 \end{bmatrix}^{-1} \begin{bmatrix} 0 & & & 1 \\ e^{-\kappa\Delta T} - \mathcal{R}(\kappa\Delta T) & 0 & & \\ & \ddots & \ddots & \\ & & e^{-\kappa\Delta T} - \mathcal{R}(\kappa\Delta T) & 0 \end{bmatrix}. \tag{15}$$

The asymptotic convergence factor of the fixed-point iteration (14) describing our new periodic Parareal algorithm (5)-(6) applied to the periodic problem (8) is therefore given by

$$\rho_{\text{asym}}(S) = \lim_{k\to\infty} \left(\|\boldsymbol{e}^{(k)}\| / \|\boldsymbol{e}^{(0)}\|\right)^{1/k}. \tag{16}$$

Theorem (Convergence estimate of the new periodic Parareal algorithm) Let $[0, T]$ be partitioned into N equal time intervals with $\Delta T = T/N$. Assume the fine propagator to be exact as in (9), and the coarse propagator to be a one-step method as in (11) satisfying (12). Then the asymptotic convergence factor (16) of the new periodic Parareal algorithm (5)-(6) is bounded for all $l \geq 1$ by

$$\rho_{\text{asym}}(S) < x_l, \text{ with } x_l = \left(|\mathcal{R}(\kappa\Delta T)|\, x_{l-1} + \left|e^{-\kappa\Delta T} - \mathcal{R}(\kappa\Delta T)\right|\right)^{\frac{N}{N+1}} \text{ and } x_0 = 1. \tag{17}$$

Proof Since the errors of the new periodic Parareal algorithm satisfy the same relation (14) as in [4], the proof follows by the same arguments as in [4]. □

We note that under the assumption (12), the operator S is a contraction [4], which ensures convergence of the new periodic Parareal algorithm (5)-(6).

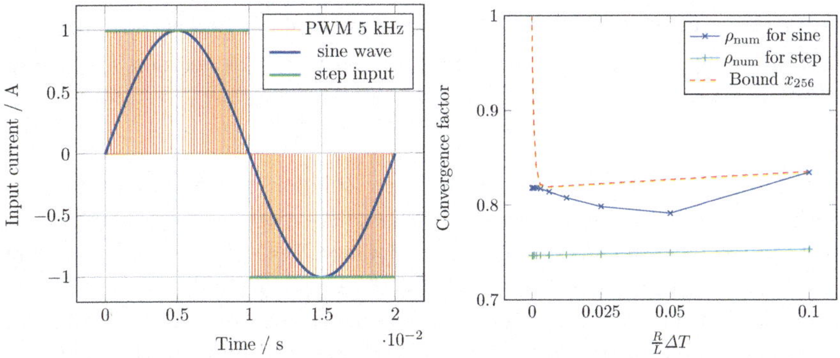

Fig. 1: Left: PWM excitation (19) of 5 kHz ($m = 100$) and two coarse inputs (20), (21). Right: convergence factor of the new periodic Parareal algorithm (5)-(6) with reduced coarse dynamics: sinusoidal waveform (20) and step function (21), together with our theoretical bound.

4 Numerical experiments for a model problem

In this section we illustrate our convergence theory for the new periodic Parareal algorithm with a periodic problem given by an RL-circuit model, namely

$$R^{-1}\phi'(t) + L^{-1}\phi(t) = f_m(t), \quad t \in (0, T), \qquad \phi(0) = \phi(T), \tag{18}$$

with the resistance $R = 0.01\ \Omega$, inductance $L = 0.001$ H, period $T = 0.02$ s, and f_m denoting the supplied PWM current source (in A) of 20 kHz, defined by

$$f_m(t) = \begin{cases} \text{sign}\left[\sin\left(\frac{2\pi}{T}t\right)\right], & s_m(t) - \left|\sin\left(\frac{2\pi}{T}t\right)\right| < 0, \\ 0, & \text{otherwise}, \end{cases} \tag{19}$$

where $s_m(t) = \frac{m}{T}t - \left\lfloor \frac{m}{T}t \right\rfloor$, $t \in [0, T]$ is the common sawtooth pattern with $m = 400$ teeth. An example of the PWM signal of 5 kHz is shown in Fig. 1 on the left. This figure also illustrates the following two choices for the coarse excitation:

$$\bar{f}_{\text{sine}}(t) = \sin\left(\frac{2\pi}{T}t\right), \quad t \in [0, T] \tag{20}$$

$$\bar{f}_{\text{step}}(t) = \begin{cases} 1, & t \in [0, T/2], \\ -1, & t \in (T/2, T]. \end{cases} \tag{21}$$

We note that the step function (21) is discontinuous only at $t = T/2$. This does not lead to any difficulties, since we use Backward Euler for the time discretization and we choose the discontinuity to be located exactly at a synchronization point.

The coarse propagator $\bar{\mathcal{G}}$ then solves an IVP for the equation $R^{-1}\phi'(t)+L^{-1}\phi(t) = \bar{f}(t)$, $t \in (0, T]$, where the RHS $\bar{f}$ is one of the functions in (20) or (21). We illustrate the estimate (17) by calculating the numerical convergence factor $\rho_{\text{num}} := \left(\|e^{(K)}\|/\|e^{(0)}\|\right)^{1/K}$ with the l^∞-norm of the error $e^{(k)}$ at iteration $k \in \{0, K\}$ defined as

$$\|e^{(k)}\| = \max_{0 \le n \le N} |\phi(T_n) - \Phi_n^{(k)}|. \tag{22}$$

Here ϕ denotes the time-periodic steady-state solution of (18) having the same accuracy as the fine propagator, and $\Phi_n^{(K)}$ is the solution obtained at the Kth iteration when (5)-(6) converged up to a prescribed tolerance. The stability function used for x_l in (17) in case of Backward Euler is $\mathcal{R}\left(\frac{R}{L}\Delta T\right) = \left(1 + \frac{R}{L}\Delta T\right)^{-1}$.

On the right in Fig. 1, we show the measured convergence factor of the new Parareal iteration (5)-(6) for the two choices of the coarse excitation (20) and (21). The fine step size is chosen to be $\delta T = T/2^{18} \sim 7.63^{-8}$, while the coarse step varies as $\Delta T = T/2^p$, $p = 1, 2, \ldots, 17$. We also show on the right in Fig. 1 the value of x_{256} to be the bound in (17). The graphs show that the theoretical estimate is indeed an upper bound for the numerical convergence factor for both coarse inputs (sine and step). However, one can observe that x_{256} gives a sharper estimate in the case of the sinusoidal RHS (20), compared to the one defined in (21). We also noticed that the number of iterations required till convergence of (5)-(6) was the same (9 iterations on average for the values of ΔT considered) for both choices of the coarse input, while the initial error $\|e^{(0)}\|$ was bigger with the step coarse input (21) than with the sinusoidal waveform (20). This led to a slightly smaller convergence factor in case of the step coarse input due to the definition of ρ_{num}.

5 Numerical experiments for an induction machine

We now test the performance of our new periodic Parareal algorithm with reduced coarse dynamics for the simulation of a four-pole squirrel-cage induction motor, excited by a three-phase PWM voltage source switching at 20 kHz. The model of this induction machine was introduced in [8]. We consider the no-load condition, when the motor operates with synchronous speed.

The spatial discretization of the two-dimensional cross-section of the machine with $n = 4400$ degrees of freedom leads to a time-periodic problem represented by the system of differential-algebraic equations (DAEs)

$$\mathbf{M}\mathrm{d}_t\mathbf{u}(t) + \mathbf{K}\left(\mathbf{u}(t)\right)\mathbf{u}(t) = \mathbf{f}(t), \quad t \in (0, T), \tag{23}$$

$$\mathbf{u}(0) = \mathbf{u}(T), \tag{24}$$

with unknown $\mathbf{u} : [0, T] \to \mathbb{R}^n$, (singular) mass matrix $\mathbf{M} \in \mathbb{R}^{n \times n}$, nonlinear stiffness matrix $\mathbf{K}(\cdot) : \mathbb{R}^n \to \mathbb{R}^{n \times n}$, and the T-periodic RHS $\mathbf{f} : [0, T] \to \mathbb{R}^n$, $T = 0.02$ s. The three-phase PWM excitation of period T in the stator under the no-load operation causes the T-periodic dynamics in $\mathbf{u}$ which allows the imposition of the periodic constraint (24). For more details regarding the mathematical model we refer to [5]. We would like to note that equation (23) is a DAE of index-1,

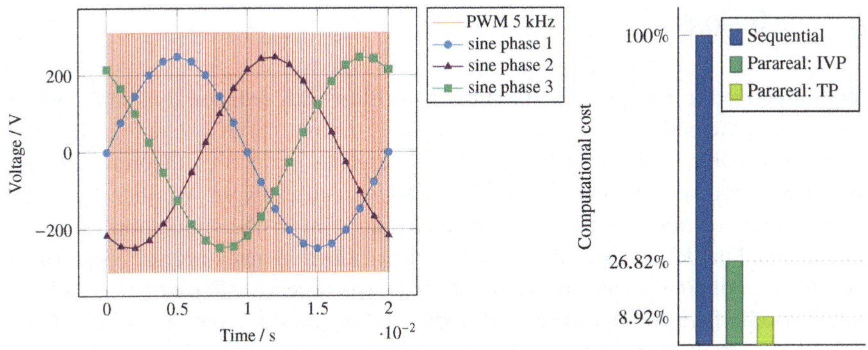

Fig. 2: Left: PWM excitation (19) of 5 kHz and three-phase sinusoidal voltage source of 50 Hz, used as the coarse input in our new periodic Parareal algorithm (5)-(6). Right: comparison of the computational costs calculated in terms of the effective number of linear algebraic systems solved for different approaches to obtain the periodic steady-state solution of the induction machine model.

which in case of discretization with Backward Euler can be treated essentially like an ODE [12].

We now use our new periodic Parareal algorithm (5)-(6) to find the solution of (23)-(24). The fine propagator $\mathcal{F}$ is then applied to (23) with the original three-phase PWM excitation of 20 kHz, discretized with the time step $\delta T = 10^{-6}$ s. The coarse solver $\bar{\mathcal{G}}$ uses a three-phase sinusoidal voltage source with frequency 50 Hz of the form (20), discretized with the time step $\Delta T = 10^{-3}$ s. Phase 1 of the PWM signal switching at 5 kHz as well as the applied periodic coarse excitation on $[0, 0.02]$ s are shown on the left in Fig. 2.

Both coarse and fine propagators solve the IVPs for (23) using Backward Euler, implemented within the GetDP library [7]. We used $N = 20$ time subintervals for the simulation of the induction machine with our new periodic Parareal algorithm, which converged after 14 iterations. Within these calculations 194 038 solutions of linearized systems of equations were performed effectively, i.e, when considering the fine solution cost only on one subinterval (due to parallelization) together with the sequential coarse solves.

On the other hand, a classical way to obtain the periodic steady-state solution is to apply a time integrator sequentially, starting from a zero initial value at $t = 0$. This computation reached the steady state after 9 periods, thereby requiring 2 176 179 linear system solves. Alternatively, one could apply the Parareal algorithm with reduced coarse dynamics, introduced in [5], to the IVP for (23) on $[0, 9T]$. In this case the simulation needed effectively 583 707 sequential linear solutions due to parallelization. However, in practice one would not know the number of periods beforehand and one could not optimally distribute the time intervals. We visualize this data on the right in Fig. 2. These results show that our new periodic Parareal algorithm (5)-(6) with reduced coarse dynamics (Parareal: TP) directly delivers the periodic steady-state solution about 11 times faster than the standard time integration (Sequential), and 3 times faster than the application of Parareal with the reduced dynamics (Parareal: IVP) to an IVP on $[0, 9T]$.

6 Conclusions

We introduced a new periodic Parareal algorithm with reduced dynamics, which is able to efficiently handle quickly-switching discontinuous excitations in time-periodic problems. We investigated its convergence properties theoretically, and illustrated them via application to a linear RL-circuit example. We then tested the performance of our new periodic Parareal algorithm in the simulation of a two-dimensional model of a four-pole squirrel-cage induction machine, and a significant acceleration of convergence to the steady state was observed. In particular, with our new periodic Parareal algorithm with reduced dynamics it is possible to obtain the periodic solution 11 times faster than when performing the classical time stepping.

References

1. Bermúdez, A., Domínguez, O., Gómez, D., Salgado, P.: Finite element approximation of nonlinear transient magnetic problems involving periodic potential drop excitations. Comput. Math. Appl. **65**(8), 1200–1219 (2013). DOI:10.1016/j.camwa.2013.02.019
2. Bose, B.K.: Power Electronics And Motor Drives. Academic Press, Burlington (2006)
3. Gander, M.J., Hairer, E.: Nonlinear convergence analysis for the parareal algorithm. In: Domain decomposition methods in science and engineering XVII, *Lect. Notes Comput. Sci. Eng.*, vol. 60, pp. 45–56. Springer, Berlin (2008). DOI:10.1007/978-3-540-75199-1_4
4. Gander, M.J., Jiang, Y.L., Song, B., Zhang, H.: Analysis of two parareal algorithms for time-periodic problems. SIAM J. Sci. Comput. **35**(5), A2393–A2415 (2013)
5. Gander, M.J., Kulchytska-Ruchka, I., Niyonzima, I., Schöps, S.: A new parareal algorithm for problems with discontinuous sources. SIAM J. Sci. Comput. **41**(2), B375–B395 (2019)
6. Gander, M.J., Vandewalle, S.: Analysis of the parareal time-parallel time-integration method. SIAM J. Sci. Comput. **29**(2), 556–578 (2007). DOI:10.1137/05064607X
7. Geuzaine, C.: GetDP: a general finite-element solver for the de Rham complex. In: PAMM, vol. 7, pp. 1010603–1010604. Wiley (2007). DOI:10.1002/pamm.200700750
8. Gyselinck, J., Vandevelde, L., Melkebeek, J.: Multi-slice FE modeling of electrical machines with skewed slots-the skew discretization error. IEEE Trans. Magn. **37**(5), 3233–3237 (2001)
9. Katagiri, H., Kawase, Y., Yamaguchi, T., Tsuji, T., Shibayama, Y.: Improvement of convergence characteristics for steady-state analysis of motors with simplified singularity decomposition-explicit error correction method. IEEE Trans. Magn. **47**(5), 1458–1461 (2011)
10. Lions, J.L., Maday, Y., Turinici, G.: Résolution d'EDP par un schéma en temps "pararéel". C. R. Acad. Sci. Paris Sér. I Math. **332**(7), 661–668 (2001). DOI:10.1016/S0764-4442(00)01793-6
11. Niyomsatian, K., Vanassche, P., Sabariego, R.V., Gyselinck, J.: Systematic control design for half-bridge converters with LCL output filters through virtual circuit similarity transformations. In: 2017 IEEE Energy Conversion Congress and Exposition (ECCE), pp. 2895–2902 (2017)
12. Schöps, S., Niyonzima, I., Clemens, M.: Parallel-in-time simulation of eddy current problems using parareal. IEEE Trans. Magn. **54**(3), 1–4 (2018). DOI:10.1109/TMAG.2017.2763090

Asymptotic Analysis for Different Partitionings of RLC Transmission Lines

Martin J. Gander, Pratik M. Kumbhar, and Albert E. Ruehli

1 Introduction

Among many applications of parallel computing, solving large systems of ordinary differential equations (ODEs) which arise from large scale electronic circuits, or discretizations of partial differential equations (PDEs), form an important part. A systematic approach to their parallel solution are Waveform Relaxation (WR) techniques, which were introduced in 1982 as a tool for circuit solvers (see [5]). These techniques are based on partitioning large circuits into smaller sub-circuits, which are then solved separately over multiple time steps, and the overall solution is obtained by an iteration between the sub-circuits. However, these techniques can lead to non-uniform and potentially slow convergence over large time windows. To overcome this issue, optimized waveform relaxation techniques (OWR) were introduced, which are based on optimizing parameters. The application of OWR to RC circuits and its asymptotic analysis can be found in [2]. We introduce overlap and analyze these methods for an RLCG transmission line type circuits with $G = 0$, which corresponds to no current loss in the dielectric medium. For the one node overlapping case, see [1]. We show that these circuit equations represent Yee scheme discretizations of the well known Maxwell equations in 1D, and give some asymptotic results.

Martin J. Gander and Pratik M. Kumbhar
Section de Mathématiques, Université de Genève, Switzerland, e-mail: martin.gander@unige.ch, pratik.kumbhar@unige.ch

Albert E. Ruehli
EMC Laboratory, Missouri University of Science And Technology, U.S, e-mail: ruehlia@mst.edu

R. Haynes et al. (eds.), *Domain Decomposition Methods in Science and Engineering XXV*,
Lecture Notes in Computational Science and Engineering 138,
https://doi.org/10.1007/978-3-030-56750-7_28

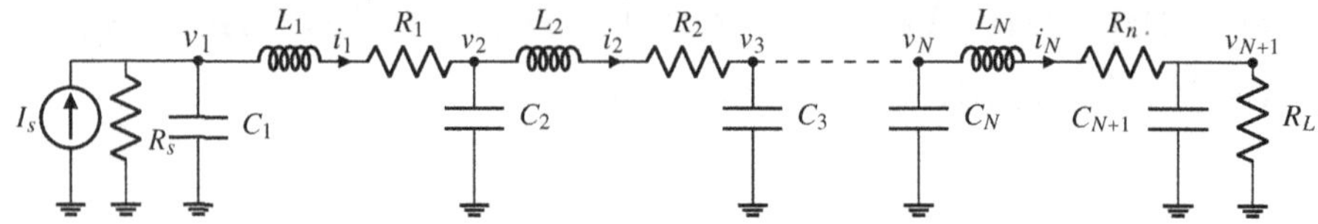

Fig. 1: RLC Transmission Line of length N.

2 Circuit Equations

We consider an infinitely long RLC transmission line where the constants R, L, C represent resistance, inductance, and capacitance per unit length of the line. The circuit equations are specified using Modified Nodal Analysis [4], whose principal element is Kirchhoff's circuit law. These circuit equations for an RLC transmission line (see Fig 1), with the length of the circuit, N, going to infinity, lead to a system of differential equations in time,

$$\dot{\mathbf{x}} = \begin{bmatrix} \ddots & \ddots & \ddots & & & & \\ & a & b & -a & & & \\ & & -c & 0 & c & & \\ & & & a & b & -a & \\ & & & & -c & 0 & c \\ & & & & \ddots & \ddots & \ddots \end{bmatrix} \mathbf{x} + \mathbf{f}, \tag{1}$$

where the entries in the above tridiagonal matrix are

$$a = \frac{1}{L}, \qquad b = -\frac{R}{L}, \qquad c = -\frac{1}{C},$$

with an unknown vector $\mathbf{x}(t) := (\ldots, x_{-1}(t), x_0(t), x_1(t), \ldots)^T$ and $\mathbf{f}(t) = (I_s(t)/C, 0, \ldots, 0)^T$. The unknowns in $\mathbf{x}(t)$ are the voltages $v(t)$ and currents $i(t)$ at the nodes aligned in a systematic way, $x_{2j}(t) = i_j(t)$ and $x_{2j-1} = v_j(t)$ for $j \in \mathbb{Z}$. Thus the even index rows, which have a and b elements correspond to current unknowns while the odd index rows correspond to voltage unknowns. We assume that all the constants R, L, C are bounded to have a well posed problem.

Before analyzing the WR algorithm, we link these circuit equations to the well known Maxwell's equations in 1D. The coupled differential equations of system (1) can be explicitly written as

$$\frac{\partial x_{2m}}{\partial t} = a x_{2m-1} + b x_{2m} - a x_{2m+1} \qquad \text{and} \qquad \frac{\partial x_{2m+1}}{\partial t} = -c x_{2m} + c x_{2m+2},$$

for $m \in \mathbb{Z}$. The parameters R, C, L are defined per unit length, $R = R_T \Delta x$, $C = C_T \Delta x$ and $L = L_T \Delta x$. Hence, substituting the values of the constants a, b, c and interpreting the differences as derivatives, we arrive at

$$\frac{\partial i}{\partial t} + \frac{1}{L_T}\frac{\partial v}{\partial x} = -\frac{R_T}{L_T} i \qquad \text{and} \qquad \frac{\partial v}{\partial t} + \frac{1}{C_T}\frac{\partial i}{\partial x} = 0.$$

Comparing with the Maxwell's equations in 1D,

$$\frac{\partial E}{\partial t} + \frac{1}{\epsilon}\frac{\partial H}{\partial x} = -\frac{\sigma}{\epsilon} E \qquad \text{and} \qquad \frac{\partial H}{\partial t} + \frac{1}{\mu}\frac{\partial E}{\partial x} = 0,$$

we see that $i \sim E$, $v \sim H$, $L_T \sim \epsilon$, $C_T \sim \mu$ and $R_T \sim \sigma$.

3 The Classical WR Algorithm

In this section, we apply the classical waveform relaxation algorithm to an RLC transmission line of infinite length and analyze its convergence. To start with this algorithm, we divide system (1) into two subsystems with unknowns $\mathbf{x}(s_1)$ and $\mathbf{x}(s_2)$, where both unknowns depend on time t but for simplicity, we have removed t from the notation. Since the system (1) consists of two different equations, one for current and the other for voltage, the type of partitioning is interesting. We first partition the system at an odd row, say at $x_{-1}(t)$, and overlap n nodes of the circuit (which corresponds to an overlap of $2n$ nodes of the two subsystems in (2) below). Thus, initially, both the subsystems have equal length and then we increase the size of $\mathbf{x}(s_1)$ by $2n-1$ to include the overlap while the size of $\mathbf{x}(s_2)$ remains unchanged. This leads to two new subsystems of differential equations

$$\begin{aligned}
\dot{\mathbf{x}}^{k+1}(s_1) &= \begin{bmatrix} \ddots & \ddots & \ddots & \\ & a & b & -a \\ & & -c & 0 \end{bmatrix} \mathbf{x}^{k+1}(s_1) + \begin{bmatrix} \vdots \\ f_{2n-4} \\ f_{2n-3} \end{bmatrix} + \begin{bmatrix} \vdots \\ 0 \\ c x_{2n-2}^{k+1}(s_1) \end{bmatrix}, \\
\dot{\mathbf{x}}^{k+1}(s_2) &= \begin{bmatrix} 0 & c & & \\ a & b & -a & \\ & \ddots & \ddots & \ddots \end{bmatrix} \mathbf{x}^{k+1}(s_2) + \begin{bmatrix} f_{-1} \\ f_0 \\ \vdots \end{bmatrix} + \begin{bmatrix} -c x_{-2}^{k+1}(s_2) \\ 0 \\ \vdots \end{bmatrix},
\end{aligned} \tag{2}$$

where k is the iteration index and the unknowns $x_{2n-2}^{k+1}(s_1)$ and $x_{-2}^{k+1}(s_2)$ are given by transmission conditions,

$$x_{2n-2}^{k+1}(s_1) = x_{2n-2}^{k}(s_2), \qquad x_{-2}^{k+1}(s_2) = x_{-2}^{k}(s_1), \tag{3}$$

which exchange only current at the interfaces. For the convergence study, we consider the homogeneous problem $\mathbf{f} = 0$ and zero initial conditions $\mathbf{x}(0) = 0$. The Laplace transform with $s \in \mathbb{C}$ for the subsystems (2) yields

$$
\begin{aligned}
s\hat{\mathbf{x}}^{k+1}(s_1) &= \begin{bmatrix} \ddots & \ddots & \ddots & \\ & a & b & -a \\ & & -c & 0 \end{bmatrix} \begin{bmatrix} \vdots \\ \hat{x}^{k+1}_{2n-4}(s_1) \\ \hat{x}^{k+1}_{2n-3}(s_1) \end{bmatrix} + \begin{bmatrix} \vdots \\ 0 \\ c\hat{x}^{k}_{2n-2}(s_2) \end{bmatrix}, \\
s\hat{\mathbf{x}}^{k+1}(s_2) &= \begin{bmatrix} 0 & c & & \\ a & b & -a & \\ & \ddots & \ddots & \ddots \end{bmatrix} \begin{bmatrix} \hat{x}^{k+1}_{-1}(s_2) \\ \hat{x}^{k+1}_{0}(s_2) \\ \vdots \end{bmatrix} + \begin{bmatrix} -c\hat{x}^{k}_{-2}(s_1) \\ 0 \\ \vdots \end{bmatrix}.
\end{aligned}
\tag{4}
$$

Theorem 1 *The convergence factor of the classical algorithm for an RLC transmission line of infinite length with n nodes overlap is*

$$
\rho_{cla}(s,a,b,c) = \begin{cases} (\lambda_1)^{2n} &, \quad |\lambda_1| < 1, \\ (\lambda_2)^{2n} &, \quad |\lambda_1| > 1, \end{cases} \tag{5}
$$

where $\lambda_{1,2} := \frac{2ac - s(s-b) \pm \sqrt{(2ac - s(s-b))^2 - 4a^2c^2}}{2ac}$ *with the property* $\lambda_1\lambda_2 = 1$.

Proof Solving the first subsystem of (4) corresponds to solving coupled recurrence equations, for $j = n-2, \dots, 0, -1, -2, \dots$

$$
\begin{aligned}
-a\hat{x}^{k+1}_{2j-1}(s_1) + (s-b)\hat{x}^{k+1}_{2j}(s_1) + a\hat{x}^{k+1}_{2j+1}(s_1) &= 0, \\
c\hat{x}^{k+1}_{2j}(s_1) + s\hat{x}^{k+1}_{2j+1}(s_1) - c\hat{x}^{k+1}_{2j+2}(s_1) &= 0.
\end{aligned}
$$

To simplify, we introduce the new notations $\hat{p}^{k+1}_j := \hat{x}^{k+1}_{2j}(s_1)$ and $\hat{q}^{k+1}_j := \hat{x}^{k+1}_{2j+1}(s_1)$ for $j = n-2, \dots, 0, -1, \dots$ to get

$$
-a\hat{q}^{k+1}_{j-1} + (s-b)\hat{p}^{k+1}_j + a\hat{q}^{k+1}_j = 0 \quad \text{and} \quad c\hat{p}^{k+1}_j + s\hat{q}^{k+1}_j - c\hat{p}^{k+1}_{j+1} = 0. \tag{6}
$$

Solving the first equation for $\hat{p}^{k+1}_j$ and substituting it into the second equation yields $ac\hat{q}^{k+1}_{j-1} + [s(s-b) - 2ac]\hat{q}^{k+1}_j + ac\hat{q}^{k+1}_{j+1} = 0$. The general solution of this recurrence equation is

$$
\hat{q}^{k+1}_j = A^{k+1}\lambda_1^j + B^{k+1}\lambda_2^j,
$$

where $\lambda_{1,2} := \frac{2ac - s(s-b) \pm \sqrt{(2ac - s(s-b))^2 - 4a^2c^2}}{2ac}$ are the roots of the characteristic equation and A^{k+1}, B^{k+1} are constants to be determined. We first consider the case $|\lambda_1| < 1$. Since $|\lambda_1^{2j-1}| \to \infty$ as $j \to -\infty$ and $\hat{q}^{k+1}_j$ are bounded, we have $A^{k+1} = 0$. The coupled equations (6) gives $\hat{q}^{k+1}_j = B^{k+1}\lambda_2^j$ and $\hat{p}^{k+1}_j = \frac{aB^{k+1}}{s-b}[\lambda_2^{j-1} - \lambda_2^j]$.

Similarly, from the second subsystem of (4), for $j = 0, 1, 2, \dots$, we define $\hat{u}^{k+1}_j := \hat{x}^{k+1}_{2j}(s_2)$ and $\hat{w}^{k+1}_j := \hat{x}^{k+1}_{2j-1}(s_2)$ to arrive at $\hat{w}^{k+1}_j = D^{k+1}\lambda_1^j$ and $\hat{u}^{k+1}_j = \frac{aD^{k+1}}{s-b}[\lambda_1^j - \lambda_1^{j+1}]$. To determine the constants B^{k+1} and D^{k+1}, we use transmission conditions in (3). The last equation of the first subsystem of (4) gives $c\hat{p}^{k+1}_{n-2} + s\hat{q}^{k+1}_{n-2} = c\hat{u}^{k}_{n-1}$. Using the properties $\lambda_1\lambda_2 = 1$ and $\lambda_1 + \lambda_2 = 2 - \frac{s(s-b)}{ac}$, we have $B^{k+1} = -D^k(\lambda_1^2)^{n-1}$. Similarly, the first equation of the second subsystem of (4) gives $D^{k+1} = -B^k\lambda_1^2$. Therefore, we have $B^{k+1} = (\lambda_1)^{2n}B^{k-1}$ and $D^{k+1} = (\lambda_1)^{2n}D^{k-1}$ which implies

$\hat{x}_j^{k+1}(s_1) = (\lambda_1)^{2n}\hat{x}_j^{k-1}(s_1)$, and $\hat{x}_j^{k+1}(s_2) = (\lambda_1)^{2n}\hat{x}_j^{k-1}(s_2)$. Similarly, we have the same convergence factor when $|\lambda_1| > 1$. □

We observe that the convergence factor is the same for all the nodes irrespective of which subsystem they belong to. Also, the convergence becomes faster by increasing the number of nodes in the overlap. Note also that, if we partition the system at an even row corresponding to a current equation, we still obtain the same convergence factor.

4 Optimized WR Algorithm

It has been observed that increasing the number of nodes in the overlap does not increase the convergence speed very much, especially for large time windows. This forces us to look for better transmission conditions to make the exchange of information between the subsystems more effective. Thus, we propose general transmission conditions for splitting the circuit at a voltage node,

$$\begin{aligned} x^{k+1}_{2n-2}(s_1) + \alpha x^{k+1}_{2n-3}(s_1) &= x^{k}_{2n-2}(s_2) + \alpha x^{k}_{2n-3}(s_2), \\ x^{k+1}_{-1}(s_2) + \beta x^{k+1}_{-2}(s_2) &= x^{k}_{-1}(s_1) + \beta x^{k}_{-2}(s_1), \end{aligned} \tag{7}$$

where α and β are weighting factors. We can have similar transmission conditions for splitting at a current node. These transmission conditions can be viewed as Robin transmission conditions which transfer both current and voltage at the boundary. Under the condition, $\alpha = 0$, and $\beta = \infty$, we recover the classical transmission conditions (3).

Theorem 2 *The convergence factor of the OWR algorithm for an RLC transmission line of infinite length with n nodes overlap and with splitting at a voltage node is given by*

$$\rho_n^v(s,a,b,c,\alpha,\beta) = \begin{cases} \left(\frac{s-\alpha c(\lambda_2-1)}{s+\alpha c(1-\lambda_1)}\right)\left(\frac{\beta s+c(\lambda_2-1)}{\beta s-c(1-\lambda_1)}\right)\left(\lambda_1\right)^{2n}, & |\lambda_1| < 1, \\ \left(\frac{s-\alpha c(\lambda_1-1)}{s+\alpha c(1-\lambda_2)}\right)\left(\frac{\beta s+c(\lambda_1-1)}{\beta s-c(1-\lambda_2)}\right)\left(\lambda_2\right)^{2n}, & |\lambda_1| > 1. \end{cases} \tag{8}$$

Similarly, for the splitting at a current node, the convergence factor $\rho_n^c(s,a,b,c,\alpha,\beta)$ is

$$\rho_n^c(s,a,b,c,\alpha,\beta) = \begin{cases} \left(\frac{s-b+a\alpha(\lambda_2-1)}{s-b-a\alpha(1-\lambda_1)}\right)\left(\frac{\beta(s-b)-a(\lambda_2-1)}{\beta(s-b)+a(1-\lambda_1)}\right)\left(\lambda_1\right)^{2n}, & |\lambda_1| < 1, \\ \left(\frac{s-b+a\alpha(\lambda_1-1)}{s-b-a\alpha(1-\lambda_2)}\right)\left(\frac{\beta(s-b)-a(\lambda_1-1)}{\beta(s-b)+a(1-\lambda_2)}\right)\left(\lambda_2\right)^{2n}, & |\lambda_1| > 1. \end{cases} \tag{9}$$

Proof The proof is similar to the proof of Theorem 1, with the change in the transmission conditions which are now given by the new transmission conditions (7). For $\beta \neq 0$,

$$x_{2n-2}^{k+1}(s_1) = x_{2n-2}^{k}(s_2) + \alpha x_{2n-3}^{k}(s_2) - \alpha x_{2n-3}^{k+1}(s_1),$$

$$x_{-2}^{k+1}(s_2) = x_{-2}^{k}(s_1) + x_{-1}^{k}(s_1)/\beta - x_{-1}^{k+1}(s_2)/\beta.$$

Performing similar operations for both cases, $|\lambda_1| < 1$ and $|\lambda_1| > 1$, we obtain the new convergence factor (8). Similarly for splitting at a current node one can obtain the convergence factor (9). □

The analysis to find optimized α and β for both convergence factors $\rho_n^v(s, a, b, c, \alpha, \beta)$ and $\rho_n^c(s, a, b, c, \alpha, \beta)$ is similar. Hence, in this article we present the analysis for the convergence factor obtained by splitting at a voltage node, i.e for $\rho_n^v(s, a, b, c, \alpha, \beta)$.

Corollary The optimized waveform relaxation algorithm for splitting at a voltage node converges in two iterations, independently of the initial waveforms if

$$\alpha_{opt} := \frac{s}{c(\lambda_2 - 1)} \quad \text{and} \quad \beta_{opt} := \frac{c(1-\lambda_2)}{s}.$$

Proof Equating the convergence factor (8) with zero provides us the expressions for the optimal α and β. □

Note that α_{opt}, β_{opt} are complicated functions of s, which would lead to non-local transmission conditions in time. Hence one searches for the optimized α, β by approximating them by a constant. For this, we solve the min-max problem

$$\min_{\alpha,\beta}\left(\max_{s} |\rho_n^v(s, a, b, c, \alpha, \beta)|\right). \tag{10}$$

By equating the denominator of $\rho_n^v(s, a, b, c, \alpha, \beta)$ with zero, we can show, provided that $\alpha < 0$ and $\beta > 0$, that $\rho_n^v(s, a, b, c, \alpha, \beta)$ is an analytic function in the right half of the complex plane. We also prove that $\rho_n^v(s, a, b, c, \alpha, \beta) \to 0$ as $s \to \infty$. These proofs are technical and will appear in [3]. The maximum principle states that the maximum of $|\rho_n^v(s, a, b, c, \alpha, \beta)|$ lies on the imaginary axis, i.e. $s = i\omega$. Further, $\rho_n^v(s, a, b, c, \alpha, \beta)$ is an even function of ω. From Corollary 1, we observe that $\beta_{opt} = \frac{-1}{\alpha_{opt}}$. This motivates to choose $\beta = \frac{-1}{\alpha}$, which means that the current in both sub-circuits at the point of partition is equal but opposite in direction. All these results and assumptions reduce our optimization problem (10) to

$$\min_{\alpha<0}\left(\max_{\omega_{min}<\omega<\omega_{max}} |\rho_n^v(\omega, a, b, c, \alpha)|\right), \tag{11}$$

where $\omega_{min} := \frac{2\pi}{T}$ and $\omega_{max} := \frac{2\pi}{\Delta t}$ with T as the total time window we are computing and Δt as the time discretization parameter.

Theorem 3 *For splitting at a voltage node, and for small* $\omega_{min} = \epsilon > 0$, *if* $\alpha_v^* = K_v \epsilon^{1/3}$, *where* $K_v = (a^2/(2nb^2c))^{1/3}$, *then the convergence factor* ρ_n^v *satisfies*

$$|\rho_n^v(\omega, a, b, c, \alpha_v^*)| \le |\rho_n^v(\omega_{min}, a, b, c, \alpha_v^*)| \sim 1 + \frac{2\sqrt{2}a\omega_{min}^{1/6}}{K_v\sqrt{bc}} + O(\omega_{min}^{1/2}). \tag{12}$$

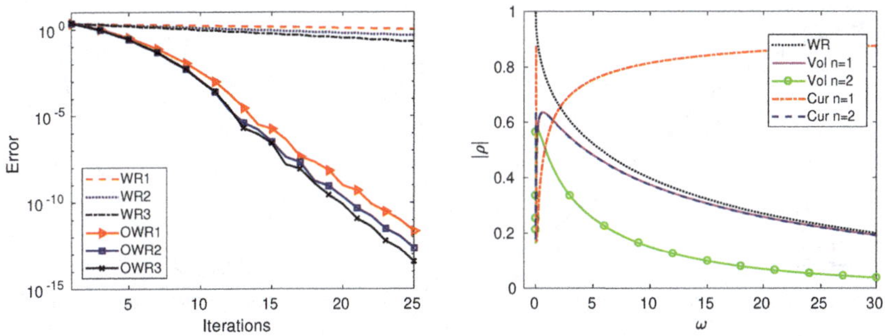

Fig. 2: Convergence for long time $T = 100$ (left) and convergence factor in Laplace space (right).

Similarly, for a splitting at a current node and with $n > 1$, if $\alpha_c^* = K_c\epsilon^{-1/3}$, where $K_c = ((2(n-1)b^2c)/a^2)^{1/3}$, then the convergence factor ρ_n^c satisfies

$$|\rho_n^c(\omega, a, b, c, \alpha_c^*)| \leq |\rho_n^c(\omega_{min}, a, b, c, \alpha_c^*)| \sim 1 + \frac{2\sqrt{2a}K_c\omega_{min}^{1/6}}{\sqrt{bc}} + O(\omega_{min}^{1/2}).$$

Proof We observe numerically (see right plot of Figure 2) that a solution of our min-max problem (11) is given by equioscillation of $|\rho_n^v(\omega, a, b, c, \alpha)|$ for $\omega = \omega_{min}$ and $\omega = \bar{\omega}$ and hence can be found by solving the coupled equations $\rho_n^v(\omega_{min} = \epsilon, a, b, c, \alpha_v^*) = \rho_n^v(\bar{\omega}_v, a, b, c, \alpha_v^*)$ and $\frac{\partial}{\partial\omega}\rho_n^v(\bar{\omega}_v, a, b, c, \alpha_v^*) = 0$, where $\omega_{min} < \bar{\omega}_v \leq \omega_{max}$. Asymptotic calculations for $\epsilon \to 0$, yield $\alpha_v^* \sim K_v\epsilon^{1/3}$ and $\bar{\omega}_v \sim \frac{2K_v c}{n}\epsilon^{1/3}$. Similar calculations yield expressions for α_c^* and $\bar{\omega}_c$. The details of this proof are complicated and too long to present in this short paper and will appear in [3]. □

Theorem 4 *The convergence of OWR is faster for the splitting at a voltage node.*

Proof We substitute the values of K_c and K_v into the expression of $\rho_n^c(\omega_{min} = \epsilon, a, b, c, \alpha_c^*)$ and $\rho_n^v(\omega_{min} = \epsilon, a, b, c, \alpha_v^*)$ respectively to prove $\rho_n^c(\omega_{min} = \epsilon, a, b, c, \alpha_c^*) > \rho_n^v(\omega_{min} = \epsilon, a, b, c, \alpha_v^*)$. The details of this proof will also appear in [3]. □

5 Numerical Results

We consider an RLC transmission line of length $N = 149$ with $R = 2K\Omega/cm$, $L = 4.95 \times 10^{-3}\mu H/cm$ and $C = 0.021pF/cm$. For the time discretization, we use backward Euler with $\Delta t = T/5000$, where T is the total time. We first compare the classical WR and OWR algorithm for large time $T = 100$. The left plot in Figure 2 clearly shows the improvement in the convergence factor when optimized transmission conditions are used. The dashed and dotted lines show the results for

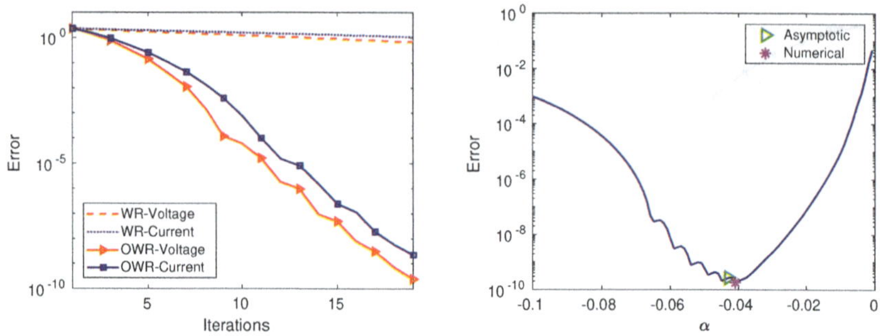

Fig. 3: Comparison of different splittings in time (left) and of values of optimized alpha (right).

classical WR while solid lines represent the OWR algorithm. We also see the effect of overlapping nodes (e.g. WR1 denotes the WR algorithm with one node overlap). Increasing the overlap increases the convergence speed. However, the gain is very small. The right plot of Figure 2 compares the convergence factor for OWR in Laplace space for both splittings, at a current node and a voltage node. The dotted black line is for WR with single node overlap while the other lines are for OWR. For OWR, the splitting at a voltage node leads to faster convergence. This is also true in the time domain, see the left plot of Figure 3. But for classical WR, splitting does not matter, see Theorem 1. Finally, the right plot of Figure 3 validates our asymptotic result (12). Both numerically computed and asymptotically derived values of the optimal α for splitting at a voltage node are very close.

6 Conclusion

This is the first analysis of WR and OWR for an RLC transmission line with overlap and with splitting either at a current or voltage node. We show that using optimized transmission conditions, we can achieve a drastic improvement in the convergence rate. Note that our analysis is in the Laplace domain since the analysis is easier and the convergence in the Laplace domain implies convergence in the time domain, see Remark 1 in [2]. We also see that overlapping nodes increase the convergence rate for both WR and OWR algorithms but the improvement (by the factor of $(\lambda_1)^{2n}$) is not large. Further, for OWR, the splitting at a voltage node leads to a little faster convergence than the splitting at a current node, while this splitting does not effect the convergence of WR. We finally compared the values of the optimized α found numerically and by asymptotic analysis and they are very close.

References

1. Gander, M.J., Al-Khaleel, M., Ruehli, A.E.: Optimized waveform relaxation methods for longitudinal partitioning of transmission lines. IEEE Trans. Circuits Syst. I. Regul. Pap. **56**(8), 1732–1743 (2009). DOI:10.1109/TCSI.2008.2008286
2. Gander, M.J., Kumbhar, P.M., Ruehli, A.E.: Analysis of overlap in waveform relaxation methods for rc circuits. In: Domain Decomposition Methods in Science and Engineering XXIV, pp. 281–289. Springer International Publishing, Cham (2018)
3. Gander, M.J., Kumbhar, P.M., Ruehli, A.E.: Optimized waveform relaxation methods applied to RLCG transmission line and their asymptotic analysis. In Preparation (2019)
4. Ho, C.W., Ruehli, A., Brennan, P.: The modified nodal approach to network analysis. IEEE Transactions on Circuits and Systems **22**(6), 504–509 (1975)
5. Lelarasmee, E., Ruehli, A.E., Sangiovanni-Vincentelli, A.L.: The waveform relaxation method for time-domain analysis of large scale integrated circuits. IEEE Transactions on Computer-Aided Design of Integrated Circuits and Systems **1**(3), 131–145 (1982)
6. Menkad, T., Dounavis, A.: Resistive coupling-based waveform relaxation algorithm for analysis of interconnect circuits. IEEE Transactions on Circuits and Systems I: Regular Papers **64**(7), 1877–1890 (2017)

Optimized Schwarz-based Nonlinear Preconditioning for Elliptic PDEs

Yaguang Gu and Felix Kwok

1 Introduction

In this paper, we consider the following nonlinear elliptic equation,

$$\begin{cases} \eta u - \nabla \cdot (a(x, u, \nabla u)\nabla u) = f \text{ in } \Omega, \\ \mathcal{B}u = h \text{ on } \partial\Omega, \end{cases} \tag{1}$$

where $\eta \geq 0$, $a(x, u, \nabla u)$ is a positive scalar function uniformly bounded away from zero, and $\mathcal{B}u$ represents boundary conditions (e.g. Dirichlet or Neumann) such that the problem is well posed. This type of equation often arises from the implicit discretization of a time-dependent problem or from a steady state calculation, for example the Forchheimer equation [5] in porous media flow.

Once the problem (1) is discretized, there are many ways to solve the large nonlinear algebraic problem by domain decomposition methods. A classical approach is to use the Newton-Krylov-Schwarz method [1]: the problem is first attacked by Newton's method, and within each Newton iteration, the linearized problem is solved using a Krylov method with a Schwarz domain decomposition preconditioner. Alternative approaches consist of applying these components in a different order. One such possibility, known as the Nested Iteration approach, was formulated in [7, 8] for nonlinear parabolic PDEs: the solution (in space and time) is first rewritten as the fixed point of a parallel Schwarz waveform relaxation iteration. Next, using the interface values as primary unknowns, one derives the fixed point equation, which is then solved using Newton's method. Within each Newton iteraiton, the Jacobian

Yaguang Gu
Department of Mathematics, Hong Kong Baptist University, Kowloon Tong, Hong Kong, e-mail: 16482980@life.hkbu.edu.hk

Felix Kwok
Department of Mathematics, Hong Kong Baptist University, Kowloon Tong, Hong Kong e-mail: felix_kwok@hkbu.edu.hk

R. Haynes et al. (eds.), *Domain Decomposition Methods in Science and Engineering XXV*,
Lecture Notes in Computational Science and Engineering 138,
https://doi.org/10.1007/978-3-030-56750-7_29

systems are solved by a Krylov method, where each matrix-vector multiplication corresponds to the solution of a linear parabolic problem.

For elliptic problems, the authors of [3] introduced the Restricted Additive Schwarz Preconditioned Exact Newton (RASPEN) method, which can be regarded as the Newton-accelerated version of the Restricted Additive Schwarz (RAS) method with classical (Dirichlet) transmission conditions. Similar to Nested Iteration, the RAS method is first written in fixed point form, and the resulting fixed point equation is solved by Newton's method, using a Krylov method as linear solver for calculating the Newton step. Unlike the ASPIN method [2], which uses approximate Jacobians, RASPEN uses exact Jacobians: it was shown in [3] that the product of the exact Jacobian matrix with an arbitrary vector can be obtained using components already computed during the subdomain solves, so the Newton corrections can be calculated cheaply. Thus, RASPEN is a true Newton method and converges quadratically close to the solution. Nonetheless, the Krylov solver within each Newton iteration converges relatively slowly, which is typical of classical RAS methods with Dirichlet transmission conditions. In this paper, we propose an optimized RASPEN (ORASPEN) method, where a zeroth order optimized (i.e. Robin) transmission condition is used to communicate information across subdomain interfaces. This allows us to take advantage of the extra Robin parameter to obtain faster convergence in the Krylov solver, just like in optimized Schwarz methods for linear problems.

2 The ORASPEN method

In this section, we derive the ORASPEN method and explain how the matrix-vector multiplication by the Jacobian can be performed by reusing components from the subdomain solves. We first recall the RASPEN method with classical transmisison conditions, as defined in [3]. Assume that the physical domain Ω is decomposed into overlapping subdomains $\Omega = \bigcup_{i=1}^{K} \Omega_i$. Then given the n-th iterate u^n, the restricted Additive Schwarz (RAS) method first calculates $u_i^{n+1} = G_i(u^n), \quad i = 1, 2, \ldots, K$, where G_i is the local solution operator which produces solutions to local subdomain problems by freezing degrees of freedom outside Ω_i. More concretely, suppose we use a finite element discretization of (1) to obtain for the ℓ-th degree of freedom

$$F_\ell(u) = \int_\Omega (\eta u \phi_\ell + a(x, u, \nabla u)\nabla u \cdot \nabla \phi_\ell)\, dx - \int_\Omega f \phi_\ell\, dx = 0, \tag{2}$$

where $\phi_\ell \in H_0^1(\Omega)$ denotes the ℓ-th finite element basis function. Let $F(u) = (F_1(u), F_2(u), \ldots)^T$ be the set of all such equations, so that the global nonlinear problem has the form $F(u) = 0$. If u^n is a finite element function whose trace on $\partial\Omega_i$ is used as Dirichlet values for the subdomain solve on Ω_i, then the subdomain solution $u_i^{n+1} = G_i(u^n) \in V_i$ can be obtained by solving the equation

$$R_i F(P_i G_i(u^n) + (I - P_i R_i)u^n) = 0 \tag{3}$$

for the unknown $G_i(u^n)$, where R_i is the restriction operator from the finite element space $V \subset H_0^1(\Omega)$ to the subspace $V_i = V \cap H_0^1(\Omega_i)$, and $P_i = R_i^T$ is the prolongation operator. Note that the subdomain solution u_i^{n+1} is none other than the solution to the following problem when we apply the parallel classical Schwarz method for (1):

$$\begin{cases} \eta u_i^{n+1} - \nabla \cdot (a(x, u_i^{n+1}, \nabla u_i^{n+1}) \nabla u_i^{n+1}) = f & \text{in } \Omega_i, \\ \mathcal{B} u_i^{n+1} = h & \text{on } \partial\Omega_i \cap \partial\Omega, \\ u_i^{n+1} = u_j^n & \text{on } \partial\Omega_i \cap \bar{\Omega}_j,\ j \in \mathcal{I}_i, \end{cases} \tag{4}$$

where $\mathcal{I}_i$ contains the indices of all the subdomains that have overlap with Ω_i.

Once the $G_i(u^n)$ are calculated for each i, the new global iterate is formed using the relation

$$u^{n+1} = \sum_{i=1}^{K} \tilde{P}_i G_i(u^n). \tag{5}$$

Here, $\tilde{P}_i$ is the *restricted* prolongation operator, formed from the P_i above and a partition of unity, so that the relation $\sum_{i=1}^K \tilde{P}_i R_i = I$ holds; see detailed definitions in [3]. When the iteration (5) converges, it does so linearly in general. The RASPEN idea consists of forming the fixed point equation

$$\tilde{\mathcal{F}}(u) = \sum_{i=1}^{K} \tilde{P}_i G_i(u) - u = 0 \tag{6}$$

and applying Newton's method to solve (6). This requires calculating the Jacobian $\tilde{\mathcal{F}}'(u)$, which in turn requires the derivative $G_i'(u)$. The latter can be obtained by differentiating (3).

We now derive the ORASPEN algorithm by showing how to incorporate optimized transmission conditions. In the ORASPEN algorithm, we still solve (6) by Newton, except that the underlying fixed point iteration (5) is replaced by the optimized RAS method of [9], so the local solution operator $G_i(u)$ is now based on Robin transmission conditions rather than Dirichlet.

Let $u_i^* = R_i u^*$ be the restriction of u^* to Ω_i, u^* being the solution to (1). Given a set of initial guesses $(u_i^0)_{i=1}^K$, the parallel optimized Schwarz method generates a sequence $(u_i^n)_{i=1}^K$, $n = 0, 1, \ldots$, that approximate $(u_i)_{i=1}^K$ by

$$\begin{cases} \eta u_i^{n+1} - \nabla \cdot (a(x, u_i^{n+1}, \nabla u_i^{n+1}) \nabla u_i^{n+1}) = f & \text{in } \Omega_i, \\ \mathcal{B} u_i^{n+1} = h & \text{on } \partial\Omega_i \cap \partial\Omega, \\ a(x, u_i^{n+1}, \nabla u_i^{n+1}) \frac{\partial u_i^{n+1}}{\partial \mathbf{n}_i} + p u_i^{n+1} = a(x, u_j^n, \nabla u_j^n) \frac{\partial u_j^n}{\partial \mathbf{n}_i} + p u_j^n & \text{on } \partial\Omega_i \cap \bar{\Omega}_j,\ j \in \mathcal{I}_i, \end{cases} \tag{7}$$

where p is the Robin parameter and $\mathbf{n}_i$ is the unit outward-pointing normal vector. If finite elements are used to discretize (7), then for each basis function ϕ_ℓ^i with support in Ω_i, the corresponding residual function becomes

$$F_\ell^i(u_i^{n+1}) = A_\ell^i(u_i^{n+1}) - \int_{\Omega_i} f \phi_\ell^i \, dx - \int_{\Gamma_i} g \phi_\ell^i \, ds, \tag{8}$$

where $g = \big(a(u_j^n, \nabla u_j^n)\frac{\partial}{\partial \mathbf{n}_i} + p\big)(u_j^n)$, $\Gamma_i = \partial\Omega_i \setminus \partial\Omega$, and

$$A_\ell^i(u_i) = \int_{\Omega_i} (\eta u_i \phi_\ell^i + a(x, u_i, \nabla u_i)\nabla u_i \cdot \nabla \phi_\ell^i)\, dx + \int_{\Gamma_i} p u_i \phi_\ell^i\, ds.$$

The evaluation of g, which involves Robin traces and must be taken in the weak sense, is non-trivial. Therefore, we mimic the approach in [4] for the linear case and exploit the equivalence between optimized parallel Schwarz and optimized RAS: we update the local solution via the full approximation scheme

$$A^i(u_i^{n+1}) - A^i(R_i u^n) = -R_i F(u^n), \tag{9}$$

where $A^i(u_i) = (A_1^i(u_i), A_2^i(u_i), \ldots)^T$, and $F(u^n) = (F_1(u^n), F_2(u^n), \ldots)^T$ is the global residual as defined in (2). Under the usual coercivity assumptions, (9) defines a mapping $G_i : u^n \mapsto u_i^{n+1}$. The fixed point iteration is completed by the update formula $u^{n+1} = \sum_{i=1}^K \tilde{P}_i G_i(u^n)$, as in (5). It is clear from (9) that if $u^n = u^*$ is the exact solution of $F(u) = 0$, then $u_i^{n+1} = R_i u^n$, so the exact solution is a fixed point of the iteration. Thus, the ORASPEN approach consists of solving (6), but with the G_i now defined by (9) instead of (3).

To calculate the Newton steps necessary for the solution of (6), one must solve linear systems involving the Jacobian matrix $\tilde{\mathcal{F}}'(u)$. Since (O)RASPEN uses Krylov methods for solving such linear systems, we need to know how to multiply $\tilde{\mathcal{F}}'(u)$ by an arbitrary vector v. Differentiating (6) with respect to u and multiplying the result by v gives

$$\tilde{\mathcal{F}}'(u^n)v = \sum_{i=1}^K \tilde{P}_i G_i'(u^n)v - v. \tag{10}$$

To evaluate $G_i'(u^n)v$, we let $u_i^{n+1} = G_i(u^n)$ in (9) and differentiate implicitly to obtain

$$\frac{\partial A^i}{\partial u}(u_i^{n+1})G_i'(u^n) - \frac{\partial A^i}{\partial u}(R_i u^n)R_i = -R_i F'(u^n).$$

Isolating $G_i'(u^n)$ in the above and substituting into (10) yields

$$\tilde{\mathcal{F}}'(u^n)v = \sum_{i=1}^K \tilde{P}_i \left(\frac{\partial A^i}{\partial u}(u_i^{n+1})\right)^{-1} \left(\frac{\partial A^i}{\partial u}(R_i u^n)R_i v - R_i F'(u^n)v\right) - v. \tag{11}$$

Note that $\frac{\partial A^i}{\partial u}(u_i^{n+1})$ is none other than the Jacobian matrix for the subdomain problem (9). If Newton's method was used to solve these subdomain problems, this Jacobian would have already been formed and factored during the calculation of u_i^{n+1}, so the multiplication by $\big(\frac{\partial A^i}{\partial u}(u_i^{n+1})\big)^{-1}$ in (11) requires only a forward-backward substitution involving the precomputed LU factors. Thus, the Krylov iterations have relatively low computational cost.

To understand the convergence of the Krylov method, it is instructive to consider the linear case, when $a(x, u, \nabla u) \equiv a(x)$ is independent of u, and $\frac{\partial A^i}{\partial u} =: J_i$ is independent of u_i^{n+1}. In that case, (11) simplifies to

$$\tilde{\mathcal{F}}'(u^n)v = -\sum_{i=1}^{K} \tilde{P}_i J_i^{-1} R_i F'(u^n)v,$$

which is identical to the preconditioned matrix for optimized RAS [9]. Therefore, for well-chosen Robin parameters, we expect ORASPEN to exhibit much faster convergence than classical RASPEN in terms of inner Krylov iterations, even when the number of outer Newton iterations remains similar. This will be verified experimentally in the next section.

3 Numerical Results

In this section, we illustrate the behaviour of ORASPEN by comparing it with classical one-level RASPEN, as defined in [3], for two model problems. All tests in this section are discretized using the P1 (conforming piecewise linear) finite element method. In the first test, we show results for the nonlinear diffusion problem

$$\begin{cases} -\nabla \cdot ((1+u^2)\nabla u) = x\sin(y) & \text{in } \Omega = [0,1]\times[0,1], \\ u = 1 & \text{on } x = 1, \\ \frac{\partial u}{\partial n} = 0 & \text{elsewhere}, \end{cases} \tag{12}$$

with the initial guess $u^0 = 1$.

We compare the linear and nonlinear iteration counts needed by ORASPEN with those needed by RASPEN for the 4×4 subdomain test case, using different Robin parameters p and mesh ratios H/h. In Table 1, we report the following numbers:

- **Nits**, the number of outer Newton iterations required for convergence to within a tolerance of 10^{-8};
- **Lits**, the number of linearized subdomain problems that must be solved. This number includes (i) all linear solves within the subdomain problems, and (ii) all multiplications by the matrix $(\partial A^i/\partial u)^{-1}$ within GMRES due to Equation (11);
- **Avg Lits**, the average number of linear iterations per Neweton step; and
- p, the Robin parameter that leads to the lowest iteration counts for each mesh ratio H/h.

We also include the number of unpreconditioned classical Newton iterations required for convergence. Although one cannot use these numbers to directly compare classical Newton with (O)RASPEN (we must also consider which preconditioner to use, and how many preconditioned GMRES iterations are required by the Jacobian solves within each Newton step), such numbers are useful for determining the difficulty of the unpreconditioned problem. Here, we observe that ORASPEN always requires the fewest nonlinear iterations, compared to classical Newton and RASPEN.

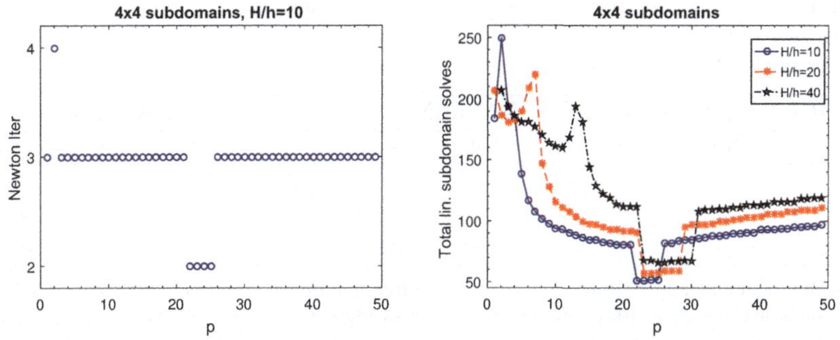

Fig. 1: Numerical results for the nonlinear diffusion problem. Left: Newton iteration counts for 4×4 subdomains with $H/h = 10$, H being the diameter of the subdomain. Right: Total linear iteration counts for the 4×4 subdomain case with different Robin parameters and mesh sizes.

We show in Figure 1 the linear and nonlinear iteration counts as a function of the Robin parameter p for different ratios H/h. We can see that the Robin parameter p has a large impact on the linear and nonlinear iteration counts. Observe that when ORASPEN is used with the optimal Robin parameter, the average number of linear iterations per Newton becomes much lower for ORASPEN than for RASPEN, and this number does not grow as quickly as for RASPEN when we refine the mesh. Moreover, with the optimal Robin parameter p, ORASPEN converges in two Newton iterations, which is slightly better than the three iterations required by RASPEN. Note that these extra savings are tolerance dependent: if we change TOL in the stopping criterion from 10^{-8} to 10^{-10}, then it will take at least three Newton iterations for ORASPEN to converge. Nevertheless, since ORASPEN needs fewer linear iterations per Newton step than RASPEN, ORASPEN will still outperform RASPEN, even when both methods take three Newton iterations to converge.

Next, we fix the ratio H/h and vary the number of subdomains; see the results in Table 2. We observe that ORASPEN again requires fewer linear iterations to converge than RASPEN, which is consistent with the linear case [6]. However, the iteration counts for both methods grow with the number of subdomains, as the inner subdomains move farther and farther away from the physical boundary.

For the second set of tests, we consider the same Forchheimer problem as in [3]:

$$\begin{cases} -\nabla \cdot \mathbf{q} = 0 & \text{in } \Omega = [0,1] \times [0,1], \\ \mathbf{q} \cdot \mathbf{n} = 0 & \text{on } \partial\Omega \setminus (\Gamma_{d0} \cup \Gamma_{d1}), \\ u = 0 \text{ on } \Gamma_{d0}, & u = 1 \quad \text{on } \Gamma_{d1}, \end{cases} \tag{13}$$

where

$$\mathbf{q} = \frac{2\Lambda(x,y)\nabla u}{1 + \sqrt{1 + 4\beta|\Lambda(x,y)\nabla u|}},$$

$\Gamma_{d0} = \{(x,y) \in \partial\Omega; x + y < 0.2\}$ and $\Gamma_{d1} = \{(x,y) \in \partial\Omega; x + y > 1.8\}$. The permeability $\Lambda(x,y)$ is equal to 1000 except in the two inclusions $[0,0.5] \times [0.2,0.4]$ and $[0.5,1] \times [0.6,0.8]$, where it is equal to 1. The nonlinearity of the Forchheimer

Table 1: Linear and nonlinear iteration counts for 4×4 subdomains for the nonlinear diffusion problem with different mesh sizes. An overlap of 4-cell widths and a stopping criterion of 10^{-8} are used for all tests.

H/h	Classical Newton	RASPEN			ORASPEN			
	Nits	Nits	Lits	Avg Lits	p	Nits	Lits	Avg Lits
10	4	3	113	37.67	22	2	51	25.50
20	4	3	152	50.67	23	2	57	28.50
40	4	3	208	69.33	25	2	66	33.00

Table 2: Linear and nonlinear iteration counts for different numbers of subdomains for the nonlinear diffusion problem with the fixed ratio $H/h = 10$. An overlap of 4-cell widths and a stopping criterion of 10^{-8} are used for all tests.

$N \times N$	Classical Newton	RASPEN			ORASPEN			
	Nits	Nits	Lits	Avg Lits	p	Nits	Lits	Avg Lits
2×2	4	3	59	19.67	12	2	34	17.00
4×4	4	3	113	37.67	22	2	51	25.50
8×8	4	3	211	70.33	28	3	133	44.33
16×16	4	3	418	139.33	31	3	247	82.33

equation is much stronger than in the first test problem, due to the appearance of ∇u in the denominator of $\mathbf{q}$ and the large contrast in $\Lambda(x, y)$. Therefore, we adopt the continuation approach, where we solve (13) first for $\beta = 0$ (which is a linear problem), then for $\beta = 0.1$ and $\beta = 1$, using the solution for the previous β as the initial guess for the next one. (Without continuation, classical Newton takes 15–20 iterations to converge, whereas (O)RASPEN takes only 4–8 in our examples.) For the *fixed* fine mesh shown in Figure 2, we vary the number of subdomains and show the iteration counts for ORASPEN and RASPEN in Table 3. We again observe significantly lower linear iteration counts in ORASPEN than in classical RASPEN. Finally, we remark that the performance of ORASPEN is sensitive to the Robin parameter p, as can be seen from Figure 3. A poor choice of the Robin parameter may lead to a higher number of nonlinear iterations compared to classical RASPEN, negating the benefits of faster linear convergence. A good parameter choice for this problem, and more generally for ORASPEN, is therefore the subject of ongoing work.

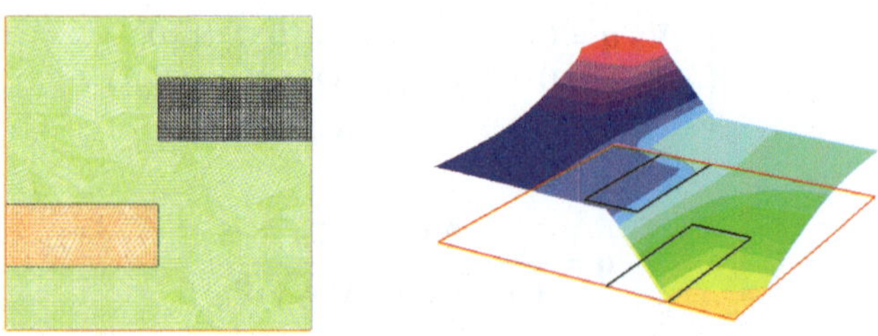

Fig. 2: Fixed grid for the 2×2, 4×4 and 8×8 subdomain test cases, and the solution profile for the Forchheimer problem.

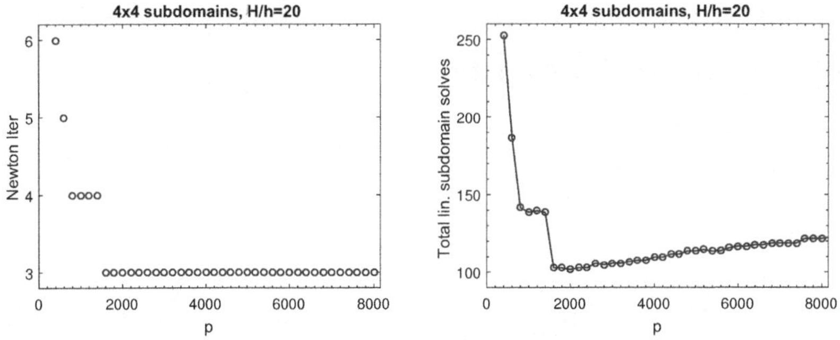

Fig. 3: Newton and linear iteration counts for the Forchheimer problem as a function of p.

Table 3: Linear and nonlinear iteration counts for the Forchheimer problem, with a continuation sequence of $\beta = 0, 0.1, 1$. An overlap of 4-cell widths and a stopping criterion of 10^{-8} are used for all tests.

β	$N \times N$	Classical Newton	RASPEN			ORASPEN			
		Nits	Nits	Lits	Avg Lits	p	Nits	Lits	Avg Lits
0.1	2×2	5	3	112	37.33	1800	3	70	23.33
	4×4	5	3	158	52.67	2000	3	103	34.33
	8×8	5	3	236	78.67	2200	4	218	54.50
1.0	2×2	4	3	109	36.33	200	2	43	21.50
	4×4	4	2	101	50.50	950	2	69	34.50
	8×8	4	3	232	77.33	1050	3	161	53.67

Acknowledgements This work is partially supported by the Hong Kong Research Grants Council (ECS-22300115) and the National Natural Science Foundation of China (RFYS-11501483).

References

1. Cai, X.C., Gropp, W.D., Keyes, D.E., Tidriri, M.D.: Newton-Krylov-Schwarz methods in CFD. In: Numerical methods for the Navier-Stokes equations, pp. 17–30. Springer (1994)
2. Cai, X.C., Keyes, D.E.: Nonlinearly preconditioned inexact Newton algorithms. SIAM J. Sci. Comput. **24**(1), 183–200 (2002)
3. Dolean, V., Gander, M.J., Kheriji, W., Kwok, F., Masson, R.: Nonlinear preconditioning: How to use a nonlinear Schwarz method to precondition Newton's method. SIAM J. Sci. Comput. **38**(6), A3357–A3380 (2016)
4. Dolean, V., Jolivet, P., Nataf, F.: An Introduction to Domain Decomposition Methods: Algorithms, Theory, and Parallel Implementation, vol. 144. SIAM (2015)
5. Forchheimer, P.: Wasserbewegung durch Boden. Z. Ver. Deutsch, Ing. **45**, 1782–1788 (1901)
6. Gander, M.J.: Optimized Schwarz methods. SIAM J. Numer. Anal. **44**(2), 699–731 (2006)
7. Haeberlein, F.: Time Space Domain Decomposition Methods for Reactive Transport — Application to CO2 Geological Storage. Ph.D. thesis, Université Paris-Nord - Paris XIII (2011). URL `https://tel.archives-ouvertes.fr/tel-00634507`
8. Haeberlein, F., Halpern, L., Michel, A.: Newton-Schwarz optimised waveform relaxation Krylov accelerators for nonlinear reactive transport. In: Domain decomposition methods in science and engineering XX, pp. 387–394. Springer (2013)
9. St-Cyr, A., Gander, M.J., Thomas, S.J.: Optimized multiplicative, additive, and restricted additive Schwarz preconditioning. SIAM J. Sci. Comput. **29**(6), 2402–2425 (2007)

Coarse Spaces for Nonlinear Schwarz Methods on Unstructured Grids

Alexander Heinlein and Martin Lanser

1 Introduction

We are concerned with the solution of nonlinear problems

$$F(u) = 0 \tag{1}$$

in some finite element space V. The function $F : V \longrightarrow V'$ is obtained by a finite element discretization of a nonlinear partial differential equation (PDE) on a domain $\Omega \subset \mathbb{R}^d$, $d = 2, 3$. To solve (1), we consider nonlinear domain decomposition methods of the Schwarz type, e.g., ASPIN (Additive Schwarz Preconditioned Inexact Newton) [1, 10] or RASPEN (Restricted Additive Schwarz Preconditioned Exact Newton) [3]. More precisely, we suggest a new approach to implement a second level or coarse level into RASPEN, which is different to FAS-RASPEN (Full Approximation Scheme - RASPEN) introduced in [3]. The coarse space is applied multiplicatively, similar to the application of multiplicative nonlinear corrections in MSPIN (Multiplicative Schwarz Preconditioned Inexact Newton); see, e.g., [9]. Therefore, we consider a standard Lagrangian coarse space as well as multiscale coarse spaces that can also be constructed for unstructured meshes and unstructured domain decompositions, e.g., decompositions obtained using METIS [8]. We compare our new approaches for the example of homogeneous and heterogeneous p-Laplace equations; see section 2. In section 3, we first describe the one level RASPEN method and our approach to implement a multiplicative second level for ASPIN and RASPEN. Second, we define three different coarse spaces - one based on a $\mathbb{P}1$ discretization on a coarse mesh and the other two based on MsFEM (Multiscale

Alexander Heinlein and Martin Lanser
Department of Mathematics and Computer Science, University of Cologne, Weyertal 86-90, 50931 Köln, Germany, e-mail: alexander.heinlein@uni-koeln.de, martin.lanser@uni-koeln.de, url: http://www.numerik.uni-koeln.de
Center for Data and Simulation Science, University of Cologne, Germany, url: http://www.cds.uni-koeln.de

R. Haynes et al. (eds.), *Domain Decomposition Methods in Science and Engineering XXV*, Lecture Notes in Computational Science and Engineering 138,
https://doi.org/10.1007/978-3-030-56750-7_30

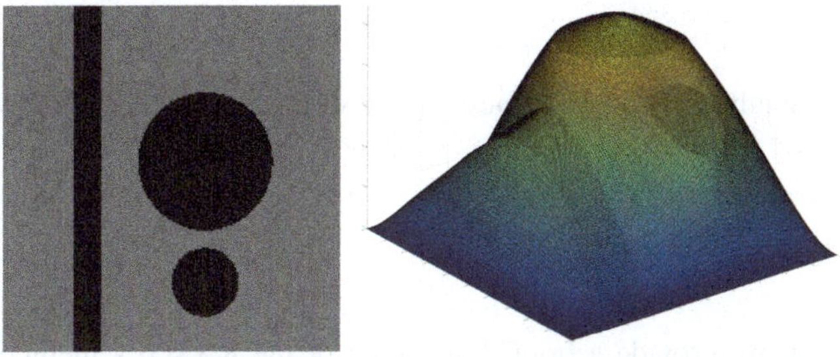

Fig. 1: Left: Definition of Ω_R (black part); **Right:** Solution of equation (2) with coefficient functions defined in (3).

Finite Element Method) [7] type discretizations on the subdomains. The MsFEM coarse spaces can easily be used in the case of unstructured decompositions and differ only in the chosen extensions from the interface to the interior parts of the nonoverlapping domain decomposition. Finally, we present numerical results considering homogeneous and heterogeneous model problems in section 4.

2 Model Problem

We consider the nonlinear model problem:

$$\begin{aligned} -\alpha\Delta_p u - \beta\Delta_2 u &= 1 \qquad \text{in } \Omega \\ u &= 0 \qquad \text{on } \partial\Omega, \end{aligned} \tag{2}$$

with the scaled p-Laplace operator $\alpha\Delta_p u := \operatorname{div}(\alpha|\nabla u|^{p-2}\nabla u)$ for $p \geq 2$ and the coefficient functions $\alpha, \beta : \Omega \to \mathbb{R}$. For all computations in this paper, we always use the unit square $\Omega = [0,1] \times [0,1]$ as the computational domain. However, our approach is not restricted to this case. We consider two different coefficient distributions: a homogeneous p-Laplace equation, i.e., $\alpha(x) = 1$ and $\beta(x) = 0$ for all $x \in \Omega$, and a heterogeneous problem with a channel and two circular inclusions carrying different coefficients than the remainder of Ω, i.e.,

$$\alpha(x) = \begin{cases} 1\,000 & \text{if } x \in \Omega_R, \\ 0 & \text{elsewhere,} \end{cases} \qquad \beta(x) = \begin{cases} 0 & \text{if } x \in \Omega_R, \\ 1 & \text{elsewhere.} \end{cases} \tag{3}$$

The set Ω_R and the solution of the corresponding heterogeneous model problem are depicted in Figure 1. If not stated otherwise, p is always chosen as 4.

With a standard finite element discretization of a variational formulation of (2), we can derive the nonlinear discrete problem

$$K(u) - f = 0 :\Leftrightarrow F(u) = 0. \tag{4}$$

Let us remark that (4) is linear for $p = 2$. We define the corresponding equation

$$K^{lin}u - f = 0, \tag{5}$$

where K^{lin} is equivalent to the stiffness matrix of the (scaled) diffusion equation.

3 The RASPEN Method

In this section, we provide a brief description of the RASPEN method, which is based on the ASPIN algorithm; see [1, 3] for a more detailed description and a local convergence analysis. As all nonlinear domain decomposition approaches, RASPEN is based on a reformulation of (1) using a decomposition of the underlying nonlinear PDE. In the case of RASPEN, a nonlinear system

$$G(F(u)) =: \mathcal{F}(u) = 0 \tag{6}$$

is derived, where the nonlinear left-preconditioner G is given implicitly. We consider a decomposition of Ω into nonoverlapping subdomains Ω_i, $i = 1, ..., N$, and, by adding layers of finite elements, we obtain overlapping subdomains Ω_i', $i = 1, ..., N$. We denote the local finite element spaces associated with the overlapping subdomains by V_i, $i = 1, ..., N$. With standard restriction operators $R_i : V \rightarrow V_i$ and corresponding prolongation operators $P_i := R_i^T$ we can define nonlinear local corrections $T_i(u)$ by

$$R_i F(u - P_i T_i(u)) = 0, \; i = 1, ..., N. \tag{7}$$

Using restricted prolongation operators $\widetilde{P}_i$, $i = 1, ..., N$, which fulfill the condition $\sum_{i=1}^N \widetilde{P}_i R_i = I$, we can define the nonlinear reformulation

$$\mathcal{F}_{RA}(u) := \sum_{i=1}^{N} \widetilde{P}_i T_i(u). \tag{8}$$

of (1). Let us remark that (8) and (1) have the same solution; see [1, 3]. In the RASPEN method, (8) is solved using Newton's method, i.e., using the iteration

$$u^{(k+1)} = u^{(k)} - \left(D\mathcal{F}_{RA}(u^{(k)})\right)^{-1} \mathcal{F}_{RA}\left(u^{(k)}\right), \tag{9}$$

with the jacobian

$$D\mathcal{F}_{RA}(u) = \sum_{i=1}^{N} \widetilde{P}_i DT_i(u) = \sum_{i=1}^{N} \widetilde{P}_i \left(R_i DF(u_i) P_i\right)^{-1} R_i DF(u_i) =: \sum_{i=1}^{N} Q_i(u_i). \tag{10}$$

Here, we have $u_i = u - P_i T_i(u)$ and $DT_i(u)$ is obtained by deriving (7). Let us remark that, in each Newton iteration and on each overlapping subdomain, the local nonlinear problem (7) has to be solved for $T_i(u^{(k)})$. This can again be done using Newton's method. The necessary local Newton iterations can be carried out in parallel. We distinguish in this paper between outer iterations, i.e., global Newton iterations as in (9), and inner iterations, i.e., local Newton iterations on the subdomain problems to compute the local nonlinear corrections $T_i(u_i)$.

3.1 A Multiplicative Coarse Space

In general, there are several approaches to implement a second level for RASPEN or ASPIN. A simple additive coarse space is suggested in [10] for ASPIN, and a multiplicative coarse space using an FAS approach is used in [3]. We choose a slightly different multiplicative approach not relying on FAS. Our coarse correction is applied after the local corrections, but different variants, i.e., applying the coarse correction before the local corrections as well as a symmetric variant doing both are suggested in [6]. All these variants can analogously be applied to ASPIN, but, for the moment, we restrict ourselves to RASPEN due to space limitations. In [6], we also discuss the differences between our proposed methods and, e.g., FAS-RASPEN, in detail. Now, let V_0 be a discrete coarse space, $R_0 : V \to V_0$ a corresponding restriction, and $P_0 := R_0^T$. Note that the columns of P_0 are just representations of the coarse basis functions on the fine mesh. The nonlinear coarse problem is given by $R_0 F(P_0 u_0)$ using a simple Galerkin approach. The nonlinear coarse correction $T_0(u)$ is then implicitly given by

$$R_0 F(u - P_0 T_0(u)) = 0. \tag{11}$$

Let us remark that the coarse correction $T_0(u)$ is computed using Newton's method in our implementation. The corresponding residual and tangential matrix of equation (11) have to be assembled on the fine grid, which can of course be done in parallel on the subdomains. Also the restriction of the residual as well as the Galerkin product necessary to form the coarse tangential matrix can be efficiently computed in parallel; see, e.g., [5, Sections 4.4 and 4.5]. There, it is also described how the coarse basis functions, i.e., the columns of P_0, can be computed in a scalable fashion; in particular, [5, Section 4.4] deals with GDSW coarse basis functions, however the coarse basis functions introduced in section 3.2 can be computed in parallel in the same way.

We can now define the two-level RASPEN method by

$$\mathcal{F}_{2l}(u) := \sum_{i=1}^{N} \widetilde{P}_i T_i(u) + P_0 T_0(u - \sum_{i=1}^{N} \widetilde{P}_i T_i(u)). \tag{12}$$

Note that the coarse correction is here applied multiplicatively after the local corrections $T_i(u_i)$. A linearization with Newton's method leads to

$$u^{(k+1)} = u^{(k)} - \left(D\mathcal{F}_{2l}(u^{(k)})\right)^{-1} \mathcal{F}_{2l}\left(u^{(k)}\right),$$

where

$$
\begin{aligned}
D\mathcal{F}_{2l}(u) &= \sum_{i=1}^{N} \widetilde{P}_i DT_i\,(u) + P_0 DT_0(u - \sum_{i=1}^{N} \widetilde{P}_i T_i(u)) \left(I - \sum_{i=1}^{N} \widetilde{P}_i DT_i(u)\right) \\
&= \sum_{i=1}^{N} Q_i(u_i) + Q_0(v_0)(I - \sum_{i=1}^{N} Q_i(u_i)) \\
&= Q_0(v_0) + (I - Q_0(v_0)) \sum_{i=1}^{N} Q_i(u_i).
\end{aligned}
\tag{13}
$$

Here, we have $v_0 = u - \sum_{i=1}^{N} P_i T_i(u) - P_0 T_0(u - \sum_{i=1}^{N} P_i T_i(u))$ and $u_i = u - P_i T_i(u)$. The projection operators $Q_i(u_i)$, $i = 1, ..., N$ are defined in (10) and

$$Q_0(v_0) := P_0 \left(R_0 DF(v_0) P_0\right)^{-1} R_0 DF(v_0)$$

is defined analogously and obtained by deriving (11). Additionally to the local Newton iterations, Newton's method is used to compute the coarse correction (11) in each outer iteration. We refer to this iterations as coarse iterations.

3.2 Different Coarse Basis Functions

We consider three different coarse spaces. The simplest one is a Lagrangian coarse space based on a coarse triangular mesh. Therefore, for a structured domain decomposition into square subdomains, each subdomain is split into two triangular finite elements. The coarse basis functions are just piecewise linear ($\mathbb{P}1$) nodal basis functions corresponding to this triangulation. In general, this coarse space relies on the availability of a suitable coarse triangulation. Therefore, we only use it for structured domain decompositions.

For arbitrary domain decompositions, we consider energy-minimizing coarse spaces of MsFEM [7] type. They are also related to reduced dimension GDSW coarse spaces [2]. As in those approaches, we use a nodal basis, i.e., containing one basis function $\Phi^{(j)}$, $j = 1, ..., N_V$, corresponding to each of the N_V vertices of the domain decomposition. Collecting the vectors $\Phi^{(j)}$ as columns in the matrix Φ, we obtain the restriction to the coarse space $R_0 := \Phi^T$. In particular, we construct the coarse basis functions such that they form a partition of unity on all subdomains which do not touch the Dirichlet boundary. This can be achieved by building a partition of unity on the interface of those subdomains and then extending the interface values to the interior in an energy-minimizing way.

To define the interface part $\Phi_\Gamma^{(j)}$ of the basis function $\Phi^{(j)T} = (\Phi_I^{(j)T}, \Phi_\Gamma^{(j)T})$ corresponding to a vertex $\mathcal{V}_j$, let $\mathcal{E}_k$ be one of the adjacent open edges and $\mathcal{V}_l$ the other vertex adjacent to $\mathcal{E}_k$. Then, we set $\Phi_\Gamma^{(j)}(\mathcal{V}_j) = 1$ and $\Phi_\Gamma^{(j)}(x) = 1 - \frac{||x-\mathcal{V}_j||}{||x-\mathcal{V}_j||+||x-\mathcal{V}_l||}$ for any $x \in \mathcal{E}_k$. We proceed equivalently with all other edges adjacent to $\mathcal{V}_j$ and define $\Phi_\Gamma^{(j)}$ as zero on the remaining interface. This results in a partition of unity on the interface, even for a METIS decomposition.

As already stated, the interior values $\Phi_I^{(j)}$ are then computed by energy-minimizing extensions. In order to do so, we propose the use of energy functionals corresponding to related linear problems. In the first alternative, we compute discrete harmonic extensions with respect to the linear operator K^{lin}; see (5). Therefore, we consider the block structure

$$K^{lin} = \begin{pmatrix} K_{II}^{lin} & K_{I\Gamma}^{lin} \\ K_{\Gamma I}^{lin} & K_{\Gamma\Gamma}^{lin} \end{pmatrix}$$

and compute the values in the interior degrees of freedom by

$$\Phi_I^{(i)} = -\left(K_{II}^{lin}\right)^{-1} K_{I\Gamma}^{lin} \Phi_\Gamma^{(i)}, \; i = 1, \ldots, N_V.$$

Alternatively, we use the tangential matrix for the initial value $u^{(0)}$, i.e.,

$$DK(u^{(0)}) = \begin{pmatrix} DK(u^{(0)})_{II} & DK(u^{(0)})_{I\Gamma} \\ DK(u^{(0)})_{\Gamma I} & DK(u^{(0)})_{\Gamma\Gamma} \end{pmatrix},$$

to compute the energy-minimizing extensions. In particular, we then define the extension to the interior of the subdomains by

$$\Phi_I^{(i)} = -\left(DK(u^{(0)})_{II}\right)^{-1} DK(u^{(0)})_{I\Gamma} \Phi_\Gamma^{(i)}, \; i = 1, \ldots, N_V.$$

In general, this is advantageous since it only depends on the nonlinear operator F and no linear Laplacian has to be assembled additionally.

Let us remark that the energy-minimizing basis functions can be computed locally by the solution of linear problems on the interior part of the nonoverlapping subdomains. Also, they are zero on all subdomains not adjacent the corresponding vertex by construction, and therefore, no extensions have to be computed on the remaining subdomains. All three coarse spaces build a partition of unity on all subdomains which do not touch the Dirichlet boundary. This property is crucial for a good linear coarse space. All coarse spaces have the same size and therefore have the same computational cost per nonlinear or linear iteration; only the costs for the construction of the energy-minimizing coarse basis functions are higher.

4 Numerical Results

For all tests and all methods, we choose the same initial value $u^{(0)}(x, y) = xy(x - 1)(y - 1)$ and the same relative stopping tolerance, i.e., we stop the outer iteration if $F(u^{(k)})/F(u^{(0)}) < 1e - 6$. All inner or, respectively, coarse iterations

Table 1: Homogeneous p-Laplace: Comparison of different coarse spaces for regular and METIS domain decompositions; best results for the largest experiment are marked in bold; *outer it.* gives the number of global Newton iterations; *inner it.* gives the number of local Newton iterations summed up over the outer Newton iterations (average over subdomains); *coarse it.* gives the number of nonlinear iterations on the second level summed up over the outer Newton iterations; *GMRES it.* gives the number of GMRES iterations summed up over the outer Newton iterations.

p-Laplace homogeneous $p = 4$; $H/h = 32$ for regular domains; overlap $\delta = 2$;									
		Regular				METIS			
N	**RASPEN Coarse Space**	**outer it.**	**inner it. (avg.)**	**coarse it.**	**GMRES it. (sum)**	**outer it.**	**inner it. (avg.)**	**coarse it.**	**GMRES it. (sum)**
9	-	5	25.9	-	99	7	41.4	-	238
	$\mathbb{P}1$	5	30.2	17	88	-	-	-	-
	$DK(u^{(0)})$ ext.	5	30.7	16	83	5	31.3	22	123
	K^{lin} ext.	5	29.9	16	83	5	30.7	19	121
16	-	14	73.8	-	358	11	62.8	-	458
	$\mathbb{P}1$	6	32.4	20	122	-	-	-	-
	$DK(u^{(0)})$ ext.	7	38.9	30	140	7	36.8	27	180
	K^{lin} ext.	5	30.6	18	99	6	32.5	21	152
25	-	6	28.4	-	201	12	57.6	-	578
	$\mathbb{P}1$	5	27.4	18	116	-	-	-	-
	$DK(u^{(0)})$ ext.	5	27.6	19	108	5	28.6	20	126
	K^{lin} ext.	5	27.2	18	108	6	31.4	22	151
36	-	15	66.9	-	563	11	53.1	-	617
	$\mathbb{P}1$	6	30.6	21	145	-	-	-	-
	$DK(u^{(0)})$ ext.	7	34.3	30	164	6	30.4	23	155
	K^{lin} ext.	5	28.7	19	117	6	30.0	21	152
49	-	6	29.0	-	268	13	60.9	-	811
	$\mathbb{P}1$	**5**	27.3	**18**	126	-	-	-	-
	$DK(u^{(0)})$ ext.	**5**	27.4	19	**121**	7	32.0	27	178
	K^{lin} ext.	**5**	**27.2**	**18**	122	**6**	**29.4**	**21**	**152**

are stopped with an equivalent relative residual criterion in the corresponding local or, respectively, coarse finite element space, after a reduction of $1e-3$ is reached. This is sufficient since the inner and coarse initial values get more and more accurate while the outer loop converges. As a linear solver for the tangential systems, we use GMRES (Generalized Minimal RESidual) iterations with a relative stopping tolerance of $1e-8$. Of course, in particular, in the first Newton steps, we might over-solve the linear systems, and choosing the forcing terms correctly could be beneficial for all methods; see [4].

We first consider a numerical scalability study for the homogeneous p-Laplace for $p = 4$; see Table 1. Here, for regular domain decompositions, we choose $H/h = 32$ and therefore 2 048 triangular finite elements per nonoverlapping subdomain. For the METIS decompositions, the global problem sizes are identical to the corresponding regularly decomposed problems. We present the number of outer or global Newton iterations, which is up to 2.5 times higher in the one level RASPEN method compared with the best of the two-level approaches. All three coarse levels show a similar performance for the regular domain decomposition and both extension based coarse spaces perform well for the METIS decompositions. In general, the two-level RASPEN method needs less inner iterations and significantly less GMRES iterations, especially for irregular domain decompositions.

Table 2: Heterogeneous p-Laplace: See Table 1 for description of column labels and Fig. 1 for the coefficient distribution.

p-Laplace heterogeneous (channel + 2 circles) $p = 4$; $H/h = 32$ for regular domains; overlap $\delta = 2$;									
		Regular				METIS			
N	RASPEN Coarse Space	outer it.	inner it. (avg.)	coarse it.	GMRES it. (sum)	outer it.	inner it. (avg.)	coarse it.	GMRES it. (sum)
36	-	5	14.3	-	321	**5**	**14.2**	-	346
	$\mathbb{P}1$	5	15.6	17	139	-	-	-	-
	$DK(u^{(0)})$ ext.	5	15.1	16	139	**5**	15.2	**18**	**125**
	K^{lin} ext.	**4**	**12.7**	**13**	**108**	**5**	15.5	**18**	128

For the chosen heterogeneous problem (see Table 2), the number of outer Newton iterations is similar for all methods. Nevertheless, the linear convergence, i.e. the number of GMRES iterations, is superior in the two-level variants. All in all, our experiments show that our multiplicative second level with the chosen coarse basis functions has a superior linear convergence and, in some cases, also a better nonlinear convergence - regardless if regular or METIS decompositions are used.

In general, the discrete extension using K_{lin} shows a slightly better performance than the extension with the tangent $DK(u^{(0)})$, but the latter one will always be available, also for different nonlinear model problems where a suitable linear operator K_{lin} cannot be found easily. Considering, e.g., nonlinear hyperlelasticity or elasto-plasticity problems, the linear elasticity model or a multi-dimensional Laplacian could be used to form K_{lin}, but for large loads or highly plastic behavior, $DK(u^{(0)})$ might be a better choice.

5 Conclusion

We have presented a new approach to implement a multiplicative coarse space for ASPIN or RASPEN, which is robust for the considered model problems. Additionally, we presented two different coarse spaces usable for irregular domain decompositions and compared both against the one level RASPEN method and, for regular domain decompositions, also against a classical $\mathbb{P}1$ coarse space. Both coarse spaces are competitive and cheap to compute.

References

1. Cai, X.C., Keyes, D.E.: Nonlinearly preconditioned inexact Newton algorithms. SIAM J. Sci. Comput. **24**(1), 183–200 (2002). DOI:10.1137/S106482750037620X
2. Dohrmann, C.R., Widlund, O.B.: On the design of small coarse spaces for domain decomposition algorithms. SIAM J. Sci. Comput. **39**(4), A1466–A1488 (2017)

3. Dolean, V., Gander, M.J., Kheriji, W., Kwok, F., Masson, R.: Nonlinear preconditioning: how to use a nonlinear Schwarz method to precondition Newton's method. SIAM J. Sci. Comput. **38**(6), A3357–A3380 (2016). DOI:10.1137/15M102887X
4. Eisenstat, S.C., Walker, H.F.: Choosing the forcing terms in an inexact Newton method. SIAM J. Sci. Comput. **17**(1), 16–32 (1996). DOI:10.1137/0917003. Special issue on iterative methods in numerical linear algebra (Breckenridge, CO, 1994)
5. Heinlein, A., Klawonn, A., Rheinbach, O.: A parallel implementation of a two-level overlapping Schwarz method with energy-minimizing coarse space based on Trilinos. SIAM J. Sci. Comput. **38**(6), C713–C747 (2016)
6. Heinlein, A., Lanser, M.: Additive and hybrid nonlinear two-level Schwarz methods and energy minimizing coarse spaces for unstructured grids. Technical report, Universität zu Köln (2019). URL https://kups.ub.uni-koeln.de/9845/
7. Hou, T.Y., Wu, X.H.: A multiscale finite element method for elliptic problems in composite materials and porous media. Journal of Computational Physics **134**(1), 169 – 189 (1997). DOI: http://dx.doi.org/10.1006/jcph.1997.5682. URL http://www.sciencedirect.com/science/article/pii/S0021999197956825
8. Karypis, G., Kumar, V.: A fast and high quality multilevel scheme for partitioning irregular graphs. SIAM J. Sci. Comput. **20**(1), 359–392 (1998). DOI:10.1137/S1064827595287997
9. Liu, L., Keyes, D.E.: Field-split preconditioned inexact Newton algorithms. SIAM J. Sci. Comput. **37**(3), A1388–A1409 (2015). DOI:10.1137/140970379
10. Marcinkowski, L., Cai, X.C.: Parallel performance of some two-level ASPIN algorithms. In: Domain decomposition methods in science and engineering, *Lect. Notes Comput. Sci. Eng.*, vol. 40, pp. 639–646. Springer, Berlin (2005). DOI:10.1007/3-540-26825-1_68

A Reynolds Number Dependent Convergence Estimate for the Parareal Algorithm

Martin J. Gander and Thibaut Lunet

1 The Parareal algorithm

Time parallel time integration has received substained attention over the last decades, for a review, see [2]. More recently, renewed interest in this area was sparked by the invention of the Parareal algorithm [5] for solving initial value problems like

$$\frac{d\boldsymbol{u}}{dt} = \mathcal{L}_h(\boldsymbol{u}(t), t), \quad \boldsymbol{u}(0) = \boldsymbol{u}_0, \quad t \in [0, T], \tag{1}$$

with $\mathcal{L}_h : \mathbb{R}^p \times \mathbb{R}^+ \to \mathbb{R}^p$, $\boldsymbol{u}(t) \in \mathbb{R}^p$, $\boldsymbol{u}_0 \in \mathbb{R}^p$, p being the total number of degrees of freedom and T a positive real value. Problem (1) often arises from the spatial discretization of a (non-)linear system of partial differential equations (PDEs) through the method-of-lines. For Parareal, one decomposes the global time interval $[0, T]$ into N time subintervals $[T_{n-1}, T_n]$ of size ΔT, $n = 1, \cdots, N$, where N is the number of processes to be considered for the time parallelization. In the following, we denote by $\boldsymbol{U}_n$ the approximation of $\boldsymbol{u}$ at time T_n, *i.e.* , $\boldsymbol{U}_n \approx \boldsymbol{u}(T_n)$. Let $\underset{T_{n-1}\to T_n}{\mathcal{F}^{\delta t}}(\boldsymbol{U}_{n-1})$ denote the result of approximately integrating (1) on the time subinterval $[T_{n-1}, T_n]$ from a given starting value $\boldsymbol{U}_{n-1}$ using a fine propagator $\mathcal{F}$ with time step δt. Similarly, Parareal also needs a coarse propagator $\mathcal{G}$ (with time step Δt), which has to be much cheaper than $\mathcal{F}$ resulting in less accuracy.

The Parareal algorithm consists of a prediction step and a correction iteration. In the prediction step, Parareal computes an inital guess of the starting values $\boldsymbol{U}_n^0$ at the beginning of each time subinterval using the coarse propagator,

Martin J. GANDER
University of Geneva, 2-4 rue du Lièvre, 1211 Genève 4, Suisse, e-mail: martin.gander@unige.ch

Thibaut LUNET
University of Geneva, 2-4 rue du Lièvre, 1211 Genève 4, Suisse, e-mail: thibaut.lunet@unige.ch

R. Haynes et al. (eds.), *Domain Decomposition Methods in Science and Engineering XXV*,
Lecture Notes in Computational Science and Engineering 138,
https://doi.org/10.1007/978-3-030-56750-7_31

$$\forall n = 1, \cdots, N, \ \boldsymbol{U}_n^0 = \underset{T_{n-1}\to T_n}{\mathcal{G}^{\Delta t}}(\boldsymbol{U}_{n-1}^0), \ \boldsymbol{U}_0^0 = \boldsymbol{u}_0. \tag{2}$$

A correction iteration is then applied in PARAREAL, using concurrently the fine propagator $\mathcal{F}$ on each time subinterval:

$$\boldsymbol{U}_n^k = \underset{T_{n-1}\to T_n}{\mathcal{F}^{\delta t}}(\boldsymbol{U}_{n-1}^{k-1}) + \underset{T_{n-1}\to T_n}{\mathcal{G}^{\Delta t}}(\boldsymbol{U}_{n-1}^k) - \underset{T_{n-1}\to T_n}{\mathcal{G}^{\Delta t}}(\boldsymbol{U}_{n-1}^{k-1}), \tag{3}$$

where $\boldsymbol{U}_n^k$ denotes the approximation of $\boldsymbol{u}$ at time T_n at the k-th iteration of PARAREAL ($k = 1, \cdots, K, n = 1, \cdots, N$). While the application of $\mathcal{F}$ can be performed independently for each time subinterval, PARAREAL remains limited by the sequential nature of the coarse integration performed by $\underset{T_{n-1}\to T_n}{\mathcal{G}^{\Delta t}}$ in (3). PARAREAL will thus reduce the total computational time compared to a direct time-serial integration only if the application of $\mathcal{G}$ is cheap enough and if the total number of iterations K of PARAREAL is small. We will use the following result, which is an extension of [4, Th. 4.9] following indications of [4, Sec. 4.5] for the Dahlquist test equation

$$\frac{du}{dt} = \lambda u, \ \ \lambda \in \mathbb{C}, \ \ u(0) = u_0 \in \mathbb{C}. \tag{4}$$

Theorem 1 (Linear convergence bound - Dahlquist test equation) *Let $\mathcal{G}$ be a one-step time-integration method, and $\mathcal{F}$ be the same time integrator, but using m time-steps instead of a single one (i.e. $\Delta t = \Delta T = m\delta t$). If $\mathcal{G}$ is used such that $\lambda\Delta t$ is in its region of absolute stability, then*

$$\sup_{n>0} |u_n^{\mathcal{F}} - U_n^k| \leq \rho(\lambda\Delta T)^k \sup_{n>0} |u_n^{\mathcal{F}} - U_n^0|, \tag{5}$$

where $u_n^{\mathcal{F}}$ is the fine solution at time T_n, and the convergence factor *is given by*

$$\rho(\lambda\Delta T) = \frac{|R(\lambda\Delta T/m)^m - R(\lambda\Delta T)|}{1 - |R(\lambda\Delta T)|}, \tag{6}$$

with R the stability function of the coarse (and fine) solver.

2 Semi-discretization of the advection-diffusion problem

We are interested in the linear advection-diffusion equation on a one-dimensional spatial domain $[0, L]$

$$\frac{\partial u}{\partial t} = -a\frac{\partial u}{\partial x} + \nu\frac{\partial^2 u}{\partial x^2} + f(x,t), \ \ u(x,0) = u_0(x), \tag{7}$$

with $a, \nu \in \mathbb{R}_+^*$ the advection and diffusion coefficients, $f : \mathbb{R} \times \mathbb{R}^+ \to \mathbb{R}$ a source term, and periodic boundary conditions for the spatial domain

$$\forall t \in [0,T], \quad u(0,t) = u(L,t). \tag{8}$$

We discretize $[0, L]$ using a uniform mesh $[x_1, ..., x_p]^T$ for p unknowns, which gives a mesh size $\delta_x = L/p$. We use centered finite differences of order 2 (FD-C2) for the diffusion operator in (7), and also use either a FD-C2 discretization for the advection operator, or a 1st order upwind scheme (FD-U1). This leads to the semi-discrete system of ODEs

$$\frac{d\boldsymbol{u}}{dt}(t) = A\boldsymbol{u}(t) + \boldsymbol{f}(t), \quad \boldsymbol{u}(0) = \boldsymbol{u}_0, \tag{9}$$

where $A \in \mathbb{R}^{p \times p}$ and $\boldsymbol{f} : \mathbb{R} \to \mathbb{R}^p$ represents the source term.

Two dimensionless numbers can be defined to characterize this problem:

$$Re := \frac{aL}{\nu}, \quad Re_{\delta_x} := \frac{a\delta_x}{\nu}, \tag{10}$$

where Re is the Reynolds number[1] and Re_{δ_x} is the mesh Reynolds number ($Re = pRe_{\delta_x}$). Re indicates by its large (resp. small) value a major influence of advection (resp. diffusion) on the solution $u(x,t)$. As Re compares this advection/diffusion ratio with the characteristic length L (that can be chosen differently for a different situation), Re_{δ_x} compares this ratio to the mesh size. Decreasing the diffusion coefficient will increase Re, as the advection becomes more dominant. It does not necessarily induce an increase of Re_{δ_x}, as δ_x can also be decreased to keep a constant value for Re_{δ_x}. This mesh refinement when Re increases is commonly done for Direct Numerical Simulation (DNS) of the Navier-Stokes equations [6, Chap. 4], or also for stationary forms of (9) with Dirichlet boundary conditions, to keep a certain accuracy in the approximate solution (solutions become qualitatively wrong when $Re_{\delta_x} \geq 2$ for FD-C2 dicretizations of the advection term [1, Sec. 2, § 5]).

The mesh Reynolds number Re_{δ_x} is useful to formulate convergence results for PARAREAL. In particular, we can use it to express the eigenvalues of A in (9):

Lemma 1 (Eigenvalues of the spatial advection-diffusion operator)
For the FD-C2 discretization of the diffusion term with periodic boundary conditions, the eigenvalues of the discrete spatial operator A with FD-C2 discretization of the advection are

$$\lambda_\kappa = -\frac{a}{\delta_x}\left[i \sin\left(\frac{2\kappa\pi}{p}\right) + \frac{2}{Re_{\delta_x}}\left(1 - \cos\left(\frac{2\kappa\pi}{p}\right)\right)\right], \quad \kappa \in \{0, 1, ..., p-1\}, \tag{11}$$

with $i := \sqrt{-1}$. *For the FD-U1 discretization of the advection, the eigenvalues are*

$$\lambda_\kappa = -\frac{a}{\delta_x}\left[1 - e^{-i\frac{2\kappa\pi}{p}} + \frac{2}{Re_{\delta_x}}\left(1 - \cos\left(\frac{2\kappa\pi}{p}\right)\right)\right], \quad \kappa \in \{0, 1, ..., p-1\}. \tag{12}$$

[1] What we call here the Reynolds number is in fact the Peclet number, since there is no non-linear advective term in (7). However, we prefer to use Reynolds number, since our analysis is a first step toward Navier-Stokes equations, and this links our results to those already in the literature.

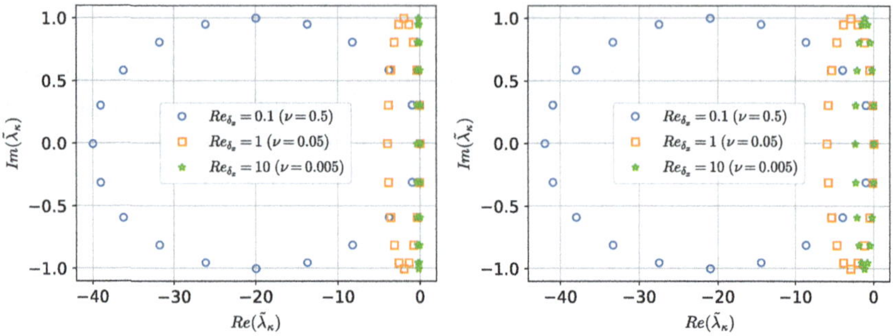

Fig. 1: Influence of Re_{δ_x} on the eigenvalues of the advection-diffusion problem when varying ν only and keeping a and δ_x fixed. For the advection term, we used FD-C2 (left) and FD-U1 (right), with $p = 20$, $a = 1$, $L = 1$, and for the axes, we used $\tilde{\lambda}_\kappa := \delta_x/a\lambda_\kappa$.

Proof For the FD-C2 discretization of the advection term, the space operator matrix is given by

$$A := -\frac{a}{2\delta_x}\begin{pmatrix} 0 & 1 & & -1 \\ -1 & \ddots & \ddots & \\ & \ddots & \ddots & 1 \\ 1 & & -1 & 0 \end{pmatrix} + \frac{\nu}{\delta_x^2}\begin{pmatrix} -2 & 1 & & 1 \\ 1 & \ddots & \ddots & \\ & \ddots & \ddots & 1 \\ 1 & & 1 & -2 \end{pmatrix}, \tag{13}$$

where the eigenvalues of each matrix are well known (see, *e.g.* [6, Chap. 3]). Each circulant matrix is diagonalized by the same Fourier basis, and hence the eigenvalues λ_κ of A are just the sum of the eigenvalues of each matrix in (13), i.e.

$$\lambda_\kappa = -i\frac{a}{2\delta_x}\sin(2\kappa\pi/p) + 2\frac{\nu}{\delta_x^2}\left(1 - \cos\left(\frac{2\kappa\pi}{p}\right)\right). \tag{14}$$

Extracting the common factor a/δ_x and using the definition of Re_{δ_x} then leads to (11). The result for FD-U1 in (12) is obtained similarly. □

In Fig. 1, we show the eigenvalues for both discretizations, FD-C2 and FD-U1, and their dependency on Re_{δ_x} when varying ν only. We see that the eigenvalues are distributed along an ellipse that flattens toward the imaginary axis when Re_{δ_x} increases. For small Re_{δ_x}, the eigenvalues are very similar, but for large Re_{δ_x}, the flattening toward the imaginary axis is more pronounced for FD-C2 than for FD-U1, which comes from the numerically more diffusive nature of FD-U1.

3 Linear bound of Parareal for advection-diffusion

Theorem 2 (Linear convergence bound - Advection-diffusion equation) *Let $\mathcal{G}$ be a one-step time-integration method, $\mathcal{F}$ be the same time integrator using m time-*

steps, and $\boldsymbol{u}_n^{\mathcal{F}}$ be the fine sequential solution at T_n. If $\lambda_\kappa \Delta T$ is in the region of absolute stability of $\mathcal{G}$ for each eigenvalue λ_κ of A, then the error in the PARAREAL algorithm satisfies the linear error bound

$$E_\infty^k := \sup_{n>0} \left\| \boldsymbol{u}_n^{\mathcal{F}} - \boldsymbol{U}_n^k \right\|_2 \leq \rho_{ad}^k C_\infty^0, \quad C_\infty^0 = \sqrt{\sum_\kappa \sup_{n>0} \left| \hat{u}_n^{\mathcal{F}}(\kappa) - \hat{U}_n^0(\kappa) \right|^2} \tag{15}$$

where $\hat{u}(\kappa)$ is the κth Fourier component of $\boldsymbol{u}$ and ρ_{ad} denotes the linear convergence factor *of PARAREAL,*

$$\rho_{ad} = \sup_{\lambda_\kappa}[\rho(\lambda_\kappa \Delta T)] = \sup_{\lambda_\kappa}\left[\frac{|R(\lambda_\kappa \Delta T/m)^m - R(\lambda_\kappa \Delta T)|}{1 - |R(\lambda_\kappa \Delta T)|}\right], \tag{16}$$

with R the stability function of the coarse (and fine) solver. In particular, PARAREAL convergence is ensured if $\rho_{ad} < 1$.

Proof As the unitary Discrete Fourier Transform (DFT) matrix transforms A into diagonal form, (9) is then a combination of decoupled Dahlquist equations in Fourier space. Using Theorem 1, we can bound the PARAREAL error of each Fourier component for all time subintervals,

$$\forall n \in \mathbb{N}, \ |\hat{u}_n^{\mathcal{F}}(\kappa) - \hat{U}_n^k(\kappa)| \leq \sup_{n>0}|\hat{u}_n^{\mathcal{F}}(\kappa) - \hat{U}_n^k(\kappa)| \leq \rho(\lambda_\kappa \Delta T)^k \sup_{n>0}|\hat{u}_n^{\mathcal{F}}(\kappa) - \hat{U}_n^0(\kappa)|. \tag{17}$$

For each κ, $\rho(\lambda_\kappa \Delta T)$ can be bounded by $\sup_{\lambda_\kappa}[\rho(\lambda_\kappa \Delta T)]$. Then, taking the power 2 of each extremal part of the inequality, summing on κ and computing the square root gives

$$\left\| \hat{\boldsymbol{u}}_n^{\mathcal{F}} - \hat{\boldsymbol{U}}_n^k \right\|_2 \leq \sup_{\lambda_\kappa}[\rho(\lambda_\kappa \Delta T)]^k C_\infty^0 \tag{18}$$

Using the Parseval-Plancherel theorem and bounding the left term for $n \in \mathbb{N}$ then leads to (15). □

As we saw previously, the eigenvalues λ_κ are fully characterized by Re_{δ_x} and a/δ_x. Hence, we can define a dimensionless number,

$$CFL_{\mathcal{P}} = \frac{a\Delta T}{\delta_x}, \tag{19}$$

which we call Courant-Friedrichs-Lewy (CFL) number for PARAREAL (as if the algorithm were simply considered as a standard time integration method with time-step ΔT). It is worth mentioning that $CFL_{\mathcal{P}}$ is the CFL number of the coarse solver, and m times the CFL number of the fine solver, and we obtain the following result:

Lemma 2 *For a given mesh with p mesh points, the linear convergence factor ρ_{ad} of PARAREAL for the advection-diffusion equation in* (16) *depends only on Re_{δ_x}, $CFL_{\mathcal{P}}$, and the coarse/fine solver settings, i.e. the stability function R and the number of time-steps m per time subinterval.*

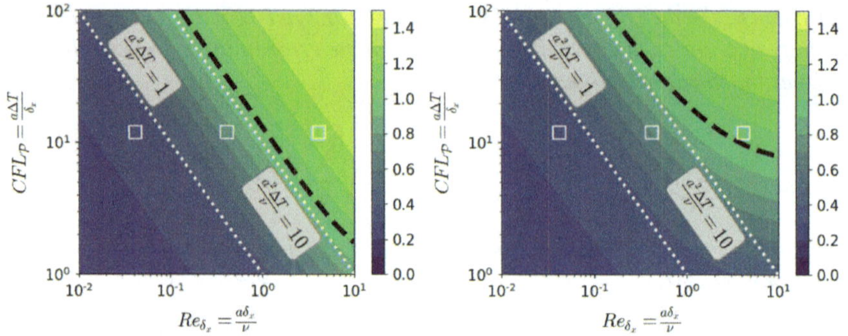

Fig. 2: Dependence of ρ_{ad} on Re_{δ_x} and $CFL_{\mathcal{P}}$, for Backward Euler ($m = 30$ for $\mathcal{F}$), and FD-C2 (left) and FD-U1 (right). Dotted black line: $\rho_{ad} = 1$. White squares are $(Re_{\delta_x}, CFL_{\mathcal{P}})$ tuples used in Sec. 4. White dotted lines are $(Re_{\delta_x}, CFL_{\mathcal{P}})$ values for some given constant ratio $a^2\Delta T/\nu$.

Proof Looking for example at the FD-C2 discretization for advection, combining (19) with (11), we get

$$\lambda_\kappa \Delta T = -CFL_{\mathcal{P}} \left[i \sin\left(\frac{2\kappa\pi}{p}\right) + \frac{2}{Re_{\delta_x}}\left(1 - \cos\left(\frac{2\kappa\pi}{p}\right)\right)\right], \quad \kappa \in \{0, 1, ..., p-1\}, \quad (20)$$

which depends only on κ, Re_{δ_x} and $CFL_{\mathcal{P}}$. As ρ_{ad} is a maximum bound over all κ, we obtain the result from (16). The proof is similar for FD-U1. □

We present the dependency proved in Lemma 2 graphically using contour plots for ρ_{ad} in Fig. 2, with a Backward Euler time integrator for $\mathcal{F}$ and $\mathcal{G}$, using $m = 30$ and $p = 5000$. For both discretizations, we observe an increase of ρ_{ad} with both Re_{δ_x} and $CFL_{\mathcal{P}}$, in agreement with numerical results in the literature (see, *e.g.* , [3]). Our analysis quantifies this convergence deterioration, and shows how ρ_{ad} depends precisely on Re_{δ_x} and $CFL_{\mathcal{P}}$. Furthermore, for sufficient space resolution (small Re_{δ_x}), ρ_{ad} is determined by $CFL_{\mathcal{P}} Re_{\delta_x} = a^2\Delta T/\nu$ (white dotted lines in Fig. 2). In particular, convergence is only ensured when $a^2\Delta T/\nu$ is less than a given value (around 10 in our case). This implies that ΔT must be in the order of ν/a^2, or in other words, the coarse time step must be small enough for $\mathcal{G}$ to capture the diffusion time-scale, requiring that $\mathcal{G}$ has an $\mathcal{F}$-like resolution.

Using the FD-U1 discretization only changes the convergence factor for large values of Re_{δ_x}, which may not be of use since $Re_{\delta_x} >> 1$ can lead to an important loss of accuracy for the numerical solution, as we will see in Sec. 4.

4 Numerical experiments

We perform now numerical experiments similar to those already in the literature (see, *e.g.* , [7, 3]), where we use a fixed number of mesh points p, and decrease the diffusion

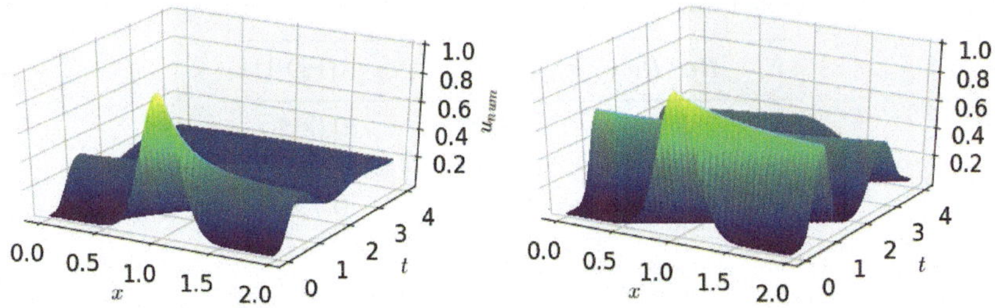

Fig. 3: Space-time numerical solution with the $\mathcal{F}$ solver using FD-C2 for the advection discretization. Left: $\nu = 0.1$ ($Re = 20$), right: $\nu = 0.01$ ($Re = 200$).

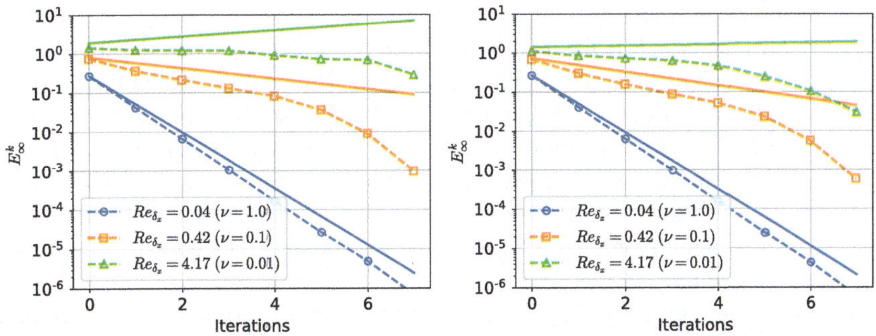

Fig. 4: Influence of an increase of Re_{δ_x} by lowering ν on PARAREAL convergence, using FD-C2 for the advection discretization (left) and FD-U1 (right), linear bounds in plain lines.

coefficient ν to obtain larger values of Re, for a fixed value $a = 1$ of the advection. Numerical simulations are done with $L = 2$ and $p = 48$, until $T = 4$. Backward Euler is used for the $\mathcal{F}$ and $\mathcal{G}$ solvers, with $m = 30$. Since we use $N = 8$ time subintervals for PARAREAL, this implies a fine time step $\delta_t = 1/60$ and $CFL_{\mathcal{P}} = 12$. We use a Gaussian as initial condition, $u_0(x) = e^{-20(x-1)^2}$, and no source term. Three different viscosity coefficients are chosen, $\nu \in \{1, 0.1, 0.01\}$. Numerical solutions are shown in Fig. 3 for the two smaller values of ν; for the largest value $\nu = 1$, the solution is almost purely diffusive, and is constant for $t > 1$. We show in Fig. 4 the error against the fine solution at each PARAREAL iteration, using the FD-C2 and FD-U1 discretizations. For each Re_{δ_x} value corresponding to the chosen Re, the linear bound is indicated by the plain lines, and the corresponding $(CFL_{\mathcal{P}}, Re_{\delta_x})$ tuples are indicated with the white squares in Fig. 2. We observe for both discretizations a degradation of the PARAREAL convergence when ν decreases and thus Re and Re_{δ_x} increase, which is well predicted by the linear convergence bound (plain lines in Fig. 4). The use of the FD-U1 discretization lessens this convergence degradation a little for high Re_{δ_x} numbers (low ν), which is due to the artificial dissipation brought by the Upwind scheme that makes the problem (wrongly) more diffusive.

This loss of accuracy is particularly visible when comparing the fine solution to the analytical solution of (7) with periodic boundary conditions, $u(x,t) =$

Table 1: Main parameters for the numerical experiments

ν	Re	Re_{δ_x}	$\epsilon_{T,S}$ (FD-C2)	$\epsilon_{T,S}$ (FD-U1)
1	2	0.042	0.006	0.005
0.1	20	0.42	0.040	0.096
0.01	200	4.2	0.321	0.724

$\frac{1}{\sqrt{4\pi\nu t}}\int_{-\infty}^{+\infty} u_0(\xi)\exp(-\frac{(x-at-\xi)^2}{4\nu t})d\xi$. We define the numerical error "in time and space" of the fine solution $\epsilon_{T,S} := \frac{1}{N_{step}}\sum_{i=1}^{N_{step}} ||\boldsymbol{u}^{ref}(i\delta_t) - \boldsymbol{u}^{\mathcal{F}}(i\delta_t)||_2$, where N_{step} is the number of time steps for the fine solver to cover $[0,T]$ (in our case, $N_{step} = 240$), and $\boldsymbol{u}^{ref}$ is the analytical solution computed at each mesh point. We give $\epsilon_{T,S}$ for each discretization and Re_{δ_x} in Tab. 1. One can see that the accuracy decreases dramatically when Re and Re_{δ_x} increase, the effect being more important for the FD-U1 discretization, compared to the FD-C2 discretization. In order to reduce this error, a mesh refinement would be necessary, which would have led to lower Re_{δ_x} values for the chosen Re, but also to corresponding higher values of $CFL_{\mathcal{P}}$, thus not changing anything in the convergence behavior of the method (*cf.* white dotted lines in Fig. 2).

In conclusion, both our theoretical results and our numerical experiments show that PARAREAL algorithm convergence deteriorates when the ratio $a^2\Delta T/\nu$ becomes large, this ratio being proportional to the Reynolds number when time-step and mesh size are kept constant. One also has to be careful when using numerically diffusive schemes not to jump to false conclusions: truly transport dominated solutions are hard to approximate effectively with the classical PARAREAL algorithm.

References

1. Birkhoff, G., Gartland Jr, E., Lynch, R.: Difference methods for solving convection-diffusion equations. Computers & Mathematics with Applications **19**(11), 147–160 (1990)
2. Gander, M.J.: 50 years of time parallel time integration. In: T. Carraro, M. Geiger, S. Körkel, R. Rannacher (eds.) Multiple Shooting and Time Domain Decomposition Methods, pp. 69–114. Springer (2015)
3. Gander, M.J.: Five decades of time parallel time integration, and a note on the degradation of the performance of the Parareal algorithm as a function of the Reynolds number. Oberwolfach Report (2017)
4. Gander, M.J., Vandewalle, S.: Analysis of the Parareal time-parallel time-integration method. SIAM J. Sci. Comput. **29**(2), 556–578 (2007)
5. Lions, J.L., Maday, Y., Turinici, G.: A "Parareal" in time discretization of PDE's. C. R. Math. Acad. Sci. Paris **332**(7), 661–668 (2001)
6. Lunet, T.: Stratégies de parallélisation espace-temps pour la simulation numérique des écoulements turbulents. Ph.D. thesis, Toulouse, ISAE (2018)
7. Steiner, J., Ruprecht, D., Speck, R., Krause, R.: Convergence of Parareal for the Navier-Stokes equations depending on the Reynolds number. In: A. Abdulle, S. Deparis, D. Kressner, F. Nobile, M. Picasso (eds.) Numerical Mathematics and Advanced Applications - ENUMATH 2013, vol. 103, pp. 195–202. Springer (2015)

The Domain Decomposition Method of Bank and Jimack as an Optimized Schwarz Method

Gabriele Ciaramella, Martin J. Gander, and Parisa Mamooler

1 Bank-Jimack Domain Decomposition Method

In 2001 Randolph E. Bank and Peter K. Jimack [1] introduced a new domain decomposition method for the adaptive solution of elliptic partial differential equations, see also [2]. The novel feature of this algorithm is that each of the subproblems is defined over the entire domain. To describe the method, we consider a linear elliptic PDE on a domain Ω, and two overlapping subdomains Ω_1 and Ω_2, $\Omega = \Omega_1 \cup \Omega_2$. Discretizing the problem on a global fine mesh leads to a linear system $K\boldsymbol{u} = \boldsymbol{f}$, where K is the stiffness matrix, $\boldsymbol{u}$ is the vector of unknown nodal values on the global fine mesh, and $\boldsymbol{f}$ is the load vector. We partition now the vector $\boldsymbol{u} = \left[\boldsymbol{u}_1, \boldsymbol{u}_s, \boldsymbol{u}_2\right]^T$, where $\boldsymbol{u}_1$ is the vector of unknowns on the nodes in $\Omega_1 \setminus \Omega_2$, $\boldsymbol{u}_s$ is the vector of unknowns on the nodes in $\Omega_1 \cap \Omega_2$, and $\boldsymbol{u}_2$ is the vector of unknowns on the nodes in $\Omega_2 \setminus \Omega_1$. We can then write the linear system in block matrix form,

$$\begin{bmatrix} A_1 & B_1 & 0 \\ B_1^T & A_s & B_2^T \\ 0 & B_2 & A_2 \end{bmatrix} \begin{bmatrix} \boldsymbol{u}_1 \\ \boldsymbol{u}_s \\ \boldsymbol{u}_2 \end{bmatrix} = \begin{bmatrix} \boldsymbol{f}_1 \\ \boldsymbol{f}_s \\ \boldsymbol{f}_2 \end{bmatrix}. \tag{1}$$

The idea of the Bank-Jimack method is to consider two further meshes on Ω, one identical to the original fine mesh in Ω_1, but coarse on $\Omega\backslash\Omega_1$, and one identical to the original fine mesh in Ω_2, but coarse on $\Omega\backslash\Omega_2$. This leads to the two further linear systems

Martin J. Gander and Parisa Mamooler
Université de Genève, Section de mathématiques, e-mail: {Martin.Gander}{Parisa.Mamooler}@unige.ch

Gabriele Ciaramella
Department of Mathematics and Statistics, University of Konstanz, e-mail: gabriele.ciaramella@uni-konstanz.de

R. Haynes et al. (eds.), *Domain Decomposition Methods in Science and Engineering XXV*,
Lecture Notes in Computational Science and Engineering 138,
https://doi.org/10.1007/978-3-030-56750-7_32

$$\begin{bmatrix} A_1 & B_1 & 0 \\ B_1^T & A_s & C_2 \\ 0 & \widetilde{B}_2 & \widetilde{A}_2 \end{bmatrix} \begin{bmatrix} \boldsymbol{v}_1 \\ \boldsymbol{v}_s \\ \boldsymbol{v}_2 \end{bmatrix} = \begin{bmatrix} \boldsymbol{f}_1 \\ \boldsymbol{f}_s \\ M_2 \boldsymbol{f}_2 \end{bmatrix}, \quad \begin{bmatrix} \widetilde{A}_1 & \widetilde{B}_1 & 0 \\ C_1 & A_s & B_2^T \\ 0 & B_2 & A_2 \end{bmatrix} \begin{bmatrix} \boldsymbol{w}_1 \\ \boldsymbol{w}_s \\ \boldsymbol{w}_2 \end{bmatrix} = \begin{bmatrix} M_1 \boldsymbol{f}_1 \\ \boldsymbol{f}_s \\ \boldsymbol{f}_2 \end{bmatrix}, \tag{2}$$

where we introduced the restriction matrices M_j to restrict $\boldsymbol{f}_j$ to the corresponding coarse meshes.The Bank-Jimack method is then performing the following iteration:

Algorithm 1: Bank-Jimack Domain Decomposition Method:

1: Set $k = 0$ and choose an initial guess $\boldsymbol{u}^0$.
2: Repeat until convergence

2.1 $$\begin{bmatrix} \boldsymbol{r}_1^k \\ \boldsymbol{r}_s^k \\ \boldsymbol{r}_2^k \end{bmatrix} := \begin{bmatrix} \boldsymbol{f}_1 \\ \boldsymbol{f}_s \\ \boldsymbol{f}_2 \end{bmatrix} - \begin{bmatrix} A_1 & B_1 & 0 \\ B_1^T & A_s & B_2^T \\ 0 & B_2 & A_2 \end{bmatrix} \begin{bmatrix} \boldsymbol{u}_1^k \\ \boldsymbol{u}_s^k \\ \boldsymbol{u}_2^k \end{bmatrix}$$

2.2 $$Solve \begin{bmatrix} A_1 & B_1 & 0 \\ B_1^T & A_s & \widetilde{B}_2^T \\ 0 & \widetilde{B}_2 & \widetilde{A}_2 \end{bmatrix} \begin{bmatrix} \boldsymbol{v}_1^{k+1} \\ \boldsymbol{v}_s^{k+1} \\ \boldsymbol{v}_2^{k+1} \end{bmatrix} = \begin{bmatrix} \boldsymbol{r}_1^k \\ \boldsymbol{r}_s^k \\ M_2 \boldsymbol{r}_2^k \end{bmatrix}, \quad \begin{bmatrix} \widetilde{A}_1 & \widetilde{B}_1 & 0 \\ \widetilde{B}_1^T & A_s & B_2^T \\ 0 & B_2 & A_2 \end{bmatrix} \begin{bmatrix} \boldsymbol{w}_1^{k+1} \\ \boldsymbol{w}_s^{k+1} \\ \boldsymbol{w}_2^{k+1} \end{bmatrix} = \begin{bmatrix} M_1 \boldsymbol{r}_1^k \\ \boldsymbol{r}_s^k \\ \boldsymbol{r}_2^k \end{bmatrix}$$

2.3 $$\begin{bmatrix} \boldsymbol{u}_1^{k+1} \\ \boldsymbol{u}_s^{k+1} \\ \boldsymbol{u}_2^{k+1} \end{bmatrix} := \begin{bmatrix} \boldsymbol{u}_1^k \\ \boldsymbol{u}_s^k \\ \boldsymbol{u}_2^k \end{bmatrix} + \begin{bmatrix} \boldsymbol{v}_1^{k+1} \\ \frac{1}{2}(\boldsymbol{v}_s^{k+1} + \boldsymbol{w}_s^{k+1}) \\ \boldsymbol{w}_2^{k+1} \end{bmatrix}$$

2.4 $k := k + 1$

To get more insight into the Bank-Jimack method, and to relate it to Schwarz methods using optimized Schwarz theory, we consider the concrete example of the 1D Poisson equation

$$-u_{xx} = f \quad \text{in } \Omega = (0, 1), \qquad u(0) = u(1) = 0. \tag{3}$$

We define a *global fine mesh* with N mesh points (see Figure 1 (top row)), and mesh size $h := \frac{1}{N+1}$. Using a finite difference discretization, we find the linear system

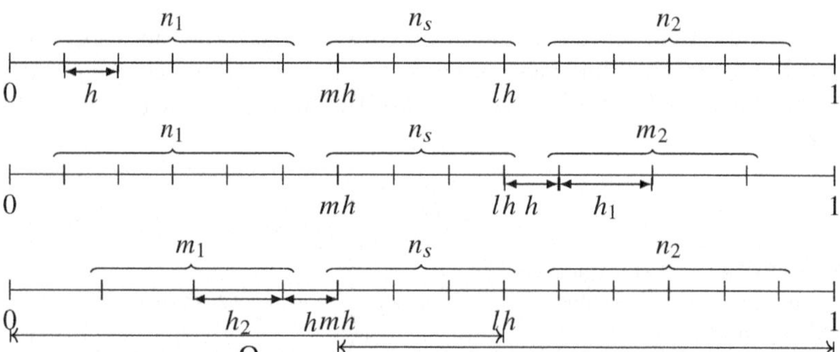

Fig. 1: Global fine mesh, and two partially coarse meshes.

$$\boldsymbol{K}\boldsymbol{u} = \boldsymbol{f}, \quad K := \begin{bmatrix} A_1 & B_1 & 0 \\ B_1^T & A_s & B_2^T \\ 0 & B_2 & A_2 \end{bmatrix} = \frac{1}{h^2}\begin{bmatrix} 2 & -1 & & \\ -1 & 2 & \ddots & \\ & \ddots & \ddots & \end{bmatrix},$$

where $A_1 \in \mathbb{R}^{n_1 \times n_1}$, $B_1 \in \mathbb{R}^{n_1 \times n_s}$, $A_s \in \mathbb{R}^{n_s \times n_s}$, $B_2 \in \mathbb{R}^{n_2 \times n_s}$ and $A_2 \in \mathbb{R}^{n_2 \times n_2}$, and $N = n_1 + n_s + n_2$ (Fig. 1). For the Bank-Jimack method, we also need the two further meshes shown in Figure 1, one with $N_1 := n_1 + n_s + m_2$ mesh points which is fine on Ω_1 with mesh size h and coarse on $\Omega\backslash\Omega_1$ with mesh size h_1, which leads to a linear system of equations of the form (2) (left), with system matrix

$$\begin{bmatrix} A_1 & B_1 & 0 \\ B_1^T & A_s & C_2 \\ 0 & \widetilde{B}_2 & \widetilde{A}_2 \end{bmatrix} := \left[\begin{array}{ccc|ccc|cccc} \frac{2}{h^2} & \frac{-1}{h^2} & & & & & & & & \\ \frac{-1}{h^2} & \frac{2}{h^2} & \ddots & & & & & & & \\ & \ddots & \ddots & \frac{-1}{h^2} & & & & & & \\ \hline & & \frac{-1}{h^2} & \frac{2}{h^2} & \frac{-1}{h^2} & & & & & \\ & & & \frac{-1}{h^2} & \frac{2}{h^2} & \ddots & & & & \\ & & & & \ddots & \ddots & \frac{-1}{h^2} & & & \\ \hline & & & & & \frac{-2}{h(h+h_1)} & \frac{2}{hh_1} & \frac{-2}{h_1(h+h_1)} & & \\ & & & & & & \frac{-1}{h_1^2} & \frac{2}{h_1^2} & \frac{-1}{h_1^2} & \\ & & & & & & & \frac{-1}{h_1^2} & \frac{2}{h_1^2} & \ddots \\ & & & & & & & & \ddots & \ddots \end{array}\right], \tag{4}$$

and one with $N_2 = m_1 + n_s + n_2$ mesh points on Ω which is fine on Ω_2 with mesh size h, and coarse on $\Omega\backslash\Omega_2$, with coarse mesh size h_2, which leads to a linear system of equations of the form (2) (right), with system matrix

$$\begin{bmatrix} \widetilde{A}_1 & \widetilde{B}_1 & 0 \\ C_1 & A_s & B_2^T \\ 0 & B_2 & A_2 \end{bmatrix} := \left[\begin{array}{cccc|ccc|ccc} \frac{2}{h_2^2} & \frac{-1}{h_2^2} & & & & & & & & \\ \frac{-1}{h_2^2} & \ddots & \ddots & & & & & & & \\ & \ddots & \ddots & -\frac{1}{h_2^2} & & & & & & \\ & & -\frac{2}{h_2(h+h_2)} & \frac{2}{hh_2} & \frac{-2}{h(h+h_2)} & & & & & \\ \hline & & & \frac{-1}{h^2} & \frac{2}{h^2} & \frac{-1}{h^2} & & & & \\ & & & & \frac{-1}{h^2} & \frac{2}{h^2} & \ddots & & & \\ & & & & & \ddots & \ddots & \frac{-1}{h^2} & & \\ \hline & & & & & & \frac{-1}{h^2} & \frac{2}{h^2} & \frac{-1}{h^2} & \\ & & & & & & & -\frac{1}{h^2} & \frac{2}{h^2} & \ddots \\ & & & & & & & & \ddots & \ddots \end{array}\right]. \tag{5}$$

For this example, M_j are the transpose of linear interpolation matrices from the fine grid (Fig. 1, top row) to the coarse grids (Fig. 1, second and third row). We find them using an algorithm which is similar to the algorithm introduced in [6] for finding the interface matrices for non-matching grids in one dimension. Running the Bank-Jimack method on this example does not lead to a convergent method[1], see Fig. 2 (left) in the numerical experiments Section 4. This is due to the averaging used in the overlap in step 2.3 of the method, and can be fixed using a specific partition of unity given by the diagonal matrices $\widetilde{D}_1$ and $\widetilde{D}_2$ such that

$$\widetilde{D}_1 = diag(1,\times,\dots,\times,0), \quad \widetilde{D}_2 = diag(0,\times,\dots,\times,1), \quad \widetilde{D}_1 + \widetilde{D}_2 = I_{n_s} \tag{6}$$

One then has to replace step 2.3 in the the method of Bank-Jimack by

$$\begin{bmatrix} \boldsymbol{u}_1^{k+1} \\ \boldsymbol{u}_s^{k+1} \\ \boldsymbol{u}_2^{k+1} \end{bmatrix} := \begin{bmatrix} \boldsymbol{u}_1^{k} \\ \boldsymbol{u}_s^{k} \\ \boldsymbol{u}_2^{k} \end{bmatrix} + \begin{bmatrix} \boldsymbol{v}_1^{k+1} \\ \widetilde{D}_1 \boldsymbol{v}_s^{k+1} + \widetilde{D}_2 \boldsymbol{w}_s^{k+1} \\ \boldsymbol{w}_2^{k+1} \end{bmatrix}. \tag{7}$$

We now present an important property of the Bank-Jimack method with (7):

Lemma 1 *The Bank-Jimack Algorithm with step 2.3 replaced by* (7) *produces for any initial guess* $\boldsymbol{u}^0$ *and arbitrary partitions of unity satisfying* (6) *for* $k = 1, 2, \dots$ *zero residual components outside the overlap,* $\boldsymbol{r}_1^k = \boldsymbol{r}_2^k = \boldsymbol{0}$.

Proof From step 2.1 in the Bank-Jimack method, we obtain

$$\begin{aligned} \boldsymbol{r}_1^k &= \boldsymbol{f}_1 - (A_1 \boldsymbol{u}_1^k + B_1 \boldsymbol{u}_s^k) \\ &= \boldsymbol{f}_1 - A_1(\boldsymbol{u}_1^{k-1} + \boldsymbol{v}_1^k) - B_1(\boldsymbol{u}_s^{k-1} + \widetilde{D}_1 \boldsymbol{v}_s^k + \widetilde{D}_2 \boldsymbol{w}_s^k) && \text{(step 2.3 at } k-1 \text{ and (7))} \\ &= \boldsymbol{f}_1 - A_1 \boldsymbol{u}_1^{k-1} - B_1 \boldsymbol{u}_s^{k-1} - A_1 \boldsymbol{v}_1^k - B_1(\widetilde{D}_1 \boldsymbol{v}_s^k + \widetilde{D}_2 \boldsymbol{w}_s^k) && \text{(rearrange)} \\ &= \boldsymbol{r}_1^{k-1} - A_1 \boldsymbol{v}_1^k - B_1(\widetilde{D}_1 \boldsymbol{v}_s^k + \widetilde{D}_2 \boldsymbol{w}_s^k) && \text{(using step 2.1)} \\ &= B_1 \boldsymbol{v}_s^k - B_1(\widetilde{D}_1 \boldsymbol{v}_s^k + \widetilde{D}_2 \boldsymbol{w}_s^k), \end{aligned}$$

since $\boldsymbol{r}_1^{k-1} - A_1 \boldsymbol{v}_1^k = B_1 \boldsymbol{v}_s^k$ because of the first system satisfied in step 2.3 at $k-1$. Now using the definition of B_1 from (4), we have

$$-B_1 \widetilde{D}_1 \boldsymbol{v}_s^k = \frac{1}{h^2} \begin{bmatrix} & & \\ & & \\ 1 & & \end{bmatrix} \begin{bmatrix} 1 & & & \\ & \times & & \\ & & \ddots & \\ & & & \times & \\ & & & & 0 \end{bmatrix} \begin{bmatrix} v_{s,1}^k \\ \vdots \\ \vdots \\ v_{s,n_s}^k \end{bmatrix} = \frac{1}{h^2} \begin{bmatrix} 0 \\ \vdots \\ 0 \\ v_{s,1}^k \end{bmatrix},$$

independently of the middle elements of $\widetilde{D}_1$, and thus $B_1 \boldsymbol{v}_s^k - B_1 \widetilde{D}_1 \boldsymbol{v}_s^k = \boldsymbol{0}$. On the other hand

[1] Bank and Jimack used the method as a preconditioner for a Krylov method.

$$-B_1\widetilde{D}_2\boldsymbol{w}_s^k = \frac{1}{h^2}\begin{bmatrix} & & \\ & & \\ 1 & & \end{bmatrix}\begin{bmatrix} 0 & & & \\ & \times & & \\ & & \ddots & \\ & & & \times \\ & & & & 1 \end{bmatrix}\begin{bmatrix} w_{s,1}^k \\ \vdots \\ \vdots \\ w_{s,n_s}^k \end{bmatrix} = \begin{bmatrix} 0 \\ \vdots \\ \vdots \\ 0 \end{bmatrix},$$

also independently of the middle elements of $\widetilde{D}_2$, which proves that $\boldsymbol{r}_1^k = \boldsymbol{0}$ for $k = 1, 2, \ldots$. The proof for $\boldsymbol{r}_2^k$ is similar. □

2 Optimized Schwarz Methods

Optimized Schwarz Methods (OSMs) use more effective transmission conditions than the classical Schwarz methods, for an introduction, see [4], and for their relation to sweeping and other more recent domain decomposition methods, see [7]. We now apply a parallel OSM with Robin transmission conditions to our Poisson equation (3) for two subdomains as shown in Fig. 1,

$$\begin{array}{llll} -\partial_{xx}u_1^k = f & \text{in } \Omega_1, & -\partial_{xx}u_2^k = f & \text{in } \Omega_2, \\ u_1^k = 0 & x = 0, & u_2^k = 0 & x = 1, \\ \frac{\partial u_1^k}{\partial n_1} + p_{12}u_1^k = \frac{\partial u_2^{k-1}}{\partial n_1} + p_{12}u_2^{k-1} & x = lh, & \frac{\partial u_2^k}{\partial n_2} + p_{21}u_2^k = \frac{\partial u_1^{k-1}}{\partial n_2} + p_{21}u_2^{k-1} & x = mh. \end{array} \tag{8}$$

Theorem 1 (Special case of Theorem 2 in [3]) *If $p_{12} = \frac{1}{1-lh}$ and $p_{21} = \frac{1}{mh}$, then the OSM* (8) *converges independently of the initial guess in 2 iterations, and is thus an optimal Schwarz method.*

Discretizing the OSM using the same mesh with N grid points as for the method of Bank-Jimack, we obtain

$$\begin{aligned} \frac{1}{h^2}\begin{bmatrix} 2 & -1 & & \\ -1 & \ddots & \ddots & \\ & \ddots & 2 & -1 \\ & & -1 & 1+p_{12}h \end{bmatrix}\begin{bmatrix} u_{1,1}^k \\ \vdots \\ \vdots \\ u_{1,l}^k \end{bmatrix} &= \begin{bmatrix} f_1 \\ \vdots \\ \vdots \\ f_l + (\frac{p_{12}}{h} - \frac{1}{h^2})u_{2,n_s}^{k-1} + \frac{1}{h^2}u_{2,n_s+1}^{k-1} \end{bmatrix}, \\ \frac{1}{h^2}\begin{bmatrix} 1+p_{21}h & -1 & & \\ -1 & 2 & \ddots & \\ & \ddots & \ddots & -1 \\ & & -1 & 2 \end{bmatrix}\begin{bmatrix} u_{2,1}^k \\ \vdots \\ \vdots \\ u_{2,N-m}^k \end{bmatrix} &= \begin{bmatrix} f_m + (\frac{p_{21}}{h} - \frac{1}{h^2})u_{1,m}^{k-1} + \frac{1}{h^2}u_{1,m-1}^{k-1} \\ \vdots \\ \vdots \\ f_N \end{bmatrix}. \end{aligned} \tag{9}$$

3 Bank-Jimack's Method as an Optimized Schwarz Method

We now prove that the method of Bank-Jimack is an optimized Schwarz method with a special choice of the Robin parameter. To do so, we reformulate the matrix systems in step 2.2 of the method: using Lemma 1, we have $M_2\boldsymbol{r}_2^1 = M_1\boldsymbol{r}_1^1 = \mathbf{0}$, and thus one can eliminate the corresponding parts from the equations to obtain

$$\begin{bmatrix} A_1 & B_1 \\ B_1^T & A_s - C_2\widetilde{A}_2^{-1}\widetilde{B}_2 \end{bmatrix} \begin{bmatrix} \boldsymbol{v}_1^1 \\ \boldsymbol{v}_s^1 \end{bmatrix} = \begin{bmatrix} \boldsymbol{r}_1^1 \\ \boldsymbol{r}_s^1 \end{bmatrix}, \quad \begin{bmatrix} A_s - C_1\widetilde{A}_1^{-1}\widetilde{B}_1 & B_2^T \\ B_2 & A_2 \end{bmatrix} \begin{bmatrix} \boldsymbol{w}_s^1 \\ \boldsymbol{w}_2^1 \end{bmatrix} = \begin{bmatrix} \boldsymbol{r}_s^1 \\ \boldsymbol{r}_2^1 \end{bmatrix}, \tag{10}$$

and we are interested in the structure of the Schur complement matrices $A_s - C_2\widetilde{A}_2^{-1}\widetilde{B}_2$ and $A_s - C_1\widetilde{A}_1^{-1}\widetilde{B}_1$.

Lemma 2 (See [9]) *The elements of the inverse of the tridiagonal matrix*

$$T = \begin{bmatrix} a_1 & b_1 & & \\ c_1 & a_2 & \ddots & \\ & \ddots & \ddots & b_{n-1} \\ & & c_{n-1} & a_n \end{bmatrix} \quad \textit{are } (T^{-1})_{ij} = \begin{cases} (-1)^{i+j} b_i \dots b_{j-1}\theta_{i-1}\phi_{j+1}/\theta_n & i < j, \\ \theta_{i-1}\phi_{j+1}/\theta_n & i = j, \\ (-1)^{i+j} c_j \dots c_{i-1}\theta_{j-1}\phi_{i+1}/\theta_n & i > j, \end{cases}$$

where $\theta_0 = 1$, $\theta_1 = a_1$, and $\theta_i = a_i\theta_{i-1} - b_{i-1}c_{i-1}\theta_{i-2}$ for $i = 2, \dots, n$, and $\phi_{n+1} = 1$, $\phi_n = a_n$, and $\phi_i = a_i\phi_{i+1} - b_i c_i \phi_{i+2}$ for $i = n-1, \dots, 1$.

Lemma 3 *The matrices $C_2\widetilde{A}_2^{-1}\widetilde{B}_2$ and $C_1\widetilde{A}_1^{-1}\widetilde{B}_1$ in the Schur complements in* (10) *are given by*

$$C_2\widetilde{A}_2^{-1}\widetilde{B}_2 = \frac{1}{h^2}\begin{bmatrix} 0 & & \\ & \ddots & \\ & & \frac{m_2 h_1}{h+m_2 h_1} \end{bmatrix}, \quad C_1\widetilde{A}_1^{-1}\widetilde{B}_1 = \frac{1}{h^2}\begin{bmatrix} \frac{m_1 h_2}{h+m_1 h_2} & & \\ & 0 & \\ & & \ddots \end{bmatrix}.$$

Proof Using the sparsity of C_2 and $\widetilde{B}_2$, we obtain

$$C_2\widetilde{A}_2^{-1}\widetilde{B}_2 = \begin{bmatrix} 0 & & \\ & \ddots & \\ \frac{-1}{h^2} & & 0 \end{bmatrix} \widetilde{A}_2^{-1} \begin{bmatrix} 0 & & \frac{-2}{h(h+h_1)} \\ & \ddots & \\ & & 0 \end{bmatrix} = \frac{1}{h^2}\begin{bmatrix} 0 & & \\ & \ddots & \\ & & \frac{2}{h(h+h_1)}(\widetilde{A}_2^{-1})_{11} \end{bmatrix},$$

and we thus need to find the first entry of $\widetilde{A}_2^{-1}$. For convenience, we find the first entry of $(h_1^2\widetilde{A}_2)^{-1}$, and then we multiply it by h_1^2. Using Lemma 2, we have $(h_1^2\widetilde{A}_2^{-1})_{11} = \frac{\theta_0\phi_2}{\theta_{m_2}}$ where $\theta_0 = 1$, and

$$\begin{aligned} \theta_{m_2} &= 2\theta_{m_2-1} - \theta_{m_2-2} = 2(2\theta_{m_2-2} - \theta_{m_2-3}) - \theta_{m_2-2} = 3\theta_{m_2-2} - 2\theta_{m_2-3} \\ &= \dots = (m_2 - 1)\left(\frac{4h_1}{h} - \frac{2h_1}{h+h_1}\right) - (m_2 - 2)\frac{2h_1}{h} = \frac{2h_1(m_2 h_1 + h)}{h(h+h_1)}, \end{aligned} \tag{11}$$

and

$$\phi_2 = 2\phi_3 - \phi_4 = 2(2\phi_4 - \phi_5 - \phi_4) = 3\phi_4 - 2\phi_5 \cdots = 2(m_2 - 1) - (m_2 - 2) = m_2.$$

We thus obtain $\frac{\theta_0 \phi_2}{\theta_{m_2}} = \frac{m_2 h(h+h_1)}{2h_1(m_2 h_1 + h)}$, which shows the first claim. The second one is proved similarly. □

Theorem 2 *The Bank-Jimack method in 1D with the partition of unity* (7) *is an optimized Schwarz method with the parameters chosen as* $p_{12} = \frac{1}{h+m_2 h_1}$ *and* $p_{21} = \frac{1}{h+m_1 h_2}$.

Proof It suffices to compare the matrix systems of the OSM (9) with the matrix systems in step 2.3 of the Bank-Jimack method, rewritten as in (10), since in stationary iterations, the standard form and the correction form are equivalent [8, Section 11.2.2]. The system matrices can be made identical by choosing p_{12} such that $1 + p_{12}h = 2 - \frac{m_2 h_1}{h+m_2 h_1}$ and p_{21} such that $1 + p_{21}h = 2 - \frac{m_1 h_2}{h+m_1 h_2}$. □

Since the parameters p_{12} and p_{21} are positive in Theorem 2, it follows from optimized Schwarz theory that the Bank-Jimack method with a partition of unity of the form (7) converges to the monodomain solution, and the convergence is independent of the particular values chosen in the partition of unity, see [5].

Corollary 1 *The Bank-Jimack method in 1D with the partition of unity* (7) *is an optimal Schwarz method: it selects the best possible Robin parameter, independently of how coarse the mesh is in the remaining parts outside of the subdomains, and thus converges in two iterations.*

Proof From Theorem 2 we can see that the Robin parameters p_{12} and p_{21} chosen by the method of Bank-Jimack are independent of the choice of the coarse grid parameters h_1 and h_2, $p_{12} = \frac{1}{h+m_2 h_1} = \frac{1}{1-lh}$ and $p_{21} = \frac{1}{h+m_1 h_2} = \frac{1}{mh}$, which are precisely the optimal choices in Theorem 1 for the OSM. □

4 Numerical Experiments

We first show numerical experiments in one spatial dimension. We discretize the Poisson equation (3) using $N = 2^i$, for $i = 4, \ldots, 7$, gridpoints on the global fine mesh (Fig.1, top row), choose $n_s = 2$ gridpoints in $\Omega_1 \cap \Omega_2$, and $m_1 = m_2 = 2$ coarse mesh points outside the subdomains (Fig. 1, middle and last rows). In Fig. 2, we show on the left that the method of Bank-Jimack using the original partition of unity is not converging. On the right, we show that the method with the new partition of unity converges in two iterations, as expected from the equivalence with the optimal Schwarz method proved in Corollary 1. In Fig. 3, we show on the left that the convergence does not depend on the number of coarse mesh points. We finally show in Fig. 3 on the right a numerical experiment in 2D, where the optimal choice of the Robin parameter in the OSM would lead to a non-local operator involving

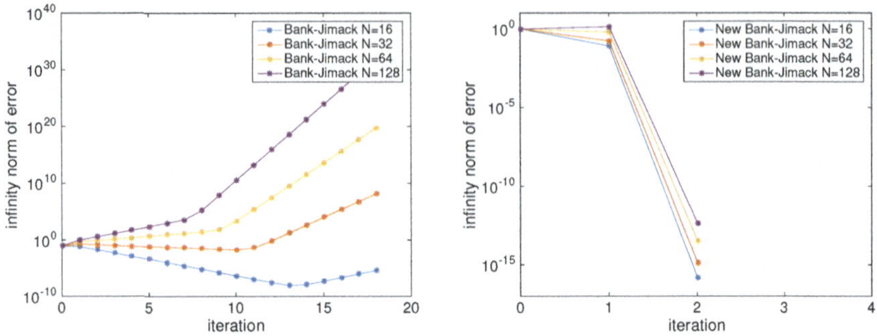

Fig. 2: Error as a function of iteration count of the method of Bank-Jimack with the original partition of unity (left) and new partition of unity (right) for various numbers of global fine mesh points.

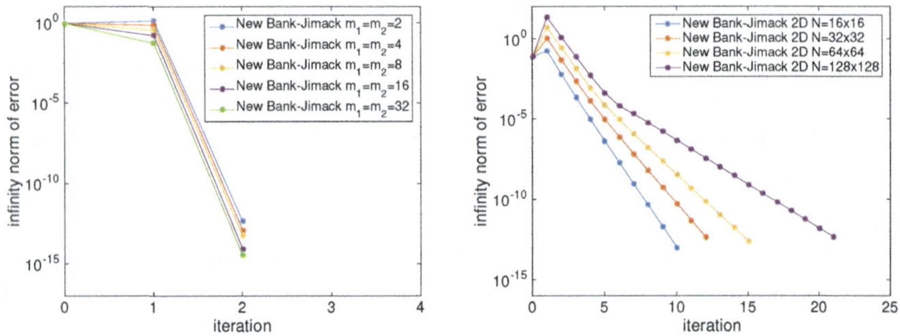

Fig. 3: Left: convergence of the method of Bank-Jimack using $N = 128$ gridpoints on the global fine mesh and various number of gridpoints on the coarse regions. Right: convergence of the method in 2D for various number of gridpoints on the global fine mesh, choosing $n_s = 2$, and $m_1 = m_2 = 2$.

a DtN map, and the method of Bank-Jimack is choosing some approximation. The study of the type of approximation chosen is our current focus of research.

References

1. Bank, R.E., Jimack, P.K.: A new parallel domain decomposition method for the adaptive finite element solution of elliptic partial differential equations. Concurrency and Computation: Practice and Experience **13**(5) (2001)
2. Bank, R.E., Vassilevski, P.S.: Convergence analysis of a domain decomposition paradigm. Computing and Visualization in Science **11**(4-6), 333–350 (2008)
3. Chaouqui, F., Ciaramella, G., Gander, M.J., Vanzan, T.: On the scalability of classical one-level domain-decomposition methods. Vietnam Journal of Mathematics **46 (4)**, 1053–1088 (2018)
4. Gander, M.J.: Optimized Schwarz methods. SIAM Journal on Numerical Analysis **44**(2) (2006)
5. Gander, M.J.: Does the partition of unity influence the convergence of schwarz methods? In: International Conference on Domain Decomposition Methods. Springer (2019). Submitted

6. Gander, M.J., Japhet, C., Maday, Y., Nataf, F.: A new cement to glue nonconforming grids with Robin interface conditions: the finite element case. In: Domain decomposition methods in science and engineering, pp. 259–266. Springer (2005)
7. Gander, M.J., Zhang, H.: A class of iterative solvers for the Helmholtz equation: Factorizations, sweeping preconditioners, source transfer, single layer potentials, polarized traces, and optimized Schwarz methods. SIAM Review **61 (1)**, 3–76 (2019)
8. Gander, W., Gander, M.J., Kwok, F.: Scientific computing-An introduction using Maple and MATLAB, vol. 11. Springer Science & Business (2014)
9. Usmani, R.A.: Inversion of a tridiagonal Jacobi matrix. Linear Algebra and its Applications **212**(213), 413–414 (1994)

Adaptive Schwarz Method for DG Multiscale Problems in 2D

Leszek Marcinkowski* and Talal Rahman

1 Introduction

In many real physical phenomena, there is heterogeneity, e.g., in some ground flow problems in heterogeneous media. When some finite element discretization method is applied to a physical model, one usually obtains a discrete problem which is very hard to solve by a preconditioned iterative method like, e.g., Preconditioned Conjugate Gradient (PCG) method. One of the most popular methods of constructing parallel preconditioners are domain decomposition methods, in particular, non-overlapping or overlapping additive Schwarz methods (ASM), cf. e.g., [16]. In Schwarz methods, a crucial role is played by carefully constructed coarse spaces. For multiscale problems with heterogeneous coefficients standard overlapping Schwarz methods with classical coarse spaces fail often to be fast and robust solvers. Therefore we need new coarse spaces which are adaptive to the jumps of the coefficients, i.e. the convergence of the ASM method is independent of the distribution and the magnitude of the coefficients of the original problem. We refer to [6], [15] and the references therein for similar earlier works on domain decomposition methods which used adaptivity in the construction of the coarse spaces.

In our paper, we consider the Symmetric Interior Penalty Galerkin (SIPG) finite element discretization, i.e., a symmetric version of the interior penalty discontinuous Galerkin (DG) method. DG methods became increasingly popular in recent years,

Leszek Marcinkowski
Faculty of Mathematics, University of Warsaw, Banacha 2, 02-097 Warszawa, Poland
Leszek.Marcinkowski@mimuw.edu.pl

Talal Rahman
Faculty of Engineering and Science, Western Norway University of Applied Sciences, Inndalsveien 28, 5063 Bergen, Norway Talal.Rahman@hvl.no

* This work was partially supported by Polish Scientific Grant: National Science Center: 2016/21/B/ST1/00350.

R. Haynes et al. (eds.), *Domain Decomposition Methods in Science and Engineering XXV*,
Lecture Notes in Computational Science and Engineering 138,
https://doi.org/10.1007/978-3-030-56750-7_33

since they allow that the finite element functions can be completely discontinuous across the element edges, cf. e.g. [14] for an introduction to DG methods.

In the case when the coefficients are discontinuous only across the interfaces between subdomains and are homogeneous inside them, then Schwarz methods with standard coarse spaces are fast and efficient, cf. e.g., [3, 16]. This is however not true in the case when the coefficients may be highly varying and discontinuous almost everywhere, what has in recent years brought many researchers' interest to the construction of new coarse spaces, cf. e.g. [11, 5, 7, 8, 9, 12, 13, 15, 10].

2 Discrete Problem

Let us consider the following elliptic second order boundary value problem in 2D: Find $u^* \in H_0^1(\Omega)$

$$\int_\Omega \alpha(x)\nabla u^*\nabla v\, dx = \int_\Omega fv\, dx, \qquad \forall v \in H_0^1(\Omega), \tag{1}$$

where Ω is a polygonal domain in $\mathbb{R}^2$, $\alpha(x) \geq \alpha_0 > 0$ is the coefficient, and $f \in L^2(\Omega)$.

We introduce $\mathcal{T}_h$ the quasi-uniform triangulation of Ω consisting of closed triangles such that $\bar{\Omega} = \bigcup_{K\in\mathcal{T}_h} K$. Further h_K denotes the diameter of K, and let $h = \max_{K\in\mathcal{T}_h} h_K$ be the mesh parameter for the triangulation.

We will further assume that α is piecewise constant on $\mathcal{T}_h$. Let be given a coarse non-overlapping partitioning of Ω into the open, connected Lipschitz polytopes Ω_i, called substructures or subdomains, such that $\overline{\Omega} = \bigcup_{i=1}^N \overline{\Omega}_i$. We also assume that those substructures are aligned with the fine triangulation, i.e. any fine triangle of $\mathcal{T}_h$ is contained in one substructure. For the simplicity of presentation, we further assume that these substructures form a coarse triangulation of the domain which is shape regular in the sense of [1]. Let Γ_{ij} denote the open edge common to subdomains Ω_i and Ω_j not in $\partial\Omega$ and let Γ be the union of all $\partial\Omega_k \setminus \partial\Omega$.

Further let us define a discrete space S_h as the piecewise linear finite element space defined on the triangulation $\mathcal{T}_h$,

$$S_h = S_h(\Omega) := \{u \in L^2(\overline{\Omega}) : u_{|K} \in P_1,\ K \in \mathcal{T}_h\}.$$

Note that the functions in S_h are multivalued on boundaries of all fine triangles of $\mathcal{T}_h$ except on $\partial\Omega$. Therefore we introduce a set of all edges of elements of $\mathcal{T}_h$ as $\mathcal{E}_h$. Let the $\mathcal{E}_h^\partial \subset \mathcal{E}_h$ be the subset of boundary edges i.e. the edges contained in $\partial\Omega$, and $\mathcal{E}_h^I = \mathcal{E}_h \setminus \mathcal{E}_h^\partial$ be the subset of interior edges, i.e. the edges interior to Ω. We define the L^2-inner products over the elements and the edges respectively as follows, $(u, v)_{\mathcal{T}_h} = \sum_{K\in\mathcal{T}_h} \int_K uv\, dx$ and $(u, v)_{\mathcal{E}_h} = \sum_{e\in\mathcal{E}_h} \int_e uv\, ds$ for $u, v \in S_h$.

The following weights, cf. e.g. [2], are introduced $\omega_+^e = \alpha_-/(\alpha_+ + \alpha_-)$ and $\omega_-^e = \alpha_+/(\alpha_+ + \alpha_-)$, $e \in \mathcal{E}_h^I$, where e is the common edge between two neighboring triangles K_+ and K_-, α_+ and α_- are the restrictions of α to K_+ and K_-, respectively. We have $\omega_+^e + \omega_-^e = 1$. We also need the following notations: $[u] = u_+ n_+ + u_- n_-$ and $\{u\} = \omega_+^e u_+ + \omega_-^e u_-$, where u_+ and u_- are the traces of $u_{|K_+}$ and $u_{|K_-}$ on $e \subset \mathcal{E}_h^I$, while n_+ and n_- are the unit outer normal to ∂K_+ and ∂K_-, respectively. On the boundary we introduce $[u] = u\, n$ and $\{u\} = u$, where n is the unit outer normal to the edge $e \subset \partial K \cap \partial \Omega$, and u is the trace of $u_{|K}$ onto e. We consider SIPG method discrete problems: (cf. [2]). Find $u_h^* \in S_h$

$$a(u_h^*, v) = f(v) \qquad \forall v \in S_h, \tag{2}$$

where $a(u, v) = (\alpha \nabla u, \nabla v)_{\mathcal{T}_h} - (\{\alpha \nabla u\}, [v])_{\mathcal{E}_h} - (\{\alpha \nabla v\}, [u])_{\mathcal{E}_h} + \gamma(\Psi_h [u], [v])_{\mathcal{E}_h}$. Here Ψ_h is a piecewise constant function over the edges of $\mathcal{E}_h$, and γ is a constant positive penalty parameter. The function Ψ_h when restricted to $e \in \mathcal{E}_h^I$, is defined as follows, cf. [2], $\Psi_{h|e} = h_e^{-1}(\omega_+^e \alpha_+ + \omega_-^e \alpha_-) = h_e^{-1} \frac{2}{\frac{1}{\alpha_+} + \frac{1}{\alpha_-}}$ on $\overline{e} = \partial K_+ \cap \partial K_-$,with h_e being the length of the edge $e \in \mathcal{E}_h$.

We have, cf. e.g. [2], $h_e^{-1} \alpha_{min} \leq \Psi_{h|e} \leq 2 h_e^{-1} \alpha_{min}$, $\alpha_{min} = \min(\alpha_+, \alpha_-)$. On a boundary edge $e \in \mathcal{E}_h^{\partial}$ we define $\Psi_{h|e} = h_e^{-1} \alpha_{|K}$. Note that ∇u_h^* is piecewise constant over the fine elements. The discrete problem has a unique solution provided the penalty parameter is sufficiently large, cf. [2]. Let us define a patch around an interface (edge) Γ_{kl}, denoted by Γ_{kl}^{δ}, as the interior of the union of all closed fine triangles having at least a vertex on Γ_{kl}. For the simplicity of the presentation let us assume that the patches cannot share a fine triangle. We divide any patch Γ_{kl}^{δ} into two disjoint open domains - subpatches, $\Gamma_{kl}^{\delta,i} = \Gamma_{kl}^{\delta} \cap \Omega_i$ for $i = k, l$.

The discrete boundary layer of Ω_k: Ω_k^{δ}, is defined as the sum of all subpatches and parts of their boundaries belonging to a subdomain Ω_k, i.e. we have $\overline{\Omega}_k^{\delta} = \bigcup_{\Gamma_{kl} \subset \partial \Omega_k \cap \Gamma} \overline{\Gamma}_{kl}^{\delta,k}$. Each subdomain inherits a local triangulation $\mathcal{T}_h(\Omega_i)$ from $\mathcal{T}_h$, thus we can define a local subspace extended by zero to the remaining substructures: $S_i := \{u \in S_h : u_{|K} = 0 \quad K \not\subset \Omega_i\}$ and its subspace S_i^{δ} formed by the functions from S_i which are also zero on the patch Ω_i^{δ}.

Since the form $a(u, v)$ is positive definite over S_i^{δ} we can introduce a local projection operator $\mathcal{P}_i : S_h \to S_i^{\delta}$: find $\mathcal{P}_i u \in S_i^{\delta}$ such that for $u \in S_h$

$$a(\mathcal{P}_i u, v) = a(u, v), \quad \forall v \in S_i^{\delta}.$$

Note that $\mathcal{P}_i u$ can be computed by solving a local problem over Ω_i.

The discrete harmonic part of $u \in S_i$ is defined as $\mathcal{H}_i u := u_{|\Omega_i} - \mathcal{P}_i u \in S_i$. We say that a function $u \in S_h$ is discrete harmonic if it is discrete harmonic in each subdomain, i.e. $u_{|\Omega_i} = \mathcal{H}_i u$ for $i = 1, \ldots, N$. Knowing the values of discrete harmonic $u \in S_i$ on the patch Ω_i^{δ} allows us to compute u over the remaining triangles contained in Ω_i by solving a local problem. We also introduce spaces related to an edge patch Γ_{kl}^{δ}. Let $S_{kl} \subset S_h$ be the space formed by all discrete harmonic functions which are zero on the all patches except Γ_{kl}^{δ}. We see that $S_{kl} \subset S_k \cup S_l$.

3 Additive Schwarz Method

In this section, we present our overlapping additive Schwarz method for solving (2). Our method is based on the abstract Additive Schwarz Method framework, cf. e.g., [16] for details.

The space S_h is decomposed into the local sub-spaces and a global coarse space. For the local spaces we take $\{S_i\}_{i=1}^N$. We have $S_h = \sum_{i=0}^N S_i$. The global coarse space S_0 is defined in (7), cf. § 3.1, below. Note that the supports of two functions $u_i \in S_i, u_j \in S_j$ for $i \neq j$ with $i, j > 0$ are disjoint, but $a(u_i, u_j)$ may be nonzero due to the edge terms in the bilinear form $a(u, v)$. Thus we see that $S_h = \sum_{i=0}^N S_i$ is a direct sum, but not an orthogonal one in terms of $a(u, v)$. We can interpret this space decomposition as an analog of a classical P_1 continuous finite element decomposition into overlapping subspaces with the minimal overlap.

Next we define the projection like operators $T_i : S_h \to S_i$ as

$$a(T_i u, v) = a(u, v), \quad \forall v \in S_i, \qquad i = 0, \ldots, N. \tag{3}$$

Note that to compute $T_i u$ $i = 1, \ldots, N$ we have to solve N independent local problems, but to get $T_0 u$ we have to solve a global one, cf. § 3.1. Let $T := T_0 + \sum_{i=1}^N T_i$, be the Additive Schwarz operator. We further replace (2) by the following equivalent problem: Find $u_h^* \in S_h$ such that

$$Tu_h^* = g, \tag{4}$$

where $g = \sum_{i=0}^N g_i$ and $g_i = T_i u_h^*$. Note that g_i may be computed without knowing the solution u_h^* of (2), cf. e.g., [16]. The following theoretical estimated of the condition number can be derived:

Theorem 1 *For all $u \in S_h$, the following holds,*

$$c \left(1 + \max_{\Gamma_{kl}} \frac{1}{\lambda_{n_{kl}+1}^{\Gamma_{kl}}} \right)^{-1} a(u, u) \leq a(Tu, u) \leq C\, a(u, u),$$

where C, c are positive constants independent of the coefficient α, the mesh parameter h and the subdomain size H and $\lambda_{n_{kl}+1}^{\Gamma_{kl}}$ is defined in (6).

Below, in § 3.2 we give a sketch of the proof.

3.1 Adaptive patch coarse space

We introduce our adaptive patch based coarse space in this section.

First, we introduce a DG analog of the classical multiscale space, see e.g. [7]. Let $S_{ms} \subset S_h$ be the space of discrete harmonic functions such that for each patch Γ_{kl}^δ a function $u \in S_{ms}$ satisfies

$$a_{kl}(u,v) = 0 \qquad \forall v \in S_{kl}^v, \tag{5}$$

where $a_{kl}(u,v) = \sum_{K\subset\Gamma_{kl}^\delta} \int_K \alpha\nabla u\cdot\nabla v\,dx + \sum_{e\subset\Gamma_{kl}^\delta\cup(\partial\Omega\cap\partial\Gamma_{kl}^\delta)} \Psi_h \int_e [u][v]\,ds$, and $S_{kl}^v \subset S_{kl}$ is formed by the functions which are zero at all degrees of freedom which are at the geometrical ends (crosspoints) of the edge Γ_{kl}. Note that the second sum in the definition of $a_{kl}(u,v)$ is over the fine edges that are either interior to the patch or are on the boundary of Ω.

We introduce the edge generalized eigenvalue problem, which is to find the eigenvalue and its eigenfunction: $(\lambda_j^{kl},\psi_j^{kl}) \in \mathbb{R}_+ \times S_{kl}^v$ such that

$$a_{kl}(\psi_j^{kl},v) = \lambda_j^{\Gamma_{kl}} b_{kl}(\psi_j^{kl},v), \qquad \forall v \in S_{kl}^v, \tag{6}$$

where $a_{kl}(u,v)$ is introduced above. The form $b_{kl}(u,v)$ may be equal to $b_{kl}^{(0)}(u,v) = a(u,v)$ or as in [4] it can be equal to $b_{kl}^{(1)}(u,v) = h^{-2}\int_{\Gamma_{kl}^\delta} \alpha uv\,dx$ or equals the scaled discrete L^2-version of the $b_{kl}^{(1)}$ form, namely, $b_{kl}^{(2)}(u,v) = \sum_{K\in\Gamma_{kl}^\delta} \alpha_{|K} \sum_{j=1}^3 |u(v_j)|^2$. Here in the last sum v_j, for $j = 1,2,3$, denote the vertices of the fine triangle K. Thus we get three different versions of the eigenproblem. Note that the last discrete form $b_{kl}^{(2)}$ can be represented by a diagonal matrix in a matrix form of the eigenproblem. Hence we see that this generalized eigenproblem can be rewritten as a standard eigenproblem, which makes the computations cheaper, cf. also § 4.3 in [9].

We order the eigenvalues in the increasing way as follows $0 < \lambda_1^{kl} \le \ldots \le \lambda_{M_{kl}}^{kl}$ for $M_{kl} = \dim(S_{kl}^v)$. We now can define the local face spectral component of the coarse space for all $\Gamma_{kl} \subset \Gamma$ and the whole coarse space V_0 as follows

$$S_{kl}^{eig} = \mathrm{Span}(\psi_j^{kl})_{j=1}^{n_{kl}}, \qquad S_0 = S_{ms} + \sum_{\Gamma_{kl}\subset\Gamma} S_{kl}^{eig}, \tag{7}$$

where $n_{kl} \le M_{kl}$ is the number of eigenfunctions ψ_j^{kl} chosen by us, e.g. in such a way that the eigenvalue $\lambda_{n_{kl}}^{kl}$, is below a given threshold.

3.2 The sketch of the proof of Theorem 1

The proof follows the lines of the proof of Theorem 3 in [4] and is based on the abstract framework of Additive Schwarz Method, cf. e.g. § 2.3 in [16]. Below C denotes a generic constant independent of the mesh parameters and the problem coefficients. We have to check three key assumptions. The latter two ones, namely, the Strengthened Cauchy Inequalities and Local Stability are verified in a standard way with constants independent of coefficients or mesh parameters. It remains to verify the Stable Splitting assumption. Let $u \in S_h$ and we first define $u_0 \in S_0$ as $u_{ms} + \sum_{\Gamma_{kl}\subset\Gamma} u_{kl}$ where $u_{ms} \in S_{ms}$ takes the values of u at all DOFs at crosspoints. Next on any patch Γ_{kl}^δ let u_{kl} be the b_{kl}-orthogonal projection of $u - u_{ms}$ onto S_{kl}^{eig}, i.e.

$$u_{kl} = \sum_{j \le n_{kl}} \frac{b_{kl}(\psi_j^{kl}, u - u_{ms})}{b_{kl}(\psi_j^{kl}, \psi_j^{kl})} \psi_j^{kl} \in S_{kl}^{eig}.$$

Finally, we define

$$u_j := (u - u_0)_{|\overline{\Omega}_j} \in S_j \qquad j = 1, \ldots, N,$$

what gives us the splitting: $u = u_0 + \sum_{j=1}^{N} u_j$.

Then we estimate the discrete harmonic part $w = \mathcal{H}(u - u_0) = \mathcal{H}u - u_0$, which is zero at crosspoints. Namely, we have the following splitting: $w = \sum_{\Gamma_{kl}} w_{kl}$, where $w_{kl} \in S_{kl}^{v}$ is a discrete harmonic function, which is equal to $u - u_0$ on the respective patch. Note that w_{kl} is b_{kl}-orthogonal to S_{kl}^{eig}. Next we can show that $a(w, w) \le C \sum_{\Gamma_{kl}} b_{kl}(w_{kl}, w_{kl})$ for all types of the bilinear form b_{kl}.

Using the classical theory of the eigenvalue problems, and some technical tools related to SIPG discretizations we can show the stable splitting

$$a(u_0, u_0) + \sum_{k=1}^{N} a(u_k, u_k) \le C \left(1 + \max_{\Gamma_{kl}} \frac{1}{\lambda_{n_{kl}+1}^{\Gamma_{kl}}} \right) a(u, u).$$

The statement in Theorem 1 follows from the abstract ASM theory.

4 Numerical tests

In the tests, our model problem is defined on the unit square with zero Dirichlet boundary condition and a constant force function. We solve it using the SIPG discretization, and the PCG iteration with our additive Schwarz preconditioner. The RHS form in the eigenvalue problem is the scaled L^2 one, i.e., $b_{kl}^{(1)}$. We decompose the domain into 8×8 non-overlapping square sub-domains. We have $H/h = 16$. The penalty parameter γ is equal to four, and the iterations are stopped when the relative residual norm became less than 10^{-6}.

Fig. 1: The coefficient is equal to one on the background and α_0 on the channels. A domain with $8x8$ subdomains. The channels are crossing each other.

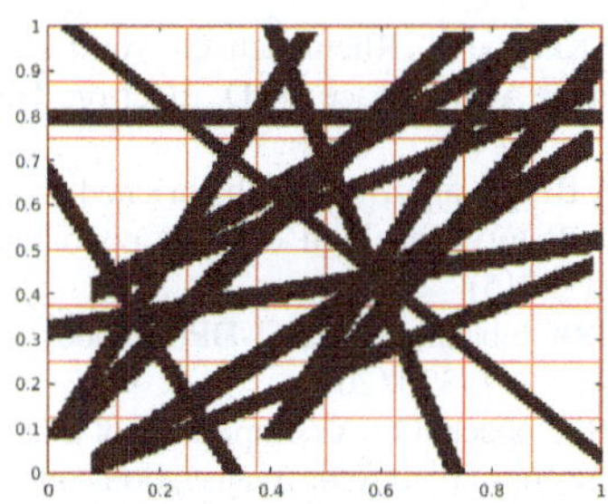

For the adaptive coarse space the threshold for including an eigenfunction is $\lambda \leq 0.18$.

	DG on distribution Fig. 1			
	#Enrichments= 0	#Enrichments= 2	#Enrichments= 4	Adaptive
α_0	Cond.	Cond.	Cond.	Cond.
10^0	57.31(53)	15.65(31)	9.64(24)	15.65(31)
10^2	1.41×10^2(83)	27.03(44)	12.01(31)	26.77(44)
10^4	2.12×10^2(97)	46.71(57)	12.12(32)	27.05(45)
10^6	2.13×10^2(102)	46.78(59)	11.39(35)	26.98(48)

Table 1: Numerical results showing condition number estimates and iteration counts (in parentheses). #Enrichments is per patch (edge).

References

1. Brenner, S.C., Sung, L.Y.: Balancing domain decomposition for nonconforming plate elements. Numer. Math. **83**(1), 25–52 (1999)
2. Cai, Z., Ye, X., Zhang, S.: Discontinuous Galerkin finite element methods for interface problems: a priori and a posteriori error estimations. SIAM J. Numer. Anal. **49**(5), 1761–1787 (2011)
3. Dohrmann, C.R., Widlund, O.B.: An overlapping Schwarz algorithm for almost incompressible elasticity. SIAM J. Numer. Anal. **47**(4), 2897–2923 (2009)
4. Eikeland, E., Marcinkowski, L., Rahman, T.: Adaptively enriched coarse space for the discontinuous Galerkin multiscale problems. Eprint, arXiv:1706.02325v3 [math.NA] (2018). URL `http://arxiv.org/abs/1706.02325v3`
5. Eikeland, E., Marcinkowski, L., Rahman, T.: Overlapping schwarz methods with adaptive coarse spaces for multiscale problems in 3d. Numerische Mathematik **142**(1), 103–128 (2019)
6. Galvis, J., Efendiev, Y.: Domain decomposition preconditioners for multiscale flows in high-contrast media. Multiscale Model. Simul. **8**(4), 1461–1483 (2010)
7. Gander, M.J., Loneland, A., Rahman, T.: Analysis of a new harmonically enriched multiscale coarse space for domain decomposition methods. arXiv preprint arXiv:1512.05285 (2015)
8. Heinlein, A., Klawonn, A., Knepper, J., Rheinbach, O.: An adaptive gdsw coarse space for two-level overlapping schwarz methods in two dimensions. In: P.E. Bjørstad, S.C. Brenner, L. Halpern, H.H. Kim, R. Kornhuber, T. Rahman, O.B. Widlund (eds.) Domain Decomposition Methods in Science and Engineering XXIV, pp. 373–382. Springer International Publishing, Cham (2018)
9. Heinlein, A., Klawonn, A., Knepper, J., Rheinbach, O.: Multiscale coarse spaces for overlapping Schwarz methods based on the ACMS space in 2D. Electron. Trans. Numer. Anal. **48**, 156–182 (2018)
10. Kim, H.H., Chung, E.T.: A BDDC algorithm with enriched coarse spaces for two-dimensional elliptic problems with oscillatory and high contrast coefficients. Multiscale Modeling & Simulation **13**(2), 571–593 (2015)
11. Klawonn, A., Radtke, P., Rheinbach, O.: FETI-DP methods with an adaptive coarse space. SIAM J. Numer. Anal. **53**(1), 297–320 (2015)
12. Klawonn, A., Radtke, P., Rheinbach, O.: A comparison of adaptive coarse spaces for iterative substructuring in two dimensions. Electronic Transactions on Numerical Analysis **45**, 75–106 (2016)

13. Nataf, F., Xiang, H., Dolean, V., Spillane, N.: A coarse space construction based on local Dirichlet-to-Neumann maps. SIAM J. Sci. Comput. **33**(4), 1623–1642 (2011)
14. Rivière, B.: Discontinuous Galerkin methods for solving elliptic and parabolic equations, *Frontiers in Applied Mathematics*, vol. 35. Society for Industrial and Applied Mathematics (SIAM), Philadelphia, PA (2008). Theory and implementation
15. Spillane, N., Dolean, V., Hauret, P., Nataf, F., Pechstein, C., Scheichl, R.: Abstract robust coarse spaces for systems of PDEs via generalized eigenproblems in the overlaps. Numer. Math. **126**, 741–770 (2014)
16. Toselli, A., Widlund, O.: Domain decomposition methods—algorithms and theory, *Springer Series in Computational Mathematics*, vol. 34. Springer-Verlag, Berlin (2005)

Domain Decomposition Coupling of FV4 and DDFV for Numerical Weather Prediction

Oliver Fuhrer, Martin J. Gander, and Sandie Moody

1 Introduction

In the context of Numerical Weather Prediction (NWP) and more precisely in the context of regional weather prediction models, the spatial domains considered usually are non-convex, because of the orography representing mountain ranges. Moreover, the grids used are highly constrained: mesh cells for solving the prognostic equations numerically are much longer and larger than high, e.g. $1.1km \times 1.1km \times 10m$ in COSMO-1. A common practice in NWP is to use terrain-following grids defined such that the distance between the levels grows with altitude (see Figure 1 left). Most weather prediction models use a coordinate change in order to solve the modified prediction equations in a computational domain which uses an equidistant grid (see Figure 1 middle and right). This has the advantage that simple numerical methods such as the finite difference method can be used. However, this also leads to metric terms in the equations due to the mapping, which can cause numerical difficulties.

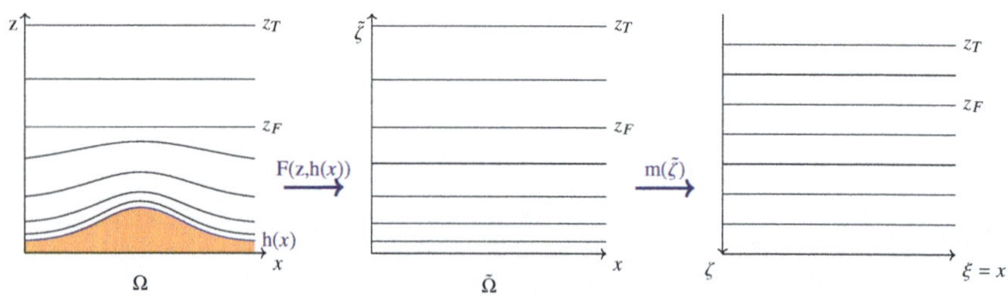

Fig. 1: Mapping of an irregular terrain-following grid to a regular equidistant grid.

Oliver Fuhrer
Vulcan Inc. E-mail: oliverf@vulcan.com

Martin J. Gander and Sandie Moody
Université de Genève. E-mail: {Martin.Gander,Sandie.Moody}@unige.ch

R. Haynes et al. (eds.), *Domain Decomposition Methods in Science and Engineering XXV*, Lecture Notes in Computational Science and Engineering 138,
https://doi.org/10.1007/978-3-030-56750-7_34

In a terrain-following coordinate system, the lowest surface of constant vertical coordinate is conformal to the orography. Any monotonic function can then be used to define the vertical coordinate, denoted by ζ. The COSMO local model [1] offers three options for the terrain-following coordinate. The first one is a pressure-based vertical coordinate, the second one is a height based coordinate, and the third one is a height based SLEVE (Smooth Level Vertical) coordinate. Both height based coordinates are similar to the *Gal-Chen* coordinate [3]. Figure 1 illustrates the height based hybrid coordinate and its mapping to a regular grid.

Definition 1 Let $h(x)$ denote the height of the local topography. The height based hybrid coordinate is defined by

$$\zeta = \begin{cases} \dfrac{z - h(x)}{1 - \frac{h(x)}{z_F}} & \text{if } z < z_F, \\ z & \text{if } z_F \leq z \leq z_T, \end{cases}$$

where z_T is the model top.

Numerical weather prediction models are based on a set of seven governing equations. They comprise the equations of motion, the thermodynamic equation, the continuity equation, the equation of state and the water vapour equation. These equations contain diffusion and advection terms which are treated, in the COSMO model, using a time-splitting method. The diffusion is treated implicitly which implies the solution of a Poisson equation of the form

$$\Delta\phi = f, \tag{1}$$

where ϕ can represent wind components, temperature or pressure.

In order to solve a Poisson equation (1) on the original irregular terrain-following grid Ω, the coordinate transformation described above mapping Ω to a regular equidistant grid is used (see Figure 1). The new coordinates are denoted by (ξ, ζ) and we need to compute the transformed Laplace operator in the new coordinate system ($\tilde{\Delta}_{(\xi,\zeta)}$). The derivatives of the new coordinates with respect to the original ones are expressed using subscripts and are called the *metric terms* of the coordinate change.

Proposition 1 *Let F be a mapping from $\Omega_{(x,z)}$ to $\tilde{\Omega}_{(\xi,\zeta)}$. Let $u = u(x, z)$ be a function defined on Ω and $F(u) = \tilde{u}(\xi, \zeta)$ be a function defined on $\tilde{\Omega}$. The transformed Laplace operator on $\tilde{\Omega}$ when $\xi(x, z) = x$ is given by*

$$\tilde{\Delta}\tilde{u} = \frac{\partial^2 \tilde{u}}{\partial \xi^2} + 2\zeta_x \frac{\partial^2 \tilde{u}}{\partial \xi \partial \zeta} + \frac{\partial^2 \tilde{u}}{\partial \zeta^2}\left(\zeta_x^2 + \zeta_z^2\right) + \frac{\partial \tilde{u}}{\partial \zeta}\left(\zeta_{xx} + \zeta_{zz}\right). \tag{2}$$

The normal derivative on $\partial\Omega$ is expressed by

$$\frac{\partial u}{\partial \boldsymbol{n}} = \boldsymbol{n}^T \begin{pmatrix} \xi_x & \zeta_x \\ \xi_z & \zeta_z \end{pmatrix} \begin{pmatrix} \frac{\partial \tilde{u}}{\partial \xi} \\ \frac{\partial \tilde{u}}{\partial \zeta} \end{pmatrix} = F(\phi) = \tilde{\phi}(\xi, \zeta). \tag{3}$$

Proof Using the chain rule, we find that the second order derivatives on Ω can be expressed by derivatives taken in $\tilde{\Omega}$ by

$$\frac{\partial^2 u}{\partial x^2} = \frac{\partial^2 \tilde{u}}{\partial \xi^2}\xi_x^2 + 2\frac{\partial^2 \tilde{u}}{\partial \xi \partial \zeta}\xi_x\zeta_x + \frac{\partial^2 \tilde{u}}{\partial \zeta^2}\zeta_x^2 + \frac{\partial \tilde{u}}{\partial \xi}\xi_{xx} + \frac{\partial \tilde{u}}{\partial \zeta}\zeta_{xx},$$
$$\frac{\partial^2 u}{\partial z^2} = \frac{\partial^2 \tilde{u}}{\partial \zeta^2}\xi_z^2 + 2\frac{\partial^2 \tilde{u}}{\partial \xi \partial \zeta}\xi_z\zeta_z + \frac{\partial^2 \tilde{u}}{\partial \zeta^2}\zeta_z^2 + \frac{\partial \tilde{u}}{\partial \xi}\xi_{zz} + \frac{\partial \tilde{u}}{\partial \zeta}\zeta_{zz}.$$

Since $\xi = x$, we have $\xi_x = 1$, $\xi_{xx} = \xi_z = \xi_{zz} = 0$ so the second order derivatives reduce to

$$\frac{\partial^2 u}{\partial x^2} = \frac{\partial^2 \tilde{u}}{\partial \xi^2} + 2\frac{\partial^2 \tilde{u}}{\partial \xi \partial \zeta}\zeta_x + \frac{\partial^2 \tilde{u}}{\partial \zeta^2}\zeta_x^2 + \frac{\partial \tilde{u}}{\partial \zeta}\zeta_{xx}, \quad \frac{\partial^2 u}{\partial z^2} = \frac{\partial^2 \tilde{u}}{\partial \zeta^2}\zeta_z^2 + \frac{\partial \tilde{u}}{\partial \zeta}\zeta_{zz}, \tag{4}$$

which when summed give equation (2). In order to prove (3), we simply need to write the gradient using the chain rule which leads to

$$\mathbf{n}^T \nabla_{(x,z)} u = \mathbf{n}^T \begin{pmatrix} \frac{\partial \tilde{u}}{\partial \xi}\xi_x + \frac{\partial \tilde{u}}{\partial \zeta}\zeta_x \\ \frac{\partial \tilde{u}}{\partial \xi}\xi_z + \frac{\partial \tilde{u}}{\partial \zeta}\zeta_z \end{pmatrix} = \mathbf{n}^T \begin{pmatrix} \xi_x & \zeta_x \\ \xi_z & \zeta_z \end{pmatrix} \begin{pmatrix} \frac{\partial \tilde{u}}{\partial \xi} \\ \frac{\partial \tilde{u}}{\partial \zeta} \end{pmatrix} = \tilde{\phi}(\xi, \zeta). \tag{5}$$

The first disadvantage of this method is that the metric terms ζ_x, ζ_{xx}, ζ_z and ζ_{zz} have to be approximated which leads to instabilities when the mesh size of the grid is very small in the vertical direction in comparison with the horizontal direction, which is typically the case in numerical weather prediction, as we have seen. Moreover, when it is used to solve a time-dependent problem, its CFL condition is quite restrictive. The second disadvantage is that the topography in weather prediction models is represented by the grid as a polygon in contrast to the smooth drawing in Fig. 1 (left). This has as an effect that the first and higher order derivatives of the solution expressed in the new set of coordinates (2) lack continuity in general and so the convergence of the scheme is hampered, as we will see in Section 3. We propose a new method to solve the diffusion equation on such domains and grids; the Discrete Duality Finite Volume (**DDFV**) method.

2 Discrete Duality Finite Volume Method

The DDFV method was introduced by K. Domelevo and P. Omnes in 2005 (see [2]). F. Hermeline introduced a finite volume method in 2000 which turned out to be equivalent but the construction had less inherent properties (see [6]). DDFV has the advantage that it is adapted to almost arbitrary meshes and geometries.

We now give the notations which we use to define the DDFV method and which are exemplified in Figure 2. The *primal mesh* forms a partition of Ω and is composed of I elements T_i. With each element T_i we associate a *primal node* G_i located inside T_i. The function θ_i^T is the *characteristic function* of the cell T_i. We denote by J the total

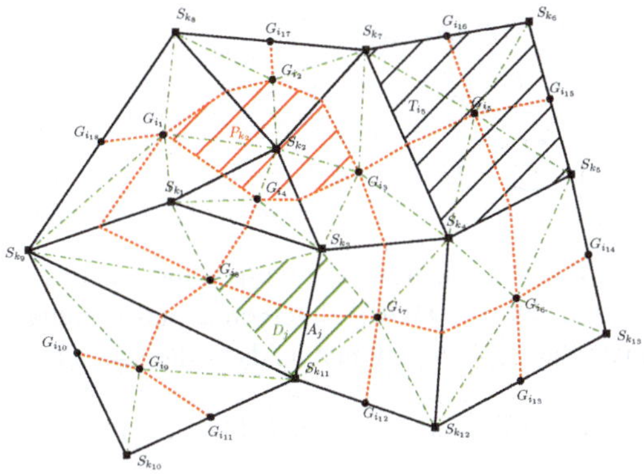

Fig. 2: Notations for the DDFV method.

number of sides of the primal mesh, and by J^Γ the number of these sides which are located on the boundary. We denote the *sides of the primal mesh* by A_j, and assume that they are ordered so that $A_j \subset \Gamma \Leftrightarrow j \in \{J - J^\Gamma + 1, J\}$. We introduce additional primal nodes to each boundary A_j, denoted by G_i with $i \in \{I+1, ..., I+J^\Gamma\}$. The nodes of the primal mesh, the *dual nodes* are denoted by S_k with $k \in \{1, .., K\}$. To each S_k, we associate a *dual cell* P_k obtained by joining the points G_i associated with the elements of the primal mesh of which S_k is a node. The dual mesh also forms a partition of Ω and its sides are denoted by A'_j. We assume that $S_k \in \Gamma$ if and only if $k \in \{K - J^\Gamma + 1, ..., K\}$.

To each A_j we associate a *diamond-cell* obtained by joining the nodes of A_j with the primal nodes associated with the primal cells which share the side A_j (see Figure 2). The unit vector normal to A_j is denoted by $\mathbf{n}_j$ and is oriented such that $\langle G_{i_2(j)} - G_{i_1(j)}, \mathbf{n}_j\rangle \geq 0$. Similarly, the unit vector normal to A'_j is denoted by $\mathbf{n}'_j$ and is oriented so that $\langle S_{k_2(j)} - S_{k_1(j)}, \mathbf{n}'_j\rangle \geq 0$. For all $i \in \{1, \ldots, I\}$, $j \in \mathcal{V}(i)$ (resp. $k \in \{1, \ldots, K\}$, $j \in \mathcal{E}(k)$) we define s_{ji} (resp. s'_{jk}) to be 1 if $\mathbf{n}_j$ points outward of T_i and -1 otherwise (resp. 1 if $\mathbf{n}'_j$ points outward of P_k and -1 otherwise). We thus can define the *outward pointing unit normal vectors* $\mathbf{n}_{ji} = s_{ji}\mathbf{n}_j$ and $\mathbf{n}'_{jk} = s'_{jk}\mathbf{n}'_j$. We define $\mathcal{V}(i) := \{j \in \{1, \ldots, J\} \mid A_j \subset T_i\}$ and $\mathcal{E}(k) := \{j \in \{1, \ldots, J\} \mid S_k \in A_j\}$.

Definition 2 Let ϕ be defined on Ω. The *discrete gradient* is defined on each diamond-cell by

$$(\nabla_h \phi)_j = \frac{1}{2|D_j|}\Big((\phi^P_{k_2} - \phi^P_{k_1})|A'_j|\mathbf{n}'_j + (\phi^T_{i_2} - \phi^T_{i_1})|A_j|\mathbf{n}_j\Big)$$

where $\phi^T_{i_\gamma} = \phi(G_{i_\gamma})$ and $\phi^P_{k_\gamma} = \phi(S_{k_\gamma})$ for $\gamma \in \{1, 2\}$.

Definition 3 The *discrete divergence* $\nabla_h \cdot$ is defined by its values over the primal and dual cells

$$(\nabla_h \cdot \phi)_i = \frac{1}{|T_i|} \sum_{j \in \mathcal{V}(i)} |A_j| \phi_j \cdot \mathbf{n}_{ji},$$

$$(\nabla_h \cdot \phi)_k = \frac{1}{|P_k|} \left(\sum_{j \in \mathcal{E}(j)} |A'_j| \phi_j \cdot \mathbf{n}'_{jk} + \sum_{j \in \mathcal{E}(j) \cap \{J-J^\Gamma+1,\ldots,J\}} \frac{1}{2} |A_j| \phi_j \cdot \mathbf{n}_j \right).$$

Let us consider the Poisson equation (1) with homogeneous Dirichlet boundary conditions. We use the discrete DDFV divergence and gradient operators defined above to approximate the Laplacian which leads to the scheme

$$\begin{cases} -(\nabla_h^T \cdot (\nabla_h \phi))_i = f_i^T & \forall i \in \{1, \ldots, I\} \\ -(\nabla_h^P \cdot (\nabla_h \phi))_k = f_k^P & \forall k \in \{1, \ldots, K - J^\Gamma\} \\ \phi_i^T = 0 & \forall i \in \{I+1, \ldots, I + J^\Gamma\}, \\ \phi_k^P = 0 & \forall k \in \{K - J^\Gamma + 1, \ldots, K\}, \end{cases} \tag{6}$$

where

$$f_i^T = \frac{1}{|T_i|} \int_{T_i} f(x)\, dx, \quad f_k^P = \frac{1}{|P_k|} \int_{P_k} f(x)\, dx.$$

Proposition 2 ([2], Proposition 3.2.) *The linear system given by (6) possesses a unique solution in V where V is defined by*

$$V := \Big\{ \phi = ((\phi_i^T), (\phi_k^P)) \in \mathbb{R}^{I+J^\Gamma} \times \mathbb{R}^K \mid \phi_i^T = 0\, \forall i \in \{I+1, \ldots, I+J^\Gamma\}$$
$$\text{and } \phi_k^P = 0\, \forall k \in \{K - J^\Gamma + 1, \ldots, K\} \Big\}.$$

3 Coupling of DDFV and FV4

One of the main concerns of weather prediction services is computational costs. Due to the fact that the DDFV method introduces additional nodes, the size of the linear system which has to be solved is roughly twice as large as the one associated with the classical Finite Volume (FV4) method. A coupling of FV4 and DDFV allows to reduce the size of this linear system considerably. Such a coupling could be achieved using optimized Schwarz techniques, see for example [7, 5, 4], but we propose here a different approach using interpolation. Let us consider a rectangular domain Ω which has a mountain at its center with slope α, see Figure 3. All cells which are not directly above the mountain are rectangular, so the standard FV4 scheme can be applied on those cells. To the cells which are irregular quadrilaterals, we apply the DDFV method (see Figure 3 , left). The points at the interface ("black diamonds") are dual points which were needed for the DDFV equations associated with primal points ("white squares") and the dual points ("black squares") at the interface. However, they are not associated with a dual cell (see Figure 3, right), so we need to define

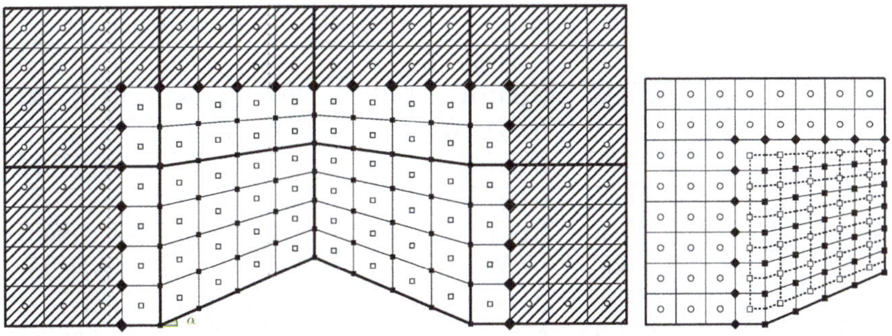

Fig. 3: (Left) Hatched area: Finite volume. White area: DDFV. (Right) FV4 point, primal DDFV points, dual DDFV points and interface points.

a coupling equation. An intuitive way to define the coupling is to set the value of the interface dual points to be the weighted average of its four primal neighboring points, which defines our DDFV-FV4 coupled scheme. For testing purposes, let us consider the problem

$$\Delta u = -5\pi^2 \sin(2\pi x)\sin(\pi y) \text{ on } \Omega,$$

with Dirichlet boundary conditions on the left and right of the domain and Neumann boundary conditions at the top and the bottom of the domain. The order of convergence of both the DDFV method and the DDFV-FV4 coupled method for this problem is 2 (see Figure 4). As for the error, which we define to be the infinity-norm of the difference between the exact solution and the numerical solution, it has a stronger dependence on the mountain angle α for the DDFV method. We then compare the time in seconds needed to solve the linear system associated with the coupled scheme and the DDFV scheme alone. We consider different domains which induce different percentages of the domain to be covered by the DDFV method i.e. different percentages of cells which are not rectangular (column "DDFV-FV4" in

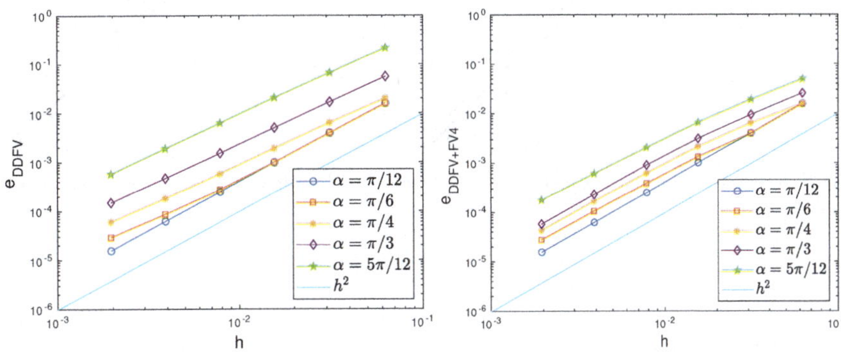

Fig. 4: Error of the DDFV method on the left and error of the DDFV-FV4 coupled scheme on the right.

DDFV-FV4	n	time in sec.	error	COSMO	n	time in sec.	error
14%	128	0.092837	0.0035973	14%	128	0.07799	0.038027
	256	0.47006	0.0010113		256	0.42181	0.022181
18%	128	0.10192	0.0036404	18%	128	0.083277	0.037821
	256	0.58811	0.0010218		256	0.43485	0.021982
37%	128	0.13081	0.004225	37%	128	0.088778	0.10018
	256	0.73254	0.00111		256	0.50752	0.048907
56%	128	0.20006	0.0038666	56%	128	0.10705	0.085701
	256	1.0613	0.00096342		256	0.53349	0.051025
79%	128	0.23772	0.0048132	79%	128	0.12153	0.12076
	256	1.1817	0.0012051		256	0.60488	0.05706
79% DDFV	128	0.19662	0.0044965				
	256	0.98724	0.0011279				

Table 1: Computational time and error of the DDFV-FV4 coupled scheme.

Table 1). We also compute the time and error obtained when using the scheme based on the coordinate transformation described in Section 1 (column "COSMO" in Table 1). We see that the coupled scheme leads to excellent accuracy, even when only a small percentage of DDFV is needed. We note however that even though the coupling method has less degrees of freedom, it is not always faster than the DDFV method. This is due to the fact that our linear system associated with the coupled method is non-symmetric, see Figure 5, whereas the DDFV method gives a symmetric matrix, which is inverted more efficiently by the Matlab solver we use.

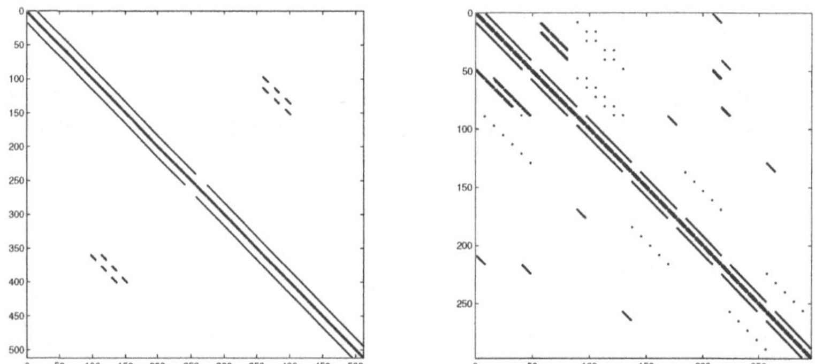

Fig. 5: Structure of the linear system associated to the DDFV method (left) and DDFV-FV4 method (right).

4 Conclusion

We presented a DDFV scheme which does not need a mapping to a regular grid on a rectangular domain for a faithful discretization of diffusion operators on the high aspect ratio grids typical in numerical weather prediction. Moreover, the scheme presented converges on domains which lead to discontinuities in the derivatives of the solution when a mapping to a regular grid is used. Since DDFV uses twice as many unknowns than a standard FV4, we also introduced a coupled DDFV-FV4 scheme which only uses DDFV where it is needed due to the mountain orography. We observed second order convergence for both DDFV and DDFV-FV4. When measuring computing times, the coupled scheme is only faster when less than half the domain is treated by DDFV, even though it always has less unknowns than the DDFV method. We identified the reason for this to be the non-symmetry introduced by our coupling through interpolation between DDFV and FV4. It is currently an open question if a symmetric coupling of these two schemes is possible.

References

1. COSMO-model Documentation. http://www.cosmo-model.org/content/model/documentation/core/default.htm (2019). [Online; accessed 11-March-2019]
2. Domelevo, K., Omnes, P.: A Finite Volume Method for the Laplace Equation on Almost Arbitrary Two-dimensional Grids. ESAIM: Mathematical Modelling and Numerical Analysis **39**(06), 1203–1249 (2005)
3. Gal-Chen, T., Somerville, R.C.J.: On the Use of a Coordinate Transformation for the Solution of the Navier-Stokes Equations. J. Comput. Phys. **17**, 209–228 (1975). DOI: 10.1016/0021-9991(75)90037-6
4. Gander, M.J., Halpern, L., Hubert, F., Krell, S.: DDFV Ventcell Schwarz algorithms. In: Domain Decomposition Methods in Science and Engineering XXII, pp. 481–489. Springer (2016)
5. Gander, M.J., Hubert, F., Krell, S.: Optimized Schwarz algorithms in the framework of DDFV schemes. In: Domain Decomposition Methods in Science and Engineering XXI, pp. 457–466. Springer (2014)
6. Hermeline, F.: A Finite Volume Method for the Approximation of Diffusion Operators on Distorted Meshes. Journal of Computational Physics **160**(2), 481 – 499 (2000). DOI:https://doi.org/10.1006/jcph.2000.6466. URL http://www.sciencedirect.com/science/article/pii/S0021999100964660
7. Krell, S.: Finite Volume Schemes for Complex Fluid Mechanics. Thèse, Université de Provence - Aix-Marseille I (2010). URL https://tel.archives-ouvertes.fr/tel-00524509

A Discrete Domain Decomposition Method for Acoustics with Uniform Exponential Rate of Convergence Using Non-local Impedance Operators

Xavier Claeys, Francis Collino, Patrick Joly, and Emile Parolin

1 Introduction

We consider the Helmholtz equation in harmonic regime in a domain $\Omega \subset \mathbb{R}^d$, $d = 2$ or 3, and a first order absorbing condition on its boundary Γ with unit outward normal vector $\mathbf{n}$. Let $k \in \mathbb{R}$ be a constant wave number and $f \in L^2(\Omega)$, we seek $u \in H^1(\Omega)$ such that

$$\begin{cases} -\Delta u - k^2 u = f, & \text{in } \Omega, \\ (\partial_{\mathbf{n}} + ik)\, u = 0, & \text{on } \Gamma. \end{cases} \tag{1}$$

In previous works [2, 3, 5], a domain decomposition method (DDM) using non-local transmission operator with suitable properties was described. The relaxed Jacobi algorithm written at the continuous level was proven to converge exponentially. However, it was only a conjecture, hinted at by numerical experiments in [5, Section 8], that the discretized algorithm using finite elements has a rate of convergence *uniformly* bounded with respect to the discretization parameter, hence does not deteriorate when the mesh is refined. In this work we prove this conjecture for the case of Lagrange finite elements. Numerical experiments in [5, Section 8.3] highlighted that this important property is not shared by DDM based on local operators [4] or rational fractions of local operators [1].

Xavier Claeys
Sorbonne Université, Université Paris-Diderot SPC, CNRS, INRIA, Laboratoire Jacques-Louis Lions, équipe Alpines, 75005 Paris, France, e-mail: claeys@ann.jussieu.fr

Francis Collino, Patrick Joly, Emile Parolin
POEMS, CNRS, INRIA, ENSTA Paris, Institut Polytechnique de Paris, 91120 Palaiseau, France, e-mail: francis.collino@orange.fr;patrick.joly@inria.fr;emile.parolin@inria.fr

R. Haynes et al. (eds.), *Domain Decomposition Methods in Science and Engineering XXV*,
Lecture Notes in Computational Science and Engineering 138,
https://doi.org/10.1007/978-3-030-56750-7_35

2 DDM algorithm: the continuous case

Impedance based transmission problem. We suppose that the domain is partitioned into two non-overlapping subdomains $\overline{\Omega} = \overline{\Omega}_- \cup \overline{\Omega}_+$. The transmission interface between the subdomains is noted Σ, with a unit normal vector $\mathbf{n}$ oriented from Ω_+ to Ω_-, and we suppose[1] that Σ does not intersect Γ. We then consider the following transmission problem

$$\begin{cases} -\Delta u_\pm - k^2 u_\pm = f|_{\Omega_\pm}, & \text{in } \Omega_\pm, \\ (\pm\partial_\mathbf{n} + ik\mathrm{T})\, u_\pm = (\pm\partial_\mathbf{n} + ik\mathrm{T})\, u_\mp, & \text{on } \Sigma, \end{cases} \tag{2}$$

with $(\partial_\mathbf{n} + ik)\, u_\pm = 0$ on $\Gamma \cap \partial\Omega_\pm$. T is a suitable impedance operator supposed to be injective, positive and self-adjoint so that the coupled problems (2) are well posed and equivalent to the model problem (1), see [2, Th. 3] and [5, Lem. 1].

Reformulation at the interface. Let $V_\pm = H^1(\Omega_\pm)$, $V = V_+ \times V_-$ and $V_\Sigma = H^{-1/2}(\Sigma)$. We define the lifting operator R

$$\mathrm{R} \, : \, V_\Sigma^2 \ni (x_+, x_-) \mapsto (\mathrm{R}_+ x_+, \mathrm{R}_- x_-) \in V. \tag{3}$$

where $u_\pm = \mathrm{R}_\pm x_\pm$ are solutions of the following decoupled boundary value problems

$$-\Delta u_\pm - k^2 u_\pm = 0, \text{ in } \Omega_\pm, \qquad (\pm\partial_\mathbf{n} + ik\mathrm{T})\, u_\pm = x_\pm, \text{ on } \Sigma, \tag{4}$$

and $(\partial_\mathbf{n} + ik)\, u_\pm = 0$ on $\Gamma \cap \partial\Omega_\pm$. We define the scattering operator S

$$\mathrm{S} \, : \, V_\Sigma^2 \ni (x_+, x_-) \mapsto (\mathrm{S}_+ x_+, \mathrm{S}_- x_-) \in V_\Sigma^2, \tag{5}$$

with $\mathrm{S}_\pm x = -x + 2ik\mathrm{T}(\mathrm{R}_\pm x)|_\Sigma$ for $x \in V_\Sigma$. We finally define the operator $\mathrm{A} = \Pi\mathrm{S}$ on V_Σ^2, where Π is an exchange operator: $\Pi\,(x_+, x_-) = (x_-, x_+)$ for a couple of traces $(x_+, x_-) \in V_\Sigma^2$. The following result provides equivalence between the decomposed problem (2) and a problem at the interface (6), see [2, Th. 5] and [5, Prop. 3].

Theorem 1 *If $u = (u_+, u_-) \in V$ is solution of (2) then the trace $x = (x_+, x_-) \in V_\Sigma^2$ defined as $x_\pm := (\pm\partial_\mathbf{n} + ik\mathrm{T})\, u_\mp|_\Sigma$ is solution of the interface problem*

$$x = Ax + b, \qquad on\ \Sigma, \tag{6}$$

where $b = 2ik\,(\mathrm{T}F_-|_\Sigma, \mathrm{T}F_+|_\Sigma) \in V_\Sigma^2$ and $F = (F_+, F_-) \in V$ is such that $-\Delta F_\pm - k^2 F_\pm = f|_{\Omega_\pm}$ in $\Omega_\pm$, $(\pm\partial_\mathbf{n} + ik\mathrm{T})\, F_\pm = 0$ on Σ and $(\partial_\mathbf{n} + ik)\, F_\pm = 0$ on $\Gamma \cap \partial\Omega_\pm$.

Reciprocally, if $x \in V_\Sigma^2$ is solution of (6), then $u = (u_+, u_-) \in V$ defined as $u = Rx + F$ is solution of (2).

Continuous DDM algorithm. The solution of (2) is computed iteratively using a relaxed Jacobi algorithm on the interface problem (6). From an initial trace $x^0 \in V_\Sigma^2$ and a relaxation parameter $r \in (0, 1)$, iteration n writes,

[1] In the presence of such intersections, the proof fails and as a matter of fact the exponential convergence is not observed numerically.

$$x^n = (1-r)x^{n-1} + r\mathrm{A}x^{n-1} + b. \tag{7}$$

Note that the application of A involves solving the decoupled local problems (4) which can be done in parallel. The previous theorem guarantees that the solution of (7) satisfies (2) at convergence. In the following we assume in addition that

$$\mathrm{T} : H^{1/2}(\Sigma) \to H^{-1/2}(\Sigma) \quad \text{is a self-adjoint isomorphism.} \tag{8}$$

Only non-local operators, constructed in practice using integral operators with appropriate singular kernels, can fit in this framework. Under those additional assumptions the algorithm (7) converges exponentially, see [2, Th. 7] and [5, Th. 1].

3 DDM algorithm: the discrete setting

We consider two series $(V_{\pm,h})_h$ of finite dimensional subspaces $V_{\pm,h} \subset V_\pm$ conformal at the interface i.e. $V_{\Sigma,h} = \{u_{\pm,h}|_\Sigma \mid u_{\pm,h} \in V_{\pm,h}\} \subset V_\Sigma$. Let $V_h = V_{+,h} \times V_{-,h} \subset V$. We define the sesquilinear form $a_{\widetilde{\Omega}}$ for a domain $\widetilde{\Omega} \in \{\Omega, \Omega_+, \Omega_-\}$: for all $u, u' \in H^1(\widetilde{\Omega})$,

$$a_{\widetilde{\Omega}}(u,u') = (\nabla u, \nabla u')_{L^2(\widetilde{\Omega})} - k^2 (u,u')_{L^2(\widetilde{\Omega})} + ik(u,u')_{L^2(\Gamma\cap\partial\widetilde{\Omega})}. \tag{9}$$

By Assumption (8), the transmission operator T induces a continuous and coercive sesquilinear form t on $H^{1/2}(\Sigma) \times H^{1/2}(\Sigma)$ such that

$$t(z,z') = \langle \mathrm{T}z, z'\rangle_\Sigma, \qquad \forall z, z' \in H^{1/2}(\Sigma). \tag{10}$$

Reformulation at the interface. We follow the approach of the continuous setting and define the discrete version R_h of the lifting operator R given in (3) by

$$\mathrm{R}_h : V_{\Sigma,h}^2 \ni (x_{+,h}, x_{-,h}) \mapsto (\mathrm{R}_{+,h}x_{+,h}, \mathrm{R}_{-,h}x_{-,h}) \in V_h. \tag{11}$$

with $\mathrm{R}_{\pm,h}$, the discrete versions of $\mathrm{R}_\pm$ given in (4), such that $u_{\pm,h} = \mathrm{R}_{\pm,h}x_{\pm,h}$ satisfies

$$a_{\Omega_\pm}(u_{\pm,h}, u'_{\pm,h}) + ik\, t(u_{\pm,h}, u'_{\pm,h}) = \langle x_{\pm,h}, u'_{\pm,h}\rangle_\Sigma, \qquad \forall u'_{\pm,h} \in V_{\pm,h}. \tag{12}$$

Similarly, the discrete version S_h of the scattering operator S defined in (5) is

$$\mathrm{S}_h : V_{\Sigma,h}^2 \ni (x_{+,h}, x_{-,h}) \mapsto (\mathrm{S}_{+,h}x_{+,h}, \mathrm{S}_{-,h}x_{-,h}) \in V_{\Sigma,h}^2, \tag{13}$$

with the discrete versions $\mathrm{S}_{\pm,h}$ of $\mathrm{S}_\pm$ are such that: for all $w'_{\pm,h} \in V_{\Sigma,h}$,

$$\langle \mathrm{S}_{\pm,h}x_{\pm,h}, w'_{\pm,h}\rangle_\Sigma = -\langle x_{\pm,h}, w'_{\pm,h}\rangle_\Sigma + 2ik\, t\left(\mathrm{R}_{\pm,h}x_{\pm,h}, w'_{\pm,h}\right). \tag{14}$$

We finally define the discrete operator $\mathrm{A}_h = \Pi \mathrm{S}_h$ on $V_{\Sigma,h}^2$. It can then be proven, in a similar fashion as for the continuous case, that the discretization of the problem (2) is equivalent to a discrete counterpart of the interface problem (6): find $x_h \in V^2_{\Sigma,h}$ such that $x_h = \mathrm{A}_h x_h + b_h$, where b_h is the discrete counterpart of b.

Discrete DDM algorithm. In the following we analyse the convergence of the discretization of the DDM algorithm (7): from an initial trace $x^0_h \in V^2_{\Sigma,h}$ and for a relaxation parameter $r \in (0, 1)$, iteration n writes

$$x^n_h = (1 - r)x^{n-1}_h + r\mathrm{A}_h x^{n-1}_h + b_h. \tag{15}$$

4 An abstract uniform exponential convergence result

We now state an abstract result specifying the conditions under which uniform exponential convergence is achieved.

Theorem 2 *If* A_h *is contractant in* $V^2_{\Sigma,h}$ *and* $\mathrm{I} - \mathrm{A}_h$ *is an isomorphism in* $V^2_{\Sigma,h}$ *with* uniformly bounded inverse, *then the relaxed Jacobi algorithm* (15) *with* $r \in (0, 1)$ *converges exponentially* uniformly *(*C *and* τ *are independent of* h *below):*

$$\exists\,\tau \in (0, 1),\ C > 0,\ h_0 > 0, \quad \forall h < h_0,\ n \in \mathbb{N}, \quad \|u^n_h - u_h\|_V \le C\tau^n. \tag{16}$$

Proof At each iteration n, the surface error $\varepsilon^n_h = x^n_h - x_h$ satisfies

$$\varepsilon^{n+1}_h = (1 - r)\varepsilon^n_h + r\mathrm{A}_h\varepsilon^n_h. \tag{17}$$

By hypothesis we have (for some $\delta \in (0, 2]$ independent of h)

$$\|\mathrm{A}_h\varepsilon^n_h\|_{V^2_\Sigma} \le \|\varepsilon^n_h\|_{V^2_\Sigma}, \qquad \text{and} \qquad \|(\mathrm{I} - \mathrm{A}_h)\varepsilon^n_h\|_{V^2_\Sigma} \ge \delta\|\varepsilon^n_h\|_{V^2_\Sigma}. \tag{18}$$

We have the identity for $r \in (0, 1)$ and $a, b \in V^2_\Sigma$

$$\|(1 - r)a + rb\|^2_{V^2_\Sigma} = (1 - r)\|a\|^2_{V^2_\Sigma} + r\|b\|^2_{V^2_\Sigma} - r(1 - r)\|a - b\|^2_{V^2_\Sigma}. \tag{19}$$

Using this identity (take $a = \varepsilon^n_h$ and $b = \mathrm{A}_h\varepsilon^n_h$) together with (17) and (18) we get

$$\|\varepsilon^{n+1}_h\|_{V^2_\Sigma} \le \tau\ \|\varepsilon^n_h\|_{V^2_\Sigma}, \qquad \text{with } \tau = \sqrt{1 - r(1 - r)\delta^2}, \tag{20}$$

and where τ is well defined in $\mathbb{R}$ since $\delta \in (0, 2]$. Since we have $u^n_h - u_h = \mathrm{R}_h\varepsilon^n_h$, the well-posedness of the local problems yields the existence of a constant $c > 0$ independent of h such that, for h sufficiently small, $\|u^n_h - u_h\|_V \le c\|\varepsilon^n_h\|_{V^2_\Sigma}$. □

Since T is assumed to be a self-adjoint isomorphism from $H^{1/2}(\Sigma)$ to $H^{-1/2}(\Sigma)$, the contractive nature of A_h and the fact that $\mathrm{I} - \mathrm{A}_h$ is an isomorphism can be proven, see [2, Th. 3 and Lem. 6] and [5, Lem. 2 and 3]. However, the *uniform* boundedness of the inverse of $\mathrm{I} - \mathrm{A}_h$ was recognized as an open question in [5, Rem. 3]. The previous proof highlights that this property is essential to prevent the convergence rate from potentially degenerating (tending to 1 as h goes to 0).

5 An abstract sufficient condition for exponential convergence

The next theorem states that a sufficient condition for the operator $\mathrm{I} - \mathrm{A}_h$ to be an isomorphism with uniformly continuous inverse relies on the existence of two liftings with suitable properties.

Theorem 3 *Assume that there exists two liftings $L_{\pm,h}$ from $V_{\Sigma,h}$ to $V_{\pm,h}$* uniformly *continuous and preserving Dirichlet boundary conditions: namely there exists $c > 0$, independent of h, such that for all $x_{\pm,h} \in V_{\Sigma,h}$,*

$$\left(L_{\pm,h} x_{\pm,h}\right)|_{\Sigma} = x_{\pm,h}, \quad \text{and} \quad \|L_{\pm,h} x_{\pm,h}\|_{V_\pm} \le c \|x_{\pm,h}\|_{V_\Sigma}. \tag{21}$$

Then $\mathrm{I} - \mathrm{A}_h$ is an isomorphism in $V^2_{\Sigma,h}$ with uniformly *bounded inverse (C is independent of h below):*

$$\exists C > 0,\ h_0 > 0, \quad \forall h < h_0,\ x_h \in V^2_{\Sigma,h}, \quad \|x_h\|_{V^2_\Sigma} \le C \,\|(\mathrm{I} - \mathrm{A}_h) x_h\|_{V^2_\Sigma}, \tag{22}$$

To prove this result, we closely follow the lines of the proof in the continuous case, which we recall below. Let $y = (y_+, y_-) \in V^2_\Sigma$ we aim at finding $x = (x_+, x_-) \in V^2_\Sigma$ such that $(\mathrm{I} - \mathrm{A})x = y$. From the definitions of Section 2, this is equivalent to finding $u_\pm \in V_\pm$ and $x_\pm \in V_\Sigma$ such that (omitting the boundary condition on $\Gamma \cap \partial\Omega_\pm$ here and in the following for brevity)

$$\begin{cases} -\Delta u_\pm - k^2 u_\pm = 0, \quad \text{in } \Omega_\pm, \qquad (\pm\partial_{\mathbf{n}} + ik\,T)\, u_\pm = x_\pm, \quad \text{on } \Sigma, \\ x_\pm - (-x_\mp + 2ik\,T u_\mp) = y_\pm, \quad \text{on } \Sigma. \end{cases} \tag{23}$$

Step 1: Definition of two jumps. A key point is to recognize that the property (8) of T allows to define the Dirichlet and Neumann jumps u_D and u_N such that

$$u_D = (ik\,T)^{-1}\,\frac{y_+ - y_-}{2} \in H^{1/2}(\Sigma), \qquad u_N = \frac{y_- + y_+}{2} \in H^{-1/2}(\Sigma). \tag{24}$$

Step 2: Transmission problem. It is then straightforward to check that the system of equations (23) is equivalent to compute directly $x_\pm = (\pm\partial_{\mathbf{n}} + ik\,T)\, u_\pm$ where $(u_+, u_-) \in V$ is solution of the transmission problem

$$-\Delta u_\pm - k^2 u_\pm = 0, \text{ in } \Omega_\pm, \qquad u_+ - u_- = u_D, \quad \partial_{\mathbf{n}} u_+ - \partial_{\mathbf{n}} u_- = u_N, \text{ on } \Sigma. \tag{25}$$

Step 3: Construction of the solution. The solution $u_\pm$ of (25) is sought in the form $u_\pm = u^c|_{\Omega_\pm} + u^d_\pm$ with $(u^d_+, u^d_-) \in V$ (discontinuous across Σ) and $u^c \in H^1(\Omega)$ (continuous across Σ), constructed as follows. We first construct $u^d_\pm$ as the result of two liftings $L_\pm$ from V_Σ to $V_\pm$ such that $u^d_\pm = \pm\frac{1}{2}\mathrm{L}_\pm u_D$. The liftings $L_\pm$ can be obtained for instance by solving a modified (coercive) Helmholtz equation in the local domains. Having found such a $u^d_\pm$ which satisfies by construction $u^d_+ - u^d_- = u_D$, it is clear that $u_\pm = u^c|_{\Omega_\pm} + u^d_\pm$ solves (25) if $u^c \in H^1(\Omega)$ satisfies (writing $u^c_\pm = u^c|_{\Omega_\pm}$)

$$\begin{cases} -\Delta u_{\pm}^{c} - k^2 u_{\pm}^{c} = \Delta u_{\pm}^{d} + k^2 u_{\pm}^{d}, & \text{in } \Omega_{\pm}, \\ \partial_{\mathbf{n}} u_{+}^{c} - \partial_{\mathbf{n}} u_{-}^{c} = u_N - \partial_{\mathbf{n}} u_{+}^{d} + \partial_{\mathbf{n}} u_{-}^{d}, & \text{on } \Sigma. \end{cases} \tag{26}$$

The problem (26) is well-posed in $H^1(\Omega)$ by application of Fredholm's alternative. The solution $x = (x_+, x_-) \in V_\Sigma^2$ of (23) can finally be computed directly as $x_\pm = (\pm\partial_{\mathbf{n}} + ikT)\, u_\pm$.

The proof at the discrete level mimics this procedure but we need to systematically verify at each step that uniform bounds hold. In the following, C denotes a constant possibly taking different values from one inequality to another.

Proof (of Theorem 3) Let $y_h \in V_{\Sigma,h}^2$, using the definitions of Section 3 the problem of finding $x_h \in V_{\Sigma,h}^2$ such that $(\mathrm{I} - \mathrm{A}_h)x_h = y_h$ writes: find $u_{\pm,h} \in V_{\pm,h}$ and $x_{\pm,h} \in V_{\Sigma,h}$ such that, for all $u'_{\pm,h} \in V_{\pm,h}$ and $z'_{\pm,h} \in V_{\Sigma,h}$,

$$\begin{cases} a_{\Omega_\pm}(u_{\pm,h}, u'_{\pm,h}) + ik\, t(u_{\pm,h}, u'_{\pm,h}) = \langle x_{\pm,h}, u'_{\pm,h}\rangle_\Sigma, \\ \langle x_{\pm,h}, z'_{\pm,h}\rangle_\Sigma - \left(-\langle x_{\mp,h}, z'_{\pm,h}\rangle_\Sigma + 2ik\, t(u_{\mp,h}, z'_{\pm,h})\right) = \langle y_{\pm,h}, z'_{\pm,h}\rangle_\Sigma. \end{cases} \tag{27}$$

Step 1: Definition of two jumps. Let $v_{D,h}$ and $u_{N,h}$ be such that

$$v_{D,h} := (ik)^{-1}\, \frac{y_{+,h} - y_{-,h}}{2}, \qquad u_{N,h} := \frac{y_{-,h} + y_{+,h}}{2}. \tag{28}$$

Both quantities belong to $V_{\Sigma,h}$ and we have, with C independent of h,

$$\|v_{D,h}\|_{V_\Sigma} \le C\, \|y_h\|_{V_\Sigma^2}, \qquad \|u_{N,h}\|_{V_\Sigma} \le C\, \|y_h\|_{V_\Sigma^2}. \tag{29}$$

Note that $v_{D,h}$ is not the discrete counterpart of u_D. A good candidate would be $u_{D,h} = T_h^{-1} v_{D,h}$ where T_h is a discrete version of T. This leads us to the definition

$$u_{D,h} = T_h^{-1} v_{D,h} \qquad \Leftrightarrow \qquad t(u_{D,h}, z'_h) = \langle v_{D,h}, z'_h\rangle_\Sigma, \quad \forall z'_h \in V_{\Sigma,h}. \tag{30}$$

Since t is supposed to be strictly coercive, such a $u_{D,h}$ exists and it holds, with C independent of h,

$$\|u_{D,h}\|_{H^{1/2}(\Sigma)} \le C\, \|v_{D,h}\|_{V_\Sigma}. \tag{31}$$

Step 2: Transmission problem. The solutions $x_{\pm,h} \in V_{\Sigma,h}$ of (27) must satisfy

$$\langle x_{\pm,h}, u'_{\pm,h}\rangle_\Sigma = a_{\Omega_\pm}(u_{\pm,h}, u'_{\pm,h}) + ik\, t(u_{\pm,h}, u'_{\pm,h}), \quad \forall u'_{\pm,h} \in V_{\pm,h}, \tag{32}$$

where $u_{\pm,h} \in V_{\pm,h}$ must satisfy a discrete version of the transmission problem (25)

$$\begin{cases} t(u_{+,h} - u_{-,h}, z'_h) = \langle v_{D,h}, z'_h\rangle_\Sigma, & \forall z'_h \in V_{\Sigma,h}, \\ a_{\Omega_+}(u_{+,h}, u'_{+,h}) + a_{\Omega_-}(u_{-,h}, u'_{-,h}) = \langle u_{N,h}, u'_h\rangle_\Sigma, & \forall (u'_{+,h}, u'_{-,h}) \in V_h \cap H^1(\Omega). \end{cases} \tag{33}$$

where the equation on the first line is obtained by taking the difference of the two equations in the second line of (27) and the equation on the second line is obtained by summing all equations in (27) with a test function in $V_h \cap H^1(\Omega)$.

Reciprocally, let $u_{\pm,h} \in V_{\pm,h}$ and $x_{\pm,h} \in V_{\Sigma,h}$ be solutions of (33) and (32). For any $z'_h \in V_{\Sigma,h}$, there exists by assumption $u'_h = (u'_{+,h}, u'_{-,h}) \in V_h \cap H^1(\Omega)$ such that $u'_{\pm,h}|_\Sigma = z'_h$. By taking linear combinations of equations of (33) with these test functions z'_h and u'_h one obtains the two equations on the second line of (27).

Step 3: Construction of the solution. The solution $u_{\pm,h}$ of (33) is sought in the form $u_{\pm,h} = u^c_h|_{\Omega_\pm} + u^d_{\pm,h}$ with $(u^d_{+,h}, u^d_{-,h}) \in V_h$ and $u^c_h \in V_h \cap H^1(\Omega)$ constructed as follows. We first construct $u^d_{\pm,h} = \pm\frac{1}{2}\mathrm{L}_{\pm,h}u_{D,h}$, by hypothesis on the liftings we have

$$\|u^d_{\pm,h}\|_{V_\pm} \leq C\,\|u_{D,h}\|_{H^{1/2}(\Sigma)}, \tag{34}$$

with C independent of h. By construction $t(u^d_{+,h} - u^d_{-,h}, z'_h) = \langle v_{D,h}, z'_h\rangle_\Sigma$ for all $z'_h \in V_{\Sigma,h}$. Hence, using the last equation in (33), $u_{\pm,h} = u^c_h|_{\Omega_\pm} + u^d_{\pm,h}$ will be solution of (33) if $u^c_h \in V_h \cap H^1(\Omega)$ is such that, for all $(u'_{+,h}, u'_{-,h}) \in V_h \cap H^1(\Omega)$,

$$a_\Omega(u^c_h, u'_h) = \langle u_{N,h}, u'_h\rangle_\Sigma - a_{\Omega_-}(u^d_{-,h}, u'_{-,h}) - a_{\Omega_+}(u^d_{+,h}, u'_{+,h}). \tag{35}$$

Since a_Ω is $H^1(\Omega)$-coercive, it is well known from the theory of Galerkin approximation of Fredholm type problem that for h sufficiently small, such a u^c_h exists and it holds, with C independent of h,

$$\|u^c_h\|_V \leq C\left(\|u_{N,h}\|_{V_\Sigma} + \|u^d_{-,h}\|_{V_-} + \|u^d_{+,h}\|_{V_+}\right). \tag{36}$$

From $u_{\pm,h} = u^d_{\pm,h} + u^c_h|_{\Omega_\pm}$ in $V_{\pm,h}$ we have, with C independent of h,

$$\|u_{\pm,h}\|_{V_\pm} \leq C\left(\|u^d_{\pm,h}\|_{V_\pm} + \|u^c_h|_{\Omega_\pm}\|_{V_\pm}\right). \tag{37}$$

The solution $x_{\pm,h} \in V_{\Sigma,h}$ of (32) hence (27) are computed using (33) hence satisfy, with C independent of h,

$$\|x_{\pm,h}\|_{V_\Sigma} \leq C\,\|u_{\pm,h}\|_{V_\pm}. \tag{38}$$

Since all the quantities computed at each step are bounded uniformly by the data used for their construction, see (29), (31) , (34), (36) and (37), the uniform bound of Theorem 3 with respect to h is established. □

6 Application to finite element approximations

In this section we assume that $\Omega_\pm$ are bounded open polyhedral Lipchitz domains discretized using conforming simplicial mesh elements and consider classical Lagrange finite element spaces. The previous proof relies on the existence of two uniformly stable liftings $L_{\pm,h}$ from $V_{\Sigma,h}$ to $V_{\pm,h}$ which must preserve the Dirichlet trace on Σ. A theoretical construction of such liftings $L_{\pm,h}$ can be obtained as $L_{\pm,h} = P_h \circ L_\pm$ where $P_h : V_\pm \rightarrow V_{\pm,h}$ is an interpolator and $L_\pm : V_\Sigma \rightarrow V_\pm$ are two continuous liftings. The construction of $L_{\pm,h}$ is hence reduced to the construction of P_h. The classical Lagrange interpolator fails to provide a practical answer because it lacks the continuity property for non-smooth functions (point-wise function evaluations). The Clément interpolator is continuous but fails to preserve the prescribed trace on the boundary. An interpolator featuring the suitable properties have been proposed by Scott and Zhang [6] for general conforming Lagrange finite elements of any order in $\mathbb{R}^d$, $d = 2, 3$. For the sake of illustration, we briefly recall below the construction of this operator for $\mathbb{P}_1$ Lagrange finite elements on triangles.

For each vertex M_i of the mesh, choose arbitrarily σ_i an edge connected to M_i. The application $v \in \mathbb{P}_1(\sigma_i) \mapsto v(M_i) \in \mathbb{R}$ is a continuous linear form on $\mathbb{P}_1(\sigma_i) \subset L^2(\sigma_i)$. From Riesz theorem, there exists a unique $\psi_i \in \mathbb{P}_1(\sigma_i)$ such that, for all $v \in \mathbb{P}_1(\sigma_i)$, we have $v(M_i) = (\psi_i, v)_{L^2(\sigma_i)}$. Let w_i be the $\mathbb{P}_1$ Lagrange basis function associated to the vertex M_i. There is a natural definition of an interpolation operator P_h on $H^1(\Omega)$ such that: for all $v \in H^1(\Omega)$,

$$P_h v := \sum_i (\psi_i, v)_{L^2(\sigma_i)} w_i. \tag{39}$$

From the trace theorem, P_h is a continuous linear mapping from $H^1(\Omega)$ to V_h and is invariant on V_h. To preserve the trace on the boundary, we require in addition that for all vertices M_i on the boundary of Ω, the edge σ_i is chosen to belong to the boundary. This operator P_h is the Scott-Zhang operator and satisfies Hypothesis (21), see [6, Th. 2.1 and Cor. 4.1].

Acknowledgements This work was supported by the research Grant ANR–15–CE23–0017–01.

References

1. Y. Boubendir, X. Antoine, and C. Geuzaine. A Quasi-Optimal Non-Overlapping Domain Decomposition Algorithm for the Helmholtz Equation. *J. Comp. Phys.*, 213(2):262–280, 2012.
2. F. Collino, S. Ghanemi, and P. Joly. Domain decomposition method for harmonic wave propagation: a general presentation. *CMAME*, 184(24):171–211, 2000.
3. F. Collino, P. Joly, and M. Lecouvez. Optimized quasi-local transmission conditions for non overlapping domain decomposition method. Submitted.
4. B. Després. *Méthodes de décomposition de domaine pour la propagation d'ondes en régime harmonique. Le théorème de Borg pour l'équation de Hill vectorielle.* PhD thesis, Université Paris IX Dauphine, 1991.
5. M. Lecouvez. *Méthodes itératives de décomposition de domaine sans recouvrement avec convergence géométrique pour l'équation de Helmholtz.* PhD thesis, Ecole polytechnique, 2015.
6. L. R. Scott and S. Zhang. Finite element interpolation of nonsmooth functions satisfying boundary conditions. *Math. Comp.*, 54(190):483–493, 1990.

Optimized Schwarz Methods for Linear Elasticity and Overlumping

Kévin Santugini-Repiquet

1 Introduction

Linear Elasticity models how elastic solids deform in the presence of surface and volume forces. The model of Linear Elasticity is valid for small deformations. For large deformations, the nonlinear theory of elasticity should be used instead. For an introduction to Linear Elasticity, we refer the reader to [3]. Linear Elasticity is commonly discretized using Finite Element Methods, see [1, Chap. 11].

Domain Decomposition Methods (DDMs) have previously been applied to Linear Elasticity [2, 6, 5, 7, 10, 11, 14, 15, 13, 19, 18, 16]. However, we found no reference on applying OSMs to the equations of Linear Elasticity.

Optimized Schwarz methods(OSMs) are a family of Domain Decompositions Methods. In iterative OSMs, at each iteration, the interior equation is solved inside each subdomain with artificial conditions on each subdomain boundary. Then, data is exchanged between neighboring subdomains to update those boundary conditions. The process is reiterated until convergence. See [8] for a full analysis of OSMs. The most common transmission conditions are Robin transmission conditions and Ventcell transmission conditions. In [9], the authors showed that we should lump (and even overlump) Robin transmission conditions when applying OSMs to a FEM(Finite Element Method) discretization of Poisson Equations.

In this paper, our main goal is to apply one-level Optimized Schwarz Methods (OSM) to the Finite Element Discretization of the Linear Elasticity problems. We first present some basic definitions on Linear Elasticity in §2. To this end, we derive transmission conditions applicable to Linear Elasticity, obtain an OSM for Linear Elasticity, and establish convergence in §3 using energy estimates. Finally, in §4, we present numerical results, and observe that numerically, overlumping transmission conditions at the discrete level yields a better convergence rate.

Kévin Santugini
Bordeaux INP, IMB, UMR 5251, F-33400,Talence, France, e-mail: Kevin.Santugini-Repiquet@bordeaux-inp.fr

R. Haynes et al. (eds.), *Domain Decomposition Methods in Science and Engineering XXV*,
Lecture Notes in Computational Science and Engineering 138,
https://doi.org/10.1007/978-3-030-56750-7_36

2 Basic Linear Elasticity Definitions

Let Ω be a domain of $\mathbb{R}^3$. Let $\boldsymbol{u} : \Omega \to \mathbb{R}^3$ be a vector field called small displacements. The strain tensor ε is defined as $\varepsilon_{ij}(\boldsymbol{u}) = \frac{1}{2}\left(\frac{\partial u_i}{\partial x_j} + \frac{\partial u_j}{\partial x_i}\right)$, and the stress tensor σ is defined as $\sigma_{ij}(\boldsymbol{u}) = \sum_{k\ell} \mathsf{C}_{ijk\ell}\varepsilon_{k\ell}(\boldsymbol{u})$. The tensor $\mathsf{C}_{ijk\ell}$ is called the stiffness tensor, depends on the material, and satisfy $\mathsf{C}_{ijk\ell} = \mathsf{C}_{k\ell ij}$ and $\mathsf{C}_{ijk\ell} = \mathsf{C}_{jik\ell}$. In addition, the stiffness tensor is positive definite, *i.e.*, there exists $\alpha > 0$ such that

$$\sum_{i,j,k,\ell} \mathsf{C}_{ijk\ell}(\boldsymbol{x})\varepsilon_{ij}\varepsilon_{k\ell} \geq \alpha \sum_{ij} |\varepsilon_{ij}|^2.$$

In this paper, we only consider homogenous isotropic materials. For isotropic material,

$$\mathsf{C}_{ijk\ell} = \frac{E}{1+\nu}\delta_i^k\delta_j^\ell + \frac{E\nu}{(1+\nu)(1-2\nu)}\delta_i^j\delta_k^\ell,$$

where E is the Young modulus, and ν is the Poisson coefficient.

Let $f_v : \Omega \to \mathbb{R}^3$ be the vector field of volume forces applied to the solid body. Let $\Gamma_d \subset \partial\Omega$. Let $f_s : \Gamma_d \to \mathbb{R}^3$ be the vector field of surface forces applied to $\Gamma_f \subset \partial\Omega$. And let $\boldsymbol{d}$ be the known displacements on $\Gamma_d = \partial\Omega \setminus \Gamma_f$. In the variational formulation of Linear Elasticity, a weak solution is defined as a $\boldsymbol{u}$ in V such that for all $\boldsymbol{v}$ in V_ℓ

$$\int_\Omega \sigma(\boldsymbol{u}) : \varepsilon(\boldsymbol{v})\mathrm{d}\boldsymbol{x} = \int_\Omega f_v \boldsymbol{v}\mathrm{d}\boldsymbol{x} + \int_{\Gamma_f} f_s \boldsymbol{v}\mathrm{d}S(\boldsymbol{x}. \qquad (1)$$

where

$$V = \{\boldsymbol{u} \subset H^1(\Omega;\mathbb{R}^3) : \boldsymbol{u} = \boldsymbol{d} \text{ on } \Gamma_d\}, \qquad V_\ell = \{\boldsymbol{v} \in H^1(\Omega) : \boldsymbol{v} = 0 \text{ on } \Gamma_d\}.$$

3 Optimized Schwarz Methods for Linear Elasticity

3.1 At the continuous level

In iterative OSMs, at each iteration, the interior equation is solved inside each subdomain with artificial transmission conditions at the interface between subdomains. Then, in order to update these conditions, data is exchanged between neighboring subdomains.

In order to apply OSMs to Linear Elasticity, adequate transmission conditions are needed. For Poisson equations, the simplest transmission conditions are Robin transmission conditions. Robin conditions are a linear combination of Dirichlet and Neumann boundary conditions. The Neumann conditions originates from the following integral equality:

$$\int_{\Omega} \nabla\phi \cdot \nabla\psi \mathrm{d}\boldsymbol{x} - \int_{\Omega} (-\Delta\phi)\psi \mathrm{d}\boldsymbol{x} = \int_{\partial\Omega} \frac{\partial\phi}{\partial\boldsymbol{n}} \psi \mathrm{d}S(\boldsymbol{x}).$$

for all $\phi\colon \Omega \to \mathbb{R}$, and $\psi : \Omega \to \mathbb{R}$ regular enough and where $\boldsymbol{n}$ is the outer-pointing normal to Ω. Likewise, from the variational formulations of Linear Elasticity (1), we get

$$\int_{\Omega} \sigma(\boldsymbol{u}) : \varepsilon(\boldsymbol{v}) \mathrm{d}\boldsymbol{x} - \int_{\Omega} (-\mathrm{div}(\sigma(\boldsymbol{u})))\boldsymbol{v} \mathrm{d}\boldsymbol{x} = \int_{\partial\Omega} (\sigma(\boldsymbol{u})\boldsymbol{n}) \cdot \boldsymbol{v} \mathrm{d}S(\boldsymbol{x}).$$

Hence, the equivalent to the Neumann boundary condition for Linear Elasticity is $\sigma(\boldsymbol{u})\boldsymbol{n}$.

To define OSMs on the equations of linear elasticity, we consider a domain Ω divided in N subdomains Ω_i $\Gamma_{ij} := \partial\Omega_i \cap \partial\Omega_j$. Let $\boldsymbol{n}_{ij}$ be the normal to Γ_{ij} pointing from Ω_i to Ω_j. Let S_{ij} be operators on some functional space defined over $\Gamma_{ij} := \partial\Omega_i \cap \partial\Omega_j$. Transmission conditions for Linear Elasticity are:

$$\sigma(\boldsymbol{u}_i^{n+1})\boldsymbol{n}_{ij} + S_{ij}\boldsymbol{u}_i^{n+1} = \sigma(\boldsymbol{u}_j^n)\boldsymbol{n}_{ij} + S_{ij}\boldsymbol{u}_j^n$$

In particular, Robin transmission conditions for Linear Elasticity are obtained when $S_{ij}(\boldsymbol{u}) = p\boldsymbol{u}$ with $p \in \mathbb{R}^+$ being the Robin parameter. In this paper, we always suppose $S_{ij} = S_{ji}$. The Optimized Schwarz Algorithm for the equations of Linear Elasticity at the continuous level is given in Algorithm 1.

Algorithm 1: (Optimized Schwarz for Linear Elasticity)

Initialize $g_{ij}^0 : \Gamma_{ij} \to \mathbb{R}^3$, to some initial guess in $L^2(\Gamma_{ij})$.
for $n \geq 0$ and until convergence **do**
In each subdomain Ω_i, compute the iterates $\boldsymbol{u}_i^n$ in parallel as the solutions in Ω_i to the variational formulation of:

$$\begin{cases} \mathrm{div}(\sigma(\boldsymbol{u}_i^n)) + \boldsymbol{f}_v = 0 \text{ in } \Omega_i, \\ \sigma(\boldsymbol{u}_i^n)\boldsymbol{n}_{ij} + S_{ij}\boldsymbol{u}_i^n = g_{ij}^n \text{ on } \Gamma_{ij}, \\ \boldsymbol{u}_i^n = \boldsymbol{d} \text{ on } \partial\Omega_i \cap \Gamma_d, \\ \sigma(\boldsymbol{u}_i^n)\boldsymbol{n} = \boldsymbol{f}_s \text{ on } \partial\Omega_i \cap \Gamma_f. \end{cases}$$

For all neighboring subdomains Ω_i and Ω_j, set $g_{ij}^{n+1} := -g_{ji}^n + (S_{ij} + S_{ji})u_j^{n+1}$.
end for

Using Energy Estimates introduced in [17, 4] for the Poisson equation, we can prove the convergence of OSMs applied to Linear Elasticity at the continuous level.

Theorem 1 *If $S_{ij}^h = S_{ji}^h$, if each S_{ij} is symmetric positive definite, and if there is one subdomain where $\Gamma_d \cap \partial\Omega_i$ is of nonzero surface measure, then the Optimized Schwarz Method (1) at the continuous level is convergent.*

Proof Due to the linearity of the equations, we can without loss of generality suppose that the volume forces $\boldsymbol{f}_s$, surface forces $\boldsymbol{f}_v$ and known displacements $\boldsymbol{d}$ are null.

For each subdomain Ω_i, we multiply the interior equation satisfied by $\boldsymbol{u}_i^n$ by $\boldsymbol{u}_i^n$ then integrate over Ω_i. After applying Green's formulas, we get:

$$\int_{\Omega_i} \sigma(\boldsymbol{u}_i^n) : \varepsilon(\boldsymbol{u}_i^n)\mathrm{d}\boldsymbol{x} = \sum_j \int_{\Gamma_{ij}} (\sigma(\boldsymbol{u}_i^n)\boldsymbol{n}_{ij}) \cdot \boldsymbol{u}_i^n \mathrm{d}S(\boldsymbol{x}). \tag{2}$$

By [12, Theorem 3.35], S_{ij} has a symmetric definite positive square root which we denote by $M_{ij} := S_{ij}^{1/2}$. And we have:

$$\begin{aligned}
&\int_{\Gamma_{ij}} (\sigma(\boldsymbol{u}_i^n)\boldsymbol{n}_{ij}) \cdot \boldsymbol{u}_i^n \mathrm{d}S(\boldsymbol{x}) = \int_{\Gamma_{ij}} (M_{ij}^{-1}\sigma(\boldsymbol{u}_i^n)\boldsymbol{n}_{ij}) \cdot M_{ij}\boldsymbol{u}_i^n \mathrm{d}S(\boldsymbol{x}) \\
&= \frac{1}{4}\left(\int_{\Gamma_{ij}} |M_{ij}^{-1}(\sigma(\boldsymbol{u}_i^n)\boldsymbol{n}_{ij} + S_{ij}\boldsymbol{u}_i^n)|^2 \mathrm{d}S(\boldsymbol{x}) - \int_{\Gamma_{ij}} |M_{ij}^{-1}(\sigma(\boldsymbol{u}_i^n)\boldsymbol{n}_{ij} - S_{ij}\boldsymbol{u}_i^n)|^2 \mathrm{d}S(\boldsymbol{x})\right), \\
&= \frac{1}{4}\left(\int_{\Gamma_{ij}} |M^{-1} g_{ij}^n|^2 \mathrm{d}S(\boldsymbol{x}) - \int_{\Gamma_{ij}} |M^{-1} g_{ij}^{n+1}|^2 \mathrm{d}S(\boldsymbol{x})\right)
\end{aligned}$$

Combining this equality with (2), and summing over the subdomain index i, and over the iteration index n, we get

$$\sum_{n=0}^{+\infty}\sum_{i=1}^{N} \int_{\Omega_i} \sigma(\boldsymbol{u}_i^n) : \varepsilon(\boldsymbol{u}_i^n)\mathrm{d}\boldsymbol{x} \leq \frac{1}{4}\sum_{ij} \int_{\Gamma_{ij}} |M^{-1} g_{ij}^0|^2 \mathrm{d}S(\boldsymbol{x}) < +\infty.$$

Since the stiffness tensor $\mathsf{C}_{ijk\ell}$ is positive definite, this implies $\varepsilon(\boldsymbol{u}_i^{n+1})$ converges to 0 as n goes to infinity. This proves that inside each subdomain, the iterates converges to an equiprojective vector field. This implies the limit is zero on the subdomain where $\Gamma_d \cap \partial\Omega_i$ is of nonzero measure. Since a domain is always connected by definition, and using the transmission condition, one gets the limit is also zero on the other subdomains. □

3.2 FEM Discretization of OSMs for Linear Elasticity

In this section, we describe how to discretize OSMs for Linear Elasticity with Finite Element Methods. Let's consider a tetrahedral mesh $\mathcal{T}^h$ of Ω compatible with the domain decomposition of Ω in N subdomains $(\Omega_i)_{1\leq i\leq N}$. Let $\mathcal{T}_i$ be the restriction of mesh $\mathcal{T}$ to subdomain Ω_i. We use $\mathcal{P}^1$ elements for each component of the small displacements. So at most three degrees (one per component) of freedom per node.

Let M be the number of degree of freedoms. Let the $\boldsymbol{\phi}_k$ be the elementary basis functions of the finite element space. For any k in $[\![1, M]\!]$, $\boldsymbol{\phi}_k$ is null on every node of the mesh except one. And on this node, $\boldsymbol{\phi}_k$ belongs to the canonical basis of $\mathbb{R}^3$. Let $\mathcal{I}_i$ be the subset of $[\![1, M]\!]$ of indices corresponding to degrees of freedoms located on a node of $\mathcal{T}_i$. The $\mathcal{I}_i$ are not disjoint. For all k, ℓ in $\mathcal{I}_i$, we set:

$$(A_i^h)_{k,\ell} := \int_{\Omega_i} \sigma(\boldsymbol{\phi}_k) : \varepsilon(\boldsymbol{\phi}_\ell), \qquad (\boldsymbol{f}_i^h)_k = \int_{\Omega_i} f \cdot \boldsymbol{\phi}_k.$$

There are multiple ways to discretize the transmission condition. For a consistent discretization, we set:

$$(S_{i,j}^{h,\text{cons}})_{k,\ell} := \int_{\Gamma_{ij}} (\sigma(\boldsymbol{\phi}_k)\boldsymbol{n}) \cdot \boldsymbol{\phi}_\ell,$$

for all k, ℓ in $\mathcal{I}_i \cap \mathcal{I}_j$. Alternatively, we can use a lump discretization, and define $S_{i,j}^{h,\text{lump}}$ as the diagonal matrix obtained by lumping $S_{i,j}^{h,\text{cons}}$. We also set the overlumped matrix $S_{ij}^{h,\omega} := (1-\omega)S_i^{h,\text{cons}} + \omega S_{ij}^{h,\text{lump}}$. For Poisson equation, overlumping has been shown to be beneficial in [9].

The main issue is deciding how transmission conditions should be updated, especially near cross-points (or cross-edges). This is especially true when cross-points (or cross-edges in 3d) are present. When using a FEM discretization of Linear Elasticity, the discrete value of $\sigma(\boldsymbol{u}_i)\boldsymbol{n}$ is only known as a variational quantity, as an integral over the boundary of $\partial\Omega_i$. Near cross-point, this variational quantity represents an integral over multiple surfaces each shared by $\partial\Omega_i$ with another subdomain. Ideally, this quantity must be split before being sent to the neighboring subdomains. Unfortunately, near cross-points, there is no canonical way to do so. See [9], for an explanation on how to discretize OSMs near cross-points for Poisson Equation, including the "Auxiliary Variable Method". When there are cross-points, at the discrete level, the g_{ij}^{n+1} cannot be derived from the discrete u_i^n. However, using (3b), they can be derived from both the g_{ij}^n and the u_i^n. Hence, in the Auxiliary Variable Method, the unknowns are not the discrete u_i^n, but the discrete g_{ij}^{n+1}.

The OSM iteration can be written at the discrete level as:

$$A_i^h \boldsymbol{u}_i^n = \boldsymbol{f}_i^h + \sum_j S_{ij}^h g_{ij}^n, \tag{3a}$$

$$g_{ij}^{n+1} := -g_{ji}^n + (S_{ij}^h + S_{ji}^h)u_j^n. \tag{3b}$$

Theorem 2 *If $S_{ij} = S_{ji}$, if each S_{ij} is symmetric positive definite, and if there is one subdomain where $\Gamma_d \cap \partial\Omega_i$ is of nonzero surface measure, then the Auxiliary Variable Method OSM, Eq.* (3)*, applied to a FEM-discretization of Linear Elasticity is convergent. I.E., if $(u_i)_{1\le i\le N}$ represents the discrete mono-domain solution, $u_i - u_i^n$ converges to* 0.

Proof There exists a finite sequence of $(g_{ij})_{ij}$ that is a fixed point of the (3b) iterate. Hence, we can suppose $\boldsymbol{f}^h$ null. Using u_i^n as the test function, we get

$$\int_{\Omega_i} \sigma(\boldsymbol{u}_i^n) : \varepsilon(\boldsymbol{u}_i^n)\mathrm{d}\boldsymbol{x} = \sum_j \int_{\Gamma_{ij}} (g_{ij}^n - S_{ij}u_{ij}^n) \cdot u_{ij}^n \mathrm{d}\boldsymbol{x}$$

Then, we can reuse the end of the proof of Theorem 1 almost verbatim. □

4 Numerical Results

We consider a cylindrical domain with a diameter of 1 and a heigh of 3.2. We subdivide this domain in two identical cylindrical subdomains. The domain is meshed using 4144 tetrahedrons. We set the Young Modulus $E = 1$ and the Poisson coefficient to either $\nu = 0.1$ or $\nu = 0.49$. We tested various values of the Robin parameter p and of the lump parameter ω. We found the best convergence for $p = 0.4$. As for Poisson equations, we found that overlumping the transmission condition substantially improves convergence, see convergence curves in Figures 1 and 2. Convergence is slower when the Poisson coefficient is near 1/2.

We also did a similar test by subdividing the same cylindrical domain into ten identical cylindrical subdomains. We set the Young Modulus $E = 1$ and the Poisson coefficient $\nu = 0.1$. See convergence curves in Figure 3. As expected in the absence of coarse spaces, the convergence of the Optimized Schwarz Method is considerably slower with ten subdomains.

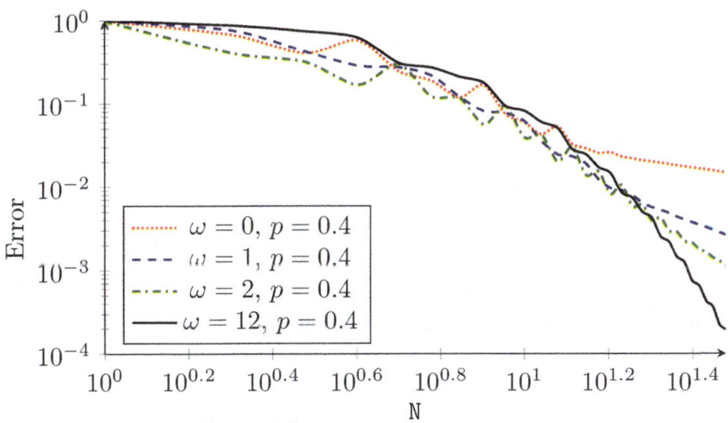

Fig. 1: Numerical Results for two subdomains with $\nu = 0.1$

5 Conclusion

In this paper, we showed how to to derive the equivalent of Robin boundary transmission for the equations of linear elasticity. Using overlumping, we improved these boundary transmission condition without the need to discretize higher order transmission conditions. We proved the theoretical convergence of Non Overlapping Optimized Schwarz Methods for linear elasticity.

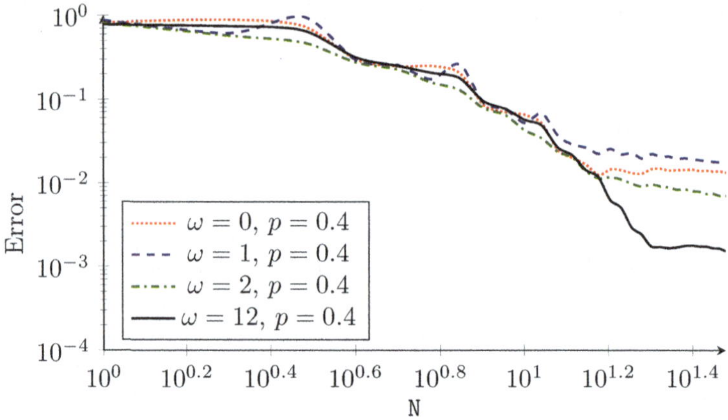

Fig. 2: Numerical Results for two cylindrical subdomains with $E = 1.0$ and $\nu = 0.49$

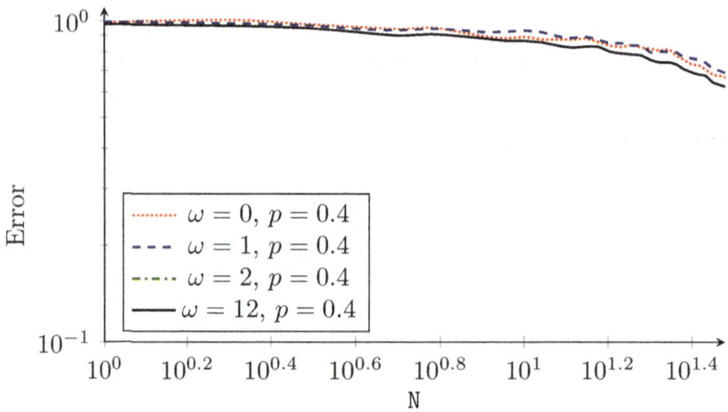

Fig. 3: Numerical Results for ten cylindrical subdomains with $E = 1.0$ and $\nu = 0.1$

As future works, we currently see three ways to expand upon this work. First, we will further study how to discretize the OSMs method for linear elasticity when cross-points are present. Then, we will generalize the Robin boundary condition for linear elasticity by replacing the scalar Robin parameter p with a 3 by 3 matrix. Finally, we are planning to add a coarse space to OSMs for linear elasticity.

References

1. Brenner, S.C., Scott, L.R.: The Mathematical Theory of Finite Element Methods. Springer New York (2008). DOI:10.1007/978-0-387-75934-0
2. Cai, M., Pavarino, L.F., Widlund, O.B.: Overlapping Schwarz methods with a standard coarse space for almost incompressible linear elasticity. SIAM Journal on Scientific Computing **37**(2),

A811–A830 (2015)
3. Ciarlet, P.G.: Mathematical Elasticity: Three-Dimensional Elasticity, *Studies in Mathematics and its Applications*, vol. 1. Elsevier, Academic Press (1988)
4. Després, B.: Domain decomposition method and the Helmholtz problem. In: G.C. Cohen, L. Halpern, P. Joly (eds.) Mathematical and numerical aspects of wave propagation phenomena, *Proceedings in Applied Mathematics Series*, vol. 50, pp. 44–52. Society for Industrial and Applied Mathematics (1991)
5. Dohrmann, C.R., Widlund, O.B.: An overlapping Schwarz algorithm for almost incompressible elasticity. SIAM Journal on Numerical Analysis **47**(4), 2897–2923 (2009)
6. Dohrmann, C.R., Widlund, O.B.: Hybrid domain decomposition algorithms for compressible and almost incompressible elasticity. International Journal for Numerical Methods in Engineering **82**(2), 157–183 (2010)
7. Farhat, C., Mandel, J.: The two-level FETI method for static and dynamic plate problems part i: An optimal iterative solver for biharmonic systems. Computer methods in applied mechanics and engineering **155**(1-2), 129–151 (1998)
8. Gander, M.J.: Optimized Schwarz methods. SIAM J. Numer. Anal. **44**(2), 699–731 (2006)
9. Gander, M.J., Santugini, K.: Cross-points in domain decomposition methods with a finite element discretization. Electron. Trans. Numer. Anal. **45**, 219–240 (2016)
10. Goldfeld, P., Pavarino, L.F., Widlund, O.B.: Balancing Neumann-Neumann methods for mixed approximations of linear elasticity. In: Lecture Notes in Computational Science and Engineering, pp. 53–76. Springer Berlin Heidelberg (2002)
11. Goldfeld, P., Pavarino, L.F., Widlund, O.B.: Balancing Neumann-Neumann preconditioners for mixed approximations of heterogeneous problems in linear elasticity. Numerische Mathematik **95**(2), 283–324 (2003). DOI:10.1007/s00211-002-0450-9
12. Kato, T.: Perturbation theory for linear operators, *Die Grundlehren der Mathematischen Wissenschaften in Einzeldarstellungen*, vol. 132. Springer-Verlag (1966)
13. Klawonn, A., Neff, P., Rheinbach, O., Vanis, S.: FETI-DP domain decomposition methods for elasticity with structural changes:p-elasticity. ESAIM: Mathematical Modelling and Numerical Analysis **45**(3), 563–602 (2010). DOI:10.1051/m2an/2010067
14. Klawonn, A., Pavarino, L.F.: An overlapping additive Schwarz method for a saddle point problem from linear elasticity. In: J. Wang, M.A. III, B. Chen, T. Mathew (eds.) Iterative Methods in Scientific Computation, pp. 219–224. IMACS Series in Computational and Applied Mathematics (1998)
15. Klawonn, A., Widlund, O.B.: A domain decomposition method with Lagrange multipliers and inexact solvers for linear elasticity. SIAM J. Sci. Comput. **22**(4), 1199–1219 (2000)
16. Le Tallec, P., De Roeck, Y.H., Vidrascu, M.: Domain-decomposition methods for large linearly elliptic three dimensional problems. J. Comput. Appl. Math. **34** (1991)
17. Lions, P.L.: On the Schwarz alternating method. III: a variant for nonoverlapping subdomains. In: T.F. Chan, R. Glowinski, J. Périaux, O. Widlund (eds.) Third International Symposium on Domain Decomposition Methods for Partial Differential Equations , held in Houston, Texas, March 20-22, 1989, pp. 202–223. SIAM, Philadelphia, PA (1990)
18. Pavarino, L.F., Widlund, O.B., Zampini, S.: BDDC preconditioners for spectral element discretizations of almost incompressible elasticity in three dimensions. SIAM Journal on Scientific Computing **32**(6), 3604–3626 (2010)
19. Smith, B.F.: An optimal domain decomposition preconditioner for the finite element solution of linear elasticity problems. SIAM J. Sci. Stat. Comput. **13**(1), 364–378 (1992)

Coupling of Navier-Stokes Equations and Their Hydrostatic Versions for Ocean Flows: A Discussion on Algorithm and Implementation

Hansong Tang and Yingjie Liu

1 Introduction

Now it is necessary to advance our capabilities to direct simulation of many emerging problems of coastal ocean flows. Two examples of such flow problems are the 2010 Gulf of Mexico oil spill and the 2011 Japan tsunami. The two examples come from different backgrounds, however, they present a same challenge to our modeling capacity; the both examples involve distinct types of physical phenomena at vastly different scales, and they are multiscale and multiphysics flows in nature. In particular, at the bottom of ocean, the spill appeared as high-speed, three dimensional (3D) jets at scales of $O(10)$ m, whereas on the ocean surface, it became two dimensional (2D) patches of oil film at horizontal sizes of $O(100)$ km [1]. The tsunami started as surface waves with tiny amplitude in a deep ocean, then evolved into walls of water as high as 39 m near seashore, and finally impacted coastal structures such as bridges at scales of $O(10)$ m [5]. These phenomena take place at different scales, and they are better and more efficiently simulated using different governing equations and numerical methods. Currently there is lack of appropriate computational methods and corresponding computer software packages that can directly and integrally simulate those multiple physics phenomena.

A natural and most feasible approach to simulation of multiscale and multiphysics coastal ocean flows is coupling of the Navier-Stokes (NS) equations and hydrostatic versions of the Navier-Stokes (HNS) equations. In the past few decades, various computational fluid dynamics (CFD) models (i.e., computer software packages) have been built on the NS equations for fully 3D fluid dynamics at complicated,

Hansong Tang
Department of Civil Engineering, City College, City University of New York, NY 10031, USA. e-mail: htang@ccny.cuny.edu

Yingjie Liu
School of Mathematics, Georgia Institute of Technology, Atlanta, GA 30332, USA. e-mail: yingjie@math.gatech.edu

R. Haynes et al. (eds.), *Domain Decomposition Methods in Science and Engineering XXV*,
Lecture Notes in Computational Science and Engineering 138,
https://doi.org/10.1007/978-3-030-56750-7_37

small scales ($\mathcal{O}(1)$ cm – $\mathcal{O}(10)$ km), such as jet flows similar to those of above oil spill [7]. At the same time, a number of CFD models have also been designed on the basis of HNS equations for geophysical fluid dynamics at large scales ($\mathcal{O}(10)$-$\mathcal{O}(10,000)$ km), such as the ocean currents carrying above oil patches [3]. Since the NS equations and the HNS equations are better bases for simulation of ocean flows at small and large scales, respectively, and coupling of them will enable us to conduct simulation of phenomena at larger or even full ranges of scales.

2 Governing Equations

The NS equations describe motion of flows, and they consist of the continuity equation and the momentum equation:

$$\begin{aligned} &\nabla \cdot \mathbf{u} = 0 \\ &\mathbf{u}_t + \nabla \cdot \mathbf{u}\mathbf{u} = \nabla \cdot (\nu \nabla \mathbf{u}) - \nabla p/\rho - g\mathbf{k} \end{aligned} \tag{1}$$

Here, $\mathbf{u}$ is the velocity vector, with u and v as the components in x and y direction, respectively, on the horizontal plane, and w as the component in z direction, or, the vertical direction, $\mathbf{k}$. ν is the viscosity, ρ the density, p the pressure, and g the gravity.

HNS equations are widely used for coastal ocean flows, and they are simplified from above NS equations; with the hydrostatic assumption, only the gravity and pressure terms are kept and all others are ignored in the vertical component of the momentum equation. As a result, the governing equations of the HNS equations consist of the continuity equation and the simplified momentum equation, which read as

$$\begin{aligned} &\nabla \cdot \mathbf{u} = 0 \\ &\mathbf{v}_t + \nabla \cdot \mathbf{u}\mathbf{v} = \nabla \cdot (\nu \nabla \mathbf{v}) - \nabla_H p/\rho \\ &p = \rho g(\eta - z) \end{aligned} \tag{2}$$

where $\mathbf{v} = (u, v)$, η is the elevation of water surface, and ∇_H is the gradient in the horizontal plane.

In view of the third equation for pressure in (2), its momentum equation in the horizontal plane can be rewritten as

$$\mathbf{v}_t + \nabla \cdot \mathbf{u}\mathbf{v} = \nabla \cdot (\nu \nabla \mathbf{u}) - g \nabla_H \eta \tag{3}$$

Additionally, by pressure splitting $p = p_d + \rho g(\eta - z)$, where p_d is the dynamic pressure, the momentum equation in (1) becomes

$$\mathbf{u}_t + \nabla \cdot \mathbf{u}\mathbf{u} = \nabla \cdot (\nu \nabla \mathbf{u}) - \nabla p_d/\rho - g \nabla_H \eta \tag{4}$$

3 Computational Methods

3.1 Transmision Condition

Let a flow field ω be divided into subdomains of NS and HNS by their interface γ, as shown in Fig. 1. Consider the weak solution of the continuity equation in (1) and (2) that satisfies

$$\int_{\omega} \mathbf{u} \cdot \nabla \phi d\omega = 0 \qquad (5)$$

for any $\phi \in C_0^\infty$. Let ω be an arbitrarily selected a region across the interface, and $\omega = \omega_1 \cup \omega_2$, with ω_1 and ω_2 falling in the regions of NS and HNS, respectively (Fig. 1). In view that

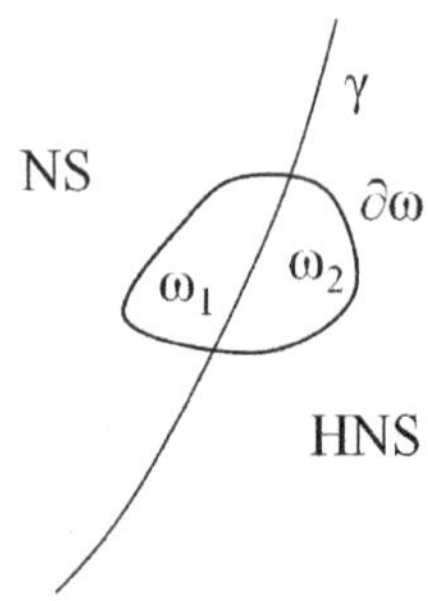

Fig. 1: Division of a flow region into a NS region abd a HNS region.

$$\int_{\omega} \mathbf{u} \cdot \nabla \phi d\omega = \int_{\gamma} \phi(\mathbf{u}_n|_{\gamma_-} - \mathbf{u}_n|_{\gamma_+}) d\omega - \int_{\omega_1 \cup \omega_2} \phi \nabla \cdot \mathbf{u} d\omega \qquad (6)$$

it is readily seen that, under the divergence-free condition (i.e., the first equation in (1) or (2)), continuity of normal velocity across the interface

$$\mathbf{u}_n|_{\gamma_-} = \mathbf{u}_n|_{\gamma_+} \qquad (7)$$

is a sufficient and necessary condition for $\mathbf{u}$ to be a weak solution. Here n means the normal direction of γ pointing from ω_1 to ω_2, $\mathbf{u}_n = \mathbf{u} \cdot \mathbf{n}$, and γ_- and γ_+ indicate the ω_1- and ω_2-side of interface γ, respectively. Therefore, condition (7) can be a transmission condition.

Similar analysis may be made for the momentum equations in the horizontal plane in (1) and (2), and it leads to the following transmission condition:

$$(\mathbf{u}_n \mathbf{v} + p_{n'}/\rho - \nu \partial \mathbf{v}/\partial n)_{\gamma_-} = (\mathbf{u}_n \mathbf{v} + p_{n'}/\rho - \nu \partial \mathbf{v}/\partial n)_{\gamma_+} \qquad (8)$$

here n' refers to the normal direction, pointing from ω_1 to ω_2, of the interface's projection onto the horizontal plane. It is noted that since p is a scalar, $p_{n'}$ may be replaced by p in (8). Also, a condition similar to (8) has been proposed in [2]. In correspondence to Eqs. (3) and (4) and also in view that surface elevation can be determined by HNS equations (see discussion in Sect. 4), its values on γ_- and γ_+ cancel each other as long as the elevation is continuous across the interface. As a result, interface condition (8) becomes

$$\left(\mathbf{u}_n\mathbf{v} + p_{d_{n'}}/\rho - \nu\partial\mathbf{v}/\partial n\right)_{\gamma_-} = \left(\mathbf{u}_n\mathbf{v} - \nu\partial\mathbf{v}/\partial n\right)_{\gamma_+} \tag{9}$$

where, for a similar reason, $p_{d_{n'}}$ may be replaced by p_d, and its value is zero in the hydrostatic region.

It is noted that, instead of those described as above, different interface conditions may be used:

$$\mathbf{u}|_{\gamma_-} = \mathbf{u}|_{\gamma_+},\ \partial p_d/\partial n|_{\gamma_-} = 0. \tag{10}$$

Here, the continuity of whole velocity is required across the interface. Interface condition (10) is commonly used to solve NS equations, and it has been used in coupling of NS and HNS equations, see Sect. 4.

3.2 Schwarz Iteration

Let discretization of NS equations and HNS equations be written as

$$\mathbf{F}(\mathbf{f}) = \mathbf{0},\ \ \mathbf{H}(\mathbf{h}) = \mathbf{0}, \tag{11}$$

in which $\mathbf{f} = (\mathbf{u}, p_d)$, and $\mathbf{h} = (\mathbf{u}, \eta)$, with the former and the latter being the solution for NS and HNS, respectively. Since the discretization is nonlinear, an iteration within each of the two equations in (11), named as the internal iteration, is needed for their solutions. Also, because the two equations in (11) are coupled with each other, another iteration between them, referred to as the external iteration, is also necessary.

From time level n to $n+1$, a Schwartz waveform relaxation approach is used to compute the discretization and exchange solution at the interfaces:

$\bar{\mathbf{f}}^{0,K_1} = \mathbf{f}^n,\ \bar{\mathbf{h}}^{0,K_2} = \mathbf{h}^n$

Do 1 $m = 1, M$

$$\begin{cases} \mathbf{F}(\bar{\mathbf{f}}^{m,k_1}) = \mathbf{0}, & \\ k_1 = 1, 2, \ldots, K_1, & \mathbf{x} \in \omega_1 \\ \bar{\mathbf{f}}^{m,k_1} = \hat{\mathbf{f}}^m, & \mathbf{x} \in \gamma_1 \end{cases} \qquad \begin{cases} \mathbf{H}(\bar{\mathbf{h}}^{m,k_2}) = \mathbf{0}, & \\ k_2 = 1, 2, \ldots, K_2, & \mathbf{x} \in \omega_2 \\ \bar{\mathbf{h}}^{m,k_2} = \hat{\mathbf{h}}^m, & \mathbf{x} \in \gamma_2 \end{cases} \tag{12}$$

1 *End Do*

$\mathbf{f}^{n+1} = \bar{\mathbf{f}}^{M,K_1},\ \bar{\mathbf{h}}^{n+1} = \bar{\mathbf{h}}^{M,K_2}$

in which $\hat{\mathbf{f}}^m = \kappa_1(\bar{\mathbf{h}}^{m-1,K_2}),\ \hat{\mathbf{h}}^m = \kappa_2(\bar{\mathbf{f}}^{m-1,K_1})$, being operators for solution exchange between NS and HNS equations on their interfaces γ_1 and γ_2, respectively (γ_1 and γ_2 overlap when the subdomains of NS and HNS patch with each other). M is a prescribed external iteration number, and K_1 and K_2 are prescribed internal iteration numbers.

An issue is how to compute (12) efficiently. A possible approach to speed up its convergence is to introduce relaxation to solution exchange at the interfaces, or, to the external iteration:

$$
\begin{aligned}
\bar{\mathbf{f}}^{m,k_1} &= \hat{\mathbf{f}}^{m-1} + \alpha(\hat{\mathbf{f}}^{m} - \hat{\mathbf{f}}^{m-1}), \quad \mathbf{x} \in \gamma_1 \\
\bar{\mathbf{h}}^{m,k_2} &= \hat{\mathbf{h}}^{m-1} + \alpha(\hat{\mathbf{h}}^{m} - \hat{\mathbf{h}}^{m-1}), \ \mathbf{x} \in \gamma_2
\end{aligned}
\tag{13}
$$

As $\alpha < 1$ and > 1, the iteration "under-relaxation" and "over-relaxation", respectively. An optimal value for α may be determined as the one that leads to a quick reduction of the residual of Eq. (11). For instance, let

$$
Q(\mathbf{f}, \mathbf{h}) = \langle \mathbf{F}, \mathbf{F} \rangle + \langle \mathbf{H}, \mathbf{H} \rangle \geq 0 \tag{14}
$$

By Taylor expansion and Eq. (13), one has

$$
\begin{aligned}
& Q(\bar{\mathbf{f}}^{m,K_1}, \bar{\mathbf{h}}^{m,K_2}) \\
= {} & Q(\bar{\mathbf{f}}^{m-1,K_1}, \bar{\mathbf{h}}^{m-1,K_2}) \\
& + \langle \partial Q / \partial \mathbf{f}|^{m} (\bar{\mathbf{f}}^{m,K_1} - \bar{\mathbf{f}}^{m-1,K_1}) \rangle_{\omega_1 \setminus \gamma_1} + \langle \partial Q / \partial \mathbf{h}|^{m} (\bar{\mathbf{h}}^{m,K_2} - \bar{\mathbf{h}}^{m-1,K_2}) \rangle_{\omega_2 \setminus \gamma_2} \\
& + \alpha \left(\langle \partial Q / \partial \mathbf{f}|^{m} (\hat{\mathbf{f}}^{m} - \hat{\mathbf{f}}^{m-1}) \rangle_{\gamma_1} + \langle \partial Q / \partial \mathbf{h}|^{m} (\hat{\mathbf{h}}^{m} - \hat{\mathbf{h}}^{m-1}) \rangle_{\gamma_2} \right) \\
& + \alpha^2 (\cdots) + \cdots
\end{aligned}
\tag{15}
$$

An expression for an optimal α can be derived from above equation by, say, letting $Q(\bar{\mathbf{f}}^{m,K_1}, \bar{\mathbf{h}}^{m,K_2}) = 0$ or $\partial Q / \partial \alpha = 0$, hoping that $\bar{\mathbf{f}}^{m,K_1}$ and $\bar{\mathbf{h}}^{m,K_2}$ are the fixed point of the iteration. It is noted that in pursuing above iteration with relaxation, it may be important to enforce the divergence-free condition.

Another approach to speed up the convergence in computation of (12) is via an optimal combination of the internal and external iterations. A natural arrangement of them is that the external iteration marches forwards only after the internal iterations converge, i.e., at sufficiently large K_1 and K_2. However, it is expected that an optimal combination of K_1 and K_2 is possible in terms of fast convergence to solutions at time level $n + 1$. With such an optimal combination, a new external iteration may start before the full convergence of the two internal iterations, and this could be an interesting topic.

4 Implementation of Model Coupling

The Solver of Incompressible Flow on Overset Meshes (SIFOM) is developed to compute NS equations (e.g., [6]), and its governing equations are

$$
\begin{aligned}
& \nabla \cdot \mathbf{u} = 0 \\
& \mathbf{u}_t + \nabla \cdot \mathbf{u}\mathbf{u} = \nabla \cdot ((\nu + \nu_t) \nabla \mathbf{u}) - \nabla p'_d / \rho - g \nabla_H \eta
\end{aligned}
\tag{16}
$$

in which ν_t is the turbulence viscosity. SIFOM discretizes above equations in curvilinear coordinates using a finite difference method [6].

An HNS solver is the Finite Volume Method Coastal Ocean Model (FVCOM), and its hydrostatic version consists of an external and an internal mode [4]. The governing equations for the external mode are the vertically averaged continuity and momentum equations:

$$\begin{aligned} &\eta_t + \nabla_H \cdot (\mathbf{V}D) = 0 \\ &(\mathbf{V}D)_t + \nabla_H \cdot (\mathbf{V}\mathbf{V}D) = -gD\, \nabla_H\, \eta + (\tau_s - \tau_b)/\rho + \mathbf{E}. \end{aligned} \tag{17}$$

The governing equations of the internal mode are the 3D continuity and momentum equations associated with the hydrostatic assumption:

$$\begin{aligned} &\eta_t + \nabla_H \cdot (\mathbf{v}D) + \omega_\sigma = 0, \\ &(\mathbf{v}D)_t + \nabla_H \cdot (\mathbf{v}\mathbf{v}D) + (\mathbf{v}\omega)_\sigma = -gD\, \nabla_H\, \eta + \nabla_H \cdot (\kappa \mathbf{e}) \\ &\qquad\qquad +(\lambda \mathbf{v}_\sigma)_\sigma / D + \mathbf{I}, \end{aligned} \tag{18}$$

In the external mode, $\mathbf{V}$ is the depth-averaged velocity vector, D is the water depth, and τ_s and τ_b are the shear stress on water surface and seabed, respectively. $\mathbf{E}$ includes the other terms such as the Coriolis force. In the internal mode, σ is the vertical coordinate, ω the vertical velocity in the σ-coordinate, $\mathbf{e}$ the strain rate, subscript σ the derivative over σ, and $\mathbf{I}$ the other terms. κ, and λ are coefficients. FVCOM solves Eqs. (17) and (18) on a triangular grid in the horizontal plane and a σ-grid in the vertical direction using a finite volume method.

An approach to integrate SIFOM and FVCOM is to couple Eqs. (16) and (18). The integration follows the algorithm (12). Interface transmission condition (10) is used for both SIFOM and FVCOM at their interfaces. It is noted that water surface elevation in SIFOM, or Eq. (16), is computed by FVCOM, or Eq. (17). Also, SIFOM and FVCOM are models for complicated, realistic flow problems, and their governing equations are not exactly but approximately same to the NS and HNS equations, respectively. For instance, Eq. (18) is a form transformed from Eq. (2). More details on the interface treatments and numerical algorithms can be found in [8].

As an example on performance of the SIFOM-FVCOM system, simulation has been made for a flow over a sill in a channel, see Fig. 2. In the simulation, SIFOM occupies the contraction section of the channel, and FVCOM covers all the channel, except a blanked region within the zone of SIFOM. Here, two regions for the SIFOM's are used; one is bigger and the other is smaller, and they lead to different interface locations. In the figure, it is seen that the simulated flow passes the interfaces of SIFOM smoothly, and no obvious artifact is generated there. However, the simulations with the two SIFOM regions present certain difference; as illustrated by streamlines, the simulation with the larger region presents more vibrating vertical motion after the contraction section, which is anticipated because SIFOM permits strong vertical motion. More simulated results for the flow are available in [8, 9], and they show that solution presents patterns similar to that by SIFOM alone, e.g., the vortical structures after the contraction section, and this is an intention of the coupling approach.

Simulation of actual ocean flows is challenging. For instance, as seen in Fig. 2, the simulated solution with a bigger SIFOM region is somewhat different from that with

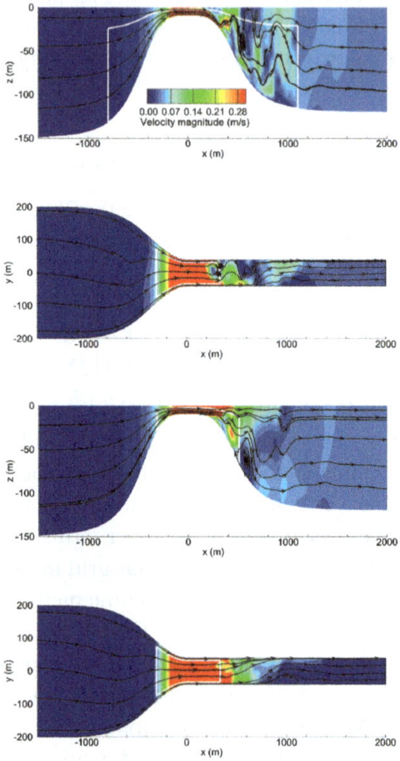

Fig. 2: Simulated instantaneous solutions of the sill flow. The solid white lines are SIFOM's interfaces, The top two panels have a bigger SIFOM region, and the bottom two panels have a smaller SIFOM region. In each set of the two panels, the first one is a side view, and the second one is a top view. In this simulation, to make it simple, M is set as 1, or, no iteration is made between SIFOM and FVCOM.

a smaller SIFOM region, indicating the influence of the size of the SIFOM's region and locations of the interfaces. Moreover, since it involves multiple times of runs of both SIFOM and FVCOM in marching from time level n to $n+1$, Schwarz iteration (12) is expensive, and it is significant to speed up the iteration. For this purpose, a preliminary effort has been made by running the SIFOM-FVCOM system with a few prescribed values for α in Eq. (13). However, no obvious speedup in convergence has been achieved, and this indicates that a deliberate design for the value of α is necessary.

Finally, it is noted that, in a recent study, another solver for the NS equations and implemented with a volume of fraction method has been coupled to FVCOM using techniques similar to those of the SIFOM-FVCOM system, and the results are encouraging [10].

5 Concluding Remarks

An successful coupling of NS and HNS equations will lead to an avenue to simulation of multiscale and multiphysics in many emerging coastal ocean flow problems. This paper presents a preliminary study on such coupling with regard to transmission conditions, Schwartz iterations, and coupling of actual models. In view that it is a relatively new topic and its realization is complicated, the coupling deserves systematic theoretical analysis and numerical experimentation, and we shall keep what discussed in this paper, in particular, the transmission conditions and the Schwarz iterations, for future study, and explore their effectiveness.

Acknowledgements This work is supported by the NSF (DMS-1622453, DMS-1622459). The example simulation in Fig. 2 is made by Mr. Wenbin Dong.

References

1. BBC.: Gulf of Mexico oil leak 'worst US environment disaster'. May 30, 2010 `http://www.bbc.co.uk/news/10194335`.
2. Blayo, E. and Rousseau, A.: About interface conditions for coupling hydrostatic and nonhydrostatic navier-stokes flows. Discrete and Continuous Dynamical Systems Series. **9**, 1565-1574 (2016).
3. Blumberg, A.F. and Mellor, G. L.: A description of a three-dimensional coastal ocean circulation model, in Three-Dimensional Coastal Models, Coastal Estuarine Ser., vol. 3, edited by N. S. Heaps, 1 – 16, AGU, Washington, D. C, 1987.
4. Chen, C. and Liu, H. and Beardsley, R.C.: An unstructured, finite-volume, three-dimensional, primitive equation ocean model: application to coastal ocean and estuaries. J. Atm. & Oceanic Tech. **20**, 159-186(2003).
5. Mimura, N. and Yasuhara, K. and Kawagoe, S. and Yokoki, H. and Kazama, S.: Damage from the great east japan earthquake and tsunami - a quick report. Mitig. Adapt. Strateg. Glob. Change. **16**, 803 – 818 (2011)
6. Tang, H.S. and Jones, S.C. and Sotiropoulos, F.: An overset-grid method for 3D unsteady incompressible flows. J. Comput. Phys. **191**, 567-600 (2003).
7. Tang, H.S. and Paik, J. and Sotiropoulos, F. and Khangaokar, T.: Three-dimensional numerical modeling of initial mixing of thermal discharges at real-life configurations. ASCE J. Hydr. Eng. **134**, 1210 – 1224 (2008)
8. Tang, H.S. and Qu, K. and Wu, X.G.: An overset grid method for integration of fully 3D fluid dynamics and geophysical fluid dynamics models to simulate multiphysics coastal ocean flows. J. Comput. Phys. **273**, 548 – 571 (2014).
9. Tang, H.S. and Qu, K. and Wu, X.G. and Zhang, Z.K.: Domain decomposition for a hybrid fully 3D fluid dynamics and geophysical fluid dynamics modeling system: A numerical experiment on a transient sill flow. Domain Decomposition Methods in Science and Engineering XXII, 407-414. Ed: Dickopf, T. and Gander, M.J. and Halpern, L. and Krause, R. and Pavarino, L.F. Springer, 2016.
10. Qu, K. and Tang, H. S. and Agrawal, A.: Integration of fully 3D fluid dynamics and geophysical fluid dynamics models for multiphysics coastal ocean flows: Simulation of local complex free-surface phenomena, Ocean Modelling, **135**, 14 –30 (2019).

BDDC for a Saddle Point Problem with an HDG Discretization

Xuemin Tu and Bin Wang

1 Introduction

The Balancing Domain Decomposition by Constraints (BDDC) algorithms, introduced in [4], are nonoverlapping domain decomposition methods. The coarse problems in the BDDC algorithms are given in terms of a set of primal constraints. An important advantage with such a coarse problem is that the Schur complements that arise in the computation will all be invertible. The BDDC algorithms have been extended to many different applications with different discretizations such as [9, 10, 13, 14, 2] and [11, 12].

In this paper, the BDDC algorithm is developed for the incompressible Stokes equation with an Hybridizable Discontinuous Galerkin (HDG) discretization. The HDG discretization for incompressible Stokes flow was introduced in [7] and analyzed in [3]. The main features of the HDG is that it reduces the globally coupled unknowns to the numerical trace of the velocity on the element boundaries and the mean of the pressure on the element. The size of the reduced saddle point problem is significantly smaller compared to the original one. In [7], the reduced saddle point problem is solved by an augmented Lagrange approach. An additional time dependent problem is introduced and solved by a backward-Euler method. Here, we solve the reduced saddle point problem directly using the BDDC methods. Similar to the earlier domain decomposition works on saddle point problems such as [8, 5, 6], and [9], we reduce the saddle point problem to a positive definite problem in a benign subspace and therefore the conjugate gradient (CG) method can be used to solve the resulting system. Due to the discontinuous pressure basis functions in this HDG

Xuemin Tu
Department of Mathematics, University of Kansas, 1460 Jayhawk Blvd, Lawrence, KS 66045, U.S.A. xuemin@ku.edu

Bin Wang
Department of Mathematical Sciences, Hood College, 401 Roesmont Ave., Frederick, MD 21701, U.S.A. wang@hood.edu

R. Haynes et al. (eds.), *Domain Decomposition Methods in Science and Engineering XXV*, Lecture Notes in Computational Science and Engineering 138,
https://doi.org/10.1007/978-3-030-56750-7_38

discretization, the complicated no-net-flux condition, which is needed to make sure all CG iterates are in the benign subspace, can be ensured by edge and face average constraints for each velocity component in two and three dimensions, respectively. These required constraints are the same as those for the elliptic problems with the HDG discretizations, cf. [11].

The rest of the paper is organized as follows. The HDG discretization for the Stokes problem are described in Section 2. In Section 3, the original system is reduced to an interface problem and a BDDC preconditioner is then introduced. The condition number estimate for the system with the BDDC preconditioner is provided in Section 4. Finally, we give some computational results in Section 5.

2 A Stokes problem and an HDG Discretization

The following Stokes problem is defined on a bounded polygonal domain Ω, in two or three dimensions, with a Dirichlet boundary condition:

$$\begin{cases} -\triangle \mathbf{u} + \nabla p = \mathbf{f}, \text{ in } \Omega, \\ \nabla \cdot \mathbf{u} = 0, \text{ in } \Omega, \\ \mathbf{u} = g, \text{ on } \partial\Omega, \end{cases} \tag{1}$$

where $\mathbf{f} \in L^2(\Omega)$ and $g \in H^{1/2}(\partial\Omega)$. Without loss of generality, we assume that $g = 0$. The solution of (1) is unique with the pressure p determined up to a constant. Here we will look for the solution with the pressure p having a zero average over the domain Ω.

We follow the approach in [7] and rewrite (1) as follows:

$$\begin{cases} \mathbf{L} - \nabla \mathbf{u} = 0, & \text{in } \Omega, \\ -\nabla \cdot \mathbf{L} + \nabla p = \mathbf{f}, & \text{in } \Omega, \\ \nabla \cdot \mathbf{u} = 0, & \text{in } \Omega, \\ \mathbf{u} = 0, & \text{in } \partial\Omega. \end{cases} \tag{2}$$

Let $P_k(D)$ be the space of polynomials of order at most k on D. We set $\mathbf{P}_k(D) = [P_k(D)]^n$ ($n = 2$ and 3 for two and three dimensions, respectively) and $\mathcal{P}_k(D) = [P_k(D)]^{n\times n}$. $\mathbf{L}$, $\mathbf{u}$, and p will be approximated by these discontinuous finite element spaces defined on a shape-regular and quasi-uniform triangulation of Ω, denoted by $\mathcal{T}_h$. Let h be the characteristic element size h of $\mathcal{T}_h$ and κ be an element in $\mathcal{T}_h$. The union of edges of elements κ is denoted by $\mathcal{E}$. $\mathcal{E}_i$ and $\mathcal{E}_\partial$ are two subsets of $\mathcal{E}$, for the edges in the interior of the domain and on its boundary, respectively. Define the following finite element spaces: $\mathbf{G}_k = \{\mathbf{G}_h \in \left[L^2(\Omega)\right]^{n\times n} : \mathbf{G}_h|_\kappa \in \mathcal{P}_k(\kappa), \forall \kappa \in \Omega\}$, $\mathbf{V}_k = \{\mathbf{v}_h \in \left[L^2(\Omega)\right]^n : \mathbf{v}_h|_\kappa \in \mathbf{P}_k(\kappa), \forall \kappa \in \Omega\}$, $W_k = \{p_h \in L^2(\Omega) : p_h|_\kappa \in P_k(\kappa), \int_\Omega p_h = 0, \forall \kappa \in \Omega\}$, $\mathbf{M}_k = \{\mu_h \in \left[L^2(\text{e})\right]^n : \mu_h|_e \in \mathbf{P}_k(e), \forall e \in \mathcal{E}\}$, and $\mathbf{\Lambda}_k = \{\mu_h \in \mathbf{M}_k : \mu_h|_e = 0, \forall e \in \partial\Omega\}$. To make our notation simple, we drop the subscript k from now on. The discrete problem resulting from the HDG discretization can be written as: to find $(\mathbf{L}_h, \mathbf{u}_h, p_h, \lambda_h) \in (\mathbf{G}, \mathbf{V}, W, \mathbf{\Lambda})$ such that for all $(\mathbf{G}_h, \mathbf{v}_h, q_h, \mu_h) \in (\mathbf{G}, \mathbf{V}, W, \mathbf{\Lambda})$

$$\begin{cases} (\mathbf{L}_h, \mathbf{G}_h)_{\mathcal{T}_h} + (\mathbf{u}_h, \nabla \cdot \mathbf{G}_h)_{\mathcal{T}_h} - \langle \lambda_h, \mathbf{G}_h \mathbf{n} \rangle_{\partial \mathcal{T}_h} & = 0, \\ (\mathbf{L}_h, \nabla \mathbf{v}_h)_{\mathcal{T}_h} - (p_h, \nabla \cdot \mathbf{v}_h)_{\mathcal{T}_h} - \langle \mathbf{L}_h \mathbf{n} - p_h \mathbf{n} - \tau_\kappa (\mathbf{u}_h - \lambda_h), \mathbf{v}_h \rangle_{\partial \mathcal{T}_h} & = (\mathbf{f}, \mathbf{v}_h)_{\mathcal{T}_h}, \\ - \langle \mathbf{L}_h \mathbf{n} - p_h \mathbf{n} - \tau_\kappa (\mathbf{u}_h - \lambda_h), \mu_h \rangle_{\partial \mathcal{T}_h} & = 0, \\ -(\mathbf{u}_h, \nabla q_h)_{\mathcal{T}_h} + \langle \lambda_h \cdot \mathbf{n}, q_h \rangle_{\partial \mathcal{T}_h} & = 0. \end{cases} \tag{3}$$

where τ_κ is a local stabilization parameter, see [7] for details.

The matrix form of (3) can be written as

$$\begin{bmatrix} A_{\mathbf{LL}} & A_{\mathbf{uL}}^T & A_{\lambda\mathbf{L}}^T & 0 \\ A_{\mathbf{uL}} & A_{\mathbf{uu}} & A_{\lambda\mathbf{u}}^T & B_{p\mathbf{u}}^T \\ A_{\lambda\mathbf{L}} & A_{\lambda\mathbf{u}} & A_{\lambda\lambda} & B_{p\lambda}^T \\ 0 & B_{p\mathbf{u}} & B_{p\lambda} & 0 \end{bmatrix} \begin{bmatrix} \mathbf{L} \\ \mathbf{u} \\ \lambda \\ p \end{bmatrix} = \begin{bmatrix} \mathbf{0} \\ \mathbf{F}_h \\ \mathbf{0} \\ 0 \end{bmatrix}, \tag{4}$$

where $\mathbf{F}_h = -(\mathbf{f}, \mathbf{v}_h)_{\mathcal{T}_h}$ and we use $\mathbf{L}$, $\mathbf{u}$, λ, and p to denote the unknowns associated with $\mathbf{L}_h$, $\mathbf{u}_h$, λ_h, and p_h, respectively. In each κ, we decompose the pressure degrees of freedom p into the element average pressure p_{0e} and the rest called the element interior pressure p_i and let $W = W_i \oplus W_{0e}$, correspondingly. We can easily eliminate $\mathbf{L}$, $\mathbf{u}$ and p_i element-wise from (4) and obtain the system for λ and p_{0e} only

$$\begin{bmatrix} A & B^T \\ B & 0 \end{bmatrix} \begin{bmatrix} \lambda \\ p_{0e} \end{bmatrix} = \begin{bmatrix} b \\ 0 \end{bmatrix}. \tag{5}$$

The global problem (4) can also be written as the following saddle point problem

$$\begin{bmatrix} A_a & B_a^T \\ B_a & 0 \end{bmatrix} \begin{bmatrix} u_a \\ p \end{bmatrix} = \begin{bmatrix} \mathbf{F}_a \\ 0 \end{bmatrix}, \tag{6}$$

where

$$A_a = \begin{bmatrix} A_{\mathbf{LL}} & A_{\mathbf{uL}}^T & A_{\lambda\mathbf{L}}^T \\ A_{\mathbf{uL}} & A_{\mathbf{uu}} & A_{\lambda\mathbf{u}}^T \\ A_{\lambda\mathbf{L}} & A_{\lambda\mathbf{u}} & A_{\lambda\lambda} \end{bmatrix}, \quad B_a^T = \begin{bmatrix} 0 \\ B_{p\mathbf{u}}^T \\ B_{p\lambda}^T \end{bmatrix}, \quad u_a = \begin{bmatrix} \mathbf{L} \\ \mathbf{u} \\ \lambda \end{bmatrix}, \quad \text{and } \mathbf{F}_a = \begin{bmatrix} \mathbf{0} \\ \mathbf{F}_h \\ \mathbf{0} \end{bmatrix}. \tag{7}$$

We note that A_a is the same as the matrix obtained using HDG discretization for elliptic problem as discussed in [11].

3 The BDDC algorithm

We decompose Ω into N nonoverlapping subdomain Ω_i with diameters H_i, $i = 1, \ldots, N$, and set $H = \max_i H_i$. We assume that each subdomain is a union of shape-regular coarse triangles and that the number of such elements forming an individual subdomain is uniformly bounded. We define edges/faces as open sets shared by two subdomains. Two nodes belong to the same edge/face when they are associated with the same pair of subdomains. Let Γ be the interface between the subdomains. The set of the interface nodes Γ_h is defined as $\Gamma_h := \left(\cup_{i\neq j}\partial\Omega_{i,h} \cap \partial\Omega_{j,h}\right) \setminus \partial\Omega_h$, where $\partial\Omega_{i,h}$ is the set of nodes on $\partial\Omega_i$ and $\partial\Omega_h$ is that of $\partial\Omega$. We assume the triangulation of each subdomain is quasi-uniform.

We decompose the velocity numerical trace $\mathbf{\Lambda} = \mathbf{\Lambda}_I \oplus \widehat{\mathbf{\Lambda}}_\Gamma$ and the element average pressure $W_{0e} = W_I \oplus W_0$, where $\widehat{\mathbf{\Lambda}}_\Gamma$ denotes the degrees of freedom associated with Γ. $\mathbf{\Lambda}_I = \prod_{i=1}^N \mathbf{\Lambda}_I^{(i)}$ and $W_I = \prod_{i=1}^N W_I^{(i)}$ are products of subdomain interior velocity numerical trace spaces $V_I^{(i)}$ and subdomain interior pressure spaces $W_I^{(i)}$, respectively. The elements of $\mathbf{\Lambda}_I^{(i)}$ are supported in the subdomain Ω_i and vanishes on its interface Γ_i, while the elements of $W_I^{(i)}$ are the restrictions of the pressure variables to Ω_i which satisfy $\int_{\Omega_i} p_I^{(i)} = 0$. $\widehat{\mathbf{\Lambda}}_\Gamma$ is the subspace of edge/face functions on Γ in $\mathbf{\Lambda}$, and W_0 is the subspace of W with constant values $p_0^{(i)}$ in the subdomain Ω_i that satisfy $\sum_{i=1}^N p_0^{(i)} m\left(\Omega_i\right) = 0$, where $m\left(\Omega_i\right)$ is the measure of the subdomain Ω_i.

We denote the space of interface velocity numerical trace variables of the subdomain Ω_i by $\mathbf{\Lambda}_\Gamma^{(i)}$, and the associated product space by $\mathbf{\Lambda}_\Gamma = \prod_{i=1}^N \mathbf{\Lambda}_\Gamma^{(i)}$; generally edge/face functions in $\mathbf{\Lambda}_\Gamma$ are discontinuous across the interface. We define the restriction operators $R_\Gamma^{(i)} : \widehat{\mathbf{\Lambda}}_\Gamma \to \mathbf{\Lambda}_\Gamma^{(i)}$ to be an operator which maps functions in the continuous global interface velocity numerical trace variable space $\widehat{\mathbf{\Lambda}}_\Gamma$ to the subdomain component space $\mathbf{\Lambda}_\Gamma^{(i)}$. Also, $R_\Gamma : \widehat{\mathbf{\Lambda}}_\Gamma \to \mathbf{\Lambda}_\Gamma$ is the direct sum of $R_\Gamma^{(i)}$.

The global interface problem is assembled from the subdomain interface problems, and can be written as: find $(\lambda_\Gamma, p_0) \in \left(\widehat{\mathbf{\Lambda}}_\Gamma, W_0\right)$ such that

$$\widehat{S}\begin{bmatrix}\lambda_\Gamma \\ p_0\end{bmatrix} = \begin{bmatrix} g_\Gamma \\ 0\end{bmatrix}, \quad \text{where } \widehat{S} = \begin{bmatrix}\widehat{S}_\Gamma & \widehat{B}_{0\Gamma}^T \\ \widehat{B}_{0\Gamma} & 0\end{bmatrix}. \tag{8}$$

Here $\widehat{S}_\Gamma$, $\widehat{B}_{0\Gamma}$, and g_Γ are assembled from the subdomain matrices.

In order to introduce the BDDC preconditioner, we first introduce a partially assembled interface space $\widetilde{\mathbf{\Lambda}}_\Gamma = \widehat{\mathbf{\Lambda}}_\Pi \oplus \mathbf{\Lambda}_\Delta = \widehat{\mathbf{\Lambda}}_\Pi \oplus \prod_{i=1}^{N} \mathbf{\Lambda}_\Delta^{(i)}$. Here, $\widehat{\mathbf{\Lambda}}_\Pi$ is the coarse level, primal interface velocity space and the space $\mathbf{\Lambda}_\Delta$ is the direct sum of the $\mathbf{\Lambda}_\Delta^{(i)}$, which are spanned by the remaining interface degrees of freedom. In the space $\widetilde{\mathbf{\Lambda}}_\Gamma$, we relax most continuity constraints across the interface but retain the continuity at the primal unknowns, which makes all the linear systems nonsingular.

We need to introduce several restriction, extension, and scaling operators between different spaces. $\overline{R}_\Gamma^{(i)} : \widetilde{\mathbf{\Lambda}}_\Gamma \to \mathbf{\Lambda}_\Gamma^{(i)}$ restricts functions in the space $\widetilde{\mathbf{\Lambda}}_\Gamma$ to the components $\mathbf{\Lambda}_\Gamma^{(i)}$ of the subdomain Ω_i. $\overline{R}_\Gamma : \widetilde{\mathbf{\Lambda}}_\Gamma \to \mathbf{\Lambda}_\Gamma$ is the direct sum of $\overline{R}_\Gamma^{(i)}$. $R_\Delta^{(i)} : \widehat{\mathbf{\Lambda}}_\Gamma \to \mathbf{\Lambda}_\Delta^{(i)}$ maps the functions from $\widehat{\mathbf{\Lambda}}_\Gamma$ to $\mathbf{\Lambda}_\Delta^{(i)}$, its dual subdomain components. $R_{\Gamma\Pi} : \widehat{\mathbf{\Lambda}}_\Gamma \to \widehat{\mathbf{\Lambda}}_\Pi$ is a restriction operator from $\widehat{\mathbf{\Lambda}}_\Gamma$ to its subspace $\widehat{\mathbf{\Lambda}}_\Pi$. $\widetilde{R}_\Gamma : \widehat{\mathbf{\Lambda}}_\Gamma \to \widetilde{\mathbf{\Lambda}}_\Gamma$ is the direct sum of $R_{\Gamma\Pi}$ and $R_\Delta^{(i)}$. We define the positive scaling factor $\delta_i^{\dagger}(x)$ as follows:

$$\delta_i^{\dagger}(x) = \frac{1}{card\,(\mathcal{I}_x)}, \qquad x \in \partial\Omega_{i,h} \cap \Gamma_h,$$

where $\mathcal{I}_x$ is the set of indices of the subdomains that have x on their boundaries, and $card\,(\mathcal{I}_x)$ counts the number of the subdomain boundaries to which x belongs. We note that $\delta_i^{\dagger}(x)$ is constant on each edge/face. Multiplying each row of $R_{\Delta}^{(i)}$ with the scaling factor gives us $R_{D,\Delta}^{(i)}$. The scaled operators $\widetilde{R}_{D,\Gamma}$ is the direct sum of $R_{\Gamma\Pi}$ and $R_{D,\Delta}^{(i)}$.

We denote the direct sum of the local interface velocity Schur complement by S_{Γ} and the partially assembled interface velocity Schur complement is defined by $\widetilde{S}_{\Gamma} = \overline{R}_{\Gamma}^{T} S_{\Gamma} \overline{R}_{\Gamma}$. Correspondingly, we define an operator $\widetilde{B}_{0\Gamma}$, which maps the partially assembled interface velocity space $\widetilde{\mathbf{\Lambda}}_{\Gamma}$ into the space of right-hand sides corresponding to W_0. $\widetilde{B}_{0\Gamma}$ is obtained from the subdomain operators by assembling them with respect to the primal interface velocity part. Using the following notation

$$\widetilde{R}_D = \begin{bmatrix} \widetilde{R}_{D,\Gamma} & \\ & I \end{bmatrix}, \qquad \widetilde{S} = \begin{bmatrix} \widetilde{S}_{\Gamma} & \widetilde{B}_{0\Gamma}^{T} \\ \widetilde{B}_{0\Gamma} & 0 \end{bmatrix}, \tag{9}$$

and the preconditioned BDDC algorithm is then of the form: find $(\lambda_{\Gamma}, p_0) \in \left(\widehat{\mathbf{\Lambda}}_{\Gamma}, W_0\right)$, such that

$$\widetilde{R}_D^{T} \widetilde{S}^{-1} \widetilde{R}_D \widehat{S} \begin{bmatrix} \lambda_{\Gamma} \\ p_0 \end{bmatrix} = \widetilde{R}_D^{T} \widetilde{S}^{-1} \widetilde{R}_D \begin{bmatrix} g_{\Gamma} \\ 0 \end{bmatrix}. \tag{10}$$

Note that $\widetilde{R}_{D,\Gamma}$ is of full rank and that the preconditioner is nonsingular.

Definition 1 (Benign Subspaces) We will call

$$\widehat{\mathbf{\Lambda}}_{\Gamma,B} = \{\lambda_{\Gamma} \in \widehat{\mathbf{\Lambda}}_{\Gamma} \mid \widehat{B}_{0\Gamma}\lambda_{\Gamma} = 0\}, \quad \widetilde{\mathbf{\Lambda}}_{\Gamma,B} = \{\lambda_{\Gamma} \in \widetilde{\mathbf{\Lambda}}_{\Gamma} \mid \widetilde{B}_{0\Gamma}\lambda_{\Gamma} = 0\}$$

the benign subspaces of $\widehat{\mathbf{\Lambda}}_{\Gamma}$ and $\widetilde{\mathbf{\Lambda}}_{\Gamma}$, respectively.

It is easy to see that the operators $\widehat{S}$ and $\widetilde{S}$, defined in (8) and (10), are symmetric positive definite on $\left(\widehat{\mathbf{\Lambda}}_{\Gamma,B}, W_0\right)$ and $\left(\widetilde{\mathbf{\Lambda}}_{\Gamma,B}, W_0\right)$, respectively. A preconditioned conjugate gradient method can then be used to solve the global BDDC preconditioned interface problem (10).

4 Condition number estimate for the BDDC preconditioner

In this section, we only consider the case that the stabilization parameter $\tau_{\kappa} = O(\frac{1}{h_{\kappa}})$, where h_{κ} the diameter of the element κ. Other choices of τ_{κ} will be considered elsewhere.

Similar to the inf-sup condition of the weak Galerkin finite element methods [15, Lemma 4.3], we have the following lemma:

Lemma 1 *There exists a positive constant β independent of h and H, such that*

$$\sup_{u_a \in (\mathbf{G},\mathbf{V},\mathbf{\Lambda})} \frac{u_a^T B_a^T p}{\left(u_a^T A_a u_a\right)^{1/2}} \geq \beta \|p\|_{L^2(\Omega)}, \tag{11}$$

for all $p \in W$. Here A_a, B_a are defined in (7). The theorem is also hold when Ω is replaced by a subdomain Ω_i.

Using Lemma 1 for each subdomain, we can prove a well-known relation between the harmonic extension and Stokes extension when the subdomain boundary velocity is given. Similar results for the standard finite element discretization can be found in [1]. Then we can prove a bound of the averaging operator E_D for the Stokes problem.

Lemma 2 *There exists a positive constant C, which is independent of H and h, such that*

$$|E_D w|_{\widetilde{S}}^2 \leq C \frac{(1+\beta)^2}{\beta^2}\left(1+\log\frac{H}{h}\right)^2 |w|_{\widetilde{S}}^2, \qquad \forall\, w = (\lambda_\Gamma,\, p_0) \in \left(\widetilde{\mathbf{\Lambda}}_{\Gamma,B}, W_0\right),$$

where β is the inf-sup stability constant.

With the help of Lemma 2, we can obtain our main result

Theorem 1 *The preconditioned operator $M^{-1}\widehat{S}$ is symmetric, positive definite with respect to the bilinear form $\langle\cdot,\cdot\rangle_{\widehat{S}}$ on the space $\left(\widehat{\mathbf{\Lambda}}_{\Gamma,B}, W_0\right)$.The condition number of $M^{-1}\widehat{S}$ is bounded by $C\frac{(1+\beta)^2}{\beta^2}\left(1+\log\left(\frac{H}{h}\right)\right)^2$, where C is a constant, which is independent of H and h, and β is the inf-sup stability constant, defined in Lemma 1.*

5 Numerical Experiments

We have applied our BDDC algorithms to the model problem (1), where $\Omega = [0,1]^2$. Zero Dirichlet boundary conditions are used. The right-hand side function $\mathbf{f}$ is chosen such that the exact solution is

$$\mathbf{u} = \begin{bmatrix} \sin^3(\pi x)\sin^2(\pi y)\cos(\pi y) \\ -\sin^2(\pi x)\sin^3(\pi y)\cos(\pi x) \end{bmatrix} \quad \text{and} \quad p = x - y.$$

We decompose the unit square into $N \times N$ subdomains with the sidelength $H = 1/N$. Equation (1) is discretized, in each subdomain, by the kth-order HDG method with an element diameter h. The preconditioned conjugate gradient iteration is stopped when the relative l_2-norm of the residual has been reduced by a factor of 10^6.

Table 1: Performance of solving (10) with HDG discretization ($\tau_K = 1/h_K$)

		$k = 0$		$k = 1$		$k = 2$	
H/h	#sub	Cond.	Iter.	Cond.	Iter.	Cond.	Iter.
8	4×4	4.21	10	4.72	12	12.72	14
	8×8	5.12	12	8.81	17	11.52	20
	16×16	5.00	13	10.43	21	13.44	24
	24×24	5.14	13	10.83	20	13.96	25
	32×32	5.14	13	10.84	20	14.09	25
#sub	H/h	Cond.	Iter.	Cond.	Iter.	Cond.	Iter.
8×8	4	2.56	9	6.23	14	8.52	17
	8	5.12	12	8.81	17	11.52	20
	16	7.59	15	11.86	20	17.86	24
	24	9.22	17	13.86	22	20.32	25
	32	10.48	19	15.37	23	22.21	26

We consider the choice of the stabilization constant $\tau_K = \frac{1}{h_K}$. We have carried out two sets of experiments to obtain iteration counts and condition number estimates. In the first set of the experiments, we fixed $\frac{H}{h} = 8$, the subdomain local problem size, and change the number of subdomains to test the scalability of the algorithms (the condition number is independent of the number of subdomains). In the second set of experiments, we fixed the number of subdomains to 64 and change $\frac{H}{h}$, the subdomain local problem size. The performance of the algorithms for the Stokes problem is similar to those for the elliptic problems. The experimental results are fully consistent with our theory.

Acknowledgements This work was supported in part by National Science Foundation Contracts No. DMS-1419069 and DMS-1723066.

References

1. Bramble, J., Pasciak, J.: A domain decomposition technique for Stokes problems. Appl. Numer. Math. **6**(4), 251–261 (1990)
2. Canuto, C., Pavarino, L.F., Pieri, A.B.: BDDC preconditioners for continuous and discontinuous Galerkin methods using spectral/hp elements with variable local polynomial degree. IMA J. Numer. Anal. **34**(3), 879–903 (2014)
3. Cockburn, B., Gopalakrishnan, J., Nguyen, N.C., Peraire, J., Sayas, F.: Analysis of HDG methods for Stokes flow. Math. Comp. **80**(274), 723–760 (2011)

4. Dohrmann, C.: A preconditioner for substructuring based on constrained energy minimization. SIAM J. Sci. Comput. **25**(1), 246–258 (2003)
5. Li, J.: A Dual-Primal FETI method for incompressible Stokes equations. Numer. Math. **102**, 257–275 (2005)
6. Li, J., Widlund, O.: BDDC algorithms for incompressible Stokes equations. SIAM J. Numer. Anal. **44**(6), 2432–2455 (2006)
7. Nguyen, N.C., Peraire, J., Cockburn, B.: A hybridizable discontinuous Galerkin method for Stokes flow. Comput. Methods Appl. Mech. Engrg. **199**(9-12), 582–597 (2010)
8. Pavarino, L., Widlund, O.: Balancing Neumann-Neumann methods for incompressible Stokes equations. Comm. Pure Appl. Math. **55**(3), 302–335 (2002)
9. Tu, X.: A BDDC algorithm for a mixed formulation of flows in porous media. Electron. Trans. Numer. Anal. **20**, 164–179 (2005)
10. Tu, X.: A BDDC algorithm for flow in porous media with a hybrid finite element discretization. Electron. Trans. Numer. Anal. **26**, 146–160 (2007)
11. Tu, X., Wang, B.: A BDDC algorithm for second-order elliptic problems with hybridizable discontinuous Galerkin discretizations. Electron. Trans. Numer. Anal. **45**, 354–370 (2016)
12. Tu, X., Wang, B.: A BDDC algorithm for the Stokes problem with weak Galerkin discretizations. Comput. Math. Appl. **76**(2), 377–392 (2018)
13. Beirão da Veiga, L., Cho, D., Pavarino, L.F., Scacchi, S.: BDDC preconditioners for isogeometric analysis. Math. Models Methods Appl. Sci. **23**(6), 1099–1142 (2013)
14. Beirão da Veiga, L., Pavarino, L., Scacchi, S., Widlund, O., Zampini, S.: Isogeometric BDDC preconditioners with deluxe scaling. SIAM J. Sci. Comput. **36**(3), A1118–A1139 (2014)
15. Wang, J., Ye, X.: A weak Galerkin finite element method for the Stokes equations. Adv. Comput. Math. **42**(1), 155–174 (2016)

A Balancing Domain Decomposition by Constraints Preconditioner for a C^0 Interior Penalty Method

Susanne C. Brenner, Eun-Hee Park, Li-Yeng Sung, and Kening Wang

1 Introduction

Consider the following weak formulation of a fourth order problem on a bounded polygonal domain Ω in $\mathbb{R}^2$:
Find $u \in H_0^2(\Omega)$ such that

$$\int_\Omega \nabla^2 u : \nabla^2 v \, dx = \int_\Omega f v \, dx \qquad \forall\, v \in H_0^2(\Omega), \tag{1}$$

where $f \in L_2(\Omega)$, and $\nabla^2 v : \nabla^2 w = \sum_{i,j=1}^2 (\partial^2 v/\partial x_i \partial x_j)(\partial^2 w/\partial x_i \partial x_j)$ is the inner product of the Hessian matrices of v and w.

For simplicity, let $\mathcal{T}_h$ be a quasi-uniform triangulation of Ω consisting of rectangles and take $V_h \subset H_0^1(\Omega)$ to be the Q_2 Lagrange finite element space associated with $\mathcal{T}_h$. (Results also hold for quadrilateral meshes.) Then the model problem (1) can be discretized by the following C^0 interior penalty Galerkin method [7, 3]:
Find $u_h \in V_h$ such that

$$a_h(u_h, v) = \int_\Omega f v \, dx \qquad v \in V_h,$$

where

Susanne C. Brenner and Li-Yeng Sung
Department of Mathematics and Center for Computation and Technology, Louisiana State University, Baton Rouge, LA 70803, USA, e-mail: brenner@math.lsu.edu, e-mail: sung@math.lsu.edu

Eun-Hee Park
School of General Studies, Kangwon National University, Samcheok, Gangwon 25913, Republic of Korea, e-mail: eh.park@kangwon.ac.kr

Kening Wang
Department of Mathematics and Statistics, University of North Florida, Jacksonville, FL 32224, USA, e-mail: kening.wang@unf.edu

R. Haynes et al. (eds.), *Domain Decomposition Methods in Science and Engineering XXV*, Lecture Notes in Computational Science and Engineering 138,
https://doi.org/10.1007/978-3-030-56750-7_39

$$a_h(v,w) = \sum_{D\in\mathcal{T}_h} \int_T \nabla^2 v : \nabla^2 w \, dx + \sum_{e\in\mathcal{E}_h} \frac{\eta}{|e|} \int_e \left[\!\!\left[\frac{\partial v}{\partial \mathbf{n}} \right]\!\!\right] \left[\!\!\left[\frac{\partial w}{\partial \mathbf{n}} \right]\!\!\right] ds$$
$$+ \sum_{e\in\mathcal{E}_h} \int_e \left(\left\{\!\!\left\{ \frac{\partial^2 v}{\partial \mathbf{n}^2} \right\}\!\!\right\} \left[\!\!\left[\frac{\partial w}{\partial \mathbf{n}} \right]\!\!\right] + \left\{\!\!\left\{ \frac{\partial^2 w}{\partial \mathbf{n}^2} \right\}\!\!\right\} \left[\!\!\left[\frac{\partial v}{\partial \mathbf{n}} \right]\!\!\right] \right) ds.$$

Here η is a positive penalty parameter, $\mathcal{E}_h$ is the set of edges of $\mathcal{T}_h$, and $|e|$ is the length of the edge e. The jump $[\![\cdot]\!]$ and the average $\{\!\{\cdot\}\!\}$ are defined as follows.

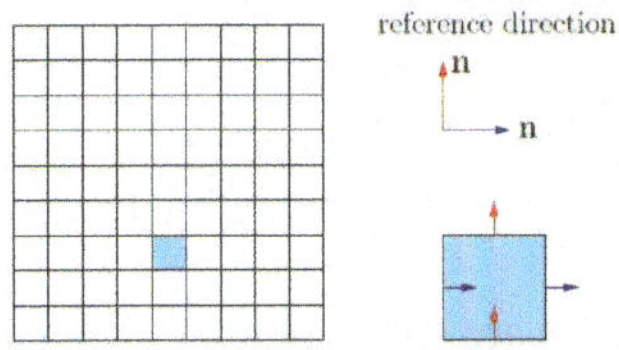

Fig. 1: (a) A triangulation of Ω. (b) A reference direction of normal vectors on the edges of $T \in \mathcal{T}_h$.

Let $\mathbf{n}_e$ be the unit normal chosen according to a reference direction shown in Fig. 1. If e is an interior edge of $\mathcal{T}_h$ shared by two elements D_- and D_+, we define on e,

$$\left[\!\!\left[\frac{\partial v}{\partial \mathbf{n}} \right]\!\!\right] = \frac{\partial v_+}{\partial \mathbf{n}_e} - \frac{\partial v_-}{\partial \mathbf{n}_e} \quad \text{and} \quad \left\{\!\!\left\{ \frac{\partial^2 v}{\partial \mathbf{n}^2} \right\}\!\!\right\} = \frac{1}{2}\left(\frac{\partial^2 v_+}{\partial \mathbf{n}_e^2} + \frac{\partial^2 v_-}{\partial \mathbf{n}_e^2} \right),$$

where $v_\pm = v|_{D_\pm}$. On an edge of $\mathcal{T}_h$ along $\partial\Omega$, we define

$$\left[\!\!\left[\frac{\partial v}{\partial \mathbf{n}} \right]\!\!\right] = \pm \frac{\partial v}{\partial \mathbf{n}_e} \quad \text{and} \quad \left\{\!\!\left\{ \frac{\partial^2 v}{\partial \mathbf{n}^2} \right\}\!\!\right\} = \frac{\partial^2 v}{\partial \mathbf{n}_e^2},$$

in which the negative sign is chosen if $\mathbf{n}_e$ points towards the outside of Ω, and the positive sign otherwise.

It is noted that for $\eta > 0$ sufficiently large (Lemma 6 in [3]), there exist positive constants C_1 and C_2 independent of h such that

$$C_1 a_h(v,v) \le |v|^2_{H^2(\Omega,\mathcal{T}_h)} \le C_2 a_h(v,v) \quad \forall v \in V_h,$$

where

$$|v|^2_{H^2(\Omega,\mathcal{T}_h)} = \sum_{D\in\mathcal{T}_h} |v|^2_{H^2(D)} + \sum_{e\in\mathcal{E}_h} \frac{1}{|e|} \left\| \left[\!\!\left[\frac{\partial v}{\partial \mathbf{n}} \right]\!\!\right] \right\|^2_{L_2(e)}.$$

Compared with classical finite element methods for fourth order problems, C^0 interior penalty methods have many advantages [3, 5, 7]. However, due to the nature of fourth order problems, the condition number of the discrete problem resulting from C^0 interior penalty methods grows at the rate of h^{-4} [8]. Thus a good preconditioner is essential for solving the discrete problem efficiently and accurately. In this paper, we develop a nonoverlapping domain decomposition preconditioner for C^0 interior penalty methods that is based on the balancing domain decomposition by constraints (BDDC) approach [6, 4, 1].

The rest of the paper is organized as follows. In Section 2 we introduce the subspace decomposition. We then design a BDDC preconditioner for the reduced problem in Section 3, followed by condition number estimates in Section 4. Finally, we report numerical results in Section 5 that illustrate the performance of the proposed preconditioner and corroborate the theoretical estimates.

2 A Subspace Decomposition

We begin with a nonoverlapping domain decomposition of Ω consisting of rectangular (open) subdomains $\Omega_1, \Omega_2, \cdots, \Omega_J$ aligned with $\mathcal{T}_h$ such that $\partial\Omega_j \bigcap \partial\Omega_\ell = \emptyset$, a vertex, or an edge, if $j \neq \ell$.

We assume the subdomains are shape regular and denote the typical diameter of the subdomains by H. Let $\Gamma = \left(\bigcup_{j=1}^J \partial\Omega_j\right) \setminus \partial\Omega$ be the interface of the subdomains, and $\mathcal{E}_{h,\Gamma}$ be the subset of $\mathcal{E}_h$ containing the edges on Γ.

Since the condition that the normal derivative of v vanishes on Γ is implicit in terms of the standard degrees of freedom (dofs) of the Q_2 finite element, it is more convenient to use the modified Q_2 finite element space (Fig. 2) as V_h. Details of the modified Q_2 finite element space can be found in [5].

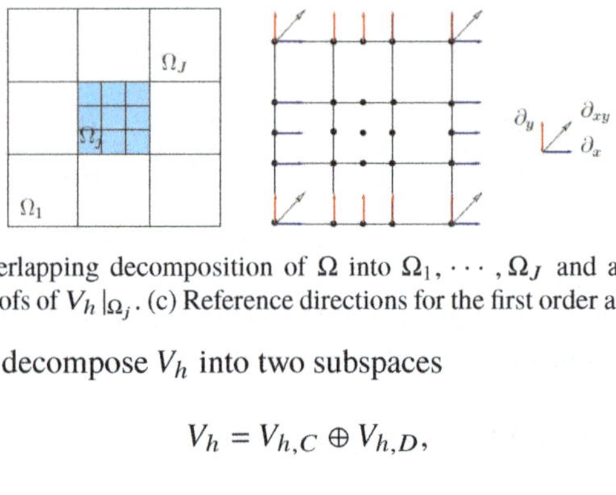

Fig. 2: (a) A nonoverlapping decomposition of Ω into $\Omega_1, \cdots, \Omega_J$ and a triangulation of the subdomain Ω_j. (b) Dofs of $V_h|_{\Omega_j}$. (c) Reference directions for the first order and mixed derivatives.

First of all, we decompose V_h into two subspaces

$$V_h = V_{h,C} \oplus V_{h,D},$$

where

$$V_{h,C} = \left\{v \in V_h : \left[\!\left[\frac{\partial v}{\partial \mathbf{n}}\right]\!\right] = 0 \text{ on the edges in } \mathcal{E}_h \text{ that are subsets of } \bigcup_{j=1}^J \partial\Omega_j\right\}$$

and

$$V_{h,D} = \left\{v \in V_h : \left\{\!\!\left\{\frac{\partial v}{\partial \mathbf{n}}\right\}\!\!\right\} = 0 \text{ on edges in } \mathcal{E}_{h,\Gamma}, \text{ and } v \text{ vanishes at all interior nodes of each subdomain}\right\}.$$

Let $A_h : V_h \to V_h'$ be the symmetric positive definite (SPD) operator defined by

$$\langle A_h v, w \rangle = a_h(v, w) \qquad \forall\, v, w \in V_h,$$

where $\langle \cdot, \cdot \rangle$ is the canonical bilinear form between a vector space and its dual. Similarly, we define $A_{h,C} : V_{h,C} \to V_{h,C}'$ and $A_{h,D} : V_{h,D} \to V_{h,D}'$ by

$$\langle A_{h,C} v, w \rangle = a_h(v, w)\ \forall\, v, w \in V_{h,C} \quad \text{and} \quad \langle A_{h,D} v, w \rangle = a_h(v, w)\ \forall\, v, w \in V_{h,D}.$$

Then we have the following lemma.

Lemma 1 *For any $v \in V_h$, there is a unique decomposition $v = v_C + v_D$, where $v_C \in V_{h,C}$ and $v_D \in V_{h,D}$. In addition, it holds that*

$$\langle A_h v, v \rangle \approx \langle A_{h,C} v_C, v_C \rangle + \langle A_{h,D} v_D, v_D \rangle \quad \forall\, v \in V_h.$$

Remark 1 Since the subspace $V_{h,D}$ only contains dofs on the boundary of subdomains, the size of the matrix $A_{h,D}$ is of order J/h. We can implement the solve $A_{h,D}^{-1}$ directly. Therefore, it is crucial to have an efficient preconditioner for $A_{h,C}$.

Because functions in $V_{h,C}$ have continuous normal derivatives on the edges in $\mathcal{E}_{h,\Gamma}$ and vanishing normal derivatives on $\partial\Omega$, it is easy to observe that

$$a_h(v, w) = \sum_{j=1}^{J} a_{h,j}(v_j, w_j) \qquad \forall\, v, w \in V_{h,C},$$

where $v_j = v\big|_{\Omega_j}, w_j = w\big|_{\Omega_j}$, and $a_{h,j}(\cdot, \cdot)$ is the analog of $a_h(\cdot, \cdot)$ defined on elements and interior edges of Ω_j. Note that $a_{h,j}(\cdot, \cdot)$ is a localized bilinear form.

Next we define

$$V_{h,C}(\Omega\backslash\Gamma) = \left\{ v \in V_{h,C} : v \text{ has vanishing derivatives up to order 1 on } \Gamma \right\}$$
$$V_{h,C}(\Gamma) = \left\{ v \in V_{h,C} : a_h(v, w) = 0,\ \forall\, w \in V_{h,C}(\Omega\backslash\Gamma) \right\}.$$

Functions in $V_{h,C}(\Gamma)$ are referred to as discrete biharmonic functions. They are uniquely determined by the dofs associated with Γ.

For any $v_C \in V_{h,C}$, there is a unique decomposition $v_C = v_{C,\Omega\backslash\Gamma} + v_{C,\Gamma}$, where $v_{C,\Omega\backslash\Gamma} \in V_{h,C}(\Omega\backslash\Gamma)$ and $v_{C,\Gamma} \in V_{h,C}(\Gamma)$. Furthermore, let $A_{h,C,\Omega\backslash\Gamma} : V_{h,C}(\Omega\backslash\Gamma) \to V_{h,C}(\Omega\backslash\Gamma)'$ and $S_h : V_{h,C}(\Gamma) \to V_{h,C}(\Gamma)'$ be SPD operators defined by

$$\langle A_{h,C,\Omega\backslash\Gamma} v, w \rangle = a_h(v, w) \quad \forall\, v, w \in V_{h,C}(\Omega\backslash\Gamma),$$
$$\langle S_h v, w \rangle = a_h(v, w) \quad \forall\, v, w \in V_{h,C}(\Gamma),$$

then it holds that for all $v_C \in V_{h,C}$ with $v_C = v_{C,\Omega\backslash\Gamma} + v_{C,\Gamma}$,

$$\langle A_{h,C} v_C, v_C \rangle = \langle A_{h,C,\Omega\backslash\Gamma} v_{C,\Omega\backslash\Gamma}, v_{C,\Omega\backslash\Gamma} \rangle + \langle S_h v_{C,\Gamma}, v_{C,\Gamma} \rangle.$$

Remark 2 It is noted that $A^{-1}_{h,C,\Omega\setminus\Gamma}$ can be implemented by solving the localized biharmonic problems on each subdomain in parallel. Hence, a preconditioner for S_h^{-1} needs to be constructed.

3 A BDDC Preconditioner

In this section a preconditioner for the Schur complement S_h is constructed by the BDDC methodology.

Let $V_{h,C,j}$, $1 \le j \le J$ be the restriction of $V_{h,C}$ on the subdomain Ω_j. We define $\mathcal{H}_j$, the space of local discrete biharmonic functions, by

$$\mathcal{H}_j = \left\{v \in V_{h,C,j} : a_{h,j}(v,w) = 0 \quad \forall\, w \in V_{h,C}(\Omega_j)\right\},$$

where $V_{h,C}(\Omega_j)$ is the subspace of $V_{h,C,j}$ whose members vanish up to order 1 on $\partial\Omega_j$. The space $\mathcal{H}_C$ is then defined by gluing the spaces $\mathcal{H}_j$ together at the cross points such that

$$\mathcal{H}_C = \left\{v \in L_2(\Omega) : v\big|_{\Omega_j} \in \mathcal{H}_j \text{ and } v \text{ has continuous dofs at subdomain corners}\right\}.$$

We equip $\mathcal{H}_C$ with the bilinear form:

$$a_h^C(v,w) = \sum_{1\le j\le J} a_{h,j}(v_j, w_j) \qquad \forall\, v, w \in \mathcal{H}_C,$$

where $v_j = v\big|_{\Omega_j}$ and $w_j = w\big|_{\Omega_j}$.

Next we introduce a decomposition of $\mathcal{H}_C$,

$$\mathcal{H}_C = \mathring{\mathcal{H}} \oplus \mathcal{H}_0$$

where

$$\mathring{\mathcal{H}} = \{v \in \mathcal{H}_C : \text{ the dofs of } v \text{ vanish at the corners of the subdomains } \Omega_1, \ldots, \Omega_J\},$$
$$\mathcal{H}_0 = \left\{v \in \mathcal{H}_C : a_h^C(v,w) = 0 \quad \forall\, w \in \mathring{\mathcal{H}}\right\}.$$

Let $\mathring{\mathcal{H}}_j$ be the restriction of $\mathring{\mathcal{H}}$ on Ω_j. We then define SPD operators $S_0 : \mathcal{H}_0 \longrightarrow \mathcal{H}_0'$ and $S_j : \mathring{\mathcal{H}}_j \longrightarrow \mathring{\mathcal{H}}_j'$ by

$$\langle S_0 v, w\rangle = a_h^C(v,w) \quad \forall\, v, w \in \mathcal{H}_0 \quad \text{and} \quad \langle S_j v, w\rangle = a_{h,j}(v,w) \quad \forall\, v, w \in \mathring{\mathcal{H}}_j.$$

Now the BDDC preconditioner B_{BDDC} for S_h is given by

$$B_{BDDC} = (P_\Gamma I_0)\, S_0^{-1}\, (P_\Gamma I_0)^t + \sum_{j=1}^{J} (P_\Gamma \mathbb{E}_j)\, S_j^{-1}\, (P_\Gamma \mathbb{E}_j)^t,$$

where $I_0 : \mathcal{H}_0 \to \mathcal{H}_C$ is the natural injection, $\mathbb{E}_j : \mathring{\mathcal{H}}_j \to \mathcal{H}_C$ is the trivial extension, and $P_\Gamma : \mathcal{H}_C \longrightarrow V_{h,C}$ is a projection defined by averaging such that for all $v \in \mathcal{H}_C$, $P_\Gamma v$ is continuous on Γ up to order 1.

Remark 3 A preconditioner $B : V_h{}' \longrightarrow V_h$ for A_h can then be constructed as follows:

$$B = I_D A_{h,D}^{-1} I_D^t + I_{h,C,\Omega\backslash\Gamma} A_{h,C,\Omega\backslash\Gamma}^{-1} I_{h,C,\Omega\backslash\Gamma}^t + I_\Gamma B_{BDDC} I_\Gamma^t,$$

where $I_D : V_{h,D} \to V_h$, $I_{h,C,\Omega\backslash\Gamma} : V_{h,C}(\Omega\backslash\Gamma) \to V_h$, and $I_\Gamma : V_{h,C}(\Gamma) \to V_h$ are natural injections.

4 Condition Number Estimates

In this section we present the condition number estimates of $B_{BDDC} S_h$. Let us begin by noting that

$$V_{h,C}(\Gamma) = P_\Gamma I_0 \mathcal{H}_0 + \sum_{j=1}^{J} P_\Gamma \mathbb{E}_j \mathring{\mathcal{H}}_j.$$

Then it follows from the theory of additive Schwarz preconditioners (see for example [10, 11, 9, 2]) that the eigenvalues of $B_{BDDC} S_h$ are positive, and the extreme eigenvalues of $B_{BDDC} S_h$ are characteristic by the following formulas

$$\lambda_{\min}(B_{BDDC} S_h) = \min_{\substack{v \in V_{h,C}(\Gamma) \\ v \neq 0}} \frac{\langle S_h v, v\rangle}{\displaystyle\min_{\substack{v = P_\Gamma I_0 v_0 + \sum_{j=1}^J P_\Gamma \mathbb{E}_j \mathring{v}_j \\ v_0 \in \mathcal{H}_0,\, \mathring{v}_j \in \mathring{\mathcal{H}}_j}} \left(\langle S_0 v_0, v_0\rangle + \sum_{j=1}^{J} \langle S_j \mathring{v}_j, \mathring{v}_j\rangle\right)},$$

$$\lambda_{\max}(B_{BDDC} S_h) = \max_{\substack{v \in V_{h,C}(\Gamma) \\ v \neq 0}} \frac{\langle S_h v, v\rangle}{\displaystyle\min_{\substack{v = P_\Gamma I_0 v_0 + \sum_{j=1}^J P_\Gamma \mathbb{E}_j \mathring{v}_j \\ v_0 \in \mathcal{H}_0,\, \mathring{v}_j \in \mathring{\mathcal{H}}_j}} \left(\langle S_0 v_0, v_0\rangle + \sum_{j=1}^{J} \langle S_j \mathring{v}_j, \mathring{v}_j\rangle\right)},$$

from which we can establish a lower bound for the minimum eigenvalue of $B_{BDDC} S_h$, an upper bound for the maximum eigenvalue of $B_{BDDC} S_h$, and then an estimate on the condition number of $B_{BDDC} S_h$.

Theorem 1 *It holds that* $\lambda_{\min}(B_{BDDC} S_h) \geq 1$ *and* $\lambda_{\max}(B_{BDDC} S_h) \leq (1 + \ln(H/h))^2/C$, *which imply*

$$\kappa(B_{BDDC}S_h) = \frac{\lambda_{\max}(B_{BDDC}S_h)}{\lambda_{\min}(B_{BDDC}S_h)} \leq C(1 + \ln(H/h))^2,$$

where the positive constant C is independent of h, H, and J.

5 Numerical Results

In this section we present some numerical results to illustrate the performance of the preconditioners B_{BDDC} and B. We consider our model problem (1) on the unit square $(0, 1) \times (0, 1)$. By taking the penalty parameter η in $a_h(\cdot, \cdot)$ and $a_{h,j}(\cdot, \cdot)$ to be 5, we compute the maximum eigenvalue, the minimum eigenvalue, and the condition number of the systems $B_{BDDC}S_h$ and BA_h for different values of H and h.

The eigenvalues and condition numbers of $B_{BDDC}S_h$ and BA_h for 16 subdomains are presented in Tables 1 and 2, respectively. They confirm our theoretical estimates. In addition, the corresponding condition numbers of A_h are provided in Table 2.

Moreover, to illustrate the practical performance of the preconditioner, we present in Table 3 the number of iterations required to reduce the relative residual error by a factor of 10^{-6} for the preconditioned system and the un-preconditioned system, from which we can observe the dramatic improvement in efficiency due to the preconditioner, especially as h gets smaller.

Table 1: Eigenvalues and condition numbers of $B_{BDDC}S_h$ for $H = 1/4$ (J = 16 subdomains)

	$\lambda_{\max}(B_{BDDC}S_h)$	$\lambda_{\min}(B_{BDDC}S_h)$	$\kappa(B_{BDDC}S_h)$
h=1/8	3.6073	1.0000	3.6073
h=1/12	2.9197	1.0000	2.9197
h=1/16	3.0908	1.0000	3.0908
h=1/20	3.2756	1.0000	3.2756
h=1/24	3.4535	1.0000	3.4535

Table 2: Eigenvalues and condition numbers of BA_h, and condition numbers of A_h for $H = 1/4$ (J = 16 subdomains)

	$\lambda_{\max}(BA_h)$	$\lambda_{\min}(BA_h)$	$\kappa(BA_h)$	$\kappa(A_h)$
h=1/8	4.0705	0.2148	18.9490	1.1064e+03
h=1/12	3.4107	0.2507	13.6054	1.3426e+04
h=1/16	3.4866	0.2578	13.5244	6.1689e+04
h=1/20	3.5947	0.2590	13.8787	1.8215e+05
h=1/24	3.7123	0.2593	14.3181	4.2288e+05

Acknowledgements The work of the first and third authors was supported in part by the National Science Foundation under Grant No. DMS-16-20273.

Table 3: Number of iterations for reducing the relative residual error by a factor of 10^{-6} for $H = 1/4$ (J = 16 subdomains)

	$Niter(A_h x = b)$	$Niter(BA_h x = Bb)$
h=1/8	95	27
h=1/12	235	23
h=1/16	434	23
h=1/20	704	23
h=1/24	1026	23

References

1. S. C. Brenner and E.-H. Park and L. -Y. Sung. A BDDC preconditioner for a symmetric interior penalty method. *Electron. Trans. Numer. Anal.*, 46, 190–214, 2017.
2. S. C. Brenner and L. R. Scott. The Mathematical Theory of Finite Element Methods (Third Edition). *Springer-Verlag*, New York, 2008.
3. S. C. Brenner and L.-Y. Sung. C^0 interior penalty methods for fourth order elliptic boundary value problems on polygonal domains. *J. Sci. Comput.*, 22/23, 83–118, 2005.
4. S. C. Brenner and L.-Y. Sung. BDDC and FETI-DP without matrices or vectors. *Comput. Methods Appl. Mech. Engrg.*, 196, 1429–1435, 2007.
5. S. C. Brenner and K. Wang. An iterative substructuring algorithm for a C^0 interior penalty method. *ETNA*, 39, 313–332, 2012.
6. C. R. Dohrmann. A preconditioner for substructuring based on constrained energy minimization. *SIAM J. Sci. Comput.*, 25, 246–258, 2003.
7. G. Engel, K. Garikipati, T. Hughes, M. Larson, L. Mazzei, and R. Taylor. Continuous/discontinuous finite element approximations of fourth order elliptic problems in structural and continuum mechanics with applications to thin beams and plates, and strain gradient elasticity. *Comput. Methods Appl. Mech. Engrg.*, 191, 3669-3750, 2002.
8. S. Li and K. Wang. Condition number estimates for C^0 interior penalty methods. *Domain Decomposition Methods in Science and Engineering XVI*, 55, 675-682, 2007.
9. T. Mathew. Domain Decomposition Methods for the Numerical Solutions of Partial Differential Equations. *Springer-Verlag*, Berlin, 2008.
10. B. Smith and P. Bjørstad and W. Gropp. Domain Decomposition. *Cambridge University Press*, Cambridge, 1996.
11. A. Toselli and O. B. Widlund. Domain Decomposition Methods - Algorithms and Theory. *Springer-Verlag*, Berlin, 2005.

Preconditioners for Isogeometric Analysis and Almost Incompressible Elasticity

Olof B. Widlund, Luca F. Pavarino, Simone Scacchi, and Stefano Zampini

1 Introduction

The aim of this work is to develop a block FETI–DP preconditioner for mixed formulations of almost incompressible elasticity discretized with mixed isogeometric analysis (IGA) methods with continuous pressure. IGA is a recent technology for the numerical approximation of Partial Differential Equations (PDEs), using the highly regular function spaces generated by B-splines and NURBS not only to describe the geometry of the computational domain but also to represent the approximate solution, see e.g. [4]. For a few previous studies, focused on effective solvers for IGA of saddle point problems, see [7, 6].

Inspired by previous work by Tu and Li [10] for finite element discretizations of the Stokes system, the proposed preconditioner is applied to a reduced positive definite system involving only the pressure interface variable and the Lagrange multipliers of the FETI–DP algorithm. A novelty of our contribution consists of using BDDC with deluxe scaling for the interface pressure block and FETI–DP with deluxe scaling for the multiplier block. The numerical results reported in this paper show the robustness of this solver with respect to jumps in the elastic coefficients and the degree of incompressibility of the material.

Olof B. Widlund
Courant Institute, 251 Mercer Street, New York, NY 10012, USA. e-mail: widlund@cims.nyu.edu

Luca F. Pavarino
Dipartimento di Matematica, Università degli Studi di Pavia, Via Ferrata 5, 27100 Pavia, Italy. e-mail: luca.pavarino@unipv.it

Simone Scacchi
Dipartimento di Matematica, Università degli Studi di Milano, Via Saldini 50, 20133 Milano, Italy. e-mail: simone.scacchi@unimi.it

Stefano Zampini
Extreme Computing Research Center, King Abdullah University of Science and Technology, Thuwal 23955, Saudi Arabia. e-mail: stefano.zampini@kaust.edu.sa

R. Haynes et al. (eds.), *Domain Decomposition Methods in Science and Engineering XXV*, Lecture Notes in Computational Science and Engineering 138,
https://doi.org/10.1007/978-3-030-56750-7_40

2 Two variational formulations of elasticity systems

Let Ω be a domain in $\mathbb{R}^3$, which can be represented exactly by the isogeometric analysis system. It is decomposed into N non-overlapping subdomains Ω_i, of diameter H_i, which are images under a geometric map $\mathbf{F}$ of a coarse element partition τ_H of a reference domain. The interface of the decomposition is given by

$$\Gamma = \left(\bigcup_{i=1}^{N} \partial\Omega_i \right) \setminus \partial\Omega.$$

The boundary $\partial\Omega$ is the union of two disjoint sets $\partial\Omega_D$ and $\partial\Omega_N$ where $\partial\Omega_D$ is of non-zero surface measure. We work with two load functions $\boldsymbol{g} \in [L^2(\Omega)]^3$ and $\boldsymbol{g}_N \in [L^2(\partial\Omega_N)]^3$, and the spaces

$$\boldsymbol{V} := \{\mathbf{v} \in H^1(\Omega)^3 : \mathbf{v}|_{\partial\Omega_D} = 0\}, \quad Q := L^2(\Omega).$$

The load functions define a linear functional

$$< \boldsymbol{f}, \mathbf{v} >:= \int_\Omega \boldsymbol{g} \cdot \mathbf{v} dx + \int_{\partial\Omega_N} \boldsymbol{g}_N \cdot \mathbf{v} dA.$$

If the material is compressible, we can use the variational formulation of the linear elasticity (LE) equations:

$$2 \int_\Omega \mu \boldsymbol{\varepsilon}(\mathbf{u}) : \boldsymbol{\varepsilon}(\mathbf{v})\, dx + \int_\Omega \lambda \text{div}\mathbf{u}\ \text{div}\mathbf{v}\, dx \ = < \mathbf{f}, \mathbf{v} > \quad \forall \mathbf{v} \in \boldsymbol{V}. \tag{1}$$

Here $\boldsymbol{\varepsilon}$ is the symmetric gradient operator and $\mu(x)$ and $\lambda(x)$ the Lamé parameters of the material that for simplicity, when developing the theory, are assumed to be constant in each subdomain Ω_i, i.e. $\mu = \mu_i$ and $\lambda = \lambda_i$ in Ω_i. These parameters can be expressed in terms of the local Poisson ratio ν_i and Young's modulus E_i as

$$\mu_i := \frac{E_i}{2(1+\nu_i)}, \qquad \lambda_i := \frac{E_i \nu_i}{(1+\nu_i)(1-2\nu_i)}. \tag{2}$$

The elastic material approaches the incompressible limit when $\nu_i \rightarrow 1/2$. Our main focus will be on a mixed formulation of linear elasticity for almost incompressible (AIE) materials as, e.g., in [2, Ch. 1]: find the material displacement $\mathbf{u} \in \boldsymbol{V}$ and pressure $p \in Q$ such that

$$\begin{cases} 2 \int_\Omega \mu\, \boldsymbol{\varepsilon}(\mathbf{u}) : \boldsymbol{\varepsilon}(\mathbf{v})\, dx - \int_\Omega \text{div}\, \mathbf{v}\, p\, dx = < \boldsymbol{f}, \mathbf{v} > & \forall \mathbf{v} \in \boldsymbol{V}, \\ -\int_\Omega \text{div}\, \mathbf{u}\, q\, dx \quad - \int_\Omega \frac{1}{\lambda} pq\, dx = \quad 0 & \forall q \in Q. \end{cases} \tag{3}$$

Factoring out the constants μ_i and $\frac{1}{\lambda_i}$, we can define local bilinear forms in terms of integrals over the subdomains Ω_i and we obtain for the almost incompressible case

$$\begin{aligned}
\mu a(\mathbf{u},\mathbf{v}) &:= \sum_{i=1}^{N} \mu_i a_i(\mathbf{u},\mathbf{v}) := \sum_{i=1}^{N} 2\mu_i \int_{\Omega_i} \boldsymbol{\varepsilon}(\mathbf{u}) : \boldsymbol{\varepsilon}(\mathbf{v})\, dx \\
b(\mathbf{v},q) &:= \sum_{i=1}^{N} b_i(\mathbf{v},q) := -\sum_{i=1}^{N} \int_{\Omega_i} \operatorname{div}\mathbf{v}\; q\, dx, \qquad (4)\\
\frac{1}{\lambda} c(p,q) &:= \sum_{i=1}^{N} \frac{1}{\lambda_i} c_i(p,q) := \sum_{i=1}^{N} \frac{1}{\lambda_i} \int_{\Omega_i} p\, q\, dx.
\end{aligned}$$

The isogeometric approximation of the mixed elasticity problem is obtained by selecting spaces for the displacements $\mathbf{u}$ and pressure p, respectively. Following Bressan and Sangalli, [3], we select mapped NURBS functions of polynomial degree $p \geq 2$ with $p-2$ continuous derivatives for the displacement and of polynomial degree $p-1$ with $p-2$ continuous derivatives for the pressure; see, e.g., [8] for details on these Taylor–Hood spaces. The resulting pair of spaces is known to be inf-sup stable, see [3]. A major difference from finite element approximations stems from the fact that except for the lowest order case, there is no nodal basis which leads to *fat interfaces*, see [11, Sec. 4.2] and [12, Sec. 3]. This fact makes the construction of small primal spaces more urgent and complicated.

The knots of the isogeometric analysis problems are partitioned into interior knots with basis functions, with support in the subdomain interiors which do not intersect the boundaries of any subdomain, and interface knots. The latter set is partitioned into equivalence classes. These equivalence classes are associated with the subdomain vertices, edges, and faces. Thus, such a vertex class is given by the knots with basis functions with a subdomain vertex in the interior of their supports. A detailed definition of the edge and face classes are given in [8, Section 3]. These equivalence classes are important in the design, analysis, and programming of BDDC and FETI–DP as well as many other domain decomposition algorithms.

3 Dual–Primal decomposition and a FETI–DP reduced system

The interface displacement variable $\mathbf{u}$ is partitioned into a *dual* part $\mathbf{u}_\Delta$ and a *primal* part $\mathbf{u}_\Pi$. To be competitive, the space of primal variables, with functions which are continuous across the interface, should be of much smaller dimension than that of the space of dual variables, for which we allow jumps across the interface. The displacement variables $\mathbf{u}$ is split into interior $\mathbf{u}_I$, dual $\mathbf{u}_\Delta$, and primal $\mathbf{u}_\Pi$ components, and the pressure p into interior p_I and interface p_Γ components, and we denote by λ_Δ the vector of Lagrange multipliers used to enforce the continuity of the dual displacements across the interface.

Following Tu and Li, [10], we reorder the variables as $\mathbf{u}_I$, p_I, $\mathbf{u}_\Delta$, $\mathbf{u}_\Pi$, p_Γ, and λ_Δ and splitting the matrices μA, B, and $\frac{1}{\lambda}C$, defined by the the bilinear forms of (4) and the mixed method, into appropriate blocks associated with this splitting. The original saddle point system resulting from (4) is equivalent to

$$\begin{bmatrix} \mu A_{II} & B_{II}^T & \mu A_{I\Delta} & \mu A_{I\Pi} & B_{\Gamma I}^T & 0 \\ B_{II} & -\frac{1}{\lambda}C_{II} & B_{I\Delta} & B_{I\Pi} & -\frac{1}{\lambda}C_{\Gamma I}^T & 0 \\ \mu A_{\Delta I} & B_{I\Delta}^T & \mu A_{\Delta\Delta} & \mu A_{\Delta\Pi} & B_{\Gamma\Delta}^T & B_{\Delta}^T \\ \mu A_{\Pi I} & B_{I\Pi}^T & \mu A_{\Pi\Delta} & \mu A_{\Pi\Pi} & B_{\Gamma\Pi}^T & 0 \\ B_{\Gamma I} & -\frac{1}{\lambda}C_{\Gamma I} & B_{\Gamma\Delta} & B_{\Gamma\Pi} & -\frac{1}{\lambda}C_{\Gamma\Gamma} & 0 \\ 0 & 0 & B_{\Delta} & 0 & 0 & 0 \end{bmatrix} \begin{bmatrix} \mathbf{u_I} \\ p_I \\ \mathbf{u_\Delta} \\ \mathbf{u_\Pi} \\ p_\Gamma \\ \lambda_\Delta \end{bmatrix} = \begin{bmatrix} \mathbf{f_I} \\ 0 \\ \mathbf{f_\Delta} \\ \mathbf{f_\Pi} \\ 0 \\ 0 \end{bmatrix}, \tag{5}$$

where $B_\Delta = \begin{bmatrix} B_\Delta^{(1)} & B_\Delta^{(2)} & \dots & B_\Delta^{(N)} \end{bmatrix}$ is a Boolean matrix which enforces continuity, $B_\Delta \mathbf{u_\Delta} = \mathbf{0}$, of the dual displacement variables $\mathbf{u_\Delta}$ shared by neighboring subdomains. If we confine ourselves to the case where λ_Δ belongs to the range of B_Δ, this matrix, although indefinite, is nonsingular under the condition that the primal space is large enough.

If the primal space is relatively small, we can, at an acceptable cost, reduce the indefinite system (5) to a symmetric, positive definite system by eliminating the $\mathbf{u_I}$, p_I, $\mathbf{u_\Delta}$, and $\mathbf{u_\Pi}$ variables and changing the sign. We obtain a Schur complement and a reduced linear system

$$G \begin{bmatrix} p_\Gamma \\ \lambda_\Delta \end{bmatrix} = g, \tag{6}$$

which is then solved by a preconditioned conjugate gradient algorithm with a block preconditioner. Here,

$$G := \widetilde{B}_C \widetilde{A}^{-1} \widetilde{B}_C^T + \frac{1}{\lambda}\widetilde{C}, \qquad g := -\widetilde{B}_C \widetilde{A}^{-1} \begin{bmatrix} \mathbf{f_I} \\ 0 \\ \mathbf{f_\Delta} \\ \mathbf{f_\Pi} \end{bmatrix}, \tag{7}$$

and where $\widetilde{A}$ is the leading 4-by-4 principal minor of the matrix of (5) and

$$\widetilde{B}_C := \begin{bmatrix} B_{\Gamma I} & -\frac{1}{\lambda}C_{\Gamma I} & B_{\Gamma\Delta} & B_{\Gamma\Pi} \\ 0 & 0 & B_\Delta & 0 \end{bmatrix} \quad \text{and } \widetilde{C} := \begin{bmatrix} C_{\Gamma\Gamma} & 0 \\ 0 & 0 \end{bmatrix}. \tag{8}$$

4 Deluxe scaling

For the Lagrange multiplier λ_Δ, we use, following Tu and Li [10], a FETI-DP preconditioner borrowed from our work on the compressible case reported in [8]. In BDDC, the average $\bar{\mathbf{u}} := E_D \mathbf{u}$ of an element in the partially discontinuous space of displacements is computed separately for the sets of interface degrees of freedom of the vertex, edge, and face equivalence classes; the operator E_D is central for both the algorithm and the analysis, see, e.g., [11]. For FETI-DP methods the complementary projection $P_D := I - E_D$ is similarly relevant. We start by defining the deluxe scaling in the simplest case of a class with only two elements, i, j, for a face $\mathcal{F}$; for more details on the fat interface and the definition of the fat equivalence classes, we refer to [11, Sec. 4.2] and [12, Sec. 3].

Let $S^{(i)}$ be the Schur interface complement of the subdomain Ω_i, and define two principal minors, $S_{\mathcal{F}}^{(i)}$ and $S_{\mathcal{F}}^{(j)}$, obtained from $S^{(i)}$ and $S^{(j)}$ by removing all rows and columns which do not belong to variables associated with $\mathcal{F}$.

With $\mathbf{u}_{\mathcal{F}}^{(i)}$ the restriction of an element in the dual space to the face $\mathcal{F}$, the deluxe average across $\mathcal{F}$ is then defined as

$$\bar{\mathbf{u}}_{\mathcal{F}} = \left(S_{\mathcal{F}}^{(i)} + S_{\mathcal{F}}^{(j)}\right)^{-1}\left(S_{\mathcal{F}}^{(i)}\mathbf{u}_{\mathcal{F}}^{(i)} + S_{\mathcal{F}}^{(j)}\mathbf{u}_{\mathcal{F}}^{(j)}\right). \tag{9}$$

We also need to define deluxe averaging operators for subdomain edges and subdomain vertices. Given the simple hexahedral subdomain geometry of the parameter space that we are considering, we find that such an equivalence class will have four and eight elements for any fat subdomain edge and vertex, respectively, in the interior of Ω. Thus, for such a fat subdomain edge $\mathcal{E}$ shared by subdomains $\Omega_i, \Omega_j, \Omega_k$, and Ω_ℓ, we use the formula

$$\bar{\mathbf{u}}_{\mathcal{E}} := \left(S_{\mathcal{E}}^{(i)} + S_{\mathcal{E}}^{(j)} + S_{\mathcal{E}}^{(k)} + S_{\mathcal{E}}^{(\ell)}\right)^{-1}\left(S_{\mathcal{E}}^{(i)}\mathbf{u}_{\mathcal{E}}^{(i)} + S_{\mathcal{E}}^{(j)}\mathbf{u}_{\mathcal{E}}^{(j)} + S_{\mathcal{E}}^{(k)}\mathbf{u}_{\mathcal{E}}^{(k)} + S_{\mathcal{E}}^{(\ell)}\mathbf{u}_{\mathcal{E}}^{(\ell)}\right).$$

An analogous formula holds for the fat vertices and involves eight operators. Edges and vertices located on the Neumann boundary of the domain will have fewer elements, depending on the number of subdomains that share them.

For each subdomain Ω_i, we then define a scaling matrix by its restriction $D_\Delta^{(i)}$ to subdomain Ω_i as the direct sum of diagonal blocks given by the deluxe scaling of the face, edge, and vertex terms belonging to the interface of Ω_i:

- for subdomain faces: $D_{\mathcal{F}}^{(i)} := S_{\mathcal{F}}^{(i)}\left(S_{\mathcal{F}}^{(i)} + S_{\mathcal{F}}^{(j)}\right)^{-1}$,
- for subdomain edges: $D_{\mathcal{E}}^{(i)} := S_{\mathcal{E}}^{(i)}\left(S_{\mathcal{E}}^{(i)} + S_{\mathcal{E}}^{(j)} + S_{\mathcal{E}}^{(k)} + S_{\mathcal{E}}^{(\ell)}\right)^{-1}$,
- for subdomain vertices: an analogous formula with eight operators.

These scaling matrices and their transposes provide factors of FETI-DP preconditioning operator. In terms of the complementary projection operator $P_D = I - E_D$, we have for a fat face of Ω_i :

$$P_D\mathbf{u}_{\mathcal{F}} = \left(S_{\mathcal{F}}^{(i)} + S_{\mathcal{F}}^{(j)}\right)^{-1} S_{\mathcal{F}}^{(j)}(\mathbf{u}_{\mathcal{F}}^{(i)} - \mathbf{u}_{\mathcal{F}}^{(j)}).$$

Similar formulas are easily developed for the other types of equivalence classes.

For the preconditioner block associated with the λ_Δ variable, we can borrow directly a successful preconditioner developed in [8] for compressible elasticity. We note that the bilinear form of (1) has a term additional to $\mu a(\cdot,\cdot)$ but that this does not have any real consequences.

In the present work, the pressure sub-solver $M_{p_\Gamma}^{-1}$ is chosen as the inverse of $\frac{1}{\mu}S_{\Gamma\Gamma}^{C}$ obtained from the subdomain mass matrices associated with the interface pressure variables p_Γ. This matrix is obtained by subassembling the local Schur complements $S_{\Gamma\Gamma}^{C^{(i)}}$ of the subdomain mass matrices $C^{(i)}$ weighted by $\frac{1}{\mu_i}$ and defined by

$$\frac{1}{\mu_i} S_{\Gamma\Gamma}^{C^{(i)}} := \frac{1}{\mu_i} C_{\Gamma\Gamma}^{(i)} - \frac{1}{\mu_i} C_{\Gamma I}^{(i)} C_{II}^{(i)^{-1}} C_{I\Gamma}^{(i)}.$$

To develop a competitive algorithm, we then replace the inverse of this Schur complement, defining $M_{p_\Gamma}^{-1}$, by a BDDC deluxe preconditioner built from the subdomain matrices $\frac{1}{\mu_i} S_{\Gamma\Gamma}^{C^{(i)}}$. In our experience, this has proven very successful even without a primal subspace. We note that such a preconditioner is quite helpful given that the mass matrices of the isogeometric Taylor–Hood elements are quite ill-conditioned.

5 Numerical results

We report results of some numerical experiments for the LE (1) and AIE (3) systems in two and three dimensions, discretized with isogeometric NURBS spaces with a uniform mesh size h, polynomial degree p, and regularity k. Results for much larger problems are reported in [13]. The boundary of a reference unit cube has a zero Dirichlet condition on one face, an inhomogeneous Neumann condition on the opposite face, and zero Neumann conditions on all the other faces. The domain Ω is decomposed into N non-overlapping subdomains of characteristic size H.

The tests have been performed using PetIGA-MF [5, 9] as a discretization package; the solvers used are available in the latest release, 3.10, of the PETSc library [1], and have been contributed by Stefano Zampini (see also [14]). In all experiments, the norm of the residual vector has been decreased by a factor 10^{-8}.

5.1 Checkerboard jumping coefficient test

This test is devoted to investigating the robustness of the proposed block FETI-DP preconditioners for the 2D and 3D AIE system with elastic coefficients configured in a checkerboard pattern. We consider jumps in both the Young modulus E and the Poisson ration ν. In Tables 1 and 2, the conjugate gradient (CG) iteration count (n_{it}) and the maximal (λ_M) and minimal (λ_m) eigenvalues of the preconditioned operator are reported. In the 2D test, we have fixed the number of subdomains to $N = 49 = 7 \times 7$ and the mesh size to $1/h = 128$. In the 3D test, the number of subdomains is $N = 27 = 3 \times 3 \times 3$ and the mesh size $1/h = 16$. The displacement field spline parameters of the Taylor-Hood pair are $p = 3$, $k = 1$; therefore the pressure spline parameters are $p = 2$, $k = 1$. The results show that the proposed solver is very robust with respect to all the jumps considered, since both the number of CG iterations and the extreme eigenvalues approach constant values when E becomes large or the material becomes incompressible ($\nu \longrightarrow 0.5$).

2D jump test for E				2D jump test for ν			
E	n_{it}	λ_M	λ_m	ν	n_{it}	λ_M	λ_m
1e+00	23	2.91e+00	2.91e-01	0.3	10	3.16e+00	6.27e-01
1e+01	25	1.91e+00	1.83e-01	0.4	11	3.12e+00	5.04e-01
1e+02	39	2.09e+00	8.30e-02	0.45	12	3.07e+00	4.17e-01
1e+03	51	2.15e+00	3.72e-02	0.49	14	3.02e+00	3.35e-01
1e+04	52	2.16e+00	3.31e-02	0.499	14	3.01e+00	3.21e-01
1e+05	49	2.16e+00	4.23e-02	0.4999	14	3.01e+00	3.20e-01

Table 1: FETI-DP for AIE on 2D checkerboard jumping coefficient tests. Conjugate gradient iteration counts (n_{it}) and extreme eigenvalues (λ_M, λ_m) of the preconditioned operator. **Jump test for E:** $E = 1$ in black subdomains, E shown in the table in red subdomains; fixed $\nu = 0.49$. **Jump test for ν:** $\nu = 0.3$ in black subdomains, ν shown in the table in red subdomains; fixed $E = 1e+06$. In both tests: $N = 49 = 7 \times 7$ subdomains; $1/h = 128$, displacement field spline parameters $p = 3$, $k = 1$.

3D jump test for E				3D jump test for ν			
E	n_{it}	λ_M	λ_m	ν	n_{it}	λ_M	λ_m
1e+00	28	2.72e+00	2.46e-01	0.3	12	3.04e+00	5.52e-01
1e+01	39	2.83e+00	1.27e-01	0.4	13	2.82e+00	4.34e-01
1e+02	62	2.92e+00	4.35e-02	0.45	14	2.76e+00	3.69e-01
1e+03	80	2.99e+00	2.39e-02	0.49	15	2.73e+00	3.19e-01
1e+04	83	3.00e+00	2.18e-02	0.499	16	2.72e+00	3.05e-01
1e+05	83	3.00e+00	2.20e-02	0.4999	16	2.72e+00	3.04e-01

Table 2: FETI-DP for AIE on 3D checkerboard jumping coefficient tests. Conjugate gradient iteration counts (n_{it}) and extreme eigenvalues (λ_M, λ_m) of the preconditioned operator. **Jump test for E:** $E = 1$ in black subdomains, E shown in the table in red subdomains; fixed $\nu = 0.49$. **Jump test for ν:** $\nu = 0.3$ in black subdomains, ν shown in the table in red subdomains; fixed $E = 1e + 06$. In both tests: $N = 27 = 3 \times 3 \times 3$ subdomains; $1/h = 16$, displacement field spline parameters $p = 3$, $k = 1$.

5.2 A comparison between FETI-DP for LE and for AIE

The aim of this test is to compare the FETI-DP preconditioner for 3D LE developed previously in [8] with the block FETI-DP solver for 3D AIE proposed in the current project, in terms of the robustness with respect to incompressibility of the material, i.e., when $\nu \to 0.5$.

In Table 3, the CG iteration count (n_{it}) and the maximal (λ_M) and minimal (λ_m) eigenvalues of the preconditioned operator are reported. The Young modulus is kept fixed to $E = 1$ in the whole domain, while ν varies as detailed in the tables. We fix the number of subdomains to $N = 27 = 3 \times 3 \times 3$ and the mesh size to $1/h = 16$. The displacement field spline parameters are $p = 3$, $k = 1$. Using a Taylor-Hood pair for the case of AIE, this results in pressure spline parameters of $p = 2$, $k = 1$.

The results show, as expected, that the FETI-DP solver for LE degenerates when the material approaches the incompressible limit, while the FETI-DP solver for

AIE is very robust in terms of both CG iterations and extreme eigenvalues of the preconditioned operator.

3D comparison						
ν	FETI-DP for LE			FETI-DP for AIE		
	n_{it}	λ_M	λ_m	n_{it}	λ_M	λ_m
0.3	16	8.73e+00	1.03e+00	20	3.04e+00	5.07e-01
0.4	19	1.30e+01	1.03e+00	23	2.73e+00	3.59e-01
0.45	25	2.02e+01	1.02e+00	25	2.73e+00	2.94e-01
0.49	43	5.49e+01	1.03e+00	28	2.72e+00	2.46e-01
0.499	100	2.48e+02	1.02e+00	28	2.72e+00	2.36e-01
0.4999	283	1.85e+03	1.02e+00	28	2.72e+00	2.35e-01

Table 3: 3D comparison between FETI-DP for LE and for AIE. Conjugate gradient iteration counts (n_{it}) and extreme eigenvalues (λ_M, λ_m) of the preconditioned operator. $E = 1$, $N = 27 = 3 \times 3 \times 3$ subdomains; $1/h = 16$, displacement field spline parameters $p = 3$, $k = 1$.

References

1. Balay, S., Abhyankar, S., Adams, M., Brown, J., Brune, P., Buschelman, K., Dalcin, L., Dener, A., Eijkhout, V., Gropp, W., Kaushik, D., Knepley, M., May, D., McInnes, L.C., Mills, R.T., Munson, T., Rupp, K., Sanan, P., Smith, B., Zampini, S., Zhang, H.: PETSc Web page. http://www.mcs.anl.gov/petsc (2018)
2. Boffi, D., Brezzi, F., Fortin, M.: Mixed finite element methods and applications, *Springer Series in Computational Mathematics*, vol. 44. Springer, Heidelberg (2013). DOI:10.1007/978-3-642-36519-5
3. Bressan, A., Sangalli, G.: Isogeometric discretizations of the Stokes problem: stability analysis by the macroelement technique. IMA J. Numer. Anal. **33**(2), 629–651 (2013). DOI:10.1093/imanum/drr056
4. Cottrell, J.A., Hughes, T.J.R., Bazilevs, Y.: Isogeometric analysis. John Wiley & Sons, Ltd., Chichester (2009). DOI:10.1002/9780470749081. Toward integration of CAD and FEA
5. Dalcin, L., Collier, N., Vignal, P., Côrtes, A., Calo, V.: PetIGA: a framework for high-performance isogeometric analysis. Comput. Methods Appl. Mech. Engrg. **308**, 151–181 (2016). DOI:10.1016/j.cma.2016.05.011
6. Montardini, M., Sangalli, G., Tani, M.: Robust isogeometric preconditioners for the Stokes system based on the fast diagonalization method. Comput. Methods Appl. Mech. Engrg. **338**, 162–185 (2018). DOI:10.1016/j.cma.2018.04.017
7. Pavarino, L.F., Scacchi, S.: Isogeometric block FETI-DP preconditioners for the Stokes and mixed linear elasticity systems. Comput. Methods Appl. Mech. Engrg. **310**, 694–710 (2016). DOI:10.1016/j.cma.2016.07.012
8. Pavarino, L.F., Scacchi, S., Widlund, O.B., Zampini, S.: Isogeometric BDDC deluxe preconditioners for linear elasticity. Math. Models Methods Appl. Sci. **28**(7), 1337–1370 (2018). DOI:10.1142/S0218202518500367
9. Sarmiento, A., Cortess, A., Garcia, D., Dalcin, L., Collier, N., Calo, V.: PetIGA-MF: a multi-field high-performance toolbox for structure-preserving B-splines spaces (2017)
10. Tu, X., Li, J.: A FETI-DP type domain decomposition algorithm for three-dimensional incompressible Stokes equations. SIAM J. Numer. Anal. **53**(2), 720–742 (2015). DOI:10.1137/13094997X
11. Beirão da Veiga, L., Pavarino, L.F., Scacchi, S., Widlund, O.B., Zampini, S.: Isogeometric BDDC preconditioners with deluxe scaling. SIAM J. Sci. Comput. **36**(3), A1118–A1139 (2014). DOI:10.1137/130917399

12. Beirão da Veiga, L., Pavarino, L.F., Scacchi, S., Widlund, O.B., Zampini, S.: Adaptive selection of primal constraints for isogeometric BDDC deluxe preconditioners. SIAM J. Sci. Comput. **39**(1), A281–A302 (2017). DOI:10.1137/15M1054675
13. Widlund, O.B., Zampini, S., Pavarino, L.F., Scacchi, S.: Block FETI–DP/BDDC preconditioners for mixed isogeometric discretizations of three-dimensional almost incompressible elasticity. Tech. Rep. TR2019-994, Department of Computer Science of the Courant Institute (2019)
14. Zampini, S.: PCBDDC: a class of robust dual-primal methods in PETSc. SIAM J. Sci. Comput. **38**(5), S282–S306 (2016). DOI:10.1137/15M1025785

Dispersion Correction for Helmholtz in 1D with Piecewise Constant Wavenumber

Pierre-Henri Cocquet, Martin J. Gander, and Xueshuang Xiang

1 Introduction

The Helmholtz equation is the simplest model for time harmonic wave propagation, and it contains already all the fundamental difficulties such problems pose when trying to compute their solution numerically. Since time harmonic wave propagation has important applications in many fields of science and engineering, the numerical solution of such problems has been the focus of intensive research efforts, see [15, 16, 20, 10, 5, 23, 28], and the review [17] and references therein for domain decomposition approaches, and [3, 11, 13, 25, 14, 7] and references therein for multigrid techniques. The main problem is that all grid based numerical methods like finite differences or finite elements are losing accuracy because of what is called the *pollution effect* [2, 1]. It is not sufficient to just choose a number of grid points large enough to resolve the wave length determined by the wave number to obtain an accurate solution; the larger the wave number, the more grid points per wave length are needed. This leads to extremely large linear systems that need to be solved when the wave number becomes large, which is hard using classical iterative methods, see [12] and references therein. The pollution effect is due to the *numerical dispersion*, a property which unfortunately all grid based methods have, see also [22, 26] and references therein. In the case of a constant wave number, to reduce the numerical dispersion of the standard 5-point finite difference scheme, a rotated 9-point FDM was proposed in [19] which minimizes the numerical dispersion, see also [4, 24, 27, 6] for more recent such approaches. In particular, in [8] a new approach was introduced which does not only modify the finite difference stencil, but also

P-H. Cocquet
Université de la Réunion, France e-mail: pierre-henri.cocquet@univ-reunion.fr

M. J. Gander
Université de Genève, Switzerland e-mail: Martin.Gander@unige.ch

X. Xiang
Qian Xuesen Laboratory of Space Technology, Beijing e-mail: xiangxueshuang@qxslab.cn

R. Haynes et al. (eds.), *Domain Decomposition Methods in Science and Engineering XXV*, Lecture Notes in Computational Science and Engineering 138,
https://doi.org/10.1007/978-3-030-56750-7_41

the wave number itself in the discrete scheme to minimize dispersion. This led to a new finite difference scheme in two spatial dimensions with much smaller dispersion error than all previous approaches, see also [18]. Minimizing numerical dispersion is also important for effective coarse grid corrections in domain decomposition and for constructing efficient multigrid solvers [25]: for a constant wave number in 1D it is even possible to obtain perfect multigrid efficiency using standard components and dispersion correction, it suffices to use a suitably modified numerical wave number on each level [13], see also [21]. Note that this is very different from [9] where a large complex shift is used in the spirit of [11].

In all previous work on dispersion correction, a main assumption is that the wave number is constant throughout the domain. We propose and study here a new dispersion correction for the Helmholtz equation in 1D in the case where the wave number is only piecewise constant, and allowed to jump in between. Using the exact solution of a transmission problem, we determine for a finite difference discretization a dispersion correction at the interface where the wave number is jumping by introducing a modified numerical wave number there. We show by numerical experiments that this dispersion correction leads to much more accurate solutions than the scheme without dispersion correction, and this at already few points per wavelength resolution. We then also show that this dispersion correction has a very good effect on a two-grid method, by studying numerically the contraction factor of a two grid scheme in the important regime where the coarse and fine mesh are far from resolving the problem. We then conclude by discussing further research directions.

2 Problem Setting

We consider the 1D Helmholtz equation[1] with source term $f \in L^2(-1,1)$,

$$-\partial_{x^2}^2 u(x) - k(x)^2 u(x) = f(x), \quad x \in (-1,1), \quad u(-1) = 0, \quad u(1) = 0, \tag{1}$$

where $k(x)$ is the wave number, which we assume to be piecewise constant, $k(x) := k_1$ if $x \leq 0$, and $k(x) := k_2$ if $x > 0$. We discretize Problem (1) with a standard 3-point centered finite difference scheme on a uniform mesh[2] with n interior mesh points and meshsize $h = 1/(n+1)$. Assuming that $x = 0$ is always a grid point, and denoting by n_1 the number of interior mesh points in $(-1,0)$ and n_2 the number of interior mesh points in $(0,1)$, the continuous problem is thus approximated by a linear system $A\mathbf{u} = \mathbf{f}$ with

$$A = \frac{1}{h^2}\text{tridiag}(-1,2,-1) - \text{diag}(k_1^2 I_{n_1}, k_0^2, k_2^2 I_{n_2}), \quad \mathbf{f} = \left(f(x_j)\right)_{j=1}^n, \tag{2}$$

[1] We only choose wave number configurations such that this problem is well posed

[2] We use for simplicity the same mesh size in both regions

where $k_0 = k(0) = k_1$, but we could also have chosen k_2 here.

3 Dispersion Correction for Piecewise Constant Wave Number

We use a modified wave number like in [13, p. 26 Eq. (3.15)] in the regions where the wave number is constant,

$$\widehat{k}_h(x) := \begin{cases} \sqrt{2h^{-2}(1-\cos(k_1 h))} & \text{if } x < 0, \\ \widehat{k}_0 & \text{at } x = 0, \\ \sqrt{2h^{-2}(1-\cos(k_2 h))} & \text{if } x > 0. \end{cases}$$

This modified wave number was obtained in [13] by making the exact solution of the homogeneous Helmholtz equation on $\mathbb{R}$ satisfy the finite difference scheme. Similarly we determine $\widehat{k}_0$ such that it satisfies the equation

$$h^{-2}(2u_e(0) - u_e(-h) - u_e(h)) - \widehat{k}_0^2 u_e(0) = 0, \tag{3}$$

where u_e is the exact solution of the Helmholtz equation with discontinuous k on $\mathbb{R}$ for the transmission problem of an incoming wave, given by

$$u_e(x) := \begin{cases} Ae^{ik_1x} + Be^{-ik_1x} & \text{if } x < 0 \\ Ce^{ik_2x} & \text{if } x \geq 0 \end{cases}, \text{ with } A = \frac{k_1 + k_2}{2k_1},\ B = \frac{k_1 - k_2}{2k_1},\ C = 1. \tag{4}$$

The matrix associated to the new FD scheme with dispersion correction is then given by (2) with k replaced by $\widehat{k}_h$.

We show in Figure 1 the great influence this dispersion correction has on the numerical quality of the solution. We used $k_1 = 3.2\pi$ and $k_2 = k_1/2$ (top) and $k_2 = k_1/4$ (bottom) and solved (2) with and without dispersion correction using a mesh size $h = \frac{1}{16}$ which implies 10 points per wavelength for $x < 0$ (left) and $h = \frac{1}{8}$ (right), which implies 5 points per wavelength for $x < 0$. As a source term, we used a linear combination of the first sine functions $\sin(\omega\frac{\pi(x+1)}{2})$, $\omega = 1, 2, \ldots, 16$ with random coefficients, and we denote by exact a numerical solution without dispersion correction using a four times finer grid. We clearly see that dispersion correction is also possible in the case of a non-constant wave number, and we next study the influence of such a correction on a two-grid method.

4 Influence of Dispersion Correction on Multigrid

A two-grid algorithm for a general linear system $A\mathbf{u} = \mathbf{f}$ is given by performing for $n = 0, 1, \ldots$

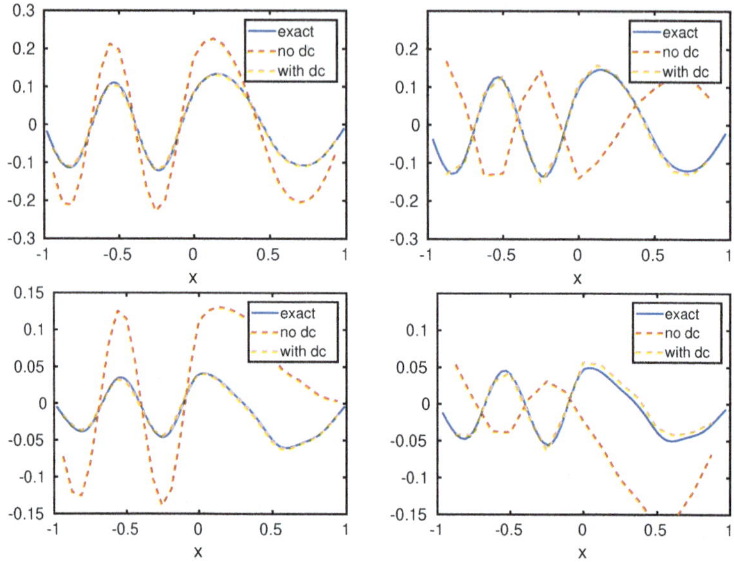

Fig. 1: Four numerical examples showing the impact of dispersion correction in the case of piecewise constant wave speed: contrast 2 (top row) and 4 (bottom row) and 10 points per wavelength (left column) and 5 points per wavelength (right column)

$$\begin{aligned} \tilde{\mathbf{u}}^n &:= S^{\nu_1}(\mathbf{u}^n, \mathbf{f}); \qquad \text{\% pre} - \text{smoothing} \\ \mathbf{r}_c^n &:= R(\mathbf{f} - A\tilde{\mathbf{u}}^n); \\ \mathbf{e}_c^{n+1} &= A_c^{-1}\mathbf{r}_c^n; \\ \tilde{\mathbf{u}}^{n+1} &= \tilde{\mathbf{u}}^n + P\mathbf{e}^{n+1}; \\ \mathbf{u}^{n+1} &= S^{\nu_2}(\tilde{\mathbf{u}}^{n+1}, \mathbf{f}); \ \text{\% post} - \text{smoothing} \end{aligned} \tag{5}$$

where R denotes a restriction operator, P a prolongation operator, S^ν represents ν iterations of a smoother, and A_c is a coarse matrix. We define the fine grid Ω^h with meshsize $h = 1/(n+1)$ by

$$\Omega^h := \{x_j = jh : \ j = 0, \cdots, n+1\}.$$

The coarse grid is defined from Ω^h with mesh width $H = 2h$ by coarsening,

$$\Omega^H := \{x_j = jH : \ j = 0, \cdots, N+1\},$$

where $n = 2N + 1$. The prolongation operator maps grid functions $\mathbf{u}^H$ defined on a coarse grid Ω^H to a function $I_H^h \mathbf{u}^H$ defined on the fine grid Ω^h using linear interpolation. Its matrix representation P can be found in [13, p.18, Eq.(3.1)]. For the restriction operator, we use the *full weighting restriction operator* whose matrix representation is $R = P^T/2$ (see [13, p.20, Eq. (3.4)]).

For the smoother, we use a damped Kacmarz smoother whose iteration matrix is given by

$$S := I_N - \omega A_h^* A_h.$$

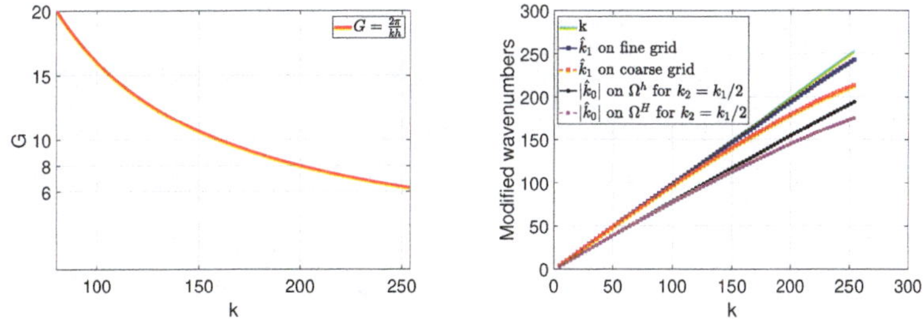

Fig. 2: Left: number of grid points per wavelength. Right: modified wavenumbers

Necessary conditions for the two-grid algorithm to converge can be found in [7, p.12 Theorem 4.1], and having $||S^\nu||_2 \leq C_S$, where $C_S > 0$ does not depend on ν, is needed. Since $S^* = S$, one has $||S||_2 = \max_{\lambda \in \sigma(S)} |\lambda|$ and one can thus chose ω to ensure that $\sigma(S) \subset [0, 1]$. This can be achieved with $\omega = \rho(A_h)^{-2}$ and it is worth noting that this gives $||S^\nu||_2 \leq 1$.

The two-grid operator with ν_1 pre- and ν_2 post-smoothing steps then reads

$$T(\nu_1, \nu_2) = S^{\nu_1} \left(I_N - P A_H^{-1} R A_h \right) S^{\nu_2}, \tag{6}$$

where now both the fine matrix A_h and the coarse matrix A_H are defined by (2) when no dispersion correction is used and with k replaced by $\hat{k}_h$ (respectively $\hat{k}_H$) when we use dispersion correction.

Note that $\rho(T(\nu_1, \nu_2)) = \rho(T(\nu_1 + \nu_2, 0))$ and thus we are going to present our numerical results using $\nu = \nu_1 + \nu_2$ smoothing steps. We use a fine grid with $n = 255$ grid points which gives $h = 1/256$, and $N = 127$ coarse grid points. We compute the spectral radius of the two-grid operator (6) for the sequence of wave numbers

$$k_{1,j} = \sqrt{\frac{2}{h^2} \left(\sin\left(\frac{(j-1)\pi h}{2} \right)^2 + \sin\left(\frac{j \pi h}{2} \right)^2 \right)}, \quad j = 1, \cdots, \widetilde{N},$$

placing $k_{1,j}^2$ exactly between two eigenvalues of the discrete Laplace operator[3]. The integer $\widetilde{N}$ is chosen so that we have a number of grid points per wavelength G satisfying $G = 2\pi/(k_{1,j}h) \geq 2\pi$ since, otherwise, the discrete dispersion relation at the coarse level is empty. The value of G satisfies $G \geq 20$ for $k \leq 80$. Figure 2 gives G for $k \geq 80$ and the modified wavenumbers as functions of k when $k_2 = k_1/2$ (similar results can be obtained for $k_2 = k_1/4$). We present in Figure 3 the spectral radius of the two grid operator without dispersion correction (left), and with dispersion correction (right), for a contrast of two in the wave number (top), and four (bottom), as a function of an increasing wave number, using various numbers of smoothing steps. These results show that without dispersion correction, $\rho(T(\nu, 0))$

[3] This choice allows us to systematically test sequences of wavenumbers for similarly conditioned Helmholtz problems as long as the contrast is not too large.

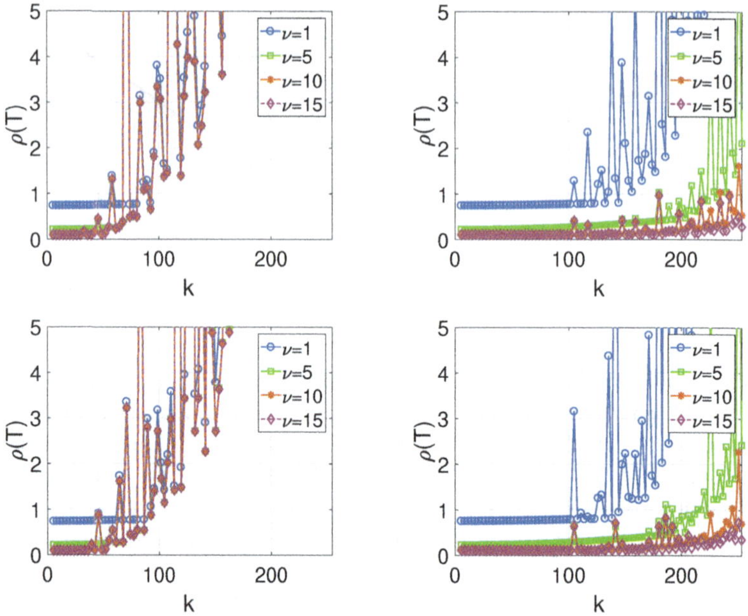

Fig. 3: Spectral radius of the two-grid operator. Left column: no dispersion correction. Right column: with dispersion correction. Top row: $k_2 = k_1/2$. Bottom row: $k_2 = k_1/4$.

is decreasing as the number of smoothing steps ν increases but the minimal ν to get $\rho(T(\nu, 0)) < 1$ is becoming too large for this method to be used in practice. In contrast, the two-grid scheme with dispersion correction is a convergent iterative method for a relatively small number of smoothing steps already.

In our approach, we computed a modified wave number $\hat{k}_h(0)$ at the interface using (3), which requires the computation of an exact solution of a transmission problem for the Helmholtz equation with piecewise constant wave number. Since it might be difficult to compute an exact solution of such a transmission problem in higher dimensions, we now test how important this dispersion correction at the interface is. The idea of this test is that since without dispersion correction we had $k(0) = k_1$, one could choose in the dispersion correction for the modified wave number $\hat{k}_h$ such that $\hat{k}_h(0) = \hat{k}_1 = \sqrt{2h^{-2}(1 - \cos(k_1 h))}$, i.e. just use the same dispersion correction at the interface as in the left region. We show in Figure 4 the spectral radius of the two-grid operator as a function of the number of smoothing steps for these two possible choices of $\hat{k}_h(0)$ for two different wave number contrasts and $k_1 = \max_j(k_{1,j}) = 253.73$. These results show that the modified wave number with $\hat{k}_h(0) = \hat{k}_1$ also yields a convergent two-grid method for a large enough number of smoothing steps, but the specific dispersion correction from the transmission problem in (3) needs a smaller number of smoothing steps to ensure that $\rho(T(\nu, 0)) < 1$ and also has a much smaller contraction factor.

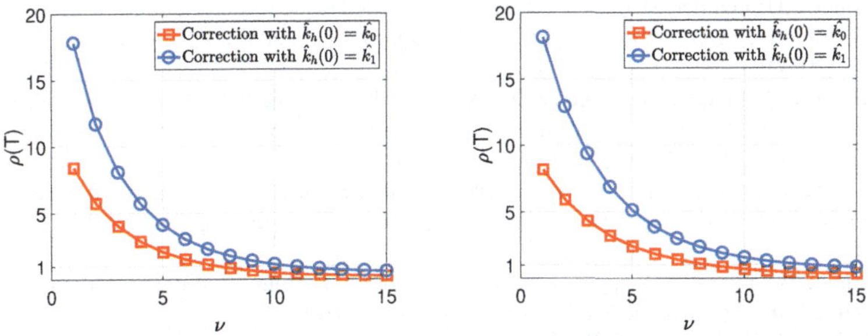

Fig. 4: Comparison of $\rho(T)$ for $k = \max_j(k_{1,j})$ using a shifted wave number $\hat{k}_h(x)$ such that $\hat{k}_h(0) = \hat{k}_1$ and $\hat{k}_h(0) = \hat{k}_0$ satisfying (3). Left: $k_2 = k_1/2$. Right: $k_2 = k_1/4$.

5 Conclusions and Outlook

We have introduced a new technique for dispersion correction for discretized Helmholtz problems in 1D for the case of piecewise constant wave numbers at the interface between regions where the wave number has a jump. The idea is to use a discrete wave number stemming from a transmission problem. We showed numerically that this dispersion correction leads to much more accurate numerical solutions, and also leads to much more efficient multigrid techniques when applied on each level of the grid hierarchy. Dispersion correction is more difficult in higher dimensions, but modifying the wave number in addition to specialized stencils has led to very good results in [8]. We are currently working on a 2D variant of the ideas presented here.

References

1. Babuška, I.M., Ihlenburg, F., Paik, E.T., Sauter, S.A.: A generalized finite element method for solving the Helmholtz equation in two dimensions with minimal pollution. Computer methods in applied mechanics and engineering **128**(3-4), 325–359 (1995)
2. Babuška, I.M., Sauter, S.A.: Is the pollution effect of the fem avoidable for the Helmholtz equation considering high wave numbers? SIAM Journal on numerical analysis **34**(6), 2392–2423 (1997)
3. Brandt, A., Livshits, I.: Wave-ray multigrid method for standing wave equations. Electron. T. Numer. Ana. **6**, 162–181 (1997)
4. Chen, Z., Cheng, D., Wu, T.: A dispersion minimizing finite difference scheme and preconditioned solver for the 3d Helmholtz equation. Journal of Computational Physics **231**(24), 8152–8175 (2012)
5. Chen, Z., Xiang, X.: A source transfer domain decomposition method for Helmholtz equations in unbounded domain. SIAM J. Numer. Anal. **51**, 2331–2356 (2013)
6. Cheng, D., Tan, X., Zeng, T.: A dispersion minimizing finite difference scheme for the Helmholtz equation based on point-weighting. Computers & Mathematics with Applications (2017)

7. Cocquet, P.H., Gander, M.J.: How large a shift is needed in the shifted Helmholtz preconditioner for its effective inversion by multigrid? SIAM Journal on Scientific Computing **39**(2), A438–A478 (2017)
8. Cocquet, P.H., Gander, M.J., Xiang, X.: A finite difference method with optimized dispersion correction for the Helmholtz equation. In: Domain Decomposition Methods in Science and Engineering XXIV, LNCSE. Springer-Verlag (2018)
9. Cools, S., Reps, B., Vanroose, W.: A new level-dependent coarse grid correction scheme for indefinite Helmholtz problems. Numerical Linear Algebra with Applications **21**(4), 513–533 (2014)
10. Engquist, B., Ying, L.: Sweeping preconditioner for the Helmholtz equation: Moving perfectly matched layers. Multiscale Model. Sim. **9**, 686–710 (2011)
11. Erlangga, Y.A., Oosterlee, C.W., Vuik, C.: A novel multigrid based preconditioner for heterogeneous Helmholtz problems. SIAM J. Sci. Comput. **27**, 1471–1492 (2006)
12. Ernst, O., Gander, M.J.: Why it is difficult to solve Helmholtz problems with classical iterative methods. In: I. Graham, T. Hou, O. Lakkis, R. Scheichl (eds.) Numerical Analysis of Multiscale Problems, pp. 325–363. Springer-Verlag, Berlin (2012)
13. Ernst, O.G., Gander, M.J.: Multigrid methods for Helmholtz problems: A convergent scheme in 1d using standard components. Direct and Inverse Problems in Wave Propagation and Applications. De Gruyer pp. 135–186 (2013)
14. Gander, M.J., Graham, I.G., Spence, E.A.: Applying GMRES to the Helmholtz equation with shifted Laplacian preconditioning: what is the largest shift for which wavenumber-independent convergence is guaranteed? Numer. Math. **131**, 567–614 (2015)
15. Gander, M.J., Magoules, F., Nataf, F.: Optimized Schwarz methods without overlap for the Helmholtz equatio n. SIAM J. Sci. Comput. **24**, 38–60 (2002)
16. Gander, M.J., Nataf, F.: An incomplete LU preconditioner for problems in acoustics. J. Comput. Acoust. **13**, 455–476 (2005)
17. Gander, M.J., Zhang, H.: A class of iterative solvers for the Helmholtz equation: Factorizations, sweeping preconditioners, source transfer, single layer potentials, polarized traces, and optimized Schwarz methods. To appear in SIAM Review (2018)
18. Harari, I., Turkel, E.: Accurate finite difference methods for time-harmonic wave propagation. Journal of Computational Physics **119**(2), 252–270 (1995)
19. Jo, C.H., Shin, C., Suh, J.H.: An optimal 9-point, finite-difference, frequency-space, 2-d scalar wave extrapolator. Geophysics **61**(2), 529–537 (1996)
20. Livshits, I.: The least squares AMG solver for the one-dimensional Helmholtz operator. Comput. Visual. Sci. **14**, 17–25 (2011)
21. Olson, L.N., Schroder, J.B.: Smoothed aggregation for Helmholtz problems. Numerical Linear Algebra with Applications **17**(2-3), 361–386 (2010)
22. Shin, C., Sohn, H.: A frequency-space 2-d scalar wave extrapolator using extended 25-point finite-difference operator. Geophysics **63**(1), 289–296 (1998)
23. Stolk, C.C.: A rapidly converging domain decomposition method for the Helmholtz equation. J. Comput. Phys. **241**, 240–252 (2013)
24. Stolk, C.C.: A dispersion minimizing scheme for the 3-d Helmholtz equation based on ray theory. Journal of computational Physics **314**, 618–646 (2016)
25. Stolk, C.C., Ahmed, M., Bhowmik, S.K.: A multigrid method for the Helmholtz equation with optimized coarse grid corrections. SIAM Journal on Scientific Computing **36**(6), A2819–A2841 (2014)
26. Turkel, E., Gordon, D., Gordon, R., Tsynkov, S.: Compact 2d and 3d sixth order schemes for the Helmholtz equation with variable wave number. Journal of Computational Physics **232**(1), 272–287 (2013)
27. Wu, T.: A dispersion minimizing compact finite difference scheme for the 2d Helmholtz equation. Journal of Computational and Applied Mathematics **311**, 497–512 (2017)
28. Zepeda-Núñez, L., Demanet, L.: The method of polarized traces for the 2D Helmholtz equation. J. Comput. Phys. **308**, 347–388 (2016)

BDDC Preconditioners for a Space-time Finite Element Discretization of Parabolic Problems

Ulrich Langer and Huidong Yang

1 Introduction

Continuous space-time finite element methods for parabolic problems have been recently studied, e.g., in [1, 9, 10, 13]. The main common features of these methods are very different from those of time-stepping methods. Time is considered to be just another spatial coordinate. The variational formulations are studied in the full space-time cylinder that is then decomposed into arbitrary admissible simplex elements. In this work, we follow the space-time finite element discretization scheme proposed in [10] for a model initial-boundary value problem, using continuous and piecewise linear finite elements in space and time simultaneously.

It is a challenging task to efficiently solve the large-scale linear system of algebraic equations arising from the space-time finite element discretization of parabolic problems. In this work, as a preliminary study, we use the balancing domain decomposition by constraints (BDDC [2, 11, 12]) preconditioned GMRES method to solve this system efficiently. We mention that robust preconditioning for space-time isogeometric analysis schemes for parabolic evolution problems has been reported in [3, 4].

The remainder of the paper is organized as follows: Sect. 2 deals with the space-time finite element discretization for a parabolic model problem. In Sect. 3, we discuss BDDC preconditioners that are used to solve the linear system of algebraic equations. Numerical results are shown and discussed in Sect. 4. Finally, some conclusions are drawn in Sect. 5.

Ulrich Langer
Johann Radon Institute, Altenberg Strasse 69, 4040 Linz, Austria, e-mail: ulrich.langer@ricam.oeaw.ac.at

Huidong Yang
Johann Radon Institute, Altenberg Strasse 69, 4040 Linz, Austria, e-mail: huidong.yang@oeaw.ac.at

R. Haynes et al. (eds.), *Domain Decomposition Methods in Science and Engineering XXV*,
Lecture Notes in Computational Science and Engineering 138,
https://doi.org/10.1007/978-3-030-56750-7_42

2 The space-time finite element discretization

The following parabolic initial-boundary value problem is considered as our model problem: Find $u : \overline{Q} \to \mathbb{R}$ such that

$$\partial_t u - \Delta_x u = f \text{ in } Q, \quad u = 0 \text{ on } \Sigma,\ u = u_0 \text{ on } \Sigma_0, \tag{1}$$

where $Q := \Omega \times (0,T)$, $\Omega \subset \mathbb{R}^2$ is a sufficiently smooth and bounded spatial computational domain, $\Sigma := \partial\Omega \times (0,T)$, $\Sigma_0 := \Omega \times \{0\}$, $\Sigma_T := \Omega \times \{T\}$.

Let us now introduce the following Sobolev spaces:

$$H_0^{1,0}(Q) = \{u \in L_2(Q) : \nabla_x u \in [L_2(Q)]^2, u = 0 \text{ on } \Sigma\},$$
$$H_{0,\bar{0}}^{1,1}(Q) = \{u \in L_2(Q) : \nabla_x u \in [L_2(Q)]^2, \partial_t u \in L_2(Q) \text{ and } u|_{\Sigma\cup\Sigma_T} = 0\},$$
$$H_{0,\underline{0}}^{1,1}(Q) = \{u \in L_2(Q) : \nabla_x u \in [L_2(Q)]^2, \partial_t u \in L_2(Q) \text{ and } u|_{\Sigma\cup\Sigma_0} = 0\}.$$

Using the classical approach [7, 8], the variational formulation for the parabolic model problem (1) reads as follows: Find $u \in H_0^{1,0}(Q)$ such that

$$a(u,v) = l(v), \quad \forall v \in H_{0,\bar{0}}^{1,1}(Q), \tag{2}$$

where

$$a(u,v) = -\int_Q u(x,t)\partial_t v(x,t)d(x,t) + \int_Q \nabla_x u(x,t) \cdot \nabla_x u(x,t)d(x,t),$$
$$l(v) = \int_Q f(x,t)v(x,t)d(x,t) + \int_\Omega u_0(x)v(x,0)dx.$$

Remark 1 (Parabolic solvability and regularity [7, 8]) If $f \in L_{2,1}(Q) := \{v : \int_0^T \|v(\cdot,t)\|_{L_2(\Omega)}dt < \infty\}$ and $u_0 \in L_2(\Omega)$, then there exists a unique generalized solution $u \in H_0^{1,0}(Q) \cap V_2^{1,0}(Q)$ of (2), where $V_2^{1,0}(Q) := \{u \in H^{1,0}(Q) : |u|_Q < \infty \text{ and } \lim_{\Delta t\to 0} \|u(\cdot,t+\Delta t) - u(\cdot,t)\|_{L_2(\Omega)} = 0, \text{ uniformly on } [0,T]\}$, and $|u|_Q := \max_{0\le\tau\le T} \|u(\cdot,\tau)\|_{L_2(\Omega)} + \|\nabla_x u\|_{L_2(\Omega\times(0,T))}$. If $f \in L_2(Q)$ and $u_0 \in H_0^1(\Omega)$, then the generalized solution u belongs to $H_0^{\Delta,1}(Q) := \{v \in H_0^{1,1}(Q) : \Delta_x u \in L_2(Q)\}$ and continuously depends on t in the norm of the space $H_0^1(\Omega)$.

To derive the space-time finite element scheme, we mainly follow the approach proposed in [10]. Let $V_h = \text{span}\{\varphi_i\}$ be the span of continuous and piecewise linear basis functions φ_i on shape regular finite elements of an admissible triangulation $\mathcal{T}_h$. Then we define $V_{0h} = V_h \cap H_{0,\underline{0}}^{1,1}(Q) = \{v_h \in V_h : v_h|_{\Sigma\cup\Sigma_0} = 0\}$. For convenience, we consider homogeneous initial conditions, i.e., $u_0 = 0$ on Ω. Multiplying the PDE $\partial_t u - \Delta_x u = f$ on $K \in \mathcal{T}_h$ by an element-wise time-upwind test function $v_h + \theta_K h_K \partial_t v_h$, $v_h \in V_{0h}$, we get

$$\int_K (\partial_t u v_h + \theta_K h_K \partial_t u \partial_t v_h - \Delta_x u(v_h + \theta_K h_K \partial_t v_h))d(x,t) =$$
$$\int_K f(v_h + \theta_K h_K \partial v_h)d(x,t),$$

where h_K refers to the diameter of an element K in the space-time triangulation $\mathcal{T}_h$ of Q. Further, θ_K denotes a stabilization parameter [10]; see Remark 3. In the space-time finite element scheme [10], the time is considered as another spatial coordinate, and the partial derivative w.r.t. time is viewed as a convection term in the time direction. Therefore, as in the classical SUPG (streamline upwind Petrov-Galerkin) scheme, we use time-upwind test functions elementwise.

Integration by parts (the first part) with respect to the space and summation yields

$$\sum_{K \in \mathcal{T}_h} \int_K (\partial_t u v_h + \theta_K h_K \partial_t u \partial_t v_h + \nabla_x u \cdot \nabla_x v_h - \theta_K h_K \Delta_x u \partial_t v_h)d(x,t)$$
$$- \sum_{K \in \mathcal{T}_h} \int_{\partial K} n_x \cdot \nabla_x u v_h ds = \sum_{K \in \mathcal{T}_h} \int_K f(v_h + \theta_K h_K \partial_t v_h)d(x,t).$$

Since $n_x \cdot \nabla_x u$ is continuous across the inner boundary ∂K of K, $n_x = 0$ on $\Sigma_0 \cup \Sigma_T$, and $v_h = 0$ on Σ, the term $-\sum_{K \in \mathcal{T}_h} \int_{\partial K} n_x \cdot \nabla_x u v_h ds$ vanishes.

If the solution u of (2) belongs to $H^{\Delta,1}_{0,\underline{0}}(\mathcal{T}_h) := \{v \in H^{1,1}_{0,\underline{0}}(Q) : \Delta_x v|_K \in L_2(K), \forall K \in \mathcal{T}_h\}$, cf. Remark 1, then the consistency identity

$$a_h(u, v_h) = l_h(v_h), \quad v_h \in V_{0h}, \tag{3}$$

holds, where

$$a_h(u, v_h) := \sum_{K \in \mathcal{T}_h} \int_K (\partial_t u v_h + \theta_K h_K \partial_t u \partial_t v_h + \nabla_x u \cdot \nabla_x v_h - \theta_K h_K \Delta_x u \partial_t v_h)d(x,t),$$
$$l_h(v_h) := \sum_{K \in \mathcal{T}_h} \int_K f(v_h + \theta_K h_K \partial_t v_h)d(x,t).$$

With the restriction of the solution to the finite-dimensional subspace V_{0h}, the space-time finite element scheme reads as follows: Find $u_h \in V_{0h}$ such that

$$a_h(u_h, v_h) = l_h(v_h), \quad v_h \in V_{0h}. \tag{4}$$

Thus, we have the Galerkin orthogonality: $a_h(u - u_h, v_h) = 0, \forall v_h \in V_{0h}$.

Remark 2 Since we use continuous and piecewise linear trial functions, the integrand $-\theta_K h_K \Delta_x u_h \partial_t v_h$ vanishes element-wise, which simplifies the implementation.

Remark 3 On fully unstructured meshes, $\theta_k = O(h_k)$ [10]; on uniform meshes, $\theta_k = \theta = O(1)$ [9]. In this work, we have used $\theta = 0.5$ and $\theta = 2.5$ on uniform meshes for testing robustness of the BDDC preconditioners. The detailed results for $\theta = 2.5$ are presented in Table 1.

It was shown in [10] that the bilinear form $a_h(\cdot,\cdot)$ is V_{0h}-coercive: $a_h(v_h,v_h) \geq \mu_c\|v_h\|_h^2$, $\forall v_h \in V_{0h}$ with respect to the norm $\|v_h\|_h^2 = \sum_{K\in\mathcal{T}_h}(\|\nabla_x v_h\|^2_{L_2(K)} + \theta_K h_K\|\partial_t v_h\|^2_{L_2(K)}) + \frac{1}{2}\|v_h\|^2_{L_2(\Sigma_T)}$. Furthermore, the bilinear form is bounded on $V_{0h,*}\times V_{0h}$: $|a_h(u,v_h)| \leq \mu_b\|u\|_{0h,*}\|v_h\|_h$, $\forall u\in V_{0h,*}$, $\forall v_h\in V_{0h}$, where $V_{0h,*} = H^{\Delta,1}_{0,\underline{0}}(\mathcal{T}_h)+V_{0h}$ equipped with the norm $\|v\|^2_{0h,*} = \|v\|_h^2 + \sum_{K\in\mathcal{T}_h}(\theta_K h_K)^{-1}\|v\|^2_{L_2(K)}$ $+\sum_{K\in\mathcal{T}_h}\theta_k h_k\|\Delta_x v\|^2_{L_2(K)}$. Let l and k be positive reals such that $l\geq k>3/2$. We now define the broken Sobolev space $H^s(\mathcal{T}_h) := \{v\in L_2(Q): v|_K\in H^s(K)\ \forall K\in\mathcal{T}_h\}$ equipped with the broken Sobolev semi-norm $|v|^2_{H^s(\mathcal{T}_h)} := \sum_{K\in\mathcal{T}_h}|v|^2_{H^s(K)}$. Using the Lagrangian interpolation operator Π_h mapping $H^{1,1}_{0,\underline{0}}(Q)\cap H^k(Q)$ to V_{0h}, we obtain $\|u-u_h\|_h \leq \|u-\Pi_h u\|_h + \|\Pi_h u - u_h\|_h$. The term $\|u-\Pi_h u\|_h$ can be bounded by means of the interpolation error estimate, and the term $\|\Pi_h u - u_h\|_h$ by using ellipticity, Galerkin orthogonality and boundedness of the bilinear form. The discretization error estimate $\|u-u_h\|_h \leq C(\sum_{K\in\mathcal{T}_h} h_K^{2(l-1)}|u|^2_{H^l(K)})^{1/2}$ holds for the solution u provided that u belongs to $H^{1,1}_{0,\underline{0}}(Q)\cap H^k(Q)\cap H^l(\mathcal{T}_h)$, and the finite element solution $u_h\in V_{0h}$, where $C>0$, independent of mesh size; see [10].

3 Two-level BDDC preconditioners

After the space-time finite element discretization of the model problem (1), the linear system of algebraic equations reads as follows:

$$Kx = f, \tag{5}$$

with $K := \begin{bmatrix} K_{II} & K_{I\Gamma} \\ K_{\Gamma I} & K_{\Gamma\Gamma}\end{bmatrix}$, $x := \begin{bmatrix} x_I \\ x_\Gamma\end{bmatrix}$, $f := \begin{bmatrix} f_I \\ f_\Gamma\end{bmatrix}$, $K_{II} = \mathrm{diag}\left[K^1_{II},\dots,K^N_{II}\right]$, where N denotes the number of polyhedral subdomains Q_i from a non-overlapping domain decomposition of Q. In system (5), we have decomposed the degrees of freedom into the ones associated with the internal (I) and interface (Γ) nodes, respectively. We aim to solve the Schur-complement system living on the interface:

$$Sx_\Gamma = g_\Gamma, \tag{6}$$

with $S := K_{\Gamma\Gamma} - K_{\Gamma I}K_{II}^{-1}K_{I\Gamma}$ and $g := f_\Gamma - K_{\Gamma I}K_{II}^{-1}f_I$.

The bilinear form $a_h(\cdot,\cdot)$ is coercive on the space-time finite element space V_{0h} like in the corresponding elliptic case. There are efficient domain decomposition preconditioners for such elliptic problems [14]. This motivated us to use such preconditioners for solving positive definite space-time finite element equations too. Following [12] (see also details in [5]), Dohrmann's (two-level) BDDC preconditioners P_{BDDC} for the interface Schur complement equation (6), originally proposed for symmetric and positive definite systems in [2, 11], can be written in the form

$$P^{-1}_{BDDC} = R^T_{D,\Gamma}(T_{sub}+T_0)R_{D,\Gamma}, \tag{7}$$

where the scaled operator $R_{D,\Gamma}$ is the direct sum of restriction operators $R^i_{D,\Gamma}$ mapping the global interface vector to its component on local interface $\Gamma_i := \partial Q_i \cap \Gamma$, with a proper scaling factor.

Here the coarse level correction operator T_0 is constructed as

$$T_0 = \Phi(\Phi^T S \Phi)^{-1}\Phi^T \tag{8}$$

with the coarse level basis function matrix $\Phi = \left[(\Phi^1)^T, \cdots, (\Phi^N)^T\right]^T$, where the basis function matrix Φ^i on each subdomain interface is obtained by solving the following augmented system:

$$\begin{bmatrix} S^i & (C^i)^T \\ C^i & 0 \end{bmatrix} \begin{bmatrix} \Phi^i \\ \Lambda^i \end{bmatrix} = \begin{bmatrix} 0 \\ R^i_{\Pi} x_{\Gamma} \end{bmatrix}. \tag{9}$$

with the given primal constraints C^i of the subdomain Q_i and the vector of Lagrange multipliers on each column of Λ^i. The number of columns of each Φ^i equals to the number of global coarse level degrees of freedom, typically living on the subdomain corners, and/or interface edges, and/or faces. Here the restriction operator R^i_{Π} maps the global interface vector in the continuous primal variable space on the coarse level to its component on Γ_i.

The subdomain correction operator T_{sub} is defined as

$$T_{sub} = \sum_{i=1}^{N} \left[(R^i_{\Gamma})^T \quad 0\right] \begin{bmatrix} S^i & (C^i)^T \\ C^i & 0 \end{bmatrix}^{-1} \begin{bmatrix} R^i_{\Gamma} \\ 0 \end{bmatrix}, \tag{10}$$

with vanishing primal variables on all the coarse levels. Here the restriction operator R^i_{Γ} maps global interface vectors to their components on Γ_i.

4 Numerical experiments

We use $u(x, y, t) = \sin(\pi x)\sin(\pi y)\sin(\pi t)$ as exact solution of (1) in $Q = (0, 1)^3$; see the left plot in Fig. 1. We perform uniform mesh refinements of Q using tetrahedral elements. By using Metis [6], the domain is decomposed into $N = 2^k$, $k = 3, 4, ..., 9$, non-overlapping subdomains Q_i with their own tetrahedral elements; see the right plot in Fig. 1. The total number of degrees of freedom is $(2^k + 1)^3$, $k = 4, 5, 6, 7$. We run BDDC preconditioned GMRES iterations until the relative residual error reaches 10^{-9}. The experiments are performed on 64 nodes each with 8-core Intel Haswell processors (Xeon E5-2630v3, 2.4Ghz) and 128 GB of memory. Three variants of BDDC preconditioners are used with corner (C), corner/edge (CE), and corner/edge/face (CEF) constraints, respectively. The number of BDDC preconditioned GMRES iterations and the computational time measured in seconds [s] with respect to the number of subdomains (row-wise) and number of degrees of freedom (column-wise) are given in Table 1 for $\theta = 2.5$. Since the system is unsymmetric but

Fig. 1 Solution (left), space-time domain decomposition (right) with 129^3 degrees of freedom and 512 subdomains.

positive definite, the BDDC preconditioners do not show the same typical robustness and efficiency behavior when applied to the symmetric and positive definite system [14]. Nevertheless, we still observe certain scalability with respect to the number subdomains (up to 128), in particular, with corner/edge and corner/edge/face constraints. Increasing θ will improve the preformance of the BDDC preconditioners with respect to the number of GMRES iterations, computational time, and scalability with respect to the number of subdomains as well as number of degrees of freedom, whereas decreasing θ leads to a worse performance. For instance, in the case of $\theta = 0.5$, the last row of Table 1 reads as follows: 129^3 $\text{OoM}/(-)$ $\text{OoM}/(-)$ $173/(126.93s)$ $171/(109.94s)$ $185/(45.05s)$ $> 500/(-)$ $206/(33.13s)$. This behaviour is expected since larger θ makes the problem more elliptic. However, we note that θ also affects the norm $\|\cdot\|_h$ in which we measure the discretization error.

5 Conclusions

In this work, we have applied two-level BDDC preconditioned GMRES methods to the solution of finite element equations arising from the space-time discretization of a parabolic model problem. We have compared the performance of BDDC preconditioners with different coarse level constraints for such an unsymmetric, but positive definite system. The preconditioners show certain scalability provided that θ is sufficiently large. Future work will concentrate on improvement of coarse-level corrections in order to achieve robustness with respect to different choices of θ.

References

1. Bank, R.E., Vassilevski, P.S., Zikatanov, L.T.: Arbitrary dimension convection-diffusion schemes for space-time discretizations. J. Comput. Appl. Math. **310**, 19–31 (2017)
2. Dohrmann, C.R.: A preconditioner for substructuring based on constrained energy minimization. SIAM J. Sci. Comput. **25**(1), 246–258 (2003)
3. Hofer, C., Langer, U., Neumüller, M., Schneckenleitner, R.: Parallel and robust preconditioning for space-time isogeometric analysis of parabolic evolution problems. SIAM J. Sci. Comput. **41**(3), A1793–A1821 (2019)

Table 1: $\theta = 2.5$. BDDC performance using different coarse level constraints ($C/CE/CEF$), with respect to the number of subdomains (row-wise) and degrees of freedoms (column-wise).

			No preconditioner				
	8	16	32	64	128	256	512
17^3	46	51	55	61	66	> 500	> 500
	(0.02s)	(0.02s)	(0.02s)	(0.03s)	(0.04s)	–	–
33^3	72	79	87	99	108	> 500	> 500
	(0.64s)	(0.32s)	(0.15s)	(0.12s)	(0.13s)	–	–
65^3	116	126	145	163	176	191	> 500
	(27.59s)	(9.17s)	(4.07s)	(1.92s)	(1.06s)	(2.02s)	–
129^3	OoM	OoM	240	271	304	> 500	382
	(–)	(–)	(145.64s)	(58.51s)	(24.3s)	(–)	(12.41s)

			C (corner) preconditioner				
17^3	23	28	27	32	36	51	110
	(0.02s)	(0.02s)	(0.03s)	(0.03s)	(0.07s)	(0.72s)	(2.48s)
33^3	30	33	39	50	50	206	182
	(0.60s)	(0.26s)	(0.14s)	(0.10s)	(0.09s)	(3.86s)	(6.2s)
65^3	35	47	61	64	69	77	287
	(19.85s)	(7.42s)	(3.47s)	(1.44s)	(0.68s)	(1.05s)	(9.00s)
129^3	OoM	OoM	94	104	107	340	112
	(–)	(–)	(124.30s)	(46.45s)	(16.90s)	(33.45s)	(5.89s)

			CE (corner+edge) preconditioner				
17^3	21	22	22	25	27	42	80
	(0.03s)	(0.02s)	(0.01s)	(0.03s)	(1.78s)	(0.78s)	(1.65s)
33^3	27	26	32	38	32	117	132
	(0.54s)	(0.23s)	(0.11s)	(0.12s)	(0.15s)	(2.52s)	(5.98s)
65^3	33	44	51	54	54	54	235
	(19.00s)	(7.27s)	(2.98s)	(1.33s)	(0.75s)	(1.32s)	(15.13s)
129^3	OoM	OoM	82	83	90	366	94
	(–)	(–)	(109.19s)	(38.59s)	(15.21s)	(37.00s)	(7.91s)

			CEF (corner+edge+face) preconditioner				
17^3	21	21	22	22	22	38	74
	(0.02s)	(0.01s)	(0.02s)	(0.04s)	(0.13s)	(0.74s)	(1.74s)
33^3	27	26	30	35	31	115	123
	(0.68s)	(0.28s)	(0.14s)	(0.12s)	(0.25s)	(3.79s)	(7.08s)
65^3	34	44	49	52	53	51	226
	(26.64s)	(9.29s)	(3.86s)	(1.61s)	(1.11s)	(1.88s)	21.30s
129^3	OoM	OoM	82	83	88	369	92
	(–)	(–)	(145.39s)	(53.10s)	(23.04s)	(52.33)	(13.8s)

4. Hofer, C., Langer, U., Neumüller, M., Toulopoulos, I.: Time-multipatch discontinuous Galerkin space-time isogeometric analysis of parabolic evolution problems. Electron. Trans. Numer. Anal. **49**, 126–150 (2018)
5. Jing, L., Widlund, O.B.: FETI-DP, BDDC, and block Cholesky methods. Int. J. Numer. Meth. Engng. **66**(2), 250–271 (2006)
6. Karypis, G., Kumar, V.: A fast and high quality multilevel scheme for partitioning irregular graphs. SIAM Journal on Scientific Computing **20**(1), 359–392 (1998)
7. Ladyžhenskaya, O.A.: The boundary value problems of mathematical physics, *Applied Mathematical Sciences*, vol. 49. Springer-Verlag, New York (1985)
8. Ladyžhenskaya, O.A., Solonnikov, V.A., Uraltseva, N.N.: Linear and quasilinear equations of parabolic type. AMS, Providence, RI (1968)
9. Langer, U., Moore, S.E., Neumüller, M.: Space-time isogeometric analysis of parabolic evolution problems. Comput. Methods Appl. Mech. Eng. **306**, 342–363 (2016)
10. Langer, U., Neumüller, M., Schafelner, A.: Space-time finite element methods for parabolic evolution problems with variable coefficients. In: T. Apel, U. Langer, A. Meyer, O. Steinbach (eds.) Advanced finite element methods with applications - Proceedings of the 30th Chemnitz FEM symposium 2017, *Lecture Notes in Computational Science and Engineering (LNCSE)*, vol. 128, pp. 247–275. Springer, Berlin, Heidelberg, New York (2019)
11. Mandel, J., Dohrmann, C.R.: Convergence of a balancing domain decomposition by constraints and energy minimization. Numer. Linear Algebra Appl. **10**(7), 639–659 (2003)
12. Mandel, J., Dohrmann, C.R., Tezaur, R.: An algebraic theory for primal and dual substructuring methods by constraints. Appl. Numer. Math. **54**(2), 167–193 (2005)
13. Steinbach, O.: Space-time finite element methods for parabolic problems. Comput. Methods Appl. Math. **15**, 551–566 (2015)
14. Toselli, A., Widlund, O.B.: Domain decomposition methods - Algorithms and theory, *Springer Series in Computational Mathematics*, vol. 34. Springer, Berlin, Heidelberg, New York (2004)

Non-overlapping Spectral Additive Schwarz Methods

Yi Yu, Maksymilian Dryja, and Marcus Sarkis

1 Discrete Problem

For a given domain $\Omega \subset \mathbb{R}^d$, we impose homogeneous Dirichlet data on $\partial\Omega$. Let us introduce the Sobolev space $H_0^1(\Omega) := \{v \in H^1(\Omega) : v = 0 \text{ on } \partial\Omega\}$.
The continuous variational formulation is given by: Find $u \in H_0^1(\Omega)$ such that

$$a(u, v) = f(v) \qquad \text{for all} \qquad v \in H_0^1(\Omega), \tag{1}$$

where

$$a(u, v) := \int_\Omega \rho(x)\nabla u \cdot \nabla v dx \qquad f(v) := \int_\Omega f v dx,$$

where we assume $\rho(x) \geq \rho_{\min} > 0$ almost everywhere in Ω.

2 Discretization

We begin by discretizing Problem (1) in an algebraic framework. Let us consider a conforming triangulation $\mathcal{T}_h$ of Ω where $\overline{\Omega} = \bigcup_{\tau \in \mathcal{T}_h} \bar{\tau}$ and basis functions $\{\phi_k\}_{1 \leq k \leq n}$ for the finite element space $V_h(\Omega)$. We use the convention that an element $\tau \in \mathcal{T}_h$, the domain Ω, and the subdomains Ω_i are treated as open sets.

The finite element space $V_h(\Omega)$ is defined as:

Yi Yu
Worcester Polytechnic Institute, 100 Institute Rd, Worcester, USA, e-mail: yyu5@wpi.edu

Maksymilian Dryja
Warsaw University,Banacha 2, 00-097 Warsaw, Poland e-mail: dryja@mimuw.edu.pl

Marcus Sarkis
Worcester Polytechnic Institute, 100 Institute Rd, Worcester, USA, e-mail: msarkis@wpi.edu.

R. Haynes et al. (eds.), *Domain Decomposition Methods in Science and Engineering XXV*,
Lecture Notes in Computational Science and Engineering 138,
https://doi.org/10.1007/978-3-030-56750-7_43

$$V_h(\Omega) := \{v \in H_0^1(\Omega); v|_\tau \in P_1(\tau)\ \forall \tau \in \mathcal{T}_h\} = \text{Span}\{\phi_k : 1 \leq k \leq n\}.$$

The FEM matrix form associated with (1) can be written as

$$Au = b, \tag{2}$$

where

$$(A)_{kl} := a(\phi_k, \phi_l) = \sum_{\tau \in \mathcal{T}_h} a_\tau(\phi_k|_\tau, \phi_l|_\tau) \quad \text{for all } 1 \leq k, l \leq n,$$

and

$$(b)_k := f(\phi_k) = \sum_{\tau \in \mathcal{T}_h} f_\tau(\phi_k|_\tau) \qquad \text{for all } 1 \leq k \leq n.$$

2.1 Finite Element Spaces

We decompose Ω into N non-overlapping polygonal subdomains Ω_i which satisfy

$$\overline{\Omega} = \bigcup_{i=1}^{N} \overline{\Omega}_i \quad \text{and} \quad \Omega_i \cap \Omega_j = \emptyset, \quad i \neq j.$$

We require that each subdomain be a union of shape-regular triangular elements with nodes on the boundaries of neighboring subdomains matching across the interface. We define the interface of each subdomain Γ_i and the interior of each subdomain I_i, global interface $\Gamma \subset \Omega$ and global interior I as:

$$\Gamma_i := \partial\Omega_i \backslash \partial\Omega \quad \text{and} \quad \Gamma := \bigcup_{i=1}^{N} \Gamma_i \quad \text{and} \quad I = \Omega/\Gamma = \bigcup_{i=1}^{N} I_i.$$

For any finite element subset $D \subset \Omega$ let the set of degrees of freedom in D be the hat functions

$$\text{dof}(D) := \{1 \leq k \leq n; \phi_k|D \neq 0|D\},$$

where $0|D : D \to \mathbb{R}$ is identically zero. The finite element space on D is defined as

$$V_h(D) := \{u|D; u \in V_h(\Omega)\} = \text{span}\{\phi_k|D; k \in \text{dof}(D)\}.$$

2.2 Decomposition of $V_h(\Omega)$

Let us consider a family of local spaces $\{V_i, 1 \leq i \leq N\}$, where

$$V_i = V_h(\Omega_i) \cap H_0^1(\Omega_i),$$

and we define the extrapolation operators $R_i^T : V_i \to V_h(\Omega)$ where R_i^T is the extension by zero outside of Ω_i.
The coarse space V_0 is defined as the space of piecewise linear and continuous functions on Γ:

$$V_0 = V_h(\Gamma) := \{v|_\Gamma; \forall v \in V_h(\Omega)\}.$$

In Section 3, we will present different choices of the extension operator $R_0^T : V_0 \to V_h(\Omega)$. The space $V_h(\Omega)$ admits the following direct sum decomposition:

$$V_h(\Omega) = R_0^T V_0 \oplus R_1^T V_1 \oplus \cdots \oplus R_N^T V_N$$

2.3 Additive Schwarz Methods

Local solvers: For $1 \leq i \leq N$, let us introduce the exact local bilinear form

$$a_i(u, v) := a(R_i^T u, R_i^T v) \qquad u, v \in V_i,$$

and let us define $\tilde{T}_i : V_h(\Omega) \to V_i$ by

$$a_i(\tilde{T}_i u, v) = a(u, R_i^T v) \qquad v \in V_i \quad 1 \leq i \leq N, \tag{3}$$

and let $T_i : V_h(\Omega) \to V_h(\Omega)$ be given by $T_i := R_i^T \tilde{T}_i$.
Global solver: For $i = 0$ first we consider the exact global solver

$$a_0(u, v) := a(R_0^T u, R_0^T v) \qquad u, v \in V_0$$

and let us define $\tilde{T}_0 : V_h(\Omega) \to V_0$ by

$$a_0(\tilde{T}_0 u, v) = a(u, R_0^T v) \qquad v \in V_0 \tag{4}$$

and let $T_0 : V_h(\Omega) \to V_h(\Omega)$ be given by $T_0 := R_0^T \tilde{T}_0$. Note that we also will consider inexact solvers $\hat{a}_0(\cdot, \cdot)$ later in this paper. We replace (2) by the linear system

$$T_A u = g_h \quad \text{where} \quad T_A := T_0 + T_1 + \cdots + T_N, \quad g_h = \sum_{i=0}^{N} g_i$$

where g_i are obtained from (3) and (4); see [8].

3 Schur complement system

The linear system (2) can be written as

$$\begin{bmatrix} A_{\Gamma\Gamma} & A_{\Gamma I} \\ A_{I\Gamma} & A_{II} \end{bmatrix} \begin{bmatrix} u_\Gamma \\ u_I \end{bmatrix} = \sum_{i=1}^{N} R^{(i)^T} \begin{bmatrix} A_{\Gamma\Gamma}^{(i)} & A_{\Gamma I}^{(i)} \\ A_{I\Gamma}^{(i)} & A_{II}^{(i)} \end{bmatrix} R^{(i)} \begin{bmatrix} u_\Gamma \\ u_I \end{bmatrix} = \sum_{i=1}^{N} R^{(i)^T} \begin{bmatrix} b_\Gamma^{(i)} \\ b_I^{(i)} \end{bmatrix}.$$

In this equation the extrapolation operators $R^{(i)^T} : V_h(\Omega_i) \to V_h(\Omega)$ is the extension by zero at nodes outside of $\bar{\Omega}_i$. Thus we have,

$$A = \sum_{i=1}^{N} R^{(i)^T} A^{(i)} R^{(i)} = \sum_{i=1}^{N} R^{(i)^T} \begin{bmatrix} A_{\Gamma\Gamma}^{(i)} & A_{\Gamma I}^{(i)} \\ A_{I\Gamma}^{(i)} & A_{II}^{(i)} \end{bmatrix} R^{(i)} \quad \text{and} \quad b = \begin{bmatrix} b_\Gamma \\ b_I \end{bmatrix} = \sum_{i=1}^{N} R^{(i)^T} \begin{bmatrix} b_\Gamma^{(i)} \\ b_I^{(i)} \end{bmatrix},$$

where $A^{(i)}$ is the matrix corresponding to the bilinear form of

$$a^{(i)}(u_i, v_i) = \sum_{\tau \in \mathcal{T}_{h|\Omega_i}} a_\tau(u_i|_\tau, v_i|_\tau) \qquad u_i, v_i \in V_h(\Omega_i).$$

Moreover, if we label the interface nodes first and then label the interior nodes, we can decompose the Boolean matrices $R^{(i)^T}$ as:

$$R^{(i)^T} = \begin{bmatrix} R_{\Gamma_i\Gamma}^T & 0 \\ 0 & R_{I_i I}^T \end{bmatrix} \quad \text{and} \quad \begin{bmatrix} u_\Gamma^{(i)} \\ u_I^{(i)} \end{bmatrix} = R^{(i)} \begin{bmatrix} u_\Gamma \\ u_I \end{bmatrix},$$

where $R_{\Gamma_i\Gamma}^T : V_h(\Gamma_i) \to V_h(\Omega)$ and $R_{I_i I}^T : V_i \to V_h(I)$ are zero extension operators. We now rewrite (2) in terms of Schur complement system (see [8])

$$\begin{bmatrix} S & 0 \\ A_{I\Gamma} & A_{II} \end{bmatrix} \begin{bmatrix} u_\Gamma \\ u_I \end{bmatrix} = \sum_{i=1}^{N} R^{(i)^T} \begin{bmatrix} S_{\Gamma\Gamma}^{(i)} & 0 \\ A_{I\Gamma}^{(i)} & A_{II}^{(i)} \end{bmatrix} \begin{bmatrix} u_\Gamma^{(i)} \\ u_I^{(i)} \end{bmatrix} = \sum_{i=1}^{N} R^{(i)^T} \begin{bmatrix} b_\Gamma^{(i)} - A_{\Gamma I}^{(i)} A_{II}^{(i)^{-1}} b_I^{(i)} \\ b_I^{(i)} \end{bmatrix} = \begin{bmatrix} \tilde{b}_\Gamma \\ b_I \end{bmatrix},$$

where

$$S_{\Gamma\Gamma}^{(i)} = A_{\Gamma\Gamma}^{(i)} - A_{\Gamma I}^{(i)} A_{II}^{(i)^{-1}} A_{I\Gamma}^{(i)},$$

$$\tilde{b}_{\Gamma_i} = \sum_{i=1}^{N} R_{\Gamma_i\Gamma}^T (b_\Gamma^{(i)} - A_{\Gamma I}^{(i)} A_{II}^{(i)^{-1}} b_I^{(i)}) \quad \text{and} \quad S = \sum_{i=1}^{N} R_{\Gamma_i\Gamma}^T S_{\Gamma\Gamma}^{(i)} R_{\Gamma_i\Gamma} \quad \text{and} \quad S u_\Gamma = \tilde{b}_\Gamma.$$

We note that the best extension R_0^T is the a-discrete harmonic extension from Γ to I due to the orthogonality of the coarse and local problems. In this case

$$a(R_0^T v_\Gamma, R_0^T u_\Gamma) = v_\Gamma^T S u_\Gamma.$$

The motivation is to replace S by a good preconditioner S_0 of S; see [1],[3],[2],[6].

4 New Method: Spectral Schwarz methods with exact solver

In our spectral method, we define a new R_0^T extension operator. To do that, the first goal is to represent the best k_i-dimensional subspace of V_i to approximate the a-discrete harmonic extension operator inside the subdomains. We fix a threshold

$\delta < 1$, and choose the smallest k_i eigenvalues in each subdomain smaller than δ. First solve the following generalized eigenproblem in each subdomain separately:

$$S^{(i)}\xi_j^{(i)} \equiv (A_{\Gamma\Gamma}^{(i)} - A_{\Gamma I}^{(i)}(A_{II}^{(i)})^{-1}A_{I\Gamma}^{(i)})\xi_j^{(i)} = \lambda_j^{(i)} A_{\Gamma\Gamma}^{(i)}\xi_j^{(i)} \tag{5}$$

These eigenvalue problems are based on Neumann matrix associated to nonoverlapping subdomains, therefore, differ from those in GenEO [7] and AGDSW [4].

We choose the smallest k_i eigenvalues and corresponding eigenvectors:
For $j = 1 : k_i$, let $Q_j^{(i)} = \xi_j^{(i)} \qquad P_j^{(i)} = -(A_{II}^{(i)})^{-1}A_{I\Gamma}^{(i)}\xi_j^{(i)}$
And $Q^{(i)} = [Q_1^{(i)}, Q_2^{(i)}, \cdots, Q_{k_i}^{(i)}]$ and $P^{(i)} = [P_1^{(i)}, P_2^{(i)}, \cdots, P_{k_i}^{(i)}]$.
Thus we have three identities, where the left-hand sides involve operators on Γ_i only:

1. $-A_{\Gamma\Gamma}^{(i)}Q^{(i)}D^{(i)} = A_{\Gamma I}^{(i)}P^{(i)}$
2. $-D^{(i)}Q^{(i)^T}A_{\Gamma\Gamma}^{(i)} = P^{(i)^T}A_{I\Gamma}^{(i)}$
3. $D^{(i)}Q^{(i)^T}A_{\Gamma\Gamma}^{(i)}Q^{(i)} = Q^{(i)^T}A_{\Gamma\Gamma}^{(i)}Q^{(i)}D^{(i)} = P^{(i)^T}A_{II}^{(i)}P^{(i)}$, where

$$D^{(i)} = \text{diagonal}(1-\lambda_1, 1-\lambda_2, \cdots, 1-\lambda_{k_i}) = I - \Lambda^{(i)}$$

Also the $Q^{(i)}$ consist of the generalized eigenvectors from (5), and we can normalize the eigenvectors so that $Q^{(i)^T}A_{\Gamma\Gamma}^{(i)}Q^{(i)} = I^{(i)}$ and $Q^{(i)^T}S^{(i)}Q^{(i)} = \Lambda^{(i)}$ however in the implementation we do not assume normalized eigenvectors, so we keep $Q^{(i)^T}A_{\Gamma\Gamma}^{(i)}Q^{(i)}$. Define the global extension $R_0^T : V_0 \to V_h(\Omega)$ as:

$$R_0^T u_\Gamma = \begin{bmatrix} u_\Gamma \\ -\sum_{i=1}^N R_{I_i I}^T P^{(i)}(P^{(i)^T}A_{II}^{(i)}P^{(i)})^{-1}P^{(i)^T}A_{I\Gamma}^{(i)}R_{\Gamma_i\Gamma}u_\Gamma \end{bmatrix}$$
$$= \begin{bmatrix} u_\Gamma \\ \sum_{i=1}^N R_{I_i I}^T P^{(i)}(Q^{(i)^T}A_{\Gamma\Gamma}^{(i)}Q^{(i)})^{-1}Q^{(i)^T}A_{\Gamma\Gamma}^{(i)}R_{\Gamma_i\Gamma}u_\Gamma \end{bmatrix}.$$

And for $u, v \in V_0$, we define the exact coarse solver as:

$$a_0(u,v) = a(R_0^T u, R_0^T v) = v^T \sum_{i=1}^N R_{\Gamma_i\Gamma}^T (A_{\Gamma\Gamma}^{(i)} - A_{\Gamma I}^{(i)}P^{(i)}(P^{(i)^T}A_{II}^{(i)}P^{(i)})^{-1}P^{(i)^T}A_{I\Gamma}^{(i)})R_{\Gamma_i\Gamma}u$$
$$= v^T \sum_{i=1}^N R_{\Gamma_i\Gamma}^T (A_{\Gamma\Gamma}^{(i)} - A_{\Gamma\Gamma}^{(i)}Q^{(i)}D^{(i)}(Q^{(i)^T}A_{\Gamma\Gamma}^{(i)}Q^{(i)})^{-1}Q^{(i)^T}A_{\Gamma\Gamma}^{(i)})R_{\Gamma_i\Gamma}u.$$

On each subdomain, we have the following lemmas and theorem:

Lemma 1 *([5]) Let $\Pi_S^{(i)}u$ be the projection of $u \in V_h(\Gamma_i)$ onto Span of $Q^{(i)}$, that is, $\Pi_S^{(i)}u \triangleq Q^{(i)}(Q^{(i)^T}A_{\Gamma\Gamma}^{(i)}Q^{(i)})^{-1}Q^{(i)^T}A_{\Gamma\Gamma}^{(i)}u$. Define the local bilinear form $a_0^{(i)}(u,v) = v^T(A_{\Gamma\Gamma}^{(i)} - A_{\Gamma\Gamma}^{(i)}Q^{(i)}D^{(i)}(Q^{(i)^T}A_{\Gamma\Gamma}^{(i)}Q^{(i)})^{-1}Q^{(i)^T}A_{\Gamma\Gamma}^{(i)})u$ where $u, v \in V_h(\Gamma_i)$. Then:*

$$a_0^{(i)}(u,v) = (\Pi_S^{(i)}v)^T S^{(i)}(\Pi_S^{(i)}u) + (v - \Pi_S^{(i)}v)^T A_{\Gamma\Gamma}^{(i)}(u - \Pi_S^{(i)}u).$$

Lemma 2 *([5]) Let $u \in V_0$ then*

$$a_0(u,u) = \sum_{i=1}^{N} a_0^{(i)}(R_\Gamma^{(i)}u, R_\Gamma^{(i)}u) \le \sum_{i=1}^{N} \frac{1}{\delta} u^T R_\Gamma^{(i)^T} S^{(i)} R_\Gamma^{(i)} u = \frac{1}{\delta} u^T S u$$

From Lemma 1 and Lemma 2 and the classical Schwarz Theory [8] we have:

Theorem 1 *([5]) For any $u \in V_h(\Omega)$ the following holds:*

$$(2+\frac{3}{\delta})^{-1} a(u,u) \le a(T_A u, u) \le 2a(u,u) \Longrightarrow k(T_A) \le 2(2+\frac{3}{\delta})$$

5 Complexity of Spectral Schwarz Method and with inexact coarse solver

The solution $u_\Gamma = \tilde{T}_0 u$ of $a_0(u_\Gamma, v_0) = a(u, R_0^T v_0) = (R_0^T v_0)^T b$ is of the form:

$$\sum_{i=1}^{N} R_{\Gamma_i\Gamma}^T (A_{\Gamma\Gamma}^{(i)} - A_{\Gamma\Gamma}^{(i)} Q^{(i)} D^{(i)} (Q^{(i)^T} A_{\Gamma\Gamma}^{(i)} Q^{(i)})^{-1} Q^{(i)^T} A_{\Gamma\Gamma}^{(i)}) R_{\Gamma_i\Gamma} u_\Gamma$$
$$= \sum_{i=1}^{N} R_{\Gamma_i\Gamma}^T (b_\Gamma^{(i)} + A_{\Gamma\Gamma}^{(i)} Q^{(i)} (Q^{(i)^T} A_{\Gamma\Gamma}^{(i)} Q^{(i)})^{-1} P^{(i)^T} b_I^{(i)}).$$

Denote $A_{\Gamma\Gamma} = \sum_{i=1}^{N} R_{\Gamma_i\Gamma}^T A_{\Gamma\Gamma}^{(i)} R_{\Gamma_i\Gamma}$, $U = \sum_{i=1}^{N} R_{\Gamma_i\Gamma}^T A_{\Gamma\Gamma}^{(i)} Q^{(i)} R_{\lambda_i}$,

$D = \sum_{i=1}^{N} R_{\lambda_i}^T D^{(i)} R_{\lambda_i}$, $C = \sum_{i=1}^{N} R_{\lambda_i}^T (Q^{(i)^T} A_{\Gamma\Gamma}^{(i)} Q^{(i)})^{-1} R_{\lambda_i}$ and $P = \sum_{i=1}^{N} R_{I_i I}^T P^{(i)} R_{\lambda_i}$.

where R_{λ_i} is a restriction chosen $[u_{i1}, \cdots, u_{ik_i}]^T$ from $\bar{u} = [u_{11}, \cdots, u_{1k_1}, \cdots, u_{Nk_1}, \cdots, u_{Nk_N}]^T$. Here k_i is the number of eigenfunctions chosen from the i-th subdomain, and $\bar{u}$ has dimension k equals to the number of all eigenvectors chosen from all N subdomains. Then we can rewrite the coarse mesh problem as:

$$(A_{\Gamma\Gamma} - UDCU^T)u_\Gamma = b_\Gamma + UCP^T b_I,$$

and we use Woodbury identity for implementation:

$$(A_{\Gamma\Gamma} - UDCU^T)^{-1} = A_{\Gamma\Gamma}^{-1} + A_{\Gamma\Gamma}^{-1} U (C^{-1}D^{-1} - U^T A_{\Gamma\Gamma}^{-1} U)^{-1} U^T A_{\Gamma\Gamma}^{-1}.$$

Then the complexity of the method is associated with $A_{\Gamma\Gamma}^{-1}$, C^{-1} and the $k \times k$ matrix $(C^{-1}D^{-1} - U^T A_{\Gamma\Gamma}^{-1} U)^{-1}$, where k = the number of all eigenfunctions.

We can make $A_{\Gamma\Gamma}$ and C block diagonal or diagonal matrices if we replace the exact $A_{\Gamma\Gamma}^{(i)}$ on the right-hand side of the generalized eigenproblems by $\hat{A}_{\Gamma\Gamma}^{(i)}$, where the $\hat{A}_{\Gamma\Gamma}^{(i)}$ are block diagonal or diagonal versions of the $A_{\Gamma\Gamma}^{(i)}$. Note that for the block diagonal case we eliminate the connections across different faces, edges and corners of the subdomains. These inexact cases can be analyzed and given in Theorem 2.

We introduce the local generalized eigenproblems:

$$S^{(i)}\xi_i^{(i)} \equiv (A_{\Gamma\Gamma}^{(i)} - A_{\Gamma I}^{(i)}(A_{II}^{(i)})^{-1}A_{I\Gamma}^{(i)})\xi_j^{(i)} = \hat{\lambda}_j^{(i)}\hat{A}_{\Gamma\Gamma}^{(i)}\hat{\xi}_j^{(i)}.$$

And for $u, v \in V_0(\Omega)$, we define the inexact coarse solver as:

$$\hat{a}_0(u,v) = v^T\sum_{i=1}^{N} R_\Gamma^{(i)^T}(\hat{A}_{\Gamma\Gamma}^{(i)} - \hat{A}_{\Gamma\Gamma}^{(i)}\hat{Q}^{(i)}\hat{D}^{(i)}(\hat{Q}^{(i)^T}\hat{A}_{\Gamma\Gamma}^{(i)}\hat{Q}^{(i)})^{-1}\hat{Q}^{(i)^T}\hat{A}_{\Gamma\Gamma}^{(i)})R_\Gamma^{(i)}u.$$

Where $\hat{Q}^{(i)}$ are the generalized eigenvectors and $\hat{D}^{(i)} = \text{diagonal}(1-\hat{\lambda}_1, \cdots, 1-\hat{\lambda}_{k_i})$. Then, a condition number estimate for the inexact case is given by the following theorem:

Theorem 2 *([5])*

For any $u \in V_h(\Omega)$ the following holds:

$$(2+7\max\{1, \frac{1}{\delta}\})^{-1}a(u,u) \le a(\hat{T}_A u, u) \le 4a(u,u) \Longrightarrow k(\hat{T}_A) \le 4(2+7\max\{1, \frac{1}{\delta}\})$$

6 Numerical Experiments

We present results for problem (1) for $f \equiv 1$ of our Adaptive Spectral Schwarz method with highly heterogeneous coefficients in the format of stripes crossing the interface of the subdomains (see Figure 1). We divide the square domain into $H \times H$ congruent square subdomains and in each subdomain we have two horizontal stripes and two vertical stripes. The coefficient on the stripe (in grey) is $\rho(x) = 10^{-6}$ and $\rho(x) = 1$ elsewhere. Experiments show (not presented here) that the Additive Average Schwarz method can lead to a large condition number that depends on ρ_{max}/ρ_{min}. In contrast, when we use Adaptive Spectral Schwarz method with a threshold $\delta = \frac{1}{4}\frac{h}{H}$, we have a well conditioned problem with a low number of iterations; in Table 1 we see the robustness of the adaptive spectral Schwarz method with exact solver and Table 2 with inexact solver using diagonal of $A_{\Gamma\Gamma}^{(i)}$.

Length of subdomain	Iterations	Condition number	Number of eigenvectors	Complexity of problem
H=1/4	11	6.4719	84	84×84
H=1/8	12	6.4719	420	420×420
H=1/16	12	6.4719	1860	1860×1860

Table 2: Adaptive Spectral Schwarz method with diagonal inexact solver and the number of eigenvectors. We fix $H/h = 8$ and the number of iterations required to reduce the residual by 10^{-6}. The condition number is estimated by the Arnoldi matrix in the CG method.

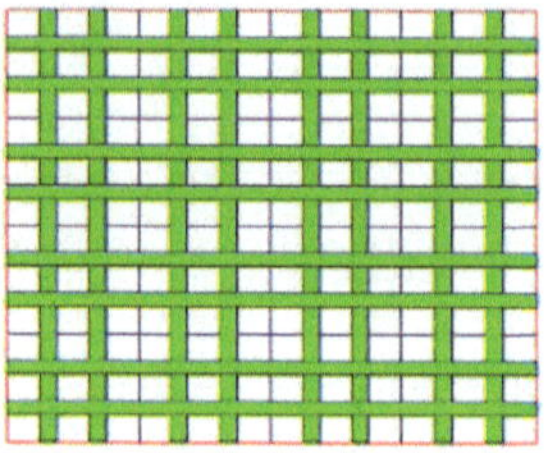

Fig. 1: In the stripe mesh, coefficient $\rho(x) = 10^{-6}$ in each stripe, and $\rho(x) = 1$ in other area.

Length of subdomain	CG Iterations	Condition number	Number of eigenvectors
$H = 1/4$	10	4.7684	84
$H = 1/8$	11	4.7684	420
$H = 1/16$	11	4.7684	1860

Table 1: Adaptive Spectral Schwarz method with exact solver and the number of eigenvectors. We fix $H/h = 8$ and the number of iterations required to reduce the residual by 10^{-6}. The condition number is estimated by the Arnoldi matrix in the CG method.

7 Conclusion

We introduced new two-dimensional and three-dimensional adaptive Schwarz methods derived from the additive average Schwarz method which are robust with respect to the jumps of coefficients with $O(H/h)$ condition number estimates. A unique feature of our methods is that our coarse space is based on generalized eigenvectors obtained in each nonoverlapping subdomain separately. One of the new methods has good parallelization properties since the global coarse matrix is sparse.

References

1. Bjørstad, P.E., Dryja, M., Vainikko, E.: Additive schwarz methods without subdomain overlap and with new coarse spaces. Domain decomposition methods in sciences and engineering (Beijing, 1995) pp. 141–157 (1997)
2. Bjørstad, P.E., Koster, J., Krzyżanowski, P.: Domain decomposition solvers for large scale industrial finite element problems. In: International Workshop on Applied Parallel Computing, pp. 373–383. Springer (2000)
3. Dryja, M., Sarkis, M.: Additive average schwarz methods for discretization of elliptic problems with highly discontinuous coefficients. Computational Methods in Applied Mathematics Comput. Methods Appl. Math. **10**(2), 164–176 (2010)
4. Heinlein, A., Klawonn, A., Knepper, J., Rheinbach, O.: Multiscale coarse spaces for overlapping schwarz methods based on the acms space in 2d. Electron. Trans. Numer. Anal **48**, 156–182 (2018)
5. Maksymilian Dryja, M.S., Yu., Y.: From additive average schwarz methods to non-overlapping spectral additive schwarz methods (2019)
6. Marcinkowski, L., Rahman, T.: Additive average schwarz with adaptive coarse spaces: scalable algorithms for multiscale problems. Electronic Transactions on Numerical Analysis **49**, 28–40 (2018)
7. Spillane, N., Rixen, D.J.: Automatic spectral coarse spaces for robust finite element tearing and interconnecting and balanced domain decomposition algorithms. International Journal for Numerical Methods in Engineering **95**(11), 953–990 (2013)
8. Toselli, A., Widlund, O.: Domain decomposition methods-algorithms and theory, vol. 34. Springer Science & Business Media (2006)

Auxiliary Space Preconditioners for Linear Virtual Element Method

Yunrong Zhu

1 Introduction

In this paper, we present the auxiliary space preconditioning techniques for solving the linear system arising from linear virtual element method (VEM) discretizations on polytopal meshes of second order elliptic problems in both 2D and 3D domains. The VEMs are generalizations of the classical finite element methods (FEMs), which permit the use of general polygonal and polyhedral meshes. Using polytopal meshes allows for more flexibility in dealing with complex computational domains or interfaces (cf. [12]). It also provides a unified treatment of different types of elements on the same mesh. In recent years, a lot of work has been devoted to the design and analysis of the discretization methods. Less attention has been paid to developing efficient solvers for the resulting linear systems. Only recently, have the balancing domain decomposition by constraint (BDDC) and the finite element tearing and interconnecting dual primal (FETI-DP) methods been studied in [6] for VEM methods. Some two-level overlapping domain decomposition preconditioners were developed and analyzed in [8, 9] for VEM in two dimensions. A p-version multigrid algorithm was proposed and analyzed in [1].

The auxiliary space preconditioners we consider here can be understood as two-level methods, with a standard smoother on the fine level and a "coarse space" correction. The fine level problem is the VEM discretization on polytopal mesh, and the coarse level problem is a standard conforming $\mathbb{P}_1$ finite element space defined on an auxiliary simplicial mesh. It is natural to choose the standard $\mathbb{P}_1$ finite element space as the coarse space for a couple of reasons: (1) the degrees of freedom of the coarse space are included in the VEM space – so asymptotically, the "coarse" space should provide a good approximation for the solution on the "fine" space; (2) there are a lot of works on developing efficient (and robust) solvers for the standard conforming

Yunrong Zhu
Department of Mathematics & Statistics, Idaho State University, 921 S. 8th Ave., Stop 8085 Pocatello, ID 83209, USA. e-mail: zhuyunr@isu.edu

R. Haynes et al. (eds.), *Domain Decomposition Methods in Science and Engineering XXV*, Lecture Notes in Computational Science and Engineering 138,
https://doi.org/10.1007/978-3-030-56750-7_44

$\mathbb{P}_1$ finite element discretization, so we can use any existing solvers/preconditioners as a coarse solver. One of the main benefits of these preconditioners is that they are easy to implement in practice. The procedure is the same as for the standard multigrid algorithms with the grid-transfer operators between the virtual element space and the conforming $\mathbb{P}_1$ finite element space. Since the same degrees of freedom are used, we can simply use the identity operator as the intergrid transfer operator between the coarse and fine spaces.

Due to page limitation, we only state the main result and provide some numerical experiments to support it. We refer to [20] for more detailed analysis and further discussion of the preconditioners. The rest of this paper is organized as follows. In Section 2, we give basic notation and the virtual element discretization. Then in Section 3, we present the auxiliary space preconditioners and discuss its convergence. Finally, in Section 4, we present several numerical experiments in both 2D and 3D to verify the theoretical result.

2 Virtual Element Methods

Let $\Omega \subset \mathbb{R}^d$ $(d = 2, 3)$ be a bounded open polygonal domain. Given $f \in L^2(\Omega)$, we consider the following model problem: Find $u \in V := H_0^1(\Omega)$ such that

$$a(u, v) := (\kappa \nabla u, \nabla v) = (f, v), \qquad \forall v \in V, \tag{1}$$

where $(\cdot,\cdot)$ denotes the $L^2(\Omega)$ inner product, $\kappa = \kappa(x) \in L^\infty(\Omega)$ is assumed to be piecewise positive constant with respect to the polytopal partition $\mathcal{T}_h$ of Ω but may have large jumps across the interface of the partition.

Let $\mathcal{T}_h$ be a partition of Ω into non-overlapping simple polytopal elements K. Here we use h_K for the diameter of the element $K \in \mathcal{T}_h$ (the greatest distance between any two vertices of K), and define $h = \max_{K\in\mathcal{T}_h} h_K$, the maximum of the diameters. Following [11], we make the following assumption on the polytopal mesh:

(**A**) Each polytopal element $K \in \mathcal{T}_h$ has a triangulation $\mathcal{T}_K$ of K such that $\mathcal{T}_K$ is uniformly shape regular and quasi-uniform. Each edge of K is an edge of certain elements in $\mathcal{T}_K$.

On each polytopal element $K \in \mathcal{T}_h$, we define the local virtual finite element space:

$$V_h^K := \{v \in H^1(K) \ : \ v|_{\partial K} \in \mathbb{B}_1(\partial K),\ \Delta v = 0\},$$

where $\mathbb{B}_1(\partial K) := \{v \in C^0(\partial K) : v|_e \in \mathbb{P}_1(e),\ \forall e \subset \partial K\}$. Note that $V_h^K \supset \mathbb{P}_1(K)$, and may contain implicitly some other non-polynomial functions. The global virtual element space V_h is then defined as:

$$V_h := \{v \in V \ : \ v|_K \in V_h^K,\ \forall K \in \mathcal{T}_h\}.$$

The VEM discretization of (1) is given by a symmetric bilinear form $a_h : V_h \times V_h \to \mathbb{R}$ such that

$$a_h(u_h, v_h) = \sum_{K \in \mathcal{T}_h} a_h^K(u_h, v_h), \qquad \forall u_h, v_h \in V_h,$$

where $a_h^K(\cdot,\cdot)$ is a computable bilinear form defined on $V_h^K \times V_h^K$. So the VEM discretization of (1) reads: Find $u_h \in V_h$ such that

$$a_h(u_h, v_h) = (f, v_h), \quad \forall v_h \in V_h. \tag{2}$$

Further details on how to construct the computable bilinear form a_h, as well as a study of the convergence and stability properties of the VEM can be found in [2, 4, 5]. We refer to [3, 15] for detailed discussion on the implementation of the methods, and refer to [7, 11] for the error estimates of the methods.

Let A be the operator induced by the bilinear form $a_h(\cdot,\cdot)$, that is,

$$(Av, w) = (v, w)_A := a_h(v, w), \qquad \forall v, w \in V_h.$$

Then solving (2) is equivalent to solving the linear system

$$Au_h = f. \tag{3}$$

It is clear that the operator A is symmetric and positive definite, and we can show that the condition number satisfies $\mathcal{K}(A) \lesssim \mathcal{J}(\kappa)h^{-2}$, where $\mathcal{J}(\kappa) = \max_x \kappa(x)/\min_x \kappa(x)$ is the variation of the discontinuous coefficient (see for example [20, Lemma 2.2]). Thus the resulting linear system of the VEM discretization (2) can be very ill-conditioned with the condition number depending on both the mesh size and the variation in the discontinuous coefficient. It is difficult to solve using the classic iterative methods such as Jacobi, Gauss-Seidel or conjugate gradient method, without effective preconditioners. In the next section, we describe efficient auxiliary space preconditioners for (3) that are robust with respect to the variation in the discontinuous coefficient and the mesh size.

3 Auxiliary Space Preconditioners

To solve the discrete system (3) efficiently, we use the auxiliary space preconditioning technique (cf. [17]). For this purpose, we need an "auxiliary space". For each polytopal element $K \in \mathcal{T}_h$, we introduce an auxiliary triangulation $\mathcal{T}_K$ of it such that each edge of K is an edge of some element in this triangulation. By Assumption (**A**), this is possible and can be done using a Delaunay triangulation. With this triangulation, we obtain a conforming quasi-uniform triangulation $\mathcal{T}_h^c := \bigcup_{K \in \mathcal{T}_h} \mathcal{T}_K$ of the whole domain Ω. Let $V_h^c \subset V$ be the standard conforming $\mathbb{P}_1$ finite element space defined on this auxiliary triangulation $\mathcal{T}_h^c$. We introduce the auxiliary problem: find $u_h^c \in V_h^c$ such that

$$a(u_h^c, v_h) = (f, v_h), \qquad \forall v_h \in V_h^c. \tag{4}$$

Similarly, let A_c be the operator induced by the bilinear form $a(\cdot,\cdot)$, that is,

$$(A_c v, w) = (v, w)_{A_c} := a(v, w), \qquad \forall v, w \in V_h^c.$$

The auxiliary space preconditioners can be understood as a two-level algorithm involving a "fine level" and a "coarse level". In this setting, the fine level problem is the VEM discretization (2) on polytopal mesh $\mathcal{T}_h$, and the coarse level problem is the standard conforming $\mathbb{P}_1$ finite element space defined on the auxiliary simplicial mesh (4). Since A_c is the standard conforming piecewise linear finite element discretization of (1) on the auxiliary quasi-uniform triangulation $\mathcal{T}_h^c$, the "coarse" problem in V_h^c can be solved by many existing efficient solvers such as the standard multigrid methods or domain decomposition methods (see, for example [18, 19] and the references cited therein). It can be either an exact solver or an approximate solver. We denote $B_c : V_h^c \to V_h^c$ to be such a "coarse" solver, that is $B_c \approx A_c^{-1}$. Next, on the fine space V_h, we define a "smoother" $R : V_h \to V_h$, which is symmetric positive definite. For example, R could be a Jacobi or symmetric Gauss-Seidel smoother. Finally, to connect the "coarse" space V_h^c with the "fine" space V_h, we need a "prolongation" operator $\Pi : V_h^c \to V_h$. The restriction operator $\Pi^t : V_h \to V_h^c$ is then defined as

$$(\Pi^t v, w) = (v, \Pi w), \quad \text{for } v \in V_h \text{ and } w \in V_h^c.$$

Note that the auxiliary space defined in this way has a natural intergrid transfer operator because the degrees of freedom for the space V_h^c are included among the degrees of freedom for the space V_h. Thus for each $v \in V_h$, we can define $\Pi^t v = v^c \in V_h^c$ such that $v^c(z_i) = v(z_i)$ for each vertex z_i in the element $K \in \mathcal{T}_h$. We can view this as a linear interpolation of v onto V_h^c. Then, the auxiliary space preconditioner $B : V_h \to V_h$ can be chosen as

$$\text{Additive} \qquad B_{\text{add}} = R + \Pi B_c \Pi^t, \tag{5}$$

$$\text{Multiplicative} \qquad I - B_{\text{mul}} A = (I - RA)(I - \Pi B_c \Pi^t)(I - RA). \tag{6}$$

For these preconditioners, we have the following theorem.

Theorem 1 *The auxiliary space preconditioner $B = B_{\text{add}}$ defined by* (5) *or $B = B_{\text{mul}}$ defined by* (6) *satisfies:*

$$\mathcal{K}(BA) \le C,$$

where the constant $C > 0$ depends only on the shape-regularity of the auxiliary triangulation, and is independent of the mesh size h and the coefficients κ.

The analysis is based on the auxiliary space framework [17], with some technical error estimates from [11]. Due to the page limitation, we refer to [20] for more detailed analysis and discussion.

Remark 1 In the auxiliary space preconditioners defined in (5) and (6), if we ignore the smoother R, the resulting preconditioner is usually called the *fictitious space preconditioner* ((cf. [14]). In this case, we denote $B_{\text{fict}} := \Pi B_c \Pi^t$. In fact, the auxiliary space preconditioners can be viewed as a generalization of the fictitious space preconditioner by a special choice of the "fictitious space". In particular, the fictitious space is defined as the product space $V_h \times V_h^c$. Including V_h as a component of the fictitious space makes it easier to construct the map from the fictitious space to the original space, which is required to be surjective. For example, there is no surjective mapping from the linear FEM space V_h^c to higher order VEM space. In this case the smoother R will play an important role in the auxiliary space preconditioners.

On the other hand, note that the operator Π defined above is surjective for linear VEM discretization. If the mesh satisfies Assumption (A), one can show that the fictitious space preconditioner is also robust with respect to the problem size and the discontinuous coefficients. We refer to [20] for more detailed discussion. However, our numerical experiments indicate that B_{fict} is more sensitive to the shape-regularity of the auxiliary triangulation, while B_{add} and B_{mul} are more stable with respect to the mesh quality.

4 Numerical Experiments

In this section, we present some numerical experiments in both 2D and 3D to verify the result in Theorem 1. In all these tests, we use 2-sweeps symmetric Gauss-Seidel smoother. The stopping criteria is $\|r_k\|/\|r_0\| < 10^{-12}$ for the PCG algorithm, where $r_k = f - Au_k$ is the residual. For the coarse solver, we use the AMG algorithm implemented in *i*FEM [10].

In the first example, we consider the model problem (1) in the unit square $\Omega = [0, 1]^2$ with constant coefficient $\kappa = 1$. Figure 1 is an example of the polytopal mesh of the unit square domain (with 100 elements) generated using PolyMesher [16], and Figure 2 is the corresponding Delaunay triangular mesh. The VEM discretization is defined on the polytopal mesh (cf. Figure 1), while the auxiliary space is the standard conforming $\mathbb{P}_1$ finite element discretization defined on the corresponding triangular mesh (cf. Figure 2).

Tables 1 shows the estimated condition number and the number of PCG iteration in parenthesis for the un-preconditioned and preconditioned systems with various preconditioners. Here and in the sequel, B_{sgs} is the (2-sweep) symmetric Gauss-Seidel preconditioner; B_{fict} is the fictitious space preconditioner defined in Remark 1; B_{add} is the additive auxiliary space preconditioner defined in (5); and B_{mul} is the multiplicative auxiliary space preconditioner defined in (6). As we can observe from

Table 1: Estimated condition number (number of PCG iteration) in 2D with constant coefficients.

# Polytopal Elements	10	10^2	10^3	10^4	10^5
$\mathcal{K}(A)$	3.45 (9)	3.86e01 (41)	3.80e02 (117)	3.88e03 (351)	4.07e04 (1100)
$\mathcal{K}(B_{\text{sgs}}A)$	1.07(6)	3.78 (15)	3.20e01 (37)	3.17e02 (104)	3.17e03 (318)
$\mathcal{K}(B_{\text{fict}}A)$	2.92 (8)	5.75 (26)	7.53 (29)	8.73 (32)	9.67(36)
$\mathcal{K}(B_{\text{add}}A)$	1.53 (9)	1.71 (14)	1.94 (14)	1.99 (14)	2.00 (13)
$\mathcal{K}(B_{\text{mul}}A)$	1.06 (8)	1.21 (10)	1.04 (7)	1.02 (6)	1.02 (6)

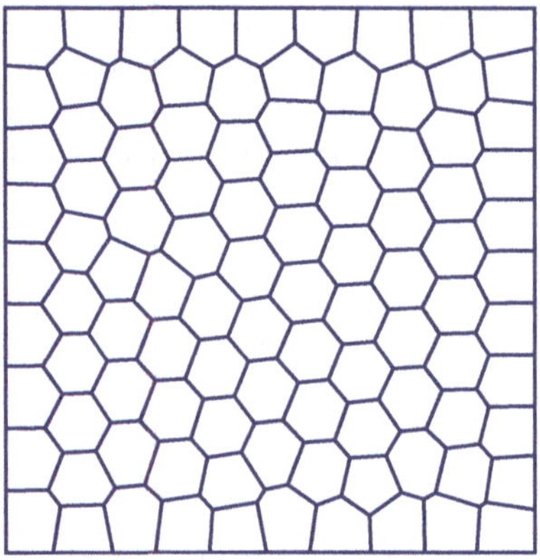

Fig. 1: Polygonal Mesh $\mathcal{T}_h$ of the Unit Square Domain (100 Elements)

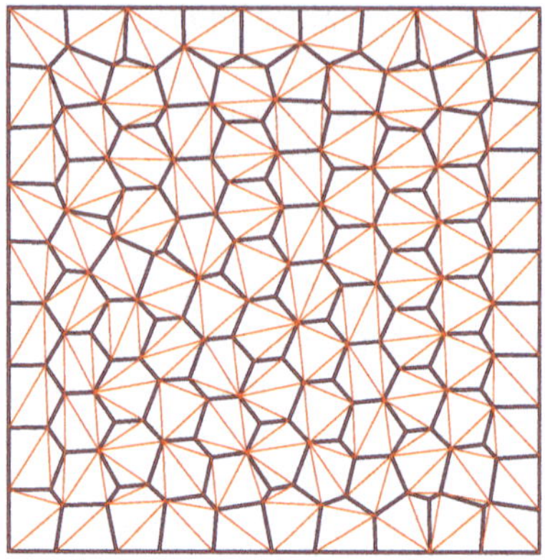

Fig. 2: The Corresponding Delaunay Triangle Mesh $\mathcal{T}_h^c$

this table, while the condition numbers $\mathcal{K}(A)$ and $\mathcal{K}(B_{\text{sgs}}A)$ increase as the mesh refined, the condition numbers $\mathcal{K}(B_{\text{fict}}A)$, $\mathcal{K}(B_{\text{add}}A)$ and $\mathcal{K}(B_{\text{mul}}A)$ are uniformly bounded.

In the second test, we consider the problem with discontinuous coefficients. The coefficient κ is generated randomly on each polygon element (see Figure 3 for an example of the coefficient distribution with 100 elements). Note that the

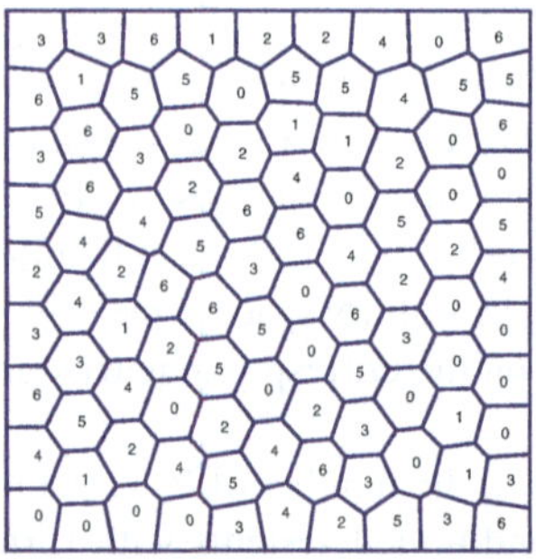

Fig. 3: Random Discontinuous Coefficients 10^k (100 Elements)

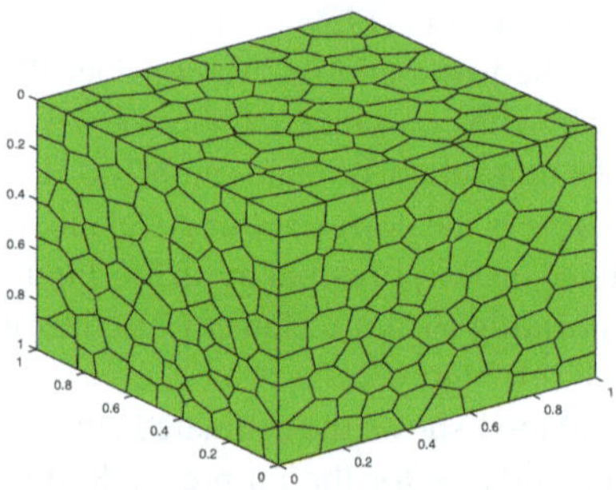

Fig. 4: Polyhedral mesh generated by CVT (9^3 Elements)

coefficient settings are different in different polytopal mesh. Tables 2 shows the estimated condition number and the number of PCG iteration in parenthesis. Here - denotes that the PCG algorithm fail to converge after 1200 iterations. As we can see from this table, while $\mathcal{K}(A)$ and $\mathcal{K}(B_{\text{sgs}}A)$ increase dramatically, the condition numbers $\mathcal{K}(B_{\text{fict}}A)$, $\mathcal{K}(B_{\text{add}}A)$ and $\mathcal{K}(B_{\text{mul}}A)$ are nearly uniformly bounded. These observations verify the conclusions given in Theorem 1 and Remark 1.

Finally, we consider the model problem on a 3D cubic domain $\Omega = [0, 1]^3$. We create a polyhedral mesh using Centroidal Voronoi tessellations (CVT, cf.[13]), see Fig 4 for an example. The VEM discretization is defined on the polyhedral mesh.

Table 2: Estimated condition number (number of PCG iteration) in 2D with discontinuous coefficients.

# Polytopal Elements	10	10^2	10^3	10^4	10^5
$\mathcal{K}(A)$	2.44 (11)	2.73e06 (578)	-	-	-
$\mathcal{K}(B_{\text{sgs}}A)$	1.18(5)	3.90e02 (26)	3.93e03 (409)	-	-
$\mathcal{K}(B_{\text{fict}}A)$	3.27 (8)	6.94 (33)	6.42 (36)	11.6 (44)	13.6 (53)
$\mathcal{K}(B_{\text{add}}A)$	1.54 (9)	3.51 (20)	3.60 (25)	3.67 (25)	3.80 (26)
$\mathcal{K}(B_{\text{mul}}A)$	1.06 (6)	1.74 (15)	1.82 (16)	1.84 (16)	1.88 (17)

Then we subdivide each polyhedron into tetrahedrons using Delaunay triangulation to define the $\mathbb{P}_1$ conforming finite element discretization on this auxiliary mesh.

Table 3: Estimated condition number (number of PCG iteration) in 3D with $\kappa \equiv 1$.

PolyElem	3^3	9^3	15^3	21^3	27^3
TetQuality	7.08e-06	4.09e-08	1.11e-09	2.98e-11	1.28e-11
$\mathcal{K}(A)$	7.91 (25)	5.94e+01 (60)	1.35e+02 (77)	3.20e+02 (104)	6.66e+02 (139)
$\mathcal{K}(B_{\text{add}}A)$	2.36 (16)	4.29 (21)	3.38 (21)	4.35 (24)	5.36 (27)
$\mathcal{K}(B_{\text{mul}}A)$	1.00 (5)	1.10 (8)	1.13 (8)	1.14 (8)	1.18 (9)

Table 3 shows the performance of the B_{add} and B_{mul}. We do not present B_{fict} here because the PCG algorithm does not converge within 200 iterations. To understand the reason, we have calculated the mesh quality of the auxiliary triangulation. Here the TetQuality is the minimum value: $\min_T \frac{r_i}{r_c}$ for all tetrahedral elements T, where r_i and r_c are the radii of the inscribed and circumscribed spheres of T, respectively. From this table, we notice that both B_{add} and B_{mul} are still robust, even in the case of poor TetQuality (which violates Assumption (A)). On the other hand, B_{fict} is sensitive to the shape-regularity of the auxiliary tetrahedral mesh.

Acknowledgements This work was partially supported by NSF DMS-1319110. The author would also like to thank the anonymous referee for carefully proofreading this manuscript and his suggestions, which greatly improved the presentation of this paper.

References

1. P. F. Antonietti, L. Mascotto, and M. Verani. A multigrid algorithm for the p-version of the virtual element method. *arXiv preprint arXiv:1703.02285*, 2017.
2. L. Beirão da Veiga, F. Brezzi, A. Cangiani, G. Manzini, L. D. Marini, and A. Russo. Basic principles of virtual element methods. *Math. Models Methods Appl. Sci.*, 23(01):199–214, 2013.
3. L. Beirão da Veiga, F. Brezzi, L. D. Marini, and A. Russo. The hitchhiker's guide to the virtual element method. *Math. Models Methods Appl. Sci.*, 24(08):1541–1573, 2014.
4. L. Beirão da Veiga, F. Brezzi, L. D. Marini, and A. Russo. Virtual element method for general second-order elliptic problems on polygonal meshes. *Math. Models Methods Appl. Sci.*, 26(04):729–750, 2016.

5. L. Beirão da Veiga, C. Lovadina, and A. Russo. Stability analysis for the virtual element method. *Math. Models Methods Appl. Sci.*, 27(13):2557–2594, 2017.
6. S. Bertoluzza, M. Pennacchio, and D. Prada. BDDC and FETI-DP for the virtual element method. *Calcolo*, 54(4):1565–1593, Dec 2017.
7. S. C. Brenner, Q. Guan, and L.-Y. Sung. Some estimates for virtual element methods. *Comput. Methods Appl. Math.*, 17(4):553–574, 2017.
8. J. G. Calvo. On the approximation of a virtual coarse space for domain decomposition methods in two dimensions. *Math. Models Methods Appl. Sci.*, 28(07):1267–1289, Mar. 2018.
9. J. G. Calvo. An overlapping Schwarz method for virtual element discretizations in two dimensions. *Comput. Math. Appl.*, Nov. 2018.
10. L. Chen. *i*FEM: an integrate finite element methods package in MATLAB. Technical report, University of California at Irvine, 2009.
11. L. Chen and J. Huang. Some error analysis on virtual element methods. *Calcolo*, 55(1):5, Feb 2018.
12. L. Chen, H. Wei, and M. Wen. An interface-fitted mesh generator and virtual element methods for elliptic interface problems. *J. Comput. Phys.*, 334:327 – 348, 2017.
13. Q. Du, V. Faber, and M. Gunzburger. Centroidal voronoi tessellations: Applications and algorithms. *SIAM Rev.*, 41(4):637–676, 1999.
14. S. V. Nepomnyaschikh. Decomposition and fictitious domains methods for elliptic boundary value problems. In *Fifth International Symposium on Domain Decomposition Methods for Partial Differential Equations (Norfolk, VA, 1991)*, pages 62–72. SIAM, Philadelphia, PA, 1992.
15. O. J. Sutton. The virtual element method in 50 lines of MATLAB. *Numer. Algorithms*, 75(4):1141–1159, Aug 2017.
16. C. Talischi, G. H. Paulino, A. Pereira, and I. F. M. Menezes. PolyMesher: a general-purpose mesh generator for polygonal elements written in Matlab. *Struct. Multidiscip. Optim.*, 45(3):309–328, Mar 2012.
17. J. Xu. The auxiliary space method and optimal multigrid preconditioning techniques for unstructured meshes. *Computing*, 56:215–235, 1996.
18. J. Xu and Y. Zhu. Uniform convergent multigrid methods for elliptic problems with strongly discontinuous coefficients. *Math. Models Methods Appl. Sci.*, 18(1):77 –105, 2008.
19. Y. Zhu. Domain decomposition preconditioners for elliptic equations with jump coefficients. *Numer. Linear Algebra Appl.*, 15(2-3):271–289, 2008.
20. Y. Zhu. Auxiliary space preconditioners for virtual element methods. *Submitted*, 2018. Available as *http://arxiv.org/abs/1812.04423*

Part III
Contributed Talks and Posters (CT)

Multi-step Variant of the Parareal Algorithm

Katia Ait-Ameur, Yvon Maday, and Marc Tajchman

1 Introduction

In the field of nuclear energy, computations of complex two-phase flows are required for the design and safety studies of nuclear reactors. System codes are dedicated to the thermal-hydraulic analysis of nuclear reactors at system scale by simulating the whole reactor. We are here interested in the Cathare code developed by CEA, [5]. Typical cases involve up to a million of numerical time iterations, computing the approximate solution during long physical simulation times. A space domain decomposition method has already been implemented. To improve the response time, we will consider a strategy of time domain decomposition, based on the *parareal method* [11]. The Cathare time discretization is based on a multi-step time scheme (see [8]). In this paper, we derive a strategy to adapt the parareal algorithm to multi-step schemes that is not implemented in the code. The paper is organized as follows. In Section 2, we recall the classical version of the parareal algorithm, and then detail the variant that allows us to use multi-step time schemes for the fine solvers. A couple of remarks on the algorithm will be discussed. The numerical convergence for a simplified test case is shown in Section 3 on a Dahlquist test equation followed by numerical results on an advection-diffusion equation and on an industrial test case with an application on the Cathare code.

Katia Ait-Ameur
Laboratoire Jacques Louis Lions (LJLL), Sorbonne Université, 75005 Paris, France
C.E.A, CEA Saclay - DEN/DANS/DM2S/STMF/LMES - 91191 Gif-Sur-Yvette Cedex, France
e-mail: aitameur.katia@gmail.com

Yvon Maday
Laboratoire Jacques Louis Lions (LJLL), Sorbonne Université, 75005 Paris, France
Institut Universitaire de France
e-mail: maday@ann.jussieu.fr

Marc Tajchman
C.E.A, CEA Saclay - DEN/DANS/DM2S/STMF/LMES - 91191 Gif-Sur-Yvette Cedex, France
e-mail: marc.tajchman@cea.fr

R. Haynes et al. (eds.), *Domain Decomposition Methods in Science and Engineering XXV*,
Lecture Notes in Computational Science and Engineering 138,
https://doi.org/10.1007/978-3-030-56750-7_45

2 Parareal algorithm and the multi-step variant

After the discretization of a PDE in space, we obtain an ODE system of the form:

$$\frac{\partial u}{\partial t} + A(t,u) = 0, \quad t \in [0,T], \quad u(t=0) = u_0 \tag{1}$$

mat where $A : \mathbb{R} \times \mathbb{R}^{\mathcal{N}} \rightarrow \mathbb{R}^{\mathcal{N}}$, and $\mathcal{N}$ denotes the number of degrees of freedom.
We here recall the classical parareal algorithm as initially proposed in [11], [2], [4]. Let G and F be two propagators such that, for any given $t \in [0,T]$, $s \in [0,T-t]$ and any function w in a Banach space, $G(t,s,w)$ (respectively $F(t,s,w)$) takes w as an initial value at time t and propagates it at time $t+s$. The full time interval is divided into N^c sub-intervals $[T^n, T^{n+1}]$ of size ΔT that will each be assigned to a processor. The algorithm is defined using two propagation operators:

- $G(T^n, \Delta T, u^n)$ computes a coarse approximation of $u(T^{n+1})$ with initial condition $u(T^n) \simeq u^n$ (low computational cost)
- $F(T^n, \Delta T, u^n)$ computes a more accurate approximation of $u(T^{n+1})$ with initial condition $u(T^n) \simeq u^n$ (high computational cost)

Starting from a coarse approximation u_0^n at times $T^0, T^1, \cdots, T^{N^c}$, obtained using G, the parareal algorithm performs for $k = 0, 1, \cdots$ the following iteration:

$$u_{k+1}^{n+1} = G(T^n, \Delta T, u_{k+1}^n) + F(T^n, \Delta T, u_k^n) - G(T^n, \Delta T, u_k^n)$$

In the parareal algorithm, the value $u(T^n)$ is approximated by u_k^n at each iteration k with an accuracy that tends rapidly to the one achieved by the fine solver, when k increases. The coarse approximation G can be chosen much less expensive than the fine solver F by the use of a scheme with a much larger time step (even $\delta T = \Delta T$) $\delta T \gg \delta t$ (time step of the fine solver) or by using a reduced model. All the fine propagations are made in parallel over the time windows and the coarse propagations are computed in a sequential way but have a low computational cost. We refer to [12] about the parallel efficiency of parareal and a recent work offering a new formulation of the algorithm to improve the parallel efficiency of the original one. The main convergence properties were studied in [7] and stability analysis was made in [14], [3].
In the sequel, we will consider the case that the coarse solver is based on a one-step time scheme and the fine solver on a two-step time scheme. Hence we will use the following notation for the fine solver that takes two initial values: $F(t,s,x,y)$, for $t \in [0,T]$, $s \in [0, T-t[$ and x, y in a Banach space.

Example 1. If one solves (1) with a multi-step time scheme as fine propagator F like the second-order BDF method:

$$\frac{3}{2}u^{j+1} - 2u^j + \frac{1}{2}u^{j-1} = -\delta t A(u^{j+1}, t^{j+1}), \quad j = 1, \cdots, N^f, t^{j+1} - t^j = \delta t$$

Here the fine solver reads as: $u^{j+1} = F(t^j, \delta t, u^{j-1}, u^j)$. Now, we apply the parareal algorithm with a coarse grid: $T^0, \cdots, T^{N^c}$ where:
$T^{n+1} - T^n = \Delta T = N^f \delta t$.
Then we can write: $u(T^n + j\delta t) \simeq u^{n,j}, j = 1, \cdots, N^f, n = 1, \cdots, N^c$.
In order to perform the fine propagation, in a given time window $[T^n, T^{n+1}]$, we only need the local initial condition u_k^n and a consistent approximation of $u(T^n - \delta t)$.

In [1], the authors propose a consistent approximation in the context of the simulation of molecular dynamics. The proposed method was linked to the nature of the model and the symplectic character of their algorithm is shown, which is an important property to verify for molecular dynamics.
In the context of our application to the thermalhydraulic code Cathare, we want to derive a multi-step variant of parareal that will not be intrusive in the software. We seek a consistent approximation of $u(T^n - \delta t)$. The only fine trajectory at our disposal is $F(T^{n-1}, \Delta T, u_k^{n-2,N^f-1}, u_k^{n-1})$. Its final value at T^n is:
$F(T^{n-1}, \Delta T, u_k^{n-2,N^f-1}, u_k^{n-1})(T^n)$ from which we compute u_{k+1}^n by the parareal correction. Hence, we translate the solution:
$F(T^{n-1}, \Delta T - \delta t, u_k^{n-2,N^f-1}, u_k^{n-1})(T^n - \delta t)$ by the same correction:
$u_{k+1}^n - F(T^{n-1}, \Delta T, u_k^{n-2,N^f-1}, u_k^{n-1})$ and obtain the so called consistent approximation u_{k+1}^{n-1,N^f-1} to initialize the fine propagation in $[T^n, T^{n+1}]$. We now detail our algorithm:

$$\begin{cases} u_0^{n+1} &= G(T^n, \Delta T, u_0^n), \quad 0 \leq n \leq N-1 \\ u_{k+1}^{n+1} &= G(T^n, \Delta T, u_{k+1}^n) + F(T^n, \Delta T, u_k^{n-1,N^f-1}, u_k^n) \\ & \quad -G(T^n, \Delta T, u_k^n), \quad 0 \leq n \leq N-1, \quad k \geq 0 \\ u_{k+1}^{n,N^f-1} &= F(T^n, \Delta T - \delta t, u_k^{n-1,N^f-1}, u_k^n) + u_{k+1}^{n+1} \\ & \quad -F(T^n, \Delta T, u_k^{n-1,N^f-1}, u_k^n), \quad 0 \leq n \leq N-1, \quad k \geq 0 \end{cases} \tag{2}$$

Another option to treat this issue is to use a one-step time scheme to initialize the fine computation or to make one iteration with a second-order Runge Kutta method. We will see from the numerical results that these choices modify the fine scheme and prevent the parareal algorithm to converge to the fine solution: even after N^c iterations (where N^c is the number of time windows), the parareal algorithm does not converge to the fine solution with the machine precision but the parareal error stagnates around 10^{-6}.
This method adds consistency with the fine scheme. Also, this strategy can be applied to multi-step time schemes involving several fine time steps preceding the time T^n by applying the same correction to terms taking the form: $u_{k+1}^{n,N^f-i}, \quad i = 1, \cdots, I$. The convergence analysis will be shown in a forthcoming work.

3 Numerical results

We now show some numerical experiments, first for an ordinary differential equation, the test equation of Dahlquist, then for a partial differential equation, namely the advection-diffusion equation and finally on an industrial application with the Cathare code.

3.1 Dahlquist equation

We first use the Dahlquist test equation :

$$u'(t) = \lambda u(t), \quad t \in (0, T), \text{ with } u(0) = 1,$$

discretized by a second-order BDF method. With $\lambda = -1, T = 5, \Delta T = T/50$, which correspond to 50 processors, and $\delta t = T/5000$, we obtain the convergence curves shown in Fig. 1. Here, the fine solver is based on a two-step time scheme where the computation of the solution $u^{n,j+1}$ at time $T^n + (j+1)\delta t$ needs the knowledge of the solutions $u^{n,j}$ and $u^{n,j-1}$ at times $T^n + j\delta t$ and $T^n + (j-1)\delta t$, respectively. We use the multi-step variant of parareal (2) to initialize the fine solver in each time window, starting from the parareal iteration $k \geq 2$. At the parareal iteration $k = 1$, we use a Backward Euler method to initialize the fine solver since we don't have the fine solution yet. The coarse solver is based on a one-step time scheme, namely the Backward Euler method. We plot the relative error in $L^\infty(0, T)$ between the fine solution computed in a sequential way and the parareal solution as a function of iteration number for the classical parareal algorithm where the Backward Euler method is used at each iteration for the initialization of the fine solver (circles), and for the multi-step variant of the parareal algorithm that we introduced in the previous section (squares). We see in the Fig. 1 (without multi-step) that starting from the fourth parareal iteration the error stagnates around 10^{-6} without recovering the fine solution at the machine precision, even after 50 iterations. On the other hand, we see in Fig. 1 (with multi-step) that the error continues to decrease after the fourth parareal iteration until reaching the machine precision at the eleventh iteration. In this case, if we don't use the multi-step variant of parareal, we loose one of the well known property of parareal: to recover the fine solution at the machine precision after N^c iterations.

3.2 Advection-diffusion equation

We now study the behavior of the multi-step parareal (2) applied to the advection-diffusion equation:

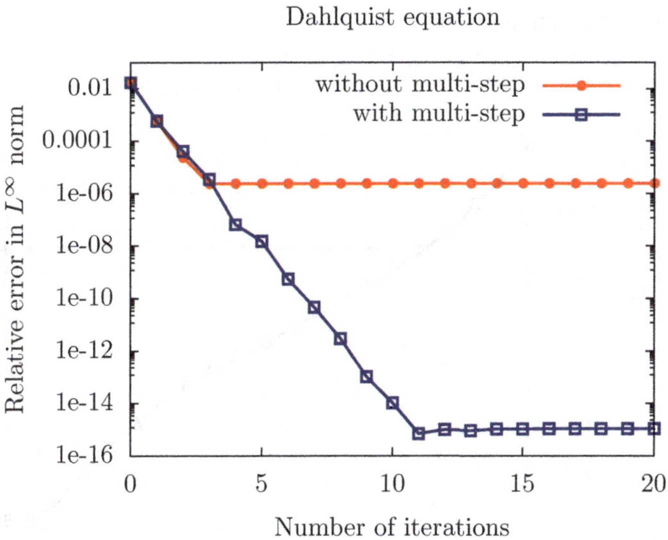

Fig. 1: Convergence of the multi-step parareal for the Dahlquist test equation

$$\begin{cases} \frac{\partial u}{\partial t} - \frac{\partial^2 u}{\partial x^2} + \frac{\partial u}{\partial x} = 0, & (x,t) \in (0,2\pi) \times (0,T) \\ u(x,0) = u_0(x), & x \in (0,2\pi) \\ u(x,t) = u(x+2\pi,t), & t \in (0,T) \end{cases} \tag{3}$$

We have chosen a spectral Fourier approximation in space (truncated series with $N = 16$) and a second-order BDF method in time for a propagation over $[0,2]$. The parareal in time algorithm is implemented with $\Delta T = 0.1$ and $\delta t = 10^{-3}$. We have chosen the following initial condition:

$$u_0(x) = \sum_{l=-N/2+1}^{l=N/2} \hat{u}_l e^{ilx}, \text{ with } \hat{u}_l = \frac{sgn(l)}{|l|^p}.$$

We choose $p = 4$, hence the initial condition is sufficiently regular.

The coarse and fine solvers are the same as in the previous numerical example and we use the same initialization of the fine solver at the first parareal iteration. We plot the following error as a function of iteration number: $E_k^n = \frac{max_n ||U_{seq}(T^n) - U_k^n||_{L^2((0,2\pi))}}{max_n ||U_{seq}(T^n)||_{L^2((0,2\pi))}}$, where U_{seq} is the fine solution computed in a sequential way. In Fig. 2 (circles), we observe a similar behavior as in the previous case, the error stagnates around 10^{-6} after the fifth parareal iteration without reaching the machine precision when the number of iterations is equal to the number of time windows (20 in this case). In Fig. 2 (squares), using the multi-step variant, the error continues to decrease after the fifth iteration until reaching the machine precision around 10^{-16} at iteration number 20.

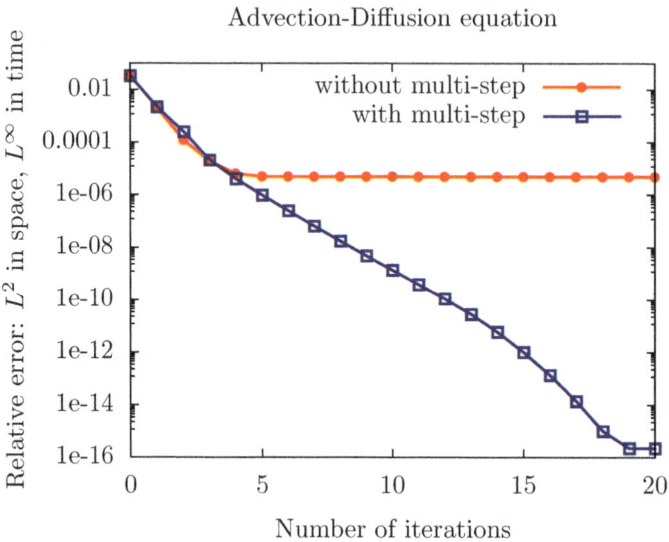

Fig. 2: Convergence of the multi-step parareal for the advection-diffusion equation

3.3 Application to the Cathare code

The Cathare code simulates two-phase flows at a macroscopic scale and the model used is the six-equation two-fluid model ([6],[10],[13]) that considers a set of balance laws (mass, momentum and energy) for each phase liquid and vapor. The unknowns are the volume fraction $\alpha_k \in [0,1]$, the enthalpies H_k, the velocity u_k of each phase and the pressure p. The density $\rho_k \geq 0$ is computed with equations of state $(k = l, g)$. The Cathare scheme is based on a finite volume method on a staggered grid (MAC scheme) and on a two-step time scheme. Here, we write the time discretization of the Cathare scheme:

$$\begin{cases} \frac{(\alpha_k\rho_k)^{n+1}-(\alpha_k\rho_k)^n}{\Delta t} + \partial_x(\alpha_k\rho_k u_k)^{n+1} = 0 \\ (\alpha_k\rho_k)^{n+1}\frac{u_k^{n+1}-u_k^n}{\Delta t} + (\alpha_k\rho_k u_k)^{n+1}\partial_x u_k^{n+1} + \alpha_k^{n+1}\partial_x p^{n+1} = (\alpha_k\rho_k)^{n+1}g \\ \qquad + F_k^{n,n+1} \\ \frac{1}{\Delta t}\left[(\alpha_k\rho_k)^{n+1}\left(H_k + \frac{u_k^2}{2}\right)^{n,n+1} - (\alpha_k\rho_k)^n\left(H_k + \frac{u_k^2}{2}\right)^{n-1,n}\right] \\ \qquad + \partial_x\left[\alpha_k\rho_k u_k\left(H_k + \frac{u_k^2}{2}\right)\right]^{n+1} = \alpha_k^{n+1}\frac{p^{n+1}-p^n}{\Delta t} + (\alpha_k\rho_k u_k)^{n+1}g \end{cases} \tag{4}$$

This choice of time discretization was made for stability purpose.
Here we apply the parareal algorithm to the solution of an oscillating manometer. This test case is proposed in [9] for system codes to test the ability of each numerical scheme to preserve system mass and to retain the gas-liquid interface.
We have used coarse and fine solvers such that $\delta t = 10^{-5}$ for F and $\Delta T = 10\delta t$ for G. All calculations have been evaluated with a stopping criterion where the tolerance is fixed to the precision of the numerical scheme, $\epsilon = 5 \cdot 10^{-2}$.
In order to perform the fine propagation, in a given time window $[T^n, T^{n+1}]$, at the first parareal iteration we need to choose a different consistent approximation of $u(T^n - \delta t)$, since we have not used the fine solver yet. In the context of the application to the Cathare code, we choose a non intrusive initialization by imposing $u_0^{n-1,N^f-1} = u_0^n, \quad 1 \leq n \leq N-1$.
In Fig. 3, we plot the evolution of the relative error in L^2 norm between the parareal solution and the fine one accross the time. This result illustrates that the multi-step variant (2) of the parareal algorithm effectively converges when applied to the problem of the oscillating manometer.

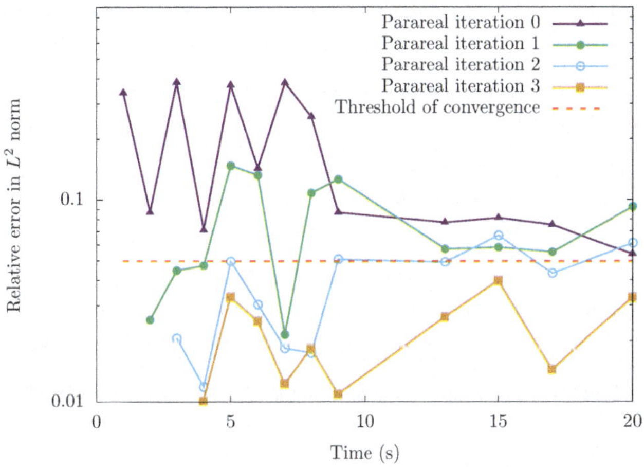

Fig. 3: Multi-step parareal for an industrial application with Cathare code

4 Conclusion

We have built a new variant of parareal algorithm allowing to overcome the issue of initializing the fine solver when the time scheme involve the numerical solution at times preceding the local initial condition in a given time window. The results of this study show that this variant converges numerically on different examples: the two simple test equations allow to see clearly the advantage of our strategy. The application on an industrial code shows its efficiency on a more realistic test case without being intrusive in the software.

The convergence analysis of this algorithm will be the subject of a forthcoming paper. The extension of this method to the use of a multi-step scheme in the coarse solver will be also investigated. This point was treated in [1] and a similar strategy will be studied.

Acknowledgements This research is partially supported by ANR project CINE-PARA (ANR-15-CE23-0019).

References

1. Audouze, C., Massot, M., Volz, S.: Symplectic multi-time step parareal algorithms applied to molecular dynamics. `http://hal.archives-ouvertes.fr/hal-00358459/fr/` (2009)
2. Baffico, L., Bernard, S., Maday, Y., Turinici, G., Zérah, G.: Parallel-in-time molecular-dynamics simulations. Physical Review E **66**, p. 057,701 (2002)
3. Bal, G.: Parallelization in time of (stochastic) ordinary differential equations. `http://www.columbia.edu/gb2030/PAPERS/paralleltime.pdf` (2003)
4. Bal, G., Maday, Y.: A "parareal" time discretization for non-linear pde's with application to the pricing of an american put. Recent developments in domain decomposition methods **23**, pp. 189–202 (2002)
5. Bestion, D.: The physical closure laws in the cathare code. Nuclear Engineering and Design **vol. 124**, pp 229–245 (1990)
6. Drew, D., Passman, S.: Theory of multicomponent fluids. Springer-Verlag, New-York (1999)
7. Gander, M.J., Vandewalle, S.: Analysis of the parareal time-parallel time-integration method. SIAM J. Sci. Comput. **29(2)**, pp. 556–578 (2007)
8. Hairer, E., Norsett, S., Wanner, G.: Solving Ordinary Differential Equations I: Nonstiff Problems. Springer-Verlag, Second Revised Edition (1993)
9. Hewitt, G., Delhaye, J., Zuber, N.: Multiphase science and technology, vol. vol. 6 (1991)
10. Ishii, M.: Thermo-fluid dynamic theory of two-phase flow. Eyrolles, Paris (1975)
11. Lions, J.L., Maday, Y., Turinici, G.: Résolution par un schéma en temps "pararéel". C. R. Acad. Sci. Paris **332(7)**, pp. 661–668 (2001)
12. Maday, Y., Mula, O.: An adaptive parareal algorithm. `https://arxiv.org/pdf/1909.08333.pdf` (2019)
13. Ndjinga, M.: Influence of interfacial pressure on the hyperbolicity of the two-fluid model. C. R. Acad. Sci. Paris **Ser. I 344**, pp. 407–412 (2007)
14. Staff, G., Ronquist, E.: Stability of the parareal algorithm. Domain Decomposition Methods in Science and Engineering, Lecture Notes in Computational Science and Engineering **vol. 40**, pp. 449–456 (2005)

A Domain Decomposition Method for a Geological Crack

O. Bodart, A. Chorfi, and J. Koko

1 Introduction

The computational cost is a key issue in crack identification or propagation problems. One of the solutions is to avoid re-meshing the domain when the crack moves by using a fictitious domain method [2]. We consider a geological crack in which the sides do not pull apart. To avoid re-meshing, we propose an approach combining the finite element method, the fictitious domain method, and a domain decomposition approach. We first extend artificially the crack to split the domain into two subdomains with a nonpenetration condition (negative relative normal displacement) on the crack, a prescribed homogeneous displacement jump condition (continuous displacement) on the fictitious crack. We obtain a convex linearly constrained minimization problem with a quadratic cost function. We use a (primal-dual) interior points method, see e.g.[7, sect 16.6],[5], for the numerical realization.

The paper is organized as follows. In Section 2 we present the model problem, followed by the domain decomposition in Section 3. In Section 4, we describe the finite element discretization and the algebraic problem. Results are presented in Section 5.

O. Bodart
The Lyon University, Université Jean Monnet Saint-Étienne, CNRS UMR 5208, Institut Camille Jordan, F-42023 Saint-Etienne, France, e-mail: olivier.bodart@univ-st-etienne.fr

A. Chorfi and J. Koko
LIMOS, Université Clermont-Auvergne – CNRS UMR 6158, F-63000 Clermont-Ferrand, France, e-mail: chorfi@isima.fr, koko@isima.fr

R. Haynes et al. (eds.), *Domain Decomposition Methods in Science and Engineering XXV*,
Lecture Notes in Computational Science and Engineering 138,
https://doi.org/10.1007/978-3-030-56750-7_46

2 Model description

Let Ω be an open and bounded domain in $\mathbb{R}^2$ with smooth boundary $\Gamma = \Gamma_D \cup \Gamma_N$, where Γ_D and Γ_N are Dirichlet and Neumann parts ($\Gamma_D \cap \Gamma_N = \emptyset$). We denote by $\boldsymbol{u}$ the displacement fields and by $\boldsymbol{f}$ the density of the external forces. The Cauchy stress tensor $\sigma(\boldsymbol{u})$ and the strain tensor $\varepsilon(\boldsymbol{u})$ are given by

$$\sigma(\boldsymbol{u}) = 2\mu\varepsilon(\boldsymbol{u}) + \lambda(\varepsilon(u))\mathbb{I}_{\mathbb{R}^2} \quad \text{and} \quad \varepsilon(\boldsymbol{u}) = (\nabla\boldsymbol{u} + \nabla\boldsymbol{u}^\top)/2,$$

where λ and μ are the Lamé constants. The top boundary (Γ_N ground surface) is subject to homogeneous Neumann boundary condition and, on Γ_D homogeneous Dirichlet boundary conditions are assumed.

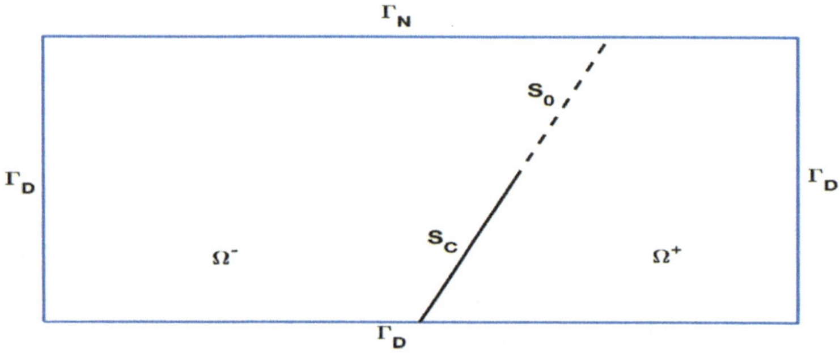

Fig. 1: Domain Ω with the crack S_C and fictitious crack S_0

We assume that Ω contains a crack S_C represented by a curve (cf. Figure 1), parametrized by an injective map. A nonpenetration condition is prescribed on S_C. Denoting by S_C^+, S_C^- the right and left sides of S_C we can set $\boldsymbol{u}^+ = \boldsymbol{u}_{|S_C^+}$ and $\boldsymbol{u}^- = \boldsymbol{u}_{|S_C^-}$, the displacement fields on the right and left sides of S_C. Then the nonpenetration condition is given by the negative relative normal displacement, i.e., $[\boldsymbol{u}_n] := (\boldsymbol{u}^+ - \boldsymbol{u}^-)\cdot\boldsymbol{n} \le 0$, assuming no normal gap in the undeformed configuration.

The linear elastostatic model with crack is governed by the following system of equations

$$-\mathrm{div}\sigma(\boldsymbol{u}) = f \text{ in } \Omega, \tag{1}$$

$$\boldsymbol{u} = 0 \text{ on } \Gamma_D, \qquad \sigma(\boldsymbol{u})\cdot n = 0 \text{ on } \Gamma_N, \tag{2}$$

$$[\boldsymbol{u}_n] \le 0, \text{ on } S_C. \tag{3}$$

In the next section we extend the crack to split the domain into two subdomains.

3 Domain Decomposition

We extend artificially the crack to split the domain into two subdmoains $\Omega^\pm$ as shown in Figure 1. Let S_0 be the fictitious crack. On S_0 we prescribed the (displacement) continuity condition $[\boldsymbol{u}] := (\boldsymbol{u}^+ - \boldsymbol{u}^-) = 0$ and the normal derivative continuity condition $[\sigma(\boldsymbol{u})n] := (\sigma(\boldsymbol{u}^+) - \sigma(\boldsymbol{u}^-)) \cdot n = 0$.

Let us introduce the functions space $V = \{v \in H^1(\Omega);\ v = 0 \text{ on } \Gamma_D\}$, and the forms

$$a(\boldsymbol{u}, \boldsymbol{v}) = \int_\Omega \sigma(\boldsymbol{u}) : \varepsilon(\boldsymbol{v})\, dx \quad \text{and} \quad f(\boldsymbol{v}) = \int_\Omega f\, \boldsymbol{v}\, dx.$$

Then the total potential energy is

$$J(\boldsymbol{v}) = \frac{1}{2} a(\boldsymbol{v}, \boldsymbol{v}) - f(\boldsymbol{v}). \tag{4}$$

The elastostatic problem with extended crack can now be formulated as the following constrained minimization problem

$$\min J(\boldsymbol{u}), \tag{5}$$

$$[\boldsymbol{u}_n] \leq 0 \text{ on } S_C, \tag{6}$$

$$[\boldsymbol{u}] = 0 \text{ on } S_0 \tag{7}$$

Since the functional (4) is strongly convex on V and constraints (6)-(7) are linear, the constrained minimization problem (5)-(7) has a unique solution.

Remark 1 The stress continuity condition is no longer taken into acount in the formulation (5)-(7). It will be ensure by the Lagrange multiplier associated with the displacement continuity condition (7).

With (5)-(7) we associate the Lagrangian functional $\mathcal{L}$ defined on $\mathbf{V} \times \mathbf{L}^2(S_C) \times \mathbf{L}^2(S_0)^2$ by:

$$\mathcal{L}(\boldsymbol{v}, \mu_C, \mu_0) = J(\boldsymbol{v}) + (\mu_C, [\boldsymbol{u}_n])_{S_C} + (\mu_0, [\boldsymbol{u}])_{S_0}, \tag{8}$$

where $\mu_C \in \mathbf{L}^2(S_C)$, $\mu_0 \in \mathbf{L}^2(S_0)^2$ are the Lagrange multipliers associated with (6) and (7), respectively . Note that the the multiplier associated with (7) must be non negative, i.e. $\mu_C \geq 0$ on S. Since (5)-(7) is linear a constrained convex minimization problem, a saddle point of $\mathcal{L}$ exists and (5)-(7) is equivalent to the saddle point problem

Find $(\boldsymbol{u}, \lambda_C, \lambda_0)$ such that

$$\mathcal{L}(\boldsymbol{u}, \mu_C, \mu_0) \leq \mathcal{L}(\boldsymbol{u}, \lambda_C, \lambda_0) \leq \mathcal{L}(\boldsymbol{v}, \lambda_C, \lambda_0), \quad \forall (\boldsymbol{v}, \mu_C, \mu_0) \tag{9}$$

Since $\mathcal{L}$ is Gateaux differentiable on $\mathbf{V} \times \mathbf{L}^2(S_C) \times \mathbf{L}^2(S_0)^2$, the solution of (9) is characterized by the saddle-point (Euler-Lagrange) equations of the primal and dual problems as follows

Find $(\boldsymbol{u}, \lambda_C, \lambda_0)$ such that

$$a(\boldsymbol{u},\boldsymbol{v}) + (\lambda_C, [\boldsymbol{v}_n])_{S_C} + (\lambda_0, [\boldsymbol{v}])_{S_0} = \boldsymbol{f}(\boldsymbol{v}), \quad \forall \boldsymbol{v} \in \mathbf{V}, \tag{10}$$

$$\lambda_C [\boldsymbol{u}_n] = 0, \quad \text{on } S_C, \tag{11}$$

$$(\mu_0, [\boldsymbol{u}])_{S_0} = 0, \quad \forall \mu_0 \in \mathbf{L}^2(S_0)^2, \tag{12}$$

where $(.,.)_{S_c}$ and $(.,.)_{S_0}$ are L^2-scalar product on S_c and S_0, respectively.
The equality (11) (i.e. the complementarity condition) is true almost everywhere, and if $\lambda_C > 0$ then $[\boldsymbol{u}_n] = 0$, and if $[\boldsymbol{u}_n] < 0$ (non contact), then $\lambda_C = 0$.

4 Finite element discretization and the algebraic problem

4.1 Finite element discretization

The saddle-point equations are suitable for a fictitious domain approach, i.e. the crack mesh is defined independently of the domain mesh, see e.g.,[2]. We use a fictitious domain method inspired by the extended finite element method (XFEM) in which basis functions are cut across the crack, e.g. [1].

We assume that the domain Ω has a polygonal shape such that it can be entirely triangulated. Let $\mathcal{T}_h$ be a triangulation of Ω.We define the finite elements space

$$V_h = \{\boldsymbol{v}_h \in C^0(\bar{\Omega}); \boldsymbol{v}_{h|T} \in P_k(T)\ \forall T \in \mathcal{T}_h; \boldsymbol{v}_h = 0 \text{ on } \Gamma\} \subset V,$$

Here, $P_k(T)$ is the space of the polynomials of degree $\leq k$ on the mesh T. We define on $S = S_C \cup S_0$ a finite elements space

$$\Lambda_h = \{\lambda^h \in C^0(S);\ \lambda_{h|I} \in P_k(I)\ \forall I \in \mathcal{I}_h\} \subset L^2(S),$$

This approach is similar to XFEM [6], except that the standard basis functions near the crack are not enriched by singular functions but only multiplied by Heaviside functions :

$$H(x) = \begin{cases} 1 \text{ if} & x \in \Omega^{\pm} \text{ (computational domain)} \\ 0 \text{ otherwise.} & \end{cases}$$

For element K containing the crack, the stiffness term $\int_K \sigma(\phi_i) : \varepsilon(\phi_j)$ is replaced by $\int_K \sigma(H(\phi_i)) : \varepsilon(H(\phi_j))$.

4.2 Algebraic problem and algorithm

Assuming that $\mathbf{u} = [\mathbf{u}^+\ \mathbf{u}^-]^\top \in \mathbb{R}^{2n}$ is the unknown vector of nodal values of the displacement fields on Ω_h. Let us define the following matrices and vectors:

- $\mathbf{A}$ the stiffness matrix ($2n \times 2n$ symmetric positive definite), $\mathbf{A} = diag(\mathbf{A}^+, \mathbf{A}^-)$.
- $\mathbf{f}$, the external forces (vector of $\mathbb{R}^{2n}$), $\mathbf{f} = [\mathbf{f}^+ \, \mathbf{f}^-]^\top$
- $\mathbf{B}_C$, the relative normal displacement matrix at the contact nodes $\mathbf{B}_C\mathbf{u} := (\mathbf{u}^+ - \mathbf{u}^-) \cdot \mathbf{n}$.
- $\mathbf{B}_0$, the displacement jump matrix across S_0, $\mathbf{B}_0\mathbf{u} := (\mathbf{u}^+ - \mathbf{u}^-)$.

We define the scalar products

$$(\lambda, \mu)_{M_C} = \lambda^\top M_C \mu \quad \text{and} \quad (\lambda, \mu)_{M_0} = \lambda^\top M_0 \mu,$$

where M_C and M_0 are the mass matrices on S_C andS_0, respectively.

With the above notations, the algebraic Lagrangian of the problem is

$$\mathcal{L}(\mathbf{u}, \lambda_C, \lambda_0) = \frac{1}{2}\mathbf{v}^\top \mathbf{A}\mathbf{v} - \mathbf{v}^\top \mathbf{f} + (\lambda_C, \mathbf{B}_C\mathbf{v})_{M_C} + (\lambda_0, \mathbf{B}_0\mathbf{v})_{M_0},$$

for which the saddle point (KKT) equation are

Find $(\mathbf{u}, \lambda_C, \lambda_0)$ such that:

$$\nabla_\mathbf{u}\mathcal{L}(\mathbf{u}, \lambda_C, \lambda_0) = 0 \tag{13}$$

$$\nabla_{\lambda_C}\mathcal{L}(\mathbf{u}, \lambda_C, \lambda_0) \le 0, \quad \lambda_C \ge 0, \; \lambda_C \cdot \nabla_{\lambda_C}\mathcal{L}(\mathbf{u}, \lambda_C, \lambda_0) = 0 \tag{14}$$

$$\nabla_{\lambda_0}\mathcal{L}(\mathbf{u}, \lambda_C, \mu_0) = 0, \tag{15}$$

where $(\cdot)$ stands for element-wise (or Hadamard) multiplication. Note that in (13), the primal problem, the unknowns $\mathbf{u}^\pm$ are uncoupled if the Lagrange multipliers $\lambda_C, \lambda_0)$ are known. Then a primal-dual algorithm is suitable for solving (13)-(15). To apply an primal-dual interior point method, we set $\mathbf{z} = -\nabla_{\lambda_C}\mathcal{L}(\mathbf{u}, \lambda_C, \lambda_0)$, such that (13)-(15) becomes

Find $(\mathbf{u}, \mathbf{z}, \lambda_C, \lambda_0)$, with $\mathbf{z} \ge 0$ and $\lambda_C \ge 0$, such that

$$\nabla_\mathbf{u}\mathcal{L}(\mathbf{u}, \lambda_C, \lambda_0) = 0 \tag{16}$$

$$\nabla_{\lambda_C}\mathcal{L}(\mathbf{u}, \lambda_C, \lambda_0) + \mathbf{z} = 0, \tag{17}$$

$$\nabla_{\lambda_0}\mathcal{L}(\mathbf{u}, \lambda_C, \lambda_0) = 0. \tag{18}$$

$$\lambda_C \cdot \mathbf{z} = 0 \tag{19}$$

Since $\mathbf{A}$ is positive definite, (16)-(19) are necessary and sufficient conditions. Then we have to solve a nonlinear system of the form

$$F(\mathbf{u}, \mathbf{z}, \lambda_C, \lambda_0) = 0, \quad \mathbf{z} \ge 0, \quad \lambda_C \ge 0. \tag{20}$$

Let us introduce the vector $\mathbf{e} = (1, \ldots, 1)^\top$ and define the complementarity measure $\mu = \lambda_C^\top \mathbf{z}/m$, where m is the dimension of $\mathbf{z}$. We then replace (20) by the following perturbed KKT conditions

$$F(\mathbf{u}, \mathbf{z}, \lambda_C, \lambda_0) = (0^\top, 0^\top, 0^\top, \tau\mu\mathbf{e}^\top)^\top, \tag{21}$$

that is

$$\mathbf{A}\mathbf{u} = \mathbf{f} - B_C^\top \lambda_C - B_0^\top \lambda_0, \tag{22}$$
$$\mathbf{B}_C\mathbf{u} + \mathbf{z} = 0, \tag{23}$$
$$\mathbf{B}_0\mathbf{u} = 0, \tag{24}$$
$$\lambda_C \cdot \mathbf{z} = \tau\mu\mathbf{e}, \tag{25}$$

where $(\tau, \mu) > 0$. Solutions of (22)-(25) for all positive values of τ and μ define a curve $C(\tau, \mu)$, called the *central path*, which is the trajectory that leads to the solution of the quadratic problem as $\tau\mu$ tends to zero. The primal-dual interior point algorithm for solving the saddle point system (13)-(15) consists of applying the damped Newton method to (22)-(25). The damped parameter, τ and μ are adjusted iteratively to ensure fast convergence (see e.g., [7, sect 16.6],[5]). Solving (21) with primal-dual interior point method consists of solving a primal-dual linear system equivalent to the optimality conditions for an equality-constrained convex quadratic program. Applying a Uzawa conjugate gradient method to the (linearized) optimality conditions leads to solving primal linear systems of the form (22) which breaks down naturally into ± sub-systems.

5 Numerical results

We have implemented the method described in the previous section in MATLAB (R2016b) on a Linux workstation equipped with a quad-core Intel Xeon E5 with 3.00GHz clock frequency and 32GB RAM. We use the mesh generation package `KMG2D` [3], and the fast FEM assembling functions package `KPDE` [4]. The test problem used is designed to illustrate the numerical behavior of the algorithm more than to model an actual geological crack.

We consider $\Omega = (0,\ 10) \times (0,\ 5)$ with the boundary partition

$$\Gamma_D = (0,\ 10) \times \{0\} \cup \{1\} \times (0,\ 5) \cup \{0\} \times (0,\ 5) \tag{26}$$
$$\Gamma_N = (0,\ 10) \times \{1\}. \tag{27}$$

The crack is given by

$$S_C = \{(x, 1.25(x-3) \mid x \in (3,\ 5.4)\}, \quad S_0 = \{(x, 1.25(x-3) \mid x \in (5.4,\ 7)\}.$$

The mesh sample is shown in Figure 2. The material constants are $E = 9 \times 10^6$ (Young's modulus) and $\nu = 0.3$ (Poisson's ratio).The applied force to the domain is the gravity with a value density of 1500.

We use the couple $P2/P1$ for the discretization: continuous $P2$ triangular element for Ω_h, continuous $P1$ segment for the crack. The choice of the finite element pair $P2/P1$ is made to ensure the inf-sup condition . We first consider a uniform discretization of Ω consisting of 561 nodes and 256 triangles. The interior point

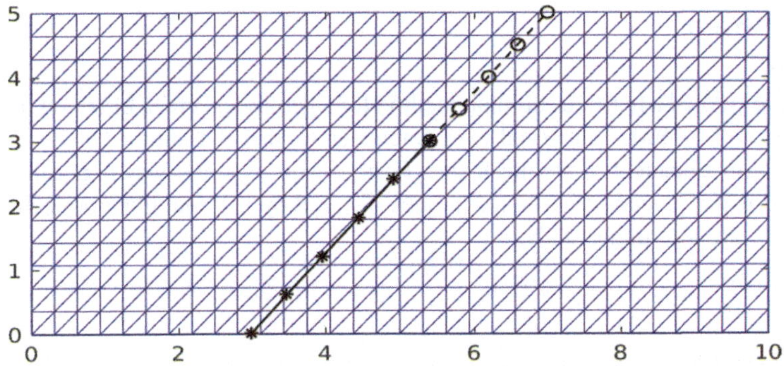

Fig. 2: Mesh sample of Ω with real crack and fictitious crack

algorithm stops after 11 iterations. The deformed configuration is shown in Figure 3 and reveals the presence of a crack.

To study the behavior of our algorithm, the initial mesh is uniformly refined to produce meshes with 2145, 8385, 33153 and 131841 nodes. The performances of the algorithm is shown in Table 1. One can observe that the number of iterations required for convergence is virtually independent of the mesh size.

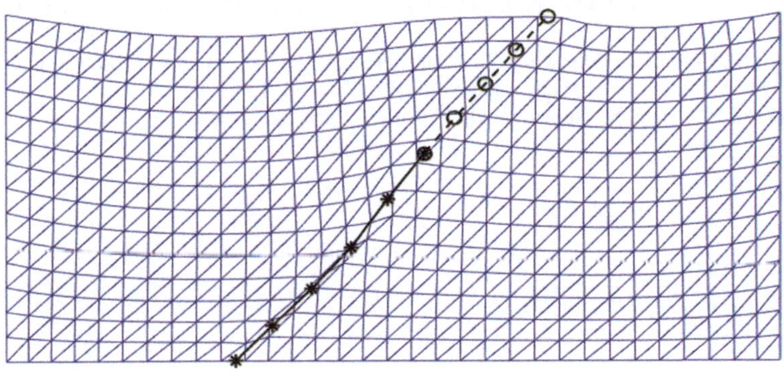

Fig. 3: Mesh sample of Ω with real crack and fictitious crack (magnification=20)

Nodes/Triangles	561/256	2145/1024	8385/4096	33153/16384	131841/65536
Iterations	11	12	15	15	14
CPU Times (Sec.)	0.21	0.47	2.37	17.64	191.92

Table 1: Number of iterations and CPU times (in Sec.) for the interior point algorithm

Conclusion

We have studied a fictitious domain method for a geological crack based on fictitious domain and XFEM. Numerical experiments show that the number of iterations is virtually independent of the mesh size. Further work is under way to accelerate the method using preconditioning techniques inspired by [5]. Stabilization techniques, as in [1], are also under study.

References

1. Bodart, O., Cayol, V., Court, S., J., K.: Xfem-based fictitious domain method for linear elastic model with crack. SIAM J. Sci. Comput. **38**, 219–246 (2016)
2. Glowinski, R., Pan, T., Periaux, J.: A fictitious domain method for dirichlet problem and applications. Comput. Methods Appl. Mech. Engrg. **111**, 283–303 (1994)
3. J., K.: A MATLAB mesh generator for the two-dimensional finite element method. Appl. Math. Comput. **250**, 650–664 (2015)
4. J., K.: Fast MATLAB assembly of fem matrices in 2d and 3d using cell array approach. Int. J. Model. Simul. Sci. Comput. **7** (2016)
5. Kučera, R., Machalová, J., Netuka, H., Zenčá, P.: An interior-point algorithm for the minimization arising from 3d contact problems with friction. Optimization Methods and Software **28**, 1195–1217 (2012)
6. Moës, N., Dolbow, J., Belytschko, T.: A finite element method for crack growth without remeshing. Int. J. Numerical Mathods Engineering **46**, 131–150 (1999)
7. Nocedal, J., Wright, S.: Numerical Optimization. Springer (2006)

Fictitious Domain Method for an Inverse Problem in Volcanoes

Oliver Bodart, Valérie Cayol, Farshid Dabaghi, and Jonas Koko

1 General framework and problem setting

Problems in volcanology often involve elasticity models in presence of cracks (see e.g. [5]). Most of the time the force exerted on the crack is unknown, and the position and shape of the crack are also frequently unknown or partially known (see e.g. [2]). The model may be approximated *via* boundary element methods.These methods are quite convenient to take into account the crack since the problem is then reformulated into an external problem where the crack is the only object to be meshed. However these methods do not allow to take the heterogeneity and/or the anisotropy of the medium into account. Another drawback is that, when it comes to identifying the shape and/or location of the crack, the variation of the latter implies a remeshing and assembling of all the matrices of the problem.

Using a domain decomposition technique then appears as the natural solution to these problems. In [1], a first step was made with the development of a direct solver implementing a domain decomposition method. The present work represents a step further with the use of such a solver, which has been improved since the publication of [1], to solve inverse problems in the field of earth sciences. To our knowledge, this is the first work using these kind of techniques in this field of application. The next step of our project will be the shape optimization problem to identify the shape and location of the crack.

Oliver Bodart
Université Jean Monnet Saint-Étienne, CNRS UMR 5208, Institut Camille Jordan, F-42023 Saint-Etienne, France, e-mail: olivier.bodart@univ-st-etienne.fr

Valérie Cayol, Farshid Dabaghi
Laboratoire Magmas et Volcans, Université Jean Monnet-CNRS-IRD, Saint-Etienne F-42023, France e-mail: v.cayol@opgc.fr, farshid.dabaghi@univ-st-etienne.fr

Jonas Koko
LIMOS, UMR 6158, Université Clermont Auvergne, BP 10448, F-63173 Aubière Cedex, France e-mail: jonas.koko@uca.fr

R. Haynes et al. (eds.), *Domain Decomposition Methods in Science and Engineering XXV*, Lecture Notes in Computational Science and Engineering 138,
https://doi.org/10.1007/978-3-030-56750-7_47

Let Ω be a bounded open set in $\mathbb{R}^d$, $d = 2, 3$ with smooth boundary $\partial\Omega := \Gamma_D \cup \overline{\Gamma_N}$ where Γ_D and Γ_N are of nonzero measure and $\Gamma_D \cap \Gamma_N = \emptyset$. We assume that Ω is occupied by an elastic solid and we denote by $\mathbf{u}$ the displacement field of the solid and the density of external forces by $\mathbf{f} \in \mathbf{L}^2(\Omega)$. The Cauchy stress $\sigma(\mathbf{u})$ and strain $\varepsilon(\mathbf{u})$ are given by

$$\sigma(\mathbf{u}) = \lambda(\mathrm{Tr}\varepsilon(\mathbf{u}))\mathbf{I}_{\mathbb{R}^d} + 2\mu\varepsilon(\mathbf{u}) \quad \text{and} \quad \varepsilon(\mathbf{u}) = \frac{1}{2}(\nabla\mathbf{u} + \nabla\mathbf{u}^\top),$$

where (λ, μ) are the Lamé coefficients, $\mathbf{I}_{\mathbb{R}^d}$ denotes the identity tensor, and $\mathrm{Tr}(\cdot)$ represents the matrix trace. Consider a crack $\Gamma_C \subset \Omega$ represented by a line ($d = 2$) or a surface ($d = 3$) parametrized by an injective mapping. Around the crack, Ω is split into Ω^- and Ω^+. The deformation field of the solid is supposed to satisfy the following elastostatic system:

$$\begin{cases} \text{Find } \mathbf{u} \in \mathbf{H}^1(\Omega \setminus \Gamma_C) \text{ such that :} & \\ -\mathrm{div}\, \sigma(\mathbf{u}) = \mathbf{f} & \text{in } \Omega \setminus \Gamma_C, \\ \mathbf{u} = 0 & \text{in } \Gamma_D, \\ \sigma(\mathbf{u}) \cdot \mathbf{n} = 0 & \text{on } \Gamma_N, \\ \sigma(\mathbf{u}) \cdot \mathbf{n}^\pm = p\mathbf{n}^\pm & \text{on } \Gamma_C. \end{cases} \tag{1}$$

where $\mathbf{n}$ is the outward unit normal to its boundaries. Typically in such a situation, Γ_N is the ground surface and free to move. Practically, the displacement field can be observed on Γ_N, whereas the pressure p exerted on the crack is unknown most of the time.

Consider the following function defined on $\mathbf{L}^2(\Gamma_C)$:

$$J(p) := \frac{1}{2} \int_{\Gamma_N} (\mathbf{u} - \mathbf{u}_d) \mathrm{C}^{-1} (\mathbf{u} - \mathbf{u}_d)^\top \, d\Gamma_N + \frac{\alpha}{2} \|p\|^2_{\mathbf{L}^2(\Gamma_C)}, \tag{2}$$

where $u_d \in \mathbf{L}^2(\Gamma_N)$ is the measured displacement field and u is the solution of (1) associated with p. Moreover, the matrix C is the covariance operator of the measurements uncertainties, and is assumed to be positive definite (see e.g. [4]), and finally $\alpha > 0$ is a regularization parameter. The aim of this work is to study the following problem, of optimal control type:

$$\min_{p \in \mathbf{L}^2(\Gamma_C)} J(p). \tag{3}$$

The paper is organized as follows : the next section will be devoted to the presentation of the domain decomposition method and its discretization. Section 3 gives the optimality conditions for the problem (3) and establishes their discrete version. A special focus will be made on the adaptation of the problem to a domain decomposition formulation. Finally we present a relevant numerical test in section 4 and discuss the next steps of our project.

2 Domain decomposition : the direct solver

To solve the direct problem (1), we use a domain decomposition method. More precisely, following [1], the domain Ω is split into two subdomains such that each point of the domain lies on one side of the crack or on the crack. Moreover, the global unknown solution $\mathbf{u}$ is decoupled in two sub–solutions for each side of the crack. For this purpose, we are using an artificial extension of the considered crack Γ_C (e.g. Γ_0 in Figure 1). Therefore, instead of the crack problem (1), we have to solve two

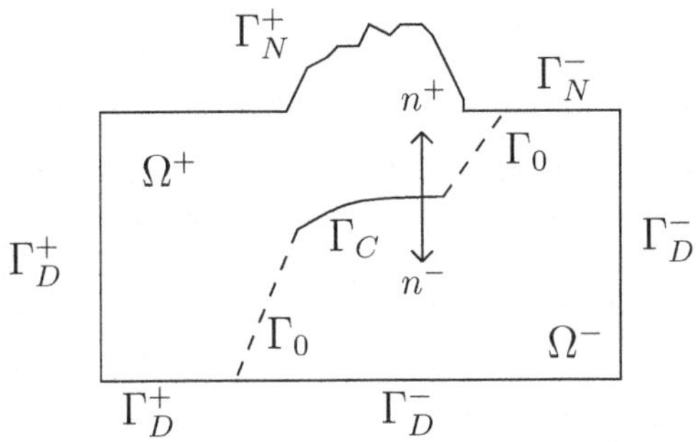

Fig. 1: Splitting the volcanic cracked domain

Neumann–type boundary problems such that for each problem we impose a pressure on Γ_C, which is more convenient from both theoretical and numerical points of view. More precisely we solve the following system:

$$\begin{cases} \text{Find } \mathbf{u} \in \mathbf{H}^1(\Omega) \text{ such that :} & \\ -\text{div}\,\sigma(\mathbf{u}^\pm) = \mathbf{f}^\pm & \text{in } \Omega^\pm, \\ \mathbf{u}^\pm = 0 & \text{on } \Gamma_D \cap \partial\Omega^\pm, \\ (\sigma(\mathbf{u}) \cdot \mathbf{n})^\pm = 0 & \text{on } \Gamma_N \cap \partial\Omega^\pm, \\ (\sigma(\mathbf{u}) \cdot \mathbf{n})^\pm = p\mathbf{n}^\pm & \text{on } \Gamma_C, \\ [\mathbf{u}] = 0 & \text{on } \Gamma_0, \\ [\sigma(\mathbf{u})] \cdot \mathbf{n}^+ = 0 & \text{on } \Gamma_0, \end{cases} \tag{4}$$

where $\mathbf{u}^+ = \mathbf{u}_{|\Omega^+}$ and $\mathbf{u}^- = \mathbf{u}_{|\Omega^-}$, and $[\mathbf{v}]$ denotes the jump of v across Γ_0. The two last conditions in (4) enforce the continuity of displacement and stress across Γ_0. Notice that the boundary conditions on Γ_0 ensure the construction of a global displacement field in $\mathbf{H}^1(\Omega \setminus \Gamma_C)$ solving the original problem (1).

Let us define the following Hilbert spaces

$$\mathbf{V}^\pm = \{v \in \mathbf{H}^1(\Omega^\pm) \mid v = 0 \text{ on } \Gamma_D \cap \partial\Omega^\pm\}, \qquad \mathbf{W} = (\mathbf{H}^{\frac{1}{2}}(\Gamma_0)).$$

and their dual spaces $\mathbf{V}'^\pm$ and $\mathbf{W}'$, endowed with their usual norms. Prescribing the continuity of displacement across the Γ_0 *via* a Lagrangian formulation, the mixed weak formulation of Problem (4) reads as follows:

$$
\begin{cases}
\text{Find } \mathbf{u}^{\pm} \in \mathbf{V}^{\pm} \text{ and } \lambda \in \mathbf{W}' \text{ such that :} & \\
a(\mathbf{u}^{\pm}, \mathbf{v}^{\pm}) \pm b(\lambda, \mathbf{v}^{\pm}) = l^{\pm}(\mathbf{v}^{\pm}) & \forall \mathbf{v}^{\pm} \in \mathbf{V}^{\pm}, \\
b(\mu, [\mathbf{u}]) = 0 & \forall \mu \in \mathbf{W}',
\end{cases}
\tag{5}
$$

with

$$
a(\mathbf{u}^{\pm}, \mathbf{v}^{\pm}) = \int_{\Omega^{\pm}} \sigma(\mathbf{u}^{\pm}) : \varepsilon(\mathbf{v}^{\pm})\, \mathrm{d}\Omega^{\pm}
$$

bilinear, symmetric, coercive and

$$
l^{\pm}(\mathbf{v}^{\pm}) = \int_{\Omega^{\pm}} f \cdot \mathbf{v}^{\pm}\, \mathrm{d}\Omega^{\pm} + \int_{\Gamma_C} (p\mathbf{n})^{\pm} \cdot \mathbf{v}^{\pm}\, \mathrm{d}\Gamma_C
$$

linear and continuous. Moreover, b is defined as the duality pairing between $\mathbf{W}'$ and $\mathbf{W}$: $b(\lambda, \mathbf{v}^{\pm}) = \langle \lambda, \mathbf{v}^{\pm} \rangle_{\mathbf{W}', \mathbf{W}}$. Therefore, it is straightforward to prove the existence and uniqueness of a solution to Problem (5) (see e.g. [1] ans references within).

Denoting then $\mathbf{f}^{\pm}$ and $\mathbf{p}$ the approximations of f and p in $\mathbf{V}^{\pm}$ and $\widehat{\mathbf{W}}_h$, setting $p_n = \mathbf{p}\mathbf{n}^+ = -\mathbf{p}\mathbf{n}^-$ and

$$
\mathrm{K} = \begin{pmatrix} A^+ & 0 & B^{+T} \\ 0 & A^- & -B^{-T} \\ B^+ & -B^- & 0 \end{pmatrix}, \qquad \mathbf{X} = \begin{pmatrix} \mathbf{u}^+ \\ \mathbf{u}^- \\ \lambda \end{pmatrix},
$$

$$
\mathbf{F} = \begin{pmatrix} \mathbf{F}^+ \\ \mathbf{F}^- \\ 0 \end{pmatrix} = \begin{pmatrix} M_{\Omega}^+ \cdot \mathbf{f}^+ \\ M_{\Omega}^- \cdot \mathbf{f}^- \\ 0 \end{pmatrix} + \begin{pmatrix} +M_c^+ \cdot p_n \\ -M_c^- \cdot p_n \\ 0 \end{pmatrix} := L_{\Omega}.\mathbf{f} + L_c.\mathbf{p},
$$

the discretized form of system (5) has the linear algebraic formulation

$$
\mathrm{K}\mathbf{X} = \mathbf{F}. \tag{6}
$$

The system (6) can be solved by a Uzawa Conjugate gradient/domain decomposition method [1]. The method can be classically stabilized and the convergence of the numerical scheme can be proved as $h \to 0$.

In what follows, we will focus on the adaptation of a crack inverse problem to this domain decomposition formulation and its application to a realistic problem.

3 The crack inverse problem

First, we have the following result.

Proposition 1 *For any* $\alpha > 0$, *the problem* (3) *admits a unique solution* p^* *in* $\mathbf{L}^2(\Gamma_C)$.

Proof The proof is classical: applying the same method as in [3], one easily shows that J is strictly convex and coercive on $\mathbf{L}^2(\Gamma_C)$. □

The objective function J being strictly convex, first order optimality conditions can be computed to implement an suitable optimization method (in our case the conjugate gradient). Let us introduce the adjoint system

$$\begin{cases} -\operatorname{div} \sigma(\boldsymbol{\phi}) = 0 & \text{in } \Omega, \\ \boldsymbol{\phi} = 0 & \text{on } \Gamma_{\mathrm{D}}, \\ \sigma(\boldsymbol{\phi}) \cdot \mathbf{n} = \mathrm{C}^{-1}(\mathbf{u} - \mathbf{u}_d) & \text{on } \Gamma_{\mathrm{N}}, \end{cases} \tag{7}$$

where $\mathbf{u}$ is a solution of system (1). Is is easy to prove that this adjoint system admits a unique solution $\boldsymbol{\phi} \in H^1(\Omega)$. We have the following

Proposition 2 *Let p^* be the solution of problem* (3) *and $(\mathbf{u}^*, \boldsymbol{\phi}^*)$ be the associated solutions of* (1) *and* (7). *Then, the following optimality condition holds:*

$$\alpha p^* + (\boldsymbol{\phi}^* \cdot \mathbf{n}^{\pm}) = 0. \tag{8}$$

This result can be proved using a classical sensitivity analysis technique. The important point here is that it gives a way to compute the gradient of the function J: for a given $p \in L^2(\Gamma_{\mathrm{C}})$, compute $(\mathbf{u}, \boldsymbol{\phi})$ which solve (1) and (7). Then, the Gâteaux derivative of J is given in $L^2(\Gamma_{\mathrm{C}})$ by

$$J'(p) = \alpha p + (\boldsymbol{\phi} \cdot \mathbf{n}^{\pm}), \tag{9}$$

For a given pressure $p \in L^2(\Gamma_{\mathrm{C}})$, the computation of the gradient $J'(p)$ then requires to solve two systems.

Since we transformed our direct problem into system (4), we now need to adapt the inverse problem to this formulation. The cost function J defined by (2) then rewrites into

$$J(p) := \frac{1}{2} \int_{\Gamma_{\mathrm{N}}^{\pm}} (\mathbf{u}^{\pm} - \mathbf{u}_d) \mathrm{C}^{-1} (\mathbf{u}^{\pm} - \mathbf{u}_d)^{\mathsf{T}} \, \mathrm{d}\Gamma_{\mathrm{N}}^{\pm} + \frac{\alpha}{2} \|p\|^2_{\mathbf{L}^2(\Gamma_{\mathrm{C}})}. \tag{10}$$

Notice that the observed data $\mathbf{u}_d$ can be interpolated on two sub–domains $\Omega^{\pm}$ to obtain $\mathbf{u}_d^{\pm}$ corresponding to $\mathbf{u}^{\pm}$.

In view of (6), denoting R the reduction matrix $R : \mathbf{X} \to \mathbf{U}$ and

$$\mathbf{U} = \begin{pmatrix} \mathbf{u}^+ \\ \mathbf{u}^- \end{pmatrix}, \quad \mathbf{U}_d = \begin{pmatrix} \mathbf{u}_d^+ \\ \mathbf{u}_d^- \end{pmatrix},$$

the discrete cost function is defined as

$$J_d(\mathbf{p}) = \frac{1}{2} (R\mathbf{X} - \mathbf{U}_d)^{\mathsf{T}} \mathrm{C}^{-1} M_N (R\mathbf{X} - \mathbf{U}_d) + \frac{\alpha}{2} (\mathbf{p}^{\mathsf{T}} M_F \mathbf{p}), \tag{11}$$

where $\mathbf{X}$ is the solution of (6), M_N and M_F are the mass matrices on Γ_N and Γ_C, respectively. This finite dimensional problem then boils down to finding the saddle point of the following Lagrangian

$$\mathcal{L}(\mathbf{X}, \mathbf{p}, \mathbf{\Phi}) = J_d(\mathbf{p}) - \langle \mathrm{K}\mathbf{X} - (L_\Omega \mathbf{f} + L_c \mathbf{p}), \mathbf{\Phi} \rangle.$$

Computing the KKT conditions for this problem allows to compute the gradient of J_d : for a given vector $\mathbf{p}$, let $\mathbf{X}$ be the solution of (6) and $\mathbf{\Phi}$ be the solution of the adjoint problem

$$\mathrm{K}^{\mathsf{T}} \mathbf{\Phi} = \mathrm{C}^{-1} M_N (R\mathbf{X} - \mathbf{U}_d). \tag{12}$$

Then, we have

$$\nabla J_d(\mathbf{p}) = \alpha M_F \mathbf{p} + L_c^{\mathsf{T}} \mathbf{\Phi}, \tag{13}$$

The system (12) and the gradient (13) are the discrete counterparts of (7) and (9). As (6), the adjoint system (12) is solved by a Uzawa conjugate gradient/domain decomposition method.

Computational aspects: the problem studied here is actually of quadratic type. Hence it is natural to use a suitable minimization technique, namely a conjugate gradient algorithm. It is important to notice that, using the underlying quadratic form, one can determine the optimal step size. Therefore no line search algorithm is necessary, which consequently reduces the computational cost.

4 Numerical experiments

Aiming at practical applications, we applied the technique to a realistic volcano, the Piton de la Fournaise, Île de la Réunion, France. The mesh was built from a digital elevation model (DEM), provided by the french institute IGN (Institut Géographique National, French National Geographic Institute). Both the boundary and volume mesh for the whole domain were generated by Gmsh software (Figure 2, left). The crack geometry is assumed to be quadrangular and intersecting the surface. It is constructed following [2] (see Figure 2, left). The crack mesh does not match the volume mesh. Moreover, it can be easily extended in order to split the domain. We assume that the crack is submitted to an initial pressure $\mathbf{p}^0$. The inverse problem will consist in determining the unknown pressure from the surface displacements (Figure 2, right). The convergence curves in Figure 3, highlight the efficiency of adapted optimization algorithm. The conjugate gradient minimization performs efficiently, even for fine meshes.

5 Conclusion

We have studied a conjugate gradient type method for an interface pressure inverse problem using a Uzawa conjugate gradient domain decomposition method (from [1]) as inner solver. Further study is underway to derive a single-loop conjugate gradient domain decomposition method by (directly) considering the constrained minimization problem (1)-(2) and using sensitivity and adjoint systems techniques.

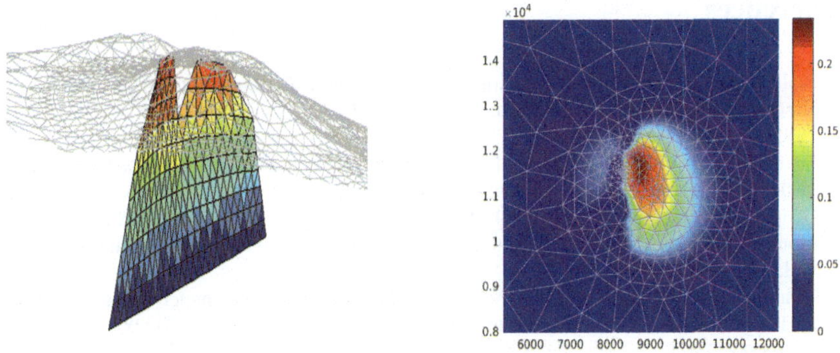

Fig. 2: Triangular surface mesh [2] representing the crack (left), amplitude of the displacement of a realistic volcano (right).

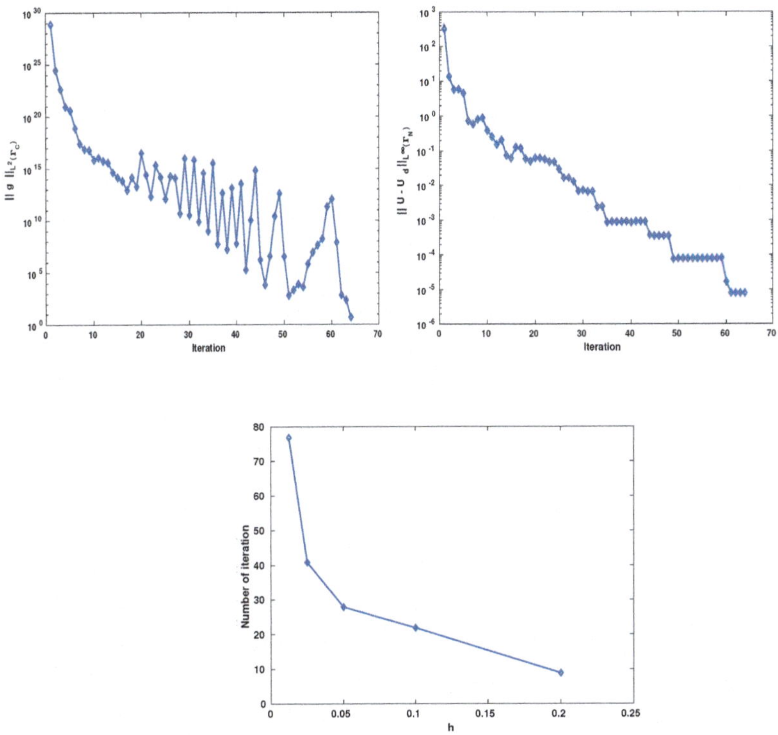

Fig. 3: Decay of the norm of gradient (left), the error of displacement on the ground in each iteration (right) and number of iteration to the converged **p** after each refinement of the mesh (bottom)

References

1. Bodart, O., Cayol, V., Court, S., Koko, J.: XFEM-based fictitious domain method for linear elasticity model with crack. SIAM J. Sci. Comput. **38**(2), B219–B246 (2016). DOI:10.1137/15M1008385
2. Fukushima, Y., Cayol, V., Durand, P.: Finding realistic dike models from interferometric synthetic aperture radar data: The february 2000 eruption at piton de la fournaise. Journal of Geophysical Research: Solid Earth **110**(B3) (2005). DOI:10.1029/2004JB003268
3. Lions, J.L.: Optimal control of systems governed by partial differential equations. Springer-Verlag, New York-Berlin (1971)
4. Tarantola, A.: Inverse problem theory and methods for model parameter estimation. Society for Industrial and Applied Mathematics (SIAM), Philadelphia, PA (2005). DOI:10.1137/1.9780898717921
5. Wauthier, C., Cayol, V., Kervyn, F., d'Oreye, N.: Magma sources involved in the 2002 nyiragongo eruption, as inferred from an insar analysis. Journal of Geophysical Research: Solid Earth **117**(B5) (2012). DOI:10.1029/2011JB008257

A Schwarz Method for the Magnetotelluric Approximation of Maxwell's Equations

Fabrizio Donzelli, Martin J. Gander, and Ronald D. Haynes

1 Introduction

Maxwell's equations can be used to model the propagation of electro-magnetic waves in the subsurface of the Earth. The interaction of such waves with the material in the subsurface produces response waves, which carry information about the physical properties of the Earth's subsurface, and their measurement allows geophysicists to detect the presence of mineral or oil deposits. Since such deposits are often found to be invariant with respect to one direction parallel to the Earth's surface, the model can be reduced to a two dimensional complex partial differential equation. Following [20], the magnetotelluric approximation is derived from the full 3D Maxwell's equations,

$$\frac{\partial B}{\partial t} + \nabla \times E = 0, \quad -\frac{\partial D}{\partial t} + \nabla \times H = J, \tag{1}$$

in the quasi-static (i.e. long wavelength, low frequency) regime, which implies that $\frac{\partial D}{\partial t}$ in (1) is neglected. Assuming a time dependence of the form $e^{i\omega t}$, where ω is the pulsation of the wave, using Ohm's law, $J = \sigma E + J^e$, where J^e denotes some exterior current source, and the constitutive relation $B = \mu H$ where μ is the permeability of free space, we obtain

$$\nabla \times E = -i\omega\mu H, \quad \nabla \times H = \sigma E + J^e. \tag{2}$$

Assuming the plane-wave source of magnetotellurics, and a two-dimensional Earth structure such that $\sigma = \sigma(x, z)$, the electric and magnetic fields can be decomposed

Fabrizio Donzelli
Memorial University of Newfoundland, Canada, e-mail: fdonzelli@mun.ca

Martin J. Gander
University of Geneva, Switzerland, e-mail: martin.gander@unige.ch

Ronald D. Haynes
Memorial University of Newfoundland, Canada, e-mail: rhaynes@mun.ca

R. Haynes et al. (eds.), *Domain Decomposition Methods in Science and Engineering XXV*, Lecture Notes in Computational Science and Engineering 138,
https://doi.org/10.1007/978-3-030-56750-7_48

into two independent modes. For the TM-, or H-polarization, mode, we have $E = (E_x, 0, E_z)$ and $H = (0, H_y, 0)$. Hence the first vector valued equation in (2) becomes a scalar equation,

$$\frac{\partial E_x}{\partial z} - \frac{\partial E_z}{\partial x} = -i\omega\mu H_y, \tag{3}$$

and the second vector valued equation in (2) gives two scalar equations,

$$-\frac{\partial H_y}{\partial z} = \sigma E_x + J_x^e \quad \text{or} \quad E_x = -\frac{1}{\sigma}\frac{\partial H_y}{\partial z} - \frac{J_x^e}{\sigma}, \tag{4}$$

and

$$\frac{\partial H_y}{\partial x} = \sigma E_z + J_z^e \quad \text{or} \quad E_z = \frac{1}{\sigma}\frac{\partial H_y}{\partial x} - \frac{J_z^e}{\sigma}. \tag{5}$$

Substituting (4) and (5) into (3) thus leads to a scalar equation for H_y,

$$-\frac{\partial}{\partial z}\left\{\frac{1}{\sigma}\frac{\partial H_y}{\partial z}\right\} - \frac{\partial}{\partial x}\left\{\frac{1}{\sigma}\frac{\partial H_y}{\partial x}\right\} + i\omega\mu H_y = \frac{\partial}{\partial z}\left\{\frac{J_x^e}{\sigma}\right\} - \frac{\partial}{\partial x}\left\{\frac{J_z^e}{\sigma}\right\}. \tag{6}$$

In geophysical applications the coefficient of conductivity σ is in general a non-constant, piece-wise continuous function. We will assume, however, for simplicity that $\sigma \equiv 1$. If we then set $u := H_y$, assume homogeneous Dirichlet boundary conditions and let $f := -\frac{\partial}{\partial z}\left\{\frac{J_x^e}{\sigma}\right\} + \frac{\partial}{\partial x}\left\{\frac{J_z^e}{\sigma}\right\}$, we obtain the magnetotelluric approximation of the Maxwell equations (cf. equation (2.86) in [20])

$$\Delta u - i\omega u = f \quad \text{in } \Omega\,, u = 0 \quad \text{on } \partial\Omega\,. \tag{7}$$

We further assume for simplicity that Ω is a domain with smooth boundary and that $f \in C^\infty(\Omega) \cap C(\overline{\Omega})$. The pulsation ω is assumed to be real and non-zero. Note that the solution u of equation (7) could also represent a component of the electric field if the model had been derived in an analogous fashion from the TE mode.

We are interested in solving the magnetotelluric approximation (7) using Schwarz methods. The alternating Schwarz method, introduced by H.A. Schwarz in 1869 [18] to prove existence and uniqueness of solutions to Laplace's equation on irregular domains, is the foundational idea of the field of domain decomposition, and has inspired work in both theoretical aspects and applications to all fields of science and engineering, see [9, 4] and references therein for more information about the historical context. Lions [16, 17] reconsidered the problem of the convergence of the method for the Poisson equation on more general configurations of overlapping subdomains. In his second paper [17], he followed the idea of Schwarz and proved convergence of the alternating Schwarz method using the maximum principle for harmonic functions. He also introduced a parallel variant of the Schwarz method, where all subdomain problems are solved simultaneously. Schwarz methods have also been introduced and studied for the original Maxwell equations (1), see [2, 1, 6, 5, 7, 8], in regimes where the maximum principle can not be used to prove convergence. We show here that for the magnetotelluric approximation of Maxwell equations in (7),

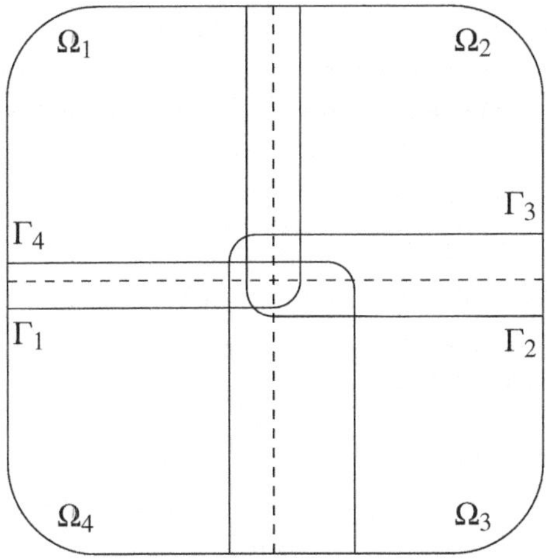

Fig. 1: Strongly overlapping subdomain decomposition obtained by enlarging a non-overlapping decomposition, indicated by the dashed lines, by a layer of strictly positive width

which also has complex solutions like the original Maxwell equations, the convergence of the parallel Schwarz method can be proved using a maximum modulus principle satisfied by complex solutions of (7).

2 Well-Posedness, Schwarz Method and Convergence

We start by establishing the well-posedness of the magnetotelluric approximation of the Maxwell equations in (7).

Theorem 1 *Let $\Omega \subset \mathbb{R}^2$ be a bounded domain with smooth boundary. Assume that $f \in L^2(\Omega)$ and ω is a non-zero constant. Then the boundary value problem* (7) *has a unique solution $u \in H_0^1(\Omega)$, depending continuously on f.*

Proof This result follows from a standard application of the Riesz Representation Theorem and the Lax-Milgram Lemma. □

We now decompose the domain $\Omega \subset \mathbb{R}^2$ first into non-overlapping subdomains, and then enlarge each subdomain by a layer of positive width to obtain the overlapping subdomains Ω_j, for $j = 1, \ldots, J$, leading to a strongly overlapping subdomain decomposition of Ω. An example is shown in Figure 1, where thc non-overlapping decomposition is indicated by the dashed lines, see also [10]. For such strongly overlapping decompositions, one can define a smooth partition of unity $\{\chi_j\}_{j=1}^J$ subordinated to the open covering $\{\Omega_j\}_{j=1}^J$, such that the support of χ_j is a set K_j contained in the open subdomain Ω_j for each $j = 1, 2, \ldots, J$, see [19, Theorem 15, Chapter 2]. The assumption of a strongly overlapping decomposition is not strictly necessary to use maximum principle arguments, see for example [11, 12], which contain even accurate convergence estimates, but we make it here since it simplifies the application of the maximum modulus principle (via Corollary 1) for studying Schwarz methods for equations with complex valued solutions. For each Ω_j, we denote by Γ_j the portion of $\partial\Omega_j$ in the interior of Ω.

The parallel Schwarz method for such a multi-subdomain decomposition starts with a global initial guess for the solution of (7), $u^0_{\text{glob}} \in C^2(\Omega) \cap C^0(\overline{\Omega})$ (less regularity would also be possible, because of the regularization provided by the equation). If at step n of the parallel Schwarz method the global approximation u^n_{glob} has been constructed, and $u^n_{\text{glob}} \in C^2(\Omega) \cap C^0(\overline{\Omega})$, then the iteration produces the next global approximation by solving, for $j = 1, \ldots, J$, the Dirichlet problems

$$\begin{aligned} \Delta u_j^{n+1} - i\omega u_j^{n+1} &= f && \text{in } \Omega_j, \\ u_j^{n+1} &= 0 && \text{on } \overline{\Omega}_j \cap \partial\Omega, \\ u_j^{n+1} &= u^n_{\text{glob}} && \text{on } \Gamma_j, \end{aligned} \tag{8}$$

and then defining the $(n+1)^{th}$ global iterate by using the partition of unity,

$$u^{n+1}_{\text{glob}} = \sum_{j=1}^{J} \chi_j u_j^{n+1} . \tag{9}$$

Since the initial guess u^0_{glob} is smooth, by induction it follows that $u^{n+1}_{\text{glob}} \in C^2(\Omega) \cap C^0(\overline{\Omega})$. This fact allows us to use the classical (i.e. non-variational) formulation of the maximum modulus principle.

Definition 1 A real valued function v of class $C^2(\Omega)$ is said to be subharmonic if $\Delta v(x) \geq 0$, $\forall x \in \Omega$, and strictly subharmonic if $\Delta v(x) > 0$, $\forall x \in \Omega$.

Note that the above definition is not the most general one, but it is suitable for the purposes of our paper. The property that we will use to prove the convergence of the parallel Schwarz method is the well-known maximum principle, which is the content of the next theorem (see [13], Theorem J-7).

Theorem 2 *Let* $v \in C^2(\Omega) \cap C(\overline{\Omega})$ *be a non-constant subharmonic function. Let* $O \subset \Omega$ *be a proper open subset. Then* v *satisfies the strong maximum principle, namely* $\max_O v < \max_{\partial\Omega} v$.

The following corollary contains the key estimate for proving the convergence of the parallel Schwarz method.

Corollary 1 *Let* K *be a closed subset of* Ω. *Then there exists a constant* $\gamma \in [0, 1)$ *such that* $\max_K u < \gamma \max_{\partial\Omega} u$, *for all non-constant subharmonic functions* $u \in C^2(\Omega) \cap C^0(\overline{\Omega})$.

Proof The result follows as an application of a Lemma originally stated by Schwarz (see [15], pp. 632-635). □

Since the solution of the magnetotelluric approximation (7) of Maxwell's equation has complex valued solutions, it is not directly possible to use the maximum principle result in Corollary 1 for proving convergence of the associated Schwarz method (8)-(9). The key additional ingredient is to prove the following property on the modulus of solutions of the magnetotelluric approximation:

Theorem 3 *Let $u \in C^2(\Omega) \cap C(\overline{\Omega})$ be a non-zero solution of the homogeneous form of equation* (7). *Then $|u|^2$ is a non-constant subharmonic function.*

Proof Taking the complex conjugate of the partial differential equation (7) with $f = 0$, gives a pair of equations, $\Delta u - i\omega u = 0$ and $\Delta \overline{u} + i\omega \overline{u} = 0$. Hence we can compute

$$\begin{aligned}\Delta |u|^2 = \Delta(u\overline{u}) = \nabla(\overline{u}\nabla u + u\nabla\overline{u}) = \nabla\overline{u}\nabla u + \overline{u}\Delta u + \nabla u \nabla \overline{u} + u\Delta\overline{u} = \\ = 2|\nabla u|^2 + i\omega|u|^2 - i\omega|u|^2 = 2|\nabla u|^2 \geq 0\,.\end{aligned}$$

Therefore, $|u|^2$ is subharmonic. If $|u|^2$ is constant, the same calculations show that $\nabla u \equiv 0$, which implies that u is a constant solution, hence it must be identically equal to zero, since the equation $\Delta u - i\omega u = 0$ has no constant non-zero solutions.□

We now prove the convergence of the parallel Schwarz method for the magnetotelluric approximation of Maxwell's equation in the infinity norm, which we denote by $||\cdot||_S$ for any function on a subdomain S.

Theorem 4 *The parallel Schwarz method* (8)-(9) *for the magnetotelluric approximation* (7) *of Maxwell's equations is convergent and satisfies the error estimate*

$$\max_{j=1,\ldots,J} ||u - u_j^n||_{\Omega_j} \leq \gamma^n \max_{j=1,\ldots,J} ||u - u_j^0||_{\Omega_j}, \tag{10}$$

where u denotes the global solution of problem (7) *and u_j^n the approximations from the parallel Schwarz method* (8)-(9), *and the constant $\gamma < 1$ comes from Corollary 1.*

Proof For $j = 1, \ldots, J$, let $K_j \subset \Omega_j$ be the support of the partition of unity function χ_j, and let $e_j^n := u - u_j^n$ be the error. Then e_j^n is solution of the homogeneous equation $\Delta e_j^n - i\omega e_j^n = 0$, and hence by Theorem 3 its modulus is a subharmonic function, and thus by Theorem 2, the modulus of the error $|e_j^n|$ satisfies the strong maximum principle. We can then estimate on each subdomain Ω_j

$$\begin{aligned}||e_j^{n+1}||_{\Omega_j} &= ||e_j^{n+1}||_{\Gamma_j} = ||\sum_{j'=1}^{J} \chi_{j'} e_{j'}^n||_{\Gamma_j} \\ &\leq \max_{j'=1,\ldots,J} ||e_{j'}^n||_{K_{j'}} \leq \gamma \max_{j'=1,\ldots,J} ||e_{j'}^n||_{\Gamma_{j'}} = \gamma \max_{j'=1,\ldots,J} ||e_{j'}^n||_{\Omega_{j'}},\end{aligned}$$

where $\gamma \in [0, 1)$ is the maximum of the factor introduced in Corollary 1 over all Ω_j and corresponding K_j. Since this holds for all j, we can take the maximum on the left and obtain

$$\max_{j=1,\ldots,J} ||e_j^{n+1}||_{\Omega_j} \leq \gamma \max_{j'=1,\ldots,J} ||e_{j'}^n||_{\Omega_{j'}},$$

which proves by induction (10). □

Remark 1 The convergence factor $\gamma < 1$ is not quantified in Theorem 4, since Corollary 1 does not provide a method to estimate the constant γ in the generality of the decomposition we used, but such an estimate is possible for specific decompositions, see for example [11, 12].

Remark 2 In [17], Lions proved the convergence of the classical Schwarz method for the Poisson equation with Dirichlet boundary conditions using a method that does not use the maximum principle. His remarkable proof is based on the method of orthogonal projections, and relies on the fact that the bilinear form associated with the weak formulation of the Poisson equation is an inner product in the solution space $H_0^1(\Omega)$. We do not see how this method can be extended to prove convergence of the classical Schwarz method applied to the magnetotelluric approximation of the Maxwell equations. In our case, the bilinear form associated to the weak formulation of (7) of the global problem is not an inner product, as it fails to be symmetric and positive-definite.

3 Numerical examples

We now present two numerical experiments. The simulations are computed on a domain Ω that consists of two squares Ω_1 and Ω_2, each of unit size 1×1. The discretization for each square consists of a uniform grid of 30×30 points. The overlap is along a vertical strip whose width is specified by the number of grid points, denoted by d.

We first compute the error $e_j^n := u - u_j^n$, as used in the proof of Theorem 4. In Figure 2 we show, from left to right, the modulus of the error on the left subdomain for iteration $n = 1$, $n = 5$ and $n = 15$, for an overlap of $d = 6$ horizontal grid points. We chose $\omega = 1$, and the initial error was produced by generating random values uniformly distributed on the range $[0, 1]$. Note how the modulus of the error clearly satisfies the maximum principle. In Figure 3, we plot the dependence of the interface residual (in the 2-norm) on the iteration number, for three different overlap sizes $d = 2, 4, 6$. As expected, the performance of the algorithm improves as we increase the size of the overlap, since increasing the overlap improves the constant γ in Corollary 1 which is the key quantity governing the convergence of the parallel Schwarz method.

4 Conclusion

We showed in this paper that even though the solutions of the magnetotelluric approximation of Maxwell's equations are complex valued, maximum principle arguments can be used to prove convergence of a parallel Schwarz method. The main new ingredient is a maximum modulus principle which is satisfied by the solutions of the magnetotelluric approximation. In a forthcoming paper, we will analyze the

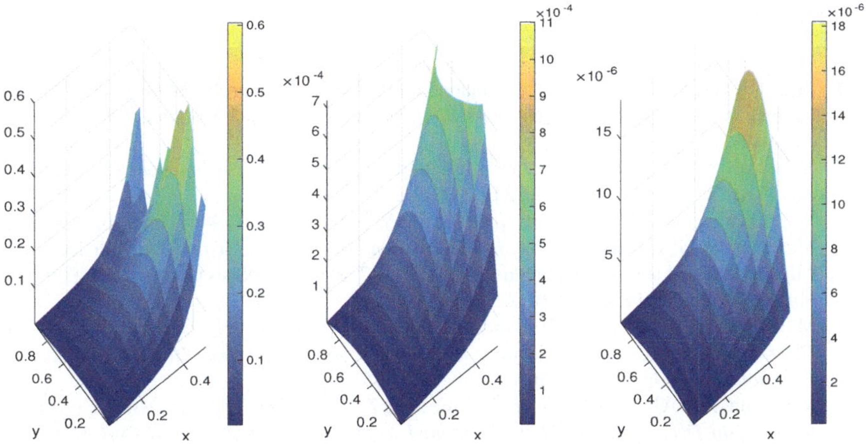

Fig. 2: Modulus of the error $e_1^n := u - u_1^n$ for $n = 1, 5, 15$ on the left subdomain when using the parallel Schwarz method for solving $\Delta u - i\omega u = 0$. Note how the modulus satisfies the maximum principle.

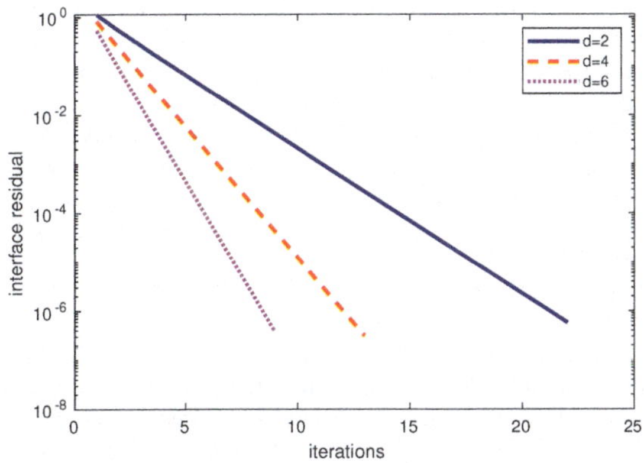

Fig. 3: Decay of the interface residual (in the 2-norm) as a function of the iteration number when using the parallel Schwarz method for solving $\Delta u - i\omega u = 0$ using different overlap sizes (d denotes the number of grid points in the overlap).

convergence rate of the parallel Schwarz method via Fourier analysis, and we will also introduce more efficient transmission conditions of Robin (or higher-order) type at the interfaces between the subdomains, which leads to optimized Schwarz methods, see [14, 3] and references therein.

Acknowledgements We would like to thank Dr. Hormoz Jandahari and Dr. Colin Farquharson for the insightful explanation of the physics behind the model discussed in this manuscript. The research of the first author was funded, in part, by the Canada Research Chairs program, the IgniteR&D program of the Research and Development Corporation of Newfoundland and Labrador (RDC), all authors were supported by the NSERC Discovery Grant program.

References

1. M. E. Bouajaji, V. Dolean, M. J. Gander, and S. Lanteri, *Optimized Schwarz methods for the time-harmonic Maxwell equations with damping*, SIAM J. Sci. Comput., **34** (2012), pp. A2048–A2071
2. V. Dolean, M. J. Gander, and L. Gerardo-Giorda, *Optimized Schwarz methods for Maxwell's equations*, SIAM J. Sci. Comput., **31** (2009), pp. 2193–2213
3. M. J. Gander, *Optimized Schwarz Methods*, SIAM J. Numer. Anal., **44** (2006), pp. 669–731
4. M. J. Gander, G. Wanner, *The Origins of the Alternating Schwarz Method*, Domain decomposition methods in science and engineering XXI, 487-495, Lect. Notes Comput. Sci. Eng., 98, Springer, Cham, 2014. 65N55 (01A55 65-03)
5. M. El Bouajaji, V. Dolean, M. J. Gander, S. Lanteri and R. Perrussel, *Discontinuous Galerkin Discretizations of Optimized Schwarz Methods for Solving the Time-Harmonic Maxwell's Equations*, Electron. Trans. Numer. Anal., **44** (2015), pp. 572–592
6. V. Dolean, M. J. Gander, S. Lanteri, J. F. Lee and Z. Peng, *Effective Transmission Conditions for Domain Decomposition Methods applied to the Time-Harmonic Curl-Curl Maxwell's equations*, J. Comput Phys, **280** (2015), pp. 232–247
7. V. Dolean, M. J. Gander, E. Veneros, *Schwarz Methods for Second Order Maxwell Equations in 3D with Coefficient Jumps*, Domain Decomposition Methods in Science and Engineering XXII, Lect. Notes Comput. Sci. Eng., Springer-Verlag, (2016), pp. 471–479
8. V. Dolean, M. J. Gander, E. Veneros, *Asymptotic Analysis of Optimized Schwarz Methods for Maxwell's Equations with Discontinuous Coefficients*, to appear in M2AN (2018)
9. M. J. Gander, *Schwarz Methods over the Course of Time*, ETNA, **31** (2008), pp. 228–155
10. M. J. Gander, H. Zhao, *Overlapping Schwarz Waveform Relaxation for the Heat Equation in n-Dimensions*, BIT (2002), Vol 42(4), No. 4, pp. 779–795
11. G. Ciaramella, M. J. Gander, *Analysis of the parallel Schwarz method for growing chains of fixed-sized subdomains: part II*, SIAM J. Numer. Anal., **56(3)** (2018), pp. 1498–1524
12. G. Ciaramella, M. J. Gander, *Happy 25th anniversary ddm! ... but how fast can the Schwarz method solve your logo?*, Domain Decomposition Methods in Science and Engineering XXV, (2018) submitted
13. R. C. Gunning, H. Rossi, *Analytic functions of several complex variables*, Englewood Cliffs, N.J. Prentice-Hall (1965)
14. C. Japhet, *Méthode de décomposition de domaine et conditions aux limites artificielles en mécanique des fluides*, Ph.D. Thesis (Université Paris 13, 1998)
15. L. V. Kantorovich, V. I. Krylov, *Approximate methods of higher analysis*, Translated from the 3rd Russian edition by C. D. Benster Interscience Publishers, Inc., New York; P. Noordhoff Ltd., Groningen 1958 xv+681 pp. 65.00
16. P.-L. Lions, *On the Schwarz alternating method I*, in The First international symposium on domain decomposition methods for partial differential equations (Paris, France, 1988), pp. 1-42
17. P.-L. Lions, *On the Schwarz alternating method II: Stochastic interpretation and order properties*, in Domain Decomposition Methods. SIAM (Philadelphia, PA, 1989), pp. 47–70
18. A. H. Schwarz, *Über einige Abbildungsaufgaben*, J. Reine Angew. Math., **70** (1869), pp. 105–120
19. M. Spivak, *A Comprehensive Introduction to Differential Geometry*, Publish or Perish Inc (Houston, TX, 1999)
20. C. J. Weiss, The two-and three-dimensional forward problems, in ed. Chave, A. D. and Alan. G. J. *The Magnetotelluric Method: Theory and Practice*, (Cambridge University Press, 2012), pp. 303–346

Can Classical Schwarz Methods for Time-harmonic Elastic Waves Converge?

Romain Brunet, Victorita Dolean, and Martin J. Gander

1 Mathematical model

The propagation of waves in elastic media is a problem of undeniable practical importance in geophysics. In several important applications - e.g. seismic exploration or earthquake prediction - one seeks to infer unknown material properties of the earth's subsurface by sending seismic waves down and measuring the scattered field which comes back, implying the solution of inverse problems. In the process of solving the inverse problem (the so-called "full-waveform inversion") one needs to iteratively solve the forward scattering problem. In practice, each step is done by solving the appropriate wave equation using explicit time stepping. However in many applications the relevant signals are band-limited and it would be more efficient to solve in the frequency domain. For this reason we are interested here in the time-harmonic counterpart of the Navier or Navier-Cauchy equation (see [6, Chapter 5.1] or [9, Chapter 9][1]), which is a linear mathematical model for elastic waves

$$-\left(\Delta^e + \omega^2\rho\right)\mathbf{u} = \mathbf{f} \quad \text{in } \Omega, \quad \Delta^e\mathbf{u} = \mu\Delta\mathbf{u} + (\lambda+\mu)\nabla(\nabla\cdot\mathbf{u}), \tag{1}$$

where $\mathbf{u}$ is the displacement field, $\mathbf{f}$ is the source term, ρ is the density that we assume real, $\mu, \lambda \in [\mathbb{R}_+^*]^2$ are the Lamé coefficients, and ω is the time-harmonic

Romain Brunet
Department of Mathematics and Statistics, University of Strathclyde, United Kingdom, e-mail: romain.brunet@strath.ac.uk

Victorita Dolean
Université Côte d'Azur, France, Department of Mathematics and Statistics, University of Strathclyde, United Kingdom, e-mail: victorita.dolean@strath.ac.uk

Martin J. Gander
Section des Mathématiques, Université de Genève, Switzerland, e-mail: martin.gander@unige.ch

[1] For the fascinating history on how Navier discovered the equation and then rapidly turned his attention to fluid dynamics, see [3].

R. Haynes et al. (eds.), *Domain Decomposition Methods in Science and Engineering XXV*,
Lecture Notes in Computational Science and Engineering 138,
https://doi.org/10.1007/978-3-030-56750-7_49

frequency for which we are interested in the solution. An example of a discretization of this equation was presented in [10]. In our case, we assumed small deformations which lead to linear equations and consider isotropic and homogeneous materials which implies that the physical coefficients are independent of the position and the direction. Due to their indefinite nature, the Navier equations in the frequency domain (1) are notoriously difficult to solve by iterative methods, especially if the frequency ω becomes large, similar to the Helmholtz equation [7], and there are further complications as we will see. We study here if the classical Schwarz method could be a candidate for solving the time harmonic Navier equations (1) iteratively.

2 Classical Schwarz Algorithm

To understand the convergence of the classical Schwarz algorithm applied to the time harmonic Navier equations (1), we study the equations on the domain $\Omega := \mathbb{R}^2$, and decompose it into two overlapping subdomains $\Omega_1 := (-\infty, \delta) \times \mathbb{R}$ and $\Omega_2 := (0, \infty) \times \mathbb{R}$, with overlap parameter $\delta \geq 0$. The classical parallel Schwarz algorithm computes for iteration index $n = 1, 2, \ldots$

$$\begin{aligned} -\left(\Delta^e + \omega^2 \rho\right) \mathbf{u}_1^n &= \mathbf{f} && \text{in } \Omega_1, \\ \mathbf{u}_1^n &= \mathbf{u}_2^{n-1} && \text{on } x = \delta, \\ -\left(\Delta^e + \omega^2 \rho\right) \mathbf{u}_2^n &= \mathbf{f} && \text{in } \Omega_2, \\ \mathbf{u}_2^n &= \mathbf{u}_1^{n-1} && \text{on } x = 0. \end{aligned} \tag{2}$$

We now study the convergence of the classical parallel Schwarz method (2) using a Fourier transform in the y direction. We denote by $k \in \mathbb{R}$ the Fourier variable and $\hat{\mathbf{u}}(x, k)$ the Fourier transformed solution,

$$\hat{\mathbf{u}}(x, k) = \int_{-\infty}^{\infty} \mathrm{e}^{-\mathrm{i}ky}\, \mathbf{u}(x, y)\, \mathrm{d}y, \quad \mathbf{u}(x, y) = \frac{1}{2\pi} \int_{-\infty}^{\infty} \mathrm{e}^{\mathrm{i}ky}\, \hat{\mathbf{u}}(x, k)\, \mathrm{d}k. \tag{3}$$

Theorem 1 (Convergence factor of the classical Schwarz algorithm) *For a given initial guess $(\mathbf{u}_1^0 \in (L^2(\Omega_1)^2)$, $(\mathbf{u}_2^0 \in (L^2(\Omega_2)^2)$, each Fourier mode k in the classical Schwarz algorithm* (2) *converges with the corresponding convergence factor*

$$\rho_{cla}\left(k, \omega, C_p, C_s, \delta\right) = \max\{|r_+|, |r_-|\},$$

where

$$r_{\pm} = \frac{X^2}{2} + e^{-\delta(\lambda_1+\lambda_2)} \pm \frac{1}{2}\sqrt{X^2\left(X^2 + 4e^{-\delta(\lambda_1+\lambda_2)}\right)},\ X = \frac{k^2 + \lambda_1\lambda_2}{k^2 - \lambda_1\lambda_2}\left(e^{-\lambda_1\delta} - e^{-\lambda_2\delta}\right). \tag{4}$$

Here, $\lambda_{1,2} \in \mathbb{C}$ are the roots of the characteristic equation of the Fourier transformed Navier equations,

$$\lambda_1 = \sqrt{k^2 - \tfrac{\omega^2}{C_s^2}}, \quad \lambda_2 = \sqrt{k^2 - \tfrac{\omega^2}{C_p^2}}, \quad C_p = \sqrt{\tfrac{\lambda+2\mu}{\rho}}, \quad C_s = \sqrt{\tfrac{\mu}{\rho}}. \tag{5}$$

Proof By linearity it suffices to consider only the case $\mathbf{f} = 0$ and analyze convergence to the zero solution, see for example [8]. After a Fourier transform in the y direction, (1) becomes

$$\begin{cases} \left[(\lambda + 2\mu)\,\partial_x^2 + \left(\rho\omega^2 - \mu k^2\right)\right]\hat{u}_x + ik(\mu+\lambda)\partial_x\hat{u}_z = 0, \\ \left[\mu\partial_x^2 + \left(\rho\omega^2 - (\lambda+2\mu)\,k^2\right)\right]\hat{u}_z + ik\,(\mu+\lambda)\,\partial_x\hat{u}_x = 0. \end{cases} \tag{6}$$

This is a system of ordinary differential equations, whose solution is obtained by computing the roots r of its characteristic equation,

$$\begin{bmatrix} (\lambda+2\mu)r^2 + \rho\omega^2 - \mu k^2 & ik(\mu+\lambda)r \\ ik(\mu+\lambda)r & \mu r^2 + \rho\omega^2 - (\lambda+2\mu)k^2 \end{bmatrix}\begin{bmatrix} \hat{u}_x \\ \hat{u}_z \end{bmatrix} = 0. \tag{7}$$

A direct computation shows that these roots are $\pm\lambda_1$ and $\pm\lambda_2$ where $\lambda_{1,2}$ are given by (5). Therefore the general form of the solution is

$$\hat{\mathbf{u}}(x,k) = \alpha_1\mathbf{v}_+e^{\lambda_1 x} + \beta_1\mathbf{v}_-e^{-\lambda_1 x} + \alpha_2\mathbf{w}_+e^{\lambda_2 x} + \beta_2\mathbf{w}_-e^{-\lambda_2 x}, \tag{8}$$

where $\mathbf{v}_\pm$ and $\mathbf{w}_\pm$ are obtained by successively inserting these roots into (7) and computing a non-trivial solution,

$$\mathbf{v}+ = \begin{pmatrix} 1 \\ \frac{i\lambda_1}{k} \end{pmatrix}, \quad \mathbf{v}_- = \begin{pmatrix} 1 \\ -\frac{i\lambda_1}{k} \end{pmatrix}, \quad \mathbf{w}_+ = \begin{pmatrix} -\frac{i\lambda_2}{k} \\ 1 \end{pmatrix}, \quad \mathbf{w}_- = \begin{pmatrix} \frac{i\lambda_2}{k} \\ 1 \end{pmatrix}. \tag{9}$$

Because the local solutions must remain bounded and outgoing at infinity, the subdomain solutions in the Fourier transformed domain are

$$\hat{\mathbf{u}}_1(x,k) = \alpha_1\mathbf{v}_+e^{\lambda_1 x} + \alpha_2\mathbf{w}_+e^{\lambda_2 x}, \qquad \hat{\mathbf{u}}_2(x,k) = \beta_1\mathbf{v}_-e^{-\lambda_1 x} + \beta_2\mathbf{w}_-e^{-\lambda_2 x}. \tag{10}$$

The coefficients $\alpha_{1,2}$ and $\beta_{1,2}$ are then uniquely determined by the transmission conditions. Before using the iteration to determine them, we rewrite the local solutions at iteration n in the form

$$\begin{aligned} \hat{\mathbf{u}}_1^n &= \alpha_1^n\mathbf{v}_+e^{\lambda_1 x} + \alpha_2^n\mathbf{w}_+e^{\lambda_2 x} = \begin{bmatrix} e^{\lambda_1 x} & -\frac{i\lambda_2}{k}e^{\lambda_2 x} \\ \frac{i\lambda_1}{k}e^{\lambda_1 x} & e^{\lambda_2 x} \end{bmatrix}\begin{pmatrix} \alpha_1^n \\ \alpha_2^n \end{pmatrix} =: M_x\boldsymbol{\alpha}^n, \\ \hat{\mathbf{u}}_2^n &= \beta_1^n\mathbf{v}_-e^{-\lambda_1 x} + \beta_2^n\mathbf{w}_-e^{-\lambda_2 x} = \begin{bmatrix} e^{-\lambda_1 x} & \frac{i\lambda_2}{k}e^{-\lambda_2 x} \\ -\frac{i\lambda_1}{k}e^{-\lambda_1 x} & e^{-\lambda_2 x} \end{bmatrix}\begin{pmatrix} \beta_1^n \\ \beta_2^n \end{pmatrix} =: N_x\boldsymbol{\beta}^n. \end{aligned} \tag{11}$$

We then insert (11) into the interface iteration of the classical Schwarz algorithm (2),

$$M_\delta\boldsymbol{\alpha}^n = N_\delta\boldsymbol{\beta}^{n-1} \Leftrightarrow \boldsymbol{\alpha}^n = M_\delta^{-1}N_\delta\boldsymbol{\beta}^{n-1}, \quad N_0\boldsymbol{\beta}^n = M_0\boldsymbol{\alpha}^{n-1} \Leftrightarrow \boldsymbol{\beta}^n = N_0^{-1}M_0\boldsymbol{\alpha}^{n-1}.$$

This leads over a double iteration to

$$\alpha^{n+1} = (M_\delta^{-1} N_\delta N_0^{-1} M_0)\alpha^{n-1} =: R_\delta^1 \alpha^{n-1}, \quad \beta^{n+1} = (N_0^{-1} M_0 M_\delta^{-1} N_\delta)\beta^{n-1} =: R_\delta^2 \beta^{n-1},$$

where $R_\delta^{1,2}$ are the iteration matrices which are spectrally equivalent. The iteration matrix R_δ^1 is given by

$$R_\delta^1 = \begin{bmatrix} \mathrm{e}^{-\delta(\lambda_1+\lambda_2)} X_2^2 \frac{\lambda_1}{\lambda_2} + \mathrm{e}^{-2\lambda_1\delta} X_1^2 & X_1 X_2 \left(\mathrm{e}^{-2\lambda_1\delta} - \mathrm{e}^{-\delta(\lambda_1+\lambda_2)}\right) \\ X_1 X_2 \frac{\lambda_1}{\lambda_2} \left(\mathrm{e}^{-\delta(\lambda_1+\lambda_2)} - \mathrm{e}^{-2\lambda_2\delta}\right) & \mathrm{e}^{-\delta(\lambda_1+\lambda_2)} X_2^2 \frac{\lambda_1}{\lambda_2} + \mathrm{e}^{-2\lambda_2\delta} X_1^2 \end{bmatrix}, \tag{12}$$

where $X_1 = \frac{k^2+\lambda_1\lambda_2}{k^2-\lambda_1\lambda_2}$ and $X_2 = -\mathrm{i}\frac{2k\lambda_2}{k^2-\lambda_1\lambda_2}$. A direct computation then leads to the eigenvalues (r_+, r_-) of R_δ^1,

$$r_\pm = \frac{X^2}{2} + \mathrm{e}^{-\delta(\lambda_1+\lambda_2)} \pm \frac{1}{2}\sqrt{X^2\left(X^2 + 4e^{-\delta(\lambda_1+\lambda_2)}\right)}, \quad X = \frac{k^2+\lambda_1\lambda_2}{k^2-\lambda_1\lambda_2}\left(\mathrm{e}^{-\lambda_1\delta} - \mathrm{e}^{-\lambda_2\delta}\right). \tag{13}$$

The convergence factor is given by the spectral radius of the matrix $R_\delta^{1,2}$,

$$\rho_{cla}\left(k, \omega, C_p, C_s, \delta\right) = \max\{|r_+|, |r_-|\}, \tag{14}$$

which concludes the proof. □

Corollary 1 (Classical Schwarz without Overlap) *In the case without overlap, $\delta = 0$, we obtain from (4) that $r_\pm = 1$, since $\left(R_\delta^1 = \mathrm{Id}\right)$. Therefore, the classical Schwarz algorithm is not convergent without overlap, it just stagnates.*

The result in Corollary 1 is consistent with the general experience that Schwarz methods without overlap do not converge, but there are important exceptions, for example for hyperbolic problems [4], and also optimized Schwarz methods can converge without overlap [8]. Unfortunately also with overlap, the Schwarz method has difficulties with the time harmonic Navier equations (1):

Corollary 2 (Classical Schwarz with Overlap)

The convergence factor of the overlapping classical Schwarz method (2) with overlap δ applied to the Navier equations (1) verifies for δ small enough

$$\rho_{cla}\left(k, \omega, C_p, C_s, \delta\right) \begin{cases} = 1, \; k \in [0, \frac{\omega}{C_p}] \cup \{\frac{\omega}{C_s}\}, \\ > 1, \; k \in (\frac{\omega}{C_p}, \frac{\omega}{C_s}), \\ < 1, \; k \in (\frac{\omega}{C_s}, \infty). \end{cases}$$

It thus converges only for high frequencies, diverges for medium frequencies, and stagnates for low frequencies.

Proof We only give here the outline of the proof, the details will appear in [2], see also [1]: for the first interval $k \in [0, \frac{\omega}{C_p})$, the proof is obtained by a direct, but long and technical calculation. For the second and third interval, $k \in (\frac{\omega}{C_p}, \frac{\omega}{C_s})$ and

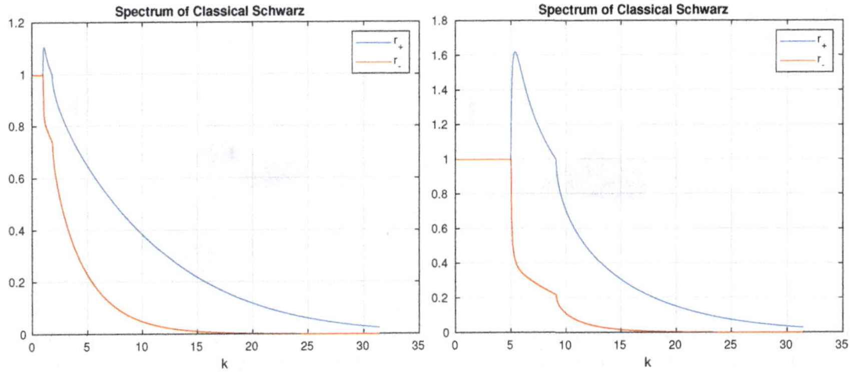

Fig. 1: Modulus of the eigenvalues of the iteration matrix as a function of Fourier frequency for the classical Schwarz method with $C_p = 1$, $C_s = 0.5$, $\rho = 1$, $\delta = 0.1$. Left: for $\omega = 1$. Right: for $\omega = 5$.

$k \in (\frac{\omega}{C_s}, \infty)$, we compute the modulus of the eigenvalues and expand them for δ small to obtain the result. At the boundary between those intervals for $k \in \{\frac{\omega}{C_p}, \frac{\omega}{C_s}\}$, the natural simplifications $\lambda_2 = 0$ or $\lambda_1 = 0$ lead directly to the result[2]. We illustrate the three zones of different convergence behavior for two examples in Figure 1.

We see from Corollary 2 that the classical Schwarz method with overlap can not be used as an iterative solver to solve the time harmonic Navier equations, since it is in general divergent on the whole interval of intermediate frequencies $(\frac{\omega}{C_p}, \frac{\omega}{C_s})$. This is even worse than for the Helmholtz or Maxwell's equations where the overlapping classical Schwarz algorithm is also convergent for high frequencies, and only stagnates for low frequencies, but is never divergent. A precise estimate of how fast the classical Schwarz method applied to the time harmonic Navier equations diverges depending on the overlap is given by the following theorem:

Theorem 2 (Asymptotic convergence factor) *The maximum of the convergence factor of the classical Schwarz method (2) applied to the Navier equations (1) behaves for small overlap δ asymptotically as*

$$\max_k(\max|r_\pm|) \sim 1 + \frac{\sqrt{2}C_s\omega\left(3C_p^2-\sqrt{C_p^4+8C_s^4}\right)\sqrt{C_p^2\sqrt{C_p^4+8C_s^4}-C_p^4-2C_s^4}}{C_p(C_p^2+C_s^2)^{\frac{3}{2}}\left(\sqrt{C_p^4+8C_s^4}-C_p^2\right)}\delta.$$

Proof According to Corollary 2, the maximum of the convergence factor is attained in the interval where the algorithm is divergent, $k \in (\frac{\omega}{C_p}, \frac{\omega}{C_s})$, and this quantity is larger than one. For a fixed k and a small overlap δ, the convergence factor

[2] The two values $k = \frac{\omega}{C_p}$ and $k = \frac{\omega}{C_s}$ correspond to points in the spectrum where the underlying Navier equations are singular, and are similar to the one resonance frequency in the Helmholtz case. They are avoided in practice either by using radiation boundary conditions on parts of the boundary of the computational domain, or by choosing domain geometries such that these frequencies are not part of the discrete spectrum of the Navier operator on the bounded domain.

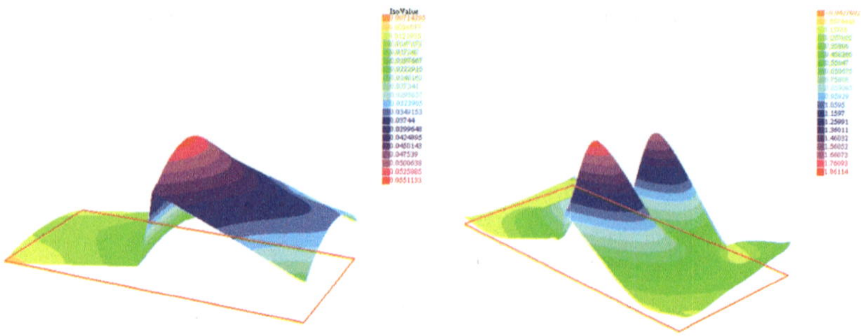

Fig. 2: Error in modulus at iteration 25 of the classical Schwarz method with 2 subdomains, where one can clearly identify the dominant mode in the error: Left: $\omega = 1$. Right: $\omega = 5$.

$\rho_{cla}(k, \omega, C_p, C_s, \delta)$ for $k \in (\frac{\omega}{C_p}, \frac{\omega}{C_s})$ is given by

$$\rho_{cla}(k, \omega, C_p, C_s, \delta) = 1 + \frac{2\omega^2 \lambda_2 \bar{\lambda}_1^2}{C_p^2 (k^4 + \bar{\lambda}_1^2 \lambda_2^2)} \delta + O\left(\delta^2\right) \in \mathbb{R}_+^*. \tag{15}$$

We denote by $C(k)$ the coefficient in front of δ. In order to compute the maximum of (15) we solve the equation $\frac{dC(k)}{dk} = 0$. We denote by k_s the only positive critical point for which $\frac{d^2C(k)}{dk^2}(k_s) < 0$. By replacing k_s into the expression of $C(k)$ we get the desired result, see [1] for more details. □

3 Numerical experiments

We illustrate now the divergence of the classical Schwarz algorithm with a numerical experiment. We choose the same parameters $C_p = 1$, $C_s = 0.5$, $\rho = 1$ and overlap $\delta = 0.1$ as in Figure 1. We discretize the time-harmonic Navier equations using $P1$ finite elements on the domain $\Omega = (-1, 1) \times (0, 1)$ and impose absorbing boundary conditions on $\partial\Omega$. We decompose the domain into two overlapping subdomains $\Omega_1 = (-1, 2h) \times (0, 1)$ and $\Omega_2 = (-2h, 1) \times (0, 1)$ with $h = \frac{1}{40}$, such that the overlap $\delta = 0.1 = 4h$. Our computations are performed with the open source software Freefem++. We show in Figure 2 the error in modulus at iteration 25 of the classical Schwarz method, on the left for $\omega = 1$ and on the right for $\omega = 5$. In the first case, $\omega = 1$, we observe very slow convergence, the error decreases from $7.89e-1$ to $5e-2$ after 25 iterations. This can be understood as follows: the lowest frequency along the interface on our domain Ω is $k = \pi$, which lies outside the interval $[\frac{\omega}{C_p}, \frac{\omega}{C_s}] = [1, 2]$ of frequencies on which the method is divergent. The method thus converges, all frequencies lie in the convergent zone in the plot in Figure 1 on the left where $\rho_{cla} < 1$. The most slowly convergent mode is $|\sin(ky)|$ with $k = \pi$, which is clearly visible in Figure 2 on the left. This is different for $\omega = 5$, where we see in Figure 2 on the right the dominant growing mode. The interval of frequencies on

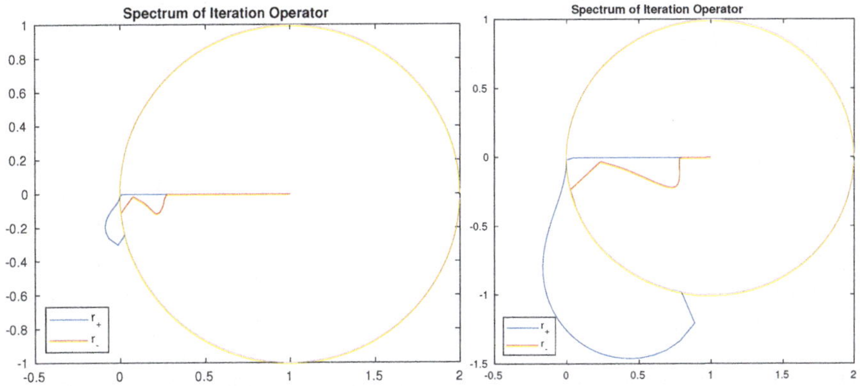

Fig. 3: Spectrum of the iteration operator for the same example as in Figure 1, together with a unit circle centered around the point $(1, 0)$. Left: $\omega = 1$. Right: $\omega = 5$

which the method is divergent is given by $[\frac{\omega}{C_p}, \frac{\omega}{C_s}] = [5, 10]$, and we clearly can identify in Figure 2 on the right a mode with two bumps along the interface, which corresponds to the mode $|\sin(ky)|$ along the interface for $k = 2\pi \approx 6$, which is the fastest diverging mode according to the analytical result shown in Figure 1 on the right.

One might wonder if the classical Schwarz method is nevertheless a good preconditioner for a Krylov method, which can happen also for divergent stationary methods, like for example the Additive Schwarz Method applied to the Laplace problem, which is also not convergent as an iterative method [5], but useful as a preconditioner. To investigate this, it suffices to plot the spectrum of the identity matrix minus the iteration operator in the complex plane, which corresponds to the preconditioned systems one would like to solve. We see in Figure 3 that the part of the spectrum that leads to a contraction factor ρ_{cla} with modulus bigger than one lies unfortunately close to zero in the complex plane, and that is where the residual polynomial of the Krylov method must equal one. Therefore we can infer that the classical Schwarz method will also not work well as a preconditioner. This is also confirmed by the numerical results shown in Figure 4, where we used first the classical Schwarz method as a solver and then as preconditioner for GMRES. We see that GMRES now makes the method converge, but convergence depends strongly on ω and slows down when ω grows.

4 Conclusion

We proved that the classical Schwarz method with overlap applied to the time harmonic Navier equations cannot be used as an iterative solver, since it is not convergent in general. This is even worse than for the Helmholtz or time harmonic Maxwell's equations, for which the classical Schwarz algorithm also stagnates for all propaga-

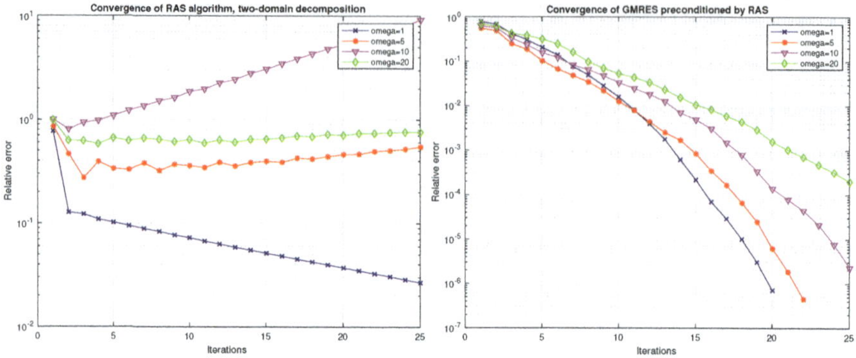

Fig. 4: Convergence history for RAS and GMRES preconditioned by RAS for different values of ω

tive modes, but at least is not divergent. We then showed that our analysis clearly identifies the problematic error modes in a numerical experiment. Using the classical Schwarz method as a preconditioner for GMRES then leads to a convergent method, which however is strongly dependent on the time-harmonic frequency parameter ω. We are currently studying better transmission conditions between subdomains, which will lead to optimized Schwarz methods for the time harmonic Navier equation.

References

1. Brunet, R.: Domain decomposition methods for time-harmonic elastic waves. Ph.D. thesis, University of Strathclyde (2018)
2. Brunet, R., Dolean, V., Gander, M.: Natural domain decomposition algorithms for the solution of time-harmonic elastic waves. submitted (2019)
3. Darrigol, O.: Between hydrodynamics and elasticity theory: the first five births of the Navier-Stokes equation. Archive for History of Exact Sciences **56**(2), 95–150 (2002)
4. Dolean, V., Gander, M.: Why classical Schwarz methods applied to certain hyperbolic systems converge even without overlap. In: Domain decomposition methods in science and engineering XVII, pp. 467–475. Springer (2008)
5. Efstathiou, E., Gander, M.: Why restricted additive Schwarz converges faster than additive Schwarz. BIT Numerical Mathematics **43**(5), 945–959 (2003)
6. Eringen, C., Suhubi, E.: Elastodynamics: Vol. 2. ACADEMIC PRESS New York San Francisco London (1977)
7. Ernst, O., Gander, M.: Why it is difficult to solve Helmholtz problems with classical iterative methods. In: Numerical analysis of multiscale problems, pp. 325–363. Springer (2012)
8. Gander, M.: Optimized Schwarz methods. SIAM Journal on Numerical Analysis **44**(2), 699–731 (2006)
9. Lautrup, B.: Physics of continuous matter: exotic and everyday phenomena in the macroscopic world. CRC press (2011)
10. T., H., P., M., F., C., J.P., K.: The ultra-weak variational formulation for elastic wave problems. SIAM Journal on Scientific Computing **25**(5), 1717–1742 (2004)

Asymptotic Analysis for the Coupling Between Subdomains in Discrete Fracture Matrix Models

Martin J. Gander, Julian Hennicker, and Roland Masson

1 Introduction

We study the behavior of solutions of PDE models on domains containing a heterogeneous layer of aperture tending to zero. We consider general second order differential operators on the outer domains and elliptic operators inside the layer. Our study is motivated by the modeling of flow through fractured porous media, when one represents the fractures as entities of co-dimension one with respect to the surrounding rock matrix. These models are called Discrete Fracture Matrix (DFM) models [2, 4, 1]. A recent study on DFM models and their discretization can be found in [3]. Our focus lies on the derivation of coupling conditions, which have to be satisfied by the traces of the solutions for the matrix domain on each side of the matrix-fracture interfaces. We emphasize that we are not only concerned with the derivation of coupling conditions that have to be fulfilled in the limit of vanishing aperture, but in particular with the derivation of coupling conditions that have to be fulfilled up to a certain order of the aperture, which in turn occurs as a model parameter. In our work flow, we first derive exact coupling conditions by means of Fourier analysis. Reduced order coupling conditions are then obtained by truncation of the exact conditions at the desired order. Our approach is very systematic and allowed us to reproduce various coupling conditions from the literature as well as assess the error of the reduced models.

Martin J. Gander
Université de Genève, 2-4 rue du Lièvre, CP 64, CH-1211 Genève, e-mail: martin.gander@unige.ch

Julian Hennicker
Université de Genève, 2-4 rue du Lièvre, CP 64, CH-1211 Genève, e-mail: julian.hennicker@unige.ch

Roland Masson
Université Côte d'Azur, CNRS, Inria team COFFEE, LJAD, France, e-mail: roland.masson@unice.fr

R. Haynes et al. (eds.), *Domain Decomposition Methods in Science and Engineering XXV*,
Lecture Notes in Computational Science and Engineering 138,
https://doi.org/10.1007/978-3-030-56750-7_50

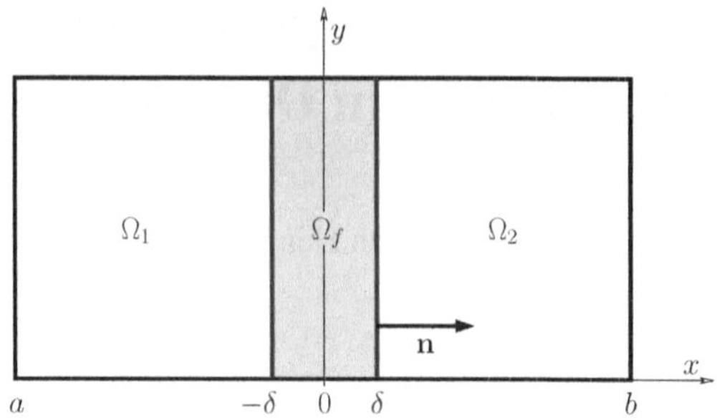

Fig. 1 Illustration of the domain under consideration. In our study, we restrict ourselves to a simple geometry, where $\Omega_1 = (a, -\delta) \times \mathbb{R}$, $\Omega_2 = (\delta, b) \times \mathbb{R}$ and $\Omega_f = (-\delta, \delta) \times \mathbb{R}$, with $a, b \in \overline{\mathbb{R}}$. $\mathbf{n}$ denotes the unit normal in x-direction.

2 Model problem

We consider the following problem on a threefold domain as illustrated in Figure 1:

$$\mathcal{L}_j(u_j, \mathbf{q}_j) = h_j \quad \text{in} \quad \Omega_j,\ j = 1, 2, f, \tag{1}$$

$$\mathbf{q}_j = \mathcal{G}_j u_j \quad \text{in} \quad \Omega_j,\ j = 1, 2, f, \tag{2}$$

$$u_j = u_f \quad \text{on} \quad \partial\Omega_j \cap \partial\Omega_f,\ j = 1, 2, \tag{3}$$

$$\mathbf{q}_j \cdot \mathbf{n} = \mathbf{q}_f \cdot \mathbf{n} \quad \text{on} \quad \partial\Omega_j \cap \partial\Omega_f,\ j = 1, 2, \tag{4}$$

where $\mathcal{L}_j, \mathcal{G}_j$ are differential operators, together with some suitable boundary conditions. Only inside the fracture domain Ω_f, we will restrict our study to the class of general elliptic models, i.e. we assume that

$$\mathcal{L}_f(u_f, \mathbf{q}_f) = -\mathrm{div}\mathbf{q}_f + \frac{\mathbf{b}}{2} \cdot \nabla u_f + (\eta - \mathrm{div}\frac{\mathbf{b}}{2})u_f \quad \text{and} \quad \mathcal{G}_f u_f = (\mathbf{A}\nabla - \frac{\mathbf{b}}{2})u_f \tag{5}$$

with $\eta \in \mathbb{R}_{\geq 0}$, $\mathbf{b} \in \mathbb{R}^2$ and coercive $\mathbf{A} \in \mathbb{R}^{2\times 2}$. For simplicity, we also assume a trivial source term inside the fracture, i.e. $h_f = 0$.

3 Derivation of the reduced models by Fourier analysis

From (1),(2),(5) the Fourier coefficients $\hat{u}_f(x, k)$ of $u_f(x, y)$ have to fulfill for all $k \in \mathbb{R}$

$$-a_{11}\partial_{xx}\hat{u}_f + \Big(b_1 - (a_{12} + a_{21})ik\Big)\partial_x\hat{u}_f + (a_{22}k^2 + b_2 ik + \eta)\hat{u}_f = 0 \quad \text{in} \quad \Omega_f. \tag{6}$$

The roots of the characteristic polynomial associated with (6) are $\lambda_{1,2} = r \pm s$, where

$$r = -\frac{1}{2a_{11}}((a_{12} + a_{21})ik - b_1) \quad \text{and} \quad s = \left(r^2 + \frac{1}{a_{11}}(a_{22}k^2 + b_2 ik + \eta)\right)^{\frac{1}{2}}.$$

The ansatz for the solution of (6),

$$\hat{u}_f(x,k) = A_f(k)e^{\lambda_1 x} + B_f(k)e^{\lambda_2 x},$$

together with (3) and (4) immediately yields for the Fourier coefficients $\hat{u}_j(x,k)$ of $u_j(x,y)$ and $\hat{\mathbf{q}}_j(x,k)$ of $\mathbf{q}_j(x,y)$, $j = 1, 2$, on the interfaces,

$$\hat{u}_1(-\delta,k) = A_f(k)e^{-\delta\lambda_1} + B_f(k)e^{-\delta\lambda_2}, \tag{7}$$

$$\hat{u}_2(\delta,k) = A_f(k)e^{\delta\lambda_1} + B_f(k)e^{\delta\lambda_2}, \tag{8}$$

$$\hat{\mathbf{q}}_1(-\delta,k)\cdot\mathbf{n} = a_{11}\lambda_1 A_f(k)e^{-\delta\lambda_1} + a_{11}\lambda_2 B_f(k)e^{-\delta\lambda_2} + (a_{12}ik - \frac{b_1}{2})\hat{u}_1(-\delta,k), \tag{9}$$

$$\hat{\mathbf{q}}_2(\delta,k)\cdot\mathbf{n} = a_{11}\lambda_1 A_f(k)e^{\delta\lambda_1} + a_{11}\lambda_2 B_f(k)e^{\delta\lambda_2} + (a_{12}ik - \frac{b_1}{2})\hat{u}_2(\delta,k). \tag{10}$$

Equations (7) and (8) are now solved for A_f and B_f, which can then be substituted into the remaining two equations (9) and (10). After some calculations, this leads to the exact coupling conditions

$$\begin{aligned} \sinh(2s\delta)\hat{\mathbf{q}}_1(-\delta)\cdot\mathbf{n} + (a_{11}s\cosh(2s\delta) + \rho\sinh(2s\delta))\hat{u}_1(-\delta) \\ = a_{11}se^{-2\delta r}\hat{u}_2(\delta), \end{aligned} \tag{11}$$

$$\begin{aligned} -\sinh(2s\delta)\hat{\mathbf{q}}_2(\delta)\cdot\mathbf{n} + (a_{11}s\cosh(2s\delta) - \rho\sinh(2s\delta))\hat{u}_2(\delta) \\ = a_{11}se^{2\delta r}\hat{u}_1(-\delta), \end{aligned} \tag{12}$$

where $\rho = \frac{a_{21}-a_{12}}{2}ik$. For the remaining part of the paper, we will drop the arguments indicating the evaluation at $x = -\delta$ for the functions living in Ω_1 and at $x = \delta$ for those living in Ω_2. Taking the sum (11) + (12) yields an expression related to the normal velocity jump across the fracture, whereas the difference (11) – (12) gives an expression related to the pressure jump accross the fracture,

$$\begin{aligned} &\sinh(2s\delta)(\hat{\mathbf{q}}_2 - \hat{\mathbf{q}}_1)\cdot\mathbf{n} \\ &= a_{11}s\Big(\cosh(2s\delta)(\hat{u}_1 + \hat{u}_2) - (e^{2\delta r}\hat{u}_1 + e^{-2\delta r}\hat{u}_2)\Big) + \rho\sinh(2s\delta)(\hat{u}_1 - \hat{u}_2), \end{aligned} \tag{13}$$

$$\begin{aligned} &a_{11}s\Big(\cosh(2s\delta)(\hat{u}_2 - \hat{u}_1) + (e^{-2\delta r}\hat{u}_2 - e^{2\delta r}\hat{u}_1)\Big) \\ &= \sinh(2s\delta)(\hat{\mathbf{q}}_1 + \hat{\mathbf{q}}_2)\cdot\mathbf{n} + \rho\sinh(2s\delta)(\hat{u}_1 + \hat{u}_2). \end{aligned} \tag{14}$$

We now expand (13), (14) into a series in δ and truncate at a given order. We then obtain the following reduced order coupling conditions at $x = \pm\delta$:

1. Truncation after the leading-order term, which we call coupling conditions of type zero (CC0 coupling conditions):

$$(\hat{\mathbf{q}}_2 - \hat{\mathbf{q}}_1)\cdot\mathbf{n} = 0 \qquad \text{and} \qquad \hat{u}_2 - \hat{u}_1 = 0.$$

2. Truncation after the next-to-leading-order term, which we call CC1 coupling conditions:

$$(\hat{\mathbf{q}}_2 - \hat{\mathbf{q}}_1) \cdot \mathbf{n} = \delta\Big(a_{22}k^2 + b_2 ik + \eta\Big)(\hat{u}_1 + \hat{u}_2) + \Big(-a_{21}ik + \frac{b_1}{2}\Big)(\hat{u}_2 - \hat{u}_1),$$
$$\delta(\hat{\mathbf{q}}_2 + \hat{\mathbf{q}}_1) \cdot \mathbf{n} = a_{11}(\hat{u}_2 - \hat{u}_1) + \delta\Big(a_{12}ik - \frac{b_1}{2}\Big)(\hat{u}_1 + \hat{u}_2).$$

Of course, we could derive higher order coupling conditions by using higher order expansions.

We now want to get back to the physical unknowns u_j and $\mathbf{q}_j$, $j = 1, 2$. To do so, we perform an inverse Fourier transform by formally applying the rules,

$$\hat{u}_j \mapsto u_j, \quad \hat{\mathbf{q}}_j \mapsto \mathbf{q}_j, \quad k^2 \mapsto -\partial_{yy}, \quad ik \mapsto \partial_y.$$

We therefore obtain as reduced order approximations of the exact coupling conditions between the subdomains Ω_1 and Ω_2

1. CC0 coupling conditions:

$$\mathbf{q}_2 \cdot \mathbf{n} - \mathbf{q}_1 \cdot \mathbf{n} = 0 \qquad \text{and} \qquad u_2 - u_1 = 0. \tag{15}$$

2. CC1 coupling conditions:

$$(\mathbf{q}_2 - \mathbf{q}_1) \cdot \mathbf{n} = \delta\Big(-a_{22}\partial_{yy} + b_2\partial_y + \eta\Big)(u_1 + u_2) + \Big(-a_{21}\partial_y + \frac{b_1}{2}\Big)(u_2 - u_1), \tag{16}$$

$$\delta(\mathbf{q}_1 + \mathbf{q}_2) \cdot \mathbf{n} = a_{11}(u_2 - u_1) + \delta\Big(a_{12}\partial_y - \frac{b_1}{2}\Big)(u_1 + u_2). \tag{17}$$

4 Comparison to the literature

DFM models are a tool for the simulation of flow through fractured porous media, where the governing equations are mass conservation and Darcy's law. The approach illustrated above covers more general problems, and in order to compare our coupling conditions to existing ones from the literature, we now let

$$\mathbf{b} := 0, \quad \eta := 0, \quad \text{and} \quad \mathbf{A} := \begin{pmatrix} a_{11} & 0 \\ 0 & a_{22} \end{pmatrix}.$$

As outlined in [4], one typically derives the reduced order coupling conditions by integrating the equations over the fracture width,

$$0 = \int_{-\delta}^{\delta} \mathrm{div}\mathbf{q}_f \,\mathrm{d}x = \mathbf{q}_f \cdot \mathbf{n}(\delta) - \mathbf{q}_f \cdot \mathbf{n}(-\delta) + \partial_y \int_{-\delta}^{\delta} \mathbf{q}_f \,\mathrm{d}x$$
$$= \mathbf{q}_2 \cdot \mathbf{n} - \mathbf{q}_1 \cdot \mathbf{n} + 2\delta a_{22}\partial_y^2 U_f, \tag{18}$$
$$\int_{-\delta}^{\delta} \mathbf{q}_f \cdot \mathbf{n}\mathrm{d}x = a_{11}(u_f(\delta) - u_f(-\delta)) = a_{11}(u_2 - u_1), \tag{19}$$

and then uses some ad-hoc approximations

$$\int_{-\delta}^{\delta} \mathbf{q}_f \cdot \mathbf{n}\mathrm{d}x \approx 2\delta \frac{\mathbf{q}_f \cdot \mathbf{n}(\delta) + \mathbf{q}_f \cdot \mathbf{n}(-\delta)}{2} = \delta(\mathbf{q}_1 \cdot \mathbf{n} + \mathbf{q}_2 \cdot \mathbf{n}), \tag{20}$$
$$u_2 + u_1 \approx 2U_f, \tag{21}$$

where $U_f := \frac{1}{2\delta}\int_{-\delta}^{\delta} u_f \,\mathrm{d}x$. Combining these equations leads to the coupling conditions

$$\delta a_{22}\partial_y^2(u_1 + u_2) + \mathbf{q}_2 \cdot \mathbf{n} - \mathbf{q}_1 \cdot \mathbf{n} = 0, \tag{22}$$
$$\delta(\mathbf{q}_2 \cdot \mathbf{n} + \mathbf{q}_1 \cdot \mathbf{n}) = a_{11}(u_2 - u_1). \tag{23}$$

Note that, by means of (21), condition (22) is equivalent to the tangential mass conservation inside the fracture together with mass exchange between the fracture and rock matrix.

Theorem 1 *The coupling conditions* (22), (23) *coincide with the coupling conditions* (16),(17) *for the diffusion equation with diagonal matrix A. Furthermore, the exact solution obeys formally* (22), (23) *with an error of order three, for* $\delta \to 0$.

Proof For the diffusion equation with diagonal matrix A, the terms in the coupling conditions (16), (17), which are related to the advection and reaction constants and the terms related to the off-diagonal entries in the diffusion matrix vanish. By direct comparison, we observe that the resulting equations coincide with the coupling conditions (22), (23), which shows the first statement of the theorem. Furthermore, for the diffusion equation with diagonal matrix A, the coupling conditions (13), after dividing by $\sinh(2s\delta)$, and (14), after dividing by $\cosh(2s\delta)$, yield

$$\mathbf{q}_1 \cdot \mathbf{n} - \mathbf{q}_2 \cdot \mathbf{n} = \left(\delta a_{22}\partial_{yy} + \frac{1}{3}\frac{\delta^3 a_{22}^2}{a_{11}}\partial_{yy}^2 + \frac{2}{15}\frac{\delta^5 a_{22}^3}{a_{11}^2}\partial_{yy}^3 + \cdots\right)(u_1 + u_2), \tag{24}$$
$$u_2 - u_1 = \left(\frac{\delta}{a_{11}} + \frac{1}{3}\frac{\delta^3 a_{22}}{a_{11}^2}\partial_{yy} + \frac{2}{15}\frac{\delta^5 a_{22}^2}{a_{11}^3}\partial_{yy}^2 + \cdots\right)(\mathbf{q}_1 + \mathbf{q}_2) \cdot \mathbf{n}. \tag{25}$$

Hence, by substitution of the exact solution into the approximate coupling conditions (22), (23), we formally obtain residuals of order three, for $\delta \to 0$, which confirms the second statement of the theorem. □

From (24),(25), we observe that the asymptotic behavior of the exact coupling conditions depends only on the asymptotic behavior of the ratio $\frac{\delta}{a_{11}}$ and of the product δa_{22}. We call these two characteristic quantities the fracture resistivity and conductivity, respectively. In [5], a rigorous asymptotic analysis for the Laplace equation is conducted, with the focus on the solution in the limit $\delta = 0$. In this context, coupling conditions (at $x = \pm 0$) are derived, for the cases $\frac{\delta}{a_{11}} \to \gamma \in \mathbb{R}$, $\frac{\delta}{a_{11}} \to \infty$, $\frac{\delta}{a_{11}} \to 0$, provided $a_{11} \to 0$, which turn out to correspond to the coupling conditions, which we derive by means of truncating (24),(25) at order δ^0 (with $a_{11} = a_{22}$ for isotropic diffusion).

1. Case $\frac{\delta}{a_{11}} \to \gamma \in \mathbb{R}$ (note that this implies $\delta a_{11} \to 0$):

$$\mathbf{q}_1 \cdot \mathbf{n} - \mathbf{q}_2 \cdot \mathbf{n} = 0 \quad \text{and} \quad u_2 - u_1 = \gamma(\mathbf{q}_1 + \mathbf{q}_2) \cdot \mathbf{n}.$$

2. Case $\frac{\delta}{a_{11}} \to \infty$ (note that this implies $\delta a_{11} \to 0$):

$$\mathbf{q}_1 \cdot \mathbf{n} - \mathbf{q}_2 \cdot \mathbf{n} = 0 \quad \text{and} \quad \mathbf{q}_1 \cdot \mathbf{n} + \mathbf{q}_2 \cdot \mathbf{n} = 0.$$

3. Case $\frac{\delta}{a_{11}} \to 0$ and $\delta a_{11} \to 0$ corresponds to (15).

We can now complete this study by considering the cases $\delta a_{11} \to \gamma \in \mathbb{R}$ or $\delta a_{11} \to \infty$ (which both imply $\frac{\delta}{a_{11}} \to 0$). We obtain

4. Case $\delta a_{11} \to \gamma \in \mathbb{R}$:

$$\mathbf{q}_1 \cdot \mathbf{n} - \mathbf{q}_2 \cdot \mathbf{n} = \gamma \partial_{yy}(u_1 + u_2) \quad \text{and} \quad u_2 - u_1 = 0.$$

5. Case $\delta a_{11} \to \infty$:

$$\partial_{yy}(u_1 + u_2) = 0 \quad \text{and} \quad u_2 - u_1 = 0.$$

5 Numerical results

We present here a series of test cases for isotropic diffusion in all of the three domains $\Omega_1 = (-10, -\delta) \times (-10, 10)$, $\Omega_2 = (\delta, 10) \times (-10, 10)$ and $\Omega_f = (-\delta, \delta) \times (-10, 10)$. The diffusion coefficients are ν in Ω_f and 1 in the domains Ω_1, Ω_2. This means that we consider the model solved on the full domain, which consists of the Laplace equation $\Delta u_j = 0$ in Ω_j, $j = 1, 2, f$, together with the coupling conditions

$$\begin{aligned} u_1(-\delta) &= u_f(-\delta) \quad \text{and} \quad u_2(\delta) = u_f(\delta), \\ \partial_x u_1(-\delta) &= \nu \partial_x u_f(-\delta) \quad \text{and} \quad \partial_x u_2(\delta) = \nu \partial_x u_f(\delta), \end{aligned}$$

and compare the solution to those obtained by the reduced models, which consist of the Laplace equation $\Delta u_j = 0$ in Ω_j, $j = 1, 2$, together with either leading order (CC0) coupling conditions,

$$u_1(-\delta) = u_2(\delta) \quad \text{and} \quad \partial_x u_1(-\delta) = \partial_x u_2(\delta),$$

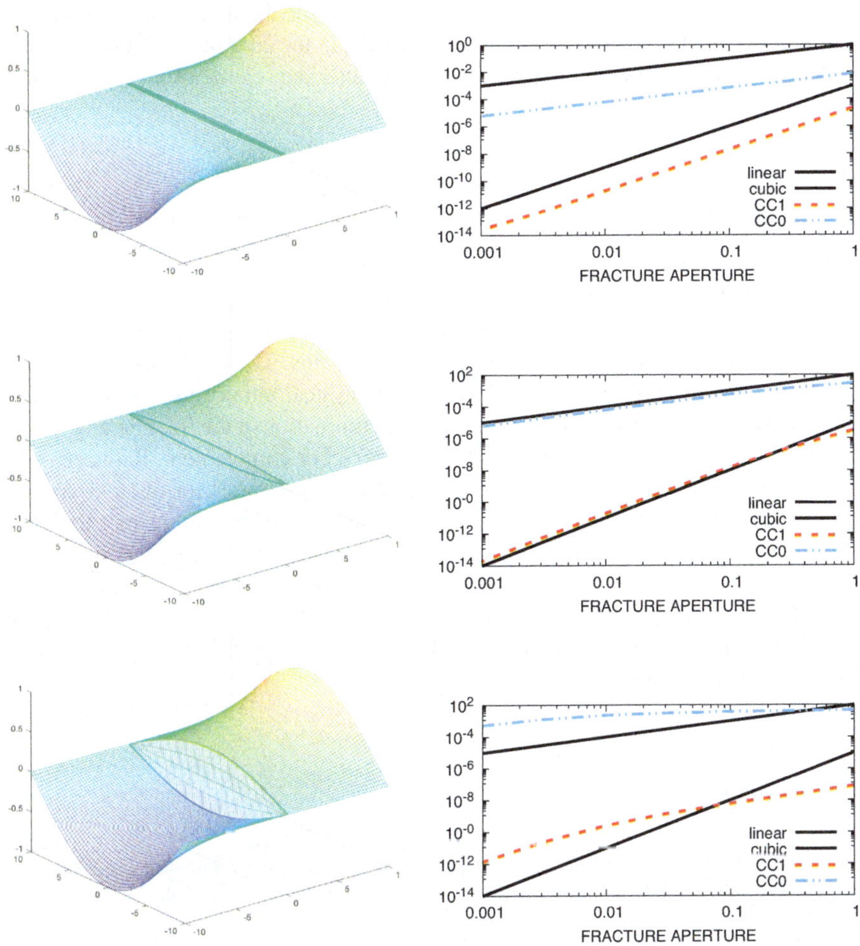

Fig. 2: The reference solution and the L^{∞}-error for the solutions of the reduced models for $\nu = 10$, $\nu = 0.1$ and $\nu = 0.001$ (from top to bottom). The error is plotted for CC0 and for CC1 coupling conditions.

or coupling conditions containing next-to-leading-order corrections (CC1),

$$\partial_x u_1(-\delta) - \partial_x u_2(\delta) = \delta\nu\partial_{yy}(u_1(-\delta) + u_2(\delta)),$$
$$u_2(\delta) - u_1(-\delta) = \delta\nu^{-1}(\partial_x u_2(\delta) + \partial_x u_1(-\delta)),$$

which have been shown to have an error of $O(\delta^3)$ compared to the exact solution, for diffusion problems with diagonal matrix A. We use homogeneous Dirichlet boundary conditions at $y = \pm 10$ and non-homogeneous Dirichlet boundary conditions with values $\pm\cos(\pi y/20)$ at $x = \pm 10$. From Figure 2, we observe an increase of the pressure jump across the fracture, when increasing the fracture resistivity, as encoded in the coupling conditions. From the error plots, we see that the theoretical order of convergence is reproduced, although we note that, in the case of $\nu = 0.001$, we need to decrease the fracture width quite severely to enter the regime of theoretical order of convergence.

6 Conclusion

We presented a rigorous derivation of coupling conditions for DFM models of very general type, i.e. advection-diffusion-reaction in the fracture and even more general second order PDEs in the surrounding matrix domains. The derivation of coupling conditions relies on a Fourier transform of the physical unknowns in direction tangential to the fracture and, subsequently, on the elimination of the fracture unknowns' Fourier coefficients by performing a continuous Schur complement. Reduced order coupling conditions are then obtained by straightforward truncation of an expansion. We compared the coupling conditions to a commonly used family of (diffusion) models from the literature and obtained correspondence for the coupling conditions truncated after the next-to-leading-order terms. We further derived coupling conditions for the fracture resistivity tending to a constant, to infinity and to zero, and found correspondence to the literature, which contains results for the special case of the Laplace equation only.

References

1. P. Angot, F. Boyer, and F. Hubert. Asymptotic and numerical modelling of flows in fractured porous media. *ESAIM: Mathematical Modelling and Numerical Analysis*, 43(2):239–275, mar 2009.
2. E. Flauraud, F. Nataf, I. Faille, and R. Masson. Domain decomposition for an asymptotic geological fault modeling. *Comptes Rendus Mécanique*, 331(12):849–855, dec 2003.
3. J. Hennicker. *Hybrid dimensional modeling of multi-phase Darcy flows in fractured porous media*. PhD thesis, Université Côte d'Azur, 2017.
4. V. Martin, J. Jaffré, and J. E. Roberts. Modeling fractures and barriers as interfaces for flow in porous media. *SIAM Journal on Scientific Computing*, 26(5):1667–1691, 2005.
5. E. Sánchez-Palencia. Problèmes de perturbations liés aux phénomènes de conduction à travers des couches minces de grande résistivité. *J. Math. Pures et Appl.*, 53(9):251–269, 1974.

A Nonlinear Elimination Preconditioned Inexact Newton Algorithm for Steady State Incompressible Flow Problems on 3D Unstructured Meshes

Li Luo, Rongliang Chen, Xiao-Chuan Cai, and David E. Keyes

1 Introduction

The Newton algorithm and its variants are frequently used to obtain the numerical solution of large nonlinear systems arising from the discretization of partial differential equations, e.g., the incompressible Navier-Stokes equations in computational fluid dynamics. Near quadratic convergence can be observed when the nonlinearities in the system are well-balanced. However, if some of the equations have stronger nonlinearities than the others in the system, a Newton-like algorithm may suffer from slow convergence in the form of a long stagnation in the residual history, or not converge at all.

Nonlinear preconditioning aims to tackle this problem by creating an inner iteration with improved balance, which can be thought of as making the residual contours more spherical (i.e., hypersphericity in high dimension). Nonlinear preconditioners require solving nonlinear subproblems in inner iterations to remove implicitly local high nonlinearities that cause Newton's method to take small updates, so that the fast convergence of global Newton iteration can be restored. A nonlinear preconditioner can be applied on the left or on the right of the nonlinear function. The idea of left preconditioning [2] is to replace the nonlinear function by a preconditioned one with

Li Luo
Extreme Computing Research Center, King Abdullah University of Science and Technology, Thuwal, Saudi Arabia, e-mail: li.luo@kaust.edu.sa

Rongliang Chen
Shenzhen Institutes of Advanced Technology, Shenzhen, China, e-mail: rl.chen@siat.ac.cn

Xiao-Chuan Cai
Department of Computer Science, University of Colorado Boulder, Boulder, USA, e-mail: cai@cs.colorado.edu

David E. Keyes
Extreme Computing Research Center, King Abdullah University of Science and Technology, Thuwal, Saudi Arabia, e-mail: david.keyes@kaust.edu.sa

R. Haynes et al. (eds.), *Domain Decomposition Methods in Science and Engineering XXV*,
Lecture Notes in Computational Science and Engineering 138,
https://doi.org/10.1007/978-3-030-56750-7_51

more balanced nonlinearities, and then solve the new system using a Newton-like algorithm. In contrast, right preconditioning such as nonlinear elimination (NE) [4, 3] does not change the nonlinear function but modifies the unknown variables of the original system. The application of NE can be viewed as a subspace correction step to provide a new starting point for the global Newton iteration, then the solution is updated in the whole space.

In this paper, we develop a nonlinear elimination preconditioned inexact Newton algorithm for steady state flow problems in 3D. It is well known that such problems are usually difficult to solve if a good initial guess is not available. The Newton-like algorithms may diverge though applied with some globalization techniques such as line search. To overcome the difficulty, we introduce an iterative restricted elimination approach based on the magnitude of the local residual, which successfully reduces the number of global Newton iterations. Numerical experiments show the value of the proposed algorithm in comparison to the classical inexact Newton method applied globally, and the impact of tuning parameters.

2 The nonlinear elimination preconditioned inexact Newton algorithm

Consider $F : R^n \to R^n$. We aim to find $x^* \in R^n$, such that

$$F(x^*) = 0, \tag{1}$$

starting from an initial guess $x^0 \in R^n$, where $F = (F_1, \ldots, F_n)^T, F_i = F_i(x_1, \ldots, x_n)$, and $x = (x_1, \ldots, x_n)^T$. We first recall the inexact Newton algorithm with backtracking (INB). Assume x^k is the current approximate solution. A new x^{k+1} can be computed via

$$x^{k+1} = x^k + \lambda^k s^k, \tag{2}$$

where λ^k is the step length, and the inexact Newton direction s^k satisfies

$$\|F'\left(x^k\right) s^k + F\left(x^k\right)\| \leq \eta^k \|F(x^k)\|. \tag{3}$$

Here $\eta^k \in [0, 1)$ is a forcing term that determines how accurately the Jacobian system needs to be solved. To enhance the robustness of INB, we adapt the choice of the forcing term based on norms that are by-products of the iteration, as suggested by Eisenstat and Walker [5].

In many practical situations, especially for nonlinear equations that have unbalanced nonlinearities, λ^k is much smaller than 1 since it is often determined by the components with the strongest nonlinearities. The objective of nonlinear elimination (NE) is to balance the overall nonlinearities of the system through subspace correction. To illustrate the algorithm, we denote by $y = L(\mathcal{F}, x)$ the operation of the subspace correction step, where $\mathcal{F}$ is a modified nonlinear function and x is an intermediate approximate solution. The basic algorithm of NE preconditioned inexact Newton method with backtracking (INB-NE) can be described as follows:

Algorithm 1: The nonlinear elimination preconditioned inexact Newton method with backtracking (INB-NE)

Step 1 Start from the initial guess x^0 and set $k = 0$, $x^{-1} = x^0$.

Step 2 Check convergence:

- If the global condition $\|F(x^k)\| \leq \gamma_r \|F(x^0)\|$ is satisfied, stop.
- If $\|F(x^k)\|/\|F(x^{k-1})\| > \rho_0$ and $k < N_{ne}$, go to **Step 3**; otherwise, go to **Step 4**.

Step 3 The NE step: perform subspace correction iteratively.

- Set $x^{(0)} = x^k$.
- For $l = 0, \ldots, N_l - 1$:
  (i) Construct the nonlinear function $\mathcal{F}(x)$.
  (ii) Evaluate $x^{(l+1)} = L\left(\mathcal{F}, x^{(l)}\right)$.
  (iii) If $\|F(x^{(l+1)})\|/\|F(x^{(0)})\| < \rho_1$, break.
- Set $x^k = x^{(l+1)}$, go to **Step 4**.

Step 4 The global INB step:

- Inexactly solve $F'(x^k)s^k = -F(x^k)$.
- Compute λ^k using the cubic backtracking technique.
- Update $x^{k+1} = x^k + \lambda^k s^k$.
- Set $k = k + 1$, go to **Step 2**.

In the algorithm, γ_r is the relative tolerance for the nonlinear solver, ρ_0 and ρ_1 are preselected factors to measure the relative reduction of the global residual, and N_{ne} is used to control the number of applications of NE.

Next, we discuss the construction of $\mathcal{F}(x)$ and $L(\mathcal{F}, x)$ in detail. In this paper, we consider a point-based elimination approach, i.e., when one variable on some particular mesh point is selected to eliminate, all other variables corresponding to that mesh point are also eliminated. Specifically, let I be an index set of M mesh points, where each index corresponds to m unknown components x_{i_c} and m nonlinear residual components F_{i_c}, $c = 0, \ldots, m-1$. At each subspace correction step, we decompose I into a "bad" subset $I_b^{(l)}$ with $M_b^{(l)}$ mesh points and a "good" subset $I_g^{(l)} = I \backslash I_b^{(l)}$ with $M - M_b^{(l)}$ mesh points, where $I_b^{(l)}$ and $I_g^{(l)}$ correspond to the variables that have strong and weak nonlinearities, respectively. In this paper we consider the bad subset of mesh points $I_b^{(l)}$ as

$$I_b^{(l)} = \left\{ i \mid \text{If } \max_c\{|F_{i_c}(x^{(l)})|\} > \beta \|F(x^{(l)})\|_\infty, c = 0, \ldots, m-1 \right\}, \tag{4}$$

where $\beta > 0$ is a preselected factor. With this subset, we define two subspaces

$$V_b^{(l)} = \left\{ v \mid v = (v_0, \ldots, v_{n-1})^T \in R^n,\ v_{i_c} = 0 \text{ if } i \notin I_b^{(l)} \right\}, \tag{5}$$

$$\text{and } V_g^{(l)} = \left\{ v \mid v = (v_0, \ldots, v_{n-1})^T \in R^n,\ v_{i_c} = 0 \text{ if } i \in I_b^{(l)} \right\}. \tag{6}$$

The corresponding restriction operators are denoted as $R_b^{(l)}$ and $R_g^{(l)}$, which map the vectors from R^n to $V_b^{(l)}$ and $V_g^{(l)}$, respectively. Then, the modified nonlinear function $\mathcal{F}$ is defined as

$$\mathcal{F}(x) = R_g^{(l)}(x - x^{(l)}) + R_b^{(l)}(F(x)). \tag{7}$$

The nonlinear system $\mathcal{F}(x) = 0$ is solved by using the classical INB algorithm with the initial guess $x^{(l)}$. $x^{(*)}$ is accepted as the approximate solution if the stopping condition $\|\mathcal{F}(x^{(*)})\| \leq \gamma_r^{ne} \|\mathcal{F}(x^{(l)})\|$ is satisfied, where γ_r^{ne} is the relative tolerance for the nonlinear solver. In practice, we replace the equations corresponding to the good components by $x_{i_c} - x_{i_c}^{(l)} = 0$ and keep the others unchanged. Therefore, the solve of $\mathcal{F}(x) = 0$ can be performed in the whole space.

To construct the operator L, we introduce a restricted bad subset

$$I_{b,\varepsilon}^{(l)} = \left\{ i \mid \text{If } \max_c \{ |F_{i_c}(x^{(l)})| \} > (\beta + \varepsilon) \|F(x^{(l)})\|_\infty, c = 0, \ldots, m-1 \right\}, \tag{8}$$

where the restricted size $\varepsilon > 0$ is a given parameter. With this subset, we define the corresponding subspaces $V_{b,\varepsilon}^{(l)}$ and $V_{g,\varepsilon}^{(l)}$ in a similar way to (5) and (6). The corresponding restriction operators are denoted as $R_{b,\varepsilon}^{(l)}$ and $R_{g,\varepsilon}^{(l)}$, respectively. Then, with the approximate solution $x^{(*)}$, we define the corrected solution $x^{(l+1)}$ for the subspace correction step as

$$x^{(l+1)} = L(\mathcal{F}, x^{(l)}) = R_{g,\varepsilon}^{(l)}(x^{(l)}) + R_{b,\varepsilon}^{(l)}(x^{(*)}). \tag{9}$$

Remark 1 In this paper, an additive Schwarz preconditioned GMRES method is employed as the linear solver to obtain the solution of the Jacobian systems in both the global INB process and the NE step.

3 Numerical experiments

Let Ω be a bounded domain in R^3. The system of interest can be described by the steady state incompressible Navier-Stokes equations, as follows:

$$\begin{cases} \rho(\mathbf{u} \cdot \nabla)\mathbf{u} - \nabla \cdot \sigma = \mathbf{0}, & \text{in } \Omega, \\ \nabla \cdot \mathbf{u} = 0, & \text{in } \Omega. \end{cases}$$

Here $\mathbf{u} = (u, v, w)^T$ is the velocity, ρ is the density, and $\sigma = -p\mathbf{I} + \mu \left(\nabla\mathbf{u} + \nabla\mathbf{u}^T \right)$ is the Cauchy stress tensor, where $\mathbf{I}$ is an identity matrix, p is the pressure, and μ is the viscosity coefficient. A P_1-P_1 stabilized finite element method is used to discretize the incompressible Navier-Stokes equations, which results in a nonlinear system $F(x) = 0$ to be solved. Here x is a vector of the velocity and pressure unknowns defined at the mesh points.

The algorithms studied in this paper are implemented in PETSc [1]. All computations are carried out on the Tianhe-2 supercomputer. The Jacobian matrices arising from both the global INB process and the NE step are computed analytically. The relative tolerances γ_r and γ_r^{ne} are set to be 10^{-10} and 10^{-3}, respectively. The restart value of GMRES is fixed at 200. In the linear Schwarz preconditioner, the size of overlap is fixed to 2. A point-block incomplete LU (ILU) factorization with 3 fill-in levels is used to apply the approximate inverse action of the subdomain matrix. For all the numerical tests, we fix the parameters $\rho_0 = 0.8$, $\rho_1 = 0.2$, and $N_{ne} = 3$. A zero vector is used as the initial guess, i.e., $x^0 = \mathbf{0}$. It is observed in our tests that only one application of NE is needed for the global Newton to converge quickly, which usually happens at the 3rd global Newton step.

We first consider a lid-driven cubical cavity flow with different Reynolds numbers. The length of the cavity is $D = 1$m. A fluid with density $\rho = 1$kg/m^3 is driven by the wall at $y = D$ which moves tangentially in the x direction with a constant velocity $U = 1$m/s. The other walls impose a no slip boundary condition. The Reynolds number is defined as $Re = \rho U D/\mu$. We vary the viscosity μ to test different Reynolds numbers 1000, 1600, 1800, and 2000. In this case, we use $\beta = 10^{-2}$, $\varepsilon = 0$, and $N_l = 1$. A tetrahedral mesh with 1,761,316 elements and 313,858 nodes is used for the test. The simulation is conducted using 240 processor cores.

For $Re = 2000$, the projections of streamlines on equidistant planes are shown in Fig. 1 (left). In Fig. 1 (right), we show the histories of the nonlinear residuals by using the classical INB and the proposed INB-NE. It is observed that, for the classical INB, the residual curve converges quickly for case $Re = 1000$, but stagnates longer for cases $Re = 1600$ and $Re = 1800$, and diverges for case $Re = 2000$. For INB-NE, the residual curves for these four cases converge within 16 global nonlinear steps. Comparatively, the proposed algorithm is more robust with respect to higher Reynolds numbers.

To study how NE removes the strongest nonlinearities, we show in Fig. 2 the residual contour of component u before and after the application of NE, and the corresponding "bad" subset colored in red, at the 3rd global Newton step for case $Re = 2000$. Table 1 shows the numbers of iterations and compute times obtained using different Re. In the table, "NI_{global}" denotes the number of global Newton iterations, "LI_{global}" denotes the averaged number of GMRES iterations per global Newton, "NI_{ne}" refers to the averaged number of Newton iterations per subspace correction in NE, "LI_{ne}" is the averaged number of GMRES iterations per Newton in NE, "T_{ne}(s)" is the compute time in second for the NE preconditioner, and "T_{total}(s)" is the total compute time in second. As Re increases, though extra compute time is spent on the nonlinear preconditioning, the total compute time of INB-NE is less than that of the classical INB due to significant decrease of the number of global Newton iterations.

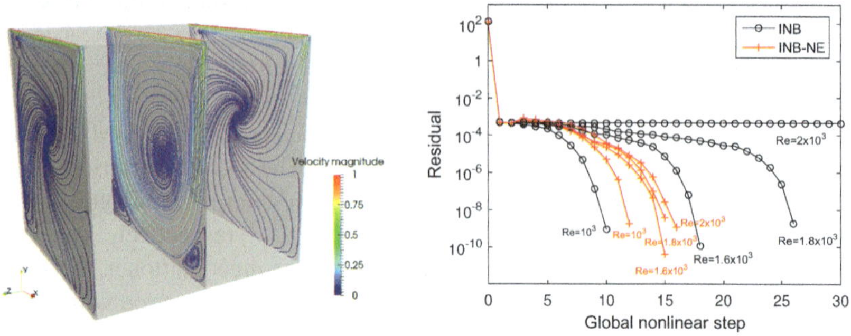

Fig. 1: Lid-driven cavity flow: (left) streamlines for case $Re = 2000$, (right) nonlinear residual history. $\beta = 10^{-2}$, $\varepsilon = 0$, and $N_l = 1$.

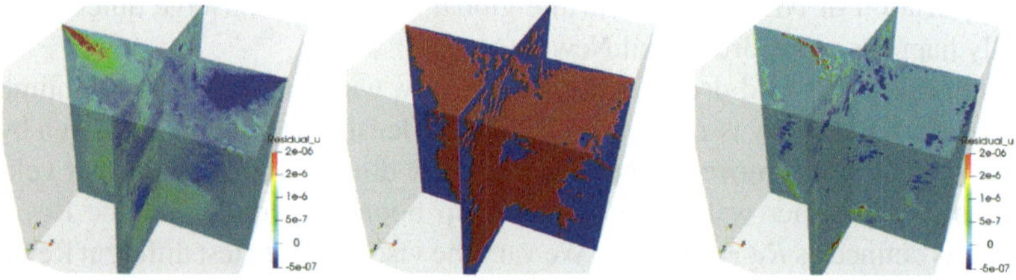

Fig. 2: Lid-driven cavity flow at the 3rd global Newton step for case $Re = 2000$: (left) residual contour of component u before NE, (middle) "bad" subset in red, (right) residual contour of component u after NE.

Table 1: The numbers of iterations and compute times obtained using different Re in the case of lid-driven cavity flow. "–" indicates that the case fails to converge.

	INB			INB-NE					
Re	NI_{global}	LI_{global}	T_{total}(s)	NI_{global}	LI_{global}	NI_{ne}	LI_{ne}	T_{ne}(s)	T_{total}(s)
1000	10	24.40	31.44	12	29.25	8	6.62	21.38	60.90
1600	18	32.29	64.72	15	28.00	8	6.87	21.86	70.82
1800	26	39.54	101.52	15	34.60	8	7.38	21.85	73.84
2000	–	–	–	16	37.93	8	7.63	21.73	78.23

We next consider another well-understood benchmark problem, flow around a cylinder, as defined in [6]. The detailed geometry can be found in the reference. The height and width of the channel is $H = 0.41$m, and the diameter of the cylinder is $D = 0.1$m. The inflow condition is $U(0, y, z) = 16U_m yz(H - y)(H - z)/H^4$, yielding $Re = \rho \bar{U} D/\mu$, where $\bar{U} = 4U_m/9$. The outlet is imposed with a natural outflow boundary condition. We fix the density to $\rho = 1$kg/m^3, the velocity to $U_m = 0.45$m/s, and vary the viscosity μ to test different Reynolds numbers $Re = 20$, 120, 170, and 200. In this case, we use $\beta = \varepsilon = 5 \times 10^{-4}$ and $N_l = 2$. A tetrahedral mesh with 4,909,056 elements and 851,024 nodes is used for the test. The simulation is conducted using 480 processor cores.

The velocity contour on the plane $z = 0.5H$ for case $Re = 20$ is shown in Fig. 3 (left). The histories of the nonlinear residuals obtained using the classical INB and the the present INB-NE are shown in Fig. 3 (right). As Re increases, the number of Newton iterations of the classical INB increases rapidly. When Re is up to 200, INB fails to converge. In contrast, INB-NE converges well for all the four cases and requires fewer nonlinear iterations than the classical INB. Table 2 shows the numbers of iterations and compute times obtained using different Re. Similar to the case of driven cavity flow, when Re becomes larger, the proposed INB-NE is more efficient than the classical INB in terms of the total compute time.

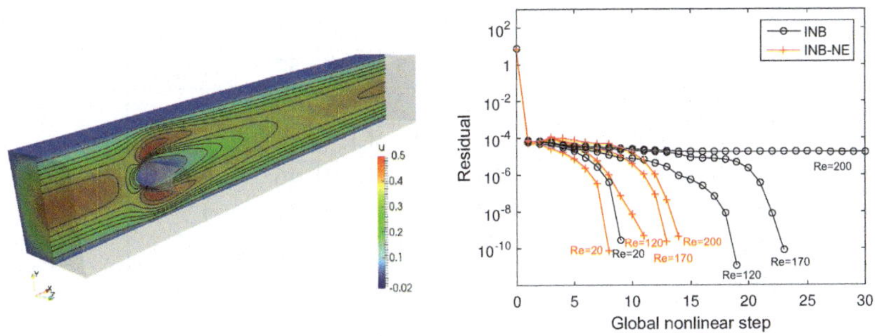

Fig. 3: Flow around a cylinder: (left) velocity contour for case $Re = 20$, (right) nonlinear residual history. $\beta = \varepsilon = 5 \times 10^{-4}$, $N_l = 2$.

Table 2: The numbers of iterations and compute times obtained using different Re in the case of flow around a cylinder. "–" indicates that the case fails to converge.

	INB			INB-NE					
Re	NI_{global}	LI_{global}	T_{total}(s)	NI_{global}	LI_{global}	NI_{ne}	LI_{ne}	T_{ne}(s)	T_{total}(s)
20	9	31.67	41.51	8	34.38	5.5	12.73	43.72	81.76
120	19	52.05	106.85	11	52.27	6.5	11.85	49.65	108.29
170	23	64.13	142.63	13	45.62	6.5	12.62	50.34	116.87
200	–	–	–	14	43.57	7.5	12.27	57.69	127.43

To study the impact of the parameters on the performance of the NE preconditioner, we test the case of flow around a cylinder at $Re = 200$ with different values of N_l, β, and ε. Results are listed in Table 3. In general, when increasing the number of subspace correction steps N_l or decreasing the preselected factor β, the number of global Newton iterations decreases, but this does not necessarily result in a better performance in terms of the total compute time. On the other hand, a suitable choice of the restricted size ε improves the convergence of the global Newton iteration. It is seen form the table that the configuration of $N_l = 2$ and $\beta = \varepsilon = 5 \times 10^{-4}$ leads to the smallest compute time for the concerned problem.

Table 3: The numbers of iterations and compute times obtained using different values of parameters for the case of flow around a cylinder at $Re = 200$.

$\beta = \varepsilon = 5 \times 10^{-4}$						
N_l	NI_{global}	LI_{global}	NI_{ne}	LI_{ne}	$\mathrm{T}_{ne}(\mathrm{s})$	$\mathrm{T}_{total}(\mathrm{s})$
1	20	54.7	8.5	12.6	66.8	181.8
2	14	43.6	7.5	12.3	57.7	127.4
3	14	49.1	6.7	9.5	73.6	146.5
$\varepsilon = \beta,\ N_l = 2$						
β	NI_{global}	LI_{global}	NI_{ne}	LI_{ne}	$\mathrm{T}_{ne}(\mathrm{s})$	$\mathrm{T}_{total}(\mathrm{s})$
2×10^{-3}	17	41.6	7.5	6.5	53.3	137.1
10^{-3}	16	46.2	7	8.2	51.3	133.8
5×10^{-4}	14	43.6	7.5	12.3	57.7	127.4
$\beta = 5 \times 10^{-4},\ N_l = 2$						
ε	NI_{global}	LI_{global}	NI_{ne}	LI_{ne}	$\mathrm{T}_{ne}(\mathrm{s})$	$\mathrm{T}_{total}(\mathrm{s})$
0	18	49.7	6	14.9	46.7	142.4
5×10^{-5}	16	43.3	6	15.0	47.3	127.9
5×10^{-4}	14	43.6	7.5	12.3	57.7	127.4
5×10^{-3}	16	42.1	8.5	11.2	67.0	149.3

4 Conclusions

We demonstrated the robustifying effect of a nonlinearly preconditioned inexact Newton algorithm for steady state incompressible flow problems in 3D. The key idea is to perform iterative subspace correction steps to remove the local high nonlinearities that cause difficulty for classical Newton-like algorithms. We tested the algorithm using two well-understood examples including a lid-driven cavity flow and the flow around a cylinder. Results of numerical experiments show that the proposed algorithm is more robust and converges faster than the classical algorithm in problems with high Reynolds numbers where globalized Newton methods may stagnate.

Acknowledgements The research was supported by the Shenzhen basic research grant JCYJ20160331193229720, JCYJ20170307165328836, JCYJ20170818153840322, and the NSFC 11701547, 61531166003.

References

1. Balay, S., Abhyankar, S., Adams, M.F., Brown, J., Brune, P., Buschelman, K., Dalcin, L., Eijkhout, V., Gropp, W.D., Kaushik, D., Knepley, M.G., May, D.A., McInnes, L.C., Mills, R.T., Munson, T., Rupp, K., Sanan, P., Smith, B.F., Zampini, S., Zhang, H., Zhang, H.: PETSc Users Manual. Tech. Rep. ANL-95/11 Revision 3.10, Argonne National Laboratory (2019)
2. Cai, X.C., Keyes, D.E.: Nonlinearly preconditioned inexact Newton algorithms. SIAM J. Sci. Comput. **24**, 183–200 (2002)

3. Cai, X.C., Li, X.: Inexact Newton methods with restricted additive Schwarz based nonlinear elimination for problems with high local nonlinearity. SIAM J. Sci. Comput. **33**, 746–762 (2011)
4. Cresta, P., Allix, O., Rey, C., Guinard, S.: Nonlinear localization strategies for domain decomposition methods: Application to post-buckling analyses. Comput. Methods Appl. Mech. Eng. **196**, 1436–1446 (2007)
5. Eisenstat, S.C., Walker, H.F.: Choosing the forcing terms in an inexact Newton method. SIAM J. Sci. Comput. **17**, 16–32 (1996)
6. Schäfer, M., Turek, S.: Benchmark computations of laminar flow around a cylinder. Notes Numer. Fluid Mech. **52**, 547–566 (1996)

A Neumann-Neumann Method for Anisotropic TDNNS Finite Elements in 3D Linear Elasticity

Lukas Maly and Dalibor Lukas

1 Introduction

We are interested in solving a problem of linear elasticity in three dimensions. Let $\Omega \in \mathbb{R}^3$ be a bounded connected domain with the Lipschitz boundary $\partial\Omega$. An elastostatic problem is described by the equilibrium equation (2) and Hooke's law (1), which couples the strain and stress tensors for linear elastic materials. We seek for the displacement vector $\mathbf{u}\colon \Omega \to \mathbb{R}^3$ and the stress field $\underline{\sigma}\colon \Omega \to \mathbb{R}^{3\times 3}_{\mathrm{sym}}$ subject to volume forces $\mathbf{f}$ and the boundary conditions (3) and (4) $\partial\Omega = \overline{\Gamma_D} \cap \overline{\Gamma_N}$. Therefore, we solve the problem

$$\underline{\mathbf{C}}^{-1}\,\underline{\sigma} - \underline{\varepsilon}(\mathbf{u}) = 0 \qquad \text{in } \Omega, \tag{1}$$

$$-\operatorname{\mathbf{div}} \underline{\sigma} = \mathbf{f} \qquad \text{in } \Omega, \tag{2}$$

$$\mathbf{u} = \mathbf{u}_D \qquad \text{on } \Gamma_D, \tag{3}$$

$$\sigma_n = \mathbf{t}_N \qquad \text{on } \Gamma_N, \tag{4}$$

where $\mathbf{u}_D$ and $\mathbf{t}_N$ are the prescribed displacement and surface traction, respectively. The tensor $\underline{\varepsilon}(\mathbf{u}) = \frac{1}{2}\left(\nabla\mathbf{u} + (\nabla\mathbf{u})^\top\right)$ is a symmetric strain tensor, and by $\underline{\mathbf{C}}^{-1}$ we denote the compliance tensor, which implements Hooke's law for a given Young modulus and Poisson ratio.

Let $\mathbf{n}$ be an outer unit normal vector. Then the normal component v_n and the tangential component $\mathbf{v}_\tau$ of a vector field $\mathbf{v}$ on the boundary are given by

Lukas Maly
IT4Innovations, VSBB – Technical University of Ostrava,
17. listopadu 2172/15, 708 00 Ostrava-Poruba, Czech Republic
e-mail: lukas.maly@vsb.cz

Lukas Maly, Dalibor Lukas
Department of Applied Mathematics, VSB – Technical University of Ostrava,
17. listopadu 2172/15, 708 00 Ostrava-Poruba, Czech Republic
e-mail: lukas.maly@vsb.cz, dalibor.lukas@vsb.cz

R. Haynes et al. (eds.), *Domain Decomposition Methods in Science and Engineering XXV*,
Lecture Notes in Computational Science and Engineering 138,
https://doi.org/10.1007/978-3-030-56750-7_52

$$v_n = \mathbf{v} \cdot \mathbf{n}, \qquad \mathbf{v}_\tau = \mathbf{v} - v_n \mathbf{n},$$

where the dot symbol stands for the inner product between vector fields. The vector-valued normal component $\boldsymbol{\sigma}_n$ of a tensor $\underline{\sigma}$ can be split into a normal-normal component σ_{nn} and a normal-tangential component $\boldsymbol{\sigma}_{n\tau}$ by

$$\boldsymbol{\sigma}_n = \underline{\sigma}\mathbf{n}, \qquad \sigma_{nn} = \boldsymbol{\sigma}_n \cdot \mathbf{n}, \qquad \boldsymbol{\sigma}_{n\tau} = \boldsymbol{\sigma}_n - \sigma_{nn}\mathbf{n}.$$

2 TDNNS Formulation

We want to solve the stated problem on thin plate-type domains, therefore we use the tangential displacement normal-normal stress (TDNNS) formulation as introduced in [13] and published in a series of papers [11, 10, 9, 8]. The authors developed a new mixed method for the Hellinger-Reissner formulation of elasticity, where the displacement $\mathbf{u}$ is sought in the $\mathbf{H}(\mathbf{curl})$ Sobolev space, i.e., continuity of the tangential components of the displacements is preserved. Meanwhile the stresses live in a new Sobolev space $\underline{\mathbf{H}}(\operatorname{div}\mathbf{div})$, which can be approximated with a symmetric stress tensor preserving continuity of the normal part of their normal component.

The TDNNS elements are applicable for nearly incompressible materials and for structurally anisotropic discretization of slim domains. Here we assume that the Ω is a polyhedral Lipschitz domain (possibly a thin layer in one direction). Let $\mathcal{T}_h = \bigcup_{k=1}^{m}\{T_k\}$, $T_k = T^{\mathrm{x}} \times T^t : T^{\mathrm{x}} \in \mathcal{T}_h^{\mathrm{x}}, T^t \in \mathcal{T}_{h^t}^t$ be a tensor product triangulation of Ω. Since we want to incorporate anisotropic geometrical elements, we need to distinguish sizes of mesh elements in plane- (isotropy-) and thickness- (anisotropy-) directions. We denote them by h and h^t, respectively. Then for the displacements we use the second family of Nédélec space $\mathbf{V}_h$ with a continuous tangential component, and for the stresses, we use a normal-normal continuous space $\underline{\Sigma}_h$. Correct definitions of the appropriate tensor product finite element spaces require more technical details, therefore we leave the spaces undefined here, and refer the interested reader to [10, Chapter 6] or [11, 6] for their correct definitions.

The discrete mixed TDNNS formulation of the original problem (1)–(4) reads as: find $\mathbf{u} \in \mathbf{V}_h$ and $\underline{\sigma} \in \underline{\Sigma}_h$ such that

$$\int_\Omega \left(\underline{\mathbf{C}}^{-1}\underline{\sigma}\right) : \underline{\tau}\,\mathrm{d}\mathbf{x} + \langle \mathbf{div}\,\underline{\tau}, \mathbf{u}\rangle_V = \int_{\Gamma_D} u_{D,n}\tau_{nn}\,\mathrm{d}s \qquad \forall \underline{\tau} \in \underline{\Sigma}_h, \quad (5)$$

$$\langle \mathbf{div}\,\underline{\sigma}, \mathbf{v}\rangle_V = -\int_\Omega \mathbf{f}\cdot\mathbf{v}\,\mathrm{d}\mathbf{x} + \int_{\Gamma_N} \mathbf{t}_{N,\tau}\cdot\mathbf{v}_\tau\,\mathrm{d}s \quad \forall \mathbf{v} \in \mathbf{V}_h, \quad (6)$$

with duality pairing that can be evaluated by element-wise volume and boundary integrals

$$\langle \mathbf{div}\, \underline{\tau}, \mathbf{v}\rangle_V = \sum_{T\in\mathcal{T}_h}\left[-\int_T \underline{\tau} : \underline{\varepsilon}\, \mathrm{d}\mathbf{x} + \int_{\partial T} \tau_{nn} v_n\, \mathrm{d}\mathbf{s}\right]. \tag{7}$$

We can identify the volume integral in (5) with matrix $\underline{\mathbf{A}}$, the duality product in (5) and (6) defined by (7) with matrix $\underline{\mathbf{B}}$, and the right hand sides in (5) and (6) with vectors $\mathbf{F}_1$ and $\mathbf{F}_2$, respectively. Similarly, the sought finite element solutions can be identified with vectors $\mathbf{S} \leftrightarrow \underline{\sigma}$ and $\mathbf{U} \leftrightarrow \mathbf{u}$. Then we can write a linear system for the discrete mixed TDNNS formulation in the following form,

$$\begin{bmatrix} \underline{\mathbf{A}} & \underline{\mathbf{B}}^\top \\ \underline{\mathbf{B}} & \underline{\mathbf{0}} \end{bmatrix} \begin{bmatrix} \mathbf{S} \\ \mathbf{U} \end{bmatrix} = \begin{bmatrix} \mathbf{F}_1 \\ \mathbf{F}_2 \end{bmatrix}. \tag{8}$$

2.1 Hybridization

The system matrix in (8) is symmetric but indefinite, as is typical when mixed formulations are considered. So far, the required continuity of tangential displacement and normal-normal stress is enforced directly by the conforming choice of the solution spaces $\mathbf{V}_h$ and $\underline{\Sigma}_h$. We can break the continuity of the stress space and re-enforce it via Langrangian multipliers. The Lagrangian multipliers will exist in the facet space $\overline{V}_n$ and will correspond to normal displacements on element interfaces. Therefore, we shall denote them as $\overline{u}_n$. To be equivalent with the normal-normal continuity condition for $\underline{\sigma}$, together with the traction condition $\sigma_{nn}|_{\Gamma_N} = 0$, functions in $\overline{V}_n$ have to fulfill the following equation,

$$\sum_{T\in\mathcal{T}_h} \int_{\partial T} \tau_{nn} \overline{v}_n\, \mathrm{d}\mathbf{s} = 0 \qquad \forall v_n \in \overline{V}_n. \tag{9}$$

This leads to an enlarged system with discontinuous stress finite elements,

$$\begin{bmatrix} \widetilde{\underline{\mathbf{A}}} & \underline{\mathbf{B}}_1^\top & \underline{\mathbf{B}}_2^\top \\ \underline{\mathbf{B}}_1 & \underline{\mathbf{0}} & \underline{\mathbf{0}} \\ \underline{\mathbf{B}}_2 & \underline{\mathbf{0}} & \underline{\mathbf{0}} \end{bmatrix} \begin{bmatrix} \widetilde{\mathbf{S}} \\ \mathbf{U} \\ \overline{\mathbf{U}} \end{bmatrix} = \begin{bmatrix} \widetilde{\mathbf{F}}_1 \\ \mathbf{F}_2 \\ \mathbf{0} \end{bmatrix}, \tag{10}$$

where all coupling degrees of freedom are connected to displacement quantities. The matrix $\widetilde{\underline{\mathbf{A}}}$ is block-diagonal with each block corresponding to one element. Such a matrix can be inverted in optimal complexity and thus, we can eliminate all stress degrees of freedom from the system by static condensation,

$$\begin{bmatrix} \underline{\mathbf{B}}_1 \widetilde{\underline{\mathbf{A}}}^{-1} \widetilde{\underline{\mathbf{B}}}^\top & \underline{\mathbf{B}}_1 \widetilde{\underline{\mathbf{A}}}^{-1} \underline{\mathbf{B}}_2^\top \\ \underline{\mathbf{B}}_2 \widetilde{\underline{\mathbf{A}}}^{-1} \underline{\mathbf{B}}_1^\top & \underline{\mathbf{B}}_2 \widetilde{\underline{\mathbf{A}}}^{-1} \underline{\mathbf{B}}_2^\top \end{bmatrix} \begin{bmatrix} \mathbf{U} \\ \overline{\mathbf{U}} \end{bmatrix} = \begin{bmatrix} \underline{\mathbf{B}}_1 \widetilde{\underline{\mathbf{A}}}^{-1} \widetilde{\mathbf{F}}_1 - \mathbf{F}_2 \\ \underline{\mathbf{B}}_2 \widetilde{\underline{\mathbf{A}}}^{-1} \widetilde{\mathbf{F}}_1 \end{bmatrix}. \tag{11}$$

The system matrix in (11) is symmetric and positive definite. The Langrange functions are identified with the vector $\overline{\mathbf{U}}$. We will abbreviate the system using the notation

$$\underline{\mathbf{K}}\widehat{\mathbf{U}} = \widehat{\mathbf{F}}. \tag{12}$$

3 Domain Decomposition

Finite elements in linear elasticity have typically relative high number of degrees of freedom. This is even worse for mixed formulations. Although the stresses were hybridized from the system, there is still significant number of degrees of freedom for displacements. The lowest order anisotropic prismatic and hexahedral finite elements have 60 and 84 degrees of freedom per element, respectively.

This aspect clearly brings a limitation in the sense of problem size. For a large number of mesh elements, the corresponding linear system becomes too large to be solved using direct solvers. Therefore, we resort to an iterative solver and substructuring domain decomposition method as a preconditioner technique [2, 3, 12, 14]. We start our research of preconditioners of mixed TDNNS elements with one which is straightforward and relatively simple to implement, to get a preliminary overview, namely the Neumann-Neumann method described in Section 3.1.

We partition the original domain Ω into N non-overlapping subdomains $\Omega^{(i)}$:

$$\overline{\Omega} = \bigcup_{i=1}^{N} \overline{\Omega}^{(i)}, \qquad \Omega^{(i)} \cap \Omega^{(j)} = \emptyset \quad \text{for } i \neq j, \qquad \Gamma := \bigcup_{i=1}^{N} \partial\Omega^{(i)},$$

such that each subdomain is a union of elements of the global mesh. Using the index (i) we indicate an association to subdomain $\Omega^{(i)}$. By the union of individual subdomain boundaries without the global boundary of Ω we define the *interface* Γ.

The degrees of freedom can be subdivided into two groups; coupling, those being associated with the interface (shared with at least one of the other subdomains, or being on the Dirichlet or Neumann boundary), and interior, which are not coupling. In our setting, the coupling degrees of freedom are associated with an edge or face. All the coupling degrees of freedom in the system are denoted by the lower index C while the interior ones are denoted by the lower index I. The system (12) can then be reordered into the following form

$$\begin{bmatrix} \underline{\mathbf{K}}_{II} & \underline{\mathbf{K}}_{IC} \\ \underline{\mathbf{K}}_{CI} & \underline{\mathbf{K}}_{CC} \end{bmatrix} \begin{bmatrix} \widehat{\mathbf{U}}_I \\ \widehat{\mathbf{U}}_C \end{bmatrix} = \begin{bmatrix} \widehat{\mathbf{F}}_I \\ \widehat{\mathbf{F}}_C \end{bmatrix}. \tag{13}$$

The interior degrees of freedom are related only to the individual subdomains and thus can easily be eliminated from the system using the same idea we used in the hybridization of stresses. This procedure leads to a classical method referred to in the literature as primal domain decomposition, the Schur complement, particular-solution, and the three-step approach [7, 5].

The main idea of using the described domain decomposition procedure resides in preconditioning of the global Schur complement matrix in such a way that the resulting method can be effectively run in parallel. In future, we plan to apply some of the known preconditioner techniques used in [2, 3, 14].

3.1 Neumann-Neumann preconditioner

One of the basic preconditioners of the Schur complement system is the Neumann-Neumann method, which is derived from its local additive construction. Since the Schur complement can be assembled subdomain-wise using local Schur complements multiplied with the restriction operator

$$\underline{\mathbf{S}} = \sum_i \underline{\mathbf{R}}^{(i)\top} \underline{\mathbf{S}}^{(i)} \underline{\mathbf{R}}^{(i)}, \qquad \underline{\mathbf{S}}^{(i)} = \underline{\mathbf{K}}_{CC}^{(i)} - \underline{\mathbf{K}}_{CI}^{(i)} \left(\underline{\mathbf{K}}_{II}^{(i)}\right)^{-1} \underline{\mathbf{K}}_{IC}^{(i)}, \tag{14}$$

a simple idea for how to obtain an approximation of $\underline{\mathbf{S}}$ is to also assemble individual inverses subdomain-wise,

$$\underline{\mathbf{S}}^{-1} \approx \sum_i \underline{\mathbf{D}}^{(i)} \underline{\mathbf{R}}^{(i)\top} \left(\underline{\mathbf{S}}^{(i)}\right)^{-1} \underline{\mathbf{R}}^{(i)} \underline{\mathbf{D}}^{(i)} =: \underline{\mathbf{M}}^{-1}. \tag{15}$$

Matrix $\underline{\mathbf{D}}^{(i)}$ in the formula is a diagonal matrix, whose entry $\underline{\mathbf{D}}_{kk}^{(i)}$ is computed as a reciprocal of a number of subdomains that share the k-th degree of freedom.

It is important to note that the Schur complement $\underline{\mathbf{S}}^{(i)}$ on a subdomain has the same null space dimension as the stiffness matrix $\underline{\mathbf{K}}^{(i)}$. Therefore, local problems on floating subdomains have to be treated with care, since the matrices are singular. Then, the application of the preconditioner corresponds to solving a problem with pure Neumann boundary conditions. For more details on Neumann-Neumann preconditioning, see [1, 4]. In the numerical experiment presented in Section 4, none of the subdomains is floating due to the Dirichlet boundary condition and the two-dimensional decomposition.

4 Numerical Experiments

We present here a simple problem of linear elasticity in three dimensions. Our domain $\Omega := (0; 1) \times (0; 1) \times (0; L_z)$ is represented by a plate with varying thickness in the z-direction. The plate is rigidly fixed on the bottom and top side. As we described above, all volume forces are reflected on the right hand side in (12) and thus they do not play any role in the study of the system matrix properties. We set Young's modulus E to be 1, and Poisson's ration ν to be 0.285. A diagram of the model problem is depicted in Fig. 1 on the left. On the right, we present a simple two-dimensional ($N \times N$) domain decomposition of the plate geometry.

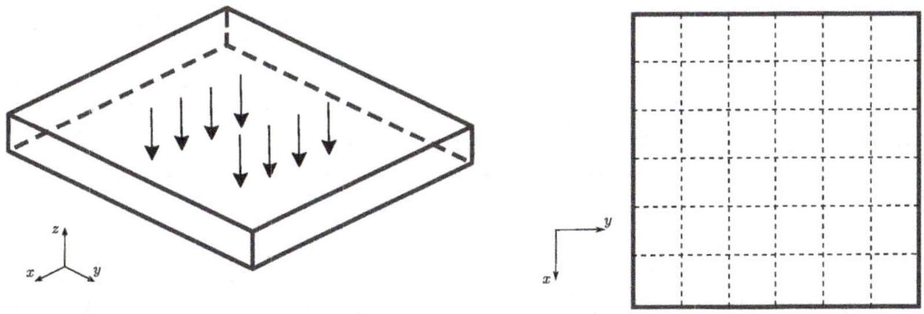

Fig. 1: A diagram of the model problem on plate geometry of thickness L_z and its two-dimensional decomposition into $N \times N$ subdomains in the xy-plane. Dashed lines on the right represent the interface.

We started the development of the scalable parallel algorithm with this simple $N \times N$ domain decomposition to study and fully understand the behavior of the presented systems within the TDNNS formulation of linear elasticity. We construct the Schur complement matrix and its Neumann-Neumann preconditioner as described above. In Table 1 and Table 2, we present condition numbers with respect to discretization size h. To discretize the domain Ω ,we use anisotropic hexahedral elements with only one element in the thickness direction, i.e. $h^t = L_z$. The number of subdomains varies from 2×2 to 8×8.

Presented numerical experiments were implemented in Matlab (version 8.5.0.19-7613 (R2015a)). The implementation uses sparse matrices. Condition numbers presented in Tables 1 and 2 were computed using the built-in *condest* function, and inverse matrices were assembled explicitly. Computations were performed on a classical portable laptop, the biggest problem consisted of 4,096 elements that in case of 2×2 decomposition translate into more than 20,000 inner, and more than 3,000 coupling degrees of freedom.

Table 1: Condition numbers of the Schur complement and preconditioned Schur complement for a plate with thickness $L_z = 0.25$.

$N \times N$	2×2		4×4		8×8	
H/h	$\kappa(\underline{\mathbf{S}})$	$\kappa(\underline{\mathbf{M}}^{-1}\underline{\mathbf{S}})$	$\kappa(\underline{\mathbf{S}})$	$\kappa(\underline{\mathbf{M}}^{-1}\underline{\mathbf{S}})$	$\kappa(\underline{\mathbf{S}})$	$\kappa(\underline{\mathbf{M}}^{-1}\underline{\mathbf{S}})$
1	$1.14 \cdot 10^4$	$6.16 \cdot 10^2$	$3.05 \cdot 10^3$	$1.61 \cdot 10^3$	$1.23 \cdot 10^3$	$1.56 \cdot 10^3$
2	$1.42 \cdot 10^3$	$8.72 \cdot 10^1$	$8.14 \cdot 10^2$	$2.54 \cdot 10^2$	$2.81 \cdot 10^3$	$4.78 \cdot 10^3$
4	$5.21 \cdot 10^2$	$8.50 \cdot 10^1$	$1.48 \cdot 10^3$	$1.00 \cdot 10^3$	$1.38 \cdot 10^4$	$3.03 \cdot 10^4$
8	$1.17 \cdot 10^3$	$5.63 \cdot 10^2$	$7.22 \cdot 10^3$	$8.31 \cdot 10^3$	$8.16 \cdot 10^4$	$2.72 \cdot 10^5$
16	$4.92 \cdot 10^3$	$3.81 \cdot 10^3$	$4.59 \cdot 10^4$	$8.22 \cdot 10^4$		
32	$3.23 \cdot 10^4$	$3.43 \cdot 10^4$				

As presented in Table 1, the preconditioner is not working very well when the thickness is 0.25. The conditioning stays more or less the same except in the case of 4 subdomains and a low H/h ratio. The situation significantly differs when the thickness is 0.01, as in Table 2. In this case, the conditioning is decreased by two orders for all decompositions regardless of the H/h ratio.

Table 2: Condition numbers of the Schur complement and preconditioned Schur complement for a plate with thickness $L_z = 0.01$.

$N \times N$	2×2		4×4		8×8	
H/h	$\kappa(\underline{\mathbf{S}})$	$\kappa(\underline{\mathbf{M}}^{-1}\underline{\mathbf{S}})$	$\kappa(\underline{\mathbf{S}})$	$\kappa(\underline{\mathbf{M}}^{-1}\underline{\mathbf{S}})$	$\kappa(\underline{\mathbf{S}})$	$\kappa(\underline{\mathbf{M}}^{-1}\underline{\mathbf{S}})$
1	$2.19 \cdot 10^9$	$5.58 \cdot 10^7$	$5.36 \cdot 10^8$	$6.32 \cdot 10^7$	$1.36 \cdot 10^8$	$1.59 \cdot 10^7$
2	$3.19 \cdot 10^8$	$3.53 \cdot 10^6$	$8.14 \cdot 10^7$	$2.24 \cdot 10^6$	$2.10 \cdot 10^7$	$6.13 \cdot 10^5$
4	$6.76 \cdot 10^7$	$7.75 \cdot 10^5$	$1.80 \cdot 10^7$	$2.23 \cdot 10^5$	$4.99 \cdot 10^6$	$8.32 \cdot 10^4$
8	$1.78 \cdot 10^7$	$1.97 \cdot 10^5$	$4.60 \cdot 10^6$	$5.82 \cdot 10^4$	$1.57 \cdot 10^6$	$1.47 \cdot 10^4$
16	$4.60 \cdot 10^6$	$5.05 \cdot 10^4$	$1.26 \cdot 10^6$	$1.29 \cdot 10^4$		
32	$1.47 \cdot 10^6$	$1.29 \cdot 10^4$				

It is well known that the efficiency of the local Neumann-Neumann preconditioner deteriorates with a growing number of subdomains, therefore we expect the same trend for a continuation of Tables 1 and 2. In order to improve the presented preconditioner, one needs an additional coarse problem, e.g. by projecting (deflating) against certain modes (yet to be found) or by using sophisticated primal constraints (yet to be found) in a BDDC framework.

5 Conclusion and outlook

We have briefly introduced the TDNNS formulation for a problem of linear elasticity in 3-dimensions, which leads to large and ill-conditioned systems. Based on our experience, we apply a primal domain decomposition procedure to get an initial overview. We try to follow similar ideas as presented in [9], where the authors introduced FETI preconditioned methods for TDNNS elements in 2-dimensions. Our $N \times N$ domain decomposition of thin plate geometry demonstrates the limited efficiency of the Neumann-Neumann preconditioner, and moves us to further research. We aim to end up with a parallel and scalable method, therefore, next we plan to implement some of the modern methods that achieve a bound for the condition number of order $C(1 + \log(H/h))^2$, as discussed in [14].

Acknowledgements The research was supported by The Ministry of Education, Youth and Sports from the National Programme of Sustainability (NPS II) project 'IT4Innovations excellence in science – LQ1602' and by the IT4Innovations infrastructure which is supported from the Large Infrastructures for Research, Experimental Development and Innovations project 'IT4Innovations National Supercomputing Center – LM2015070'.

References

1. Bourgat, J.F., Glowinski, R., Le Tallec, P., Vidrascu, M.: Variational formulation and algorithm for trace operation in domain decomposition calculations. Research Report RR-0804, INRIA (1988). URL https://hal.inria.fr/inria-00075747
2. Bramble, J.H., Pasciak, J.E., Schatz, A.H.: The construction of preconditioners for elliptic problems by substructuring, I. Math. Comp. **47**(175), 103–134 (1986)
3. Bramble, J.H., Pasciak, J.E., Schatz, A.H.: The construction of preconditioners for elliptic problems by substructuring, II. Math. Comp. **49**, 1–16 (1987)
4. De Roeck, Y.H., Le Tallec, P.: Analysis and test of a local domain decomposition preconditioner. In: R. Glowinski, Y. Kuznetsov, G. Meurant, J. Périaux, O. Widlund (eds.) Fourth International Symposium on Domain Decomposition Methods for Partial Differential Equations, pp. 112–128. SIAM, Philadelphia, PA (1991)
5. Lukas, D., Bouchala, J., Vodstrcil, P., Maly, L.: 2-dimensional primal domain decomposition theory in detail. Application of Mathematics **60**(3), 265–283 (2015)
6. Lukas, D., Schöberl, J., Maly, L.: Dispersion analysis of displacement-based and TDNNS mixed finite elements for thin-walled elastodynamics. – pp. 1–13 (2019). Submitted.
7. Maly, L.: Primal domain decomposition methods and boundary elements. Diploma thesis, VSB-Technical University of Ostrava (2013)
8. Meindlhumer, M., Pechstein, A.: 3d mixed finite elements for curved, flat piezoelectric structures. International Journal of Smart and Nano Materials pp. 1–19 (2018). DOI:10.1080/19475411.2018.1556186. URL https://doi.org/10.1080/19475411.2018.1556186
9. Pechstein, A., Pechstein, C.: A FETI method for a TDNNS discretization of plane elasticity. RICAM-Report **11**, 30 (2013). URL http://www.sfb013.uni-linz.ac.at/index.php?id=reportshttp://www.ricam.oeaw.ac.at/publications/list/http://www.numa.uni-linz.ac.at/Publications/List/
10. Pechstein, A., Schöberl, J.: Tangential-displacement and normal-normal-stress continuous mixed finite elements for elasticity. Mathematical Models and Methods in Applied Sciences **21**(08), 1761–1782 (2011). DOI:10.1142/S0218202511005568. URL http://www.worldscientific.com/doi/abs/10.1142/S0218202511005568
11. Pechstein, A., Schöberl, J.: Anisotropic mixed finite elements for elasticity. International Journal for Numerical Methods in Engineering **90**(2), 196–217 (2012). DOI:10.1002/nme.3319. URL http://dx.doi.org/10.1002/nme.3319
12. Saad, Y.: Iterative Methods for Sparse Linear Systems: Second Edition. Society for Industrial and Applied Mathematics (2003). URL https://books.google.cz/books?id=ZdLeBlqYeF8C
13. Sinwel, A.: A new family of mixed finite elements for elasticity. PhD thesis, Johannes Kepler University, Institute of Computational Mathematics (2008). URL http://www.numa.uni-linz.ac.at/Teaching/PhD/Finished/sinwel
14. Toselli, A., Widlund, O.: Domain Decomposition Methods - Algorithms and Theory, *Springer Series in Computational Mathematics*, vol. 34. Springer (2004)

Domain Decomposition for the Closest Point Method

Ian May, Ronald D. Haynes, and Steven J. Ruuth

1 Introduction

The discretization of elliptic PDEs leads to large coupled systems of equations. Domain decomposition methods (DDMs) are one approach to the solution of these systems, and can split the problem in a way that allows for parallel computing. Herein, we extend two DDMs to elliptic PDEs posed intrinsic to surfaces as discretized by the Closest Point Method (CPM) [19, 16]. We consider the positive Helmholtz equation

$$(c - \Delta_{\mathcal{S}})\, u = f, \tag{1}$$

where $c \in \mathbb{R}^+$ is a constant and $\Delta_{\mathcal{S}}$ is the Laplace-Beltrami operator associated with the surface $\mathcal{S} \subset \mathbb{R}^d$. The evolution of diffusion equations by implicit time-stepping schemes and Laplace-Beltrami eigenvalue problems [14] both give rise to equations of this form. The creation of efficient, parallel, solvers for this equation would ease the investigation of reaction-diffusion equations on surfaces [15], and speed up shape classification [18], to name a couple applications.

Several methods exist for the discretization of surface intrinsic PDEs. The surface may be parametrized to allow the use of standard methods in the parameter space [10]. Unfortunately, many surfaces of interest do not have simple, or even known, parametrizations. Given a triangulation of the surface, a finite element discretization can be formed [9]. This approach leads to a sparse and symmetric system but is sensitive to the quality of the triangulation. Level set methods for surface PDEs [4] solve the problem in a higher dimensional embedding space over a narrow band

Ian May
Simon Fraser University, 8888 University Dr., Burnaby, BC V5A 1S6, e-mail: mayianm@sfu.ca

Ronald Haynes
Memorial University of Newfoundland, 230 Elizabeth Ave, St. Johns, NL A1C 5S7 e-mail: rhaynes@mun.ca

Steven Ruuth
Simon Fraser University, 8888 University Dr., Burnaby, BC V5A 1S6, e-mail: sruuth@sfu.ca

R. Haynes et al. (eds.), *Domain Decomposition Methods in Science and Engineering XXV*,
Lecture Notes in Computational Science and Engineering 138,
https://doi.org/10.1007/978-3-030-56750-7_53

containing the surface. The solution of model equation (1) by this method requires using gradient descent, as the approach was formulated only for parabolic problems. The CPM is also discretized over a narrow band in the embedding space, but has the advantage of using a direct discretization of equation (1).

The solution of the linear system arising from the CPM discretization of the model equation (1) has relied primarily on direct methods, although a multigrid method was discussed in [5]. Herein we formulate restricted additive Schwarz (RAS) and optimized restricted additive Schwarz (ORAS) solvers compatible with the CPM to step towards efficient iterative solvers and to allow for parallelism. The optimized variant of the classical RAS solver uses Robin transmission conditions (TCs) to pass additional information between the subdomains [11, 20], and can accelerate convergence dramatically. This formulation is described in Sections 4 and 5 after reviewing the CPM in Section 2 and (O)RAS solvers in Section 3. Then, we discuss a PETSc [1, 2] implementation and show some numerical examples in Section 6. A more thorough exploration of these solvers, and an initial look at their use as preconditioners, can be found in May's thesis [17].

2 The closest point method

The CPM was introduced in [19] as an embedding method for surface intrinsic PDEs. It allows the reuse of standard flat space discretizations of differential operators and provides a surface agnostic implementation. At the core of this method is the closest point mapping, $CP_S(x) = \arg\min_{y \in S} |x - y|$ for $x \in \mathbb{R}^d$, which identifies the closest point on the surface for (almost) any point in the embedding space. This mapping exists and is continuous in the subset of $\mathbb{R}^d$ consisting of all points within a distance κ_∞^{-1} of the surface, where κ_∞ is an upper bound on the principal curvatures of the surface [6].

From this mapping, an extension operator E can be defined that sends functions defined on the surface, $f : S \to \mathbb{R}$, to functions defined on the embedding space via composition with the closet point mapping, $Ef = f \circ CP_S$. The extended functions are constant in the surface normal direction and retain their original values on the surface. This extension operator can be used to define surface intrinsic differential operators from their flat space analogs [19].

Discretization typically requires a Cartesian grid on the embedding space within a narrow tube surrounding the surface. The extension operator can be defined by any suitable interpolation scheme, with tensor product barycentric Lagrangian interpolation [3] being used here. As such, the computational tube must be wide enough to contain the interpolation stencil for any point on the surface. Using degree p interpolation and a grid spacing of Δx requires that the tube contain all points within a distance of $\gamma = \Delta x(p+2)\sqrt{d}/2$ from the surface, thus limiting the acceptable grid spacings in relation to κ_∞.

The grid points within the computational tube form the set of active nodes, Σ_A. For (1), we need only discretize the regular Laplacian on $\mathbb{R}^d$. Here we consider the second order accurate centered difference approximation requiring $2d + 1$ points. Around Σ_A and lying outside the tube, a set of ghost nodes, Σ_G, is formed from any incomplete differencing stencils. With a total of N_A active nodes and N_G ghost nodes, we define the discrete Laplacian and extension operators, $\Delta^h : \mathbb{R}^{N_A+N_G} \longrightarrow \mathbb{R}^{N_A}$ and $\mathbf{E} : \mathbb{R}^{N_A} \longrightarrow \mathbb{R}^{N_A+N_G}$, where Δ^h applies the centered difference Laplacian over all active nodes, and $\mathbf{E}$ is the discretization of E. $\mathbf{E}$ extends data on the active nodes to both the active and ghost nodes, and has entries consisting of the interpolation weights for each node's closest point.

The Laplace-Beltrami operator can be directly discretized as $\Delta^h_{S,dir} = \Delta^h \mathbf{E}$, which was used successfully for parabolic equations with explicit time-stepping in [19]. However, for implicit time-stepping [16] and eigenvalue problems [14] a modified form is needed. In [16] it was recognized that there was a redundant interpolation being performed, and that its removal could stabilize the discretization. The stabilized form $\Delta^h_S = -\frac{2d}{\Delta x^2}\mathbf{I} + \left(\frac{2d}{\Delta x^2}\mathbf{I} + \Delta^h\right)\mathbf{E}$, will be used in the remainder of this work.

3 (Optimized) Restricted additive Schwarz

Both RAS and ORAS are overlapping DDMs, and can work on the same set of subdomains (given an additional overlap condition for ORAS [20]). We define these solvers from the continuous point of view and subsequently discretize, rather than defining them purely algebraically. This will ease the discussion of TCs within the context of the CPM later in Section 5.

First, the whole surface $\mathcal{S}$ is decomposed into N_S disjoint subdomains, $\widetilde{\mathcal{S}}_j$, for $j = 1, \ldots, N_S$. These disjoint subdomains are then grown to form overlapping subdomains $\mathcal{S}_j$, whose boundaries are labelled depending on where they lie in the disjoint partitioning. Taking $\Gamma_{jk} = \partial\mathcal{S}_j \cap \widetilde{\mathcal{S}}_k$ gives $\partial\mathcal{S}_j = \bigcup_k \Gamma_{jk}$ and allows the definition of the local problems

$$\begin{cases} (c - \Delta_S)\, u_j^{(n+1)} = f, & \text{in } \mathcal{S}_j, \\ \mathcal{T}_{jk} u_j^{(n+1)} = \mathcal{T}_{jk} u^{(n)}, & \text{on } \Gamma_{jk},\ k = 1, \ldots, N_S,\ k \neq j, \end{cases} \tag{2}$$

where $\mathcal{T}_{jk}$ are generally linear boundary operators defining the TCs. RAS is achieved by choosing $\mathcal{T}_{jk}$ as identity operators, corresponding to Dirichlet TCs, while ORAS uses Robin TCs, $\mathcal{T}_{jk} = \left(\frac{\partial}{\partial \hat{\mathbf{n}}_{jk}} + \alpha\right)$, where $\hat{\mathbf{n}}_{jk}$ is the outward pointing boundary normal on Γ_{jk} and $\alpha \in \mathbb{R}^+$ is a constant weight on the Dirichlet contribution.

The subproblems in equation (2) are initialized with a guess for the global solution $u^{(0)}$ (defined at least over the boundaries $\Gamma_{jk}, \forall j, k$), which is usually just taken as $u^{(0)} = 0$. After all of the subproblems have been solved a new global solution is constructed with respect to the disjoint partitioning, $u^{(n+1)} = \sum_j u_j^{(n+1)}\Big|_{\widetilde{\mathcal{S}}_j}$, where the use of the term *restricted* indicates that the portion of the local solutions in the overlap regions are discarded. From this new approximation for the global solution the local problems may be solved again with new boundary data, and the process repeats until the global solution is satisfactory.

4 Subdomain construction

To solve problems arising from the CPM we first need to decompose the global set of active nodes Σ_A. (O)RAS solvers rely on both a disjoint partitioning of the active nodes and an induced overlapping partitioning. Following the notation in Section 3, disjoint partitions will be denoted by $\widetilde{\Sigma}_j$, overlapping partitions by Σ_j, and the boundaries of the overlapping partitions by Λ_j.

To ensure the solvers work on a variety of surfaces, we seek an automated and surface agnostic partitioning scheme to generate the disjoint partitions. METIS [13] is a graph partitioner that is frequently used within the DD community to partition meshes [8]. The stencils of Δ^h and $\mathbf{E}$ may be used to induce connectivity between the active nodes and define a graph. Here we only consider nearest neighbor coupling through the stencil for Δ^h. Fig. 1 shows a portion of one such disjoint partition, in black circles, for a circular surface.

With $\widetilde{\Sigma}_j$ obtained from METIS, overlapping subdomains Σ_j can be formed. This construction proceeds in the following steps:

1. All nodes in $\widetilde{\Sigma}_j$ are added to Σ_j.
2. N_O layers of overlap nodes are added around Σ_j. Layers are added one at a time from globally active nodes neighboring Σ_j.
3. A subset of the ghost nodes, Σ_G, are placed in Σ_j^G which consists of nodes that neighbor a member of Σ_j.
4. The shapes of the disjoint and overlapping subdomains are not known in advance. The boundary ∂S_j is approximated discretely by the closest points of the final layer of overlap nodes, and held in the set Λ_j.
5. Nodes needed to complete stencils from the ambient Laplacian or extension operator, including extension from the points $x_i \in \Lambda_j$, are placed in the set Σ_j^{BC}.
6. For ORAS a layer of ghost nodes around Σ_j^{BC} are also placed in Σ_j^{BC}.

The active nodes in the jth subdomain consist of Σ_j and the active portion of Σ_j^{BC}. Σ_j^{BC} is kept separate as that is where the TCs in Section 5 are defined. Each of these sets are shown in Fig. 1, which shows a portion of one subdomain on a circle in the vicinity of the points in Λ_j at one of its boundaries.

The Robin TCs, to be defined in Section 5, need some final information about the subdomain. Every node in Σ_j^{BC} is identified with the point in Λ_j that is closest to it. This identification will be used to override the global closest point function in the following section. For each point in Λ_j we also need to know the direction that is simultaneously orthogonal to the boundary and the surface normal direction. We call this the conormal direction. It is in this direction that the Neumann component of the

Fig. 1 A portion of a subdomain from a circular problem with $N_S = 8$ subdomains and $N_O = 4$ layers of overlap is shown here. The nodes are marked according to their role in the subdomain as described in Section 4. The points belonging to the set Λ_j are shown by (nearly coincident) black stars.

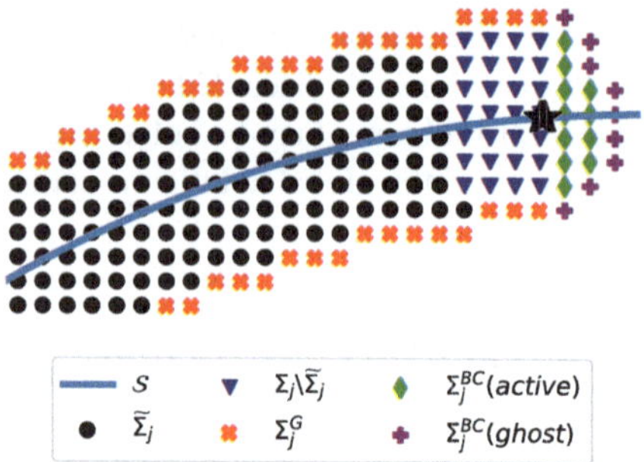

Robin condition will be enforced. However, the discrete nature of Λ_j makes this construction difficult. Instead we define the conormal vectors from the point of view of the boundary nodes. Take $x_i \in \Sigma_j^{BC}$ as a node whose associated conormal direction, $\hat{q}_i$, is sought. Let y_i be its closest point in Λ_j, and $\hat{n}_i$ be the unit surface normal there. Connecting the boundary location to the boundary node via $d_i = x_i - y_i$, we obtain a usable approximation to the conormal by computing the component of d_i that is orthogonal to $\hat{n}_i$ and normalizing, i.e., $\hat{q}_i = (d_i - (d_i \cdot \hat{n}_i)\,\hat{n}_i)/(|d_i - (d_i \cdot \hat{n}_i)\,\hat{n}_i|)$. In the unlikely event that d_i lies perfectly in the surface normal direction, we set $\hat{q}_i = 0$ which recovers the natural boundary condition on the computational tube as discussed at the end of Section 5.

5 Transmission conditions

Boundary conditions in the CPM are imposed by modifying the extension operator over the nodes Σ_j^{BC} beyond the surface boundary [14]. As such, the local operators will take the form

$$\mathbf{A}_j = \left(c + \frac{2d}{h^2}\right) - \left(\frac{2d}{h^2} + \Delta_j^h\right)\begin{bmatrix}\mathbf{E}_j \\ \mathbf{T}_j\end{bmatrix}, \tag{3}$$

where $\mathbf{E}_j$ is the extension operator for the nodes in Σ_j as inherited from the global operator and $\mathbf{T}_j$ is the modified extension operator for the nodes in Σ_j^{BC}. When solving for the local correction to the solution the right hand side of the local problem, $\mathbf{A}_j v_j = r_j$, will be the restriction of the residual to Σ_j. The final rows of the right hand side, those lying over Σ_j^{BC}, become zeros corresponding to the homogenous TCs.

Homogeneous Dirichlet TCs can be enforced to first order accuracy by extending zeros over all of Σ_j^{BC}. With the right hand side already set to zero there, the modified extension reduces to the identity mapping, $\mathbf{T}_j = \begin{bmatrix}\mathbf{0} & \mathbf{I}\end{bmatrix}$, with the zero matrix padding the columns corresponding to the interior nodes.

We discretize the Robin condition

$$\left.\frac{\partial u}{\partial \hat{q}_i}\right|_{CP_{S_j}(x_i)} + \alpha u\left(CP_{S_j}(x_i)\right) = 0, \tag{4}$$

using a forward difference in the $\hat{q}_i$ direction for each node in Σ_j^{BC} and the first order accurate Dirichlet condition from above. Taking the partial derivative $\frac{\partial u}{\partial d_i}$, and applying the change of variables $d_i = \hat{q}_i + \hat{n}_i$, allows one to write the Neumann term in equation (4) in terms of the displacement vector d_i from Section 4. Assuming for the moment that d_i and $\hat{q}_i$ are not perpendicular, the derivative in the conormal direction can be approximated by $\left.\frac{\partial u}{\partial \hat{q}_i}\right|_{CP_{S_j}(x_i)} \approx \frac{u(x_i)-u\left(CP_{S_j}(x_i)\right)}{d_i \cdot \hat{q}_i}$ where CP_{S_j} denotes the modified closest point function identifying points in Σ_j^{BC} with points in Λ_j. Combining this with (4), and applying the identity extension for the Dirichlet component, $u(x_i) = u(CP_{S_j}(x_i))$, we find that $\mathbf{T}_j$ must enforce the extension $u(x_i) = \frac{u\left(CP_{S_j}(x_i)\right)}{1+\alpha d_i \cdot \hat{q}_i}$, with $u\left(CP_{S_j}(x_i)\right)$ replaced by the same interpolation used in the global scheme discussed in Section 2.

As d_i approaches the surface normal direction, $d_i \cdot \hat{q}_i$ will tend to zero. In this event, the extension reduces to $u(x_i) = u\left(CP_{S_j}(x_i)\right)$, which is just the standard extension corresponding to the interior. Fortuitously, this case arises when the point x_i lies adjacent to the interior points where this condition would be applied anyway, and in our experience this ensures that the method remains robust.

6 Results

The solvers described in the previous sections were implemented in C++, with PETSc [1, 2] providing the linear algebra data structures and MPI parallelization, and Umfpack [7] providing the local solutions. Here we focus on evaluating the solver, though in practice one should accelerate the solver with a Krylov method. The (O)RAS solver was placed into a PETSc PCSHELL preconditioner, allowing it to be embedded in any of their Krylov methods, and we have found coupling with GMRES to be a favorable pair.

Equation (1) was solved over the Stanford Bunny [21], which has been scaled to be two units tall. The original triangulation has not been modified in any way beyond this scaling. This surface has several holes and is complicated enough to stress the solvers, making it a good test case. Our chosen grid spacing was $\Delta x = 1/120$, which paired with tri-quadratic interpolation gives $N_A = 947,964$ active nodes in the global problem. The origin was placed at the center of the bounding box containing the bunny and the right hand side $f = \phi(\pi - \phi)\sin(3\phi)(\sin\theta + \cos(10\theta))/2$ was used after extending it to be constant along the surface normals.

Table 1 shows the effects of subdomain count N_S, overlap width N_O, and Robin parameter α. For comparison, GMRES preconditioned with the standard block-Jacobi method with 64, 96, and 128 blocks requires more than 10000 iterations. The

$N_O = 4, \alpha = 16$				$N_S = 64, \alpha = 16$				$N_S = 64, N_O = 4$			
N_S	64	96	128	N_O	4	6	8	α	16	32	64
RAS	992	1237	1533		992	747	610		992	992	992
ORAS	526	672	833		526	418	393		526	707	868

Table 1: Here the iterations to convergence of the ORAS solver are gathered for various parameters. Convergence was declared when the 2−norm of the residual was reduced by a factor of 10^6.

solvers display the expected behavior with the iteration count increasing for larger subdomain counts and decreasing with larger overlap widths. ORAS consistently requires fewer iterations than RAS, though the final sub-table shows the dependence of this performance on the appropriate choice of Robin weight α. The partitioning, the initial error, and the error in the approximate solution after 10 and 550 iterations are visible in Fig. 2 for one run of the solver.

Choosing an optimal value for α is non-trivial as it depends on the value of c, the mesh width, and the geometry. Additionally, the presence of cross points in decomposition, where more than two subdomains meet, complicate the matter. From the planar case, it is known that $\alpha \sim O\left(\Delta x^{-1/2}\right)$, but determining precise values *a priori* is limited to simple splittings [11, 12]. An upcoming work from the same authors explores this in much greater detail.

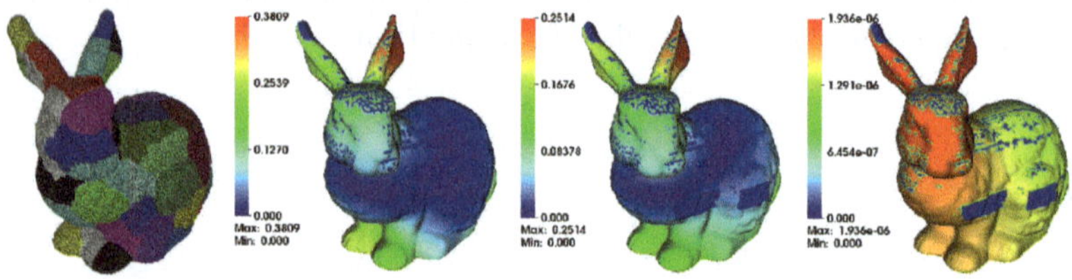

Fig. 2: The Stanford Bunny test problem solved with ORAS using $N_S = 64$, $N_O = 4$, $\alpha = 16$. The first panel shows the disjoint partitioning from METIS. The second, third, and fourth panels show the error in the solution after the 1st, 10th, and 500th iterations compared to the converged solution.

7 Conclusion

Restricted additive Schwarz and optimized restricted additive Schwarz solvers were formulated for the closest point method applied to (1). These solvers provide a solution mechanism for larger problem sizes and will allow users of the CPM to leverage large scale parallelism. Table 1 shows the dramatic reduction in iteration count when Robin TCs are used. These solvers were more completely evaluated in [17], which includes an exploration of their utility as preconditioners. The optimized conditions come at the cost of some additional complexity in the implementation, and even the standard RAS solver brings parallel capabilities to the user. Interesting extensions to this work include multiplicative methods, non-overlapping Robin schemes, two-level solvers, and inclusion of advective terms in the model equation.

Acknowledgements The authors gratefully acknowledge the financial support of NSERC Canada (RGPIN 2016-04361 and RGPIN 2018-04881), and the preliminary work of Nathan King that helped start this project.

References

1. Balay, S., Abhyankar, S., Adams, M.F., Brown, J., Brune, P., Buschelman, K., Dalcin, L., Eijkhout, V., Gropp, W.D., Kaushik, D., Knepley, M.G., May, D.A., McInnes, L.C., Mills, R.T., Munson, T., Rupp, K., Sanan, P., Smith, B.F., Zampini, S., Zhang, H., Zhang, H.: PETSc users manual. Tech. Rep. ANL-95/11 - Revision 3.9, Argonne National Laboratory (2018)
2. Balay, S., Gropp, W.D., McInnes, L.C., Smith, B.F.: Efficient management of parallelism in object oriented numerical software libraries. In: E. Arge, A.M. Bruaset, H.P. Langtangen (eds.) Modern Software Tools in Scientific Computing, pp. 163–202. Birkhäuser Press (1997)
3. Berrut, J.P., Trefethen, L.N.: Barycentric Lagrange interpolation. SIAM Rev. **46**(3), 501–517 (2004)
4. Bertalmio, M., Cheng, L.T., Osher, S., Sapiro, G.: Variational problems and partial differential equations on implicit surfaces. J. Comput. Phys. **174**(2), 759–780 (2001)
5. Chen, Y., Macdonald, C.: The closest point method and multigrid solvers for elliptic equations on surfaces. SIAM J. Sci. Comput. **37**(1), A134–A155 (2015)
6. Chu, J., Tsai, R.: Volumetric variational principles for a class of partial differential equations defined on surfaces and curves. Res. Math. Sci. **5**(2), 19 (2018)
7. Davis, T.A.: Algorithm 832: UMFPACK v4.3—an unsymmetric-pattern multifrontal method. ACM Transactions on Mathematical Software (TOMS) **30**(2), 196–199 (2004)
8. Dolean, V., Jolivet, P., Nataf, F.: An Introduction to Domain Decomposition Methods: Algorithms, Theory, and Parallel Implementation. Society for Industrial and Applied Mathematics, Philadelphia, PA, USA (2015)
9. Dziuk, G., Elliott, C.: Surface finite elements for parabolic equations. Journal of Computational Mathematics **25**(4), 385–407 (2007)
10. Floater, M.S., Hormann, K.: Surface parameterization: a tutorial and survey. Math. Vis. Advances in Multiresolution for Geometric Modelling pp. 157–186 (2005)
11. Gander, M.J.: Optimized Schwarz methods. SIAM J. Numer. Anal. **44**(2), 699–731 (2006)
12. Gander, M.J., Kwok, F.: Best Robin parameters for optimized Schwarz methods at cross points. SIAM J. Sci. Comput. **34**(4), 1849–1879 (2012)
13. Karypis, G., Kumar, V.: A fast and high quality multilevel scheme for partitioning irregular graphs. SIAM J. Sci. Compt. **20**(1), 359–392 (1998)
14. Macdonald, C.B., Brandman, J., Ruuth, S.J.: Solving eigenvalue problems on curved surfaces using the closest point method. J. Comput. Phys. **230**(22), 7944–7956 (2011)
15. Macdonald, C.B., Merriman, B., Ruuth, S.J.: Simple computation of reaction diffusion processes on point clouds. Proceedings of the National Academy of Sciences **110**(23) (2013)
16. Macdonald, C.B., Ruuth, S.J.: The implicit closest point method for the numerical solution of partial differential equations on surfaces. SIAM J. Sci. Comput. **31**(6), 4330–4350 (2010)
17. May, I.: Domain decomposition solvers and preconditioners for the implicit closest point method. Master's thesis, Simon Fraser University (2018)
18. Reuter, M., Wolter, F.E., Peinecke, N.: Laplace–Beltrami spectra as 'Shape-DNA' of surfaces and solids. Computer-Aided Design **38**(4), 342–366 (2006)
19. Ruuth, S.J., Merriman, B.: A simple embedding method for solving partial differential equations on surfaces. J. Comput. Phys. **227**(3), 1943–1961 (2008)
20. St-Cyr, A., Gander, M.J., Thomas, S.J.: Optimized multiplicative, additive, and restricted additive Schwarz preconditioning. SIAM J. Sci. Comput. **29**(6), 2402–2425 (2007)
21. Turk, G., Levoy, M.: Zippered polygon meshes from range images. In: Proceedings of the 21st annual conference on computer graphics and interactive techniques, SIGGRAPH '94, pp. 311–318. ACM (1994)

Towards a Time Adaptive Neumann-Neumann Waveform Relaxation Method for Thermal Fluid-Structure Interaction

Azahar Monge and Philipp Birken

1 Introduction

Our prime motivation is thermal fluid-structure interaction (FSI) where two domains with jumps in the material coefficients are connected through an interface. There exist two main strategies to simulate FSI models: the monolithic approach where a new code is tailored for the coupled equations and the partitioned approach that allows to reuse existing software for each sub-problem. Here we want to develop multirate methods that contribute to the time parallelization of the sub-problems for the partitioned simulation of FSI problems.

We suggest here a parallel, time adaptive multirate method to solve two heterogeneous coupled heat equations which could be applied to FSI problems. The work to be presented is the time adaptive extension of the parallel multirate method in [12]. Some work has already been done regarding time adaptive multirate methods for the simulation of FSI problems. A time adaptive partitioned approach based on the Dirichlet-Neumann iteration for thermal FSI was presented in [4, 5]. However, the Neumann-Neumann method is inherently parallel. In [10], two new iterative partitioned coupling methods that allow for the simultaneous execution of flow and structure solvers were introduced.

A new method that at each iteration solves the two subproblems simultaneously in parallel before exchanging information across the interfaces for the coupling of two parabolic problems was introduced in [9, 8, 6]. There, the Neumann-Neumann waveform relaxation (NNWR) method, which is a waveform relaxation (WR) method based on the classical Neumann-Neumann iteration, is described. It allows the use

Azahar Monge
Centre for Mathematical Sciences, Lund University, Box 118, 22100, Lund, Sweden.
DeustoTech, University of Deusto, Avenida Universidades 24, 48007, Bilbao, Spain, e-mail: azahar.monge@deusto.es

Philipp Birken
Centre for Mathematical Sciences, Lund University, Box 118, 22100, Lund, Sweden, e-mail: philipp.birken@na.lu.se

R. Haynes et al. (eds.), *Domain Decomposition Methods in Science and Engineering XXV*,
Lecture Notes in Computational Science and Engineering 138,
https://doi.org/10.1007/978-3-030-56750-7_54

of different spatial and time discretizations for each subdomain. In [13], a pipeline implementation of the NNWR method together with its parallel efficiency is analyzed for the coupling of homogeneous materials. However, parallelization in time for the coupling of heterogeneous materials was not yet considered.

In a previous article [12], we proposed and analyzed a parallel multirate partitioned approach based on the NNWR algorithm [9, 8, 6] for two coupled parabolic problems with heterogeneous material coefficients. In this work, time adaptivity is added to the multirate approach resulting in a partitioned coupled scheme that allows at each iteration to find the local solutions of the subproblems over a certain time window using different time step controllers. In this setting, one does not need to exchange information across the interface after each time step. The numerical results show the advantages of the time adaptive method over the previous multirate approach.

2 Model problem

The unsteady transmission problem reads as follows, where we consider a domain $\Omega \subset \mathbb{R}^d$ which is cut into two subdomains $\Omega = \Omega_1 \cup \Omega_2$ with transmission conditions at the interface $\Gamma = \partial\Omega_1 \cap \partial\Omega_2$:

$$\begin{cases} \alpha_m \frac{\partial u_m(\mathbf{x},t)}{\partial t} - \nabla \cdot (\lambda_m \nabla u_m(\mathbf{x},t)) = 0, \quad \mathbf{x} \in \Omega_m \subset \mathbb{R}^d, \ m = 1,2, \\ u_m(\mathbf{x},t) = 0, \quad \mathbf{x} \in \partial\Omega_m \backslash \Gamma, \\ u_1(\mathbf{x},t) = u_2(\mathbf{x},t), \quad \mathbf{x} \in \Gamma, \\ \lambda_2 \frac{\partial u_2(\mathbf{x},t)}{\partial \mathbf{n}_2} = -\lambda_1 \frac{\partial u_1(\mathbf{x},t)}{\partial \mathbf{n}_1}, \quad \mathbf{x} \in \Gamma, \\ u_m(\mathbf{x},0) = u_m^0(\mathbf{x}), \quad \mathbf{x} \in \Omega_m, \end{cases} \tag{1}$$

where $t \in [T_0, T_f]$ and $\mathbf{n}_m$ is the outward normal to Ω_m for $m = 1, 2$.

The constants λ_1 and λ_2 describe the thermal conductivities of the materials on Ω_1 and Ω_2 respectively. D_1 and D_2 represent the thermal diffusivities of the materials and they are defined by

$$D_m = \frac{\lambda_m}{\alpha_m}, \quad \text{with } \alpha_m = \rho_m c_{p_m} \tag{2}$$

where ρ_m represents the density and c_{p_m} the specific heat capacity of the material placed in Ω_m, $m = 1, 2$.

3 The Neumann-Neumann waveform relaxation algorithm

We now describe the Neumann-Neumann waveform relaxation (NNWR) algorithm [9, 8]. The main advantage of the NNWR method is that it allows to find the solution on the subdomains in parallel.

The NNWR algorithm starts by imposing continuity of the solution across the interface (i.e, given a common initial guess $g^0(\mathbf{x},t)$ on $\Gamma \times (T_0, T_f)$). One can then find the local solutions $u_m^{k+1}(\mathbf{x},t)$ on Ω_m, $m = 1,2$ through the following Dirichlet problems:

$$\begin{cases} \alpha_m \frac{\partial u_m^{k+1}(\mathbf{x},t)}{\partial t} - \nabla \cdot (\lambda_m \nabla u_m^{k+1}(\mathbf{x},t)) = 0, & \mathbf{x} \in \Omega_m, \\ u_m^{k+1}(\mathbf{x},t) = 0, & \mathbf{x} \in \partial\Omega_m \backslash \Gamma, \\ u_m^{k+1}(\mathbf{x},t) = g^k(\mathbf{x},t), & \mathbf{x} \in \Gamma, \\ u_m^{k+1}(\mathbf{x},0) = u_m^0(\mathbf{x}), & \mathbf{x} \in \Omega_m. \end{cases} \tag{3}$$

Now the second coupling condition which is the continuity of the heat fluxes is added. To this end, one solves two simultaneous Neumann problems to get the correction functions $\psi_m^{k+1}(\mathbf{x},t)$ on Ω_m, $m = 1,2$ where the Neumann boundary condition at the interface $\Gamma \times (T_0, T_f)$ is prescribed by the addition of the heat fluxes of the solutions $u_m^{k+1}(\mathbf{x},t)$ given by the Dirichlet problems:

$$\begin{cases} \alpha_m \frac{\partial \psi_m^{k+1}(\mathbf{x},t)}{\partial t} - \nabla \cdot (\lambda_m \nabla \psi_m^{k+1}(\mathbf{x},t)) = 0, & \mathbf{x} \in \Omega_m, \\ \psi_m^{k+1}(\mathbf{x},t) = 0, & \mathbf{x} \in \partial\Omega_m \backslash \Gamma, \\ \lambda_m \frac{\partial \psi_m^{k+1}(\mathbf{x},t)}{\partial \mathbf{n}_m} = \lambda_1 \frac{\partial u_1^{k+1}(\mathbf{x},t)}{\partial \mathbf{n}_1} + \lambda_2 \frac{\partial u_2^{k+1}(\mathbf{x},t)}{\partial \mathbf{n}_2}, & \mathbf{x} \in \Gamma, \\ \psi_m^{k+1}(\mathbf{x},0) = 0, & \mathbf{x} \in \Omega_m. \end{cases} \tag{4}$$

Finally, the interface values are updated with

$$g^{k+1}(\mathbf{x},t) = g^k(\mathbf{x},t) - \Theta(\psi_1^{k+1}(\mathbf{x},t) + \psi_2^{k+1}(\mathbf{x},t)), \quad \mathbf{x} \in \Gamma, \tag{5}$$

where $\Theta \in (0,1]$ is the relaxation parameter. Note that if one uses the optimal relaxation parameter, we obtain a direct solver instead of an iterative method [6, 12].

In [12, 11], we presented a multirate method for two heterogeneous coupled heat equations based on the NNWR algorithm. There, an interface interpolation that preserves a second order numerical solution of the coupled problem when using SDIRK2 was described to communicate data between the subdomains through the space-time interface in the multirate case. Furthermore, we performed a fully discrete one-dimensional analysis of the NNWR algorithm in (3)-(5). By making use of properties of Toeplitz matrices, we found the optimal relaxation parameter Θ_{opt} in 1D assuming implicit Euler in time, structured spatial grids and conforming time grids on both subdomains. Θ_{opt} then depends on the material coefficients α_1, α_2,

λ_1, λ_2, the spatial resolution Δx and the time resolution Δt. In the limits of $\Delta t/\Delta x^2$ to zero and to infinity, respectively, the optimal relaxation parameter is given by

$$\Theta_{opt}^0 = \frac{\alpha_1\alpha_2}{(\alpha_1+\alpha_2)^2}, \quad \Theta_{opt}^\infty = \frac{\lambda_1\lambda_2}{(\lambda_1+\lambda_2)^2}. \tag{6}$$

Using Θ_{opt}, we get the exact solution at the interface after one iteration, leaving only to solve the two Dirichlet problems once. We then showed numerically that the nonmultirate 1D Θ_{opt} gives excellent estimates for the multirate case using SDIRK2 both in 1D and 2D.

4 Time adaptive method

We now introduce a new adaptive scheme that, in contrast to the multirate method in [12], increases efficiency by allowing larger step sizes without increasing the error of the numerical solution. We build our partitioned time adaptive approach on the SDIRK2-NNWR algorithm introduced in [11, chap. 6] and in [12]. To that end, an error estimate at each time step is needed to be able to choose a new step size. In particular, we use an embedded technique [7, chap. IV.8].

In our approach, time adaptive integrators for the two Dirichlet problems (3) build two independent time grids τ_1 and τ_2. The Neumann problems (4) and the update step (5) then use these grids.

As our time adaptive SDIRK2-NNWR algorithm contains two time adaptive Dirichlet solvers, the corresponding local errors are given by the difference

$$\mathbf{r}_m^{n+1} = \mathbf{u}_I^{(m),n+1} - \hat{\mathbf{u}}_I^{(m),n+1}, \tag{7}$$

where $\mathbf{u}_I^{(m),n+1}$ and $\hat{\mathbf{u}}_I^{(m),n+1}$ are the two solutions of the embedded SDIRK2 method for $m = 1, 2$ and n is the index of the time recursion. Taking the Euclidean norm throughout we consider the error estimate at each time step given by $\|\mathbf{r}_m^{n+1}\|_2$, $m = 1, 2$. We then use a proportional-integral controller (PI controller) for implicit Runge-Kutta methods of order p introduced by [14, 15],

$$\Delta t_m^{n+1} = \Delta t_m^n \left(\frac{tol}{\|\mathbf{r}_m^{n+1}\|_2}\right)^{1/6p} \left(\frac{tol}{\|\mathbf{r}_m^{n}\|_2}\right)^{1/6p}, \tag{8}$$

on the subdomain Ω_m for $m = 1, 2$ respectively and $p = 2$ for SDIRK2. In the first step, the estimate of the previous local error $\mathbf{r}_m^0$ is not available and then we use $\mathbf{r}_m^0 = tol$.

In order to start the integration, one also needs to pick an initial step size. We use the following formula suggested by Gustaf Söderlind and inspired by [1, pp. 682-683] which is dependent on the rhs of the ODE evaluated at t_0, i.e, $f(\mathbf{u}_0)$:

$$\Delta t_m^0 = \frac{|T_f - T_0| \cdot tol^{1/2}}{100 \cdot (1 + \|f(\mathbf{u}_0)\|_2)} = \frac{|T_f - T_0| \cdot tol^{1/2}}{100 \cdot \left(1 + \|\mathbf{M}_{II}^{(m)^{-1}} \mathbf{A}_{II}^{(m)} \mathbf{u}_I^{(m),0}\|_2\right)}, \tag{9}$$

where $\mathbf{M}_{II}^{(m)}$ and $\mathbf{A}_{II}^{(m)}$ for $m = 1, 2$ correspond to the mass and stiffness matrices of the finite element (FE) discretization of the first equation in (3) respectively.

We choose the inner time adaptive tolerance finer than the outer tolerance TOL used to terminate the iteration. Specifically, we take $tol = TOL/5$ for $m = 1, 2$. This choice is motivated by [16] and already used in a similar context in [3, sec. 6].

4.1 Relaxation parameter in the time adaptive case

The aim here is to adapt the formula derived for Θ_{opt} for a fixed step size Δt in [12] to the variable step size context. We propose to start the algorithm with an initial guess for Θ and update the value after each iteration once the time grids τ_1 and τ_2 have already been computed.

For the non adaptive SDIRK2-NNWR, it was observed in [12] that the optimal relaxation parameter moves between the spatial and the temporal limits (6) of Θ_{opt} in terms of $\Delta t/\Delta x^2$. Therefore, we suggest to take an intermediate value between the two limits for the first iteration. Although other options were tried as the geometric mean between the limits, the minimum or the maximum, the arithmetic mean was found to be the most efficient.

To update the relaxation parameter after each iteration, we average all obtained variable step sizes getting the means $\bar{\Delta t}_1$ and $\bar{\Delta t}_2$ for each space-time subdomain $\Omega_1 \times [T_0, T_f]$ and $\Omega_2 \times [T_0, T_f]$. Once we have the values $\bar{\Delta t}_1$ and $\bar{\Delta t}_2$ we choose Θ by inserting the the larger of the averaged time steps into the formula from [12] for the fixed time step multirate SDIRK2-NNWR algorithm.

5 Numerical results

All the results in this section have been produced by implementing the algorithm in Python using the classical 1D or 2D linear FE discretization on equidistant and identical triangular meshes on both subdomains and using as a initial condition the smooth function $g(x) = -1668x^4 + 5652x^3 - 5553x^2 + 1842x$ in 1D or $g(x, y) = 2\sin(\pi y^2)\sin((\pi x^2)/2)$ in 2D on the domain $\Omega = \Omega_1 \cup \Omega_2 = [0, 1] \cup [1, 2]$ or $\Omega = \Omega_1 \cup \Omega_2 = [0, 1] \times [0, 1] \cup [1, 2] \times [0, 1]$ respectively. Physical properties of the materials are shown in table 1.

Figure 1 shows the global error of the overall solution on Ω with respect to the tolerance for the coupling between air and steel in 1D and 2D. They have been

Table 1: Physical properties of the materials. λ is the thermal conductivity, ρ the density, c_p the specific heat capacity and $\alpha = \rho c_p$.

Material	λ (W/mK)	ρ (kg/m^3)	c_p (J/kgK)	α (J/K m^3)
Air	0.0243	1.293	1005	1299.5
Water	0.58	999.7	4192.1	4.1908e6
Steel	48.9	7836	443	3471348

calculated with respect to a reference solution u_{ref} that has been computed using the time adaptive SDIRK2-NNWR algorithm for a very fine tolerance. One observes in Figure 1 how the error decreases proportionally to the tolerance as expected in a time adaptive numerical method.

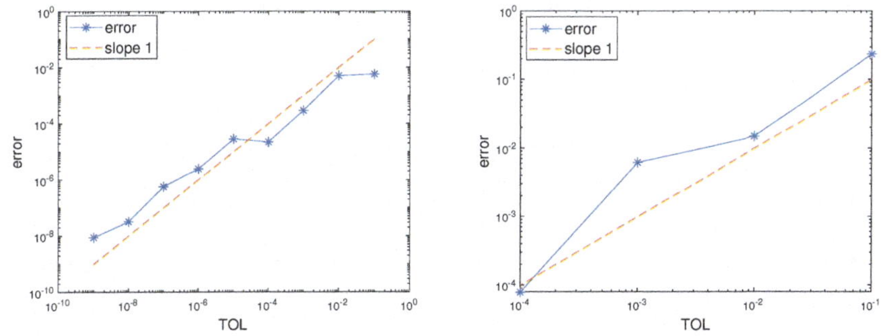

(a) Air-water 1D

(b) Air-steel 2D

Fig. 1: Global error as a function of the tolerance of the time adaptive SDIRK2-NNWR algorithm for different couplings in 1D and 2D. Figure (a): $\Delta x = 1/50$, $[T_0, T_f] = [0, 1]$ and $TOL = 1e-9, 1e-8, .., 1e-1$. Figure (b): $\Delta x = 1/10$, $[T_0, T_f] = [0, 100]$ and $TOL = 1e-4, 1e-3, .., 1e-1$.

Finally, we compare the performance of the multirate SDIRK2-NNWR algorithm with fixed time steps in [12] to the time adaptive SDIRK2-NNWR algorithm introduced in this paper. Figure 2 shows the global error as a function of work for both variants. To compute the work we added together all timesteps performed on both subdomains over all iterations. The stepsizes Δt_m, $m = 1, 2$ for the multirate case are the minimum stepsizes chosen by the time adaptive algorithm on each subdomain. This way, the methods produce almost the same error. In order to get the relation between the number of timesteps and the global error, we measure both magnitudes for a decreasing sequence of tolerances. Number of iterations are specified in table 2. Figure 2 shows that the time adaptive curve is below the multirate curve meaning that less work is employed to reach the same accuracy of the solution using the time adaptive scheme. This difference increases when the tolerance decreases. In 1D this results in 100 times less time steps and in the more relevant 2D case, in 10 times less time steps.

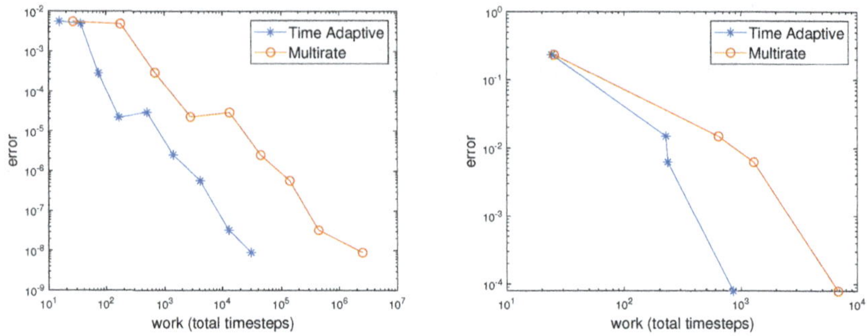

(a) Air-water 1D **(b)** Air-steel 2D

Fig. 2: Comparison between time adaptive and multirate SDIRK2-NNWR algorithm. Global error as a function of work with respect to the total number of timesteps for different couplings in 1D and 2D. Figure (a): $\Delta x = 1/50$, $[T_0, T_f] = [0, 1]$ and $TOL = 1e-9, 1e-8, .., 1e-1$. Figure (b): $\Delta x = 1/10$, $[T_0, T_f] = [0, 100]$ and $TOL = 1e-4, 1e-3, .., 1e-1$.

Table 2: Total number of fixed point iterations (FPI) and total number of timesteps over all FPI (Work) of the time adaptive SDIRK2-NNWR algorithm for different tolerances.

TOL	$1e-1$	$1e-2$	$1e-3$	$1e-4$	$1e-5$	$1e-6$	$1e-7$	$1e-8$	$1e-9$
FPI 1D	4	6	8	10	14	15	15	15	11
FPI 2D	3	15	10	16	19	-	-	-	-
Work 1D	15	36	74	164	496	1397	4135	12796	29996
Work 2D	24	225	233	849	1023	-	-	-	-

However, the method is not as robust in 2D as in 1D and fails for tolerances smaller than $1e-5$. This is because the convergence rate is extremely sensitive to the relaxation parameter. Due to lack of better choices, we use the optimal parameter from 1D in 2D, and combined with the adaptive time step this apparently leads to decreased robustness.

6 Conclusions and Further Work

We have introduced a time adaptive extension of the multirate SDIRK2-NNWR method in [12]. We inserted two different controllers in the Dirichlet solvers to build two independent time grids τ_1 and τ_2 increasing the efficiency of the algorithm. The new algorithm achieves the same solution as the multirate SDIRK2-NNWR algorithm in [12] while optimizing the number of time steps. Numerical results show that the time adaptive method uses 100 times less time steps than the multirate method in 1D and 10 times less time steps in 2D. However, the 2D extension of the time adaptive SDIRK2-NNWR algorithm is not as robust as the 1D version.

Many aspects of the time adaptive approach are left for further research. The extension of the approach to 3D, investigate alternatives adding time step controllers on the Neumann problems as well, implement time adaptivity with respect to macrosteps or study the influence of the initial condition on the performance of the method. Another future direction would be to apply the time adaptive multirate approach explained in this paper to nonlinear thermal FSI cases.

References

1. Arévalo, C. and Söderlind, G.: Grid-independent construction of multistep methods. J. Comp. Math. **35(5)**, 670–690 (2017)
2. Birken, P. and Gleim, T. and Kuhl, D. and Meister, A.: Fast Solvers for Unsteady Thermal Fluid Structure Interaction. Int. J. Numer. Meth. Fluids. **79(1)**, 16–29 (2015)
3. Birken, P. and Monge, A.: Numerical Methods for Unsteady Thermal Fluid Structure Interaction. In: Fluid-Structure Interaction. Modeling, Adaptive Discretisations and Solvers, Contributions in Mathematical and Computational Sciences, Springer, Berlin, 129–168 (2017)
4. Birken, P. and Quint, K.J. and Hartmann, S. and Meister, A.: Choosing norms in adaptive FSI calculations. PAMM, 555–556 (2010)
5. Birken, P. and Quint, K.J. and Hartmann, S. and Meister, A.: A time-adaptive fluid-structure interaction method for thermal coupling. Comp. Vis. in Science. **13(7)**, 331–340 (2011)
6. Gander, M.J. and Kwok, F. and Mandal, B.C.: Dirichlet-Neumann and Neumann-Neumann waveform relaxation algorithms for parabolic problems. ETNA. **45**, 424–456 (2016)
7. Hairer, E. and Wanner, G.: Solving Ordinary Differential Equations II - Stiff and Differential-Algebraic Problems. Springer-Verlag (1996)
8. Kwok, F.: Neumann-Neumann waveform relaxation for the time-dependent heat equation. In: Domain Decomposition Methods in Science and Engineering XXI. Lecture Notes in Computer Science and Engineering. **98**, 189–198. Springer, Berlin (2014)
9. Mandal, B.: Convergence analysis of substructuring Waveform Relaxation methods for space-time problems and their application to Optimal Control Problems. PhD thesis, University of Geneva, Geneva, Switzerland (2014)
10. Mehl, M. and Uekermann, B. and Bijl, H. and Blom, D. and Gatzhammer, B. and van Zuijlen, A.: Parallel coupling numerics for partitioned fluid-structure interaction simulations. Comput. Math. Appl. **71(4)**, 869–891 (2016)
11. Monge, A.: Partitioned methods for time-dependent thermal fluid-structure interaction. PhD thesis, Lund University, Lund, Sweden (2018)
12. Monge, A. and Birken, P.: A multirate Neumann-Neumann waveform relaxation method for heterogeneous coupled heat equations, arXiv: 1805.04336, submitted to SISC (2018)
13. Ong, B.W. and Mandal, B.C.: Pipeline implementations of Neumann-Neumann and Dirichlet-Neumann waveform relaxation methods. Numer. Algor. **78**, 1–20 (2018)
14. Söderlind, G.: Automatic Control and Adaptive Time-Stepping. Numer. Algor. **31**, 281–310 (2002)
15. Söderlind, G.: Digital filters in adaptive time-stepping. ACM Trans. Math. Software. **29**, 1–26 (2003)
16. Söderlind, G. and Wang, L.: Adaptive time-stepping and computational stability. J. Comp. Appl. Math. **185**, 225–243 (2006)

Localization of Nonlinearities and Recycling in Dual Domain Decomposition

Andreas S. Seibold, Michael C. Leistner, and Daniel J. Rixen

1 Introduction

Newton-Krylov domain decomposition methods are well suited for solving nonlinear structural mechanics problems in parallel, especially due to their scalability properties. A Newton-Raphson method in combination with a dual domain decomposition technique, such as a FETI method, takes advantage of the quadratic convergence behaviour of the Newton-Raphson algorithm and the scalabality and high parallelizability of FETI methods. In order to reduce expensive communication between computing cores and thus Newton-iterations, a localization step for nonlinearities was proposed for FETI2, FETI-DP and BDDC solvers [12, 8]. Further methodologies on nonlinear preconditioning of a global Newton method for cases with high local nonlinearities can be found in literature as well [2]. To further improve the efficiency of FETI2-solvers methods have been developed, such as adaptive multipreconditioning [15], derived from simultaneous FETI [7], and reuse techniques of Krylov subspaces [6]. These reuse techniques are rather memory-intensive. More efficient recycling strategies based on Ritz-vectors were therefore developed [9]. In this contribution, we combine those recycling methods with localization of nonlinearities and apply them to static and dynamic structural mechanics problems. We start with the introduction of the model problems and the solution strategy in Sec. 2.1. Then we introduce the localization technique in Sec. 2.2, the adaptive multipreconditioning in Sec. 3 and the used recycling methods in Sec. 4. Finally, we present our numerical results in Sec. 5.

Andreas S. Seibold, e-mail: andreas.seibold@tum.de
Michael C. Leistner, e-mail: m.leistner@tum.de
Daniel J. Rixen, e-mail: rixen@tum.de
Technical University of Munich, Department of Mechanical Engineering, Boltzmannstr. 15 D - 85748 Garching

R. Haynes et al. (eds.), *Domain Decomposition Methods in Science and Engineering XXV*,
Lecture Notes in Computational Science and Engineering 138,
https://doi.org/10.1007/978-3-030-56750-7_55

2 Localized nonlinearities in dual domain decomposition

2.1 Modelproblem and nonlinear solution strategy

We consider a static structural mechanics problem with nonlinear material behavior, discretized with Finite Elements and decomposed into N_s substructures s of the form

$$\mathbf{f}_{int}^{(s)}(\mathbf{u}^{(s)}) + \mathbf{B}^{(s)^T}\lambda - \mathbf{f}_{ext}^{(s)} = \mathbf{0}, \qquad \sum_{s=1}^{N_s} \mathbf{B}^{(s)}\mathbf{u}^{(s)} = \mathbf{0}, \tag{1}$$

where $\mathbf{u}$ describes the displacements of the elastic structure and the primary solution of the problem. The substructures are coupled with Lagrange-multipliers λ imposed on the boundary of each substructure by a signed Boolean matrix $\mathbf{B}$ [5]. Accelerations $\ddot{\mathbf{u}}$ and velocities $\dot{\mathbf{u}}$ are added with the related mass $\mathbf{M}$ and damping $\mathbf{D}$ for a structural mechanics problem, which results in the dynamic nonlinear system of equations

$$\mathbf{M}^{(s)}\ddot{\mathbf{u}}^{(s)} + \mathbf{D}^{(s)}\dot{\mathbf{u}}^{(s)} + \mathbf{f}_{int}^{(s)}(\mathbf{u}^{(s)}) + \mathbf{B}^{(s)^T}\lambda - \mathbf{f}_{ext}^{(s)} = \mathbf{0}, \qquad \sum_{s=1}^{N_s} \mathbf{B}^{(s)}\ddot{\mathbf{u}}^{(s)} = \mathbf{0}.$$

This dynamic system can now be integrated by a suitable time-integration scheme and handled as subsequently described for the static system. For our experiments we use a generalized-α scheme [3]. These systems are solved by a Newton-Raphson scheme and the resulting linearized system by a FETI-method [5] at each time or load step. By linearizing the system of equations (1) at Newton-iteration n and resolving it for the incremental displacements, we get the tangent interface problem

$$\begin{bmatrix} \mathbf{F}_n & -\mathbf{G}_n \\ \mathbf{G}_n^T & \mathbf{0} \end{bmatrix} \begin{bmatrix} \delta\lambda \\ \delta\alpha \end{bmatrix} = \begin{bmatrix} \mathbf{0} \\ \mathbf{e}_n \end{bmatrix} - \begin{bmatrix} \mathbf{d}_n \\ \mathbf{G}_n^T\lambda_n \end{bmatrix} \qquad \mathbf{F}_n = \sum_{s=1}^{N_s} \mathbf{B}^{(s)}\mathbf{K}_{T,n}^{(s)^+}\mathbf{B}^{(s)^T}$$

$$\mathbf{G}_n = \begin{bmatrix} \mathbf{B}^{(1)}\mathbf{R}_n^{(1)} & \dots & \mathbf{B}^{(N_s)}\mathbf{R}_n^{(N_s)} \end{bmatrix} \qquad \mathbf{e}_n = \begin{bmatrix} \mathbf{R}_n^{(1)^T}\mathbf{f}_{ext}^{(1)} & \dots & \mathbf{R}_n^{(N_s)^T}\mathbf{f}_{ext}^{(N_s)} \end{bmatrix}^T$$

$$\mathbf{d}_n = -\sum_{s=1}^{N_s} \mathbf{B}^{(s)}\mathbf{K}_{T,n}^{(s)^+}\left(\mathbf{f}_{int}(\mathbf{u}_n^{(s)}) - \mathbf{f}_{ext} + \mathbf{B}^{(s)^T}\lambda_n\right) + \mathbf{B}^{(s)}\mathbf{u}_n^{(s)}$$

and the local linear solves for the incremental displacements

$$\delta\mathbf{u}^{(s)} = -\mathbf{K}_{T,n}^{(s)^+}\mathbf{B}^{(s)^T}\delta\lambda - \mathbf{K}_{T,n}^{(s)^+}\left(\mathbf{f}_{int}^{(s)}(\mathbf{u}_n^{(s)}) - \mathbf{f}_{ext}^{(s)} + \mathbf{B}^{(s)^T}\lambda_n)\right) + \mathbf{R}_n^{(s)}\delta\alpha$$

Here, $\mathbf{K}_{T,n}^{(s)}$ is the tangent stiffness, $\mathbf{R}_n^{(s)}$ its null space and $\mathbf{F}_n$ the tangent interface operator with the superscript + denoting a pseudoinverse. The null space $\mathbf{R}_n^{(s)}$ and its corresponding additional unknowns $\delta\alpha$ can be seen as rigid body modes of floating substructures and are needed for solvability [5]. This isn't needed in structural dynamics due to the additional mass matrix. The interface problem is then solved by a preconditioned conjugate gradient.

2.2 Localization of nonlinearities

In order to reduce Newton-iterations and hence iterations of the conjugate gradient method, one can solve local nonlinear problems as some kind of preconditioning step for the global linear solution step within the Newton algorithm [12, 11, 8]. In case of a FETI2-solver this is achieved by solving local nonlinear Neumann-problems while keeping the Lagrange-multipliers constant, whereas for FETI-DP and BDDC far more options are available [8]. Thus, the displacements

$$\delta \mathbf{u}^{(s)} = -\mathbf{K}^{(s)^{+}}_{T,ng,nl} \left(\mathbf{f}^{(s)}_{int}(\mathbf{u}^{(s)}_{ng,nl}) - \mathbf{f}^{(s)}_{ext} + \mathbf{B}^{(s)^{T}} \lambda_{ng-1} \right), \quad \mathbf{u}^{(s)}_{ng,nl+1} = \mathbf{u}^{(s)}_{ng,nl} + \delta \mathbf{u}^{(s)}$$

are calculated within local Newton iterations nl after using the displacements of the previous global Newton-iteration $ng - 1$ as an initialization. To ensure local solvability, the Lagrange-multipliers have to be initialized with the natural coarse grid [12, 11]

$$\mathbf{G}_0^T \lambda_0 = \mathbf{e}_0 \qquad \lambda_0 = \mathbf{G}_0 \left(\mathbf{G}_0^T \mathbf{G}_0 \right)^{-1} \mathbf{e}_0 \tag{2}$$

3 Adaptive Multipreconditioning

A preconditioner $\mathbf{H} = \sum_{s=1}^{N_s} \mathbf{B}^{(s)} \mathbf{S}^{(s)} \mathbf{B}^{(s)}$ is commonly used for an efficient solution of the interface problem, here a Dirichlet-preconditioner with the Schur-complement $\mathbf{S}$ [13]. Due to the summation of the local preconditioners, some local information gets lost. Hence, multipreconditioning, also known as *simultaneous FETI* (S-FETI) [7], has been proposed using separated preconditioners $\mathbf{H}^{(s)}$ leading to independent search directions $\mathbf{z}_i^{(s)} = \mathbf{H}^{(s)} \mathbf{r}_i$ in each FETI-iteration i for the residual $\mathbf{r}$. To avoid large search spaces, a τ-criterion has been introduced to modify the S-FETI to an *adaptive multipreconditioned FETI* (AMP-FETI) method [16, 9]. The τ-criterion controls which substructures are chosen for multipreconditioning. To this end the expression

$$\Theta_i^{(s)} = \frac{\gamma_i^T \mathbf{W}_i^T \mathbf{F}^{(s)} \mathbf{W}_i \gamma_i}{\mathbf{r}_{i+1}^T \mathbf{H}^{(s)} \mathbf{r}_{i+1}}, \qquad \mathbf{W}_i = \mathbf{P} \mathbf{Z}_i$$

is used with γ_i being step-lengths from the CG iteration i and the natural coarse-grid projector $\mathbf{P}$. Only the substructures that fulfill the criterion $\Theta_i^{(s)} < \tau$ are chosen. The parameter τ can be set by the user and $\tau = 0.1$ leads to robust behavior in most cases and has been used in this paper [15, 1]. The search space is constructed with such $J = (j_1, j_2, \dots)$ chosen substructures as

$$\mathbf{Z}_i = \left[\sum_{k \notin J} \mathbf{z}_i^{(k)} \mid \mathbf{z}_i^{(j_1)} \mid \mathbf{z}_i^{(j_2)} \mid \ldots\right]$$

4 Recycling methods for dual solutions

In order to further increase the FETI-solver's efficiency and render it scalable, we introduce a deflation or coarse space $\mathbf{C}$ for the search directions $\mathbf{W}_i$, which leads to a two-level FETI (FETI2) solver [4]. A coarse-problem is solved during the initialization and iterations of the FETI-solver. The remaining search space has to be $\mathbf{F}$-conjugate, which is ensured by the projector

$$\mathbf{P}_C = \mathbf{I} - \mathbf{C}(\mathbf{C}^T\mathbf{F}\mathbf{C})^{-1}\mathbf{C}^T\mathbf{F}.$$

In the so-called *total reuse of Krylov subspaces* (TRKS), proposed in [6], all the previous solutions are reused to build the coarse grid

$$\mathbf{C}_{ng} = \left[\mathbf{C}_{ng-1}\ \mathbf{w}_{ng,i=1}\ \ldots\ \mathbf{w}_{ng,i=i_{end}}\cdot\right]$$

In order to improve memory-efficiency, people have investigated the convergence behavior of a preconditioned conjugate gradient algorithm. This is mainly governed by the eigenspectrum of the preconditioned operator $\mathbf{HF}$. High, well-separated eigenvalues might slow down convergence according to studies in [14]. These high eigenvalues are usually captured during the first few iterations of the FETI-solver. Hence, by first solving the eigenvalue-problem

$$\mathbf{S}^{(s)}\mathbf{y}^{(s)} = \Phi^{(s)}\mathbf{B}^{(s)^T}\mathbf{H}\mathbf{B}^{(s)}\mathbf{y}^{(s)}, \tag{3}$$

called *generalized eigenvalues in the overlaps* (GenEO), the high eigenmodes are precomputed separately [16]. Here $\Phi^{(s)}$ are the eigenvalues and $\mathbf{y}^{(s)}$ the corresponding eigenvectors. To reduce the high initial cost of the Schur-complements, a local Ritz Ansatz has been applied in [9], approximating the GenEO eigenvectors and resulting in a smaller eigenproblem. The Ritz space of substructure s is then constructed as

$$\mathbf{V}^{(s)} = \mathbf{S}^{(s)^{-1}}\mathbf{B}^{(s)^T}\mathbf{V}_W^{(s)}, \qquad \mathbf{V}_W^{(s)} = \left[\mathbf{W}_0\gamma_0 \ldots \mathbf{W}_{n^s-1}\gamma_{n^s-1}\right], \quad n^s \leq i_{end},$$

where the solution space of the first n^s iterations is considered and n_s limits the Ritz space size. With such a Ritz space follows the approximation of (3)

$$\mathbf{V}^{(s)^T}\mathbf{S}^{(s)}\mathbf{V}^{(s)}\mathbf{q}^{(s)} = \Phi^{(s)}\mathbf{V}^{(s)^T}\mathbf{B}^{(s)^T}\mathbf{H}\mathbf{B}^{(s)}\mathbf{V}^{(s)}\mathbf{q}^{(s)},$$

which can be rewritten as

$$\mathbf{V}_W^{(s)^T}\mathbf{F}^{(s)}\mathbf{V}_W^{(s)}\mathbf{q}^{(s)} = \Phi^{(s)}\mathbf{V}_W^{(s)^T}\mathbf{F}^{(s)}\mathbf{H}\mathbf{F}^{(s)}\mathbf{V}_W^{(s)}\mathbf{q}^{(s)}, \quad \mathbf{F}^{(s)} = \mathbf{B}^{(s)}\mathbf{S}^{(s)^{-1}}\mathbf{B}^{(s)^T}.$$

The resulting coarse space with the first k^s local Ritz vectors is

$$\mathbf{C}^{(s)} = \left[\mathbf{HF}^{(s)}\mathbf{V}_W^{(s)}\mathbf{q}_1^{(s)} \ \dots \ \mathbf{HF}^{(s)}\mathbf{V}_W^{(s)}\mathbf{q}_{k^s}^{(s)},\right]$$

where k^s has to fulfill $k^s \leq n^s$. This method is subsequently called *local Ritz* (LRitz) approach. It may be even reasonable to build the coarse space directly out of Ritz spaces only, without solving an eigenproblem [10]

$$\mathbf{C} = \left[\mathbf{HF}^{(1)}\mathbf{V}_W^{(1)} \ \dots \ \mathbf{HF}^{(N_s)}\mathbf{V}_W^{(N_s)}.\right]$$

This method is referred to as *local Ritz direct* (LRitzDir) below.

5 Numerical results

5.1 Recycling methods applied to static mechanical problems

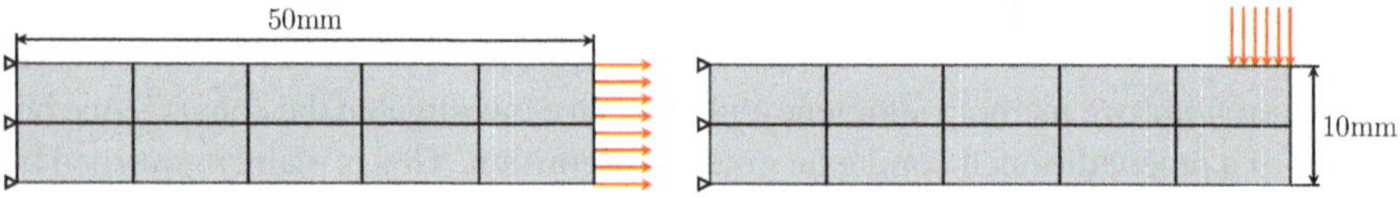

Fig. 1: Left clamped cantilever beam partitioned to 10 rectangular substructures under pull load (left) and bending load (right). Mooney-Rivlin-material (Invariant-parameters: $A_{10} = 0.4N/mm^2$, $A_{01} = 0.1N/mm^2$, $K = 1 \cdot 10^2 N/mm^2$); pull-load: $5N$, bending-load: $1.5 \cdot 10^{-3}N$

Table 1: FETI-iterations cumulated over Newton-iterations and loadsteps and normalized to the NLF-method without recycling. (NLF: classic nonlinear FETI, LoNo: FETI with localized nonlinearities) Load case: pull, 10 loadsteps; Absolute cumulated number of iterations for NLF None: 224

NL-Method	NLF	NLF	LoNo	NLF	NLF	LoNo	LoNo	LoNo
Recycling	None	plReuse	None	LRitzDir	LRitz	plReuse	LRitz	LRitzDir
rel. Iter	1	0.75	0.7366	0.6964	0.5804	0.5268	0.4955	0.4420

We apply the methods introduced above to a homogeneous, nonlinear cantilever beam (Mooney-Rivlin material model and geometrical nonlinearity without damping) under static pull and static bending load and rectangular substructuring, as shown in Fig. 1. The cumulated numbers of FETI-iterations are normalized to the classic nonlinear FETI method (NLF) without recycling in Table 1 since that is the reference we want to compare the performance gain to. The TRKS approach is renamed *plain Reuse* (plReuse) as we no longer have Krylov-subspaces due to multipreconditioning [9]. The coarse spaces are limited to a fixed global size to get compareable results. The combination of localizations and LRitzDir resulted in a reduction of global iterations by 55%. Hence, localizations combine well with recycling methods. The LRitzDir method in particular performs better with localizations than in combination with the classic nonlinear FETI. The LRitz approach suffers from a slower build up

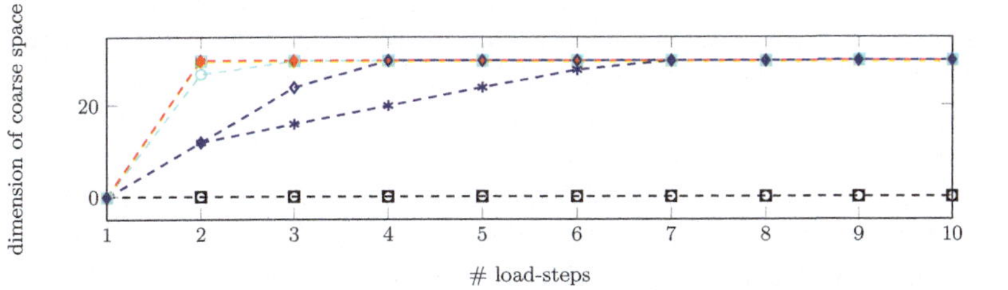

Fig. 2: Coarse grid dimension over load steps in static pull case (Colours are the same as in Fig. 3).

Table 2: Over Newton-iterations and load steps cumulated numbers of FETI-iterations normalized to the NLF-method without recycling. Load case: bending; Absolute cumulated number of iterations for NLF None: 519

NL-Method	LoNo	NLF	LoNo	NLF	LoNo	NLF	NLF	LoNo
Recycling	None	None	plReuse	LRitzDir	LRitz	LRitz	plReuse	LRitzDir
rel. Iter	1.0617	1	0.9422	0.8112	0.7380	0.4566	0.3988	0.3738

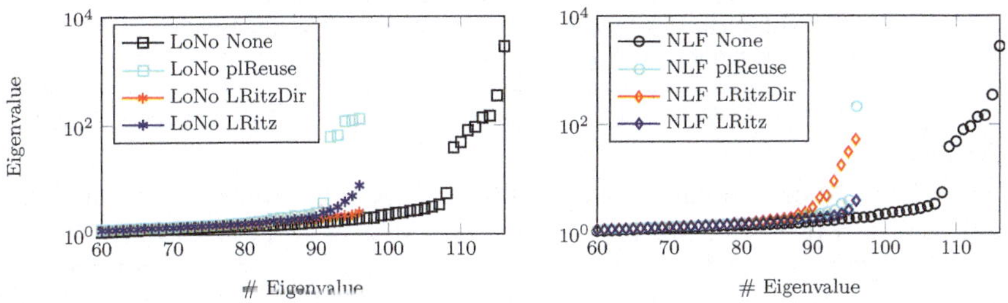

Fig. 3: Eigenvalue spectrum of interface operator $\mathbf{HP}_C^T\mathbf{F}$ sorted from lowest to highest in last load step 40 with localizations (left) and classical nonlinear FETI (right). Loadcase: bending

Table 3: Global Newton iterations of first 3 (of 40) loadsteps. Loadcase: bending

NL-Method	LoNo	LoNo	NLF	NLF	LoNo	LoNo	NLF	NLF
Recycling	None	plReuse	None	LRitzDir	LRitz	LRitzDir	LRitz	plReuse
LoadStep1	14	19	4	3	10	13	3	3
LoadStep2	4	4	4	3	2	2	2	2
LoadStep3	4	4	4	3	2	2	2	2

of the coarse grid in Fig. 2. Due to the small chosen limit of the Ritz-space by $n_s = 4$ and a reduction of global Newton-iterations, the solver is unable to capture the high modes fast enough. In the bending case, the localizations lead to worse performance than the NLF method without recycling due to instabilities of local rotational modes, mentioned as non-physical nonlinearities in literature [12, 11]. The combined localization and LRitzDir outperforms the NLF though. In Fig. 3, it is able to capture all the bad modes better than in the NLF. This doesn't apply for the LRitz method. The coarse grid is filled up within the first load step of LoNo-method due to many global Newton iterations, but with unfavorable modes. In NLF, it takes more load steps, but apparently better modes are chosen here, which accelerates the solution process. In the case of LRitzDir, the higher number of global Newton iterations in the first load step is well compensated by fewer Newton iterations compared to NLF in later load steps. Anyway, the high number of load steps has been chosen to obtain a stable convergence of the algorithm with localizations. Fewer load steps would have been needed for the classic nonlinear method. Moreover, one has to bear in mind the cost of more local solves for the localization method.

5.2 Recycling methods applied to dynamic mechanical problems

We also apply the localization and recycling methods to a dynamic mechanical bending problem, meshed with Gmsh 3.0.6 and partitioned with its Metis partitioner. Here with the plain reuse technique more iterations are needed than with Ritz approximations due to persistent high modes. The application of localizations leads to slightly more iterations, even with recycling methods. The influence of nonlinear material is rather low due to time stepping and localization won't be able to reduce global iterations significantly.

Table 4: Number of FETI-iterations cumulated over time steps and global Newton-iterations and normalized to the NLF-method without recycling. Load case: dynamic bending beam. Absolute number of cumulated iterations for NLF None: 1473

NL-Method	LoNo	LoNo	NLF	NLF	LoNo	NLF	LoNo	NLF
Recycling	None	plReuse	None	plReuse	LRitz	LRitz	LRitzDir	LRitzDir
rel. Iter	1.0930	1.0088	1	0.8771	0.8629	0.7916	0.7461	0.7264

6 Conclusions

In this work, we applied recent recycling methods and adaptive multipreconditioning for a FETI2-method together with nonlinear localization to static and dynamic structural mechanics problems. We were able to reduce global iterations by up to 62% with this combination, even for homogeneous material properties in the static bending case. This is counterbalanced by very low load step-sizes though, as otherwise the localized method would not converge due to instabilities in rotational rigid body modes. However, the static case under pull load shows quite promising results and localization combines well with recycling techniques. Hence, if the stability issues could be fixed, these methods would be a reasonable technique to reduce communication, but at a cost of additional local solves. We were unable to test these methods in parallel due to our current implementation limitations. So it still has to be evaluated, whether the increased local solves are compensated by the reduced global iterations. Moreover, we applied these methods to dynamic structural mechanics problems, where we don't encounter the stability issues due to the present mass-matrix. Localizations didn't provide any reduction of iterations either due to limited nonlinear influences caused by time stepping. Hence, it might be different for a model with local, highly nonlinear phenomena, such as cracks and damaging, which will be supported by our implementation in the future.

References

1. Bovet, C., Parret-Fréaud, A., Spillane, N., Gosselet, P.: Adaptive multipreconditioned feti: scalability results and robustness assessment. Computers and Structures **193**, 1–20 (2017). DOI:10.1016/j.compstruc.2017.07.010
2. Cai, X.C., Li, X.: Inexact Newton methods with restricted additive Schwarz based nonlinear elimination for problems with high local nonlinearity. SIAM J. Sci. Comput. **33**(2), 746–762 (2011). DOI:10.1137/080736272
3. Chung, J., Hulbert, G.M.: A time integration algorithm for structural dynamics with improved numerical dissipation: the generalized-α method. Trans. ASME J. Appl. Mech. **60**(2), 371–375 (1993). DOI:10.1115/1.2900803
4. Farhat, C., Pierson, K., Lesoinne, M.: The second generation feti methods and their application to the parallel solution of large-scale linear and geometrically non-linear structural analysis problems. Computer methods in applied mechanics and engineering **184**, 333–374 (2000)
5. Farhat, C., Roux, F.X.: A method of finite element tearing and interconnecting and its parallel solution algorithm. Internat. J. Numer. Methods Engrg. **32**(6), 1205–1227 (1991). DOI:10.1002/nme.1620320604
6. Gosselet, P., Rey, C., Pebrel, J.: Total and selective reuse of Krylov subspaces for the resolution of sequences of nonlinear structural problems. Internat. J. Numer. Methods Engrg. **94**(1), 60–83 (2013). DOI:10.1002/nme.4441
7. Gosselet, P., Rixen, D., Roux, F.X., Spillane, N.: Simultaneous FETI and block FETI: robust domain decomposition with multiple search directions. Internat. J. Numer. Methods Engrg. **104**(10), 905–927 (2015). DOI:10.1002/nme.4946
8. Klawonn, A., Lanser, M., Rheinbach, O., Uran, M.: Nonlinear FETI-DP and BDDC methods: a unified framework and parallel results. SIAM J. Sci. Comput. **39**(6), C417–C451 (2017). DOI:10.1137/16M1102495
9. Leistner, M.C., Gosselet, P., Rixen, D.J.: Recycling of solution spaces in multipreconditioned FETI methods applied to structural dynamics. Internat. J. Numer. Methods Engrg. **116**(2), 141–160 (2018). DOI:10.1002/nme.5918
10. Leistner, M.C., Gosselet, P., Rixen, D.J.: A simplification of the ritzgeneo recycling strategy for adaptive multi-preconditioned feti applied to multiple right-hand sides. PAMM Preprint pp. 1–2 (2018). DOI:hal-01815402
11. Negrello, C., Gosselet, P., Rey, C., Pebrel, J.: Substructured formulations of nonlinear structure problems—influence of the interface condition. Internat. J. Numer. Methods Engrg. **107**(13), 1083–1105 (2016). DOI:10.1002/nme.5195
12. Pebrel, J., Rey C. Gosselet, P.: A nonlinear dual domain decomposition method: application to structural problems with damage. Internat. J. Multiscale Comput. Engrg. **6**(3), 251–262 (2008). DOI:10.1615/IntJMultCompEng.v6.i3.50

13. Rixen, D.J., Farhat, C.: A simple and efficient extension of a class of substructure based preconditioners to heterogeneous structural mechanics problems. Internat. J. Numer. Methods Engrg. **44**(4), 489–516 (1999). DOI:10.1002/(SICI)1097-0207(19990210)44:4<489::AID-NME514>3.3.CO;2-Q
14. Roux, F.X.: Spectral analysis of the interface operators associated with the preconditioned saddle-point principle domain decomposition method. In: Fifth International Symposium on Domain Decomposition Methods for Partial Differential Equations (Norfolk, VA, 1991), pp. 73–90. SIAM, Philadelphia, PA (1992)
15. Spillane, N.: An adaptive multipreconditioned conjugate gradient algorithm. SIAM Journal on Scientific Computing **38**(3), 1896–1918 (2016). DOI:10.1137/15M1028534
16. Spillane, N., Rixen, D.J.: Automatic spectral coarse spaces for robust finite element tearing and interconnecting and balanced domain decomposition algorithms. Internat. J. Numer. Methods Engrg. **95**(11), 953–990 (2013). DOI:10.1002/nme.4534

New Coarse Corrections for Optimized Restricted Additive Schwarz Using PETSc

Martin J. Gander and Serge Van Criekingen

1 Introduction

Additive Schwarz Methods (ASM) are implemented in the PETSc library [2, 1, 3] within its PCASM preconditioning option. By default this applies the Restricted Additive Schwarz (RAS) method of Cai and Sarkis [4]. We here present the implementation, using PETSc tools, of two further improvements for this method: a new and more effective coarse correction, as well as optimized transmission conditions, resulting in an Optimized two-level Restricted Additive Schwarz (or ORAS2) method.

It is well known that domain decomposition methods applied to elliptic problems need a coarse correction to be scalable, since without it, information is only transferred from each subdomain to its direct neighbors which makes the number of iterations grow with the number of subdomains; for exceptions, see [6, 7]. Scalability is achieved by introducing a coarse grid on which a reduced-size calculation is performed to compute a coarse correction at each iteration of the solution process, yielding a two-level method. Our choice of the coarse grid points follows the method introduced in [11]: the coarse grid points are chosen in 1D to be the extreme grid points of the non-overlapping subdomains used to define RAS, and for a rectangular decomposition in 2D, four coarse grid points are placed around each cross point of the non-overlapping decomposition. This choice of placing the coarse grid nodes leads to substantially faster convergence than the classical option of equally distributing the coarse grid points within each subdomain.

As for optimized transmission conditions, we consider Robin transmission conditions instead of the classical Dirichlet ones, i.e., a well-chosen combination of Dirichlet and Neumann values at subdomain interfaces such as to minimize the

Martin J. Gander
University of Geneva, e-mail: martin.gander@unige.ch

Serge Van Criekingen
CNRS/IDRIS e-mail: serge.van.criekingen@idris.fr

R. Haynes et al. (eds.), *Domain Decomposition Methods in Science and Engineering XXV*,
Lecture Notes in Computational Science and Engineering 138,
https://doi.org/10.1007/978-3-030-56750-7_56

number of iterations. We follow here the method described in [8] which only requires modifying the diagonal entries of interface nodes in the subdomain matrices. A good choice of these diagonal entries leads to a much faster convergence of the associated domain decomposition method than using the standard diagonal entries from RAS.

We present weak scaling numerical results on a 2-D Laplace test case using up to 16384 CPU cores. Combining coarse correction and optimized transmission conditions, we obtain substantially improved computation times with the new optimized two-level RAS method which, despite a larger memory footprint, proves to be competitive with the multigrid library HYPRE (with the default options of the PETSc interface to this library).

2 Coarse Correction and the two-level RAS method

We consider the solution of $A\mathbf{x} = \mathbf{b}$ on a domain Ω decomposed into a set of possibly overlapping subdomains Ω_j. Introducing a restriction operator R_j onto each subdomain Ω_j, local matrices can be built as $A_j = R_j\, A\, R_j^T$. To obtain the Restricted Additive Schwarz (RAS) method, we need also to introduce a partition of Ω into non-overlapping subdomains $\tilde{\Omega}_j$, as well as the corresponding restriction operators $\tilde{R}_j$. Then, the RAS method is defined by the iterations [4]

$$\mathbf{x}^{n+1} = \mathbf{x}^n + \sum_{j=1}^{J} \tilde{R}_j^T\, A_j^{-1}\, R_j\, (\mathbf{b} - A\mathbf{x}^n). \tag{1}$$

The RAS method has the drawback of yielding a non-symmetric system even for symmetric problems, but was shown to converge faster than the Additive Schwarz method because it remedies its non-convergent behavior in the overlaps [9].

To obtain a two-level method through coarse correction, we introduce a restriction operator R_c to the coarse space, such that the coarse system matrix reads $A_c = R_c\, A\, R_c^T$. In turn, the two-level RAS method with multiplicative coarse correction (denoted RAS2 in what follows) can be written as

$$\mathbf{x}^{n+1/2} = \mathbf{x}^n + \sum_{j=1}^{J} \tilde{R}_j^T\, A_j^{-1}\, R_j\, (\mathbf{b} - A\mathbf{x}^n), \tag{2}$$

$$\mathbf{x}^{n+1} = \mathbf{x}^{n+1/2} + R_c^T\, A_c^{-1}\, R_c\, (\mathbf{b} - A\mathbf{x}^{n+1/2}). \tag{3}$$

The definition of the coarse space, that is, the choice of the coarse grid nodes, is critical to obtain an efficient two-level method. Two possible choices are shown in Fig. 1. Compared to the classical approach (circles), the new approach (squares) introduced in [11] shows superior performance since it resolves the residual location along the interfaces well (see also [8, 12]), and is therefore the choice made here (- we however compare the iteration counts for the two methods in Section 4). For the

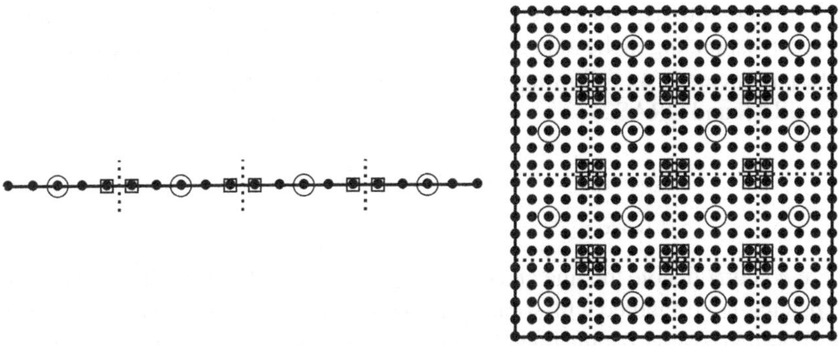

Fig. 1: Two choices of the coarse grid nodes in 1-D and 2-D: 1) the middle of each subdomain (circles) or 2) one node on each side of the (non-overlapping) subdomain interfaces (squares) in 1-D, or in 2-D four nodes around each cross point of the (non-overlapping) decomposition.

1-D case, it was actually shown in [11] that, for the Laplace equation, the new coarse correction yields convergence in two iterations, which is because the new coarse basis functions are harmonic within the subdomains.

In PETSc, the coarse correction was implemented using the `PCSHELL` preconditioning tool, which gives the possibility to entirely define the preconditioner. This self-defined preconditioner was then (multiplicatively) composed with the built-in `PCASM` (i.e., RAS) preconditioner using the `PCCOMPOSITE` composition tool [2].

3 Optimized Interface Conditions and the ORAS2 method

In the RAS iterations (1), each local A_j matrix corresponds to a discretized local problem with homogeneous Dirichlet boundary conditions. Optimized interface conditions are introduced by modifying these matrices into $\hat{A}_j$ matrices, each corresponding to a discretized local problem with homogeneous Robin boundary conditions of the type

$$\frac{\partial u_j}{\partial n_j} + p\, u_j = 0 \qquad \text{on } \partial\Omega_j \setminus \partial\Omega. \tag{4}$$

The resulting optimized RAS method will be denoted by ORAS, and a good choice of the parameter p in (4) is important for good performance.

Starting from the RAS2 iterations (2)-(3), the optimized two-level RAS method, denoted by ORAS2, is obtained as in the one-level case by modifying the local A_j matrices into $\hat{A}_j$ matrices to express Robin interface conditions.

In the numerical experiments below, we consider the 2-D Laplace problem on the unit square, discretized using the 5-point finite difference stencil. Following [8], we obtain $\hat{A}_j$ using only a first-order accurate discretization of the normal derivative in the Robin conditions, which requires modifying only diagonal entries of A_j, namely those corresponding to the unknowns on the interfaces. As for the optimized value p^* of the parameter p, we follow again [8] and take, for the one- and two-level methods (i.e., ORAS and ORAS2)

$$p^*_{\text{one-level}} = 2^{-1/3}\pi^{2/3}(ovlp \cdot h)^{-1/3}, \tag{5}$$

$$p^*_{\text{two-level}} = 2^{-1/3}\pi^{2/3}(ovlp \cdot h)^{-1/3}(H_{x,y})^{-2/3}, \tag{6}$$

where h and H denote the fine and coarse mesh sizes. As for the value of the overlap $ovlp$, it has to be handled with some care: in the formulas (5)-(6), it is the geometrical (i.e., physical) overlap of the method, while the overlap value defined in PETSc is the number of extra mesh layers per subdomain at interfaces. An overlap of 1 in PETSc implies one extra mesh layer for both subdomains at an interface, thus an algebraic overlap of 2 (- an algebraic overlap of 0 corresponds to Block Jacobi). An algebraic overlap of 2 means a geometrical overlap of $3h$ for the RAS method and h for the (one- or two-level) ORAS method [10, 14], and thus $ovlp = 1$ in the above formulas. Similarly, an overlap of 2 in PETSc corresponds to an algebraic overlap of 4, that is a geometrical overlap of $5h$ for RAS and $3h$ for ORAS, and thus $ovlp = 3$.

To end this section, note that PETSc provides the `PCSetModifySubMatrices` tool to modify the diagonal values of the local matrices.

4 Weak Scalability Results

As stated earlier, we perform numerical experiments on the 2-D Laplace problem on the unit square discretized using the 5-point finite difference stencil. We perform a weak scalability analysis, that is, increase the size of the problem while maintaining constant the workload per processor. Each subdomain of the decomposition is handled by one CPU core (corresponding to one MPI rank). We increase the number J of subdomains/cores following the list $J = 4, 16, 64, 256, 1024, \ldots$ with decomposition into $J = \frac{1}{H} \times \frac{1}{H}$ subdomains on the unit square (H being the coarse mesh size as before). To maintain the workload per CPU core constant, the fine mesh size h is decreased proportionally, such that the ratio h/H remains constant as well as, in turn, the local problem size within each subdomain. We consider two workloads, the first one with a 256×256 fine mesh within each subdomain, the second (heavier) one with 512×512 local meshes, yielding a h/H ratio of .004 and .002, respectively.

Three different supercomputers were used to perform our tests: `Ada` and `Turing` at the Institute for Development and Resources in Intensive Scientific Computing (CNRS/IDRIS), and `Occigen` at the National Computing Center for Higher Education (CINES). The `Ada` and `Occigen` machines are meant for a wide-ranging usage and are composed of large memory SMP nodes interconnected by a high-speed InfiniBand network, for a cumulated peak performance of 233 Tflop/s and 3.5 Pflop/s, respectively. The `Turing` machine is an IBM Blue Gene/Q massively parallel architecture with a cumulated performance of 1.258 Pflop/s.

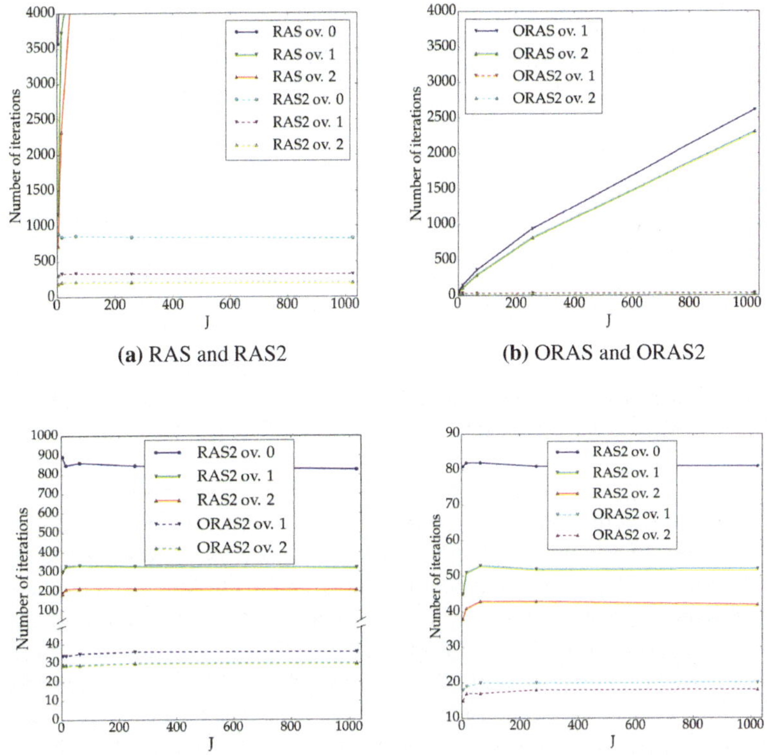

(a) RAS and RAS2 (b) ORAS and ORAS2

(c) RAS2 and ORAS2 (zoom) (d) RAS2 and ORAS2 (GMRES zoom)

Fig. 2: Number of iterations in the weak scaling experiment with $h/H = .004$ (last panel with GMRES acceleration).

Fig. 2 shows in the first three pannels the number of stationary iterations obtained using the one- and two-level (O)RAS methods up to 1024 CPU cores. As can be seen on Figs. 2a and 2b, the one-level methods do not scale (here in terms of iterations), while the two-level methods do. Fig. 2c zooms on the two-level results of the previous plots, showing the superiority of the optimized methods. In Fig. 2d we show that using GMRES acceleration to the experiments in Fig. 2c lowers the iteration counts for all methods, but does not change the relative superior performance of the optimized methods compared to the classical ones. The equivalent zoomed plot obtained (with stationary iterations) using the classical approach consisting in choosing coarse grid nodes in the middle of each subdomain (circled points in Fig. 1) is visible on Fig. 5a. As expected, these results confirm the lower iteration count of the new approach already observed in [11].

Fig. 3 shows Ada timings for the two workloads ($h/H = .004$ and $h/H = .002$) with stationary iterations, using up to 1024 cores. As above in terms of iterations,

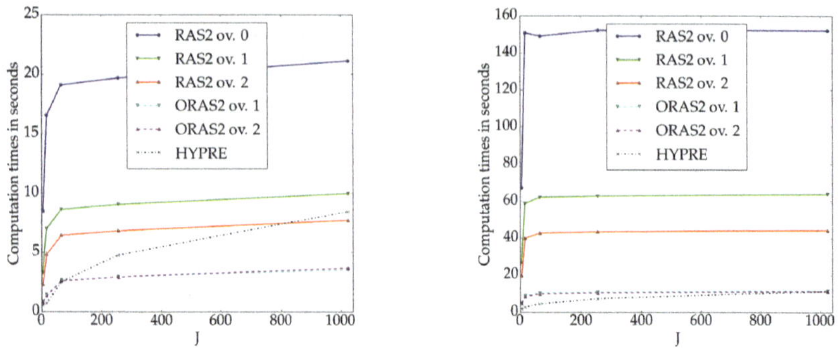

(a) $h/H = .004$ (b) $h/H = .002$

Fig. 3: Computation times (s.) for the weak scaling experiment up to 1024 cores on `Ada`, for the two different workloads. `HYPRE/BoomerAMG` is used with the default PETSc settings.

we here observe that the RAS2 and ORAS2 scale well in terms of computing time, with the optimized methods again showing superior performances. The use of a second layer of overlap does not appear beneficial in the ORAS2 method. On these plots also appear the corresponding results obtained using the multigrid library `HYPRE` as interfaced by PETSc, with the default settings. This amounts to using the `BoomerAMG` [13] component of `HYPRE`, for which the default settings are meant to work fairly well for two-dimensional diffusion problems [5]. The `HYPRE` results exhibit a scalability curve that is not as flat as the (O)RAS2 ones within this range of number of processors, with comparable computing times.

Numerical tests were pursued up to 16384 cores using the `Occigen` and `Turing` machines, as shown in Fig. 4. The scalability properties of the RAS2 and ORAS2

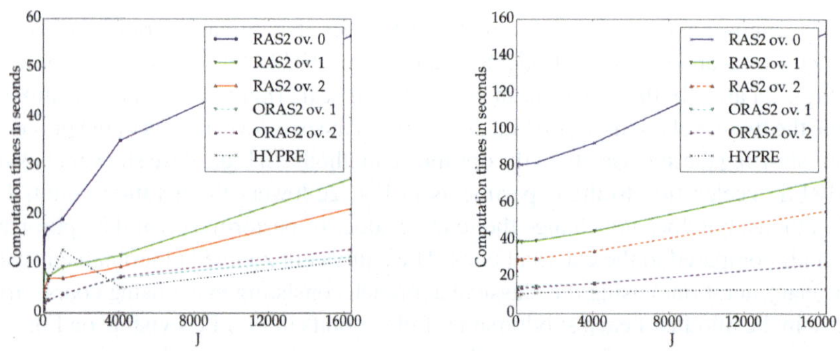

(a) `Occigen` timings (b) `Turing` timings

Fig. 4: Computation times (s.) for the weak scaling experiment up to 16384 cores on `Occigen` and `Turing` with $h/H = .004$.

methods remain decent, with the latter again performing better. As for the HYPRE results, they exhibit on Occigen (Fig. 4a) the expected scalability above 4092 cores, but not up to 1024 cores, as already observed above on Ada. This behavior remains unexplained to us, and has been observed repeatedly on these two machines of similar architecture. Changing architecture and running on Turing (Fig. 4b) however yields a flat scalability curve for HYPRE already below one thousand cores. The computing times on Turing are noticeably slower than on Occigen due to slower processors.

Finally, Fig. 5b shows the memory footprints of the different methods measured on Occigen. The overlapping RAS2 and ORAS2 methods yield very close footprints,

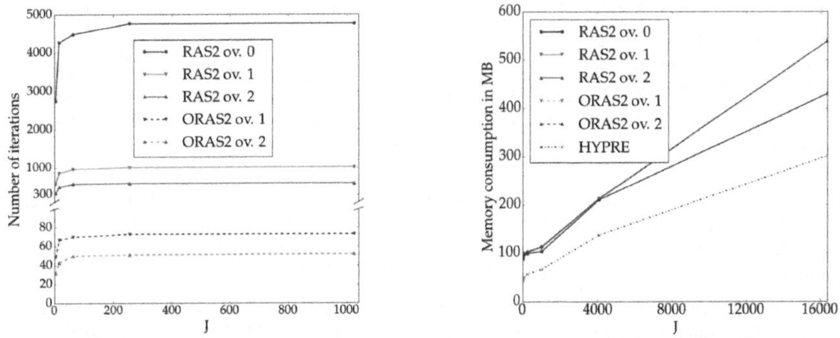

(a) RAS2 and ORAS2 (zoom, coarse nodes in the middle of each subdomain).

(b) Average (on all the MPI tasks) of the maximal physical memory consumption.

Fig. 5: Number of iterations with the classical choice of coarse grid nodes (left) and memory footprint (right) in the weak scaling experiment with $h/H = .004$.

which differ significantly from the non-overlapping ones only at 16384 cores, probably because of MPI scalability effects. Fig. 5b also shows that the HYPRE method yields the lowest memory footprint and it is unclear to us wether this comes from a better implementation or if it has a theoretical explanation.

5 Conclusions

We implemented two improvements to the RAS method built in the PETSc library, namely a new coarse correction to obtain a (scalable) two-level method, as well as optimized interface conditions. This implementation was done using only existing PETSc tools, mainly preconditioner composition and submatrix modification.

We showed numerically that combining these two improvements yields substantial improvement on the standard RAS and, on a 2-D Laplace problem, the resulting ORAS2 method appears competitive with the multigrid HYPRE library up to 16k cores, despite a larger memory footprint.

Acknowledgements The authors wish to acknowledge Olivier Dubois from John Abbot Colllege (Quebec, Canada) for providing a first PETSc implementation of the methods presented here. The second author also wishes to acknowledge the help of his fellow CNRS/IDRIS team members. This work was performed using HPC resources from GENCI-CINES and GENCI-IDRIS.

References

1. Balay, S., Abhyankar, S., Adams, M.F., Brown, J., Brune, P., Buschelman, K., Dalcin, L., Dener, A., Eijkhout, V., Gropp, W.D., Kaushik, D., Knepley, M.G., May, D.A., McInnes, L.C., Mills, R.T., Munson, T., Rupp, K., Sanan, P., Smith, B.F., Zampini, S., Zhang, H.: PETSc users manual. Tech. Rep. ANL-95/11 - Revision 3.10, Argonne National Laboratory (2018). URL `http://www.mcs.anl.gov/petsc`
2. Balay, S., Abhyankar, S., Adams, M.F., Brown, J., Brune, P., Buschelman, K., Dalcin, L., Dener, A., Eijkhout, V., Gropp, W.D., Kaushik, D., Knepley, M.G., May, D.A., McInnes, L.C., Mills, R.T., Munson, T., Rupp, K., Sanan, P., Smith, B.F., Zampini, S., Zhang, H.: PETSc Web page. `http://www.mcs.anl.gov/petsc` (2018). URL `http://www.mcs.anl.gov/petsc`
3. Balay, S., Gropp, W.D., McInnes, L.C., Smith, B.F.: Efficient management of parallelism in object oriented numerical software libraries. In: E. Arge, A.M. Bruaset, H.P. Langtangen (eds.) Modern Software Tools in Scientific Computing, pp. 163–202. Birkhäuser Press (1997)
4. Cai, X.C., Sarkis, M.: A restricted additive Schwarz preconditioner for general sparse linear systems. SIAM J. Sci. Comp. **21**(2), 239–247 (1999)
5. Center for Applied Scientific Computing: Hypre user's manual. Tech. Rep. Software Version: 2.11.2, Lawrence Livermore National Laboratory (2017). URL `https://computation.llnl.gov/projects/hypre-scalable-linear-solvers-multigrid-methods`
6. Ciaramella, G., Gander, M.J.: Analysis of the parallel Schwarz method for growing chains of fixed-sized subdomains: Part I. SIAM Journal on Numerical Analysis **55**(3), 1330–1356 (2017)
7. Ciaramella, G., Gander, M.J.: Analysis of the parallel Schwarz method for growing chains of fixed-sized subdomains: Part II. SIAM Journal on Numerical Analysis **56**(3), 1498–1524 (2018)
8. Dubois, O., Gander, M.J., Loisel, S., St-Cyr, A., Szyld, D.B.: The optimized Schwarz methods with a coarse grid correction. SIAM J. Sci. Comp. **34**(1), A421–A458 (2012)
9. Efstathiou, E., Gander, M.J.: Why restricted additive Schwarz converges faster than additive Schwarz. BIT Numerical Mathematics **43**(5), 945–959 (2003)
10. Gander, M.J.: Schwarz methods over the course of time. ETNA **31**, 228–255 (2008)
11. Gander, M.J., Halpern, L., Santugini, K.: A new coarse grid correction for ras/as. In: Domain Decomposition Methods in Science and Engineering XXI, Lecture Notes in Computational Science and Engineering, pp. 275–284. Springer-Verlag (2014)
12. Gander, M.J., Loneland, A., Rahman, T.: Analysis of a new harmonically enriched multiscale coarse space for domain decomposition methods. arXiv preprint arXiv:1512.05285 (2015)
13. Henson, V.E., Yang, U.M.: Boomeramg: a parallel algebraic multigrid solver and preconditioner. Applied Numerical Mathematics **41**, 155–177 (2002)
14. St-Cyr, A., Gander, M.J., Thomas, S.J.: Optimized multiplicative, additive, and restricted additive Schwarz preconditioning. SIAM J. Sci. Comp. **29**(6), 2402–2425 (2007)

On the Derivation of Optimized Transmission Conditions for the Stokes-Darcy Coupling

Martin J. Gander[1] and Tommaso Vanzan[1]

1 Introduction

Recently a lot of attention has been devoted to the Stokes-Darcy coupling which is a system of equations used to model the flow of fluids in porous media. In [2, 1] a non standard behaviour of the optimized Schwarz method (OSM) has been observed: the optimized parameters obtained solving the classical min-max problems do not lead to an optimized convergence. The authors in [2, 1] proposed to consider a different optimization problem and they claimed that the unexpected behaviour is due to the Krylov acceleration. In this manuscript, we study OSM as an iterative method and as a preconditioner for GMRES and we show that the discrepancy is not due to the Krylov acceleration but to a limitation in the derived convergence factor.

2 The Stokes-Darcy model

We consider a domain Ω divided by an interface Γ into two subdomains, Ω_1 and Ω_2. In Ω_1, a Newtonian fluid is present described by the Stokes equations whose unknowns are the velocity field $\mathbf{u}_f = (u, v)^\top$ and the pressure field p_f,

$$\begin{aligned} -\nabla \cdot \mathbb{T} &= f \quad \text{in} \quad \Omega_1, \\ \nabla \cdot \mathbf{u}_f &= 0 \quad \text{in} \quad \Omega_1, \end{aligned} \tag{1}$$

where $\mathbb{T} = 2\mu_f(\nabla^s \mathbf{u}_f) - p_f \mathbb{I}$ is the stress tensor, with $\nabla^s \mathbf{u}_f$ the symmetrized gradient, and μ_f is the fluid viscosity. The motion of the fluid in the porous media is modelled through the Darcy equations whose unknowns are the velocity and pressure fields in the porous media domain $\mathbf{u}_d$, p_d,

[1] Section de mathématiques, Université de Genève, 2-4 rue du Lièvre, Genève, e-mail: {martin.gander},{tommaso.vanzan}@unige.ch.

R. Haynes et al. (eds.), *Domain Decomposition Methods in Science and Engineering XXV*, Lecture Notes in Computational Science and Engineering 138,
https://doi.org/10.1007/978-3-030-56750-7_57

$$\mathbf{u}_d = -\mathbb{K}\nabla p_d + \mathbf{g}, \quad \nabla \cdot \mathbf{u}_d = 0 \quad \text{in} \quad \Omega_2, \tag{2}$$

where $\mathbb{K}$ is the permeability tensor and $\mathbf{g}$ is a body force vector. Equation (2) can be simplified taking the divergence of the first equation to obtain a second order elliptic PDE only for the pressure field,

$$-\nabla \cdot \mathbb{K}\nabla p_d = -\nabla \cdot \mathbf{g} \quad \text{in} \quad \Omega_2. \tag{3}$$

Both (1) and (3) are closed by Dirichlet boundary conditions on the external boundary $\partial\Omega \setminus \Gamma$, i.e. $\mathbf{u}_f = \mathbf{h}_f$, $p_d = h_d$ on $\partial\Omega \setminus \Gamma$. However the Stokes and Darcy equations still need to be coupled along the common interface Γ and there are many possible choices, see Paragraph 3 of [3]. In the following we prescribe the continuity of the normal velocities and of the normal stresses and the so called Beaver-Joseph-Saffman (BJS) condition,

$$\begin{aligned} \mathbf{u}_f \cdot \mathbf{n} &= -(\mathbb{K}\nabla p_d) \cdot \mathbf{n} + \mathbf{g} \cdot \mathbf{n}, \\ -\mathbf{n} \cdot (2\mu_f \nabla^s \mathbf{u}_f - p_f \mathbb{I}) \cdot \mathbf{n} &= p_d, \\ -\tau \cdot (2\mu_f \nabla^s \mathbf{u}_f - p_f \mathbb{I}) \cdot \mathbf{n} &= \chi_s (\mathbf{u}_f)_\tau. \end{aligned} \tag{4}$$

We remark that the BJS condition $(4)_3$ is not a coupling condition but only a closure condition for the Stokes equations. OSMs use enhanced transmission conditions on the interface, thus we take a linear combination of the coupling conditions $(4)_{1,2}$ introducing the real parameters s_1 and s_2 which are chosen to optimize the convergence. The OSM for the Stokes-Darcy system (1)-(3)-(4) then computes for iterations $n = 1, 2 \dots$

$$\begin{aligned} -\nabla \cdot (2\mu_f \nabla^s \mathbf{u}_f^n - p_f^n \mathbb{I}) &= \mathbf{f}, \quad \text{in} \quad \Omega_1, \\ \nabla \cdot \mathbf{u}_f^n &= 0, \quad \text{in} \quad \Omega_1 \\ -\nabla \cdot \mathbb{K}\nabla p_d^n &= -\nabla \cdot \mathbf{g}, \quad \text{in} \quad \Omega_2, \\ p_d^n - s_1 \left(\mathbb{K}\nabla p_d^n \cdot \mathbf{n} - \mathbf{g} \cdot \mathbf{n}\right) &= -\mathbf{n} \cdot (2\mu_f \nabla^s \mathbf{u}_f^{n-1} - p_f^{n-1} \mathbb{I}) \cdot \mathbf{n} \\ &\quad + s_1 \mathbf{u}_f^{n-1} \cdot \mathbf{n} \quad \text{on} \, \Gamma, \\ -\mathbf{n} \cdot (2\mu_f \nabla^s \mathbf{u}_f^n - p_f^n \mathbb{I}) \cdot \mathbf{n} - s_2 \mathbf{u}_f^n \cdot \mathbf{n} &= p_d^{n-1} + s_2 \left(\mathbb{K}\nabla p_d^{n-1} \cdot \mathbf{n} - \mathbf{g} \cdot \mathbf{n}\right) \quad \text{on} \, \Gamma, \\ -\tau \cdot (2\mu_f \nabla^s \mathbf{u}_f^n - p_f^n \mathbb{I}) \cdot \mathbf{n} &= \chi_s (\mathbf{u}_f^n)_\tau \quad \text{on} \, \Gamma. \end{aligned} \tag{5}$$

In [2], the authors perform a Fourier analysis of the OSM (5). Their analysis follows one of the standard approaches in the literature, i.e. the problem of interest is posed in a simplified setting where one can exploit the Fourier transform for unbounded domains or separation of variables for bounded domains. Unfortunately this last approach is not possible here since no analytical expression is available for the eigenvectors of the Stokes operator in bounded domains with Dirichlet boundary conditions. Furthermore, to simplify the calculations they assume that $\mathbb{K} = \text{diag}(\eta_1, \eta_2)$ with $\eta_j > 0, j = 1, 2$. They finally obtain that the convergence factor of algorithm (5) for all the Fourier frequencies $k \in \mathbb{R}$ is

$$\rho(k,s_1,s_2) = \left|\frac{2\mu_f|k| - s_1}{2\mu_f|k| + s_2} \cdot \frac{1 - s_2\sqrt{\eta_1\eta_2}|k|}{1 + s_1\sqrt{\eta_1\eta_2}|k|}\right|. \tag{6}$$

The optimal choice $s_1 = 2\mu_f|k|$ and $s_2 = \frac{1}{\sqrt{\eta_1\eta_2}|k|}$ would lead to a direct method which converges in just two iterations, however this choice corresponds to non-local operators once backtransformed. Therefore a more practical choice is to set $s_1 = 2\mu_f p$ and $s_2 = \frac{1}{\sqrt{\eta_1\eta_2}p}$ for some $p \in \mathbb{R}$. An equivalent choice of optimized parameters has been treated in [2] where the authors obtain the following result:

Theorem 1 (Proposition 3.3 in [2]) *The unique solution of the min-max problem*

$$\min_{p} \max_{k\in[k_{\min},k_{\max}]} \rho(k,p), \tag{7}$$

is given by the unique root of the non linear equation $\rho(k_{\min},p) = \rho(k_{\max},p)$.

A possible improvement consists in considering two free parameters, choosing $s_1 = 2\mu_f p$ and $s_2 = \frac{1}{\sqrt{\eta_1\eta_2}q}$ with $p,q \in \mathbb{R}$. In [1], the authors propose to choose the couple p,q such that $\rho(k_{\min},p,q) = \rho(\widehat{k},p,q) = \rho(k_{\max},p,q)$, i.e. they impose equioscillation to obtain the optimized parameters. Even though often the solution of such min-max problems is indeed given by equioscillation, a priori there is no reason why this should be the case also for the Stokes-Darcy coupling. In fact for heterogenous problems, it has been observed that there can exist a couple of parameters which satisfies the equioscillation property, but leads to a non optimized convergence or even to a divergence method, see [6, 4, 7]. In Theorem 2 we refine Proposition 1 of [1].

Theorem 2 *The solutions of the min-max problem*

$$\min_{p,q\in\mathbb{R}} \max_{k\in[k_{\min},k_{\max}]} \rho(k,p,q), \tag{8}$$

where $\rho(k,p,q) = 2\mu_f\sqrt{\eta_1\eta_2}\left|\frac{k-p}{1+2\mu_f\sqrt{\eta_1\eta_2}kp} \cdot \frac{k-q}{1+2\mu_f\sqrt{\eta_1\eta_2}kq}\right|$, *are given by two pairs* (p_i^*,q_i^*), $i=1,2$ *which satisfy the non linear equations* $|\rho(k_{\min},p_i^*,q_i^*)| = |\rho(\widehat{k},p_i^*,q_i^*)| = |\rho(k_{\max},p_i^*,q_i^*)|$, $\widehat{k}$ *being an interior maximum. Moreover* $p_2^* = q_1^*$ *and* $q_2^* = p_1^*$.

Proof The proof is based on arguments presented in [4, 8, 7] and we outline the main steps. We first observe that $\rho(k,p,q)$ is invariant under $p \leftrightarrow q$, hence we consider only $p < q$ and moreover $\rho(k,p,q) = 0$ for $k = q$ and $k = p$. The partial derivatives with respect to the parameters satisfy $\text{sign}(\partial_p\rho) = \text{sign}(p-k)$ and $\text{sign}(\partial_q\rho) = \text{sign}(q-k)$, therefore at optimality we conclude that p,q lie in $[k_{\min},k_{\max}]$, see the proof of Theorem 1 in [8]. Solving $\partial_k\rho = 0$, we get that there exists a unique interior maximum $\widehat{k}$, with $p < \widehat{k} < q$, so that we can restrict $\max_{k\in[k_{\min},k_{\max}]}\rho(k,p,q) = \max\{\rho(k_{\min},p,q),\rho(\widehat{k},p,q),\rho(k_{\max},p,q)\}$. Repeating the same arguments of Lemma 2.9 in [7], we obtain that at the optimum we must have $\rho(k_{\min},p,q) = \rho(k_{\max},p,q)$, so that we can express q as function of p and we can restrict the study to $\min_p \max\{\rho(k_{\min},p,q(p)),\rho(\widehat{k},p,q(p))\}$. Defining $\delta := 2\mu_f\sqrt{\eta_1\eta_2}$, the equioscillation constraint is equivalent to

$$l(p) := \frac{k_{\min} - p}{1 + \delta k_{\min} p} \frac{1 + \delta k_{\max} p}{k_{\max} - p} = \frac{k_{\max} - q(p)}{1 + \delta q(p) k_{\max}} \frac{1 + \delta q(p) k_{\min}}{k_{\min} - q(p)} =: g(p). \quad (9)$$

Since $\partial_p l(p) < 0$ and $\partial_p g(p) > 0$, $q(p)$ must be a decreasing function of p so that eq (9) is satisfied. Then using the sign of the derivatives of ρ with respect to p and q and the explicit expression of $q(p)$, we have $\frac{d\rho(k_{\min},p)}{dp} > 0$ and $\frac{d\rho(\widehat{k},p)}{dp} < 0$ for $k_{\min} < p < q(p)$. These observations are sufficient to conclude, see Theorem 1 in [8], that the solution of $\min_p \max\{\rho(k_{\min}, p, q(p)), \rho(\widehat{k}, p, q(p))\}$ is given by the unique p_1^*, such that $\rho(k_{\min}, p_1^*, q(p_1^*)) = \rho(\widehat{k}, p_1^*, q(p_1^*))$ and q_1^* given by $q_1^* = q(p_1^*)$. Due to the invariance $p \leftrightarrow q$, we get the same results in the case $q < p$ and we conclude that the other couple satisfies $p_2^* = q_1^*$ and $q_2^* = p_1^*$. □

In [2, 1], the authors studied extensively the methods obtained from Theorems 1-2 as preconditioners for GMRES. They observed that these optimized parameters do not lead to an optimized convergence and they proposed to minimize the L^1 norm instead of the maximum of the convergence factor,

$$\min_p \frac{1}{k_{\max} - k_{\min}} \int_{k_{\min}}^{k_{\max}} \rho(k, p) dk. \quad (10)$$

The reason behind this choice lies in the assumption that the Krylov method can take care of isolated slow frequencies, and therefore it would be better to have a convergence factor that is very small for a large set of frequencies with possibly high peaks. This approach was first discussed in [5] for the Helmholtz problem, with the significant difference that the OSM does not converge for the Helmholtz frequency ω, and thus the authors proposed to minimize $\min_p \max_{k \in [k_{\min}, \omega^-] \cup [\omega^+, k_{\max}]} \rho(k, p)$. Since such a bad performance of the optimized parameters obtained from a min-max problem in combination with a Krylov method does not have comparison in the literature, we investigate it in details in the next Section.

3 Numerical study of the optimized Schwarz method

We consider the domains $\Omega_1 = (0, 1) \times (0, 1)$, $\Omega_2 = (0, 1) \times (-1, 0)$ and a uniform structured mesh with mesh size $h = 0.02$, so that $k_{\min} = \pi$ and $k_{\max} = \pi/h$. We discretize the corresponding error equations of (5) with Taylor-Hood finite elements $\mathbb{P}_2^2 - \mathbb{P}_1$ for the Stokes unknowns and $\mathbb{P}_2$ elements for the Darcy pressure. The physical parameters are set equal to $\mu_f = 0.1$, $\eta_1 = \eta_2 = 1$. The stopping criterion for the iterative method is $\|u^n\|_{H^1} + \|v^n\|_{H^1} + \|p_f^n\|_{L^2} + \|p_d^n\|_{H^1} < 10^{-9}$ and similarly for GMRES the tolerance is 10^{-9}. Figure 1 shows the number of iterations to reach convergence. On the left panel we show with a circle the optimized parameter p obtained from Theorem 1 and with a square the optimized p obtained solving (10). We observe that indeed the solution of (10) leads to a faster convergence than the classical approach of Theorem 1 for the preconditioned GMRES. This is in accordance with the results proposed in [2, 1], where it has been shown that the

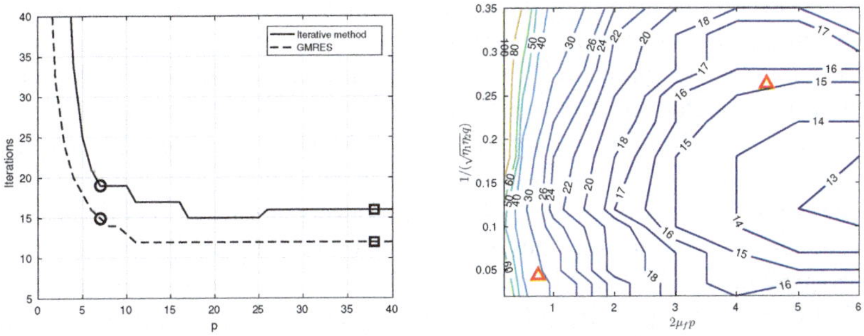

Fig. 1: Number of iterations to reach the tolerance 10^{-9} for different optimized parameters. On the left, the circle represents the solution of Theorem 1, the square corresponds to the solution of (10). On the right the triangles correspond to the double solutions of Theorem 2 and the contour plot refers to the iterative method.

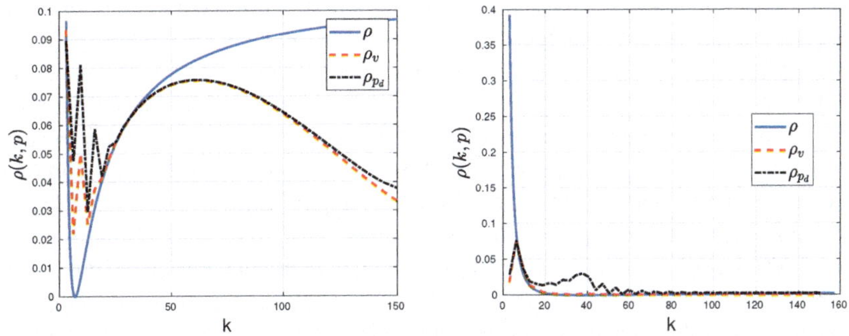

Fig. 2: Comparison of the theoretical and numerical convergence factors. On the left, optimized parameter from Theorem 1 and on the right, optimized parameter from (10).

solution of (10) leads to an equivalent or faster convergence than Theorem 1 for a wide range of parameters. However, we remark that (10) leads to a faster method than (7) also for the iterative method and not only under Krylov acceleration! On the right panel of Fig. 1 we observe that also Theorem 2 does not lead to an optimized convergence and the symmetry of the parameters has disappeared. To understand better the behaviour of the method, we initialize it setting as initial condition one by one the sine functions which correspond to the restriction of the Fourier basis $\{e^{-ikx}\}_k$ on bounded domains with Dirichlet boundary conditions. We then compute numerically an approximation of the convergence factor defining $\rho_v(k,p) = \left(\frac{\|v^3\|_{H^1}}{\|v^1\|_{H^1}}\right)$, $\rho_{p_d}(k,p) = \left(\frac{\|p_d^3\|_{H^1}}{\|p_d^1\|_{H^1}}\right)$, where v^n is the Stokes velocity in the y direction at iteration n and p_d^n is the Darcy pressure at iteration n. From the results presented in Figure 2, we observe two major issues: the first one is a very poor approximation of high frequencies. This is due to the fact that the chosen finite element spaces $\mathbb{P}_2^2 - \mathbb{P}_1 - \mathbb{P}_2$ are not capable of representing properly the exponential boundary layer of the high frequencies near the interface. We propose two remedies which can also be combined. We could first raise the order of the approximation of the finite element spaces to $\mathbb{P}_3^2 - \mathbb{P}_2 - \mathbb{P}_3$ and/or refine the mesh in the normal direction to the interface. Both remedies improve the representation of the high frequencies and in the following we only consider the first one. The second issue lies in a unusual oscillatory behaviour of the low, odd frequencies. This is due to the fact that the unbounded analysis used to obtain the convergence factor is not transferable to the bounded case, since the sines do not form a separated variable solution for the Stokes operator with Dirichlet boundary conditions. Hence, for instance in the right panel of Figure 2, the first frequency $\sin(\pi x)$ is transformed after one iteration into a complicated combination of higher frequencies so that actually the parameter p makes the method much faster than the theory predicts. Therefore it is not possible to diagonalize the iteration as the formula of the convergence factor (6) assumes. This phenomeon was first discussed in [8, 7] where the authors show that for the coupling of the Laplace equation with an advection-diffusion equation with tangential advection, the unbounded analysis leads to inefficient optimized parameters since the two equations lack a common eigenbasis. We consider now the Stokes-Darcy system (5) with periodic boundary conditions on the vertical edges in order to make the bounded problem as similar as possible to the unbounded case. In this setting there exists a separated variable solution for the Stokes problem involving the Fourier basis $\{e^{-ikx}\}_k$, see [9]. In Figure 3 we show both the numerical and theoretical convergence factors computed for even frequencies $\{\sin(2k\pi x)\}_k$. The same results are obtained using the other periodic frequencies $\{\cos(2k\pi x)\}_k$. Comparing with Figure 2, we observe that now we have an excellent agreement between the numerical and theoretical convergence factors and thus we would expect that the optimized parameters from the min-max theorems provide optimized convergence. We thus start the OSM method (5) with initial guesses given by a linear combination of periodic sine and cosine functions multiplied by random coefficients. Figure 4 shows that both Theorem 1 and 2 now lead to optimized convergence for the iterative method (5) and we also observe the symmetry of the optimized parameters in the right panel as Theorem 2 predicts. However concerning GMRES, we note that the optimized parameter from Theorem 1 is still a bit too small. This can be understood studying the eigenvalues of the preconditioned matrix system which are shown in Figure 5. Analyzing the large real eigenvalue, we have observed that the corresponding eigenvector is given by a zero velocity field $\mathbf{u}_f$, a constant pressure p_f and a linear Darcy pressure p_d. This constant mode is actually not treated by the unbounded Fourier analysis and it is not present in our initial guess for the iterative method. Defining the functions $p_d^n = D^n(y+L)$ and $p_f = P^n$ with $P, D \in \mathbb{R}$ and L is the vertical length of the subdomains, and inserting them into the OSM algorithm (5), we obtain a convergence factor $\rho(k=0,p) := \frac{1-s_2}{1+s_1}$. Solving numerically the min-max problem $\min_p \max_{k\in\{0\}\cup[k_{\min},k_{\max}]} \rho(k,p)$ we obtain the equioscillation between $\rho(0,p)$ and $\rho(k_{\min},p)$ and a numerical value of $p \approx 48$. In the right panel of Figure 5 we start the method with a totally random initial guess and this shows that taking into account the constant mode actually makes our analysis exact.

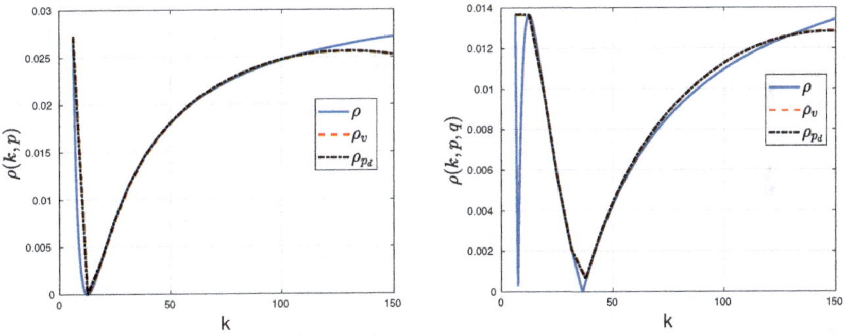

Fig. 3: Comparison of the theoretical and numerical convergence factors. On the left for the single sided optimized parameter from Theorem 1 and on the right one for the double sided parameters of Theorem 2. The minimum frequency is now $k_{\min} = 2\pi$.

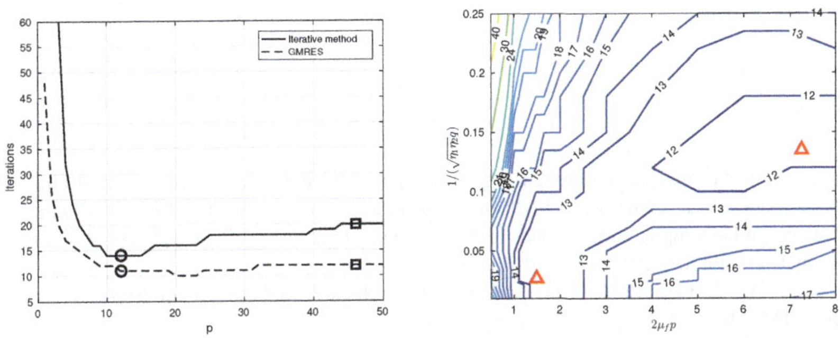

Fig. 4: Number of iterations to reach the tolerance 10^{-9} for different optimized parameters. On the left, the circle represents the solution of Theorem 1, the square corresponds to the approach of (10). On the right the triangles correspond to the double solutions of Theorem 2 and the contour plot refers to the iterative method.

4 Conclusions

In this manuscript we showed that the bad performance of the optimized parameters of the min-max problems for the Stokes-Darcy coupling is not due to the Krylov acceleration but to the difficulty of transferring the unbounded Fourier analysis to the bounded case. For Dirichlet boundary conditions, the problem lies in the odd frequencies which mix among them during the iterations and therefore the convergence factor (6) loses its accuracy. For periodic boundary conditions, we recover a perfect agreement between the unbounded analysis and the numerical simulations for periodic frequencies, however the Fourier analysis does not deal with the constant mode which is present in the bounded case. Including the constant mode in the analysis we recover the optimality of the min-max optimized parameters for periodic boundary conditions.

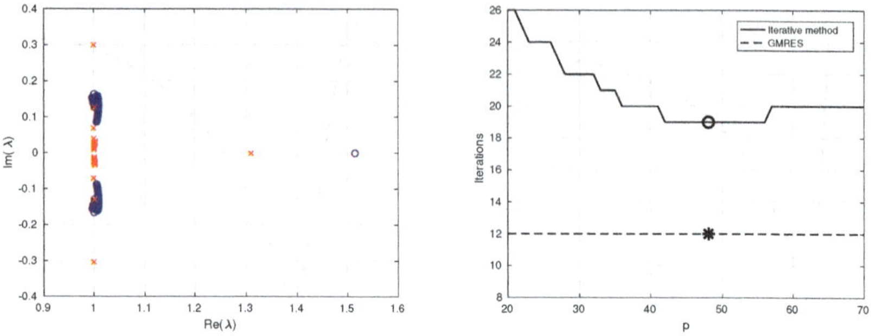

Fig. 5: On the left panel, the blue circles correspond to first 100 eigenvalues of the preconditioned volume matrix in the case with the optimized parameter of Theorem 1 and the red crosses in the case using the solution of (10). On the right panel we show the number of iterations to reach convergence with periodic boundary conditions and with a random initial guess. The circle corresponds to the solution of Theorem 1 and the star to the value of p such that we have the minimal residual of GMRES.

References

1. Discacciati, M., Gerardo-Giorda, L.: Is minimising the convergence rate a good choice for efficient optimized Schwarz preconditioning in heterogeneous coupling? the Stokes-Darcy case. In: Domain Decomposition Methods in Science and Engineering XXIV, pp. 233–241. Springer International Publishing, Cham (2018)
2. Discacciati, M., Gerardo-Giorda, L.: Optimized Schwarz methods for the Stokes-Darcy coupling. IMA Journal of Numerical Analysis (2018)
3. Discacciati, M., Quarteroni, A.: Navier-Stokes/Darcy coupling: modeling, analysis, and numerical approximation. Revista Matematica Complutense (2009)
4. Gander, M.J., Dubois, O.: Optimized Schwarz methods for a diffusion problem with discontinuous coefficient. Numerical Algorithms **69**(1), 109–144 (2015)
5. Gander, M.J., Magoules, F., Nataf, F.: Optimized Schwarz methods without overlap for the Helmholtz equation. SIAM Journal on Scientific Computing **24**(1), 38–60 (2002)
6. Gander, M.J., Vanzan, T.: Heterogeneous optimized Schwarz methods for coupling Helmholtz and Laplace equations. In: Domain Decomposition Methods in Science and Engineering XXIV, pp. 311–320. Springer International Publishing, Cham (2018)
7. Gander, M.J., Vanzan, T.: Heterogeneous optimized Schwarz methods for second order elliptic PDEs. to appear in SIAM Journal of Scientific Computing (2019)
8. Gander, M.J., Vanzan, T.: Optimized Schwarz methods for advection diffusion equations in bounded domains. In: Numerical Mathematics and Advanced Applications ENUMATH 2017, pp. 921–929. Springer International Publishing, Cham (2019)
9. Rummler, B.: The eigenfunctions of the Stokes operator in special domains. ii. ZAMM - Journal of Applied Mathematics and Mechanics **77**(9), 669–675 (1997)

Editorial Policy

1. Volumes in the following three categories will be published in LNCSE:

i) Research monographs
ii) Tutorials
iii) Conference proceedings

Those considering a book which might be suitable for the series are strongly advised to contact the publisher or the series editors at an early stage.

2. Categories i) and ii). Tutorials are lecture notes typically arising via summer schools or similar events, which are used to teach graduate students. These categories will be emphasized by Lecture Notes in Computational Science and Engineering. **Submissions by interdisciplinary teams of authors are encouraged.** The goal is to report new developments – quickly, informally, and in a way that will make them accessible to non-specialists. In the evaluation of submissions timeliness of the work is an important criterion. Texts should be well-rounded, well-written and reasonably self-contained. In most cases the work will contain results of others as well as those of the author(s). In each case the author(s) should provide sufficient motivation, examples, and applications. In this respect, Ph.D. theses will usually be deemed unsuitable for the Lecture Notes series. Proposals for volumes in these categories should be submitted either to one of the series editors or to Springer-Verlag, Heidelberg, and will be refereed. A provisional judgement on the acceptability of a project can be based on partial information about the work: a detailed outline describing the contents of each chapter, the estimated length, a bibliography, and one or two sample chapters – or a first draft. A final decision whether to accept will rest on an evaluation of the completed work which should include

- at least 100 pages of text;
- a table of contents;
- an informative introduction perhaps with some historical remarks which should be accessible to readers unfamiliar with the topic treated;
- a subject index.

3. Category iii). Conference proceedings will be considered for publication provided that they are both of exceptional interest and devoted to a single topic. One (or more) expert participants will act as the scientific editor(s) of the volume. They select the papers which are suitable for inclusion and have them individually refereed as for a journal. Papers not closely related to the central topic are to be excluded. Organizers should contact the Editor for CSE at Springer at the planning stage, see *Addresses* below.

In exceptional cases some other multi-author-volumes may be considered in this category.

4. Only works in English will be considered. For evaluation purposes, manuscripts may be submitted in print or electronic form, in the latter case, preferably as pdf- or zipped ps-files. Authors are requested to use the LaTeX style files available from Springer at http://www.springer.com/gp/authors-editors/book-authors-editors/manuscript-preparation/5636 (Click on LaTeX Template → monographs or contributed books).

For categories ii) and iii) we strongly recommend that all contributions in a volume be written in the same LaTeX version, preferably LaTeX2e. Electronic material can be included if appropriate. Please contact the publisher.

Careful preparation of the manuscripts will help keep production time short besides ensuring satisfactory appearance of the finished book in print and online.

5. The following terms and conditions hold. Categories i), ii) and iii):

Authors receive 50 free copies of their book. No royalty is paid.
Volume editors receive a total of 50 free copies of their volume to be shared with authors, but no royalties.

Authors and volume editors are entitled to a discount of 40 % on the price of Springer books purchased for their personal use, if ordering directly from Springer.

6. Springer secures the copyright for each volume.

Addresses:

Timothy J. Barth
NASA Ames Research Center
NAS Division
Moffett Field, CA 94035, USA
barth@nas.nasa.gov

Michael Griebel
Institut für Numerische Simulation
der Universität Bonn
Wegelerstr. 6
53115 Bonn, Germany
griebel@ins.uni-bonn.de

David E. Keyes
Mathematical and Computer Sciences
and Engineering
King Abdullah University of Science
and Technology
P.O. Box 55455
Jeddah 21534, Saudi Arabia
david.keyes@kaust.edu.sa

and

Department of Applied Physics
and Applied Mathematics
Columbia University
500 W. 120 th Street
New York, NY 10027, USA
kd2112@columbia.edu

Risto M. Nieminen
Department of Applied Physics
Aalto University School of Science
and Technology
00076 Aalto, Finland
risto.nieminen@aalto.fi

Dirk Roose
Department of Computer Science
Katholieke Universiteit Leuven
Celestijnenlaan 200A
3001 Leuven-Heverlee, Belgium
dirk.roose@cs.kuleuven.be

Tamar Schlick
Department of Chemistry
and Courant Institute
of Mathematical Sciences
New York University
251 Mercer Street
New York, NY 10012, USA
schlick@nyu.edu

Editor for Computational Science
and Engineering at Springer:

Martin Peters
Springer-Verlag
Mathematics Editorial IV
Tiergartenstrasse 17
69121 Heidelberg, Germany
martin.peters@springer.com

Lecture Notes in Computational Science and Engineering

1. D. Funaro, *Spectral Elements for Transport-Dominated Equations.*
2. H.P. Langtangen, *Computational Partial Differential Equations.* Numerical Methods and Diffpack Programming.
3. W. Hackbusch, G. Wittum (eds.), *Multigrid Methods V.*
4. P. Deuflhard, J. Hermans, B. Leimkuhler, A.E. Mark, S. Reich, R.D. Skeel (eds.), *Computational Molecular Dynamics: Challenges, Methods, Ideas.*
5. D. Kröner, M. Ohlberger, C. Rohde (eds.), *An Introduction to Recent Developments in Theory and Numerics for Conservation Laws.*
6. S. Turek, *Efficient Solvers for Incompressible Flow Problems.* An Algorithmic and Computational Approach.
7. R. von Schwerin, ***M**ulti **B**ody **S**ystem **SIM**ulation.* Numerical Methods, Algorithms, and Software.
8. H.-J. Bungartz, F. Durst, C. Zenger (eds.), *High Performance Scientific and Engineering Computing.*
9. T.J. Barth, H. Deconinck (eds.), *High-Order Methods for Computational Physics.*
10. H.P. Langtangen, A.M. Bruaset, E. Quak (eds.), *Advances in Software Tools for Scientific Computing.*
11. B. Cockburn, G.E. Karniadakis, C.-W. Shu (eds.), *Discontinuous Galerkin Methods.* Theory, Computation and Applications.
12. U. van Rienen, *Numerical Methods in Computational Electrodynamics.* Linear Systems in Practical Applications.
13. B. Engquist, L. Johnsson, M. Hammill, F. Short (eds.), *Simulation and Visualization on the Grid.*
14. E. Dick, K. Riemslagh, J. Vierendeels (eds.), *Multigrid Methods VI.*
15. A. Frommer, T. Lippert, B. Medeke, K. Schilling (eds.), *Numerical Challenges in Lattice Quantum Chromodynamics.*
16. J. Lang, *Adaptive Multilevel Solution of Nonlinear Parabolic PDE Systems.* Theory, Algorithm, and Applications.
17. B.I. Wohlmuth, *Discretization Methods and Iterative Solvers Based on Domain Decomposition.*
18. U. van Rienen, M. Günther, D. Hecht (eds.), *Scientific Computing in Electrical Engineering.*
19. I. Babuška, P.G. Ciarlet, T. Miyoshi (eds.), *Mathematical Modeling and Numerical Simulation in Continuum Mechanics.*
20. T.J. Barth, T. Chan, R. Haimes (eds.), *Multiscale and Multiresolution Methods.* Theory and Applications.
21. M. Breuer, F. Durst, C. Zenger (eds.), *High Performance Scientific and Engineering Computing.*
22. K. Urban, *Wavelets in Numerical Simulation.* Problem Adapted Construction and Applications.
23. L.F. Pavarino, A. Toselli (eds.), *Recent Developments in Domain Decomposition Methods.*

24. T. Schlick, H.H. Gan (eds.), *Computational Methods for Macromolecules: Challenges and Applications.*

25. T.J. Barth, H. Deconinck (eds.), *Error Estimation and Adaptive Discretization Methods in Computational Fluid Dynamics.*

26. M. Griebel, M.A. Schweitzer (eds.), *Meshfree Methods for Partial Differential Equations.*

27. S. Müller, *Adaptive Multiscale Schemes for Conservation Laws.*

28. C. Carstensen, S. Funken, W. Hackbusch, R.H.W. Hoppe, P. Monk (eds.), *Computational Electromagnetics.*

29. M.A. Schweitzer, *A Parallel Multilevel Partition of Unity Method for Elliptic Partial Differential Equations.*

30. T. Biegler, O. Ghattas, M. Heinkenschloss, B. van Bloemen Waanders (eds.), *Large-Scale PDE-Constrained Optimization.*

31. M. Ainsworth, P. Davies, D. Duncan, P. Martin, B. Rynne (eds.), *Topics in Computational Wave Propagation.* Direct and Inverse Problems.

32. H. Emmerich, B. Nestler, M. Schreckenberg (eds.), *Interface and Transport Dynamics.* Computational Modelling.

33. H.P. Langtangen, A. Tveito (eds.), *Advanced Topics in Computational Partial Differential Equations.* Numerical Methods and Diffpack Programming.

34. V. John, *Large Eddy Simulation of Turbulent Incompressible Flows.* Analytical and Numerical Results for a Class of LES Models.

35. E. Bänsch (ed.), *Challenges in Scientific Computing - CISC 2002.*

36. B.N. Khoromskij, G. Wittum, *Numerical Solution of Elliptic Differential Equations by Reduction to the Interface.*

37. A. Iske, *Multiresolution Methods in Scattered Data Modelling.*

38. S.-I. Niculescu, K. Gu (eds.), *Advances in Time-Delay Systems.*

39. S. Attinger, P. Koumoutsakos (eds.), *Multiscale Modelling and Simulation.*

40. R. Kornhuber, R. Hoppe, J. Périaux, O. Pironneau, O. Wildlund, J. Xu (eds.), *Domain Decomposition Methods in Science and Engineering.*

41. T. Plewa, T. Linde, V.G. Weirs (eds.), *Adaptive Mesh Refinement – Theory and Applications.*

42. A. Schmidt, K.G. Siebert, *Design of Adaptive Finite Element Software.* The Finite Element Toolbox ALBERTA.

43. M. Griebel, M.A. Schweitzer (eds.), *Meshfree Methods for Partial Differential Equations II.*

44. B. Engquist, P. Lötstedt, O. Runborg (eds.), *Multiscale Methods in Science and Engineering.*

45. P. Benner, V. Mehrmann, D.C. Sorensen (eds.), *Dimension Reduction of Large-Scale Systems.*

46. D. Kressner, *Numerical Methods for General and Structured Eigenvalue Problems.*

47. A. Boriçi, A. Frommer, B. Joó, A. Kennedy, B. Pendleton (eds.), *QCD and Numerical Analysis III.*

48. F. Graziani (ed.), *Computational Methods in Transport.*

49. B. Leimkuhler, C. Chipot, R. Elber, A. Laaksonen, A. Mark, T. Schlick, C. Schütte, R. Skeel (eds.), *New Algorithms for Macromolecular Simulation.*

50. M. Bücker, G. Corliss, P. Hovland, U. Naumann, B. Norris (eds.), *Automatic Differentiation: Applications, Theory, and Implementations.*

51. A.M. Bruaset, A. Tveito (eds.), *Numerical Solution of Partial Differential Equations on Parallel Computers.*

52. K.H. Hoffmann, A. Meyer (eds.), *Parallel Algorithms and Cluster Computing.*

53. H.-J. Bungartz, M. Schäfer (eds.), *Fluid-Structure Interaction.*

54. J. Behrens, *Adaptive Atmospheric Modeling.*

55. O. Widlund, D. Keyes (eds.), *Domain Decomposition Methods in Science and Engineering XVI.*

56. S. Kassinos, C. Langer, G. Iaccarino, P. Moin (eds.), *Complex Effects in Large Eddy Simulations.*

57. M. Griebel, M.A Schweitzer (eds.), *Meshfree Methods for Partial Differential Equations III.*

58. A.N. Gorban, B. Kégl, D.C. Wunsch, A. Zinovyev (eds.), *Principal Manifolds for Data Visualization and Dimension Reduction.*

59. H. Ammari (ed.), *Modeling and Computations in Electromagnetics: A Volume Dedicated to Jean-Claude Nédélec.*

60. U. Langer, M. Discacciati, D. Keyes, O. Widlund, W. Zulehner (eds.), *Domain Decomposition Methods in Science and Engineering XVII.*

61. T. Mathew, *Domain Decomposition Methods for the Numerical Solution of Partial Differential Equations.*

62. F. Graziani (ed.), *Computational Methods in Transport: Verification and Validation.*

63. M. Bebendorf, *Hierarchical Matrices.* A Means to Efficiently Solve Elliptic Boundary Value Problems.

64. C.H. Bischof, H.M. Bücker, P. Hovland, U. Naumann, J. Utke (eds.), *Advances in Automatic Differentiation.*

65. M. Griebel, M.A. Schweitzer (eds.), *Meshfree Methods for Partial Differential Equations IV.*

66. B. Engquist, P. Lötstedt, O. Runborg (eds.), *Multiscale Modeling and Simulation in Science.*

67. I.H. Tuncer, Ü. Gülcat, D.R. Emerson, K. Matsuno (eds.), *Parallel Computational Fluid Dynamics 2007.*

68. S. Yip, T. Diaz de la Rubia (eds.), *Scientific Modeling and Simulations.*

69. A. Hegarty, N. Kopteva, E. O'Riordan, M. Stynes (eds.), *BAIL* 2008 – *Boundary and Interior Layers.*

70. M. Bercovier, M.J. Gander, R. Kornhuber, O. Widlund (eds.), *Domain Decomposition Methods in Science and Engineering XVIII.*

71. B. Koren, C. Vuik (eds.), *Advanced Computational Methods in Science and Engineering.*

72. M. Peters (ed.), *Computational Fluid Dynamics for Sport Simulation.*

73. H.-J. Bungartz, M. Mehl, M. Schäfer (eds.), *Fluid Structure Interaction II - Modelling, Simulation, Optimization.*

74. D. Tromeur-Dervout, G. Brenner, D.R. Emerson, J. Erhel (eds.), *Parallel Computational Fluid Dynamics 2008.*

75. A.N. Gorban, D. Roose (eds.), *Coping with Complexity: Model Reduction and Data Analysis.*

76. J.S. Hesthaven, E.M. Rønquist (eds.), *Spectral and High Order Methods for Partial Differential Equations.*

77. M. Holtz, *Sparse Grid Quadrature in High Dimensions with Applications in Finance and Insurance.*

78. Y. Huang, R. Kornhuber, O.Widlund, J. Xu (eds.), *Domain Decomposition Methods in Science and Engineering XIX.*

79. M. Griebel, M.A. Schweitzer (eds.), *Meshfree Methods for Partial Differential Equations V.*

80. P.H. Lauritzen, C. Jablonowski, M.A. Taylor, R.D. Nair (eds.), *Numerical Techniques for Global Atmospheric Models.*

81. C. Clavero, J.L. Gracia, F.J. Lisbona (eds.), *BAIL 2010 – Boundary and Interior Layers, Computational and Asymptotic Methods.*

82. B. Engquist, O. Runborg, Y.R. Tsai (eds.), *Numerical Analysis and Multiscale Computations.*

83. I.G. Graham, T.Y. Hou, O. Lakkis, R. Scheichl (eds.), *Numerical Analysis of Multiscale Problems.*

84. A. Logg, K.-A. Mardal, G. Wells (eds.), *Automated Solution of Differential Equations by the Finite Element Method.*

85. J. Blowey, M. Jensen (eds.), *Frontiers in Numerical Analysis - Durham 2010.*

86. O. Kolditz, U.-J. Gorke, H. Shao, W. Wang (eds.), *Thermo-Hydro-Mechanical-Chemical Processes in Fractured Porous Media - Benchmarks and Examples.*

87. S. Forth, P. Hovland, E. Phipps, J. Utke, A. Walther (eds.), *Recent Advances in Algorithmic Differentiation.*

88. J. Garcke, M. Griebel (eds.), *Sparse Grids and Applications.*

89. M. Griebel, M.A. Schweitzer (eds.), *Meshfree Methods for Partial Differential Equations VI.*

90. C. Pechstein, *Finite and Boundary Element Tearing and Interconnecting Solvers for Multiscale Problems.*

91. R. Bank, M. Holst, O. Widlund, J. Xu (eds.), *Domain Decomposition Methods in Science and Engineering XX.*

92. H. Bijl, D. Lucor, S. Mishra, C. Schwab (eds.), *Uncertainty Quantification in Computational Fluid Dynamics.*

93. M. Bader, H.-J. Bungartz, T. Weinzierl (eds.), *Advanced Computing.*

94. M. Ehrhardt, T. Koprucki (eds.), *Advanced Mathematical Models and Numerical Techniques for Multi-Band Effective Mass Approximations.*

95. M. Azaïez, H. El Fekih, J.S. Hesthaven (eds.), *Spectral and High Order Methods for Partial Differential Equations ICOSAHOM 2012.*

96. F. Graziani, M.P. Desjarlais, R. Redmer, S.B. Trickey (eds.), *Frontiers and Challenges in Warm Dense Matter.*

97. J. Garcke, D. Pflüger (eds.), *Sparse Grids and Applications – Munich 2012.*

98. J. Erhel, M. Gander, L. Halpern, G. Pichot, T. Sassi, O. Widlund (eds.), *Domain Decomposition Methods in Science and Engineering XXI.*

99. R. Abgrall, H. Beaugendre, P.M. Congedo, C. Dobrzynski, V. Perrier, M. Ricchiuto (eds.), *High Order Nonlinear Numerical Methods for Evolutionary PDEs - HONOM 2013.*

100. M. Griebel, M.A. Schweitzer (eds.), *Meshfree Methods for Partial Differential Equations VII.*

101. R. Hoppe (ed.), *Optimization with PDE Constraints - OPTPDE 2014.*

102. S. Dahlke, W. Dahmen, M. Griebel, W. Hackbusch, K. Ritter, R. Schneider, C. Schwab, H. Yserentant (eds.), *Extraction of Quantifiable Information from Complex Systems.*

103. A. Abdulle, S. Deparis, D. Kressner, F. Nobile, M. Picasso (eds.), *Numerical Mathematics and Advanced Applications - ENUMATH 2013.*

104. T. Dickopf, M.J. Gander, L. Halpern, R. Krause, L.F. Pavarino (eds.), *Domain Decomposition Methods in Science and Engineering XXII.*

105. M. Mehl, M. Bischoff, M. Schäfer (eds.), *Recent Trends in Computational Engineering - CE2014.* Optimization, Uncertainty, Parallel Algorithms, Coupled and Complex Problems.

106. R.M. Kirby, M. Berzins, J.S. Hesthaven (eds.), *Spectral and High Order Methods for Partial Differential Equations - ICOSAHOM'14.*

107. B. Jüttler, B. Simeon (eds.), *Isogeometric Analysis and Applications 2014.*

108. P. Knobloch (ed.), *Boundary and Interior Layers, Computational and Asymptotic Methods – BAIL 2014.*

109. J. Garcke, D. Pflüger (eds.), *Sparse Grids and Applications – Stuttgart 2014.*

110. H. P. Langtangen, *Finite Difference Computing with Exponential Decay Models.*

111. A. Tveito, G.T. Lines, *Computing Characterizations of Drugs for Ion Channels and Receptors Using Markov Models.*

112. B. Karazösen, M. Manguoğlu, M. Tezer-Sezgin, S. Göktepe, Ö. Uğur (eds.), *Numerical Mathematics and Advanced Applications - ENUMATH 2015.*

113. H.-J. Bungartz, P. Neumann, W.E. Nagel (eds.), *Software for Exascale Computing - SPPEXA 2013-2015.*

114. G.R. Barrenechea, F. Brezzi, A. Cangiani, E.H. Georgoulis (eds.), *Building Bridges: Connections and Challenges in Modern Approaches to Numerical Partial Differential Equations.*

115. M. Griebel, M.A. Schweitzer (eds.), *Meshfree Methods for Partial Differential Equations VIII.*

116. C.-O. Lee, X.-C. Cai, D.E. Keyes, H.H. Kim, A. Klawonn, E.-J. Park, O.B. Widlund (eds.), *Domain Decomposition Methods in Science and Engineering XXIII.*

117. T. Sakurai, S.-L. Zhang, T. Imamura, Y. Yamamoto, Y. Kuramashi, T. Hoshi (eds.), *Eigenvalue Problems: Algorithms, Software and Applications in Petascale Computing.* EPASA 2015, Tsukuba, Japan, September 2015.

118. T. Richter (ed.), *Fluid-structure Interactions.* Models, Analysis and Finite Elements.

119. M.L. Bittencourt, N.A. Dumont, J.S. Hesthaven (eds.), *Spectral and High Order Methods for Partial Differential Equations ICOSAHOM 2016.* Selected Papers from the ICOSAHOM Conference, June 27-July 1, 2016, Rio de Janeiro, Brazil.

120. Z. Huang, M. Stynes, Z. Zhang (eds.), *Boundary and Interior Layers, Computational and Asymptotic Methods BAIL 2016.*

121. S.P.A. Bordas, E.N. Burman, M.G. Larson, M.A. Olshanskii (eds.), *Geometrically Unfitted Finite Element Methods and Applications.* Proceedings of the UCL Workshop 2016.

122. A. Gerisch, R. Penta, J. Lang (eds.), *Multiscale Models in Mechano and Tumor Biology*. Modeling, Homogenization, and Applications.

123. J. Garcke, D. Pflüger, C.G. Webster, G. Zhang (eds.), *Sparse Grids and Applications - Miami 2016*.

124. M. Schäfer, M. Behr, M. Mehl, B. Wohlmuth (eds.), *Recent Advances in Computational Engineering*. Proceedings of the 4th International Conference on Computational Engineering (ICCE 2017) in Darmstadt.

125. P.E. Bjørstad, S.C. Brenner, L. Halpern, R. Kornhuber, H.H. Kim, T. Rahman, O.B. Widlund (eds.), *Domain Decomposition Methods in Science and Engineering XXIV*. 24th International Conference on Domain Decomposition Methods, Svalbard, Norway, February 6–10, 2017.

126. F.A. Radu, K. Kumar, I. Berre, J.M. Nordbotten, I.S. Pop (eds.), *Numerical Mathematics and Advanced Applications – ENUMATH 2017*.

127. X. Roca, A. Loseille (eds.), *27th International Meshing Roundtable*.

128. Th. Apel, U. Langer, A. Meyer, O. Steinbach (eds.), *Advanced Finite Element Methods with Applications*. Selected Papers from the 30th Chemnitz Finite Element Symposium 2017.

129. M. Griebel, M.A. Schweitzer (eds.), *Meshfree Methods for Partial Differencial Equations IX*.

130. S. Weißer, BEM-based Finite Element *Approaches on Polytopal Meshes*.

131. V.A. Garanzha, L. Kamenski, H. Si (eds.), *Numerical Geometry, Grid Generation and Scientific Computing*. Proceedings of the 9th International Conference, NUMGRID2018/Voronoi 150, Celebrating the 150th Anniversary of G. F. Voronoi, Moscow, Russia, December 2018.

132. H. van Brummelen, A. Corsini, S. Perotto, G. Rozza (eds.), *Numerical Methods for Flows*.

133. H. van Brummelen, C. Vuik, M. Möller, C. Verhoosel, B. Simeon, B. Jüttler (eds.), *Isogeometric Analysis and Applications 2018*.

134. S.J. Sherwin, D. Moxey, J. Peiro, P.E. Vincent, C. Schwab (eds.), *Spectral and High Order Methods for Partial Differential Equations ICOSAHOM 2018*.

135. G.R. Barrenechea, J. Mackenzie (eds.), *Boundary and Interior Layers, Computational and Asymptotic Methods BAIL 2018*.

136. H.-J. Bungartz, S. Reiz, B. Uekermann, P. Neumann, W.E. Nagel (eds.), *Software for Exascale Computing - SPPEXA 2016–2019*.

137. M. D'Elia, M. Gunzburger, G. Rozza (eds.), *Quantification of Uncertainty: Improving Efficiency and Technology*.

138. R. Haynes, S. MacLachlan, X.-C. Cai, L. Halpern, H.H. Kim, A. Klawonn, O. Widlund (eds.), *Domain Decomposition Methods in Science and Engineering XXV*.

139. F.J. Vermolen, C. Vuik (eds.), *Numerical Mathematics and Advanced Applications ENUMATH 2019*.

For further information on these books please have a look at our mathematics catalogue at the following URL: www.springer.com/series/3527

Monographs in Computational Science and Engineering

1. J. Sundnes, G.T. Lines, X. Cai, B.F. Nielsen, K.-A. Mardal, A. Tveito, *Computing the Electrical Activity in the Heart.*

For further information on this book, please have a look at our mathematics catalogue at the following URL: `www.springer.com/series/7417`

Texts in Computational Science and Engineering

1. H. P. Langtangen, *Computational Partial Differential Equations.* Numerical Methods and Diffpack Programming. 2nd Edition
2. A. Quarteroni, F. Saleri, P. Gervasio, *Scientific Computing with MATLAB and Octave.* 4th Edition
3. H. P. Langtangen, *Python Scripting for Computational Science.* 3rd Edition
4. H. Gardner, G. Manduchi, *Design Patterns for e-Science.*
5. M. Griebel, S. Knapek, G. Zumbusch, *Numerical Simulation in Molecular Dynamics.*
6. H. P. Langtangen, *A Primer on Scientific Programming with Python.* 5th Edition
7. A. Tveito, H. P. Langtangen, B. F. Nielsen, X. Cai, *Elements of Scientific Computing.*
8. B. Gustafsson, *Fundamentals of Scientific Computing.*
9. M. Bader, *Space-Filling Curves.*
10. M. Larson, F. Bengzon, *The Finite Element Method: Theory, Implementation and Applications.*
11. W. Gander, M. Gander, F. Kwok, *Scientific Computing: An Introduction using Maple and MATLAB.*
12. P. Deuflhard, S. Röblitz, *A Guide to Numerical Modelling in Systems Biology.*
13. M. H. Holmes, *Introduction to Scientific Computing and Data Analysis.*
14. S. Linge, H. P. Langtangen, *Programming for Computations* - A Gentle Introduction to Numerical Simulations with MATLAB/Octave.
15. S. Linge, H. P. Langtangen, *Programming for Computations* - A Gentle Introduction to Numerical Simulations with Python.
16. H.P. Langtangen, S. Linge, *Finite Difference Computing with PDEs* - A Modern Software Approach.
17. B. Gustafsson, *Scientific Computing from a Historical Perspective.*
18. J. A. Trangenstein, *Scientific Computing.* Volume I - Linear and Nonlinear Equations.

19. J. A. Trangenstein, *Scientific Computing*. Volume II - Eigenvalues and Optimization.

20. J. A. Trangenstein, *Scientific Computing*. Volume III - Approximation and Integration.

21. H. P. Langtangen, K.-A. Mardal, *Introduction to Numerical Methods for Variational Problems*.

22. T. Lyche, *Numerical Linear Algebra and Matrix Factorizations*.

For further information on these books please have a look at our mathematics catalogue at the following URL: `www.springer.com/series/5151`

The manufacturer's authorised representative in the EU is Springer
Nature Customer Service Centre GmbH, Europaplatz 3, 69115 Heidelberg,
Germany. If you have any concerns regarding our products, please
contact ProductSafety@springernature.com

Printed and bound by CPI Group (UK) Ltd, Croydon, CR0 4YY
28/09/2026
02224062-0009